PROCEEDINGS

OF THE FIFTH INTERNATIONAL CONGRESS

OF

MATHEMATICIANS

CAMBRIDGE UNIVERSITY PRESS
London: FETTER LANE, E.C.
C. F. CLAY, Manager

Edinburgh: 100, PRINCES STREET
Berlin: A. ASHER AND CO.
Leipzig: F. A. BROCKHAUS
New York: G. P. PUTNAM'S SONS
Bombay and Calcutta: MACMILLAN AND CO., Ltd.

PROCEEDINGS

OF THE FIFTH INTERNATIONAL CONGRESS

OF

MATHEMATICIANS

(Cambridge, 22—28 August 1912)

EDITED BY THE GENERAL SECRETARIES OF THE CONGRESS

E. W. HOBSON

SADLEIRIAN PROFESSOR OF PURE MATHEMATICS IN THE UNIVERSITY OF CAMBRIDGE

AND

A. E. H. LOVE

SEDLEIAN PROFESSOR OF NATURAL PHILOSOPHY IN THE UNIVERSITY OF OXFORD

VOL. II.

COMMUNICATIONS TO SECTIONS II—IV

Cambridge:

at the University Press

1913

Cambridge:
PRINTED BY JOHN CLAY, M.A.
AT THE UNIVERSITY PRESS

COMMUNICATIONS

SECTION II

(GEOMETRY)

SUR LA NOTION DE "CLASSE" DE TRANSFORMATIONS D'UNE MULTIPLICITÉ

PAR L. E. J. BROUWER.

Soient μ et μ' des multiplicités fermées: nous dirons de deux représentations univoques et continues de μ sur μ' qu'elles appartiennent à la même *classe*, si nous pouvons arriver de l'une à l'autre au moyen d'une modification continue. Les représentations de la même classe possèdent toutes le même *degré* (voir mon travail "Ueber Abbildung von Mannigfaltigkeiten," *Mathem. Annalen*, Bd. 71), mais ce qui est remarquable, c'est qu'en un grand nombre de cas l'inverse de ce théorème est encore vrai, propriété dont j'indiquerai brièvement la démonstration pour le cas que μ et μ' sont toutes les deux des sphères.

Nous commencerons par démontrer la propriété pour certaines catégories spéciales de représentations, d'où nous nous élèverons par étapes aux représentations les plus générales.

Considérons d'abord les *représentations riemanniennes*, c'est-à-dire les représentations qui au sens de l'analysis situs sont identiques aux représentations des surfaces de Riemann de genre zéro sur le plan complexe. D'une part M. Klein a demontré que toutes les représentations de degré n à ramifications simples peuvent être transformées d'une manière continue les unes en les autres en faisant mouvoir convenablement les points de ramification (propriété découlant du théorème de Lüroth-Clebsch sur l'arrangement canonique des coupures de ramification), d'autre part on peut, pour les représentations à ramifications multiples, séparer ces ramifications en un nombre équivalent de ramifications simples, de sorte que pour les représentations riemanniennes notre théorème se trouve démontré.

Considérons ensuite les *représentations canoniques*, pour lesquelles $n-1$ courbes simples fermées, situées dans μ et sans points communs, sont représentées chacune en un seul point de μ', tandis que les n régions déterminées par elles sont représentées biunivoquement, et cela ou toutes positivement, ou bien toutes négativement. Une représentation canonique pouvant être transformée, au moyen d'une modification continue arbitrairement petite, en une représentation riemannienne, notre théorème est encore démontré pour les représentations canoniques.

Passons aux *représentations normales*, ne satisfaisant qu'aux conditions suivantes: (1) elles sont *sans contraction cyclique*, c'est-à-dire il n'existe sur μ aucune courbe simple fermée représentée en un seul point de μ'; (2) elles sont *simpliciales*; (3) elles sont *sans plis**, c'est-à-dire de chaque élément de μ, dont l'image est biunivoque, l'indicatrice de l'image a le même signe. Pour les représentations normales nous démontrons notre théorème en faisant subir aux arcs de courbe de μ' images d'éléments de μ, une sorte d'extension continue les rendant images biunivoques et transformant la représentation normale en une représentation riemannienne.

Envisageons enfin les *représentations simpliciales sans plis*, pour lesquelles il existe sur μ un nombre fini d'ensembles polygonaux d'un seul tenant, représentés chacun en un seul point de μ' et divisant μ en un nombre fini de régions, pour chacune desquelles la représentation est ou normale ou nulle part dense. Pour ces dernières représentations nous démontrons notre théorème en réduisant d'abord les ensembles polygonaux à des courbes simples fermées sans points communs—ce qui nous donne une représentation simpliciale sans plis " à contraction simple "—et réduisant ensuite la représentation de chacune des régions partielles, dont l'image n'est pas nulle part dense, successivement à une représentation riemannienne et à une représentation canonique, ce qui nous donne une telle représentation simpliciale sans plis à contraction simple, pour laquelle les régions partielles de μ se divisent en trois sortes, représentées sur μ' successivement avec le degré 0, avec le degré $+1$, et avec le degré -1. Or, au moyen de nouvelles transformations continues de la représentation, on peut d'une part remplacer l'ensemble d'une région de 2de et une région adjacente de 3me sorte par une seule région de 1re sorte, d'autre part faire absorber une région de 1re sorte par une région adjacente de 2de ou de 3me sorte, procédé qui finit par nous donner une représentation canonique.

Maintenant les *représentations générales* ne présentent plus de difficulté, puisqu'en commençant par les rendre simpliciales et détruisant ensuite les plis morceau par morceau, on peut les transformer d'une manière continue en représentations simpliciales sans plis.

* Par un "*plis*" nous entendons ici une suite finie d'éléments de μ, dont chaque élément, excepté le dernier, possède en commun avec son suivant un " côté de contact," tous ces côtés de contact étant représentés sur un même arc de courbe de μ', et le premier et le dernier élément de la suite étant représentés avec des indicatrices opposées.

ON THE EXTENSION OF A THEOREM OF W. STAHL

BY F. MORLEY.

§ 1. *Counter-curves.*

Given $d+1$ binary forms $f_0 \dots f_d$ all of order m, an involution I_d^m is defined by

$$(\xi f) \equiv \xi_0 f_0 + \dots + \xi_d f_d = 0.$$

The natural geometric representation of I_d^m is the sections of a rational curve ρ_d^m of order m in a space S_d by spaces S_{d-1}.

All forms ϕ, also of order m, which are apolar to all forms f define the counter-involution (or conjugate involution)

$$(x\phi) \equiv x_0 \phi_0 + \dots + x_{m-d-1} \phi_{m-d-1} = 0,$$

which is represented by a curve $\bar{\rho}^m_{m-d-1}$, the counter-curve of ρ_d^m.

A rational curve carries an infinity of involutions; for example in the case of a planar curve all algebraic curves of given order on the flexes will cut out an involution, and all algebraic curves of given class on the stationary lines will cut out an involution by means of the common lines.

Particular interest attaches, in the case of planar curves, to involutions given by curves on the double lines. For the curve ρ_2^m, the curve of lowest class on the double lines is of class $2(m-3)$; the common lines of this and ρ give an $I_{m-3}^{4(m-3)}$; and my point is that this is *the dual form of the counter-curve.* In other words, the counter-involution (or in Stahl's phrase fundamental involution) on a curve ρ_2^m is given by a series of theorems, the first of which was given by Stahl (*Crelle*, Vol. 101), as follows:

For the quartic ρ_2^4, any conic on the double lines has with ρ 4 common lines; these give a set of the counter-involution.

For the quintic ρ_2^5 any range of quartics on the double lines contains 5 which touch ρ; the parameters of contact are in the counter-involution.

For the sextic ρ_2^6 any web of sextics on the double lines contains 6 which osculate ρ; the 6 parameters are in the counter-involution. And so on.

I shall give the proof in the case of ρ_2^5, by a method which applies to all cases.

§ 2. *Algebraic theory of the planar rational quintic**.

The argument does not require explicit expressions, but only the degrees of the coefficients in these expressions. Thus the form (ξf) which, when zero, gives the ρ_2^5 is sufficiently indicated by

$$A\xi t^5 \quad \text{.......................................(1)},$$

where A denotes a constant, ξ the coordinates of a line in the plane, t a parameter.

Regarding this as a binary quintic we have for two such quintics $A\xi t^5$, $A\eta t^5$ the combinative transvectants

$$A^2 x\tau^8, \quad A^2 x\tau^4, \quad A^2 x \text{..............(2. 1, 2. 2, 2. 3)};$$

the first being the line-equation or point-section of the curve.

Making x and ξ incident in (1) and (2) we get

$$A^3 t^3\tau^6, \quad A^3 t^5\tau^4, \quad A^3 t^5 \quad \text{...........(3. 1, 3. 2, 3. 3)},$$

where in (3. 1) we have removed from an $A^3 t^5\tau^8$ the factor $(t\tau)^2$. The remaining factor $A^3 t^3\tau^6$ expresses the incidence of point t and line τ, and may be called the *incidence form.*

The coefficients in (3) are linear in the coordinates of ρ_2^5, that is in the 3-rowed determinants formed from the constants A.

The incidence form is a special case of Gordan's involution form

$$A^3 t_1^3 t_2^3 t_3^3 \quad \text{....................................(4)},$$

which expresses that the points t_1, t_2, t_3 are collinear. By developing Gordan's form Professor Coble (On symmetrical binary forms and involutions, *Am. Jour.* Vols. XXXI. and XXXII., in particular Vol. XXXII. p. 359) obtained the 3 forms in τ alone, which are linear in the coordinates. These linear forms are:

$$A^3\tau^9, \quad A^3\tau^5, \quad A^3\tau^3 \quad \text{..............(5. 1, 5. 2, 5. 3)},$$

or say

$$(f\tau)^9, \quad (g\tau)^5, \quad (h\tau)^3,$$

the first being the *stationary form* which gives the triple roots of (1).

If, after Study, we write the equation of ρ_2^5 symbolically:

$$(a\xi)(\alpha t)^5,$$

then f is $\quad |abc|\,|\beta\gamma|\,|\gamma\alpha|\,|\alpha\beta|\,(\alpha t)^3(\beta t)^3(\gamma t)^3,$

g is $\quad |abc|\,|\beta\gamma|\,|\gamma\alpha|\,|\alpha\beta|\,|\beta\gamma|^2(\alpha t)^3(\beta t)(\gamma t),$

h is $\quad |abc|\,|\beta\gamma|\,|\gamma\alpha|\,|\alpha\beta|\,|\beta\gamma|\,(\alpha t)\,|\gamma\alpha|^2(\beta t)^2;$

and in general to write the linear forms for a ρ_2^n we write all covariants of degree 3 of a form $(\alpha x)^{n-2}$; for a ρ_3^n all covariants of degree 4 of a form $(\alpha x)^{n-3}$, and so on.

These linear forms are fundamental. All covariant line curves of ρ_2^5 are invariants of the binary forms (1) and (5). All covariant point curves are invariants of (2) and (5). All covariant connexes are invariants of (1), (2) and (5). Covariants of the binary systems (1) and (5) yield covariant linear systems of curves. And so on.

* Compare W. Gross, *Annalen*, Vol. XXXII.

§ 3. *Quartics on the double lines of a rational planar quintic.*

Consider the web of curves of class 4 on the 12 double lines of ρ_2^5. They have with ρ_2^5 32 common lines; these are the double lines counting twice and 8 variable lines. Two of these determine the rest, that is we have an involution I_2^8, which is pictured by a curve, say of lines, $\bar{r}_2^8$. When the quartic touches a stationary line, this line counts as two common lines; there are then 9 lines of $\bar{r}_2^8$ which count twice as tangents, i.e. $\bar{r}_2^8$ has nine stationary lines.

It is then, as a point curve, of order 5, say $\bar{\rho}_2^5$; and it has the same stationary form $(f\tau)^9$ as ρ_2^5.

Hence its incidence form is

$$(ft)^3 (f\tau)^6 + \dots,$$

and is linearly built from f, g, h. Denote it by

$$A_1{}^3 t_1{}^3 \tau_1{}^6 \quad \dots\dots\dots\dots\dots\dots\dots\dots\dots\dots (6).$$

The lines τ, τ_1 of ρ and $\bar{\rho}$ cut out the sections

$$A^3 t^3 \tau^6 \,|\, t\tau\,|^2 \text{ and } A_1{}^3 t_1{}^3 \tau_1{}^6 \,|\, t_1 \tau_1\,|^2.$$

These are apolar if
$$A^6 \tau^8 \tau_1{}^8 = 0 \quad \dots\dots\dots\dots\dots\dots\dots\dots\dots\dots (7).$$

Since no form of degree 2 and order 16 can be built from f, g, h, (7) vanishes if $\tau = \tau_1$. It is then

$$|\,\tau\tau_1\,|\, A^6 \tau^7 \tau_1{}^7 \dots\dots\dots\dots\dots\dots\dots\dots\dots\dots (8).$$

Now (7) shows that eight tangent-sections of ρ are apolar to a tangent-section τ_1 of $\bar{\rho}$. Either then all sections of ρ are apolar to all those of $\bar{\rho}$, in which case the curves are counter-curves, or else the eight tangent-sections are all on a point x.

Looking now at the curve ρ_2^5 only, the form $A^6 \tau^7 \tau_1{}^7$ gives for every τ_1 a point x on the line τ_1, and the values of τ are the parameters of the other lines from x. When τ_1 is a stationary parameter, $\tau = \tau_1$. Hence when $\tau = \tau_1$, the form $A^6 \tau^7 \tau_1{}^7$ contains the factor f. The other factor must be g.

Thus as τ_1 varies the locus of x crosses each stationary line once, and therefore is a line; and it meets the curve where $g = 0$. This line g is the line (2. 3), whose equation is $A^2 x = 0$.

Hence, either the form (8) vanishes identically or it is the condition that the lines τ and τ_1 meet on the line g. This condition is obtained by making two tangent-sections of ρ_2^5 apolar. Hence when the curves ρ_2^5 and $\bar{\rho}_2^5$ are projectively distinct, they are counter-curves.

It is proved then that given the forms f, g, h, there are two and but two incidence forms, linear in the coordinates; these defining the curve itself and the counter-curve.

§ 4. *The self-counter quintic.*

The curves ρ and $\bar{\rho}$ cease to be distinct when the involution set up by the quartics is also the involution set up by lines of ρ_2^5 on a point; that is when the quartics can break up into a cubic on the double lines and an arbitrary point.

Hence *the double lines of a self-counter quintic are on a cubic.*

This interesting special curve has to satisfy 3 conditions; any two line-sections are apolar, so that the line $A^2x=0$ and therefore its section $(g\tau)^5=0$ become arbitrary.

§ 5. *Quintics on the double lines.*

Passing now to quintics on the double lines, we can write after Cayley (*Works*, Vol. II. p. 469) a covariant of (1) of degree 5 and order 9 which vanishes when (1) has two double roots. Hence we have special quintics on the double lines

$$A^5\xi^5 t^9,$$

and the apolarity-relation of this with an arbitrary form $(a\tau)^9$ gives the general such quintic, $aA^5\xi^5$.

The common lines of this and ρ are

$$aA^{15}\tau^{40},$$

and removing the parameters of the double lines, an $A^{12}\tau^{24}$, we have for the variable common lines

$$aA^3\tau^{16}.$$

This can only be the Jacobian of $(a\tau)^9$ and $(f\tau)^9$. We have then in the common lines a picture of the Jacobian of the form f and an arbitrary form a, forming an involution I_8^{16}. In the breaking up of the quintic into a quartic and a point we have the breaking up of the I_8^{16} into two counter-involutions I_2^8 and $\bar{I}_2^8$. And this raises in a new form an old question: how many curves ρ_2^5 are there when the stationary form $(f\tau)^9$ is given? This is equivalent to the question: how many planes in S_5 meet 9 given planes?—which is raised in Meyer's *Apolarität.*

Professor Coble solves this problem of enumeration as follows. Consider first the inter-relations of the linear forms f, g, h.

§ 6. *The linear forms.*

Writing $|012|$ etc. for the 3-rowed determinants formed from the coefficients of three line-sections written *without* binomial coefficients we have for the stationary form

$$\begin{aligned}\frac{f}{21} &= |012|\tau^9+3|013|\tau^8+\{6|014|+3|023|\}\tau^7\\ &+\{10|015|+8|024|+|125|\}\tau^6+\{15|025|+6|034|+3|124|\}\tau^5\\ &+\{15|035|+6|125|+3|134|\}\tau^4+\{10|045|+8|135|+|234|\}\tau^3\\ &+\{6|145|+3|235|\}\tau^2+3|245|\tau+|345| \qquad \text{(9. 1)};\end{aligned}$$

and for g and h as calculated by Mr Tracey

$$\begin{aligned}\frac{g}{5} &= \{2|014|-|023|\}\tau^5+\{10|015|-|123|\}\tau^4\\ &+\{10|025|-2|124|\}\tau^3+\{10|035|-2|134|\}\tau^2\\ &+\{10|045|-|234|\}\tau+2|145|-|235| \qquad \text{(9. 2)},\end{aligned}$$

$$\frac{h}{2} = \{20\,|015| - 5\,|024| + 2\,|123|\}\,\tau^3$$
$$+ \{15\,|025| - 15\,|034| + 3\,|124|\}\,\tau^2$$
$$- \{15\,|035| - 15\,|125| + 3\,|134|\}\,\tau$$
$$- \{20\,|045| - 5\,|135| + 2\,|234|\} \qquad (9.\ 3).$$

Thus the 20 coefficients of f, g, h are 20 linearly independent combinations of the 20 linearly independent coordinates $|012|$ etc. of the curve.

Any relation among these coefficients must be due to a syzygy connecting the comitants of f, g, h. Call $\lambda + \mu + \nu - 3$ the weight of $|\lambda\mu\nu|$. Then the coefficients of f have their natural weight, those of g and h have their natural weight increased by 2 and 3 respectively.

The weights and orders of the comitants of the second degree are:

Weight	Order	Forms
0	18	ff
2	14	$\|ff\|^2$, fg
3	12	$\|fg\|$, fh
4	10	$\|ff\|^4$, $\|fh\|$, gg, $\|fg\|^2$
5	8	$\|fh\|^2$, $\|fg\|^3$, gh
6	6	$\|ff\|^6$, $\|fh\|^3$, $\|gg\|^2$, hh, $\|fg\|^4$, $\|gh\|$
7	4	$\|fg\|^5$, $\|gh\|^2$
8	2	$\|ff\|^8$, $\|gg\|^4$, $\|hh\|^2$

......(10).

Now the linearly independent relations of the second degree in the coordinates are known; there are 30 of the type

$$|012|\,|034| + |023|\,|041| + |031|\,|024|,$$

and one of the type $$|012|\,|345| + \ldots.$$

There is then one of weight 4

$$|012|\,|034| + \ldots,$$

indicating a syzygy of weight 4 and order 10; two of weight 5,

$$|012|\,|035| + \ldots \text{ and } |102|\,|134| + \ldots,$$

indicating a syzygy of weight 5 and order 8; and so on.

In any syzygy g must occur throughout to an even or an odd degree, since we pass from a curve to the counter-curve by changing the sign of g. Thus the second syzygy must be either

$$|fh|^2 = 0$$

or $$|fg|^3 = gh,$$

and it is seen from the leading terms that the former is not the case. We have then

$$|fg|^3 = gh \qquad (11),$$

where the numerical factors are selected as in (9). The equation (11) defines g as an irrational covariant of f.

§ 7. *The enumeration.*

Inserting a factor of proportionality h_∞ we get from

$$h_\infty \,|fg|^3 = gh$$

9 bilinear forms in $g_0 \dots g_5$ and $h_\infty, h_0 \dots h_3$.

Take now 10 bilinear forms in 6 g's and 5 h's

$$(\alpha g)(\beta h),$$

and first suppose these factors αg, βh not symbolic. The condition for a common solution is the product of $\binom{10}{6}$ determinants in the coefficients α, and of $\binom{10}{5}$ determinants in β. It contains the coefficients α of each form to the degree $\binom{9}{5}$ and β to the same degree $\binom{9}{4}$.

Hence when the bilinear forms are not degenerate the invariant which expresses that they have a common solution is of degree $\binom{9}{4}$ in the coefficients of each form. Since this invariant would be obtained by substituting in the tenth form the solutions of the other 9, and multiplying, it is the number of such solutions.

There are then $\binom{9}{4}$ forms g when f is given, and therefore there are 126 *pairs of counter-quintics with a given stationary form.* Or again there are 126 *pairs of planes meeting* 9 *given planes in a space* S_5.

In starting with a ρ_2^5 we have isolated one of these 126 pairs of solutions. The other 125 pairs are then represented by pairs of nets of quintics on the double lines; for the involution I_8^{16} breaks up for each solution into two independent involutions I_2^8 and $\bar{I}_2^8$.

§ 8. *Case of the* ρ_2^4.

The explicit form for the conics on the double lines of a quartic is easy to obtain.

The line-section being $$A\xi t^4,$$
its Hessian is $A^2\xi^2t^4$.

If the counter-involution be

$$(a\tau)^4 + \lambda\,(b\tau)^4,$$

the Hessian is apolar to this if

$$aA^2\xi^2 + \lambda bA^2\xi^2 = 0 \quad \dots\dots\dots\dots(12),$$

a range of conics.

Each line of the conic λ gives a section whose Hessian is apolar to the quartic λ. Hence a common line of the range gives a section whose Hessian is apolar to all quartics λ, i.e. whose Hessian is also a section. But such a section is given by a double line. Hence (12) is the conics on the double lines.

Eliminating λ we have $$A^5\xi^2t^4.$$

In terms of the linear forms $(f\tau)^6$ and $(g\tau)^2$ of the quartic this is

$$|Hf|^3 + |Hg|,$$

where H is the Hessian of a section.

In this case of the quartic, cubics on the double lines can break up in 5 ways, once into an arbitrary point and a conic on the double lines and four times into a point on a double line and a conic on the other three. Thus the 5 binary involutions with the given Jacobian f are the counter-involution of the ρ_2^4 and those of the tangents to ρ_2^4 from a point on the double line.

Similarly for ρ_3^5 the stationary form $(f\tau)^8$ is, we know, the Jacobian of 14 pencils of binary quintics. One of these pencils is the counter-involution of the curve; the rest refer to the 13 lines, each of which is in 4 planes of the curve. From a point on such a line we can draw 5 other planes, and thus we have 13 pencils of binary quintics, obviously having the given Jacobian.

CERTAIN CONTINUOUS DEFORMATIONS OF SURFACES APPLICABLE TO THE QUADRICS

BY LUTHER PFAHLER EISENHART.

Bianchi* has established an important transformation of surfaces applicable to the quadrics, such that, if S be one of these surfaces and $\bar{S}$ a transform, the lines joining corresponding points M and $\bar{M}$ on S and $\bar{S}$ form a congruence for which S and $\bar{S}$ are the focal surfaces, and the asymptotic lines on these surfaces correspond. A congruence possessing the latter property is called a W-congruence. It is a general theorem concerning W-congruences that either focal surface admits of an infinitesimal deformation such that the direction of deformation at any point is parallel to the normal to the other focal surface at the corresponding point. Hence the Bianchi transformations of surfaces applicable to the quadrics carry with them certain infinitesimal deformations of these surfaces. It is the purpose of this paper to show that with the aid of these transformations one may derive from a surface S, applicable to a quadric Q, a suite of continuous deforms of S. Such a suite we call a *system* (Q). Although there is an essential difference in the form of the equations according as Q is a paraboloid or a central quadric, the results are analogous in all cases. Unless otherwise stated the following equations refer to the case where Q is the real hyperbolic paraboloid $\frac{x^2}{p} - \frac{y^2}{q} = 2z$ and the transformation is real.

We assume that the parametric curves $u = \text{const.}$, $v = \text{const.}$ on S correspond to the generators on Q. In this case the equations of a transform $\bar{S}$ are of the form

$$\bar{x} = x + l\frac{\partial x}{\partial u} + m\frac{\partial x}{\partial v} \qquad \text{.............................(1)},$$

where l and m are the two ratios†

$$l = \frac{U}{W}, \qquad m = \frac{V}{W} \qquad \text{.............................(2)},$$

U, V, W being algebraic functions of u, v, of an essential constant κ which fixes the confocal quadric Q_κ determining the transformation, and of a function λ which must satisfy the system‡

$$\frac{\partial \lambda}{\partial u} = \frac{\sqrt{pq}}{\kappa H}V + \frac{\epsilon}{2\kappa\sqrt{H}}(DU + D'V), \quad \frac{\partial \lambda}{\partial v} = \frac{\sqrt{pq}}{\kappa H}U + \frac{\epsilon}{2\kappa\sqrt{H}}(D'U + D''V) \ldots\text{(3)},$$

* *Lezioni di Geometria Differenziale*, Vol. III. Pisa (1909). Hereafter a reference to this volume will be of the form B. p. —.

† B. p. 13.

‡ B. p. 81.

where $4H = EG - F^2$; E, F, G and D, D', D'' being the fundamental functions for S; and ϵ is $+1$ or -1 according to the system of generators used on Q_κ to determine the transformation.

In accordance with the general theorem previously referred to, an infinitesimal deformation of S is given by

$$x_1 = x + a\xi, \quad y_1 = y + a\eta, \quad z_1 = z + a\zeta \quad \ldots\ldots(4),$$

where a is an infinitesimal constant such that powers higher than the first are negligible, and ξ, η, ζ are of the form

$$\xi = e^T \left[-(Fl + Gm)\frac{\partial x}{\partial u} + (El + Fm)\frac{\partial x}{\partial v} + \frac{2\epsilon H^{\frac{3}{2}}}{\sqrt{pq}} A X_3 \right] \quad \ldots\ldots(5),$$

the functions X_3, Y_3, Z_3 being the direction-cosines of the normal to S; A a determinate function; and T a function to be determined by the well-known conditions of infinitesimal deformation, namely

$$\Sigma \frac{\partial x}{\partial u}\frac{\partial \xi}{\partial u} = 0, \quad \Sigma \frac{\partial x}{\partial v}\frac{\partial \xi}{\partial v} = 0, \quad \Sigma \frac{\partial x}{\partial u}\frac{\partial \xi}{\partial v} + \Sigma \frac{\partial x}{\partial v}\frac{\partial \xi}{\partial u} = 0 \quad \ldots\ldots(6).$$

The resulting equations are reducible by means of (3) to

$$\left.\begin{aligned} &\frac{\partial T}{\partial u} + \frac{\partial \log H}{\partial u} + \frac{d \log m}{du} + \frac{\epsilon\sqrt{H}}{\sqrt{pq}}\frac{A}{2}\frac{D}{m} = 0 \\ &\frac{\partial T}{\partial v} + \frac{\partial \log H}{\partial v} + \frac{d \log l}{dv} - \frac{\epsilon\sqrt{H}}{\sqrt{pq}}\frac{A}{2}\frac{D''}{l} = 0 \\ &l\frac{\partial T}{\partial u} - m\frac{\partial T}{\partial v} + \frac{dl}{du} - \frac{dm}{dv} - \frac{\epsilon\sqrt{H}}{\sqrt{pq}} A D' = 0 \end{aligned}\right\} \quad \ldots\ldots(7),$$

where

$$\frac{d}{du} = \frac{\partial}{\partial u} + \frac{\partial}{\partial \lambda}\cdot\frac{\partial \lambda}{\partial u}, \quad \frac{d}{dv} = \frac{\partial}{\partial v} + \frac{\partial}{\partial \lambda}\cdot\frac{\partial \lambda}{\partial v}.$$

Equations (7) are readily shown to be consistent by means of the Gauss and Codazzi equations for S.

If we are interested not in a discrete infinitesimal deformation of S, but in a continuous deformation, we must look upon the functions appearing in the above expressions as involving a third parameter, say w. The preceding results are true in this case also, but there are further conditions to be satisfied. Now in equation (5) ξ must be replaced by $\frac{\partial x}{\partial w}$, and the conditions of integrability

$$\frac{\partial}{\partial u}\left(\frac{\partial x}{\partial w}\right) = \frac{\partial}{\partial w}\left(\frac{\partial x}{\partial u}\right), \quad \frac{\partial}{\partial v}\left(\frac{\partial x}{\partial w}\right) = \frac{\partial}{\partial w}\left(\frac{\partial x}{\partial v}\right) \quad \ldots\ldots(8),$$

must be satisfied. If we put

$$\frac{\partial x}{\partial u} = \sqrt{E} X_1, \quad \frac{\partial x}{\partial v} = \sqrt{G} X_2 \quad \ldots\ldots(9),$$

conditions (8) are reducible to

$$\frac{\partial X_1}{\partial w} = e^T (F X_1 - \sqrt{EG} X_2) + \frac{B}{\sqrt{E}} X_3, \quad \frac{\partial X_2}{\partial w} = e^T (\sqrt{EG} X_1 - F X_2) + \frac{C}{\sqrt{G}} X_3 \ldots(10),$$

where B and C are determinate functions of $u, v, \lambda, T, D, D', D''$.

From (10) follows

$$\frac{\partial X_3}{\partial w} = \frac{\sqrt{EG}}{4H}\left[(FX_1 - \sqrt{EG}\,X_2)\frac{C}{\sqrt{G}} - (\sqrt{EG}\,X_1 - FX_2)\frac{B}{\sqrt{E}}\right] \quad \text{......(11).}$$

By means of the Gauss equations for S we can find the expressions for

$$\frac{\partial X_1}{\partial u},\ \frac{\partial X_1}{\partial v};\ \frac{\partial X_2}{\partial u},\ \frac{\partial X_2}{\partial v};\ \frac{\partial X_3}{\partial u},\ \frac{\partial X_3}{\partial v} \quad \text{..................(12).}$$

The necessary and sufficient condition that a system (Q) exists is that the integrability conditions of the expressions (10), (11), (12) be satisfied. These lead to a system of five independent partial differential equations in two dependent variables λ and T, and three independent variables u, v, w. Two of these equations are of the first order and the other three of the second order. These equations are such that one may readily establish the existence of sets of integrals. When a set of values of λ and T satisfying these equations is known, the functions D, D', D'' follow directly and the corresponding system (Q) is defined intrinsically.

When one has a system (Q), the transformation (1) applied to each surface of the system leads to a family of surfaces $\bar{S}$, each of which possesses the property that the deformation of a member of the first system is in the direction of the normal to the corresponding member of the second system at the corresponding point. We call the second the *conjugate system* to the original system (Q) and we show that

The system conjugate to a given system (Q) *is itself a system* (Q) *whose conjugate system is the given system.*

In order to establish this result it is necessary to give an analytical proof of the fact that the transformation (1) is reciprocal*.

Of particular interest is the case in which all of the surfaces $w = \text{const.}$ of a system (Q) are ruled. For this case we have

$$D = 0, \quad D' = -\frac{2\sqrt{pq}}{\sqrt{H}}, \quad D'' = 2\sqrt{H}\,\phi(v, w),$$

where ϕ is to be determined. If we put

$$e^T = \frac{\sigma}{Hm},$$

the set of equations determining the system reduces to

$$\frac{\partial \phi}{\partial w} + \frac{2}{\kappa}\sigma\frac{\partial \phi}{\partial v} + \sigma F(v, \lambda)\,\phi = 0, \quad \frac{1}{\sigma}\frac{\partial \sigma}{\partial v} = \frac{1}{V}\frac{\partial V}{\partial v}, \quad \frac{\partial \lambda}{\partial v} = \frac{V\phi}{\kappa} \quad \text{......(13),}$$

where F is a determinate function. The surfaces $w = \text{const.}$ of the conjugate system are likewise ruled. We have in this case a very interesting type of rectilinear congruence, consisting of a family of applicable ruled surfaces. When the locus of the lines of striction of these surfaces is taken as the surface of reference of the congruence, the Kummer functions assume a very simple form.

When the quadric Q is a hyperboloid of revolution, the lines of striction on the

* Bianchi has given a geometrical proof of this fact for the case when S is a ruled deform of Q; B. p. 36.

applicable ruled surfaces are curves of Bertrand. For a system (Q) of these ruled surfaces the intrinsic equations of these curves of Bertrand are

$$\frac{1}{\rho}=\frac{1}{2a}\left[c\,.\,\phi \sec^4\left(\frac{s}{2a}\right)+2\right],\quad \frac{1}{\tau}=\frac{1}{2}\phi \sec^4\left(\frac{s}{2a}\right) \quad \ldots\ldots\ldots\ldots(14),$$

where $\phi(s, w)$ satisfies an equation similar to (13). The corresponding lines of striction of the conjugate system are likewise curves of Bertrand. Regarding these systems we have the theorem:

As ϕ in equations (14) *varies with w, a given Bertrand curve of the family is deformed into another curve of the family, the deformation at each point being in the direction of the principal normal to the corresponding Bertrand curve of the conjugate system.*

When in particular the quadric Q is the imaginary sphere $x^2+y^2+z^2+1=0$, the equations define a system of isogonal deforms of pseudospherical surfaces of the type studied by Bianchi*.

If in (2) and (3) we replace κ by κ_1, the corresponding equation (1) defines a transformation of S into a surface S_1. When this transformation is applied to the surfaces of the original system (Q), one gets a suite of surfaces S_1 which is not in general a system (Q). When, however, these surfaces form such a system, we have thus a transformation of a system (Q) into another of the same kind. The system conjugate to a given system arises from a particular transformation of this sort. Regarding general transformations of systems (Q), we announce the theorem:

In order that the transforms of a system (Q) *form a system* (Q), *it is necessary and sufficient that λ_1 satisfy in addition to equations analogous to* (3) *also an equation of the form*

$$\frac{\partial \lambda_1}{\partial w}=\Phi,$$

where Φ is a function whose form varies with the character of the quadric Q.

We have established this theorem, from another point of view, for the paraboloids, and owing to the analogy between all the other results we feel sure that it is true also for the central quadrics†. However, we have been unable to complete the proof in the general case by the methods of this paper.

* Sopra una classe di deformazioni continue delle superficie pseudosferiche, *Annali di Matematica*, Tome XVIII. Ser. III. pp. 1—67.

† In fact, Bianchi (l.c.) has established this theorem for systems (Q) of pseudospherical surfaces.

RECENTI PROGRESSI NELLA GEOMETRIA PROIETTIVA DIFFERENZIALE DEGLI IPERSPAZI

Di Enrico Bompiani.

Dopo le classiche ricerche di Monge e della sua scuola, ove accanto a proprietà metriche differenziali di una superficie dello spazio ordinario vengono anche studiate proprietà di carattere proiettivo, passa un lungo periodo di tempo, dovuto forse al rapido ed esuberante sviluppo della geometria metrica nella Francia stessa, prima che l' indagine geometrica volga di nuovo i suoi sforzi a quel ramo di geometria che s' indica col nome piuttosto moderno di *proiettivo-differenziale.* Si raccolgono in esso quei risultati di geometria differenziale che appartengono (secondo un concetto del Klein) al gruppo proiettivo (invarianti cioè per una trasformazione di questo gruppo).

Scarsa eredità ha lasciato a noi il secolo passato in questo campo; la teoria delle tangenti coniugate di Dupin per le superficie dello spazio ordinario ne costituisce certo la parte più preziosa.

La proiettività di Chasles fra punti e piani tangenti lungo una generatrice di una rigata appartiene pure a questo gruppo. Alla ricerca delle proiettività di Chasles degeneri si riduce il problema della distribuzione in sviluppabili delle rette di una congruenza, e a questo equivale a sua volta la determinazione di un doppio sistema coniugato di linee sopra una superficie.

Più recenti (1882) sono alcuni risultati del Koenigs sul modo di comportarsi di una congruenza o di un complesso lineare rispetto ad una congruenza o ad un complesso qualsiasi nell' intorno di una generatrice.

Questi teoremi, dati da Koenigs come analoghi al teorema di Meusnier (di natura metrica, non proiettiva) non sembrano facilmente ricollegarsi agli altri ricordati, ed occorre infatti salire agli iperspazi perchè vengano in piena luce le diverse analogie*.

In Italia si veniva intanto sviluppando la geometria proiettiva degli iperspazi; nel 1886 il Prof. Del Pezzo ottenne per via sintetica i primi resultati relativi a spazi *pluritangenti* a varietà immerse in un ambiente qualsiasi.

Poco dopo, nel 1888, il Prof. Segre estese ai sistemi ∞^{n-1} di rette di S_n le proprietà focali di una congruenza di S_3.

* Tralascio di ricordare, perchè non interviene nel seguito, una ricerca del Darboux (1880) "sur le contact des courbes et des surfaces."

Un ventennio trascorse inoperoso sopra questi risultati, fatta eccezione per le ricerche del Wilczynski (relative allo spazio ordinario) che avrò poi occasione di ricordare.

In una serie di lavori pubblicati dal 1906 al 1910 il Prof. Segre ha ripreso e condotto ad un elevato grado di perfezione questo ramo di geometria iperspaziale*. Il contributo che ad esso recano, sotto l' impulso del Segre, giovani geometri in Italia e fuori può perciò caratterizzarsi veramente italiano.

Io mi propongo di dare uno sguardo all' opera compiuta, cercando di porre in luce le idee fondamentali, e di completarne in qualche punto lo svolgimento.

* * *

Due tipi di problemi si pongono in quest' ordine di ricerche. Un primo *problema*, che può dirsi *ristretto*, ha per oggetto lo studio della costruzione dell' intorno di un punto di una varietà in un ambiente qualsiasi. Il secondo, *problema esteso*, ricerca caratteri che interessano la costruzione di tutta la varietà (per es. l' esser costituita da spazi lineari).

Vediamo come si pongano e si risolvano le quistioni del 1° tipo. Partiamo dallo spazio ordinario. All' intorno del 1° ordine di un punto di una superficie si sostituisce il piano tangente: le proprietà proiettivo-differenziali del 2° ordine risultano dal considerare le intersezioni di esso con i piani infinitamente vicini.

Per porre la questione analoga nel modo più generale ci conviene introdurre una nuova locuzione. Consideriamo sopra una varietà V_m le curve passanti per un suo punto ed aventi ivi uno stesso S_ν osculatore†; gli $S_h\ (h>\nu)$ osculatori a queste curve nel punto comune, se non riempiono l' ambiente, appartengono ad un determinato spazio che dirò *h-osculatore* alla varietà secondo lo S_ν fissato, o di indici h, ν. Se lo S_ν è il punto stesso ($\nu=0$) lo spazio definito (ove esista) si dirà semplicemente *h-osculatore* o, se si vuole, per avere una locuzione più simile all' ordinaria e a quella adottata dal Del Pezzo, *h-tangente* alla V_m nel punto‡.

Così definiti questi spazi rimane a studiare la legge di distribuzione degli spazi osculatori in essi contenuti. Fissato uno spazio S_{h-1} osculatore ($\nu=h-1$) gli S_h osculatori ivi a curve della varietà descrivono intorno ad esso un sistema lineare che riempie lo spazio h-osculatore alla V_m secondo lo S_{h-1}.

Ma le cose si presentano diversamente quando lo S_ν osculatore fissato abbia dimensione minore di $h-1$. Alcuni dei teoremi relativi sono stati studiati dal Prof. Segre e la ricerca può essere agevolmente continuata servendosi degli stessi metodi da lui usati.

* Di questi lavori citerò, come più importanti, i due seguenti: "Su una classe di superficie degli iperspazi legate colle equazioni lineari alle derivate parziali di 2° ordine," *Atti Acc. Scienze di Torino*, vol. XLII, 1906–07.

"Preliminari di una teoria delle varietà luoghi di spazi," *Rendic. Cir. Matem. Palermo*, t. XXX, 1910.

Gli altri si trovano negli *Atti della R. Acc. delle Scienze di Torino*, nel periodo di tempo ricordato. Tralascio nel seguito le indicazioni particolari relative alle memorie del Prof. Segre.

† Intendo con ciò di fissare anche gli spazi osculatori di dimensione inferiore, quando essi non vengano determinati dallo S_ν; cfr. "Sopra alcune estensioni dei teoremi di Meusnier e di Eulero" (in corso di stampa negli *Atti dell' Accad. d. Scienze di Torino*, 1913).

‡ Cf. P. Del Pezzo "Sugli spazi tangenti ad una superficie o ad una varietà immersa in uno spazio di più dimensioni," *Rend. Acc. di scienze Fis. Mat. di Napoli*, fasc. 8°, 1886.

Ad assicurarne l' interesse basterà ricordare che, seguendo le idee di Pluecker e di Klein, in questa geometria può ritenersi inclusa la geometria proiettiva di un qualsiasi elemento generatore dello spazio, e la stessa geometria metrica di un ambiente qualsiasi. E basta infatti una semplice interpretazione di linguaggio per ritrovare i teoremi ricordati di Koenigs sui sistemi di rette in S_3*, alcuni recenti teoremi di C. L. E. Moore† sui sistemi di rette in S_4 o di cerchi in S_3, e infine un' estesissima generalizzazione dei teoremi di Meusnier e di Eulero per una qualsiasi varietà immersa in un ambiente di dimensione arbitraria.

La nozione di coniugio potrà estendersi in diverso modo (in relazione alla dimensione della varietà e dell' ambiente) imponendo speciali condizioni d' incidenza a spazi osculatori a curve uscenti dai punti successivi di una curva assegnata. Le direzioni che danno luogo all' incidenza voluta possono dirsi coniugate alle tangenti alla curva data negli stessi punti, ma la corrispondenza non ha in generale carattere involutorio.

Volendo generalizzare la nozione di asintotica, si cercherà se esistono sulla varietà curve, o, in generale, varietà di dimensione inferiore, tali che alcuni loro spazi osculatori siano legati da speciali condizioni d' appartenenza con spazi di dimensione conveniente, osculatori alla varietà data. La determinazione di queste curve sopra una varietà costituisce un problema del 2° tipo.

Per portare un esempio, sopra una superficie generale di S_4 si può definire un sistema ∞^2 di curve tali che lo S_3 osculatore ad una di esse in un punto riesca ivi tangente alla superficie (mentre non esistono in generale sulla superficie asintotiche nel senso ordinario). Interpretando questo risultato nello spazio ordinario si ottengono teoremi noti sulle congruenze di rette e di sfere‡.

A queste proprietà di carattere generale ne andranno sostituite altre per quelle varietà tali che gli intorni di tutti i loro punti siano costruiti in modo particolare. Può darsi infatti che alcuni degli spazi h-osculatori sopra definiti siano di dimensione minore di quella prevista dal caso generale.

Questo fatto si esprime analiticamente ponendo dei legami lineari omogenei fra le coordinate proiettive dei punti che definiscono quegli spazi h-osculatori; cioè scrivendo che le coordinate proiettive omogenee di un punto della varietà e le loro derivate fino a quelle di un certo ordine sono fra loro legate linearmente; o, più in breve, che le coordinate di un punto della varietà sono soluzioni di un sistema di equazioni alle derivate parziali, lineari ed omogenee.

Alle superficie che rappresentano un' equazione di Laplace ha dedicato un' importante memoria il Prof. Segre.

Dall' esistenza di un doppio sistema coniugato di linee sulla superficie si deduce, come fa il Darboux per lo spazio ordinario, la trasformazione di Laplace per l' equazione ricordata. Servendosi di questa imagine geometrica riesce facile

* "Sur les propriétés infinitésimales de l'espace reglé," *Annales de l'École normale*, 1882 (2), t. 11.

† "Infinitesimal properties of lines in S_4, with applications to circles in S_3," *Proceedings of the American Academy of Arts and Sciences*, vol. XLVI, n° 5, 1911.

‡ Cf. la mia nota "Sopra una trasformazione classica di Sophus Lie," *Atti R. Accad. dei Lincei*, vol. XXI, serie 5, 1912, f. 11.

assegnare dei criteri perchè l' equazione data sia integrabile col metodo di Laplace*.

Come già accennava il Prof. Segre nel suo lavoro, le ricerche ivi contenute sono suscettibili di diverse estensioni. In luogo di una sola equazione del 2° ordine a 2 variabili si può per es. considerare un sistema di più equazioni del 2° ordine a k variabili indipendenti; il Terracini, in una memoria sulle V_k che rappresentano più di $\frac{k(k-1)}{2}$ equazioni di Laplace linearmente indipendenti†, assegna loro una proprietà (non caratteristica però) che le riavvicina alle sviluppabili dello spazio ordinario. Il caso $k=3$ era già stato ampiamente trattato in una memoria del Sig. C. H. Sisam‡.

Un' altra estensione delle ricerche del Segre si ha invece considerando una o più equazioni di ordine superiore al secondo. In particolare, l' equazione a due variabili, di qualsiasi ordine, a *caratteristica completa,* ha un' imagine geometrica del tutto simile a quella dell' equazione di Laplace: si può costruire per essa una successione di trasformate e assegnare condizioni sufficienti affinchè la sua integrazione sia ricondotta a quella di equazioni a derivate ordinarie.

* * *

Come si è visto il principio del metodo per lo studio di proprietà proiettivo-differenziali consiste nel sostituire all' intorno di un punto uno spazio lineare di dimensione conveniente. Si presenta quindi come particolarmente interessante, sotto un doppio punto di vista, lo studio delle varietà luoghi di spazî: sia per la costruzione della teoria generale, sia per vedere come si semplifichino per esse le proprietà osservate per una varietà qualsiasi.

I fondamenti di una teoria delle varietà luoghi di spazî sono stati posti dal Prof. Segre nella Memoria citata del 1910. In essa sono trattate le questioni essenziali relative agli spazi tangenti, agli spazî caratteristici, alla nozione di sviluppabilità e alle sue diverse estensioni nonchè molte delle questioni generali prima accennate per una varietà qualsiasi.

I metodi svolti in questa memoria sono serviti di modello nelle ricerche successive che già abbiamo avuto occasione di ricordare.

Un indirizzo completamente differente per la geometria proiettivo-differenziale nello spazio ordinario veniva sviluppato dal Wilczynski in una serie di Memorie la cui pubblicazione incomincia nel 1901: ne formano oggetto lo studio delle curve piane e sghembe, delle rigate, delle superficie curve, delle congruenze di rette e dei sistemi di curve piane.

Il Wilczynski, mettendosi dal punto di vista già adottato dall' Halphen nella sua Tesi e nei lavori successivi (1875—1881) per la rappresentazione di curve piane e sghembe, si è servito di un sistema di equazioni differenziali lineari opportunamente scelte per rappresentare gli enti ricordati. Dalle proprietà del sistema scelto risulta che il corrispondente ente geometrico è definito a meno di una trasformazione

* Cf. la mia memoria sull' equazione di Laplace (*Rendiconti del Circ. Mat. di Palermo*, t. XXXIV, fasc. III, 1912).

† *Circolo Matematico di Palermo*, t. XXXIII, 1912.

‡ "On Three Spreads satisfying four or more homogeneous linear partial differential equations of the second order," *American Journal of Mathematics*, vol. XXXII, n° 2, 1911.

proiettiva: gli invarianti del sistema sono dunque atti a definire le proprietà proiettive dell' ente in esame. Ed il Wilczynski è riuscito infatti a dare forma invariantiva alla teoria delle rigate, delle superficie, e delle congruenze di rette, arricchendola in molti punti di nuovi risultati*.

Nulla di più naturale che tentarne l' estensione agli iperspazi. E l' estensione è certo ricca d' interesse, perchè se da un lato la rappresentazione analitica fornisce un nuovo mezzo per lo studio degli enti geometrici, dall' altro il modello geometrico sussidia dell' intuizione ad esso propria la ricerca analitica. Ne valga ad esempio la memoria "Sugli invarianti differenziali proiettivi delle curve in un iperspazio" † in cui il Prof. Berzolari ha esteso i resultati dell' Halphen.

Consideriamo le superficie rigate di uno spazio qualsiasi S_n.

Servendosi della nozione generale di spazio h-osculatore ad una varietà in un punto, si riesce a stabilire per queste superficie una teoria delle curve tracciate su di esse e definite da proprietà proiettive, del tutto analoghe alle asintotiche sulle rigate di S_3. Insieme a questo sistema di curve, che potremmo chiamare *quasi-asintotiche*, vien definito un altro sistema di curve *associate* ad esse, che in generale non appartengono alla rigata: se vi appartengono coincidono con le curve prima definite. Questo accade per es. nello spazio ordinario. Se la rigata è algebrica rientrano in questa classe generale di curve quasi-asintotiche le direttrici minime già studiate dal Prof. Segre‡.

Dall' esistenza di queste curve si può dedurre una classificazione *proiettiva* delle rigate in un iperspazio; la natura proiettiva dell' intorno di una generatrice generica può caratterizzarsi mediante alcuni numeri interi che dirò *indici di sviluppabilità,* dei quali il più grande rappresenta il massimo numero di generatrici consecutive linearmente indipendenti.

Una rigata di S_n si dirà del tipo generale quando i suoi indici di sviluppabilità hanno i massimi valori possibili.

Una rigata qualsiasi si può rappresentare per mezzo di un sistema di 2 equazioni differenziali lineari ed omogenee e ad esse può riportarsi la nozione degli indici di sviluppabilità.

Queste considerazioni possono estendersi in direzioni diverse. Possono considerarsi sistemi semplicemente infiniti di spazî lineari S_{k-1}; una loro rappresentazione si otterrà servendosi di un sistema di k equazioni differenziali lineari omogenee. Oppure potranno considerarsi sistemi più volte infiniti di rette o di spazi qualsiasi rappresentati da sistemi di equazioni alle derivate parziali. Se, per limitarci a quanto s' è detto sopra, abbiamo in vista un sistema più volte infinito di rette, il problema analogo a quello della ricerca delle sviluppabili in una congruenza di S_3 consisterà nel ricercare entro il sistema le rigate aventi i minimi indici di sviluppabilità

* Per i lavori del Wilczynski anteriori al 1906 (che si riferiscono in modo particolare ai sistemi di rette) si può vedere il suo libro *Projective differential geometry of curves and ruled surfaces* (Teubner, 1906). Gli altri si trovano in *Transactions of the American Mathematical Society* nel periodo indicato.

† *Annali di Matematica*, s. II, t. XXVI (1897).

‡ "Sulle rigate razionali in uno spazio lineare qualunque" (*Atti dell' Accad. d. Scienze di Torino*, vol. XIX, 1884), ed. altre note sullo stesso argomento (ibid., 1885 e 1886).

possibili. Queste sono sviluppabili nel senso ordinario solo, in generale, per i sistemi ∞^{n-1} di S_n.

Non è forse illecito sperare che le considerazioni geometriche di Fano e di Enriques sulle varietà algebriche che ammettono infinite trasformazioni proiettive*, alle quali si debbono progressi notevoli nella teoria della equazione differenziale lineare le cui soluzioni sono legate da relazioni algebriche, porteranno anche aiuto nel campo più vasto al quale ho cercato ora d' accennare†.

* Cf. per es. l' esposizione del Fano nei *Math. Ann.* Bd 53 (1899), p. 493.

† Nell' intervallo di tempo trascorso fra la lettura di questa comunicazione e la stampa degli Atti (Febbraio 1913) sono apparsi i seguenti lavori sull' argomento che c' interessa:

A. Ranum: "*On the Projective Differential Geometry of N-dimensional Spreads generated by ∞^1 Flats*" (Ann. di Matem. pura ed applicata, s. III, t. XIX, p. 205).

E. Cairo: "*Sopra un sistema Σ di superficie P di S_n*" (Periodico di Matematica, anno XXVII, fasc. VI; anno XXVIII, fasc. III).

E. Artom: "*Ricerche proiettive sulle linee tracciate in una superficie immersa in uno spazio a più dimensioni*" (Periodico di Matematica, anno XXVIII, fasc. II).

THE GENERAL THEORY OF MOVING AXES

By Eric H. Neville.

For more than fifty years mathematicians both pure and applied have recognized in a moving frame of mutually perpendicular axes one of the most effective weapons in their arsenal. Geometers use it, Darboux's great work being the most astounding example of its power; in attacking particular problems in rigid dynamics it is indispensable; and if the applied mathematician when he concerns himself with curvilinear coordinates all but confines his attention to such systems as are triply orthogonal, that is chiefly, I think, because the axis-frame which corresponds to such a system is of the kind to which he is accustomed. It has been my good fortune to succeed in obtaining a general theory of moving axes, where the frame is subject to no conditions whatever of orthogonality or rigidity, to mount, as it were, the old weapon in such a way that it can be trained immediately on many objects hitherto not within its range, and brought to bear directly on many points which could be attacked in the past only after a more or less tedious detour.

In the time now available it might be possible to give the steps of the general theory, but I could then mention not even one of the numerous applications. But if I confine myself to a mere statement of results, I may succeed in shewing you something of the power of the method and of the variety of uses to which I believe we shall some day see it put.

In dealing with oblique axes, it is of the greatest importance always to be ready to use both the direction-cosines and the direction-ratios of any line, and in the case of any vector to deal not only with its three components but also with its three projections, the projection of a vector along an axis being of course the numerical magnitude of the vector multiplied by the cosine of the angle between the vector and the axis. We shall use throughout this paper λ, μ, ν for the angles between the axes of reference, A, B, C for the angles between the planes of reference, and Υ for the sine of the solid angle of the frame, the magnitude, that is, which may be defined indifferently as $\sin\mu \sin\nu \sin A$, as $\sin\nu \sin\lambda \sin B$, or as $\sin\lambda \sin\mu \sin C$; also we shall denote the components of a typical vector by x, y, z, and its projections by p, q, r, and when the vector changes continuously and there exists another vector which is the rate of change of the first, we shall denote the components of the derived vector by Dx/Dt, Dy/Dt, Dz/Dt, and its projections by Dp/Dt, Dq/Dt, Dr/Dt.

The simplest change possible to an oblique frame of reference occurs when two of the axes remain fixed in direction and the third rotates round one of these two, tracing out part of a right circular cone with that fixed line as axis. When the angle of rotation is infinitesimal we may call such a motion an elementary spin, and we realise at once that there are six such spins possible to the frame, that the frame may be brought from its initial position to any adjacent position by one and only one set of six elementary spins, and that the measures of these spins will not depend, when first order magnitudes only are considered, on the order in which the spins take place. The first spin, that of the y-axis round the x-axis, is measured by the angle through which the xy-plane has revolved in the positive direction (that is, for this spin, towards the xz-plane), and is denoted by θ_2, the other five spins similarly measured being denoted by θ_3, ϕ_3, ϕ_1, ψ_1, ψ_2; and it is a matter of no great difficulty to ascertain the effect of an elementary spin on the angles of the frame and on the components and projections of a fixed vector. When the frame changes continuously we have, instead of the six spins described above, six rates of spin which we denote by θ_2', θ_3', ϕ_3', ϕ_1', ψ_1', ψ_2'; and from the results relating to elementary spins—results which are of course only approximate—we can write down at once the following accurate and fundamental expressions,

$$\frac{d(\cos\lambda)}{dt} = \Upsilon(\theta_2' - \theta_3'),$$

$$\frac{Dx}{Dt} = \frac{dx}{dt} - (y\psi_2' - z\phi_3')\frac{\sin^2\lambda}{\Upsilon} + (z\theta_3' - x\psi_1')\cot C + (x\phi_1' - y\theta_2')\cot B,$$

$$\frac{Dp}{Dt} = \frac{dp}{dt} - \Upsilon(y\psi_1' - z\phi_1'),$$

each typical of a set of three.

It is the last of these which shews how vital it is to make use of the projections as well as of the components of our vectors, and which also leads us to expect that the increase of difficulty of manipulation as we pass from orthogonal frames to oblique ones will prove not in the slightest degree comparable with the increase of freedom; but even in the expressions for the components of the rate of change we find no unfamiliar coefficients, but merely those which occur in the equations connecting components with projections, for we have always three equations typified by

$$\Upsilon x = p\frac{\sin^2\lambda}{\Upsilon} - q\cot C - r\cot B.$$

The general theory requires for its completion the study of the features which arise when the motion of the frame depends on more variables than one. The investigation, following the lines of work by Kirchhoff and Combescure on moving frames of the simplest kind, presents no difficulty, but, anxious to say something of the applications of the abstract theory, I cannot spare time even to state so much of the notation as would enable me to write down typical results in this part of the subject.

The most evident, and it may be the most important, application of the theory now outlined is to the theory of curvilinear coordinates in space. Taking for coordinates u, v, w, let us write the square of the line-element in the form

$$ds^2 = (L, M, N, P, Q, R \between du, dv, dw)^2,$$

and let us also put

$$L = U^2, \quad M = V^2, \quad N = W^2,$$

to prevent the appearance of radicals in what follows, and again

$$L = S^{11}, \quad P = S^{23} = S^{32},$$

and so on, the double-affix notation enabling us often to include a number of investigations in one piece of work, and enabling us also to write for brevity the expression for the square of the line-element in the form

$$ds^2 = (S \between du, dv, dw)^2.$$

Then at each point of space there is associated with the coordinate system a reference frame, one axis being the tangent to the curve along which v and w are constant and having its positive direction so chosen that for a small movement in the positive direction $ds = U du$, and the other axes being similarly determined. If we suppose any curve drawn in space, and the origin of a moving frame to describe this curve while at every point the axes are those of the reference frame at that point, then the six rates of spin of the moving frame are linear functions of u', v', w', the rates of change of the coordinates. It transpires that all the coefficients in the expressions for these rates of spin can be found in terms of L, P, and so on, and their first derivatives, by means of equations which occur in the work on frames whose motion depends on several variables, and we have only to substitute in the expressions for the components and the projections of the rate of change of a given vector the values so obtained for the rates of spin, to obtain the values of such magnitudes as $\frac{Dx}{Dt} - \frac{dx}{dt}$ and $\frac{Dp}{Dt} - \frac{dp}{dt}$. The expression so found for $\frac{Dx}{Dt} - \frac{dx}{dt}$ is necessarily lineo-linear in the two sets of variables x, y, z and u', v', w', and we might expect to have to deal with its nine coefficients, and so in all with twenty-seven combinations of the first derivatives of the coefficients such as L and P; but inspection reveals that though there is no symmetry in the expression for $\frac{Dx}{Dt} - \frac{dx}{dt}$, the expression for $\frac{1}{U}\frac{Dx}{Dt} - \frac{d(x/U)}{dt}$ is symmetrical in the two sets of variables x/U, y/V, z/W and u', v', w', and contains therefore only six coefficients. We write A^{11} for the coefficient of $\frac{x}{U}u'$, and A^{23} or A^{32}, since it proves convenient to retain both forms, for the coefficient of $\frac{y}{V}w' + \frac{z}{W}v'$, and so on, and we may write compactly

$$\frac{1}{U}\frac{Dx}{Dt} = \frac{d(x/U)}{dt} + \left(A \between \frac{x}{U}, \frac{y}{V}, \frac{z}{W} \between u', v', w'\right).$$

Similarly six magnitudes B^{rs} are introduced as coefficients in the expression for $\frac{1}{V}\frac{Dy}{Dt}$, and six magnitudes C^{rs} as coefficients in the expression for $\frac{1}{W}\frac{Dz}{Dt}$, and so we have in all eighteen functions of the eighteen first derivatives of the original coefficients S^{rs}. These eighteen functions are in fact independent linear functions of the derivatives, and the expression of the derivatives in terms of them proves simple enough, for if we denote differentiation with respect to the variables u, v, w by suffixes 1, 2, 3, we find the single formula

$$S_t^{rs} = S^{r1}A^{st} + S^{r2}B^{st} + S^{r3}C^{st} + S^{s1}A^{rt} + S^{s2}B^{rt} + S^{s3}C^{rt},$$

for $r, s, t = 1, 2, 3$, to cover every case.

Already we have in outline all that is essential to enable us to write out in terms of curvilinear coordinates of any kind the majority of the equations of applied mathematics, and to deal with many problems relating to curves and surfaces; but a word must be said of the expression of rates of rates of change, not only because rates of rates of change are often needed, but also because the analysis itself is related to results of the greatest importance.

When we come to discuss rates of rates of change, we find three equations of the form

$$\frac{1}{U}\frac{D^2x}{Dt^2}=\frac{d^2(x/U)}{dt^2}+2\left(A\between\frac{d(x/U)}{dt},\ \frac{d(y/V)}{dt},\ \frac{d(z/W)}{dt}\between u',\ v',\ w'\right)$$
$$+\left(A\between\frac{x}{U},\ \frac{y}{V},\ \frac{z}{W}\between u'',\ v'',\ w''\right)+\left(X\between\frac{x}{U},\ \frac{y}{V},\ \frac{z}{W}\between u',\ v',\ w'\right)^2,$$

each containing only sixteen coefficients although we might have feared to find forty-five. Of the sixteen coefficients six are already familiar, and thus the three equations together introduce us to thirty functions of the second-order derivatives of L, P, and so on, functions which we denote by X^{rst}, Y^{rst}, Z^{rst}, r, s, t having the values 1, 2, 3, and the affixes being permutable. In terms of these thirty functions all thirty-six of the second-order derivatives are expressible, and so we come across the six relations which Cayley discovered must exist between these derivatives if the quadratic form $(S\between du,\ dv,\ dw)^2$ is to represent the square of the line-element in Euclidean space. These relations come to us however in a form far more compact than that which Cayley gives, for the two typical equations as we find them are

$$M_{33}-2P_{23}+N_{22}=2\,(S\between A^{23},\ B^{23},\ C^{23})^2-2\,(S\between A^{22},\ B^{22},\ C^{22}\between A^{33},\ B^{33},\ C^{33})$$

and

$$-L_{23}-P_{11}+Q_{12}+R_{13}=2(S\between A^{11},\ B^{11},\ C^{11}\between A^{23},\ B^{23},\ C^{23})-2(S\between A^{12},\ B^{12},\ C^{12}\between A^{13},\ B^{13},\ C^{13}).$$

In order to illustrate both the facility with which curvilinear coordinates can be manipulated and the value of the kinematical point of view in differential geometry, let me mention a result which has not, as far as I know, hitherto been given even in terms of Cartesian coordinates. Associated with a family of surfaces in a region of space there is at every point a frame which we may call the fundamental frame, formed by the tangents to the two lines of curvature of the member of the family passing through that point and the normal to this surface. This frame is completely rectangular and consequently its rotation as we pass along an orthogonal trajectory of the family requires only three rates of spin, i, j, and k, for its description. Of these, i and j, the angular velocities about the tangents to the lines of curvature, though simple enough in form, shall not detain us now; but to k, the angular velocity of the fundamental frame about the normal as we pass from surface to surface along the orthogonal trajectory, I wish to draw your attention. For k is a magnitude of the greatest importance in the study of the family, being that which will vanish when and only when the family is a Lamé family. Hence having found the analytical expression for k, namely

$$\frac{G^3}{2J^3I^2}\begin{vmatrix} G_{11}-G_1A^{11}-G_2B^{11}-G_3C^{11}, & \Phi_{11}-\Phi_1A^{11}-\Phi_2B^{11}-\Phi_3C^{11}, & 2\Phi_1, & 0, & 0, & L \\ G_{22}-G_1A^{22}-G_2B^{22}-G_3C^{22}, & \Phi_{22}-\Phi_1A^{22}-\Phi_2B^{22}-\Phi_3C^{22}, & 0, & 2\Phi_2, & 0, & M \\ G_{33}-G_1A^{33}-G_2B^{33}-G_3C^{33}, & \Phi_{33}-\Phi_1A^{33}-\Phi_2B^{33}-\Phi_3C^{33}, & 0, & 0, & 2\Phi_3, & N \\ G_{23}-G_1A^{23}-G_2B^{23}-G_3C^{23}, & \Phi_{23}-\Phi_1A^{23}-\Phi_2B^{23}-\Phi_3C^{23}, & 0, & \Phi_3, & \Phi_2, & P \\ G_{31}-G_1A^{31}-G_2B^{31}-G_3C^{31}, & \Phi_{31}-\Phi_1A^{31}-\Phi_2B^{31}-\Phi_3C^{31}, & \Phi_3, & 0, & \Phi_1, & Q \\ G_{12}-G_1A^{12}-G_2B^{12}-G_3C^{12}, & \Phi_{12}-\Phi_1A^{12}-\Phi_2B^{12}-\Phi_3C^{12}, & \Phi_2, & \Phi_1, & 0, & R \end{vmatrix},$$

where $\Phi(u, v, w) =$ constant is the equation of the family, I is the difference between the principal curvatures of the surface, suffixes denote differentiation, $J\,du\,dv\,dw$ is the element of volume bounded by six reference surfaces, and $1/G$ is $d\Phi/dn$, the rate of change of the defining function of the family as we pass along the orthogonal trajectory, so that analytically

$$J^2 = \begin{vmatrix} L, & R, & Q \\ R, & M, & P \\ Q, & P, & N \end{vmatrix}, \quad \frac{J^2}{G^2} = -\begin{vmatrix} L, & R, & Q, & \Phi_1 \\ R, & M, & P, & \Phi_2 \\ Q, & P, & N, & \Phi_3 \\ \Phi_1, & \Phi_2, & \Phi_3, & 0 \end{vmatrix},$$

while the magnitudes A^{rs}, B^{rs}, C^{rs} are those we have met earlier in this paper, we can write down Darboux's famous equation, not merely expressed in terms of coordinates the most general, but also with its meaning, its necessity, and its sufficiency, made evident.

Scant as is the best justice I can do to my subject here, I can at least mention one other and most valuable development, that of its application to the theory of a single surface. In dealing with a single surface, striated by two families of curves of reference, the frame which we have to use has two of its axes tangents to curves of reference, and the third, the normal to the surface, is at right angles to the other two. The theory of such a frame is of course only a special case of the general theory, but these frames, which may be called birectals, are of so frequent occurrence that the results relating to them are of particular interest; and indeed it was the desire to have these frames available for use in differential geometry that prompted the whole research of which I am speaking, the value of the quite general theory becoming evident only as the work proceeded. In a moving birectal the constancy of the right angles involves that θ_2 and θ_3 are equal, and also that ϕ_3 and ϕ_1 are equal, and we therefore drop the suffixes from these symbols and speak of only four spins, θ, ϕ, ψ_1, and ψ_2.

The simplicity of the whole geometrical theory will be best appreciated from the mere statement that for a curve on the surface, dividing the angle between the reference curves into angles α, β, if the rates of spin of the birectal already described, as the origin moves along the curve, are θ', ϕ', ψ_1', ψ_2', then the normal curvature κ_n, the geodesic torsion ς_g, and the geodesic curvature κ_g, are* given by the expressions

$$\kappa_n = \theta' \sin\alpha - \phi' \sin\beta,$$
$$\varsigma_g = \theta' \cos\alpha + \phi' \cos\beta,$$
$$\kappa_g = \alpha' + \psi_1' = -\beta' + \psi_2',$$

expressions which simplify not at all when the curves of reference happen to be orthogonal.

But this is not all. The analysis connected with coordinates on a surface can be deduced from the theory of the birectal just as the analysis connected with

* In Darboux's notation these three magnitudes are respectively

$$\frac{\cos\varpi}{\rho}, \quad \frac{1}{\tau} - \frac{d\varpi}{ds}, \quad \frac{\sin\varpi}{\rho}.$$

coordinates in space is obtained from the general theory, and if we write the square of the line-element in the two forms

$$ds^2 = E du^2 + 2F du dv + G dv^2$$
$$= U^2 du^2 + 2UV \cos\omega\, du dv + V^2 dv^2,$$

we are led to introduce six combinations of the first derivatives of E, F, G, and also to retain three other magnitudes describing the motion of the frame. These nine magnitudes are identical with certain of those which Forsyth has found it necessary to use in his purely algebraic work, and, adopting a notation differing as little as possible from his, we denote the last three by L, M, N, and of the first six, three by Γ^{rs} and three by Δ^{rs}, r, s having the values 1, 2 and being permutable.

Then the fundamental formulæ are

$$\frac{1}{U}\frac{Dx}{Dt} = \frac{d(x/U)}{dt} + \left(\Gamma \middle)\!\!\left(\frac{x}{U},\ \frac{y}{V} \middle)\!\!\left(u',\ v'\right) - z\left(L,\ M,\ N\middle)\!\!\left(\frac{G}{EG-F^2},\ \frac{-F}{EG-F^2}\middle)\!\!\left(u',\ v'\right),$$

$$\frac{1}{V}\frac{Dy}{Dt} = \frac{d(y/V)}{dt} + \left(\Delta \middle)\!\!\left(\frac{x}{U},\ \frac{y}{V} \middle)\!\!\left(u',\ v'\right) - z\left(L,\ M,\ N\middle)\!\!\left(\frac{-F}{EG-F^2},\ \frac{E}{EG-F^2}\middle)\!\!\left(u',\ v'\right),$$

$$\frac{Dz}{Dt} = \frac{dz}{dt} + \left(L,\ M,\ N\middle)\!\!\left(\frac{x}{U},\ \frac{y}{V}\middle)\!\!\left(u',\ v'\right),$$

from the last of which it is soon seen that the magnitudes L, M, N are the coefficients in the quadratic form representing the normal curvature of the surface, that is, are multiples by $1/UV\sin\omega$ of those magnitudes introduced by Gauss and denoted in Darboux's treatise by D, D', D''.

I cannot hope to give you even a faint indication of the range of application of these last formulæ. Let it suffice to say that from them the equation of Gauss and the equations of Codazzi and Mainardi follow immediately, that they enable the actual analytical expressions to be given for two functions of direction discovered by Laguerre and Darboux respectively, that the determination from them of the invariants connected with a family of curves on the surface is immediate, and that they have led to a geometrical interpretation of the equations of Codazzi and Mainardi which is to the best of my belief new, and which I should like to give here.

Let $\Phi =$ constant represent any family of curves on a surface, let df/ds, df/dg denote the rates of change of any scalar function f along the curve of the family and along the orthogonal trajectory respectively, and let the normal curvature, the geodesic torsion, and the geodesic curvature, of the curve of the family be κ_n, ς_g, κ_g and the mean curvature of the surface itself be H. Then the two functions

$$\frac{d(\kappa_n - \frac{1}{2}H)}{ds} - 2\varsigma_g\kappa_g, \qquad \frac{d\varsigma_g}{ds} + 2(\kappa_n - \tfrac{1}{2}H)\kappa_g$$

are virtually those which Laguerre and Darboux respectively discovered to be functions only of direction on the surface; and the two functions

$$\frac{d(\kappa_n - \frac{1}{2}H)}{dg} + 2\varsigma_g\frac{d\{\log(d\Phi/dg)\}}{ds}, \qquad \frac{d\varsigma_g}{dg} - 2(\kappa_n - \tfrac{1}{2}H)\frac{d\{\log(d\Phi/dg)\}}{ds},$$

in which the logarithmic derivative will be recognized to be the geodesic curvature of the orthogonal trajectory of the family, are analogous functions of direction which make their appearance as soon as we come to deal with a family of curves. As far as I know, these four functions, which we denote for the moment by L_s, D_s, L_g, and D_g, might well be supposed, until the actual expressions for them are forthcoming, to be distinct; but on evaluating them we find them to be connected by the two simple relations

$$L_s + D_g = \frac{1}{2}\frac{dH}{ds}, \qquad D_s - L_g = \frac{1}{2}\frac{dH}{dg},$$

which constitute the required geometrical interpretation.

I should have liked to say something of the relation of the analysis deduced from the theory of moving axes to the process of covariant differentiation, or to write down in terms of curvilinear coordinates some of the most important equations of mathematical physics, but my real desire was to arouse your interest in a subject which I believe to be worthy of some attention, and if in spite of the limits within which I have had to confine myself I have succeeded in that, I am more than satisfied.

UEBER DIE TEILUNG DES RAUMES DURCH 6 EBENEN UND DIE SECHSFLACHE

Von M. Brückner.

In dem auf dem Kongresse in Rom gehaltenen Vortrage hatte ich die Methoden der Ableitung der aussergewöhnlichen Polyeder angegeben und sie durch die Bestimmung der Sechsflache erläutert, wobei ich am Schlusse auf eine spätere ausführliche Darstellung hinwies. Bei der Ausführung des damals skizzierten Programms zeigte sich, dass die Untersuchungen einen vorher nicht vermuteten Umfang annahmen, so dass es mehrjähriger Arbeit bedurfte, das hier vorliegende Manuskript fertigzustellen. Ueberdies nahmen die Ergebnisse höchst merkwürdige, zum Teil von der früheren abweichende Gestalt an. Gestatten Sie mir, m.H., von den Endresultaten Ihnen heute einen kurzen Ueberblick zu geben, wobei ich im Interesse der Klarheit auf die einleitenden Betrachtungen näher eingehen möchte. Zu der gewünschten Ableitung der aussergewöhnlichen Polyeder im Möbiusschen Sinne waren zwei Wege vorgezeichnet. Entweder man benutzt die erweiterten Fundamentalkonstruktionen Eberhards, das Abschneiden von Ecken, Kanten und Kantenpaaren, um die $n+1$-Flache aus den als bekannt vorausgesetzten n-Flachen zu erhalten, oder man setzt die Polyeder direkt aus den sie bildenden Zellen zusammen, in die der Raum durch die betreffende Zahl von Ebenen geteilt wird. Die Anwendung beider Methoden zur Auffindung der Fünfflache ist leicht, aber an sich von geringem Interesse, da einseitige Polyeder und zweiseitige von höherem Zusammenhange noch nicht existieren. Doch bildet ein eingehendes Studium der Raumfigur aus 5 Ebenen die Voraussetzung für die Möglichkeit, die komplizierten Figuren aus 6 Ebenen übersehen zu können. Es teilen bekanntlich n Ebenen in allgemeinster Lage den projektiven Raum in $n+\frac{n(n-1)(n-2)}{1\,.\,2\,.\,3}$ Zellen, von denen $\frac{(n-1)(n-2)(n-3)}{1\,.\,2\,.\,3}$ geschlossen sind, die übrigen, transgredienten aus 2 Teilen bestehen, die im Unendlichen zusammenhängen. Für $n=5$ ergibt sich nur eine einzige Figur mit 4 endlichen und 11 transgredienten Zellen, wie auch die 5 Ebenen im Raume gelegen sein mögen. Die Linearfiguren, die in jeder Ebene durch die übrigen erzeugt werden, sind vollständige Vierseite des einen nur vorhandenen Typus und alle 5 möglichen transgredienten konvexen Fünfflache sind in diesem einen Gebilde vorhanden. Es ist vorteilhaft, ein für allemal das vollständige Fünfflach in einer bestimmten Form in der Vorstellung festzuhalten. Bezeichnet man das endliche Tetraeder mit T, das Pentaeder mit P und die transgredienten Tetraeder und Pentaeder mit T_{i+j} und P_{i+j}, wo die Indices i und j die Zahlen der Ecken auf den beiden im Unendlichen zusammenhängenden Teilen anzeigen, so besteht der durch 5 Ebenen geteilte Raum aus $2T$, $2P$, $2T_{3+1}$, $1T_{2+2}$, $2P_{5+1}$, $2P_{4+2}$, $2P'_{4+2}$, $1P_{3+3}$ und $1P'_{3+3}$, wobei die Strichelung anzeigt, dass die betreffenden Zellen mit den vorhergehenden allomorph sind. Fügt man nun eine sechste Ebene hinzu, so entstehen laut obiger Formel 10 geschlossene und 16 transgrediente Zellen, unter denen sich nun auch die beiden von einander verschiedenen Sechsflache H^1 und H^2 befinden, von denen das erste als Grenzflächen 2 Fünfecke, 2 Vierecke und 2 Dreiecke, das zweite aber 6 Vierecke hat. Da die analytische Behandlung des Problems zeigt, dass von den 26 Zellen des durch 6 Ebenen geteilten Raumes stets 6 Tetraeder, 12 Pentaeder, 6 Hexaeder H^1 und 2 Hexaeder H^2 sein müssen, es aber, wie die Untersuchung zeigt, 20 verschiedene transgrediente konvexe Sechsflache $H^1{}_{i+j}$ und 6 dergleichen $H^2{}_{i+j}$ gibt, so können nicht sämtliche transgrediente Sechsflache in einer vorliegenden Sechsteilung des Raumes zugleich auftreten. Es sind die verschiedenen Teilungen des Raumes durch 6 Ebenen demnach durch den Charakter der transgredienten Zellen ebensowohl wie durch die Art und Anordnung der endlichen Zellen zu unterscheiden. In jeder der 6 Ebenen besteht nun die durch die übrigen fünf erzeugte Linearfigur aus einem Fünfeck, 5 Vierecken und 5 Dreiecken, von denen 6 geschlossene Polygone, die übrigen transgredient sind, und es existieren 6 von einander unterschiedene Teilungen der Ebene durch 5 Gerade, wenn man beachtet, wie sich die transgredienten und geschlossenen Zellen unter die verschiedenen Polygone verteilen. Die Einführung einer sechsten Ebene in das vollständige Fünfflach ergibt demnach in jener eine der 6 Linearfiguren, deren einziges Fünfeck nur als Schnitt eines geschlossenen oder transgredienten Pentaeders P bezw. P_{i+j} entstanden sein kann. Um demnach alle von einander verschiedenen Sechsteilungen des Raumes zu erhalten, sind sämtliche möglichen Fünfeckschnitte der Pentaeder des vollständigen Fünfflaches auszuführen. Dies zeigte sich auf 152 Weisen möglich, aber die erhaltenen Sechsteilungen des Raumes sind zum grossen Teile isomorph und es ist eine eingehende Untersuchung nötig, um die Anzahl der von einander verschiedenen Typen festzustellen. Das Resultat ist: Es existieren 42 verschiedene Teilungen des Raumes durch 6 Ebenen, die nach der Art der endlichen Zellen zunächst in 9 Ordnungen zerfallen, von denen jede mehrere Gruppen enthält, die durch die Beschaffenheit und Verteilung der transgredienten Zellen unterschieden sind, wie die folgende Uebersicht zeigt:

Die endlichen Zellen	Zahl der Gruppen
$3H^1$, $4P$, $3T$	2
$3H^1$, $3P$, $4T$	4
$3H^1$, $2P$, $5T$	3
$2H^1$, $5P$, $3T$	4
$2H^1$, $4P$, $4T$	6
$2H^1$, $3P$, $5T$	2
$1H^1$, $1H^2$, $5P$, $3T$	5
$1H^1$, $1H^2$, $4P$, $4T$	3
$1H^2$, $6P$, $3T$	13

Verschiedene Kontrollen lassen, abgesehen von der Eindeutigkeit der Konstruktionsmethode die Behauptung berechtigt erscheinen, dass keine Figur übersehen ist. Es lassen sich diese 42 Sechsflache noch in verschiedener anderer Weise gruppieren, sei es, dass man ihre Schnittfiguren in der unendlichweiten Ebene untersucht, wonach sie in 4 Klassen zu bringen sind, da 6 Gerade in einer Ebene vier von einander durch die Anzahl der i-eckigen Zellen unterschiedene Teilungen ergeben, sei es, dass man die vollständigen Siebenflache betrachtet, die durch Hinzufügung der unendlichweiten Ebene zu den sechs im Endlichen liegenden entstehen, wonach die Sechsflache in 12 Klassen zerfallen würden.

Der zweite Hauptteil meiner Untersuchungen begreift die Auffindung der endlichen d. h. geschlossenen aussergewöhnlichen Sechsflache. Wie die Betrachtung der von 5 Geraden gebildeten Linearfigur zeigt, können Achtecke als Grenzflächen von Sechsflachen nicht auftreten, wohl aber Sechsecke, bei denen 2 Kanten in einer Schnittgeraden und Siebenecke, bei denen 2 Paare Kanten in 2 Schnittgeraden zu liegen kommen. Demnach sind für die zweiseitigen allgemeinen Sechsflache vom Geschlechte $p=0$ und $p=1$ die Anzahlen für die Flächen f_3, f_4, f_5, f_6, f_7 die Lösungen der Gleichungen:

$$3f_3+2f_4+f_5=12\ldots\ldots\ldots(1) \quad \text{bezw.} \quad 3f_3+2f_4+f_5=f_7\ldots\ldots\ldots(2).$$

Für die Sechsflache vom Geschlecht $p=0$ verweise ich auf die in den Akten des Romkongresses S. 294 gemachten Angaben. Die daselbst für die Ableitung dieser Gebilde erklärten Fundamentalkonstruktionen sind für die Sechsflache vom Geschlecht $p=1$ durch die gleichfalls dort schon erläuterte Methode der Kombination der Zellen des vollständigen Sechsflaches zu ersetzen. Natürlich ist diese zweite Methode die allgemeinere, die *sämtliche* ein- und zweiseitigen Sechsflache abzuleiten gestattet. Man hat nur die 10 geschlossenen Zellen der durch 6 Ebenen gebildeten Raumteilung zu 2, 3, 4, …, 9, 10 in jeder zulässigen Weise zusammenzufügen, was mit Berücksichtigung der Zahl 42 der vollständigen Figuren ein zwar mühsames, aber eine bestimmte endliche Anzahl von Sechsflachen ergebendes Verfahren ist, das die Geduld insofern auf eine harte Probe stellt, als sich natürlich jedes Sechsflach wiederholt vorfindet. Ich habe mich schliesslich, wegen der ungeheuren Anzahl existierender Sechsflache, begnügt, von allen Lösungen der obigen Gleichungen typische Vertreter zu konstruieren. Für $p=1$ existieren die folgenden Lösungen der Gleichung (2), wobei die Zahl der Kanten stets 18, die der Ecken 12 ist:

(α) $f_7=4,\ f_4=2$, (6); (β) $f_7=2,\ f_6=2,\ f_5=2$, (7); (γ) $f_6=6$, (2)*.

Diese Polyeder für $p=1$ haben zum Teil die Gestalt eines Ringes†. Einseitige Sechsflache treten für die Zusammenhangszahl $\sigma=1$ und $\sigma=2$ auf (vergl. hierzu Dehn, *Math. Encykl.* III$_1$, S. 202) und es sind die Zahlen für die i-kantigen Grenzflächen an die Gleichungen gebunden:

$$3f_3+2f_4+f_5=6+f_7; \quad \text{für } \sigma=1.(e=10,\ k=15)\ldots\ldots\ldots\ldots(3)$$

bezw.

$$3f_3+2f_4+f_5=f_7; \quad \text{für } \sigma=2.(e=12,\ k=18)\ldots\ldots\ldots\ldots(4).$$

* Die eingeklammerten Zahlen geben hier wie im folgenden die Anzahl der konstruierten und im Manuskript beschriebenen Sechsflache.

† Es wurde die Art der Ableitung im Vortrage hier an der Figur eines vollständigen Sechsflaches erläutert; eine Reihe Modelle wurde vorgelegt und ein zweiseitiges vom Geschlecht $p=1$ ausführlich besprochen.

Die erste Gleichung besitzt die konstruierbaren Lösungen:

$$(\alpha)\ f_7=3,\ f_3=3,\ (7);\qquad (\beta)\ f_7=1,\ f_6=2,\ f_4=2,\ f_3=1,\ (29)$$

$$(\gamma)\ f_6=2,\ f_5=2,\ f_4=2,\ (31);\qquad (\delta)\ f_5=6,\ (9).$$

Da die Gleichung (4) mit der früher unter (2) für die zweiseitigen Polyeder für $p=1$ aufgeführten übereinstimmt, haben die einseitigen Sechsflache für $\sigma=2$ die gleichen Grenzflächen wie oben angegeben und es bedurfte in jedem Falle noch hier wie dort der Untersuchung, ob das Möbiussche Kantengesetz gültig war oder nicht, um die betr. Sechsflache richtig einzuordnen. Es sind 33 solche einseitige Sechsflache für $\sigma=2$ aufgefunden worden*.

Die Zahl aller beschriebenen zwei- und einseitigen allgemeinen Sechsflache beträgt 228 bezw. 109. Auf die aus den allgemeinen sich ergebenden singulären vollständigen Sechsflache, so wie auf die in ihnen enthaltenen singulären geschlossenen Polyeder, möchte ich heute nicht eingehen, doch ist auch hier die Untersuchung abgeschlossen† und ich hoffe bald die ausführliche Darstellung an anderer Stelle zu veröffentlichen.

* Ein solches wurde im Modell im Anschlusse an die Zeichnung der vollständigen Figur erläutert.

† Es wurden 286 zweiseitige und 98 einseitige singuläre Sechsflache beschrieben, womit freilich deren Anzahl sicher noch nicht erschöpft ist.

SUR L'ÉQUIVALENT ANALYTIQUE DU PROBLÈME DES PRINCIPES DE LA GÉOMÉTRIE MÉTRIQUE

PAR C. STÉPHANOS.

L'ensemble des formules de la géométrie analytique Euclidienne ou non Euclidienne dans un espace à m dimensions est complètement déterminé lorsqu'on donne l'expression de la longueur ds de l'élément linéaire joignant deux points infiniment voisins $(x_1, x_2, \ldots x_m)$ et $(x_1+dx_1, \ldots x_m+dx_m)$ de cet espace. Il en est ainsi non seulement lorsqu'on fait usage de coordonnées projectives (caracterisées par ce fait que les plans sont representés par des équations linéaires), mais aussi quand on emploie des coordonnées (curvilignes) quelconques.

L'expression de ds

$$ds = f(x_1, x_2, \ldots x_m, dx_1, \ldots dx_m) = f(x, dx)$$

en coordonnées quelconques, étant la transformée de l'expression de l'élément linéaire Euclidien ou non Euclidien en coordonnées projectives, doit satisfaire à un certain nombre de conditions *nécessaires et suffisantes* pour qu'il puisse par un changement convenable de variables prendre la forme de l'élément linéaire (Euclidien ou non Euclidien) en coordonnées projectives.

Le carré de $f(x, dx)$ doit ainsi être une forme entière et homogène du second degré par rapport aux différentielles $dx_1, dx_2, \ldots dx_m$, soumise à un certain système d'équations différentielles aux dérivées partielles.

Ces conditions nécessaires et suffisantes auxquelles doit satisfaire l'expression de la longueur de l'élément linéaire en coordonnées quelconques peuvent être considérées comme l'équivalent analytique de l'ensemble des principes propres à la géométrie Euclidienne ou non Euclidienne considérée comme examinant les propriétés des figures de l'espace qui restent invariables par un certain groupe de transformations (déplacements) de l'espace Euclidien ou non Euclidien.

La considération de ces conditions nécessaires et suffisantes est indispensable lorsqu'on veut traiter en coordonnées quelconques des problèmes de géométrie différentielle, de mécanique et de physique mathématique.

Les formules auxquelles on est conduit ainsi ont la propriété de conserver leur caractère d'invariance par rapport à l'ensemble de toutes les transformations ponctuelles de l'espace à m dimensions.

Un grand nombre de questions intéressantes, pouvant conduire à des résultats bien remarquables, sont liées à cette manière de considérer le problème des principes de la géométrie.

ON RATIONAL RIGHT-ANGLED TRIANGLES

By Artemas Martin.

—"and are you such fools,
To Square for this?"—*Titus Andronicus*, ii, 1.

In a former paper the writer gave three methods of proof of the celebrated proposition that the square on the hypotenuse of a right-angled triangle is equal to the sum of the squares on the legs (sides including the right angle). Other proofs are presented here which may be of interest.

IX. Let ABC be any right-angled triangle (Fig. 1), B the right angle, O the centre of the inscribed circle.

Put OD, OE, $OF = r$ = radius of inscribed circle, $AC = a$, $AB = b$, $BC = c$.

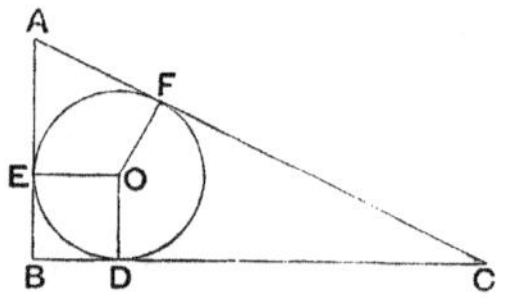

Then

area of triangle $ABC = \frac{1}{2}bc = \frac{1}{2}r(a + b + c)$...(1).

From the figure we have

$$BF + BD = b + c - a = 2r \text{..............(2)},$$

since $AE = AF$, $DC = FC$, and $BD = BE = r$.

Fig. 1.

Hence from (2) we have

$$r = \frac{1}{2}(b + c - a) \text{ ....(3)}.$$

Substituting this value in (1) and reducing we get

$$a^2 = b^2 + c^2 \text{ ....(4)}.$$

This proof was communicated in May, 1891, by Dr L. A. Bauer, then of the U.S. Coast and Geodetic Survey, now of the Carnegie Institution of Washington.

Another method of the same proof.

Let $CD = CF = m$, $AE = AF = n$, $OD = OE = OF = r$, in Fig. 1; then

$$AC = a = m + n,\ AB = b = n + r,\ BC = c = m + r.$$

Now $(m + r)(n + r) = r(2m + 2n + 2r)$ ....(1),

since each expression is double the area of the triangle ABC.

Therefore $mn = mr + nr + r^2$,

and $2mn = 2mr + 2nr + 2r^2$ ....(2).

Adding $m^2 + n^2$ to each member of (2),

$$(m+n)^2 = (m+r)^2 + (n+r)^2,$$

or

$$a^2 = b^2 + c^2.$$

This method was communicated by Lucius Brown, Hudson, Mass.

The following is not new, but is of interest because of its simplicity.

Let ABC (Fig. 2) be any right-angled triangle (right-angled at B), $DEAC$ the square on the hypotenuse AC. Produce BC to H, making $CH = AB$; produce BA to F, making $AF = BC$; draw EH perpendicular to BH, and produce HE to G, making $EG = CH$; draw FG through D; then

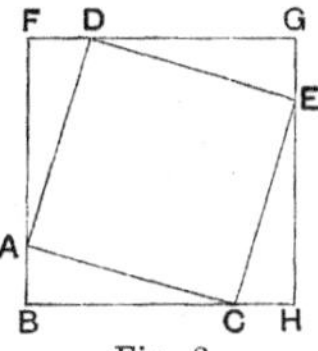

Fig. 2.

$$FD = AB = CH = GE,$$

and

$$DG = EH = BC = AF.$$

Let $AC = a$, $AB = b$, $BC = c$; then

$$BH = BF = FG = GH = b + c;\ \square DEAC = c^2,$$

$$\triangle ABC = \triangle DFA = \triangle GED = \triangle CHE = \tfrac{1}{2}bc.$$

$$\square FGBH = \square DEAC + 4\triangle ABC;$$

therefore

$$(b+c)^2 = a^2 + 2bc,$$

or

$$b^2 + 2bc + c^2 = a^2 + 2bc.$$

Dropping $2bc$ from each side we have left

$$b^2 + c^2 = a^2.$$

X. Every prime number of the form $4m + 1$ is the hypotenuse of a prime rational right-angled triangle.

The hypotenuse of a prime rational right-angled triangle is either a prime number of the form $4m + 1$ or the product of two or more such primes.

A number that is the product of n different prime numbers of the form $4m + 1$ is the common hypotenuse of 2^{n-1} different prime rational right-angled triangles.

Every number which contains a prime factor of the form $4m + 1$ is the hypotenuse of a rational (but not prime) right-angled triangle.

The hypotenuse of any rational right-angled triangle is n times a prime number of the form $4m + 1$, or n times the product of two or more such primes, where n may be any whole number. The triangle will be prime when n is a prime of the form $4m + 1$.

A prime number of the form $4n + 3$ can not be the hypotenuse of a rational right-angled triangle.

A number which does not contain a prime factor of the form $4m + 1$ can not be the hypotenuse of a rational right-angled triangle.

XI. In *Miscellaneous Notes and Queries*, edited by the late S. C. Gould, Manchester, N.H., Vol. IX, No. 6 (June, 1892), p. 141, in a paper on "Rational Right-Angled Triangles," the late Benjamin F. Burleson of Oneida Castle, N.Y., in his solution of Problem 4, "Write out all the rational right-angled triangles there are whose hypotenuses do not exceed 100," stated that there are 55 such triangles; but he inadvertently included in his second series of triangles three that are in the first series, so that three triangles (10, 24, 26), (14, 48, 50), (54, 72, 90), were counted twice. Deducting 3 from 55 we have left 52, which is the number of rational right-angled triangles whose hypotenuses do not exceed 100.

See *Mathematical Magazine*, Vol. II, No. 12 (Sept. 1910), pp. 319—20; *Miscellaneous Notes and Queries*, Vol. X, No. 3 (Sept. 1892), p. 230; *Educational Times Reprint*, Vol. XXXIX (London, 1883), pp. 37—8, Quest. 6966.

The prime rational right-angled triangles whose hypotenuses are less than 3000 are given in the appended table for reference, and are arranged in the order of the magnitude of the hypotenuses for convenience of use in this paper. I have extended this table, in manuscript, so as to include all prime rational right-angled triangles whose hypotenuses are less than 10000. For triangles whose hypotenuses exceed 3000, consult the table on pp. 301—8 of No. 12, Vol. II, of the *Mathematical Magazine*, beginning with p. 304. I have extended this table in manuscript to $p = 101$, $q = 100$.

There are 158 prime rational right-angled triangles whose hypotenuses are less than 1000; 161 whose hypotenuses are between 1000 and 2000; 158 whose hypotenuses are between 2000 and 3000; or 477 whose hypotenuses are less than 3000.

I have checked the number of these triangles in the appended table with the table by Edward Sang (*Transactions of the Royal Society of Edinburgh*, Vol. XIII, 1864, pp. 757—58), and both agree to the extent of Sang's table, except that he omitted the following two triangles (576, 943, 1105), (744, 817, 1105), at the end of his table.

XII. To find rational right-angled triangles whose sides are whole numbers and hypotenuses consecutive whole numbers.

Examples. 1. To find pairs or couples of right-angled triangles whose hypotenuses are consecutive integers.

Let u, v, w; $x, y, w+1$ denote the sides of a pair of such triangles, and, to get a general solution, we must solve the equations

$$u^2 + v^2 = w^2 \quad \text{..........................(1)},$$

$$x^2 + y^2 = (w+1)^2 \quad \text{..........................(2)};$$

but it is much easier to find the triangles wanted by trial from the appended table with the aid of Table I in Barlow's Tables (London, 1814), when the hypotenuses are not greater than 3000; when the hypotenuses are greater than 3000, use the table of right-angled triangles in the *Mathematical Magazine* on pp. 301—8, beginning with p. 304.

If we multiply the sides of the first triangle in the table at the end of this paper by 5, we have a triangle whose sides are

15, 20, 25.

Multiplying the sides of the second triangle in the table by 2, we get the triangle whose sides are

10, 24, 26;

and we have two triangles whose hypotenuses differ by unity or are consecutive numbers.

In a similar manner we find from the first and fifth triangles in the table the pair

20, 21, 29;
18, 24, 30.

From the first and third triangles we get the pair

16, 30, 34;
21, 28, 35.

The following pairs, and many others, are easily found:

15,	36,	39;	14,	48,	50;	20,	48,	52;
24,	32,	40.	24,	45,	51.	28,	45,	53.
48,	55,	73;	60,	80,	100;	40,	96,	104;
24,	70,	74.	20,	99,	101.	63,	84,	105.
24,	143,	145;	119,	120,	169;	65,	156,	169;
96,	110,	146.	102,	136,	170.	80,	150,	170.
252,	844,	900;	696,	697,	985;	324,	945,	999;
451,	780,	901.	680,	714,	986.	352,	936,	1000.

2. To find groups of three rational right-angled triangles whose hypotenuses are consecutive numbers.

Multiplying the second triangle in the appended table by 3, the first triangle by 8, we have, with the seventh triangle,

15, 36, 39;
24, 32, 40;
9, 40, 41;

a group of three triangles whose hypotenuses are consecutive numbers.

In a similar manner the following groups of three such triangles have been found, and many other groups may be found with hypotenuses not exceeding 1000.

48,	55,	73;	60,	80,	100;	40,	96,	104;
24,	70,	74;	20,	99,	101;	63,	84,	105;
45,	60,	75.	48,	90,	102.	56,	90,	106.
69,	92,	115;	81,	108,	135;	48,	140,	148;
80,	84,	116;	64,	120,	136;	51,	140,	149;
45,	108,	117.	88,	105,	137.	90,	120,	150.
52,	165,	173;	95,	168,	193;	160,	168,	232;
120,	126,	174;	130,	144,	194;	105,	208,	233;
105,	140,	175.	117,	156,	195.	90,	216,	234.

44, 240, 244;	84, 245, 259;	128, 240, 272;
147, 196, 245;	100, 240, 260;	105, 252, 273;
54, 240, 246.	180, 189, 261.	176, 210, 274.
102, 280, 298;	140, 336, 364;	81, 360, 369;
115, 276, 299;	240, 275, 365;	222, 296, 370;
180, 240, 300.	66, 360, 366.	196, 315, 371.
155, 372, 403;	280, 294, 406;	120, 391, 409;
80, 396, 404;	132, 385, 407;	168, 374, 410;
243, 324, 405.	192, 360, 408.	264, 315, 411.

3. To find groups of four rational right-angled triangles whose hypotenuses are consecutive numbers less than 1000.

In the *School Messenger*, edited by G. H. Harvill, Vol. II, No. 6 (June, 1885), p. 218, the late B. F. Burleson, in a paper on "Complementary Squares" (right-angled triangles), said: "With hypotenuses less than 1000 I have found 21 sets of series, 4 in each, in which the hypotenuses differ by unity"; but he did not give the triangles.

The present writer has found 45 such sets or groups of four triangles which are exhibited below. It is to be regretted that Mr Burleson did not publish the groups he found so that it could be seen whether any of them are missing here.

30, 40, 50;	108, 144, 180;	40, 198, 202;
24, 45, 51;	19, 180, 181;	140, 147, 203;
20, 48, 52;	70, 168, 182;	96, 180, 204;
28, 45, 53.	33, 180, 183.	45, 200, 205.
120, 182, 218;	160, 168, 232;	44, 240, 244;
144, 165, 219;	105, 208, 233;	147, 196, 245;
132, 176, 220;	90, 206, 234;	54, 240, 246;
21, 220, 221.	120, 225, 235.	95, 228, 247.
128, 240, 272;	136, 255, 289;	120, 288, 312;
105, 252, 273;	200, 210, 290;	25, 312, 313;
176, 210, 274;	195, 216, 291;	170, 264, 314;
77, 264, 275.	192, 220, 292.	189, 252, 315.
75, 308, 317;	175, 288, 337;	240, 252, 348;
168, 270, 318;	130, 312, 338;	180, 299, 349;
220, 231, 319;	45, 336, 339;	210, 280, 350;
192, 256, 320.	160, 300, 340.	135, 324, 351.
260, 288, 388;	155, 372, 403;	132, 385, 407;
189, 340, 389;	80, 396, 404;	192, 360, 408;
150, 360, 390;	243, 324, 405;	120, 391, 409;
184, 345, 391.	280, 294, 406.	168, 374, 410.

280, 351, 449;	168, 425, 457;	108, 480, 492;
126, 432, 450;	120, 442, 458;	340, 357, 493;
99, 440, 451;	216, 405, 459;	190, 456, 494;
60, 448, 452.	276, 368, 460.	287, 396, 495.
168, 490, 518;	57, 540, 543;	352, 420, 548;
156, 495, 519;	256, 480, 544;	99, 540, 549;
312, 416, 520;	100, 455, 545;	154, 528, 550;
279, 440, 521.	210, 504, 546.	380, 399, 551.
215, 516, 559;	48, 575, 577;	390, 432, 582;
336, 448, 560;	272, 511, 578;	408, 495, 583;
264, 495, 561;	285, 504, 579;	384, 440, 584;
320, 462, 562.	400, 420, 580.	225, 540, 585.
354, 472, 590;	420, 441, 609;	273, 560, 623;
84, 585, 591;	110, 600, 610;	440, 576, 624;
192, 560, 592;	235, 564, 611;	175, 600, 625;
368, 465, 593.	288, 540, 612.	50, 624, 626.
150, 616, 634;	315, 572, 653;	385, 552, 673;
381, 508, 635;	360, 546, 654;	350, 576, 674;
336, 540, 636;	393, 524, 655;	189, 648, 675;
245, 588, 637.	144, 640, 656.	260, 624, 676.
52, 675, 677;	480, 504, 696;	196, 672, 700;
90, 672, 678;	153, 680, 697;	260, 651, 701;
455, 504, 679;	360, 598, 698;	270, 648, 702;
104, 672, 680.	315, 624, 699.	228, 665, 703.
480, 550, 730;	481, 600, 769;	217, 744, 775;
344, 645, 731;	462, 616, 770;	520, 576, 776;
132, 720, 732;	96, 765, 771;	252, 735, 777;
108, 725, 733.	380, 672, 772.	378, 640, 778.
324, 768, 800;	310, 744, 806;	315, 756, 819;
351, 720, 801;	207, 780, 807;	180, 800, 820;
80, 798, 802;	160, 792, 808;	421, 700, 821;
498, 605, 803.	280, 759, 809.	528, 630, 822.
540, 629, 829;	504, 672, 840;	600, 630, 870;
498, 664, 830;	580, 609, 841;	335, 804, 871;
345, 756, 831;	58, 840, 842;	480, 728, 872;
320, 768, 832.	480, 693, 843.	585, 648, 873.

345, 828, 897;	75, 936, 939;	365, 876, 949;
560, 702, 898;	564, 752, 940;	266, 912, 950;
620, 651, 899;	580, 741, 941;	225, 924, 951;
252, 864, 900.	510, 792, 942.	448, 840, 952.

These groups of four triangles are distinct and independent; no group contains a triangle found in any other group.

4. To find groups of five triangles whose hypotenuses are consecutive whole numbers less than 1000.

I have found the following 15 groups of five such triangles whose hypotenuses are less than 1000.

120, 182, 218;	136, 255, 289;	155, 372, 403;
144, 165, 219;	200, 210, 290;	80, 396, 404;
132, 176, 220;	195, 216, 291;	243, 324, 405;
21, 220, 221;	192, 220, 292;	280, 294, 406;
72, 210, 222.	68, 285, 293.	132, 385, 407.
168, 490, 518;	390, 432, 582;	385, 552, 673;
156, 495, 519;	408, 495, 583;	350, 576, 674;
312, 416, 520;	384, 440, 584;	189, 648, 675;
279, 440, 521;	225, 540, 585;	260, 624, 676;
360, 378, 522.	136, 570, 586.	52, 675, 677.
480, 504, 696;	481, 600, 769;	217, 744, 775;
153, 680, 697;	462, 616, 770;	520, 576, 776;
360, 598, 698;	96, 765, 771;	252, 735, 777;
315, 624, 699;	380, 672, 772;	378, 640, 778;
196, 672, 700.	195, 748, 773.	171, 760, 779.
483, 644, 805;	240, 782, 818;	540, 629, 829;
310, 744, 806;	315, 756, 819;	498, 664, 830;
207, 780, 807;	180, 800, 820;	345, 756, 831;
160, 792, 808;	421, 700, 821;	320, 768, 832;
280, 759, 809.	528, 630, 822.	392, 735, 833.
345, 828, 897;	75, 936, 939;	365, 876, 949;
560, 702, 898;	564, 752, 940;	266, 912, 950;
620, 651, 899;	580, 741, 941;	225, 924, 951;
252, 864, 900;	510, 792, 942;	448, 840, 952;
451, 780, 901.	207, 920, 943.	615, 728, 953.

5. To find groups of six triangles whose hypotenuses are consecutive whole numbers less than 1000.

I have found the following seven groups of six such triangles:

155, 372, 403;	385, 552, 673;	417, 556, 695;
80, 396, 404;	350, 576, 674;	480, 504, 696;
243, 324, 405;	189, 648, 675;	153, 680, 697;
280, 294, 406;	260, 624, 676;	360, 598, 698;
132, 385, 407;	52, 675, 677;	315, 624, 699;
192, 360, 408.	90, 672, 678.	196, 672, 700.
217, 744, 775;	483, 644, 805;	345, 828, 897;
520, 576, 776;	310, 744, 806;	560, 702, 898;
252, 735, 777;	207, 780, 807;	620, 651, 899;
378, 640, 778;	160, 792, 808;	252, 864, 900;
171, 760, 779;	280, 759, 809;	451, 780, 901;
396, 672, 780.	486, 648, 810.	198, 880, 902.
365, 876, 949;	225, 924, 951;	615, 728, 953;
266, 912, 950;	448, 840, 952;	504, 810, 954.

6. To find groups of seven right-angled triangles whose hypotenuses are consecutive whole numbers less than 1000.

I have found the following four groups of seven such right-angled triangles:

155, 372, 403;	385, 552, 673;
80, 396, 404;	350, 576, 674;
243, 324, 405;	189, 648, 675;
280, 294, 406;	260, 624, 676;
132, 385, 407;	52, 675, 677;
192, 360, 408;	90, 672, 678;
120, 391, 409.	455, 504, 679.
417, 556, 695;	365, 876, 949;
480, 504, 696;	266, 912, 950;
153, 680, 697;	225, 924, 951;
360, 598, 698;	448, 840, 952;
315, 624, 699;	615, 728, 953;
196, 672, 700;	504, 810, 954;
260, 651, 701.	573, 764, 955.

7. To find groups of eight right-angled triangles whose hypotenuses are consecutive whole numbers.

I have found the following seven groups of eight such triangles. The hypotenuses of the first three are less than 1000.

155, 372, 403;	385, 552, 673;	417, 556, 695;
80, 396, 404;	350, 576, 674;	480, 504, 696;

243, 324, 405;	189, 648, 675;	153, 680, 697;
280, 294, 406;	260, 624, 676;	360, 598, 698;
132, 385, 407;	52, 675, 677;	315, 624, 699;
192, 360, 408;	90, 672, 678;	196, 672, 700;
120, 391, 409;	455, 504, 679;	260, 651, 701;
168, 374, 410.	104, 672, 680.	270, 648, 702.

528, 1025, 1153;	715, 1428, 1597;	1008, 1344, 1680;
96, 1150, 1154;	752, 1410, 1598;	720, 1519, 1681;
693, 924, 1155;	351, 1560, 1599;	82, 1680, 1682;
644, 960, 1156;	448, 1536, 1600;	792, 1485, 1683;
765, 868, 1157;	80, 1599, 1601;	116, 1680, 1684;
570, 1008, 1158;	702, 1440, 1602;	164, 1677, 1685;
209, 1140, 1159;	420, 1547, 1603;	960, 1386, 1686;
800, 840, 1160.	160, 1596, 1604.	840, 1463, 1687.

1240, 1302, 1798;
224, 1785, 1799;
504, 1728, 1800;
649, 1680, 1801;
952, 1530, 1802;
720, 1653, 1803;
396, 1760, 1804;
1083, 1444, 1805.

The late B. F. Burleson gave the first and second of the foregoing groups of eight right-angled triangles whose hypotenuses are consecutive numbers in the late S. C. Gould's *Notes and Queries*, Vol. III, No. 12 (December, 1886), p. 201, and remarked: "These are the only 2 sets of 8 triangles in a series in which the hypotenuses are in regular sequence ever obtained. They were found by Charles Kriele, of Pennsylvania, when aged, infirm, and nearly blind, and sent by him to the writer of this article."

Mr Burleson gave these two groups in Harvill's *School Messenger*, Vol. II, No. 6 (June, 1885), p. 219, saying he had found "2 sets of series, 8 in each, in which the hypotenuses differ only by unity," and made no mention there of Kriele.

I have added five other groups of 8 triangles, so *seven* such groups are now known. I have also added another triangle to the first, third and fifth groups, making three groups of *nine* triangles whose hypotenuses are consecutive whole numbers, which are exhibited below:

155, 372, 403;	417, 556, 695;	715, 1428, 1597;
80, 396, 404;	480, 504, 696;	752, 1410, 1598;
243, 324, 405;	153, 680, 697;	351, 1560, 1599;
280, 294, 406;	360, 598, 698;	448, 1536, 1600;

132, 385, 407;	315, 624, 699;	80, 1599, 1601;
192, 360, 408;	196, 672, 700;	702, 1440, 1602;
120, 391, 409;	260, 651, 701;	420, 1547, 1603;
168, 374, 410;	270, 648, 702;	160, 1596, 1604;
264, 315, 411.	420, 665, 703.	963, 1284, 1605.

I have added another triangle to the last of the groups above and thus have *one* group of 10 rational right-angled triangles whose hypotenuses are consecutive whole numbers which is given below:

715, 1428, 1597;	702, 1440, 1602;
752, 1410, 1598;	420, 1547, 1603;
351, 1560, 1599;	160, 1596, 1604;
448, 1536, 1600;	963, 1284, 1605;
80, 1599, 1601;	1056, 1210, 1606.

Since what precedes was written, Mr B. O. M. De Beck of Cincinnati, Ohio, has sent me, from notes he made from Barlow's Tables more than 60 years ago, eight groups of 10 triangles whose hypotenuses are consecutive numbers; eight groups of 11 such triangles; six groups of 12 triangles; two groups of 13 triangles, and one group of 14 triangles, the hypotenuses of all the groups being between 1000 and 10000.

He also sent me later the following groups with hypotenuses between 10000 and 20000: thirteen groups of 10 triangles; nine groups of 11 triangles; ten groups of 12 triangles; three groups of 13 triangles; one group of 14 triangles, and one group of 15 triangles.

Mr De Beck also contributed one group of 15 triangles with hypotenuses from 85478 to 85492, inclusive.

Space considerations prevent me from giving here more than one of each of the different groups whose hypotenuses are between 1000 and 10000. I give also the group of 15 triangles whose hypotenuses are between 10000 and 20000.

1247, 2140, 3103;	1440, 1650, 2190;	1692, 2256, 2820;
2080, 2304, 3104;	175, 2184, 2191;	1085, 2604, 2821;
1863, 2484, 3105;	1408, 1680, 2192;	1328, 2490, 2822;
990, 2944, 3106;	1032, 1935, 2193;	1740, 2223, 2823;
1195, 2868, 3107;	1170, 1856, 2194;	1800, 2176, 2824;
1008, 2940, 3108;	1317, 1756, 2195;	1695, 2260, 2825;
1309, 2820, 3109;	396, 2160, 2196;	1530, 2376, 2826;
1866, 2488, 3110;	845, 2028, 2197;	352, 2805, 2827;
561, 3060, 3111;	1190, 1848, 2198;	560, 2772, 2828;
1512, 2720, 3112.	324, 2175, 2199;	621, 2760, 2829;
	1320, 1760, 2200.	1698, 2264, 2830;
		969, 2660, 2831.

3630, 4840, 6050;	5535, 6552, 8577;	2460, 12177, 12423;
2376, 5565, 6051;	2080, 8322, 8578;	3960, 11776, 12424;
2848, 5340, 6052;	5796, 6325, 8579;	3479, 11928, 12425;
1635, 5828, 6053;	5148, 6864, 8580;	6840, 10374, 12426;
3354, 5040, 6054;	131, 8580, 8581;	5848, 10965, 12427;
3633, 4844, 6055;	490, 8568, 8582;	4780, 11472, 12428;
3744, 4760, 6056;	5700, 6417, 8583;	8379, 9180, 12429;
3465, 4968, 6057;	3584, 7800, 8584;	1650, 12320, 12430;
3142, 4600, 6058;	5151, 6868, 8585;	1240, 12369, 12431;
3984, 4565, 6059;	4536, 7290, 8586;	4032, 11760, 12432;
3636, 4848, 6060;	3565, 7812, 8587;	4495, 11592, 12433;
4180, 4389, 6061;	1140, 8512, 8588;	6384, 10670, 12434;
2030, 5712, 6062.	2520, 8211, 8589;	8100, 9435, 12435;
	5154, 6872, 8590.	5236, 11280, 12436;
		2355, 12212, 12437.

XIII. In each of the following nine groups of three right-angled triangles, the three *hypotenuses* of each group are the sides of a right-angled triangle.

21, 72, 75;	40, 96. 104;	56, 105, 119;
60, 80, 100;	72, 135, 153;	72, 96, 120;
35, 120, 125.	57, 176, 185.	65, 156, 169.
64, 120, 136;	60, 63, 87;	60, 144, 156;
60, 252, 273;	160, 384, 416;	460, 483, 667;
55, 300, 305.	243, 324, 425.	440, 525, 685.
240, 320, 400;	480, 504, 696;	387, 516, 645;
264, 495, 561;	455, 528, 697;	560, 588, 812;
111, 680, 689.	473, 864, 985.	315, 988, 1037.

Such groups of right-angled triangles could be obtained by solving the following simultaneous equations, but it is easier to find them from the table of right-angled triangles, with the aid of Barlow's Table I, as I have done.

Let p, q, r be the legs and hypotenuse of the first triangle; s, t, u the legs and hypotenuse of the second triangle; v, w, x the legs and hypotenuse of the third triangle; then we shall have

$$p^2 + q^2 = r^2,$$

$$s^2 + t^2 = u^2,$$

$$v^2 + w^2 = x^2,$$

$$r^2 + u^2 = x^2.$$

XIV. In the following six groups of three rational right-angled triangles, the *hypotenuses* of each group form a rational scalene triangle.

5, 12, 13;	15, 36, 39;	8, 15, 17;
24, 32, 40;	9, 40, 41;	16, 63, 65;
27, 36, 45.	14, 48, 50.	48, 64, 80.
15, 112, 113;	27, 120, 123;	88, 105, 137;
24, 143, 145;	48, 140, 148;	96, 110, 146;
130, 144, 194.	21, 220, 221.	140, 225, 265.

In each of the groups below the three *hypotenuses* of each group are the sides of a rational scalene triangle whose sides are consecutive numbers.

24, 45, 51;	95, 168, 193;	360, 627, 723;
20, 48, 52;	130, 144, 194;	76, 720, 724;
28, 45, 53.	117, 156, 195.	120, 715, 725.

1776, 2035, 2701;	5040, 8733, 10083;
1330, 2352, 2702;	280, 10080, 10084;
1428, 2295, 2703.	3960, 9275, 10085.

XV. As stated on a preceding page, a number which is the product of n different prime factors of the form $4m+1$ is the hypotenuse of 2^{n-1} (incorrectly given 2^n in Barlow's *Theory of Numbers*, p. 177) different prime right-angled triangles.

See Matteson's *Diophantine Problems with Solutions* (Washington, 1888), p. 10.

Examples. 1. The least two primes of the form $4m+1$ are 5 and 13, and their product $= 65 = 8^2 + 1^2 = 7^2 + 4^2$, is the hypotenuse of each of the two prime triangles

16, 63, 65; 33, 56, 65.

There are also two other triangles, not prime triangles, with hypotenuses = 65, obtained as follows:

$$(3,\ 4,\ 5) \times 13 = 39,\ 52,\ 65;$$
$$(5,\ 12,\ 13) \times 5 = 25,\ 60,\ 65.$$

The product of 5 and $17 = 85 = 9^2 + 2^2 = 7^2 + 6^2$, and therefore 85 is the common hypotenuse of the two prime triangles

13, 84, 85; 36, 77, 85.

The other two triangles in this case are

$$(3,\ 4,\ 5) \times 17 = 51,\ 68,\ 85;$$
$$(8,\ 15,\ 17) \times 5 = 40,\ 75,\ 85.$$

The product of 13 and $17 = 221 = 14^2 + 5^2 = 11^2 + 10^2$, and 221 is the common hypotenuse of the two prime triangles

21, 220, 221; 140, 171, 221.

The other two triangles with hypotenuses = 221 are

$$(5,\ 12,\ 13) \times 17 = 85,\ 204,\ 221;$$
$$(8,\ 15,\ 17) \times 13 = 104,\ 195,\ 221.$$

And, generally,

$$(a^2 + b^2)(c^2 + d^2) = (ac + bd)^2 + (ad - bc)^2 = (ad + bc)^2 + (ac - bd)^2.$$

There are 36 couples or pairs of prime right-angled triangles having a common hypotenuse less than 1000, 41 such pairs of triangles whose hypotenuses are between 1000 and 2000, 39 pairs whose hypotenuses are between 2000 and 3000; or, 116 pairs with hypotenuses less than 3000. See table of prime rational right-angled triangles at end of paper.

2. The product of three prime numbers of the form $4m + 1$ is the common hypotenuse of $2^2 = 4$ prime right-angled triangles. The product of the least three primes of this form is

$$5 \times 13 \times 17 = 1105 = 33^2 + 4^2 = 32^2 + 9^2 = 31^2 + 12^2 = 24^2 + 23^2,$$

which is, therefore, the hypotenuse of each of the following four prime right-angled triangles, viz.:

47, 1104, 1105;
264, 1073, 1105;
576, 943, 1105;
744, 817, 1105.

The last two of these triangles are omitted at the end of Edward Sang's table of right-angled triangles referred to on a preceding page.

There are nine other triangles having 1105 for hypotenuse which are not prime triangles, but are multiples of prime triangles. They are given below with the method of obtaining them.

$$(3,\ 4,\ 5) \times 13 \times 17 = 663,\ 884,\ 1105;$$
$$(5,\ 12,\ 13) \times 5 \times 17 = 425,\ 1020,\ 1105;$$
$$(8,\ 15,\ 17) \times 5 \times 13 = 520,\ 975,\ 1105;$$
$$(16,\ 63,\ 65) \times 17 = 272,\ 1071,\ 1105;$$
$$(33,\ 56,\ 65) \times 17 = 561,\ 952,\ 1105;$$
$$(36,\ 77,\ 85) \times 13 = 468,\ 1001,\ 1105;$$
$$(13,\ 84,\ 85) \times 13 = 169,\ 1092,\ 1105;$$
$$(21,\ 220,\ 221) \times 5 = 105,\ 1100,\ 1105;$$
$$(140,\ 171,\ 221) \times 5 = 700,\ 885,\ 1105.$$

In general,

$$\begin{aligned}(a^2 + b^2)(c^2 + d^2)(e^2 + f^2) &= [e(ad + bc) + f(ac - bd)]^2 + [e(ac - bd) - f(ad + bc)]^2, \\ &= [e(ac - bd) + f(ad + bc)]^2 + [e(ad + bc) - f(ac - bd)]^2, \\ &= [e(ad - bc) + f(ac + bd)]^2 + [e(ac + bd) - f(ad - bc)]^2, \\ &= [e(ac + bd) + f(ad - bc)]^2 + [e(ad - bc) - f(ac + bd)]^2.\end{aligned}$$

Take $a = 2$, $b = 1$; $c = 3$, $d = 2$; $e = 4$, $f = 1$; then

$$5 \times 13 \times 17 = 1105 = 32^2 + 9^2 = 24^2 + 23^2 = 31^2 + 12^2 = 33^2 + 4^2,$$

as found before.

$$5 \times 13 \times 29 = 1885 = 43^2 + 6^2 = 42^2 + 11^2 = 38^2 + 21^2 = 34^2 + 27^2;$$

hence 1885 is the common hypotenuse of the four prime right-angled triangles

427, 1836, 1885;

516, 1813, 1885;

924, 1643, 1885;

1003, 1596, 1885.

There are also nine other right-angled triangles having 1885 for a common hypotenuse that are not prime triangles, which are given below:

221, 1872, 1885;	675, 1760, 1885;	957, 1624, 1885;
312, 1859, 1885;	725, 1740, 1885;	1131, 1508, 1885;
464, 1787, 1885;	760, 1725, 1885;	1300, 1365, 1885.

There are only two groups of four prime rational right-angled triangles having a common hypotenuse less than 2000, and only three groups of such triangles with hypotenuses between 2000 and 3000, viz.:

$$5 \times 13 \times 37 = 2405,\ 5 \times 17 \times 29 = 2465,\ 5 \times 13 \times 41 = 2665;$$

$$2405 = 49^2 + 2^2 = 47^2 + 14^2 = 46^2 + 17^2 = 38^2 + 31^2;$$

$$2465 = 49^2 + 8^2 = 47^2 + 16^2 = 44^2 + 23^2 = 41^2 + 28^2;$$

$$2665 = 51^2 + 8^2 = 48^2 + 19^2 = 44^2 + 27^2 = 37^2 + 36^2.$$

Hence the three groups of 4 prime triangles are

196, 2397, 2405;	784, 2337, 2465;	73, 2664, 2665;
483, 2356, 2405;	897, 2296, 2465;	816, 2537, 2665;
1316, 2013, 2405;	1407, 2024, 2465;	1207, 2376, 2665;
1564, 1827, 2405.	1504, 1953, 2465.	1824, 1943, 2665.

The following curious theorem is given without proof in Matteson's *Diophantine Problems with Solutions*, p. 11.

If a number N is the product of 1, 2, 3, 4, 5, ... or n different prime factors of the form $4m+1$, the terms of the following series,

$$1_1,\ 4_2,\ 13_3,\ 40_4,\ 121_5,\ 364_6,\ 1093_7,\ 3280_8, \text{ etc.},$$

will represent the number of ways N^2 can be the sum of two squares, where the subscripts, 1, 2, 3, 4, ... n, denote the number of prime factors composing N. Any term of this series is equal to three times the preceding term, + 1. Thus the square of a prime number of the form $4m+1$ is the sum of two squares in *one* way only; the square of a number which is the product of two different primes of the form $4m+1$ is the sum of two squares in $1 \times 3 + 1 = 4$ different ways; the square of a number which is the product of three different primes of the form $4m+1$ is the sum of two squares in $4 \times 3 + 1 = 13$ different ways; the square of a number which is the product of four different primes of the form $4m+1$ is the sum of two squares in $13 \times 3 + 1 = 40$ different ways, etc.

If we put u_n for the nth term in the above series we have

$$u_{n+1} = 3u_n + 1.$$

The solution of this Finite-Difference equation gives

$$u_n = \frac{1}{2}(3^n - 1).$$

See Boole's *Calculus of Finite Differences*, second edition (London, 1872), p. 165.

Hence if a number is the product of n different prime numbers of the form $4m+1$, it is the common hypotenuse of $\frac{1}{2}(3^n - 1)$ different rational integral-sided right-angled triangles, 2^{n-1} of which are *prime* triangles and $\frac{1}{2}(3^n - 2^n - 1)$ are *not* prime triangles.

Examples. 1. If $n = 1$, then $\frac{1}{2}(3^1 - 1) = 1$, and a prime number of the form $4m+1$ can be the hypotenuse of but *one* rational right-angled triangle.

2. If $n = 2$, then $\frac{1}{2}(3^2 - 1) = 4$, and any number which is the product of two different primes of the form $4m+1$ is the common hypotenuse of four different rational right-angled triangles.

3. If $n = 3$, then $\frac{1}{2}(3^3 - 1) = 13$, and any number which is the product of three different primes of the form $4m+1$ is the common hypotenuse of 13 different rational right-angled triangles.

4. If $n = 4$, then $\frac{1}{2}(3^4 - 1) = 40$, and any number which is the product of four different primes of the form $4m+1$ is the common hypotenuse of 40 different rational right-angled triangles.

5. If $n = 5$, then $\frac{1}{2}(3^5 - 1) = 121$, and any number which is the product of five different primes of the form $4m+1$ is the common hypotenuse of 121 different rational right-angled triangles.

And so on, to any value of n.

Find all the rational right-angled triangles whose hypotenuses are 32045.

Solution.

$$32045 = 5 \times 13 \times 17 \times 29$$
$$= 179^2 + 2^2 = 178^2 + 19^2 = 173^2 + 46^2 = 166^2 + 67^2$$
$$= 163^2 + 74^2 = 157^2 + 86^2 = 144^2 + 109^2 = 131^2 + 122^2.$$

Hence the eight *prime* right-angled triangles whose hypotenuses are 32045 are as follows:

716, 32037, 32045; 15916, 27813, 32045;
2277, 31964, 32045; 17253, 27004, 32045;
6764, 31323, 32045; 21093, 24124, 32045;
8283, 30956, 32045; 22244, 23067, 32045.

The other thirty-two triangles are found as follows:

$(3, 4, 5) \times 13 \times 17 \times 29 = 19227, 25636, 32045;$
$(5, 12, 13) \times 5 \times 17 \times 29 = 12325, 29580, 32045;$
$(8, 15, 17) \times 5 \times 13 \times 29 = 15080, 28275, 32045;$
$(20, 21, 29) \times 5 \times 13 \times 17 = 22100, 23205, 32045;$
$(16, 63, 65) \times 17 \times 29 = 7888, 31059, 32045;$
$(33, 56, 65) \times 17 \times 29 = 16269, 27608, 32045;$
$(13, 84, 85) \times 13 \times 29 = 4901, 31668, 32045;$

$(36, 77, 85) \times 13 \times 29 = 13572, 29029, 32045;$
$(17, 144, 145) \times 13 \times 17 = 3757, 31824, 32045;$
$(24, 143, 145) \times 13 \times 17 = 5304, 31603, 32045;$
$(21, 220, 221) \times 5 \times 29 = 3045, 31900, 32045;$
$(140, 171, 221) \times 5 \times 29 = 20300, 24795, 32045;$
$(135, 352, 377) \times 5 \times 17 = 11475, 29920, 32045;$
$(152, 345, 377) \times 5 \times 17 = 12920, 29325, 32045;$
$(132, 425, 493) \times 5 \times 13 = 8580, 30875, 32045;$
$(155, 468, 493) \times 5 \times 13 = 10075, 30420, 32045;$
$(47, 1104, 1105) \times 29 = 1363, 32016, 32045;$
$(264, 1073, 1105) \times 29 = 7656, 31117, 32045;$
$(576, 943, 1105) \times 29 = 16704, 27347, 32045;$
$(744, 817, 1105) \times 29 = 21576, 23693, 32045;$
$(427, 1836, 1885) \times 17 = 7259, 31212, 32045;$
$(516, 1813, 1885) \times 17 = 8772, 30821, 32045;$
$(924, 1643, 1885) \times 17 = 15708, 27931, 32045;$
$(1003, 1596, 1885) \times 17 = 17051, 27132, 32045;$
$(784, 2337, 2465) \times 13 = 10192, 30381, 32045;$
$(897, 2296, 2465) \times 13 = 11661, 29848, 32045;$
$(1407, 2024, 2465) \times 13 = 18291, 26312, 32045;$
$(1504, 1953, 2465) \times 13 = 19552, 25389, 32045;$
$(480, 6391, 6409) \times 5 = 2400, 31955, 32045;$
$(791, 6360, 6409) \times 5 = 3955, 31800, 32045;$
$(3959, 5040, 6409) \times 5 = 19795, 25200, 32045;$
$(4200, 4841, 6409) \times 5 = 21000, 24205, 32045.$

See Matteson's *Diophantine Problems with Solutions*, p. 10; also, *Miscellaneous Notes and Queries*, Vol. x, No. 3 (Sept., 1892), p. 225.

XVI. As stated elsewhere, a prime number of the form $4m + 1$ is the sum of two squares in one way only, and therefore is the hypotenuse of only one rational right-angled triangle; but the *square* of such a prime is the sum of two squares in two ways, and therefore is the common hypotenuse of two different right-angled triangles; the *cube* of such a prime is the sum of two squares in three ways, and so is the common hypotenuse of three different right-angled triangles; and, generally, the *n*th *power* of a prime number of the form $4m + 1$ is the sum of two squares in n ways, and is the common hypotenuse of n right-angled triangles, but only *one* of the triangles is prime.

Examples. 1. $$5^2 = 3^2 + 4^2,$$
and there is *one* triangle whose hypotenuse is 5.
$$25^2 = 7^2 + 24^2 = 15^2 + 20^2,$$

and 25 is the common hypotenuse of the two triangles

7, 24, 25;

15, 20, 25.

$$125^2 = (5^3)^2 = 44^2 + 117^2 = 35^2 + 120^2 = 75^2 + 100^2,$$

and we have the three triangles

44, 117, 125;

35, 120, 125;

75, 100, 125.

$$(5^4)^2 = 625^2 = 336^2 + 527^2 = 220^2 + 585^2 = 175^2 + 600^2 = 375^2 + 500^2;$$

hence we have the four triangles

175, 600, 625;

220, 585, 625;

336, 527, 625;

375, 500, 625.

2. The next prime of the form $4m+1$ being 13, we have

$$13^2 = 5^2 + 12^2,$$

and there is only *one* triangle whose hypotenuse is 13.

$$169^2 = 119^2 + 120^2 = 65^2 + 156^2,$$

and the two triangles with hypotenuse 169 are

65, 156, 169;

119, 120, 169.

$$13^3 = 2197, \text{ and } 2197^2 = 828^2 + 2035^2 = 1547^2 + 1560^2 = 845^2 + 2028^2;$$

therefore the three triangles having hypotenuse = 2197 are

828, 2035, 2197;

845, 2028, 2197;

1547, 1560, 2197.

$13^4 = 28561$ is beyond the limit of the table of right-angled triangles given at the end of this paper, but the triangles can be computed by the aid of Barlow's Tables, before-mentioned, Table 1, p. 3. I find

$$28561^2 = 239^2 + 28560^2 = 10764^2 + 26455^2 = 20111^2 + 20280^2 = 10895^2 + 26364^2,$$

and therefore the four triangles are

239, 28560, 28561;

10764, 26455, 28561;

10895, 26364, 28561;

20111, 20280, 28561.

Table of Prime Rational Right-Angled Triangles whose Hypotenuses are less than 3000.

3	4	5	152	345	377	39	760	761	423	1064	1145
5	12	13	189	340	389	481	600	769	704	903	1145
8	15	17	228	325	397	195	748	773	528	1025	1153
7	24	25	40	399	401	56	783	785	68	1155	1157
20	21	29	120	391	409	273	736	785	765	868	1157
12	35	37	29	420	421	168	775	793	204	1147	1165
9	40	41	87	416	425	432	665	793	517	1044	1165
28	45	53	297	304	425	555	572	797	340	1131	1181
11	60	61	145	408	433	280	759	809	611	1020	1189
16	63	65	84	437	445	429	700	821	660	989	1189
33	56	65	203	396	445	540	629	829	832	855	1193
48	55	73	280	351	449	41	840	841	49	1200	1201
13	84	85	168	425	457	116	837	845	147	1196	1205
36	77	85	261	380	461	123	836	845	476	1107	1205
39	80	89	31	480	481	205	828	853	245	1188	1213
65	72	97	319	360	481	232	825	857	705	992	1217
20	99	101	44	483	485	287	816	865	140	1221	1229
60	91	109	93	476	485	504	703	865	612	1075	1237
15	112	113	132	475	493	348	805	877	280	1209	1241
44	117	125	155	468	493	369	800	881	441	1160	1241
88	105	137	217	456	505	60	899	901	799	960	1249
17	144	145	336	377	505	451	780	901	420	1189	1261
24	143	145	220	459	509	464	777	905	539	1140	1261
51	140	149	279	440	521	616	663	905	748	1035	1277
85	132	157	92	525	533	43	924	925	637	1116	1285
119	120	169	308	435	533	533	756	925	893	924	1285
52	165	173	341	420	541	129	920	929	560	1161	1289
19	180	181	33	544	545	215	912	937	72	1295	1297
57	176	185	184	513	545	580	741	941	51	1300	1301
104	153	185	165	532	557	301	900	949	255	1288	1313
95	168	193	276	493	565	420	851	949	735	1088	1313
28	195	197	396	403	565	615	728	953	360	1271	1321
84	187	205	231	520	569	124	957	965	357	1276	1325
133	156	205	48	575	577	387	884	965	884	987	1325
21	220	221	368	465	593	248	945	977	504	1247	1345
140	171	221	240	551	601	473	864	985	833	1056	1345
60	221	229	35	612	613	696	697	985	561	1240	1361
105	208	233	105	608	617	372	925	997	840	1081	1369
120	209	241	336	527	625	559	840	1009	148	1365	1373
32	255	257	100	621	629	45	1012	1013	931	1020	1381
23	264	265	429	460	629	660	779	1021	296	1353	1385
96	247	265	200	609	641	64	1023	1025	663	1216	1385
69	260	269	315	572	653	496	897	1025	53	1404	1405
115	252	277	300	589	661	192	1015	1033	444	1333	1405
160	231	281	385	552	673	315	988	1037	159	1400	1409
161	240	289	52	675	677	645	812	1037	265	1392	1417
68	285	293	37	684	685	320	999	1049	792	1175	1417
136	273	305	156	667	685	620	861	1061	371	1380	1429
207	224	305	111	680	689	731	780	1069	592	1305	1433
25	312	313	400	561	689	448	975	1073	76	1443	1445
75	308	317	185	672	697	495	952	1073	477	1364	1445
36	323	325	455	528	697	132	1085	1093	228	1435	1453
204	253	325	260	651	701	585	928	1097	583	1344	1465
175	288	337	259	660	709	47	1104	1105	936	1127	1465
180	299	349	333	644	725	264	1073	1105	380	1419	1469
225	272	353	364	627	725	576	943	1105	740	1269	1469
27	364	365	108	725	733	744	817	1105	969	1120	1481
76	357	365	216	713	745	141	1100	1109	689	1320	1489
252	275	373	407	624	745	235	1092	1117	532	1395	1493
135	352	377	468	595	757	329	1080	1129	55	1512	1513

Table of Prime Rational Right-Angled Triangles whose Hypotenuses are less than 3000 (*cont.*)

888	1225	1513	1003	1596	1885	469	2220	2269	1065	2408	2633
165	1508	1517	1311	1360	1889	752	2145	2273	1568	2145	2657
795	1292	1517	549	1820	1901	1440	1769	2281	73	2664	2665
156	1517	1525	688	1785	1913	603	2204	2285	816	2537	2665
684	1363	1525	671	1800	1921	1196	1947	2285	1207	2376	2665
312	1505	1537	1121	1560	1921	1235	1932	2293	1824	1943	2665
385	1488	1537	1092	1595	1933	1575	1672	2297	219	2660	2669
901	1260	1549	88	1935	1937	96	2303	2305	1300	2331	2669
495	1472	1553	1312	1425	1937	737	2184	2305	365	2652	2677
836	1323	1565	264	1927	1945	940	2109	2309	511	2640	2689
1036	1173	1565	793	1776	1945	480	2279	2329	1725	2068	2693
624	1457	1585	860	1749	1949	871	2160	2329	1020	2501	2701
1007	1224	1585	440	1911	1961	1365	1892	2333	1349	2340	2701
715	1428	1597	1239	1520	1961	1380	1891	2341	104	2703	2705
80	1599	1601	915	1748	1973	672	2255	2353	657	2624	2705
240	1591	1609	63	1984	1985	1128	2065	2353	312	2695	2713
988	1275	1613	616	1887	1985	1005	2132	2357	803	2604	2725
780	1421	1621	1032	1705	1993	1495	1848	2377	1764	2077	2725
57	1624	1625	315	1972	1997	69	2380	2381	520	2679	2729
1113	1184	1625	1037	1716	2005	1139	2100	2389	1491	2300	2741
285	1612	1637	1357	1476	2005	345	2368	2393	949	2580	2749
399	1600	1649	792	1855	2017	196	2397	2405	728	2655	2753
560	1551	1649	180	2021	2029	483	2356	2405	1095	2552	2777
935	1368	1657	360	2009	2041	1316	2013	2405	936	2623	2785
1140	1219	1669	1159	1680	2041	1564	1827	2405	1633	2256	2785
720	1519	1681	693	1924	2045	392	2385	2417	1700	2211	2789
164	1677	1685	1204	1653	2045	1056	2183	2425	1428	2405	2797
627	1564	1685	1428	1475	2053	1273	2064	2425	1960	2001	2801
1045	1332	1693	819	1900	2069	588	2365	2437	1241	2520	2809
328	1665	1697	1281	1640	2081	759	2320	2441	75	2812	2813
741	1540	1709	720	1961	2089	784	2337	2465	212	2805	2813
492	1645	1717	1144	1767	2105	897	2296	2465	424	2793	2825
1092	1325	1717	1376	1593	2105	1407	2024	2465	1144	2583	2825
880	1749	1721	65	2112	2113	1504	1953	2465	1775	2208	2833
1155	1292	1733	92	2115	2117	1248	2135	2473	525	2788	2837
59	1740	1741	195	2188	2117	1748	1755	2477	636	2773	2845
177	1736	1745	276	2107	2125	100	2499	2501	1387	2484	2845
656	1617	1745	1403	1596	2125	980	2301	2501	1632	2345	2857
295	1728	1753	1071	1840	2129	300	2491	2509	1900	2139	2861
84	1763	1765	455	2088	2137	1541	1980	2509	825	2752	2873
413	1716	1765	460	2091	2141	71	2520	2521	848	2745	2873
969	1480	1769	585	2072	2153	213	2516	2525	1533	2444	2885
1040	1431	1769	1320	1711	2161	1173	2236	2525	1917	2156	2885
1248	1265	1777	644	2067	2165	355	2508	2533	975	2728	2897
531	1700	1781	1197	1804	2165	1692	1885	2533	1060	2709	2909
820	1581	1781	715	2052	2173	497	2496	2545	108	2915	2917
420	1739	1789	1525	1548	2173	1176	2257	2545	1560	2479	2929
649	1680	1801	828	2035	2197	700	2451	2549	1679	2400	2929
767	1656	1825	188	2205	2213	1675	1932	2557	540	2891	2941
984	1537	1825	1260	1829	2221	639	2480	2561	2059	2100	2941
172	1845	1853	376	2193	2225	1311	2200	2561	1272	2665	2953
885	1628	1853	1496	1647	2225	781	2460	2581	1275	2668	2957
61	1860	1861	1012	1995	2237	900	2419	2581	77	2964	2965
183	1856	1865	67	2244	2245	1632	2015	2593	756	2867	2965
344	1833	1865	564	2173	2245	204	2597	2605	231	2960	2969
305	1848	1873	201	2240	2249	923	2436	2605	385	2952	2977
1148	1485	1877	1449	1720	2249	1809	1880	2609	1825	2352	2977
427	1836	1885	335	2232	2257	408	2585	2617	1425	2632	2993
516	1813	1885	1105	1968	2257	1100	2379	2621	1768	2415	2993
924	1643	1885									

THE CHARACTERS OF PLANE CURVES

By W. Esson.

In a paper published in the *Proceedings of the London Mathematical Society*, Vol. XXVIII, 1897, I took the view that an n-ic considered as an envelope consists of an m-an, 2δ points united in pairs at each of the δ double points or cuts and 3κ points united in triads at each of the κ cusps of the n-ic.

This view I developed in papers read before the Oxford Mathematical Society at various dates since 1897.

I propose to bring this view before the members of the Congress of Mathematicians for their consideration.

1. There are two kinds of joins of a point to an n-ic. (1) The tangents from the point to the curve proper. (2) The joins of the point to the double points or cuts of the curve and to the cusps of the curve. These joins are called tangent-joins, cut-joins and cusp-joins.

2. According to the view taken in this paper the total number of joins of a point to an n-ic is the sum of the number of tangent-joins, of cut-joins and of cusp-joins, and this total number depends only on the degree n of the n-ic.

3. In a pencil of n-ics each member of which is determined by the position of a point on a given curve, there are in general $3(n-1)^2$ positions of the point, each of which determines an n-ic with a double point. As the point on the given curve approaches a position which determines an n-ic with a double point two of the tangent-joins of a given point to the n-ic continuously approach each other and are ultimately united on the join of the given point to the double point. Thus a pair of tangent-joins are changed into a united pair of cut-joins. See fig. 1, I, p. 64.

4. In a pencil of n-ics each member of which has a double point in a given position there are two members each of which has a cusp at the place of the double point. As the point on the given curve approaches a position which determines an n-ic with a cusp one of the tangent-joins of a given point to the n-ic continuously approaches the join of the given point to the cusp and is ultimately united with that join. See fig. 1, II, p. 64.

5. There is thus a continuous change effected of tangent-joins into cut-joins and cusp-joins whilst the total number of joins of a given point to the curve remains the same throughout the change.

6. If this total number is M; the number of tangent-joins m; the number of cut-joins 2δ and the number of cusp-joins 3κ, then

$$m + 2\delta + 3\kappa = M,$$

and M is a number depending only upon n the degree of the pencil of n-ics.

7. If finally the n-ic takes the form of n straight lines for which m and κ are zeros and $2\delta = n(n-1)$, then

$$m + 2\delta + 3\kappa = n(n-1).$$

8. Thus the n-ic considered as an envelope consists of an m-an 2δ points united in pairs at the double points and 3κ points united in triads at the cusps and the total number of joins of a given point to the curve is $n(n-1)$.

9. The total number of joins to a curve at a double point is six, of which four are tangent-joins, two united on each tangent at the double point, and two are cut-joins each united to one of these tangents.

Let the cut be considered as the point A on the branch to which a is the tangent at A and as the point B on the branch to which b is the tangent at B. The join of A to A is the tangent a which counts as two united tangents; the join of A to B is united to the tangent b; similarly for the join of B to B and to A.

10. The total number of joins to a curve at a cusp is six, of which three are tangent-joins and three are cusp-joins all united on the tangent to the curve at the cusp.

11. When the double point A, B becomes a cusp C the tangents a, b and the cut-joins AB, BA are all united on the tangent c at C. As in the case of the joins to the curve of any point one of the four tangents now united on c becomes a cusp-join thus reducing the number of tangent-joins to three and increasing the number of cusp-joins to three.

12. The view that an n-ic considered as an envelope consists of an m-an, 2δ points united in pairs at the δ cuts and 3κ points united in triads at the κ cusps leads to the following relation between the joins of an n-ic with itself,

$$\tau + 2\delta(m-4) + 3\kappa(m-3) + 2\delta(\delta-1) + \tfrac{9}{2}\kappa(\kappa-1) + 6\delta\kappa = T,$$

τ being the number of double tangents or tangent-joins of the curve and T the whole number of joins which depends only on n the degree of the n-ic.

In addition to the number τ of tangent-joins of the curve proper there is to be reckoned:

(1) The tangent-joins to the curve of the points united in pairs at each of the δ cuts. The number of such joins from a cut is $m-4$, excluding the tangents at the cut, and the whole number is $2\delta(m-4)$.

(2) The tangent-joins to the curve of the points united in triads at each of the κ cusps. The number of such joins from a cusp is $m-3$, excluding the tangents at the cusp, and the whole number is $3\kappa(m-3)$.

(3) The joins of each pair of points at a cut to the pair at another cut. The number of such joins is $2\delta(\delta-1)$.

(4) The joins of each triad of points at a cusp to the triad at another cusp. The number of such joins is $\frac{9}{2}\kappa(\kappa-1)$.

(5) The joins of each pair of points at a cut to each triad of points at a cusp. The number of such joins is $6\delta\kappa$.

The value of T is obtained from its value when the n-ic becomes n straight lines and is $\frac{1}{2}n(n-2)(n^2-9)$ as will be shown later (§ 16). See fig. 2, p. 65.

13. If the view taken here of the n-ic considered as an envelope is correct it follows by the principle of duality that an m-an with τ double tangents and ι inflexions consists of an n-ic, 2τ straight lines united in pairs on each double tangent and 3ι straight lines united in triads on each inflexion tangent.

The cuts of a straight line with the m-an are

(1) the n cuts with the n-ic,

(2) the $2\tau+3\iota$ cuts with the straight lines united in pairs on the double tangents and in triads on the inflexion tangents.

It is not necessary to state in detail the consequences of the dual theory.

A. L. Dixon, F.R.S., has given an analytic proof of the foregoing theory. He has shown that when the lineal $f(x, y, z)$ expresses an n-ic with δ cuts and κ cusps the corresponding pointal is

$$\pi^{\delta}{}_1\alpha_r{}^2 \cdot \pi^{\kappa}{}_1\beta_s{}^3 \cdot \phi_m(\xi, \eta, \zeta),$$

α_r, β_s representing a cut and a cusp respectively, and

$$m+2\delta+3\kappa=n(n-1).$$

14. In the paper to the London Mathematical Society quoted above the characters of a curve in terms of its degree n, class m and deficiency D were stated to be

$$\kappa=2(n-1)-m+2D,$$
$$\iota=2(m-1)-n+2D,$$
$$\delta=\tfrac{1}{2}(n-1)(n-6)+m-3D,$$
$$\tau=\tfrac{1}{2}(m-1)(m-6)+n-3D.$$

The number of conditions c which determine the curve is $m+n+1-D$.

15. Assuming that a straight line is of the first degree and of class zero and is determined by two conditions, it follows that its deficiency is zero. From these data the characters of a straight line are $\kappa=0$, $\iota=-3$, $\delta=0$, $\tau=4$.

Similarly the characters of a point are $\kappa=-3$, $\iota=0$, $\delta=4$, $\tau=0$.

16. The further relation, deduced from the characters given above,

$$\iota+6\delta+8\kappa=I=3n(n-2),$$

implies that a cut absorbs six inflexions and a cusp eight inflexions.

Consider now the simplest form of an n-ic, viz. n straight lines, which is of class zero, has $\frac{1}{2}n(n-1)$ cuts and is determined by $2n$ conditions.

The deficiency obtained from $2n = n + 1 - D$ is $D = -(n-1)$.

The characters are $\kappa = 0$, $\iota = -3n$, $\delta = \frac{1}{2}n(n-1)$, $\tau = 4n$. Hence from the relation $\tau + 2\delta(m-4) + 2\delta(\delta - 1) = T$ (§ 12),

$$\begin{aligned} T &= 4n - 4n(n-1) + n(n-1)\{\tfrac{1}{2}n(n-1) - 1\} \\ &= \tfrac{1}{2}n(n-2)(n^2-9), \end{aligned}$$

the result quoted above (§ 12).

17. Similar results are obtained from the consideration of the simplest form of an m-an, viz. m points.

18. The classification of n-ics based upon their characters should include the cases of n-ics which are composed of curves of lower degrees and have zero or negative deficiency. A table of such curves will commence with

$$m = n(n-1), \quad D = \tfrac{1}{2}(n-1)(n-2),$$
$$\kappa = 0, \quad \iota = 3n(n-2), \quad \delta = 0,$$
$$\tau = \tfrac{1}{2}n(n-2)(n^2-9),$$

and end with

$$m = 0, \quad D = -(n-1), \quad \kappa = 0,$$
$$\iota = -3n, \quad \delta = \tfrac{1}{2}n(n-1), \quad \tau = 4n.$$

19. A few results are added on the characters of curves which are the loci of cuts of two (m, n) linear pencils.

20. When an $(m+n)$-ic so described has no cuts except those at the centres of the pencils and no cusps, the class of the curve is $2mn$ and its deficiency $(m-1)(n-1)$,

$$\begin{aligned} &\kappa = 0, \\ &\iota = 3(2mn - m - n), \\ &\delta = \tfrac{1}{2}m(m-1) + \tfrac{1}{2}n(n-1), \\ &\tau = 2(mn-1)(mn-2) - 4(m-1)(n-1), \\ &c = mn + 2(m+n). \end{aligned}$$

21. When $m-1$ cuts at the centre of one pencil and $n-1$ cuts at the centre of the other pencil are changed into cusps, the class of the curve is $2mn - m - n + 2$ and the deficiency is as before $(m-1)(n-1)$,

$$\begin{aligned} \kappa &= m + n - 2, \\ \iota &= 6(m-1)(n-1) + m + n - 2, \\ \delta + \kappa &= \tfrac{1}{2}(m+n-1)(m+n-2) - (m-1)(n-1), \\ \tau + \iota &= \tfrac{1}{2}(2mn - m - n + 1)(2mn - m - n) - (m-1)(n-1), \\ c &= (m+1)(n+1) + 1. \end{aligned}$$

22. The simplest form of the $(m+n)$-ic consists of an $(m+n)$-an, $(m+n)(m-1)$ points united at the centre of one pencil and $(m+n)(n-1)$ points united at the other centre. The joins of a point to the locus are $m+n$ tangent-joins and $(m+n)(m+n-2)$ cut and cusp-joins. The deficiency is zero,

$$\kappa = m+n-2,$$

$$\iota = m+n-2,$$

$$\delta+\kappa = \tfrac{1}{2}(m+n-1)(m+n-2),$$

$$\tau+\iota = \tfrac{1}{2}(m+n-1)(m+n-2),$$

$$c = 2(m+n)+1.$$

23. The analytic expression for the simplest form of the $(m+n)$-ic is the lineal

$$y^{m+n} - x^m z^n.$$

The pointal of this is

$$\xi^{(m+n)(n-1)}\zeta^{(m+n)(m-1)}\{m^m n^n \eta^{m+n} - [-(m+n)]^{m+n}\xi^m\zeta^n\},$$

which exhibits the nature of simplest form of the $(m+n)$-ic as explained in the preceding paragraph.

24. The shape of the simplest form of the $(m+n)$-ic is, when m and n are both odd numbers, an oval having tangents at the centres of the pencils meeting the curve in $m+1$, $n+1$ points respectively and, when one of the numbers m, n is even and the other odd, a curve with an inflexion at one centre and a cusp at the other centre of the pencils.

25. The following results of a research into the characters of a proper curve of degree n and least class m_0 for that degree were communicated to the Oxford Mathematical Society in 1896.

26. The least class m_0 of a curve of degree $(p-1)p$ is p and of a curve of degree intermediate to $(p-1)p$ and $p(p+1)+1$ is $p+1$, except in the case of a curve of degree $p(p+1)-1$ the least class of which is $p+2$.

27. The least deficiency D_0 of a curve of degree n and of least class m_0 for that degree is $\frac{1}{2}(n+i)-m_0+1$, i being 1 or 0 according as n is odd or even.

28. The least class m_0' of a curve of degree n and of no deficiency is $\frac{1}{2}(n+i)+1$, i being 1 or 0 according as n is odd or even.

29. If α is the curve of degree n of least class m_0 and of least deficiency D_0 for that degree and class, and β the curve of degree n, of no deficiency and of least class m_0' for that degree and deficiency, the class m_0' of β exceeds the class m_0 of α by D_0, the number of cusps κ_0 of α exceeds the number of cusps κ_0' of β by $3D_0$, the number of cuts δ_0' of β exceeds the number of cuts δ_0 of α by $4D_0$ and the number of conditions which determine β exceeds the number of conditions which determine α by $2D_0$.

30. The following tables give the values of the characters of α and β for curves of degrees from 6 to 12.

α

n	m_0	D_0	δ_0	κ_0	τ_0	ι_0	c_0
6	3	1	0	9	0	0	9
7	4	1	4	10	1	1	11
8	4	1	8	12	2	0	12
9	4	2	10	16	0	1	12
10	4	2	16	18	1	0	13
11	5	2	24	19	3	1	15
12	4	3	28	24	0	0	14

β

n	m_0'	D_0'	δ_0'	κ_0'	τ_0'	ι_0'	c_0'
6	4	0	4	6	3	0	11
7	5	0	8	7	5	1	13
8	5	0	12	9	6	0	14
9	6	0	18	10	9	1	16
10	6	0	24	12	10	0	17
11	7	0	32	13	14	1	19
12	7	0	40	15	15	0	20

There is a similar theory of the least degree of a proper curve of the mth class.

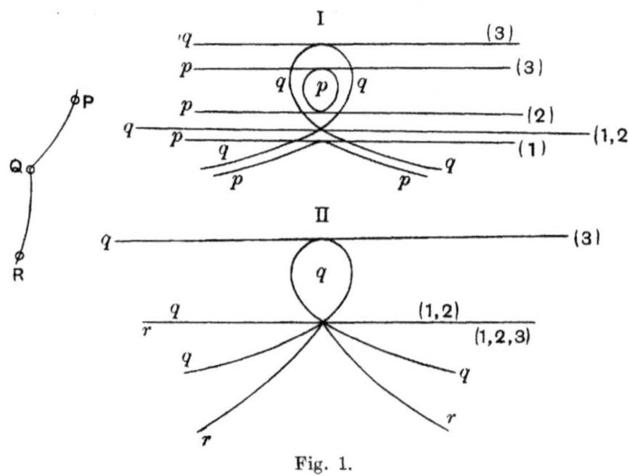

Fig. 1.

I. P determines p, one of a pencil of 4-ics which are 12-ans. Q determines a 4-ic q with a cut. The tangents (1), (2) to p become united on the join (1, 2) to the cut of q. The 12-an now consists of a 10-an and a 2-an, viz. two points united at the cut.

II. Q determines q, one of a pencil of 4-ics each consisting of a 10-an and 2-an. R determines r with a cusp. The tangent (3) is united to the join (1, 2) to the cut which is now a cusp. The 10-an and 2-an is changed into a 9-an and 3-an, viz. three points united at the cusp.

In the three cases the number of joins of a point to the curve is 12.

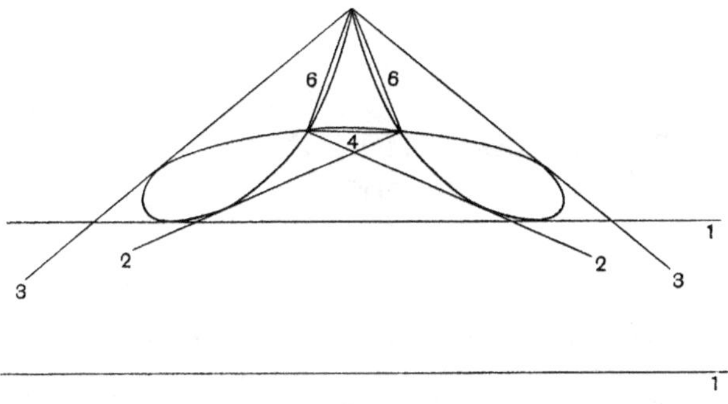

Fig. 2.

A 4-ic with 2 cuts and 1 cusp is a 5-an and 7-an, viz. 2 points at each cut and 3 points at the cusp. There are 2 tangent-joins of the curve, 4 cut-joins to the curve, 6 cusp-joins to the curve, 4 joins of cut to cut, 12 joins of cuts to cusp, in all 28 joins.

DIE ZENTRALPROJEKTION IN DER ABSOLUTEN GEOMETRIE

Von Marcel Grossmann.

Bekanntlich wird in der Zentralprojektion die Lage des Zentrums gegen die Bildebene bestimmt durch den *Distanzkreis*. Legt man die euklidische Geometrie zugrunde, so spielt dieser Kreis bei den Konstruktionen eine doppelte Rolle: er nimmt die Umlegungen des Zentrums mit Ebenen auf, die zur Bildebene normal sind, und er definiert das Orthogonalsystem des Zentrums, da er konjugiert ist zu dem imaginären Direktrixkreis des Polarsystems, welches die Bildebene aus diesem Orthogonalsystem schneidet.

Will man die Methode der Zentralprojektion anwenden auf die beiden *nichteuklidischen Geometrieen*, so hat man zu beachten, dass dem Distanzkreis diese doppelte Bedeutung nicht mehr zukommt. Ist nämlich AOB ein Durchmesser des Distanzkreises, Z das Projektionszentrum, so ist der Winkel AZB kleiner oder grösser als ein rechter Winkel, je nachdem man die Geometrie von *Lobatschefsky* oder von *Riemann* voraussetzt. Trägt man also in der Ebene AZB in Z einen halben rechten Winkel an die Distanz OZ an, so wird dessen zweiter Schenkel die Bildebene ausserhalb oder innerhalb des Distanzkreises schneiden, je nachdem man die erste bezw. zweite der genannten Geometrien voraussetzt. Das Orthogonalsystem des Zentrums wird also von der Bildebene in einem Polarsystem geschnitten, für welches der reelle Repräsentant ein dem Distanzkreis konzentrischer Kreis ist, dessen Radius grösser oder kleiner ist als die Distanz, je nachdem man die Geometrie von Lobatschefsky bezw. von Riemann annimmt. Dieser Kreis möge der *Orthogonalkreis* des Zentrums genannt werden.

Die Geometrie des Masses ist bestimmt, wenn man in einer gegebenen Bildebene den Distanzkreis D und den Orthogonalkreis K eines bestimmten Punktes Z, der nicht in der Bildebene liegt, als gegeben voraussetzt.

In Fig. 1 sei der Orthogonalkreis K grösser als der Distanzkreis D, so dass die dadurch bestimmte Geometrie hyperbolisch wird. Dann kann man leicht Geraden bestimmen, die zueinander parallel sind, d. h. Punkte des absoluten Kegelschnittes ω der Bildebene konstruieren.

Es seien nämlich AB und CD zwei zueinander rechtwinklige Durchmesser des Distanzkreises. Der Halbstrahl OC schneide den Orthogonalkreis in E. Die Tangenten, die den Distanzkreis in A bezw. D berühren, schneiden sich in einem eigentlichen Punkte F, sofern die Distanz d, was immer möglich ist, so gewählt wird, dass der Orthogonalkreis ein eigentlicher Kreis ist. Denn da die Winkel FOD und EBO je gleich $\frac{\pi}{4}$ sind, so sind die rechtwinkligen Dreiecke FDO und EOB kongruent, weil $OD = BO = d$ ist. Also ist auch $DF = OE = a$, wenn der Radius des Orthogonalkreises mit a bezeichnet wird. Da nun das Viereck $AODF$ drei rechte Winkel hat, so schneidet der Orthogonalkreis—nach *Lobatschefsky*—die Seite AF in einem Punkte G der Parallelen durch O zu DF, so dass der Schnittpunkt U von OG und DF ein Punkt des absoluten Kegelschnittes ist. Der Winkel DOG ist daher der zur Distanz d gehörige Parallelwinkel $\Pi(d)$.

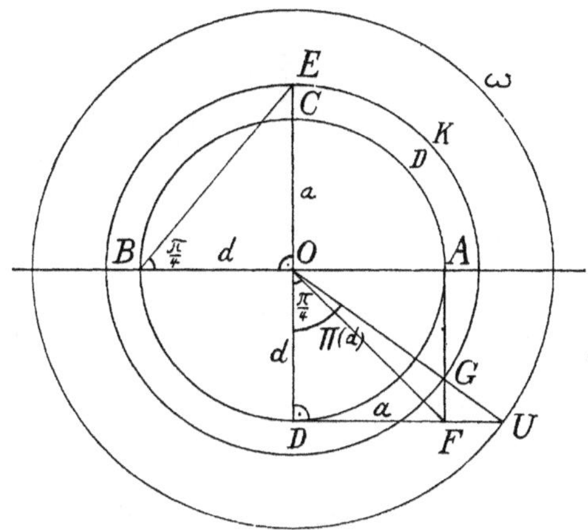

Fig. 1.

In der Zentralprojektion der euklidischen Geometrie werden die Geraden und Ebenen des Raumes durch ihre Spur- und Fluchtelemente bestimmt. Die fundamentale Bedeutung, die der unendlich-fernen Ebene bei dieser Bestimmung zukommt, ist in der Zentralprojektion der nichteuklidischen Geometrie einer Ebene zuzuweisen, deren Auswahl nach Zweckmässigkeitsgründen zu erfolgen hat. Als besonders geeignet empfiehlt sich die *absolute Polarebene des Projektionszentrums.* Es bildet für die Durchführbarkeit der Konstruktionen kein Hindernis, dass diese Ebene in der hyperbolischen Geometrie uneigentlich ist. Für den Spezialfall der euklidischen Geometrie geht die genannte Ebene in die unendlich-ferne Ebene über.

Die Zentralprojektion des Schnittpunktes einer Geraden mit der absoluten Polarebene des Zentrums möge der *Fixpunkt* der Geraden genannt werden.

Die Zentralprojektion der Schnittgeraden einer Ebene mit der absoluten Polarebene des Zentrums möge die *Fixlinie* der Ebene genannt werden. Die gemeinsame Normale der Spur- und der Fixlinie einer Ebene geht durch den Mittelpunkt des Distanzkreises (Hauptpunkt).

Die bisher getroffenen Festsetzungen ermöglichen die Entwicklung der Konstruktionen der Zentralprojektion nach einheitlicher Methode für alle drei Geometrien, d. h. unabhängig von der besonderen Form des Parallelenpostulates. In diesem Sinne möge von einer *absoluten Zentralprojektion* gesprochen werden.

Die *descriptiven* Aufgaben der absoluten Zentralprojektion werden ebenso gelöst wie in der Zentralprojektion der euklidischen Geometrie.

Als *Beispiel* für die Behandlung *metrischer* Aufgaben diene das *Normalenproblem*. In Fig. 2 sei der Distanzkreis D grösser angenommen als der zugehörige Orthogonalkreis K, so dass eine elliptische Massbestimmung definiert wird, bei welcher eine Unterscheidung von eigentlichen und uneigentlichen Elementen wegfällt. Eine Gerade g sei gegeben durch ihren Spurpunkt S_g und ihren Fixpunkt Q_g', so dass

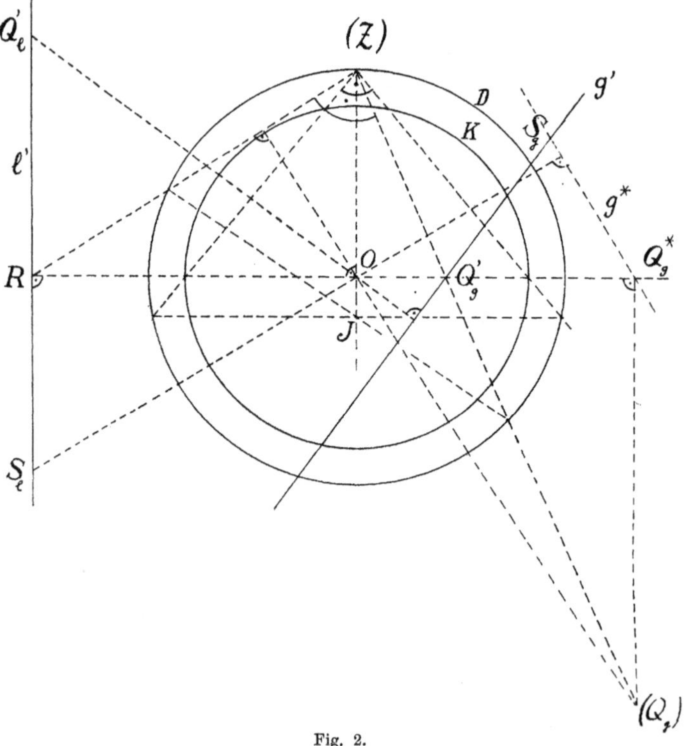

Fig. 2.

$g' \equiv S_g Q_g'$ ihre Zentralprojektion ist. Alle Ebenen, die zur Geraden g normal sind, bilden ein Ebenenbüschel, dessen Scheitelkante die zur Geraden g im absoluten Polarsystem konjugierte Gerade l ist. Die projizierende Ebene von l, d. h. die Normalebene durch Z zu g, entspricht im absoluten Polarsystem dem Punkte Q_g, ist also normal zum Fixstrahl ZQ_g'. Man lege daher die Ebene ZOQ_g' in die Bildebene um, übertrage die Rechtwinkelinvolution um (Z) auf den Distanzkreis, bestimme den Pol J dieser Involution, und konstruiere die Normale $(Z)\,R$ zum Fixstrahl $(Z)\,Q_g'$. Die Spur der gesuchten projizierenden Ebene geht durch R normal zu OR und ist die Zentralprojektion der konjugierten Geraden l.

Da nun aber die Beziehung zwischen den beiden konjugierten Geraden g und l eine gegenseitige ist, so liegt der Fixpunkt Q_e' der Geraden l in der Normalen zu g' durch den Hauptpunkt O. Zur Bestimmung der Geraden l ist nur noch der Spurpunkt S_l zu konstruieren. Die absolute Polarebene dieses Spurpunktes ist aber die normalprojizierende Ebene der Geraden g. Zur Bestimmung der Spur dieser Ebene, d. h. der Normalprojektion der Geraden g, bestimmt man die Normalprojektion des Punktes Q_g, indem man in der vorhin benützten Umlegung die Normale von O auf $(Z)R$ zieht, die den umgelegten Fixstrahl $(Z)Q_g'$ in der Umlegung (Q_g) schneidet. Die Normalprojektion Q_g^* von Q_g liegt in der Normalen von (Q_g) auf OQ_g'. Die Gerade $S_gQ_g^*$ ist die Normalprojektion g^* der Geraden g, und der Spurpunkt S_l von l liegt in der Normalen von O auf diese Gerade.

Damit ist die Gerade l durch Spur- und Fixpunkt bestimmt und das Normalenproblem bis auf deskriptive Konstruktionen erledigt.

Die Durchführung dieses Beispieles lässt erkennen, dass die metrischen Aufgaben der absoluten Zentralprojektion nur lösbar sind bei ausgiebiger Verwendung projektiver Methoden.

Eine ausführlichere Darstellung der auf die Zentralprojektion gegründeten Konstruktionen der absoluten Geometrie gedenke ich in Bälde zu veröffentlichen.

ON THE CHARACTERISTIC NUMBERS OF THE POLYTOPES $e_1 e_2 \ldots e_{n-2} e_{n-1} S(n+1)$ AND $e_1 e_2 \ldots e_{n-2} e_{n-1} M_n$ OF SPACE S_n.

By P. H. Schoute.

Introduction.

In my "Analytical treatment of the polytopes regularly derived from the regular polytopes" (*Verhandelingen* of Amsterdam, Vol. XI, No. 3) I show how it is always possible to deduce the characteristic numbers of any expansion or expansion and contraction form derived from one of the three regular polytopes, simplex, measure polytope and cross polytope, of space S_n, by passing first to the symbol of coordinates. But I am quite well aware of the fact that this way is an indirect one, and that it is to be considered as a desideratum to find a short cut leading from the expansion and contraction symbols to the characteristic numbers immediately. As to the tracing of this short cut I have tried what I could do myself with respect to the case of the measure polytope M_n; but my results are so unsatisfactory that I cannot advise anyone who is in a hurry to be at home in time for dinner to follow that short cut. If $a_{p,n}$ represents in general the number of limits $(l)_p$ of p dimensions of a polytope $(P)_n$ of n dimensions I find in the case of *successive* expansions for

$e_1 M_n$

$$a_{0,n} = 2\,(n+1)_2$$

$$a_{p,n} = \{(n-p)+1\}\,(n+1)_{p+1}$$

$$p = 1, 2, \ldots, n$$

$e_1 e_2 M_n$

$$a_{0,n} = 6\,(n+1)_3$$

$$a_{1,n} = 3n\,(n+1)_3$$

$$a_{p,n} = \{(n-p)^2 + \tfrac{1}{2}(p+1)(n-p)+1\}\,(n+1)_{p+1}$$

$$p = 2, 3, \ldots, n$$

$e_1 e_2 e_3 M_n$

$$a_{0,n} = 24\,(n+1)_4$$

$$a_{1,n} = 12n\,(n+1)_4$$

$$a_{2,n} = 2\,(2n^2 - 6n + 7)\,(n+1)_4$$

$$a_{p,n} = \{(n-p)^3 + (p-1)(n-p)^2 + \tfrac{1}{6}(p^2 - 2p + 9)(n-p) + 1\}\,(n+1)_{p+1}$$

$$p = 3, 4, \ldots, n$$

$e_1 e_2 e_3 e_4 M_n$

$$a_{0,n} = 120\,(n+1)_5$$

$$a_{1,n} = 60n\,(n+1)_5$$

$$a_{2,n} = 10\,(2n^2 - 7n + 11)\,(n+1)_5$$

$$a_{3,n} = 5\,(n^3 - 8n^2 + 23n - 22)\,(n+1)_5$$

$$a_{p,n} = \{(n-p)^4 + \tfrac{1}{2}(3p-7)(n-p)^3 + \tfrac{1}{12}(7p^2-35p+66)(n-p)^2 + \tfrac{1}{24}(p^3-10p^2+49p-36)(n-p)+1\}(n+1)_{p+1}$$
$$p = 4, 5, \ldots, n$$

from which results it is not yet an easy task to deduce the general laws.

In the hope that another mathematician may bring this general investigation to a close I will confine myself here to the case that *all* the expansion operations are applied to the *regular* cell, and try to find the characteristic numbers of these polytopes in function of the number of dimensions n, though also here the easiest way to success is the indirect one which passes by the symbol of coordinates. As the result of the application of all the expansion operations to the polarly related measure polytope and cross polytope is the same we have to consider two polytopes only, viz. $e_1 e_2 \ldots e_{n-2} e_{n-1} S(n+1)$ and $e_1 e_2 \ldots e_{n-2} e_{n-1} M_n$.

1. By means of the method indicated in the memoir quoted above we find the results deposed in the two following tables:

$e_1 e_2 \ldots e_{n-2} e_{n-1} S(n+1)$

n	a_0	a_1	a_2	a_3	a_4	a_5	a_6	a_7	a_8
2	6	6	1						
3	24	36	14	1					
4	120	240	150	30	1				
5	720	1800	1560	540	62	1			
6	5040	15120	16800	8400	1806	126	1		
7	40320	141120	191520	126000	40824	5796	254	1	
8	362880	1451520	2328480	1905120	834120	186480	18150	510	1

$e_1 e_2 \ldots e_{n-2} e_{n-1} M_n$

n	a_0	a_1	a_2	a_3	a_4	a_5	a_6	a_7	a_8
2	8	8	1						
3	48	72	26	1					
4	384	768	464	80	1				
5	3840	9600	8160	2640	242	1			
6	46080	138240	151680	72960	14168	728	1		
7	645120	2257920	3037440	1948800	595728	73752	2186	1	
8	10321920	41287680	65802240	52899840	22305024	4612608	377504	6560	1

In these tables the unit always indicates the polytope itself.

2. From the results contained in these tables we can deduce a simple relation between the characteristic numbers placed in a horizontal row and those placed in the immediately preceding one, a relation remembering the principal property of the number triangle of Pascal. So we deduce in both cases the results for $n=4$ from those of $n=3$, beginning with the unit at the right-hand side, as follows:

$S(n+1)$	M_n
$1 = 1$	$1 = 1$
$30 = 2(14+1)$	$80 = 3 \,.\, 26 + 2 \,.\, 1$
$150 = 3(36+14)$	$464 = 5 \,.\, 72 + 4 \,.\, 1$
$240 = 4(24+36)$	$768 = 7 \,.\, 48 + 6 \,.\, 72$
$120 = 5(\quad 24)$	$384 = \quad 8 \,.\, 48$

or in general:

$$\text{for } S(n+1) \ldots a_{p+1,n} = (n-p)(a_{p,n-1} + a_{p+1,n-1}) \quad \ldots\ldots (1),$$

$$\text{for } M_n \ldots a_{p+1,n} = \{2(n-p)-1\} a_{p,n-1} + 2(n-p-1) a_{p+1,n-1} \ldots (2).$$

It is the chief aim of this communication to prove these relations, hitherto of an experimental character.

3. After having communicated the experimental laws (1) and (2) to Mrs A. Boole Stott, entreating her to send me a general geometrical proof of them, I received within a month her considerations dealing with the deduction of the characteristic numbers for $n = 2$, 3 and 4 from those for $n = 1$, 2 and 3 respectively. As her study forms the basis of my general analytic proof I feel myself bound to communicate her results first. I will do so by treating in detail the transition of hexagon to tO in the case of relation (1) and that of octagon to tCO in the case of relation (2).

Deduction of tO. If on a hexagon $ABC\ldots$ (Fig. 1) situated in a horizontal plane α we construct a tO—or, if one likes, if we place a tO with a limiting hexagon

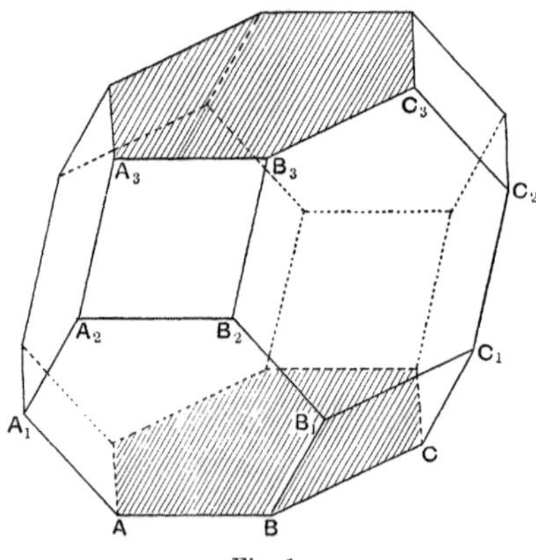

Fig. 1.

$ABC\ldots$ on the table—it is at once evident that the vertices of this polyhedron lie in four horizontal planes α, α_1, α_2, α_3, and that the last plane which is uppermost contains a limiting hexagon $A_3B_3C_3\ldots$ equipollent to $ABC\ldots$, whilst the intermediate ones divide the *tO* into three slices of the same thickness. Now from this diagram we can deduce immediately the following relations

$$\left.\begin{array}{ll}\text{number of faces} & = 2\,(1+6)\\ \text{number of edges} & = 3\,(6+6)\\ \text{number of vertices} & = 4\,.\,6\end{array}\right\},$$

as the relation (1) demands. We treat each of these limits for itself.

Faces. The original face $ABC\ldots$ presents itself in two parallel positions $ABC\ldots$ and $A_3B_3C_3\ldots$, giving 2×1. Moreover between each edge of $ABC\ldots$ and the corresponding edge of $A_3B_3C_3\ldots$ stand two inclined faces, so between AB and A_3B_3 the hexagon $ABB_1B_2A_2A_1$ and the square $A_2B_2B_3A_3$, between BC and B_3C_3 the square BCC_1B_1 and the hexagon $B_1C_1C_2C_3B_3B_2$, giving 2×6. So we find $2\,(1+6)$.

Edges. Each edge of $ABC\ldots$ presents itself in three parallel positions, so AB in the positions AB, A_2B_2, A_3B_3 and BC in the positions BC, B_1C_1, B_3C_3, giving 3×6. Moreover from each vertex of $ABC\ldots$ to the corresponding vertex of $A_3B_3C_3\ldots$ leads a broken line consisting of three inclined edges, giving 3×6. So we find $3\,(6+6)$.

Vertices. Each vertex of $ABC\ldots$ presents itself four times, giving 4×6.

Deduction of tCO. If on an octagon $ABC\ldots$ (Fig. 2) situated in a horizontal

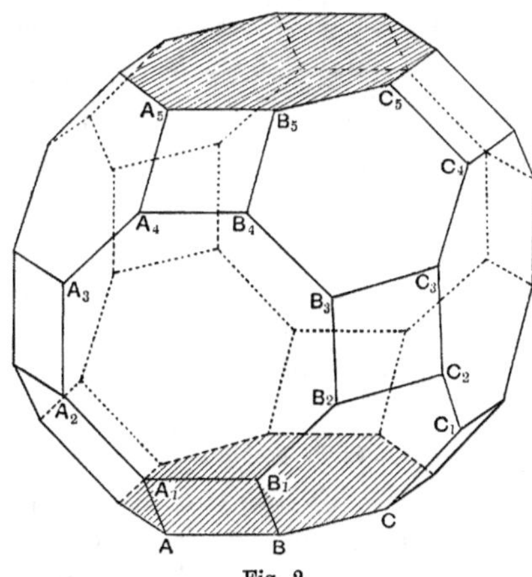

Fig. 2.

plane α we construct a *tCO*—or if we place a *tCO* with a limiting octagon $ABC\ldots$ on the table—it is at once evident that the vertices of this polyhedron lie in six horizontal planes α, α_1, α_2, α_3, α_4, α_5, that the last plane which is uppermost contains an octagon $A_5B_5C_5\ldots$ equipollent to $ABC\ldots$, whilst the four intermediate ones divide the *tCO* into five slices, all but the middle one of the same thickness. Now from this diagram we derive the relations

$$\left.\begin{array}{ll}\text{number of faces} & = 2\,.\,1 + 3\,.\,8\\ \text{number of edges} & = 4\,.\,8 + 5\,.\,8\\ \text{number of vertices} & = 6\,.\,8\end{array}\right\},$$

in accordance with relation (2).

Faces. The original face $ABC\ldots$ presents itself in two positions $ABC\ldots$ and $A_5B_5C_5\ldots$, giving 2×1. Moreover, between each edge of $ABC\ldots$ and the corresponding edge of $A_5B_5C_5\ldots$ stand three inclined faces, so between AB and A_5B_5 the square ABB_1A_1, the octagon $A_1B_1B_2B_3B_4A_4A_3A_2$ and the square $A_4B_4B_5A_5$, between BC and B_5C_5 the hexagon $BCC_1C_2B_2B_1$, the square $B_2C_2C_3B_3$ and the hexagon $B_3C_3C_4C_5B_5B_4$, giving 3×8. So we find $2.1+3.8$.

Edges. Each edge of $ABC\ldots$ presents itself in four parallel positions, so AB in the positions AB, A_1B_1, A_4B_4, A_5B_5 and BC in the positions BC, B_2C_2, B_3C_3, B_5C_5, giving 4×8. Moreover, from each vertex of $ABC\ldots$ to the corresponding vertex of $A_5B_5C_5\ldots$ leads a broken line of five inclined edges, giving 5×8. So we find $4.8+5.8$.

Vertices. Each vertex of $ABC\ldots$ presents itself six times, giving 6×8.

4. *Analytic proof of relation* (1). The symbol of coordinates of the polytope $e_1e_2\ldots e_{n-1}S(n+1)$ of space S_n is $(n, n-1, \ldots, 1, 0)$. This symbol shows that the vertices of the polytope lie in the $n+1$ parallel spaces $x_{n+1}=0$, $x_{n+1}=1, \ldots, x_{n+1}=n$, and that these spaces S_{n-1} divide the polytope into n slices of the same thickness $=1$. In each of these spaces lie $n!$ vertices, so in $x_{n+1}=k$ the vertices for which moreover $(n, n-1, \ldots, k+1, k-1, k-2, \ldots, 0)=x_1, x_2, \ldots, x_n$. Only for the extreme spaces $x_{n+1}=0$ and $x_{n+1}=n$ these vertices are the vertices of a limiting polytope

$$(n, n-1, \ldots, 1)=(n-1, n-2, \ldots, 0)=e_1e_2\ldots e_{n-2}S(n).$$

For shortness we indicate the polytope $e_1e_2\ldots e_{n-1}S(n+1)$ itself by the symbol $(P)_n$ and its limits situated in the spaces $x_{n+1}=0$ and $x_{n+1}=n$ by $(P)^{(0)}_{n-1}$ and $(P)^{(n)}_{n-1}$ respectively.

In connection with the geometrical considerations of Art. 3 we suppose, in order to facilitate the reasoning of the analytic proof of relation (1), that the limits $(P)^{(0)}_{n-1}$ and $(P)^{(n)}_{n-1}$ of $(P)_n$ are horizontal, $(P)^{(0)}_{n-1}$ being at the bottom and $(P)^{(n)}_{n-1}$ at the top, which includes that all the other limits $(l)_{n-1}$ of $(P)_n$ are inclined, and now we prove the following lemma:

"In $(P)_n$ every limit $(l)^{(0)}_p$ of $(P)^{(0)}_{n-1}$ is joined to the corresponding limit $(l)^{(n)}_p$ of $(P)^{(n)}_{n-1}$ by a chain of $n-p$ inclined limits $(l)_{p+1}$ with the property that any $(l)_{p+1}$ of this chain is in contact with each of the two immediately adjacent ones by a horizontal limit $(l)_p$ equipollent to $(l)^{(0)}_p$ and $(l)^{(n)}_p$, which enables us to consider the $n-p$ limits $(l)_{p+1}$ of this chain and the $n-p+1$ limits $(l)_p$ formed by $(l)^{(0)}_p$, the $n-p-1$ intermediate p-dimensional limits of contact and $(l)^{(n)}_p$ as the limits $(l)_{p+1}$ and $(l)_p$ of $(P)_n$ deduced from the chosen limit $(l)^{(0)}_p$ of $(P)^{(0)}_{n-1}$."

"We find *all* the limits $(l)_{p+1}$ of $(P)_n$ by joining together the system of the sets of $n-p$ *inclined* limits $(l)_{p+1}$ deduced from every limit $(l)^{(0)}_p$ of $(P)^{(0)}_{n-1}$, and the system of the sets of $n-p$ *horizontal* limits $(l)_{p+1}$ deduced from every limit $(l)^{(0)}_{p+1}$ of $P^{(0)}_{n-1}$."

In order to prove this lemma we select any limit $(l)^{(0)}_p$ of the polytope $(P)^{(0)}_{n-1}$, represented itself by

$$x_{n+1}=0, \qquad (x_1, x_2, \ldots, x_n)=1, 2, \ldots, n,$$

by following Theorem III of our memoir quoted above, i.e. by writing its extended symbol with the $n-p+1$ syllables in the form

$$x_{n+1}=0-,\ (x_1, x_2, \dots, x_{k_1})=1, 2, \dots, k_1-,$$
$$(x_{k_1+1}, x_{k_1+2}, \dots, x_{k_1+k_2})=k_1+1, k_1+2, \dots, k_1+k_2-,$$
$$\dots -,\ (x_{k_1+k_2+\dots+k_{n-p-1}+1}, x_{k_1+k_2+\dots+k_{n-p-1}+2}, \dots, x_n)$$
$$=k_1+k_2+\dots+k_{n-p-1}+1, k_1+k_2+\dots+k_{n-p-1}+2, \dots, n,$$

where the indices i of the coordinates x_i have been chosen so as to suit this arrangement. If this symbol is abridged to

$$0\,(1, \dots, k_1)\,(k_1+1, \dots, k_1+k_2)\dots(k_1+k_2+\dots+k_{n-p-1}+1, \dots, n),$$

and we maintain tacitly the order of succession $x_{n+1}, x_1, \dots, x_n$ of the coordinates, the symbol of the corresponding $(l)_p^{(n)}$ of $(P)_{n-1}^{(n)}$ becomes

$$n\,(0, \dots, k_1-1)\,(k_1, \dots, k_1+k_2-1)\dots(k_1+k_2+\dots+k_{n-p-1}, \dots, n-1),$$

and now, in accordance with the first part of the lemma, we have to look out for a limit $(l)_{p+1}$ not entirely situated in $x_{n+1}=0$ of which $(l)_p^{(0)}$ is a limit. Evidently we get this $(l)_{p+1}$ by putting the digit 0 inside the brackets of the syllable $(1, \dots, k_1)$, giving us

$$(0, 1, \dots, k_1)\,(k_1+1, \dots, k_1+k_2)\dots(k_1+k_2+\dots+k_{n-p-1}+1, \dots, n),$$

of which $(l)_{p+1}$ the limit $(l)_p$ opposite to $(l)_p^{(0)}$ is represented by

$$k_1\,(0, 1, \dots, k_1-1)\,(k_1+1, \dots, k_1+k_2)\dots(k_1+k_2+\dots+k_{n-p-1}+1, \dots, n);$$

we represent this $(l)_p$ by the symbol $(l)_p^{(k_1)}$ after the value k_1 of the horizontal space $x_{n+1}=k_1$ bearing it. So, in order to pass from $(l)_p^{(0)}$ to the first $(l)_p$ equipollent to it on the way towards $(l)_p^{(n)}$ mentioned in the lemma, we have to leap from $x_{n+1}=0$ to $x_{n+1}=k_1$, and now it is once more evident that the second limit $(l)_{p+1}$ passing through $(l)_p^{(k_1)}$ is obtained by putting the new digit k_1 of x_{n+1} inside the brackets of the syllable $(k_1+1, \dots, k_1+k_2)$, giving us for that new $(l)_{p+1}$

$$(0, \dots, k_1-1)\,(k_1, k_1+1, \dots, k_1+k_2)\dots(k_1+k_2+\dots+k_{n-p-1}+1, \dots, n),$$

and for its $(l)_p$ opposite to $(l)_p^{(k_1)}$

$$k_1+k_2\,(0, \dots, k_1-1)\,(k_1, \dots, k_1+k_2-1)\dots(k_1+k_2+\dots+k_{n-p-1}+1, \dots, n),$$

which may be represented by the symbol $(l)_p^{(k_1+k_2)}$. Going on in this manner we find the $n-p+1$ equipollent horizontal limits

$$(l)_p^{(0)}, (l)_p^{(k_1)}, (l)_p^{(k_1+k_2)}, \dots, (l)_p^{(k_1+k_2+\dots+k_{n-p-2})}, (l)_p^{(k_1+k_2+\dots+k_{n-p-1})}, (l)_p^{(n)}$$

leading to the $n-p$ inclined limits $(l)_{p+1}$ mentioned in the first part of the lemma; so this part is proved.

Any limit $(l)_{p+1}$ of $(P)_n$ is either horizontal or inclined. In the first case it lies in the space $x_{n+1}=q$, where q represents one of the digits from 0 to n, which proves that it belongs to one of the systems of $n-p$ horizontal limits deduced from a certain limit $(l)_{p+1}^{(0)}$ of $(P)_{n-1}^{(0)}$. In the second case x_{n+1} is contained under the coordinates corresponding to a certain syllable $(r, r+1, \dots, s-1, s)$ of the extended symbol of $(l)_{p+1}$, and then this $(l)_{p+1}$ admits two horizontal limits $(l)_p$ which can be got, by putting either the minimum digit r or the maximum digit s outside the brackets in the form

$$r(r+1, \ldots, s-1, s), \qquad s(r, r+1, \ldots, s-1)$$

and taking r in the first and s in the second case for the value of x_{n+1}; so any inclined limit $(l)_{p+1}$ of $(P)_n$ belongs to one of the systems of $n-p$ inclined limits deduced from a certain limit $(l)_p^{(0)}$ of $(P)_{n-1}^0$.

Evidently the lemma proved now includes the relation (1); so this relation is proved now too.

5. We now prove the following theorem:

"If $a_{p,n}$ represents the number of the limits $(l)_p$ of $(P)_n$ we have

$$a_{n-1,n} = 2(2^n - 1),$$
$$a_{n-2,n} = 3(3^n - 2.2^n + 1),$$
$$a_{n-3,n} = 4(4^n - 3.3^n + 3.2^n - 1),$$
$$a_{n-4,n} = 5(5^n - 4.4^n + 6.3^n - 4.2^n + 1),$$

....

and in general

$$a_{n-p,n} = (p+1)\sum_{k=0}^{p}(-1)^k (p)_k (p-k+1)^n \ldots\ldots\ldots\ldots\ldots\ldots (\mathrm{I}),$$

for the values of 1, 2, ..., n of p."

We prove by means of (1) that the relation (I) holds for n as soon as it holds for $n-1$. But as (I) deals with $a_{n-p,n}$ instead of $a_{p,n}$ we transcribe (1) by substituting $n-q$ for $p+1$ in the form

$$a_{n-q,n} = (q+1)(a_{n-q-1,n-1} + a_{n-q,n-1}),$$

or if we replace q by p

$$a_{n-p,n} = (p+1)(a_{n-p-1,n-1} + a_{n-p,n-1}) \ldots\ldots\ldots\ldots\ldots\ldots (1').$$

Now, when (I) holds for $n-1$ we have

$$a_{(n-1)-p,n-1} = (p+1)\sum_{k=0}^{p}(-1)^k (p)_k (p-k+1)^{n-1},$$

$$a_{(n-1)-(p-1),n-1} = p\sum_{k=0}^{p-1}(-1)^k (p-1)_k (p-k)^{n-1}.$$

In order to facilitate the addition of these two equations we transform the latter by the substitution $k = k' - 1$, after which we drop the dash of k' and replace the expression $p(p-1)_{k-1}$ then making its appearance by $k(p)_k$; so the second equation becomes

$$a_{n-p,n-1} = -p\sum_{k=1}^{p}(-1)^k k (p)_k (p-k+1)^{n-1}.$$

Then by combining the terms with $(p)_k(p-k+1)^{n-1}$ of both forms, which compels us to take the term with $k=0$ of the first equation by itself, we find

$$a_{n-p-1,n-1} + a_{n-p,n-1} = \sum_{k=1}^{p}(-1)^k \{(p+1)-k\}(p)_k (p-k+1)^{n-1} + (p+1)(p+1)^{n-1}$$
$$= \sum_{k=1}^{p}(-1)^k (p)_k (p-k+1)^n + (p+1)^n$$
$$= \sum_{k=0}^{p}(-1)^k (p)_k (p-k+1)^n.$$

So by means of (1') we get

$$a_{n-p,n} = (p+1) \sum_{k=0}^{p} (-1)^k (p)_k (p-k+1)^n,$$

which is the equation (I) of the theorem for n.

As the relation (I) holds for $n=3$ we find by application of the result deduced just now that it holds in general for any positive integer value of n.

Remark 1. By means of the general results $a_{0,n} = (n+1)!$ and $a_{1,n} = \frac{1}{2}n(n+1)!$ we find by the way the relations

$$\sum_{k=0}^{n} (-1)^k (n)_k (n-k+1)^n = n!,$$

$$\sum_{k=0}^{n-1} (-1)^k (n-1)_k (n-k)^n = \tfrac{1}{2}(n+1)!.$$

Remark 2. According to the preceding developments the determination of $a_{n-p,n}$ is a problem of the theory of permutations and combinations. If we call a row of subsequent integers a group of subsequent numbers this problem can be stated as follows:

"Given the $n+1$ numbers $1, 2, \dots, n+1$. To find all the possible ways of splitting up these $n+1$ numbers into $p+1$ groups of subsequent numbers, to determine how many times τ the numbers of each of these sets of $(p+1)$ groups can be distributed differently over $n+1$ quantities $x_1, x_2, \dots, x_{n+1}$ and to evaluate $\Sigma\tau$ extended over the different sets of the total system."

6. *Analytic proof of relation* (2). The symbol of coordinates of the polytope $e_1 e_2 \dots e_{n-1} M_n$ of space S_n is $[1+(n-1)\sqrt{2}, 1+(n-2)\sqrt{2}, \dots, 1+\sqrt{2}, 1]$. This symbol shows that the vertices of the polytope lie in the $2n$ parallel spaces $x_n = \pm(1+k\sqrt{2})$, where k assumes one of the values $n-1, n-2, \dots, 0$, and that these spaces S_{n-1} divide the polytope into $2n-1$ slices, all of the same thickness $\sqrt{2}$ but the middle one comprised between $x = \pm 1$ which admits the thickness 2. In each of these spaces lie $2^{n-1}(n-1)!$ vertices, so in $x_n = 1+k\sqrt{2}$ the vertices for which moreover

$$[1+(n-1)\sqrt{2}, \dots, 1+(k+1)\sqrt{2}, 1+(k-1)\sqrt{2}, \dots, 1+\sqrt{2}, 1] = x_1, x_2, \dots, x_{n-2}, x_{n-1}.$$

Only for the extreme spaces $x_n = \pm\{1+(n-1)\sqrt{2}\}$ these vertices are the vertices of a limiting polytope $[1+(n-2)\sqrt{2}, \dots, 1] = e_1 e_2 \dots e_{n-2} M_{n-1}$. For shortness we indicate the $e_1 e_2 \dots e_{n-1} M_n$ itself by the symbol $[M]_n$ and its limits situated in the spaces $x_n = \pm\{1+(n-1)\sqrt{2}\}$ by $[M]_{n-1}^{+}$ and $[M]_{n-1}^{-}$ respectively. We also here suppose that the spaces $x_n =$ constant are horizontal, $[M]_{n-1}^{-1}$ lying at the bottom and $[M]_{n-1}^{+}$ at the top. Now we prove the lemma:

"In $[M]_n$ every limit $(l)_p^{+}$ of $[M]_{n-1}^{+}$ is joined to the corresponding limit $(l)_p^{-}$ of $[M]_{n-1}^{-}$ by a chain of $2(n-p)+1$ inclined limits $(l)_{p+1}$ with the property that any $(l)_{p+1}$ of this chain is in contact with each of the two immediately adjacent ones by a horizontal limit $(l)_p$ equipollent to $(l)_p^{+}$ and $(l)_p^{-}$, which enables us to consider the $2(n-p)+1$ limits $(l)_{p+1}$ of this chain and the $2(n-p+1)$ limits $(l)_p$ formed by $(l)_p^{+}$, the $2(n-p)$ intermediate p-dimensional limits of contact and $(l)_p^-$ as the limits $(l)_{p+1}$ and $(l)_p$ of $[M]_n$ deduced from the chosen limit $(l)_p^+$ of $[M]_{n-1}^+$."

"We find *all* the limits $(l)_{p+1}$ of $[M]_n$ by joining together the system of the sets of $2(n-p)+1$ *inclined* limits $(l)_{p+1}$ deduced from every limit $(l)_p^+$ of $[M]_{n-1}^+$ and the system of the sets of $2(n-p)$ *horizontal* limits $(l)_{p+1}$ deduced from every limit $(l)_{p+1}^+$ of $[M]_{n-1}^+$."

The space S_{n-1} represented by $x_n=0$ is a space of symmetry of $[M]_n$. So, if starting from the limit $(l)_p^+$ situated in $x_n=1+(n-1)\sqrt{2}$, we have determined the limits $(l)_{p+1}$ of the chain mentioned in the statement of the lemma which are situated above $x_n=0$, the mirror images of these $(l)_{p+1}$ with respect to $x_n=0$ will belong also to the chain, and these two parts of the chain, the upper and the lower one, must be found to be connected by an unpaired limit $(l)_{p+1}$ which is its own mirror image, the space S_{p+1} containing it being normal to $x_n=0$. Nevertheless we call this connecting link between the two parts of the chain *inclined*, as we use inclined here for "non horizontal" and not for "non perpendicular." So we find that the number of limits $(l)_{p+1}$ forming the chain must be odd, and all we have to do is to show that the number of these $(l)_{p+1}$ lying above $x_n=0$ is $n-p$.

Now according to Theorem XXX of the second part* of my memoir quoted above the extended symbol of the limit $(l)_p^+$ selected can present itself in two different forms, which have to be considered successively, i.e. we can split up the n digits $1, 1+\sqrt{2}, \ldots, 1+(n-1)\sqrt{2}$ either in $n-p$ or in $n-p+1$ groups of subsequent digits and place all these groups between round brackets with exception of that group of the $n-p+1$ which contains the digit 1, this digit having to be placed between square brackets. So e.g. in the example of the octagon $[1+\sqrt{2}, 1]=x_1, x_2$ in the upper plane $x_3=2+\sqrt{2}$ of the tCO—see the diagram of this polyhedron—we find edges with two different kinds of symbols, i.e. edges $(1+\sqrt{2}, 1)$ and edges $[1]\,1+\sqrt{2}$; here A_5B_5 belongs to the second kind, B_5C_5 to the first, and so on alternately.

First case (of $n-p$ groups). Writing for short $1+\overline{k_1-1}\sqrt{2}$ instead of $1+(k_1-1)\sqrt{2}$ we represent the $(l)_p^+$ selected by the extended symbol

$$(1, \ldots, 1+\overline{k_1-1}\sqrt{2})(1+k_1\sqrt{2}, \ldots, 1+\overline{k_1+k_2-1}\sqrt{2})\ldots$$
$$\ldots(1+\overline{k_1+k_2+\ldots+k_{n-p-1}}\sqrt{2}, \ldots, 1+\overline{n-2}\sqrt{2})\,1+\overline{n-1}\sqrt{2},$$

where the groups of digits refer successively to the sets of positive coordinates

$$(x_1, \ldots, x_{k_1})(x_{k_1+1}, \ldots, x_{k_1+k_2})\ldots(x_{k_1+k_2+\ldots+k_{n-p-1}+1}, \ldots, x_{n-1})\,x_n,$$

the subscripts i and the signs of the coordinates x_i having been chosen so as to suit this arrangement. Then we get the first $(l)_{p+1}$ of the chain, i.e. the $(l)_{p+1}$ passing through the selected $(l)_p^+$, by bringing the digit $1+\overline{n-1}\sqrt{2}$ inside the brackets of the syllable $(1+\overline{k_1+k_2+\ldots+k_{n-p-1}}\sqrt{2}, \ldots, 1+\overline{n-2}\sqrt{2})$ and its second horizontal

* This part will appear in the beginning of 1913.

$(l)_p$ by taking the smallest digit $1+\overline{k_1+k_2+\ldots+k_{n-p-1}}\sqrt{2}$ placed at the other side as value of x_n out of the brackets. So the first new horizontal limit is represented by the symbol

$$(1,\ldots,1+\overline{k_1-1}\sqrt{2})(1+k_1\sqrt{2},\ldots,1+\overline{k_1+k_2-1}\sqrt{2})\ldots$$
$$\ldots(1+\overline{k_1+k_2+\ldots+k_{n-p-1}+1}\sqrt{2},\ldots,1+\overline{n-1}\sqrt{2})\,1+\overline{k_1+k_2+\ldots+k_{n-p-1}}\sqrt{2},$$

the order of succession and the signs of the x_i being maintained. Continuing this process by bringing the new value $1+\overline{k_1+k_2+\ldots+k_{n-p-1}}\sqrt{2}$ inside and taking $1+\overline{k_1+k_2+\ldots+k_{n-p-2}}\sqrt{2}$ out of the brackets of the syllable

$$(1+\overline{k_1+k_2+\ldots+k_{n-p-2}}\sqrt{2},\ldots,\overline{1+k_1+k_2+\ldots+k_{n-p-1}-1}\sqrt{2}),$$

etc., we find for the values of x_n for $(l)_p^+$ and the successive limits $(l)_p$ of contact lying above $x_n=0$

$$1+\overline{n-1}\sqrt{2},\ 1+\overline{k_1+k_2+\ldots+k_{n-p-1}}\sqrt{2},\ 1+\overline{k_1+k_2+\ldots+k_{n-p-2}}\sqrt{2},\ldots,\ 1+k_1\sqrt{2},\ 1.$$

So there are on either side of $x_n=0$, $(l)_p^+$ and $(l)_p^-$ included, $n-p+1$ horizontal limits $(l)_p$, and therefore *in toto* $2(n-p)+1$ limits $(l)_{p+1}$ composing the chain, etc.

Second case (of $n-p+1$ groups). Under the same notation we start here from the extended symbol

$$[1,\ldots,1+\overline{k_1-1}\sqrt{2}](1+k_1\sqrt{2},\ldots,1+\overline{k_1+k_2-1}\sqrt{2})\ldots$$
$$\ldots(1+\overline{k_1+k_2+\ldots+k_{n-p}}\sqrt{2},\ldots,1+\overline{n-2}\sqrt{2})\,1+\overline{n-1}\sqrt{2}.$$

By following closely the same process we get above $x_n=0$ for x_n the $n-p$ values

$$1+\overline{n-1}\sqrt{2},\ 1+\overline{k_1+k_2+\ldots+k_{n-p}}\sqrt{2},\ 1+\overline{k_1+k_2+\ldots+k_{n-p-1}}\sqrt{2},\ldots,\ 1+k_1\sqrt{2},$$

also leading to $2(n-p)+1$ limits $(l)_{p+1}$, etc.

Remark. The difference between the two cases (with $n-p$ and with $n-p+1$ groups), already indicated above for the edges of an octagon, greatly influences the character of the components $(l)_{p+1}$ of the chain. In the first case of $n-p$ groups these components agree in this with the components of the chain of Art. 4 that they are prismotopes the constituents of which are exclusively polytopes $e_1e_2\ldots e_{k-1}S(k+1)$ for different values of k; on the contrary in the second case these components are prismotopes one of the constituents of which is always an $e_1e_2\ldots e_{k-1}M_k$ for a certain value of k*.

The proof of the second part of the lemma, though it also breaks up into the two different cases distinguished above, can be copied from that given in Art. 4.

7. We now prove the following theorem:

"If $a_{p,n}$ represents the number of the limits $(l)_p$ of $[M]_n$ we have

$$a_{n-1,n}=3^n-1,$$
$$a_{n-2,n}=5^n-2.3^n+1,$$
$$a_{n-3,n}=7^n-3.5^n+3.3^n-1,$$
$$a_{n-4,n}=9^n-4.7^n+6.5^n-4.3^n+1,$$

....

* In the two diagrams of tO and tCO the square partakes in all the chains of faces, as the square belongs to both categories.

and in general

$$a_{n-p,n} = \sum_{k=0}^{p} (-1)^k (p)_k \{2(p-k)+1\}^n \quad \text{..................(II)},$$

for the values 1, 2, ..., n of p."

We prove by means of (2) that (II) holds for n as soon as it holds for $n-1$. But to that end we have to transform (2) into

$$a_{n-p,n} = (2p+1)\, a_{n-p-1,n-1} + 2p a_{n-p,n-1} \quad \text{..................(2')}$$

first. Now, when (II) holds for $n-1$, we have

$$(2p+1)\, a_{(n-1)-p,n-1} = (2p+1) \sum_{k=0}^{p} (-1)^k (p)_k \{2(p-k)+1\}^{n-1},$$

$$2p a_{(n-1)-(p-1),n-1} = 2p \sum_{k=0}^{p-1} (-1)^k (p-1)_{k-1} \{2(p-k)-1\}^{n-1}.$$

In the second of these equations we put $k=k'-1$, drop afterwards the dash of k', and replace $p(p-1)_{k-1}$ by $k(p)_k$; so we get for the second equation

$$2p a_{n-p,n-1} = -2 \sum_{k=1}^{p} (-1)^k k (p)_k \{2(p-k)+1\}^{n-1}.$$

Then by combining the terms with $(p)_k \{2(p-k)+1\}^{n-1}$, and taking the term with $k=0$ of the first equation by itself, we find

$$\begin{aligned}
&(2p+1)\, a_{n-p-1,n-1} + 2p a_{n-p,n-1} \\
&= \sum_{k=1}^{p} (-1)^k \{(2p+1)-2k\} (p)_k \{2(p-k)+1\}^{n-1} + (2p+1)(2p+1)^{n-1} \\
&= \sum_{k=1}^{p} (-1)^k (p)_k \{2(p-k)+1\}^n + (2p+1)^n \\
&= \sum_{k=0}^{p} (-1)^k (p)_k \{2(p-k)+1\}^n.
\end{aligned}$$

So by means of (2') we get

$$a_{n-p,n} = \sum_{k=0}^{p} (-1)^k (p)_k \{2(p-k)+1\}^n,$$

which is the equation (II) for n.

As (II) holds for $n=3$ it holds for any positive integer value of n.

Remark 1. By means of the general results $a_{0,n} = 2^n . n!$ and $a_{1,n} = 2^{n-1} . n . n!$ we find by the way the relations

$$\sum_{k=0}^{n} (-1)^k (n)_k \{2(n-k)+1\}^n = 2^n . n!$$

$$\sum_{k=0}^{n-1} (-1)^k (n-1)_k \{2(n-k)-1\}^n = 2^{n-1} . n . n!$$

Remark 2. We leave it to the reader to interpret the determination of $a_{n-p,n}$ in this case of $e_1 e_2 \dots e_{n-1} M_n$ as a problem of the theory of permutations and combinations.

CONFORMAL GEOMETRY

By Edward Kasner.

Introduction.

The connection of conformal transformations with the theory of functions of a complex variable is so important that the geometry based on the group of conformal transformations has been studied almost exclusively as an auxiliary to the theory of functions.

Conformal geometry, that is the study of those properties of geometric configurations which are invariant under all conformal transformations, has therefore not yet been developed into a systematic theory, comparable, for example, with projective geometry. The former theory is naturally more difficult than the latter, since it is based on a larger group. It would seem that it deserves study for its own sake, and also as an example of a geometry based on an infinite continuous group, instead of (in Lie's terminology), a finite continuous group.

The simplest configurations to investigate in conformal geometry are curves and sets of curves. The curves considered will throughout be assumed to be analytic: we shall confine ourselves in fact to regular real analytic arcs. The conformal transformations are assumed to be regular in a region *including* the curves considered.

A single curve has no invariants; any curve may be transformed conformally into any other curve, in particular into a straight line.

The first configuration of real interest is composed of two curves having a common point, that is, a curvilinear angle. The conformal transformations operating on the angle are assumed to be regular in a region including the vertex of the angle. The result of such a transformation is a new curvilinear angle, having the same magnitude as the original. This common magnitude we shall denote by θ: this is the radian measure of the angle formed by the tangent lines at the vertex. It is a differential invariant of the first order since it depends only on the slopes of the curves.

If two curvilinear angles are conformally equivalent, they certainly have the same θ, but is this sufficient? If two angles are equal in magnitude, that is have the same θ, will they necessarily be conformally equivalent? This is certainly a fundamental question in conformal geometry. We shall show that the answer is not always in the affirmative. Curvilinear angles exist, which while equal in magnitude, are *not* conformally equivalent. For example: it is possible to construct a curvilinear angle formed by two analytic arcs, intersecting orthogonally, which can not be transformed conformally into an angle formed by a pair of perpendicular straight lines (the transformation to be of course regular at the vertex).

This fact indicates the existence of other invariants besides θ. The principal problem of the present paper is to find these absolute invariants of higher order. The invariants considered are differential invariants involving curvatures and derivatives of the curvatures of the sides of the curvilinear angle. Only invariants of finite order are discussed. The existence of invariants of infinite order, that is, those involving *all* the coefficients in the power series representing the sides of the angle, is not settled. It is shown that if θ is a rational part of π (in which case we shall term the angle rational), there exist higher invariants. If the angle is irrational, it is shown that no higher invariants exist.

Among the rational angles we have to include the case where θ is zero: such an angle is termed a horn angle. Each type of horn angle has a unique invariant whose order depends on the degree of contact of the sides.

The question as to when two curvilinear angles are conformally equivalent is not yet completely solved. In order that two angles shall be equivalent, it is *necessary* that not simply θ, but the higher invariants obtained in this paper shall be equal. Whether this is also *sufficient* depends on the existence of invariants of infinite order.

At the end of the paper we consider briefly the corresponding (dual) problem of equilong geometry.

Conformal Invariants of Curvilinear Angles.

THEOREM I. *An irrational curvilinear angle has no conformal invariants of higher order; that is, the magnitude θ of the angle is its only invariant.*

To prove this consider two angles of the same magnitude θ, where θ/π is irrational. Without loss of generality we may assume the angles in *reduced form*, that is, one side a straight line. Thus the angle in the plane $z = x + iy$ is formed say by the axis of reals and the curve

$$y = \alpha_1 x + \alpha_2 x^2 + \ldots \qquad (1);$$

and the angle in the plane $Z = X + iY$ is formed by the axis of reals and the curve

$$Y = A_1 X + A_2 X^2 + \ldots \qquad (2).$$

By assumption $\alpha_1 = A_1 = \tan\theta$. To shew that there is no other invariant relation involving a finite number of higher coefficients of (1) and (2), we prove that a regular conformal transformation

$$Z = c_1 z + c_2 z^2 + \ldots \qquad (3),$$

(where the c's are real and $c_1 \neq 0$), converting (1) into (2), can be (formally) found.

For the determination of the c's, we have an infinite number of equations. The first is an identity in virtue of $\alpha_1 = A_1$ and allows c_1 to be arbitrary. The second involves c_1 and c_2 and is linear in c_2. The κth involves c_1 up to c_κ and is linear in c_κ. The coefficient of c_κ can be put into the form

$$\sin\kappa\theta - \tan\theta\cos\kappa\theta \qquad (4).$$

This cannot vanish for any integer $\kappa > 1$; for if it did θ would be a rational part of π. Hence for arbitrary values of $\alpha_1, \alpha_2, \ldots, \alpha_n$ and $A_1, A_2, \ldots, A_n$ values of $c_1, c_2, \ldots, c_n$ can be determined: in fact in ∞^1 ways since c_1 is arbitrary.

For arbitrary curves (1) and (2) the series (3) can be formally calculated, the coefficients being rational functions of the parameter c_1. Probably for some value of c_1 the corresponding series (3) will be convergent. This, however, the author has not succeeded in proving. If there are actually cases where (3) is divergent for all values of c_1, this would indicate the existence of invariant relations involving all the coefficients of the curves, that is, invariants of infinite order.

If the angle is rational, $\theta/\pi = p/q$, where the fraction is supposed to be reduced to lowest terms, some of the expressions (4) will vanish, namely those for which κ is $q+1$ or $2q+1$ or $3q+1$, etc. The series (3) cannot usually be found even formally. Conditions of consistency arise. The discussion of the system of equations for the c's is complicated, but by a certain device we can prove

THEOREM II. *Every curvilinear angle of rational magnitude has a conformal invariant of higher order.*

For this purpose, we consider the process of *general symmetry*, or reflexion, with respect to an analytic curve. The function-theoretic definition of Schwarz may be stated in purely geometric language as follows: two points are symmetric with respect to a given curve provided the pairs of minimal lines determined by the points intersect on the given curve. For an analytic arc a neighbourhood (region) can be found such that each point in the neighbourhood has a unique symmetric point or image in the neighbourhood. The process is covariant under the conformal group. It is fundamental in our geometry.

Consider then an angle of magnitude θ with vertex V and curved sides a and b. The image of a with respect to b is a definite curve c passing through V. The magnitude of the angle b, c is of course θ. Take the image of b with respect to c, etc. If θ is irrational we shall obtain curves passing through V so that their tangents are everywhere dense. But if θ is rational we shall after a finite number of steps arrive at a curve having the same direction as a. *Hence every rational angle determines a unique horn angle,* i.e. angle of magnitude zero, formed by two curves in contact at the vertex. Any invariant of this angle will be an invariant of the original rational angle. It now remains to prove

THEOREM III. *Every horn angle has one and only one higher conformal invariant.*

If the order of contact of the two sides of the horn angle is $h-1$, so that they have h consecutive points in common, we shall say that the angle is of type h and denote it by H_h. Of course the integer h is invariant under conformal transformation as it is under all contact transformation. The simplest type (ordinary contact) is H_2.

Consider two angles of type h. Taking one side in each plane to be the axis of reals, and the vertex as origin, the curved sides are of the form

$$y = \alpha_h x^h + \alpha_{h+1} x^{h+1} + \dots \quad \text{.........................(5)},$$

$$Y = A_h X^h + A_{h+1} X^{h+1} + \dots \quad \text{.........................(6)}.$$

Expressing the fact that the transformation (3) converts (4) into (5), we are led to a system of equations for the determination of the coefficients $c_1, c_2, \dots$. It is sufficient to note that the κth equation involves the first κ coefficients and is linear with respect to c_κ, the coefficient of this term being

$$(\kappa - h)\,\alpha_h \quad \text{(7).}$$

By assumption α_h does not vanish, therefore expression (7) vanishes when and only when $\kappa = h$. The first h equations thus involve only $c_1, c_2, \ldots, c_{h-1}$: eliminating these we find a relation involving $\alpha_h, \ldots, \alpha_{2h-1}$ and $A_h, \ldots, A_{2h-1}$. There are no other relations since no other eliminations are possible.

The order of the unique invariant of a horn angle of type h is $2h-1$. We denote the invariant by I_{2h-1}.

For a horn angle H_2 (contact of first order), the invariant in question is

$$I_3 = \frac{3\alpha_3}{2\alpha_2^2} \quad \text{(8).}$$

This is, of course, for the reduced form, in which one side is the axis of reals. For the general form where both sides of the angle are curved, we find

$$I_3 = \frac{\dfrac{d\gamma_1}{ds_1} - \dfrac{d\gamma_2}{ds_2}}{(\gamma_1 - \gamma_2)^2} \quad \text{(9),}$$

where γ_1, γ_2 denote the curvatures of the sides (at the vertex), and s_1, s_2 denote the arc lengths. *This is the simplest example of a conformal invariant of higher order.*

Every horn angle can be reduced conformally (so far as terms of finite order are concerned) to the normal form in which the sides are $y = 0$ and $y = x^h + \lambda x^{2h-1}$. The constant λ is then essential, being equivalent to the invariant I_{2h-1}.

Return now to the discussion of rational angles. Such an angle determines a horn angle by the process of successive symmetry described above. If this angle is of type h, its invariant I_{2h-1} will be an invariant of the rational angle. This proves Theorem II. In general if the ratio of θ to π is p to q (in lowest terms) then the type h equals $q+1$ so that the invariant is of order $2q+1$. However, for some rational angles, h may have a greater value. *It is necessary to classify rational angles of a given magnitude first according to the values of the arithmetic invariant h, and then according to the values of the geometric (differential) invariant I_{2h-1}.*

A question remains. Can a rational angle have more than one higher invariant? The related horn angle has only one invariant as stated in Theorem II. The horn angle is determined by the rational angle, but is the converse true? We are led to the following general problem of conformal geometry: given two curves a and b through a point V, find $n-1$ curves through V so that the image of a in the first is the second, the image of the first in the second is the third, and so on until finally the image of the penultimate in the last is b. This may be termed the *equipartition problem for curvilinear angles.* Obviously if the solution exists the magnitudes of the angles formed by the successive curves will all be equal.

The simplest case, $n = 2$, is the problem of *bisection* of a curvilinear angle: to construct a curve c so that the given curves a and b shall be symmetric with respect to c. Judging from ordinary angles, where the sides are straight lines, we should expect two solutions, an internal and an external bisector. This is, however, not always true. We state only the following results. If the given angle is a horn angle for which the sides a and b have contact of exactly the first order, then there will be an internal bisector, but no external. If the sides of the horn angle have contact of exactly the second order, then both internal and external bisectors exist*.

Consider now a curvilinear angle of magnitude $\pi/2$. The image of the first side in the second will be a curve having contact of even order with the first side. For a *general* right angle the contact will be of second order, that is, the related horn angle is of type 3. Right angles such that the type of the horn angle is 5, 7, 9, ... are of greater and greater specialty. In the extreme exceptional case, the order of contact is infinite, that is, the right angle is such that each side is its own image with respect to the other side: this kind of right angle is conformally reducible to a pair of perpendicular straight lines.

A right angle of general type has a unique invariant; this is of the fifth order.

This follows because the related horn angle, of type 3, uniquely determines the right angle. The invariant in question is the I_5 of the related horn angle. If the original right angle is given in reduced form so that one side is $y=0$ and the other is

$$x = \beta_2 y^2 + \beta_3 y^3 + \dots,$$

the invariant is found to be

$$\frac{\beta_5 - 4\beta_3\beta_2^2}{\beta_3^2}.$$

The exceptional right angles arise when β_3 vanishes. Probably for each type of rational angle, there is only one higher invariant.

Equilong Geometry.

This theory is a sort of dual (not however the direct projective dual) of the conformal theory. It is based on the infinite group of equilong transformations introduced by Scheffers in 1905. These are related to the theory of functions of dual numbers $u+jv$, where j is an imaginary unit whose square is zero, just as the conformal transformations are related to functions of the ordinary complex numbers $x+iy$, where the square of i equals -1.

Dual numbers are interpreted geometrically by directed straight lines, using Hessian line coordinates; curves are considered as envelopes of straight lines and are also oriented. Any curve may be transformed into any other curve; in particular, any curve may be transformed into a point (of course, a point is considered as the envelope of all the straight lines passing through it). A single curve, therefore, has no invariant.

The first configuration to be studied consists of two curves having a common tangent line. This is in fact the dual of the figure discussed in the first part, namely: a curvilinear angle composed of two curves having a common point.

The configuration now to be studied has not been given any special name. It possesses an obvious invariant with respect to the equilong group, namely: the

* In general if the type of the horn angle is even, no external bisector exists. If the type is odd, an external bisector probably exists.

distance measured on the common tangent between the two points of contact. The question arises as to whether the configuration has additional invariants, that is, invariants of higher (but finite) order. A special case arises when the distance between the points of contact is zero. In this case the curves have a common tangent and a common point, and therefore form a horn angle. A horn angle arises in both theories since it is in fact a self-dual configuration. The following results are obtained:

THEOREM IV. *The figure formed by two curves having a common tangent and distinct points of contact has no higher equilong invariant.*

THEOREM V. *If the points of contact coincide, so that the figure is a horn angle, then there will be one, and only one, equilong invariant of higher order.*

In this, as well as in the previous discussion, the existence of invariants of infinite order is left unsettled. The decision as to whether such invariants exist depends on questions of convergence of certain power series which can be formally constructed.

For a horn angle of type h, the unique invariant is of order $2h-1$. It is of course different from the corresponding conformal invariant I_{2h-1} and is here denoted by J_{2h-1}.

For the simplest type, $h = 2$, the equilong invariant is

$$J_3 = \frac{\dfrac{dr_1}{d\theta_1} - \dfrac{dr_2}{d\theta_2}}{(r_1 - r_2)^2}$$

where r_1, r_2 denote the radii of curvature and θ_1, θ_2 denote the inclinations of the two curves to any fixed initial line. Of course the values of the radii and the derivatives are taken at the vertex of the angle.

CHARACTERIZATION OF THE TWO GROUPS.

From the usual points of view conformal transformations and equilong transformations are characterized in entirely different ways. Conformal transformations are singled out from all *point transformations* by the requirement that the angle between two curves at a common point shall have its magnitude preserved. Equilong transformations, on the other hand, are singled out from all *line transformations* by requiring that the distance between two curves measured on a common tangent shall be preserved. We shall show that both groups may be characterized in the domain of all *contact transformations* in terms of their behaviour with respect to horn angles.

Every contact transformation turns a horn angle into a horn angle. It is sufficient to consider the simplest type, where the contact is of first order. Conformal transformations leave unaltered a certain invariant of the third order, discussed in the first part and denoted by I_3. Equilong transformations leave unaltered the invariant denoted by J_3. Both of these quantities are differential expressions of the third order; they are in fact combinations of the curvatures of the sides of the horn angle, and of the rates of variation of the curvatures.

THEOREM VI. *Conformal transformations are the only contact transformations for which the I_3 of every horn angle is unaltered. Equilong transformations are the only contact transformations for which the J_3 of every horn angle is unaltered.*

It thus appears that the conformal group may be characterized without mentioning angular magnitude, and the equilong group may be characterized without mentioning distance.

We note that the two invariants may be put into the form

$$I_3 = \frac{\dfrac{d^2\theta_1}{ds_1^2} - \dfrac{d^2\theta_2}{ds_2^2}}{\left(\dfrac{d\theta_1}{ds_1} - \dfrac{d\theta_2}{ds_2}\right)^2},$$

$$J_3 = \frac{\dfrac{d^2s_1}{d\theta_1^2} - \dfrac{d^2s_2}{d\theta_2^2}}{\left(\dfrac{ds_1}{d\theta_1} - \dfrac{ds_2}{d\theta_2}\right)^2},$$

which differ only by the interchange of the letters s and θ.

It is to be remembered that there is no (known) automatic principle by means of which we can pass from the results of conformal geometry to those of equilong geometry; we have to deal with a general analogy rather than a strict duality. What is, for example, the analogue of isothermal systems of curves, which are so important in conformal geometry? The resulting systems are defined in terms of Hessian line coordinates by linear differential equations of the first order and their theory is essentially simpler than that of isothermal systems.

SUR LES SURFACES ISOTHERMIQUES

Par G. Tzitzéica.

1. M. Wilczynski a trouvé une nouvelle méthode pour l'étude des propriétés infinitésimales des courbes, des surfaces et des congruences de droites. En se bornant aux propriétés projectives, il a pris comme base une équation différentielle linéaire dans le cas d'une courbe plane ou gauche, un système d'équations linéaires ordinaires dans le cas des surfaces réglées, et aux dérivées partielles dans le cas des surfaces générales et des congruences.

De mon côté, indépendamment de M. Wilczynski et en partant d'un problème particulier, j'avais trouvé un système d'équations aux dérivées partielles qui définissait toutes les surfaces se déduisant de l'une d'entre elles par des transformations affines.

On est conduit d'une manière naturelle à un principe général que l'on peut énoncer de la manière suivante. *Etant donnée une classe de figures du plan ou de l'espace—courbes, surfaces, congruences—pour étudier leurs propriétés infinitésimales qui restent invariables lorsqu'on applique à ces figures toutes les transformations d'un groupe, il faut d'abord trouver un système approprié de coordonnées pour la détermination d'un élément quelconque de ces figures, former ensuite une équation ou un système d'équations différentielles vérifié par les coordonnées d'un élément variable d'une figure et de toutes les figures transformées.*

Pour étudier, par exemple, les propriétés infinitésimales affines des courbes planes, on choisit d'abord des coordonnées cartésiennes obliques x et y et si, pour plus de simplicité, on prend parmi les transformations affines celles qui laissent invariable l'origine, on aura comme équation différentielle définissant une courbe plane et ses transformées une équation de la forme

$$\theta'' + p\theta' + q\theta = 0 \qquad \text{(1)},$$

p et q étant des fonctions données du paramètre t variable sur la courbe, θ' et θ'' désignant les dérivées de la fonction inconnue θ par rapport à ce paramètre. Deux solutions $x(t)$ et $y(t)$ qui forment un système fondamental de (1) définissent une courbe intégrale. Les invariants et les covariants de l'équation (1) par rapport aux transformations qui ne changent pas les courbes intégrales, à savoir celles qui remplacent θ par const. θ et t par $\phi(t)$, conduisent aux propriétés de ces courbes.

2. Je veux faire une application du principe général que je viens d'énoncer à l'étude des surfaces isothermiques, et montrer spécialement que l'on obtient par cette voie d'une manière régulière l'équation aux dérivées partielles du quatrième ordre trouvée par MM. Rothe et Calapso.

On sait que toute surface à lignes de courbure isothermes garde cette propriété après une transformation conforme de l'espace. Le groupe de transformations que nous considérerons sera donc le groupe G des transformations conformes de l'espace. Comme les transformations de ce groupe s'expriment par des transformations linéaires des coordonnées pentasphériques, ce seront ces coordonnées que nous emploierons dans cette étude. Comme enfin les lignes de courbure d'une surface restent lignes de courbure sur la surface obtenue par une transformation du groupe G, nous prendrons comme coordonnées curvilignes u et v sur une surface isothermique ses lignes de courbure.

Cela étant posé, il résulte que le système vérifié par les coordonnées pentasphériques $x_1, x_2, \ldots, x_5$ d'un point mobile sur une surface isothermique doit être linéaire. Une des équations de ce système est connue depuis longtemps, elle est due à M. Darboux; c'est une équation de Laplace à invariants égaux que nous supposerons prise sous la forme réduite

$$\frac{\partial^2 \theta}{\partial u \partial v} = h\theta \quad \ldots\ldots\ldots\ldots\ldots\ldots\ldots (2).$$

Cependant, il est clair que cette équation ne peut suffire pour définir une surface isothermique et ses transformées. Il faut ajouter à l'équation (2) un nombre d'équations suffisant, pour que le système obtenu puisse définir seulement cinq solutions linéairement indépendantes. A cet effet nous remarquons que si les fonctions x_i $(i = 1, 2, \ldots, 5)$ de u et v sont connues, nous pouvons déterminer cinq fonctions a, b, c, d, e, de manière que l'équation linéaire

$$\frac{\partial^3 \theta}{\partial u^3} = a\frac{\partial^2 \theta}{\partial u^2} + b\frac{\partial^2 \theta}{\partial v^2} + c\frac{\partial \theta}{\partial u} + d\frac{\partial \theta}{\partial v} + e\theta \quad \ldots\ldots\ldots\ldots (3)$$

admette les x_i comme solutions particulières. En effet, en introduisant ces fonctions à la place de θ dans l'équation (3), on obtient pour déterminer a, b, c, d, e un système linéaire, qui n'est impossible ou indéterminé que dans le cas où l'on aurait

$$\left| \frac{\partial^2 x_i}{\partial u^2} \; \frac{\partial^2 x_i}{\partial v^2} \; \frac{\partial x_i}{\partial u} \; \frac{\partial x_i}{\partial v} \; x_i \right| = 0 \quad \ldots\ldots\ldots\ldots\ldots\ldots (4),$$

le premier membre étant un déterminant dont on obtient les cinq lignes, en faisant $i = 1, 2, \ldots, 5$. Or, cette relation signifierait que les x_i vérifieraient une équation de la forme

$$A\frac{\partial^2 \theta}{\partial u^2} + B\frac{\partial^2 \theta}{\partial v^2} + C\frac{\partial \theta}{\partial u} + D\frac{\partial \theta}{\partial v} + E\theta = 0 \quad \ldots\ldots\ldots\ldots (5),$$

ce qui est impossible, le système formé par les équations (2) et (5) ne pouvant avoir plus de quatre solutions linéairement indépendantes et dans ce cas la surface serait une sphère.

On démontrerait de la même manière que l'on peut déterminer cinq autres fonctions a', b', c', d' et e', de manière que les x_i vérifient aussi l'équation

$$\frac{\partial^3 \theta}{\partial v^3} = a'\frac{\partial^2 \theta}{\partial u^2} + b'\frac{\partial^2 \theta}{\partial v^2} + c'\frac{\partial \theta}{\partial u} + d'\frac{\partial \theta}{\partial v} + e'\theta \quad \ldots\ldots\ldots\ldots (6).$$

Il est aisé de voir maintenant que le système formé par les équations (2), (3) et (6) admet cinq solutions linéairement indépendantes.

On peut donc dire que une surface isothermique et toutes celles qu'on en déduit par des transformations conformes de l'espace sont définies par un système de la forme de celui formé par les équations (2), (3) et (6).

3. Cependant, il faut remarquer que les coefficients h, a, b, c, d, e, a', b', c', d', e' ne sont pas des fonctions arbitraires de u et v. Ils doivent vérifier deux sortes de conditions. D'abord celles qui résultent de la relation bien connue

$$\Sigma x^2 = x_1^2 + x_2^2 + \ldots + x_5^2 = 0 \quad \ldots\ldots\ldots\ldots\ldots\ldots\ldots\ldots(7),$$

ensuite les conditions d'intégrabilité.

En partant de (7) on en déduit d'abord

$$\Sigma x \frac{\partial x}{\partial u} = 0, \quad \Sigma x \frac{\partial x}{\partial v} = 0, \quad \ldots\ldots\ldots\ldots\ldots\ldots\ldots\ldots(8),$$

de plus à l'aide de (2) on a

$$\frac{\partial}{\partial v} \Sigma \left(\frac{\partial x}{\partial u}\right)^2 = 0, \quad \frac{\partial}{\partial u} \Sigma \left(\frac{\partial x}{\partial v}\right)^2 = 0.$$

Nous laisserons de côté le cas où l'on aurait

$$\Sigma \left(\frac{\partial x}{\partial u}\right)^2 = 0, \quad \Sigma \left(\frac{\partial x}{\partial v}\right)^2 = 0.$$

On pourra donc changer les variables u et v, sans changer la forme des équations de notre système, de manière que l'on ait

$$\Sigma \left(\frac{\partial x}{\partial u}\right)^2 = 1, \quad \Sigma \left(\frac{\partial x}{\partial v}\right)^2 = 1 \ldots\ldots\ldots\ldots\ldots\ldots\ldots\ldots(9).$$

On tire alors de (8)

$$\Sigma x \frac{\partial^2 x}{\partial u^2} = -1, \quad \Sigma x \frac{\partial^2 x}{\partial v^2} = -1, \quad \Sigma \frac{\partial x}{\partial u} \frac{\partial x}{\partial v} = 0 \quad \ldots\ldots\ldots\ldots(10)$$

et de là

$$\Sigma x \frac{\partial^3 x}{\partial u^3} = 0, \quad \Sigma x \frac{\partial^3 x}{\partial v^3} = 0$$

ou

$$a + b = 0, \quad a' + b' = 0 \ldots\ldots\ldots\ldots\ldots\ldots\ldots\ldots(11).$$

On aura de même

$$\Sigma \frac{\partial x}{\partial v} \frac{\partial^2 x}{\partial u^2} = 0, \quad \Sigma \frac{\partial x}{\partial u} \frac{\partial^2 x}{\partial v^2} = 0 \quad \ldots\ldots\ldots\ldots\ldots\ldots\ldots(12)$$

et de (9)

$$\Sigma \frac{\partial x}{\partial u} \frac{\partial^2 x}{\partial u^2} = 0, \quad \Sigma \frac{\partial x}{\partial v} \frac{\partial^2 x}{\partial v^2} = 0 \ldots\ldots\ldots\ldots\ldots\ldots\ldots(13).$$

De (12) on tire aisément

$$\Sigma \frac{\partial x}{\partial v} \frac{\partial^3 x}{\partial u^3} = h, \quad \Sigma \frac{\partial x}{\partial u} \frac{\partial^3 x}{\partial v^3} = h$$

d'où

$$d = c' = h \ldots\ldots\ldots\ldots\ldots\ldots\ldots\ldots\ldots\ldots(14).$$

On a aussi

$$\Sigma \frac{\partial^2 x}{\partial u^2} \frac{\partial^2 x}{\partial v^2} = 0, \quad \Sigma \left(\frac{\partial^2 x}{\partial u^2}\right)^2 + c = 0, \quad \Sigma \left(\frac{\partial^2 x}{\partial v^2}\right)^2 + d' = 0 \quad \ldots\ldots(15).$$

Enfin de celles-ci on obtient

$$\left.\begin{aligned}&\frac{\partial c}{\partial u}=2ac+2e, \quad \frac{\partial d'}{\partial v}=2b'd'+2e',\\ &\frac{\partial c}{\partial v}=2\frac{\partial h}{\partial u}, \quad \frac{\partial d'}{\partial u}=2\frac{\partial h}{\partial v},\\ &\frac{\partial h}{\partial v}=-bd'-e, \quad \frac{\partial h}{\partial u}=-a'c-e'\end{aligned}\right\}\ \ldots\ldots\ldots\ldots\ldots\ldots(16).$$

Nous avons trouvé le premier groupe (I) de relations entre les coefficients. Ce sont les relations (11), (14) et (16).

4. Il nous reste à former les conditions d'intégrabilité. Nous écrirons d'abord que l'on a

$$\frac{\partial}{\partial v}\left(\frac{\partial^3\theta}{\partial u^3}\right)=\frac{\partial^2}{\partial u^2}\left(\frac{\partial^2\theta}{\partial u\partial v}\right)$$

et en remplaçant les dérivées données par les équations (2), (3) et (6) on obtiendra une équation du second ordre qui doit être vérifiée identiquement. On déduit les relations

$$\left.\begin{aligned}&h=\frac{\partial a}{\partial v}+a'b, \quad \frac{\partial b}{\partial v}+bb'+d=0, \quad 2\frac{\partial h}{\partial u}=ah+bc'+\frac{\partial c}{\partial v},\\ &bd'+e+\frac{\partial d}{\partial v}=0, \quad \frac{\partial^2 h}{\partial u^2}=a\frac{\partial h}{\partial u}+be'+ch+\frac{\partial e}{\partial v}\end{aligned}\right\}\ \ldots\ldots(17).$$

De même, en écrivant que l'on a

$$\frac{\partial}{\partial u}\left(\frac{\partial^3\theta}{\partial v^3}\right)=\frac{\partial^2}{\partial v^2}\left(\frac{\partial^2\theta}{\partial u\partial v}\right),$$

on trouvera les relations

$$\left.\begin{aligned}&aa'+c'+\frac{\partial a'}{\partial u}=0, \quad h=a'b+\frac{\partial b'}{\partial u}, \quad a'c+e'+\frac{\partial c'}{\partial u}=0,\\ &a'd+b'h+\frac{\partial d'}{\partial u}=2\frac{\partial h}{\partial v}, \quad \frac{\partial^2 h}{\partial v^2}=a'e+b'\frac{\partial h}{\partial v}+d'h+\frac{\partial e'}{\partial u}\end{aligned}\right\}\ \ldots\ldots(18).$$

Les relations (17) et (18) d'intégrabilité forment le groupe (II) de relations entre les coefficients de notre système.

5. Nous pouvons simplifier ces relations en remarquant d'abord que des relations (17) et (18) on tire

$$h=\frac{\partial a}{\partial v}+a'b=\frac{\partial b'}{\partial u}+a'b\ \ldots\ldots\ldots\ldots\ldots\ldots\ldots\ldots\ldots\ldots(19).$$

On pourra donc poser

$$a=\frac{1}{\phi}\frac{\partial\phi}{\partial u}, \quad b'=\frac{1}{\phi}\frac{\partial\phi}{\partial v},$$

ϕ étant une nouvelle fonction inconnue. A l'aide de (11) on a aussi

$$b=-\frac{1}{\phi}\frac{\partial\phi}{\partial u}, \quad a'=-\frac{1}{\phi}\frac{\partial\phi}{\partial v}$$

et alors (19) donne

$$h=\frac{1}{\phi}\frac{\partial^2\phi}{\partial u\partial v}.$$

Si donc on connaissait la fonction ϕ, on connaîtrait par là même les coefficients a, b, d, h, a', b', c'. Il nous reste à déterminer les coefficients c, e et d', e'. Or on a les relations

$$\frac{\partial c}{\partial u} = \frac{2c}{\phi}\frac{\partial \phi}{\partial u} + 2e, \quad \frac{\partial c}{\partial v} = 2\frac{\partial h}{\partial u} = \frac{2c}{\phi}\frac{\partial \phi}{\partial v} - 2e',$$

$$\frac{\partial d'}{\partial u} = 2\frac{\partial h}{\partial v} = \frac{2d'}{\phi}\frac{\partial \phi}{\partial u} - 2e, \quad \frac{\partial d'}{\partial v} = \frac{2d'}{\phi}\frac{\partial \phi}{\partial v} + 2e',$$

dont on tire immédiatement

$$c + d' = 2m\phi^2, \quad m = \text{const.}$$

Prenons maintenant la dernière relation de (17) et remplaçons-y $\frac{\partial h}{\partial u}$ et $\frac{\partial e}{\partial v}$ par leurs valeurs tirées de (16). On obtient ainsi

$$\frac{\partial^2 h}{\partial u^2} = -a(a'c + e') + be' + ch - \frac{\partial^2 h}{\partial v^2} - b\frac{\partial d'}{\partial v} - d'\frac{\partial b}{\partial v},$$

ou

$$\frac{\partial^2 h}{\partial u^2} + \frac{\partial^2 h}{\partial v^2} = \left(h + \frac{1}{\phi^2}\frac{\partial \phi}{\partial u}\frac{\partial \phi}{\partial v}\right)(c + d'),$$

ou enfin

$$\frac{\partial^2}{\partial u^2}\left(\frac{1}{\phi}\frac{\partial^2 \phi}{\partial u \partial v}\right) + \frac{\partial^2}{\partial v^2}\left(\frac{1}{\phi}\frac{\partial^2 \phi}{\partial u \partial v}\right) = m\frac{\partial^2}{\partial u \partial v}(\phi^2) \quad \ldots\ldots\ldots\ldots\ldots(20)$$

qui est précisément l'équation de MM. Rothe et Calapso.

Si ϕ est une intégrale de cette équation, c est déterminé par

$$\frac{\partial c}{\partial u} = 4m\phi\frac{\partial \phi}{\partial u} - 2\frac{\partial h}{\partial v}, \quad \frac{\partial c}{\partial v} = 2\frac{\partial h}{\partial u}$$

où, à cause de (20), la condition d'intégrabilité est vérifiée.

Une fois les coefficients connus le système est déterminé et complètement intégrable.

Remarques. 1. Dans l'équation (20) on ne peut avoir $m = 0$ que dans le cas d'une sphère.

2. Les groupes (I) et (II) de relations sont nécessaires pour notre problème. Nous n'avons pas démontré qu'ils sont aussi suffisants. Je le ferai à une autre occasion.

3. Les propriétés des surfaces isothermiques s'étudieraient en formant les invariants et les covariants du système vérifié par les x_i par rapport aux transformations

$$\theta' = k\theta, \quad u' = \pm u + k', \quad v' = \pm v + k'',$$

k, k' et k'' étant des constantes arbitraires.

THE PEDAL LINE OF THE TRIANGLE IN NON-EUCLIDEAN GEOMETRY

By Duncan M. Y. Sommerville.

1. Consider a triangle $\alpha\beta\gamma$, and let the equation of the absolute referred to it be, in point coordinates,

$$(\Delta \between x, y, z)^2 \equiv \left(\begin{array}{ccc} a & h & g \\ h & b & f \\ g & f & c \end{array}\right| \between x, y, z)^2 = 0$$

and in line coordinates,

$$(\nabla \between \xi, \eta, \zeta)^2 \equiv \left(\begin{array}{ccc} A & H & G \\ H & B & F \\ G & F & C \end{array}\right| \between \xi, \eta, \zeta)^2 = 0,$$

the capital letters denoting, as usual, the cofactors of the corresponding small letters in the determinant

$$\Delta \equiv \begin{vmatrix} a & h & g \\ h & b & f \\ g & f & c \end{vmatrix}.$$

The polar of the point (x', y', z') is

$$(\Delta \between x, y, z \between x', y', z') = 0,$$

and this is also the condition that the points (x, y, z), (x', y', z') should be a quadrant distant; the pole of the line (ξ', η', ζ') is

$$(\nabla \between \xi, \eta, \zeta \between \xi', \eta', \zeta') = 0,$$

and this is also the condition that the lines (ξ, η, ζ), (ξ', η', ζ') should be at right angles.

The poles of $\beta\gamma$, $\gamma\alpha$, $\alpha\beta$ are

$$\alpha' \equiv (A, H, G), \quad \beta' \equiv (H, B, F), \quad \gamma' \equiv (G, F, C).$$

The equations of the altitudes $\alpha\alpha'$, $\beta\beta'$, $\gamma\gamma'$ of the triangle are

$$Gy = Hz, \quad Hz = Fx, \quad Fx = Gy.$$

The triangles $\alpha\beta\gamma$, $\alpha'\beta'\gamma'$ are in perspective, the centre of perspective being the common orthocentre $O \equiv \left(\frac{1}{F}, \frac{1}{G}, \frac{1}{H}\right)$. The axis of perspective LMN is the polar of O

with respect to the absolute and its coordinates are $\left(\frac{1}{f}, \frac{1}{g}, \frac{1}{h}\right)$. We may call this the *orthaxis* of the triangle since it cuts each side in a point distant a quadrant from the opposite vertex.

2. Take any point P (x_1, y_1, z_1) and let the perpendiculars $P\alpha'$, $P\beta'$, $P\gamma'$, meet the corresponding sides of the triangle $\alpha\beta\gamma$ in X, Y, Z. For certain positions of P the points X, Y, Z will be collinear and the line XYZ is called the *pedal line* of the point P. We shall call the locus of P the *pedal* or *orthocentral locus*, and the envelope of XYZ the *pedal* or *orthocentral envelope* of the triangle $\alpha\beta\gamma$.

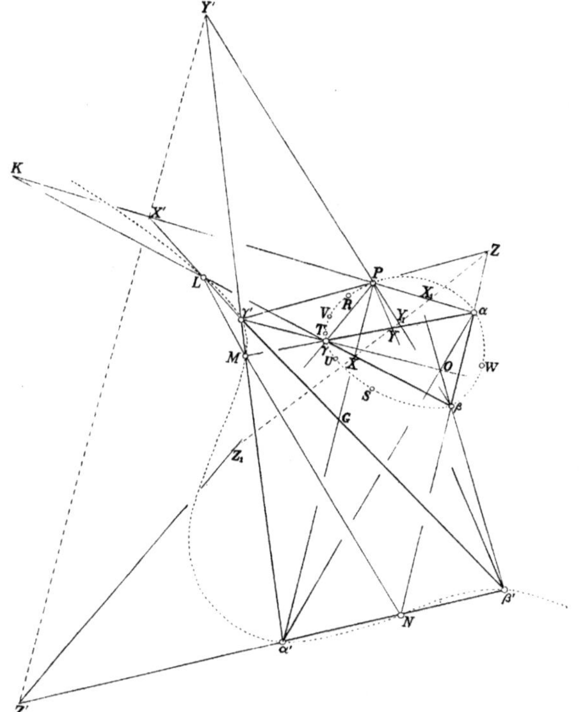

Reciprocally: take any line p (ξ_1, η_1, ζ_1) and let it cut the sides of the triangle $\alpha'\beta'\gamma'$ in X', Y', Z'. For certain positions of the line p the lines $\alpha X'$, $\beta Y'$, $\gamma Z'$ will be concurrent in a point P. We shall call the envelope of p the *orthaxal envelope* and the locus of P the *orthaxal locus* of the triangle $\alpha\beta\gamma$.

Evidently the orthaxal locus and envelope of the triangle $\alpha\beta\gamma$ are the orthocentral locus and envelope of the triangle $\alpha'\beta'\gamma'$.

If P is any point on the pedal locus, $P(\alpha\alpha'\beta\beta'\gamma\gamma')$ is a pencil in involution, and conversely.

Let these rays cut the corresponding pedal line in X_1, X, Y_1, Y, Z_1, Z, and let αX_1 cut $\beta\gamma$ in K.

Then $P(\alpha\alpha', \beta'\gamma') = (X_1X, YZ) = (KX, \gamma\beta) = P(\alpha\alpha', \gamma\beta) = P(\alpha'\alpha, \beta\gamma)$ which proves the theorem*.

Conversely, if $P(\alpha\alpha'\beta\beta'\gamma\gamma')$ is a pencil in involution and $P\alpha'$, $P\beta'$, $P\gamma'$ cut $\beta\gamma$, $\gamma\alpha$, $\alpha\beta$ in X, Y, Z, then X, Y, Z are collinear.

$$P(\alpha\alpha', \beta'\gamma') = P(\alpha'\alpha, \beta\gamma) = P(\alpha\alpha', \gamma\beta) = (KX, \gamma\beta) = \alpha(PX, \gamma\beta).$$

Now $P\alpha$ and αP coincide, and $P\alpha'$, $P\beta'$, $P\gamma'$ cut αX, $\alpha\gamma$, $\alpha\beta$ in X, Y, Z, therefore X, Y, Z are collinear. In exactly the same way it is proved that if $P\alpha$, $P\beta$, $P\gamma$ cut the corresponding sides of the triangle $\alpha'\beta'\gamma'$ in X', Y', Z', then X', Y', Z' are collinear.

Hence if P is any point on the orthocentral locus it is also on the orthaxal locus, and conversely.

Hence *the orthaxal and orthocentral loci coincide*, and so also do the envelopes.

3. To find the equation of the pedal locus.

The equation of $P\alpha'$ is

$$x(Gy_1 - Hz_1) + y(Az_1 - Gx_1) + z(Hx_1 - Ay_1) = 0,$$

and this cuts $\beta\gamma$ in X, whose coordinates are

$$0 : -(Hx_1 - Ay_1) : (Az_1 - Gx_1).$$

The equation of the locus of P is therefore

$$(Cy - Fz)(Az - Gx)(Bx - Hy) = (Fy - Bz)(Gz - Cx)(Hx - Ay),$$

or

$$\Sigma Ayz(hy + gz) = 2(fgh - abc)xyz,$$

which represents a cubic curve passing through the nine points of intersection of the lines $\alpha\gamma'$, $\beta\alpha'$, $\gamma\beta'$ with $\alpha\beta'$, $\beta\gamma'$, $\gamma\alpha'$, i.e. through the points

$$\alpha,\ \beta,\ \gamma,\ \alpha',\ \beta',\ \gamma',\ \binom{\beta\gamma'}{\beta'\gamma},\ \binom{\gamma\alpha'}{\gamma'\alpha},\ \binom{\alpha\beta'}{\alpha'\beta}.$$

Call the last three points U, V, W. Then αU, βV, γW are concurrent in the point whose coordinates are (AF, BG, CH). The following points of the cubic which are in the same row or the same column are collinear:

$$\begin{matrix} V & W & \\ \gamma' & \alpha' & M \\ \alpha & \beta & N \end{matrix}$$

Therefore LVW, and similarly MWU and NUV are sets of collinear points.

The equation of the cubic may also be written

$$\left(\frac{x}{f} + \frac{y}{g} + \frac{z}{h}\right)\left(\frac{A}{f}yz + \frac{B}{g}zx + \frac{C}{h}xy\right) = \frac{ABC - \Delta abc}{f^2g^2h^2}xyz.$$

The curve therefore passes also through L, M, N.

* It is otherwise obvious since $\alpha\beta\gamma XYZ$ is a complete quadrilateral.

If $ABC = \Delta abc$ the cubic breaks up into a straight line (the orthaxis), and a conic circumscribing the triangle $\alpha\beta\gamma$. This condition is also the condition that the triangles $\alpha\beta\gamma$, $\alpha'\beta'\gamma'$ should be inscribed in the same conic. This conic, which exists even when the cubic does not degenerate, touches the cubic at α, β, γ.

4. *The tangents to the cubic at α, α', U, L meet in a point on the cubic.*

The following points of the cubic which are in the same row or column are collinear:

$$\begin{matrix} \alpha & \alpha & \\ \beta & \gamma & L \\ N & M & L \end{matrix} \qquad \begin{matrix} \alpha' & \alpha' & \\ \beta' & \gamma' & L \\ N & M & L \end{matrix}$$

therefore the tangents at α, α' and L cut the cubic in a point L', the tangential of L.

Again we have

$$\begin{matrix} \alpha & L & \\ N & \beta' & \alpha' \\ \beta & \gamma' & U \end{matrix}$$

therefore αL and $\alpha' U$ cut in a point, R say, on the cubic. Similarly βM and $\beta' V$ cut the cubic in S, and γN and $\gamma' W$ cut the cubic in T.

Again

$$\begin{matrix} S & T & \\ \beta & \gamma & L \\ M & N & L \end{matrix} \qquad \text{and} \qquad \begin{matrix} S & T & \\ \beta' & \gamma & U \\ V & N & U \end{matrix}$$

therefore the tangent at U meets the cubic also in L'.

Since therefore four real tangents can be drawn from L' *the cubic is bipartite, and is of the sixth class; it has nine points of inflexion, of which three are real and collinear.*

5. To find the equation of the pedal envelope.

Let the coordinates of the line XYZ be ξ, η, ζ. Where this cuts $\beta\gamma$ we have $x, y, z = 0, -\zeta, \eta$, and the equation of the line through this point perpendicular to $\beta\gamma$, i.e. through α', is

$$x(H\eta + G\zeta) - yA\eta - zA\zeta = 0.$$

The three lines will be concurrent if

$$\begin{vmatrix} H\eta + G\zeta & -A\eta & -A\zeta \\ -B\xi & F\zeta + H\xi & -B\zeta \\ -C\xi & -C\eta & G\xi + F\eta \end{vmatrix} = 0,$$

that is, the line-equation of the pedal envelope is

$$\Delta \Sigma a\eta\zeta(H\eta + G\zeta) = 2(FGH - ABC)\,\xi\eta\zeta,$$

or $$(c\eta - f\zeta)(a\zeta - g\xi)(b\xi - h\eta) = (f\eta - b\zeta)(g\zeta - c\xi)(h\xi - a\eta),$$

or $$\left(\frac{\xi}{F} + \frac{\eta}{G} + \frac{\zeta}{H}\right)\left(\frac{a}{F}\eta\zeta + \frac{b}{G}\zeta\xi + \frac{c}{H}\xi\eta\right) = \frac{\Delta(\Delta abc - ABC)}{F^2G^2H^2}\,\xi\eta\zeta,$$

which represents a curve of the third class touching the lines $\beta\gamma$, $\gamma\alpha$, $\alpha\beta$, $\beta'\gamma'$, $\gamma'\alpha'$, $\alpha'\beta'$ and the polars of U, V, W with respect to the absolute; it also touches the three altitudes $\alpha\alpha'$, $\beta\beta'$, $\gamma\gamma'$.

By the transformation

$$(\xi, \eta, \zeta) = \begin{pmatrix} a & h & g \\ h & b & f \\ g & f & c \end{pmatrix}(x, y, z), \quad (x, y, z) = \begin{pmatrix} A & H & G \\ H & B & F \\ G & F & C \end{pmatrix}(\xi, \eta, \zeta),$$

$Cy - Fz$ becomes $-\Delta(h\xi - a\eta)$, etc., and the point-equation of the pedal locus passes into the line-equation of the pedal envelope, i.e. *the pedal envelope is the polar reciprocal of the pedal locus with respect to the absolute**.

The envelope is therefore a curve of the sixth degree with nine cusps of which three are real, and the three cuspidal tangents are concurrent.

If $ABC = \Delta abc$ and $\Delta \neq 0$, the pedal envelope breaks up into a point (the orthocentre) and a conic touching the sides of the triangle.

6. If $\Delta = 0$, so that the absolute degenerates to two distinct straight lines through I, and the cofactors A, B, C, F, G, H do not all vanish, the polars of α, β, γ, all pass through I and the triangle $\alpha'\beta'\gamma'$ degenerates to the point I. PI is the pedal line of any point P, so that the orthocentral locus disappears and the orthocentral envelope reduces to the point I.

The orthaxal locus and envelope are, however, definite. Let the equation of the absolute be

$$(ax + by + cz)(a'x + b'y + c'z) = 0, \text{ or shortly, } uv = 0.$$

The polar of α $(1, 0, 0)$ is $av + a'u = 0$.

Consider the line (ξ, η, ζ) or $w = 0$; let it cut the polar of α in X' and join $\alpha X'$. The equation of $\alpha X'$ is

$$2aa'w = \xi(av + a'u) \text{ or } \frac{2w}{\xi} = \frac{u}{a} + \frac{v}{a'},$$

and we get similar equations for $\beta Y'$ and $\gamma Z'$.

The condition that $\alpha X'$, $\beta Y'$, $\gamma Z'$ should be concurrent in a point P gives the line-equation of the envelope of w,

$$\begin{vmatrix} \frac{1}{\xi} & \frac{1}{a} & \frac{1}{a'} \\ \frac{1}{\eta} & \frac{1}{b} & \frac{1}{b'} \\ \frac{1}{\zeta} & \frac{1}{c} & \frac{1}{c'} \end{vmatrix} = 0,$$

which represents a conic touching the sides of the triangle $\alpha\beta\gamma$ and also touching the two lines u, v, i.e. ***the envelope is a circle inscribed in the given triangle.*** The envelope also includes the point uv.

* This is already obvious since the orthaxal envelope, which is reciprocal to the orthocentral locus, coincides with the orthocentral envelope.

7. To find the locus of $P(x', y', z')$. $P\alpha$ is $yz' - y'z = 0$ and this cuts $av + a'u = 0$, the polar of α, in

$$x, y, z = (ab' + a'b)\,y' + (ac' + a'c)\,z', \quad -2aa'y', \quad -2aa'z'.$$

The equation of the locus of P is therefore

$$\begin{vmatrix} (ab' + a'b)\,y + (ca' + c'a)\,z & -2aa'y & -2aa'z \\ -2bb'x & (bc' + b'c)\,z + (ab' + a'b)\,x & -2bb'z \\ -2cc'x & -2cc'y & (ca' + c'a)\,x + (bc' + b'c)\,y \end{vmatrix} = 0,$$

or

$$\begin{vmatrix} a'u + av - 2aa'x & -2aa'y & -2aa'z \\ -2bb'x & b'u + bv - 2bb'y & -2bb'z \\ -2cc'x & -2cc'y & c'u + cv - 2cc'z \end{vmatrix} = 0,$$

which represents a cubic passing through α, β, γ, and through the intersections of $\beta\gamma\,(x = 0)$ and the polar of $\alpha\,(av + a'u = 0)$, etc.

The equation may also be written

$$a'b'c'u^3 + abcv^3 = uv\Sigma\,(a^2b'c' + a'^2bc)\,x,$$

which shows that the cubic has a double-point at uv, u and v being the tangents at the double-point, and also $\Sigma\,(a^2b'c' + a'^2bc)\,x = 0$ is the line upon which the three points of inflexion lie.

When the double-point is a crunode, so that u, v are real, two of the points of inflexion are imaginary; when the double-point is an acnode, with u, v imaginary, the three points of inflexion are all real.

8. The triangle $\alpha\beta\gamma$ and its orthaxis form a complete quadrilateral. Take the diagonal triangle as triangle of reference and choose line-coordinates ξ, η, ζ so that the coordinates of the sides of the triangle and the orthaxis are $(-1, 1, 1)$, $(1, -1, 1)$, $(1, 1, -1)$, $(1, 1, 1)$. The point-coordinates of the vertex α are $(0, 1, 1)$. Let the line-coordinates of I be (a_0, b_0, c_0), and consider the equation

$$\frac{a_0}{\xi} + \frac{b_0}{\eta} + \frac{c_0}{\zeta} = 0.$$

This represents a conic touching the sides of the diagonal triangle. It also touches the line $\left(\frac{1}{b_0 - c_0}, -\frac{1}{a_0}, \frac{1}{a_0}\right)$ which passes through α. The harmonic conjugate of this line with respect to $\alpha\beta$, $\alpha\gamma$ is $(b_0 - c_0, -a_0, a_0)$ which is the line αI. Therefore this conic touches the sides of the diagonal triangle and the bisectors of the angles of the complete quadrilateral.

The equations of the sides of the diagonal triangle in the former coordinates are

$$\lambda X \equiv (ab' + a'b)\,y + (ca' + c'a)\,z = 0,$$
$$\mu Y \equiv (bc' + b'c)\,z + (ab' + a'b)\,x = 0,$$
$$\nu Z \equiv (ca' + c'a)\,x + (bc' + b'c)\,y = 0.$$

Taking $\lambda\,(bc' + b'c) = \mu\,(ca' + c'a) = \nu\,(ab' + a'b)$ we get

$$\frac{x}{bc' + b'c} : \frac{y}{ca' + c'a} : \frac{z}{ab' + a'b} = -X + Y + Z : X - Y + Z : X + Y - Z.$$

Then $$ax + by + cz = 2abc\left(X\frac{a'}{a} + Y\frac{b'}{b} + Z\frac{c'}{c}\right),$$

so that $$u \equiv X\frac{a'}{a} + Y\frac{b'}{b} + Z\frac{c'}{c},$$

$$v \equiv X\frac{a}{a'} + Y\frac{b}{b'} + Z\frac{c}{c'},$$

$$a_0 : b_0 : c_0 = aa'(b'^2c^2 - b^2c'^2) : bb'(c'^2a^2 - c^2a'^2) : cc'(a'^2b^2 - a^2b'^2),$$

and $$\Sigma a_0 cab' . abc' = aa'bb'cc' \Sigma a^2 (b'^2c^2 - b^2c'^2) = 0.$$

Therefore the conic touches also the lines u and v and is therefore a circle.

This corresponds in ordinary geometry to the nine-point circle, and we may therefore call it the *nine-line circle*.

9. The axis of this circle is the polar of the point I. Let us find its equation.

The pole of the line (lmn) with respect to $\frac{a_0}{\xi} + \frac{b_0}{\eta} + \frac{c_0}{\zeta} = 0$ is

$$\xi(c_0 m + b_0 n) + \eta(c_0 l + a_0 n) + \zeta(b_0 l + a_0 m) = 0,$$

therefore the polar of (xyz) is

$$l, m, n = a_0(-a_0x + b_0y + c_0z),\ b_0(a_0x - b_0y + c_0z),\ c_0(a_0x + b_0y - c_0z).$$

The polar of (a_0, b_0, c_0) is therefore

$$a_0(-a_0^2 + b_0^2 + c_0^2)X + b_0(a_0^2 - b_0^2 + c_0^2)Y + c_0(a_0^2 + b_0^2 - c_0^2)Z = 0.$$

We have to express this in terms of $x, y, z, a, b, c, a', b', c'$. We find

$$-a_0^2 + b_0^2 + c_0^2 = -(b^2c'^2 + b'^2c^2)(c^2a'^2 - c'^2a^2)(a^2b'^2 - a'^2b^2).$$

Hence we get

$$\Sigma aa'(b^2c'^2 - b'^2c^2)(b^2c'^2 + b'^2c^2)(c^2a'^2 - c'^2a^2)(a^2b'^2 - a'^2b^2)$$
$$(bc' + b'c)[(ab' + a'b)y + (ca' + c'a)z] = 0,$$

or, leaving out the symmetrical factor,

$$\Sigma x[bb'(c^2a'^2 + c'^2a^2)(ca' + c'a)(ab' + a'b) + cc'(a^2b'^2 + a'^2b^2)(ab' + a'b)(ca' + c'a)]$$
$$= \Sigma x(ca' + c'a)(ab' + a'b)[bb'(c^2a'^2 + c'^2a^2) + cc'(a^2b'^2 + a'^2b^2)]$$
$$= \Sigma x(ca' + c'a)(ab' + a'b)(bc' + b'c)(a^2b'c' + a'^2bc) = 0,$$

or, again leaving out the symmetrical factor, the equation finally reduces to

$$\Sigma(a^2b'c' + a'^2bc)x = 0,$$

and so *the line of inflexions is the axis of the nine-line circle.*

10. *The three points of inflexion are mutually equidistant.*

We may write the equation of the cubic, u and v being the tangents at the double-point and w the line of inflexions,

$$u^3 + v^3 = 3uvw,$$

or $$w^3 = u^3 + v^3 + w^3 - 3uvw$$
$$= (u + v + w)(u + \omega v + \omega^2 w)(u + \omega^2 v + \omega w).$$

The last three factors equated to zero give the equations of the inflexional tangents. The lines joining the double-point to the points of inflexion are

$$u+v=0, \quad u+\omega v=0, \quad u+\omega^2 v=0.$$

The cross-ratios $(u, v;\ u+v, u+\omega v)$, $(u, v;\ u+\omega v, u+\omega^2 v)$, $(u, v;\ u+\omega^2 v, u+v)$ are each equal to ω, hence the distance between any pair of points of inflexion is

$$\frac{i}{2}\log e^{\frac{2\pi i}{3}} = -\frac{\pi}{3} \text{ or } \frac{2\pi}{3}.$$

11. When the absolute degenerates to two coincident straight lines we have the case of Euclidean geometry.

The orthaxal locus degenerates to the line at infinity, and the orthaxal envelope disappears. For the orthocentral locus and envelope we have the well-known properties of the circumcircle, the nine-point circle and the three-cusped hypocycloid, of which those in the preceding sections are reciprocals.

12. There are other cases in which the pedal locus and envelope degenerate:

(1) When a side of the triangle touches the absolute this side forms part of the locus, and the point of contact forms part of the envelope.

(2) When a vertex of the triangle lies on the absolute the tangent at this point to the absolute forms part of the locus, and the vertex forms part of the envelope.

(3) When a side of the triangle is the polar of the opposite vertex, i.e. when the triangle has two right angles, the locus consists of this side and two lines through the opposite vertex; the envelope consists of the vertex and two points on the opposite side.

13. In a certain case the pedal locus breaks up into a straight line and a circle circumscribing the triangle.

We have first the condition $\quad ABC = \Delta abc.$

The equation of a circle with axis $lx+my+nz=0$ is

$$ax^2+by^2+cz^2+2fyz+2gzx+2hxy = k(lx+my+nz)^2.$$

This circumscribes the triangle $\alpha\beta\gamma$ if

$$a=kl^2, \quad b=km^2, \quad c=kn^2.$$

The equation then becomes

$$(f-\sqrt{bc})\,yz+(g-\sqrt{ca})\,zx+(h-\sqrt{ab})\,xy=0.$$

In order that this may be the same as

$$\frac{A}{f}yz+\frac{B}{g}zx+\frac{C}{h}xy=0,$$

we must have $\quad af^2=bg^2=ch^2=t$, say.

The equation of the cubic then becomes

$$\Sigma\frac{x}{f}\,.\,\Sigma\frac{f}{x} = -\frac{(t-p)(2t+p)}{p(t+p)},$$

where $p=fgh$.

(1) $t=p=fgh$.

The point-equation of the absolute reduces to

$$\left(\Sigma \frac{x}{f}\right)^2=0,$$

and the line-equation to

$$\Sigma f^2\xi^2-\Sigma gh\eta\zeta \equiv (f\xi+\omega g\eta+\omega^2 h\zeta)(f\xi+\omega^2 g\eta+\omega h\zeta)=0,$$

i.e. the absolute degenerates to two coincident straight lines and a pair of imaginary points, and the geometry is Euclidean.

(2) $2t+p=0$.

In this case the absolute need not degenerate, but the triangle must be equilateral.

The equation of the pair of tangents from α to the absolute is

$$Cy^2-2Fyz+Bz^2=0,$$

or

$$(p+t)\,h^2y^2+2pghyz+(p+t)\,g^2z^2=0.$$

The angle α of the triangle is therefore

$$\frac{i}{2}\log\frac{-p+\sqrt{p^2-(p+t)^2}}{-p-\sqrt{p^2-(p+t)^2}}=\cos^{-1}\frac{p}{p+t}.$$

When the cubic reduces, the angle is $\cos^{-1}2=i\cosh^{-1}2$,

the side is $\cosh^{-1}(-2)=\cosh^{-1}2+i\pi$,

and the vertices of the triangle are all outside the absolute.

The pedal envelope reduces to

$$\Sigma\frac{\xi}{f}\cdot\Sigma\frac{f}{\xi}=0,$$

i.e. the orthocentre and the circle inscribed in the triangle.

When the absolute and the degenerate cubic are represented by concentric circles on the Euclidean plane this leads to the following theorem of Euclidean geometry*:

"If $AB'CA'BC'$ is a regular hexagon inscribed in a circle, and P is any point on the circle, and if PA' cuts BC in X, PB' cuts CA in Y and PC' cuts AB in Z; then X, Y, Z are collinear on a line through the centre of the circle. And also if three parallel lines be drawn through A', B', C' cutting BC, CA, AB in X, Y, Z, then X, Y, Z are collinear and the envelope of the line XYZ is the circle inscribed in the triangle ABC."

For A', B', C' are the poles of BC, CA, AB with respect to a concentric circle of radius $\frac{1}{\sqrt{2}}$, and this circle is the absolute.

* Cf. *Educ. Times*, Sept. 1912 (Question 17308).

SUR LES HESSIENNES SUCCESSIVES D'UNE COURBE DU TROISIÈME DEGRÉ

PAR B. HOSTINSKÝ.

Soit $$x^3 + y^3 + z^3 + 6mxyz = 0 \quad \ldots\ldots\ldots\ldots\ldots\ldots\ldots\ldots(1)$$
l'équation d'une courbe du troisième degré en coordonnées homogènes x, y, z. Les invariants de la courbe sont

$$S = m(1 - m^3), \qquad T = 1 - 20m^3 - 8m^6;$$

la quantité $$r = \frac{1}{48}\frac{T^2}{S^3}$$

est son invariant absolu. A chaque valeur de r correspondent 12 valeurs de m; en général, si r est réel, deux valeurs de m (deux courbes) sont réelles et dix imaginaires.

Représentons par H_0 le premier membre de l'équation (1) et écrivons

$$H_k = \frac{1}{6}\frac{D\left(\frac{\partial H_{k-1}}{\partial x}, \frac{\partial H_{k-1}}{\partial y}, \frac{\partial H_{k-1}}{\partial z}\right)}{D(x, y, z)}.$$

Nous obtenons ainsi une suite de courbes du troisième degré

$$H_0, H_1, H_2, \ldots H_n \ldots \quad \ldots\ldots\ldots\ldots\ldots\ldots\ldots\ldots(2),$$

que nous appelerons Hessiennes successives; chaque courbe de la suite (2) est la Hessienne de la précédente et l'équation d'une quelconque de ces courbes aura toujours la forme (1). Le problème que je veux traiter maintenant consiste à rechercher les conditions pour que la suite (2) soit périodique. En d'autres mots: Un entier n étant donné, on cherche une courbe H_0 identique avec H_n.

Supposons que cette condition soit remplie par une courbe H_0.

Les courbes $$H_0, H_1, H_2 \ldots H_{n-1}$$

constituent un cycle d'ordre n; dans la suite (2) pour chaque k, les courbes H_k et H_{k+n} seront identiques. La recherche des cycles d'un ordre donné conduit à une équation algébrique dont il est aisé indiquer le mode de formation. On s'appuiera sur les formules bien connues de Salmon qui expriment les invariants S_{k+1}, T_{k+1} de H_{k+1} en fonction de ceux de H_k et sur l'équation

$$108H_{k+1} = 4S_k^2 H_k + T_k H_{k+1}.$$

On trouve ainsi $$H_n = X_n . H_0 + Y_n . H_1,$$

X_n, Y_n ne renfermant pas les coordonnées x, y, z. Pour que la courbe H_n soit identique avec H_0 il faut que

$$Y_n = 0 \quad \ldots\ldots\ldots\ldots\ldots\ldots\ldots\ldots\ldots\ldots\ldots (3).$$

Y_n est un polynôme homogène en S^3 et T^2; après avoir divisé le premier membre de (3) par une puissance convenable de S^3, on aura une équation qui ne contiendra que l'invariant absolu r de la courbe H_0. L'équation ainsi obtenue sera une équation Abélienne. Car si elle admet la racine r_k, elle doit admettre en même temps la racine r_{k+1}, r_k étant l'invariant absolu de H_k. Mais on trouve que

$$r_{k+1} = -\frac{r_k(2r_k+3)^2}{3(r_k+1)^3},$$

ce qui exprime la propriété caractéristique des équations Abéliennes.

Voici les résultats du calcul dans les cas les plus simples, où l'ordre n du cycle ne dépasse pas 4:

$n=1$. Quatre courbes dégénerées en triangles.

$n=2$. Six courbes harmoniques formant trois cycles du deuxième ordre; un seul cycle est réel. C'est à Salmon que nous devons cet exemple intéressant.

$n=3$. L'équation (3) s'écrit

$$r^2 + 3r + 3 = 0.$$

Désignons par r, r' ses racines. Tout cycle du troisième ordre est composé de courbes ayant le même invariant absolu r (ou r'). Les 12 courbes de l'invariant r (ou r') se divisent en quatre cycles; il y a donc huit cycles du troisième ordre ne comprenant que de courbes imaginaires.

$n=4$. L'équation (3) est réductible; on trouve

$$(r^2+6r+6).(r^4-3r^3-21r^2-27r-9)=0.$$

Il y a deux espèces de cycles du quatrième ordre:

(*a*) Soient r, r' les racines du facteur quadratique. Les 24 courbes dont l'invariant est r ou r' forment 6 cycles. Les quatre courbes composant un tel cycle auront resp. les invariants r, r', r, r'. Un seul cycle sera réel.

(*b*) Désignons par ρ, ρ', ρ'', ρ''' les racines du facteur biquadratique. Les 48 courbes correspondantes forment 12 cycles; les courbes d'un même cycle auront resp. les invariants ρ, ρ', ρ'', ρ'''. Deux cycles seront réels.

Si l'ordre n du cycle dépasse 4, une discussion complète exige un calcul assez long.

Toutefois on peut démontrer le théorème suivant: Il n'y a pas de cycle d'ordre n composé de courbes réelles, si n est un nombre impair.

La démonstration repose sur la remarque suivante: La Hessienne de (1) est représentée par la même équation (1) si l'on y remplace m par

$$m' = -\frac{2m^3+1}{6m^2}.$$

Une discussion de cette dernière formule nous montre que, si la courbe originale n'a pas d'ovale, la Hessienne en aura un, et si la courbe originale est douée d'un ovale, la Hessienne sera sans ovale.

Reprenons une suite de Hessiennes successives réelles. Les courbes d'indice pair H_0, H_2, $H_4 \ldots$ ont par exemple un ovale tandisque les courbes d'indice impair sont sans ovale. Si l'ordre d'un cycle n était un nombre impair, on aurait $H_0 \equiv H_n$, ce qui est impossible. Par conséquent l'ordre d'un cycle réel doit être toujours un nombre pair.

GEOMETRISCHE MAXIMA UND MINIMA MIT ANWENDUNG AUF DIE OPTIK

Von Johannes Finsterbusch.

§ 1. Ueber die Auflösungs-Methoden.

1. Geometrische Maximum- und Minimum-Aufgaben können entweder synthetisch oder analytisch behandelt werden. Die synthetische oder reingeometrische Auflösungsmethode ist durch die Anschaulichkeit und Eleganz ihrer Lösungen bemerkenswert, trägt aber vielfach den Charakter der Zufälligkeit, da sich, um mit J. Steiner zu reden, "nicht ein einziges gemeinsames Grundprinzip aufstellen lässt," während die analytische oder rechnerische Behandlung in der Anwendung der Differentialrechnung eine umfassende Methode besitzt, wie die erstere nicht ihres Gleichen hat. Damit soll aber nicht gesagt sein, dass für alle Aufgabengruppen die analytische Behandlung mit höherer Analysis allen anderen Methoden vorzuziehen sei. Selbst unter den rechnerischen Auflösungen eines Problems ist sie es bei weitem nicht immer. So führt z. B. die Anwendung des bekannten Doppelsatzes:

Bei gegebener $\left\{\begin{matrix}\textit{Summe}\\ \textit{Produkt}\end{matrix}\right\}$ *von n positiven Grössen ist deren* $\left\{\begin{matrix}\textit{Produkt}\\ \textit{Summe}\end{matrix}\right\}$ *am* $\left\{\begin{matrix}\textit{grössten}\\ \textit{kleinsten}\end{matrix}\right\}$, *wenn die n Grössen alle einander gleich sind,*

falls die Zurückführung eines Problems auf ihn gelingt, deshalb kürzer zum Ziele, weil hierdurch die Aufstellung der zu differenzierenden Funktion überflüssig wird. Auf den Beweis dieses Doppelsatzes soll hier verzichtet werden*; doch sei erwähnt, dass er für $n = 2$ und $n = 3$ auch reingeometrisch geführt werden kann.

Für $n = 2$ steht er in Euklid's *Elementen* im VI. Buche als Satz 27 und bildet nach Moritz Cantor "das erste Maximum, welches in der Geschichte der Mathematik nachgewiesen worden ist†." Ebensowenig möchte ich auf die vielen Anwendungen des Doppelsatzes näher eingehen. Nur des Zusammenhanges mit einer unten behandelten Aufgabe wegen, werde mit seiner Hilfe kurz bewiesen, dass *unter allen Sehnenvierecken von gegebenem Umfang das Quadrat den grössten Inhalt hat.* Bezeichnen a, b, c, d die Seiten eines beliebigen Sehnenvierecks vom gegebenen Umfange $2s$, so folgt aus der bekannten Heronischen Flächenformel

$$F = \sqrt{(s-a)(s-b)(s-c)(s-d)},$$

* Arithmetische Beweise finden sich z. B. in: Rudolf Sturm, *Maxima und Minima in der elementaren Geometrie*, 1910, S. 1—4.

† Moritz Cantor, *Geschichte der Mathematik*, I, S. 252.

in der die Summe der vier Faktoren $= 2s$, also konstant ist, dass ihr Produkt und daher F zum Maximum wird, wenn die Faktoren einander gleich sind, d. h. wenn das Sehnenviereck ein Quadrat ist*.

Die synthetischen Methoden verdanken ihren Ausbau in erster Linie Jacob Steiner. "Zwei der bedeutendsten Abhandlungen, in denen er die Ergebnisse seiner langjährigen Untersuchungen *über Maximum und Minimum bei den Figuren in der Ebene, auf der Kugelfläche und im Raume überhaupt*† niedergelegt hat," sind bis heute der Ausgangspunkt ähnlicher, wenn auch nicht eben zahlreicher Abhandlungen geblieben.

2. Im Folgenden will ich aus meinen Untersuchungen über denselben Gegenstand zunächst einige bekannte Aufgaben aus der *Geometrie der Vielecke* in neuer Weise synthetisch behandeln. Meine Beweise, die die bisher vorhandenen an Einfachheit und Anschaulichkeit übertreffen dürften, enthalten zugleich den Schlüssel zur Lösung anderer derartiger Aufgaben, z. B. der Steiner'schen *Schliessungs-Sätze über gespiegelte Lichtstrahlen.* Als Hauptaufgabe behandle ich die *Verallgemeinerung des Problems der Minimal-Ablenkung, die ein Lichtstrahl beim Durchgang durch ein Prisma erfährt.*

Ehe ich mich zu den Aufgaben selbst wende, möchte ich einige kurze *Bemerkungen über die synthetischen Methoden,* deren ich mich bediene, vorausschicken. Die zu lösende Aufgabe kann als *Theorem* oder als *Problem* gestellt sein. Während die analytische Auflösungsart in beiden Fällen durch das Verschwinden des ersten Differentialquotienten die Bedingungsgleichung des Extremums gewinnt und ferner durch das positive oder negative Vorzeichen des zweiten Differentialquotienten erkennt, ob ein Minimum oder Maximum vorliegt, behandelt die synthetische Methode beide Fälle verschieden:

I. Beim Beweise eines Theorems, wenn also die das Extremum kennzeichnende Eigenschaft schon bekannt ist, gehen wir von der Figur des Extremums aus und verändern diese derart, dass alle Eigenschaften der Figur, abgesehen von der besonderen des Extremums, erhalten bleiben. Durch Vergleichung beider Figuren, deren einzelne entsprechende Grössen um Endliches von einander abweichen, wird die Figur des Extremums als solche nachgewiesen. Dabei wird zugleich entschieden, ob ein Minimum oder Maximum vorliegt.

II. Handelt es sich dagegen um ein Problem, soll also erst die noch unbekannte, das Extremum bedingende Eigenschaft gefunden werden, so sind es zwei Wege, die zum Ziele führen:

II *a.* *Methode der gleichen, endlich getrennten Werte,* wie ich sie nennen möchte. Die in der Aufgabe aufgeführten Eigenschaften reichen zur vollständigen Bestimmung der Figur nicht hin. Wir nehmen deshalb als noch fehlendes Bestimmungsstück einen beliebigen Wert der veränderlichen Grösse, die zum Extremum werden soll, hinzu. Dadurch geht, wenn ein Extremum existiert, die bisher unbestimmte Aufgabe in eine immer mehrdeutig bestimmte über. Aus zwei verschiedenen Lösungen, in geeigneter Weise auf einander bezogen, ist nun abzuleiten, welche Beziehung

* Rudolf Sturm, a. a. O., S. 25.

† Jacob Steiner, *Werke*, II, S. 177—308.

zwischen den Grössen bestehen muss, wenn die beiden endlich verschiedenen Figuren zur Deckung gelangen sollen. Diese Bedingung ist die gesuchte Eigenschaft des Extremums. Damit ist das Problem zum Theorem geworden. Die Art des Extremums wird nach I entschieden.

II *b*. *Methode der gleichen, zusammenfallenden Werte* oder *Methode des Unendlich-Kleinen.* Man geht von der Analysis-Figur des Extremums aus und denkt sich diese nur unendlich wenig verändert. Die Beziehungen beider Figuren drücken sich in einer Gleichung aus, die die unendlich kleine Veränderung der veränderlichen Grösse enthält, deren Extremum gesucht wird. Wird diese unendlich kleine Veränderung gleich Null gesetzt, so erhält man die Bedingungsgleichung des Extremums. Die Art des Extremums wird auch hier nach I entschieden.

Diese Methode II *b*, eine geometrische Vorläuferin der Differentialrechnung, macht demnach wie diese von dem Verschwinden der Aenderung einer veränderlichen Grösse in der Nähe des Extremums Gebrauch. Decken sich also hierin ihrem Wesen nach beide Methoden, die synthetische und die analytische, so ist jedoch bei reingeometrischen Problemen die erstere der höheren Analysis weit vorzuziehen, da sie vielmehr als diese geeignet ist, "das eigentliche Wesen oder die wahre Ursache des Maximums und Minimums anzugeben" (Steiner, a. a. O., II, 179). Das erscheint ganz natürlich und einleuchtend, wenn man bedenkt, dass bei der synthetischen Methode die Bedingung des Extremums gleichsam organisch aus der Figur hervorgeht und so ihre wesentlichen Grössen direkt zu einander in Beziehung bringt; während bei der Differentialrechnung die zu Grunde gelegten Grössen (Koordinaten) meist recht fremdartig oder nur in losem Zusammenhange zu den wesentlichen Grössen der Aufgabe stehen, sodass es oft sehr schwer ist, aus der erhaltenen Bedingungsgleichung eine einfache Eigenschaft des Extremums herauszulesen. "Die synthetische Methode hat noch in neuester Zeit zu einzelnen schönen Ergebnissen geführt, welche durch die Anwendung der Differentialrechnung nicht gefunden worden waren" (Fiedler). Auch von den nun zu behandelnden Aufgaben hätte ich wohl schwerlich so einfache Beweise und Konstruktionen gefunden, wenn ich mich nicht synthetischer Methoden bedient hätte.

§ 2. Einige Aufgaben über Vielecke.

3. *Im spitzwinkligen Dreieck ABC hat das Dreieck der Höhenfusspunkte DEF den kleinsten Umfang.*

Von diesem Satze hat Schwarz einen schönen Beweis gegeben, der fälschlicherweise Jacob Steiner zugeschrieben worden ist*. Mein ebenfalls reingeometrischer Beweis hat nicht wie jener 6 auf einander folgende Umklappungen oder Spiegelungen des Dreiecks ABC nötig, sondern nur je eine um die Seiten AB und AC (Fig. 1).

Geht man von einem beliebigen Punkte X der Seite BC aus und sind X_1 und X_2 die ihm entsprechenden Spiegelpunkte, so hat jedes beliebige eingeschriebene Dreieck XYZ den geknickten Streckenzug X_1YZX_2 zum Umfang, also einen grösseren als das Dreieck XY_0Z_0, dessen Seite Y_0Z_0 auf der Strecke X_1X_2 liegt. Da je zwei Seiten des Höhenfusspunktdreiecks DEF mit einer Seite des gegebenen

* Vergleiche: J. Steiner, *Ges. W.*, II, Anmerkung auf Seite 728 und H. A. Schwarz, *Ges. W.*, II, 349 (nach R. Sturm, a. a. O., Seite 90).

Dreiecks gleiche Winkel einschliessen, also symmetrisch zu ihr liegen, was beim vorigen Dreieck nur an den Ecken Y_0 und Z_0 der Fall war, so ist dessen Umfang D_1EFD_2 gleich der Strecke D_1D_2, wenn D_1 und D_2 die Spiegelpunkte vom Höhenfusspunkt D bedeuten.

Werden nun die Punkte X_1 und X_2 auf D_1D_2 orthogonal projiziert, so folgt aus der Gleichheit der Strecken $DX = D_1X_1 = D_2X_2$ und der Gleichheit der Winkel bei D_1 und D_2 mit den Winkeln α bei D die Gleichheit ihrer Projektionen $X_1'D_1 = X_2'D_2$ auf D_1D_2, und also $X_1'X_2' = D_1D_2$. Da nun die Projektion $X_1'X_2'$ kleiner ist als die projizierte Strecke X_1X_2 oder der projizierte Streckenzug X_1YZX_2, so hat das Höhenfusspunktdreieck den kleinsten Umfang*.

4. Vorstehendes Theorem kann auch als "Problem" gelöst werden: Da $X_1S + X_2S$ = Umfang des Dreiecks CBS, also konstant ist, kann unsere Aufgabe auf die einfachere zurückgeführt werden: *Welches unter allen Dreiecken X_1SX_2, die in dem Winkel S an der Spitze und der Schenkelsumme übereinstimmen, hat die kleinste Grundlinie X_1X_2?* Dreieck X_1SX_2 und sein symmetrisches in Bezug auf die Winkelhalbierende von S stimmen in der Grösse von X_1X_2, dessen Extremum gesucht wird, überein. Nach Methode II a müssen im Falle des Extremums beide Dreiecke sich decken, folglich ist die Basis D_1D_2 des gleichschenkligen unter den Dreiecken X_1SX_2 das gesuchte Extremum. Seine Art wird wie oben nach Methode I bestimmt.

5. *Unter allen konvexen Vielecken V von gegebenen Seiten hat das einem Kreise eingeschriebene oder Sehnen-Vieleck V_0 die grösste Fläche.*

Da meine Beweismethode für das n-Eck genau dieselbe bleibt, wie für das Viereck, behandeln wir das letztere. Wir gehen vom Sehnen-Viereck V_0 aus (Fig. 2) und zerschneiden es vom Mittelpunkte M des umgeschriebenen Kreises aus durch die 4 nach den Ecken gehenden Radien in gleichschenklige Dreiecke D_i, deren algebraische Summe gleich der Fläche des Sehnenvierecks V_0 ist. Liegt der Mittelpunkt M innerhalb des Vierecks, so sind alle 4 Dreiecke positiv; nur wenn M ausserhalb des Vierecks liegt, ist das gleichschenklige Dreieck mit der grössten Vierecksseite als Basis negativ zu nehmen. Werden die Winkel des Sehnenvierecks (als Gelenkviereck aufgefasst) verändert (Fig. 3), so beschreiben die nach den Ecken gezogenen Radien vier Sektoren S_i, die wir als positiv oder negativ in Rechnung bringen, je nachdem durch sie der betreffende Winkel des Vierecks vergrössert oder verkleinert wird. Der Mittelpunkt M beschreibt ein Kreisbogenviereck $M_1M_2M_3M_4 \equiv M$. Dann folgt für die neue Figur des Vierecks V

$$\Sigma D_i + \Sigma S_i = M + V.$$

Nun ist aber $\Sigma D_i = V_0$ und $\Sigma S_i = 0$, weil die algebraische Summe der Zentriwinkel = 0 sein muss, damit die Winkelsumme des Vierecks konstant = 4 Rechte bleibt. Also erhalten wir

$$V_0 = V + M,$$

womit der Satz bewiesen ist.

* Laut *Jahresbericht* XXXVI—XXXIX, 1906–1909 *des Vereins für Naturkunde zu Zwickau* habe ich den Beweis in der Sitzung vom 13./I. 1908 vorgetragen. In **Weber** and **Wellstein**, *Enzykl. d. Elementarmath.*, III[1], zweite Aufl. 1910 steht ein Beweis, der mit meinem übereinstimmt und nur am Ende die Aehnlichkeit der gleichschenkligen Dreiecke X_1AX_2 und D_1AD_2 heranzieht, um $D_1D_2 < X_1X_2$ nachzuweisen. In der ersten Auflage III, 1907, S. 309, stand dieser Beweis noch nicht.

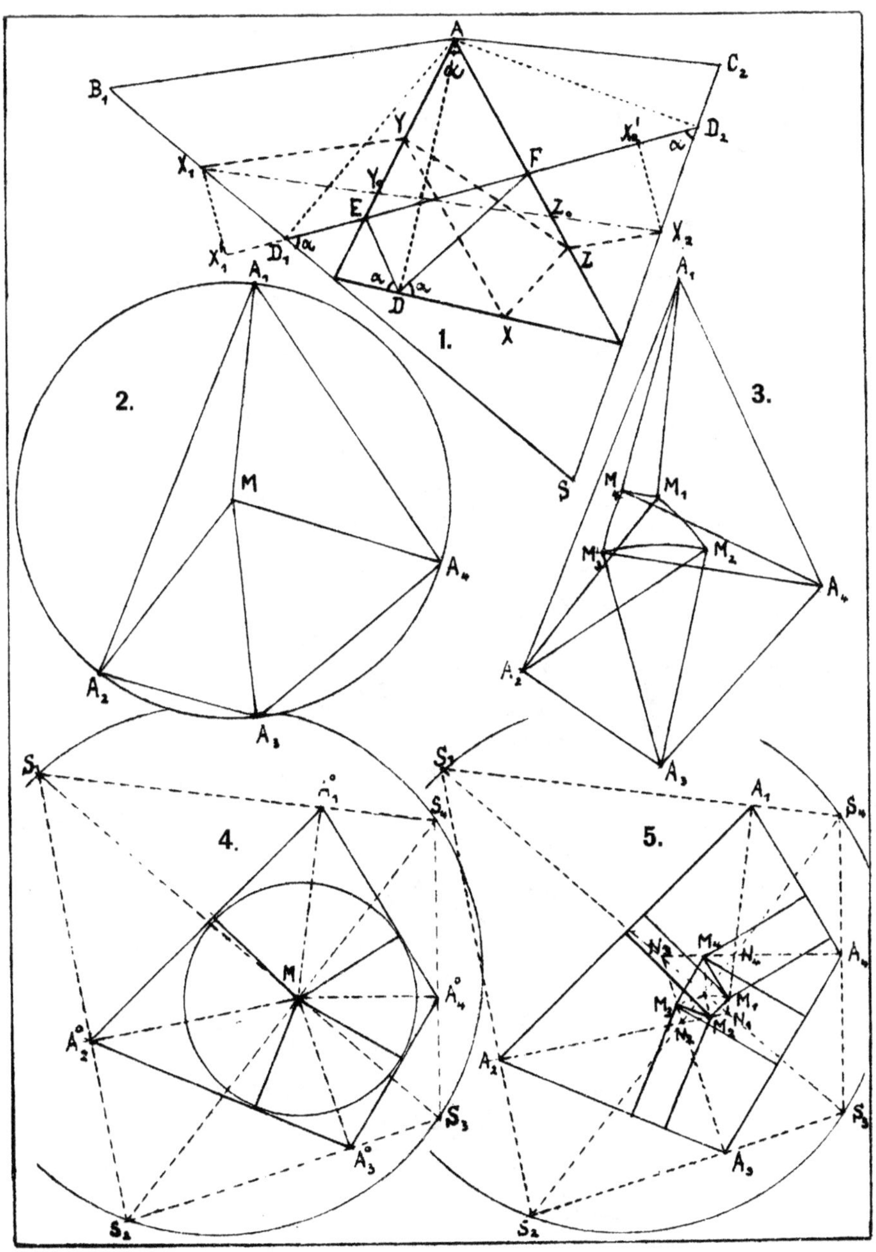
1.
2.
3.
4.
5.
A
B₁
C₂
D₂
X₁
Y
F
E
Y₀
Z₀
X₂
D₁
Z
D
X
S
α
M
A₁
A₂
A₃
A₄
M₁
M₂
M₃
M₄
S₁
S₂
S₃
S₄
N₁
N₂
N₃
N₄

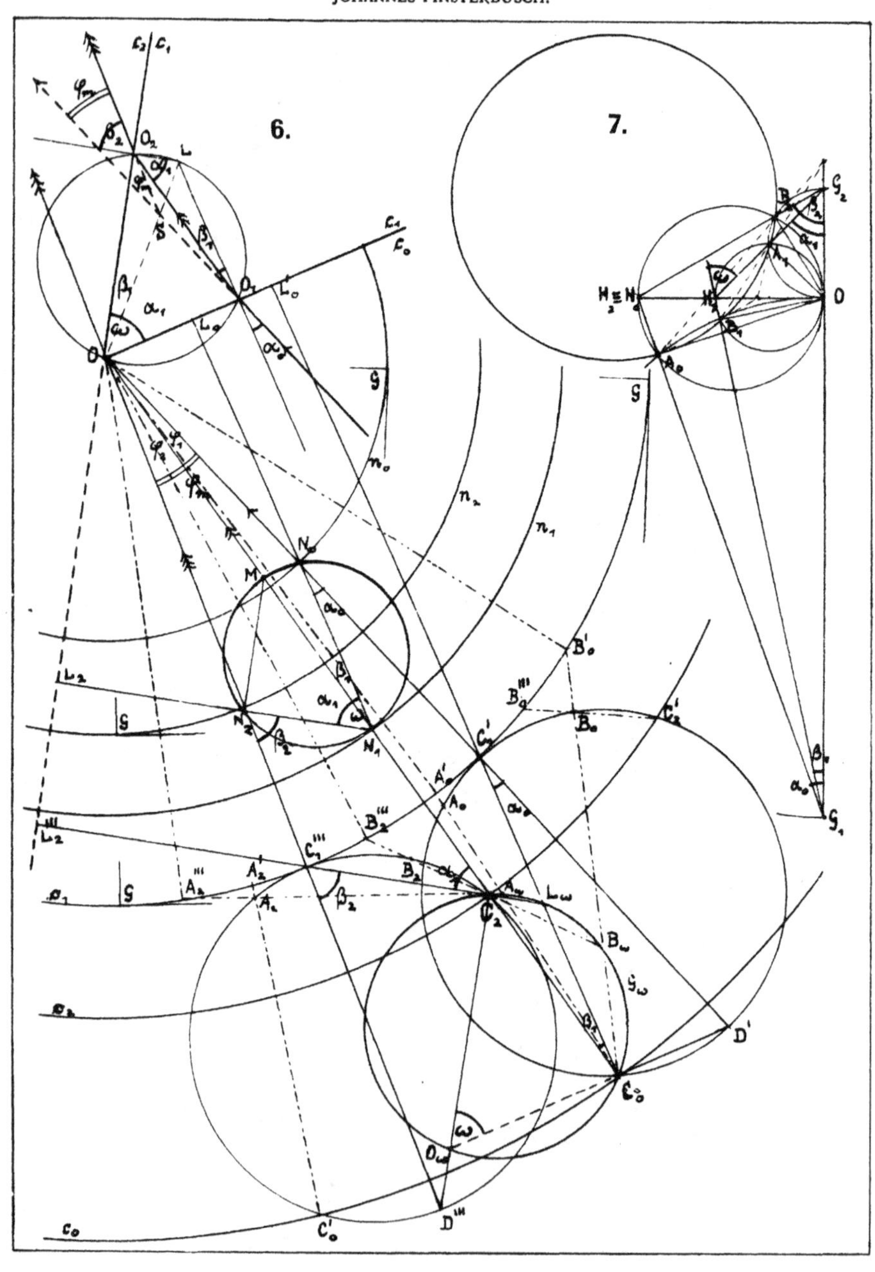
6.
7.

Die Verwandlung eines konvexen Vielecks in ein Sehnenvieleck von denselben Seiten ist immer möglich und eindeutig. Alle Sehnenvielecke von denselben Seiten, aber in verschiedener Reihenfolge, sind flächengleich.

6. $\left.\begin{matrix}(a)\\(b)\end{matrix}\right\}$ *Unter allen konvexen Vielecken V von gegebenen Winkeln und* $\left\{\begin{matrix}\textit{Umfang}\\\textit{Flächeninh.}\end{matrix}\right.$ *hat das einem Kreise umgeschriebene oder Tangentenvieleck* V_0 *den* $\left\{\begin{matrix}\textit{grössten Inhalt.}\\\textit{kleinsten Umfang.}\end{matrix}\right.$

Da auch hier die Beweisführung für das n-Eck genau dieselbe ist wie beim Viereck, geben wir wieder die letztere. Wir gehen vom Tangentenviereck V_0 aus (Fig. 4) und zerschneiden es vom Mittelpunkte M des eingeschriebenen Kreises aus durch die nach den 4 Berührungspunkten gehenden Radien in Teilvierecke, sogenannte rechtwinklige Deltoide D_i. Durch (beim Viereck abwechselnde) Verlängerungen oder Verkürzungen der Seiten, wodurch die Winkel des Vierecks erhalten bleiben, (Fig. 5) beschreiben die 4 Radien Rechtecke R_i, die wir als positiv oder negativ in Rechnung stellen, je nachdem es sich um Verlängerung oder Verkürzung der Vierecksseiten handelt. Der Mittelpunkt M beschreibt im allgemeinen ein geradliniges Viereck $M_1M_2M_3M_4 \equiv M$. Dann ergibt die neue Figur des Vierecks V

$$\Sigma D_i + \Sigma R_i = V + M.$$

Im Falle (a) ist nun, da der Umfang erhalten bleibt, die algebraische Summe der Verlängerungen und also auch die algebraische Summe der Rechtecke $\Sigma R_i = 0$ und folglich

$$V_0 = V + M,$$

womit der Satz (a) bewiesen ist.

Im Falle (b) ist, da der Inhalt nicht verändert wird, $\Sigma D_i = V_0 = V$, und mithin $\Sigma R_i = M$ oder $\Sigma R_i > 0$. Ist aber die algebraische Summe der Rechtecke grösser als Null, so ist es auch die algebraische Summe der Verlängerungen der Vierecksseiten, womit der Satz (b) bewiesen ist.

7. Nur andeutungsweise werde für $2n$-Ecke ein Zusammenhang der Aufgabe (a) mit den Steiner'schen *Schliessungssätzen über gespiegelte Lichtstrahlen* gegeben. Werden im Viereck $V \equiv A_1A_2A_3A_4$ die Winkelhalbierenden gezogen und auf ihnen in den je 4 Punkten A_i und M_i Lote errichtet, so schneiden sich die letzteren in einem Punkte M, der zugleich der Diagonalschnittpunkt von den Sehnenvierecken $N_1N_2N_3N_4$ und $S_1S_2S_3S_4$ ist. Durch Parallelverschiebung längs der Strecken M_1M, M_2M, M_3M, M_4M schliessen sich die Deltoide D_i zum Tangentenviereck $V_0 \equiv A_1^0A_2^0A_3^0A_4^0$ zusammen. Das Tangentenviereck V_0 und das Viereck V sind also demselben Sehnenviereck $S_1S_2S_3S_4$ eingeschrieben, und man erhält die Steiner'schen Sätze, die hierdurch eine neue Beleuchtung erfahren. Ich gedenke an anderer Stelle ausführlich hierauf zurückzukommen, da ich hier auf die vielen interessanten Beziehungen, die sich allgemein bei $2n$-Ecken und $(2n + 1)$-Ecken ergeben, nicht eingehen kann.

§ 3. Die Ablenkung des Lichtes in Prismen.

8. Bekanntlich erfährt ein Lichtstrahl, der im Hauptschnitt durch ein Prisma geht, immer eine Ablenkung von der brechenden Kante des Prismas weg oder nach ihr hin, je nachdem das Prisma optisch dichter oder dünner als seine Umgebung ist,

und der Ablenkungswinkel ist bei symmetrischem Durchgang am Kleinsten. Hiervon gibt es zahlreiche Beweise, die sich auf goniometrische Rechnungen mit oder ohne Anwendung der höheren Analysis stützen. Ein reingeometrischer Beweis, der diesen symmetrischen Durchgang nicht nur als extremen Wert der Ablenkung, sondern auch als Minimum erkennen lehrt, scheint nicht vorhanden zu sein. Ich gebe im Folgenden einen solchen, den ich am 12./7. 1911 gefunden habe, und zwar behandle ich gleich die allgemeine, wie ich glaube neue Aufgabe:

Das Minimum oder Maximum der Ablenkung eines Lichtstrahls im Hauptschnitt eines Prismas zu finden, dessen Begrenzungsebenen an zwei verschiedene optische Mittel angrenzen.

Als gegebene Grössen betrachten wir die Verhältnisse der Lichtgeschwindigkeiten in den drei optischen Mitteln $c_0 : c_1 : c_2$ oder die relativen Brechungsexponenten n_0, n_1, n und den brechenden Winkel ω des Prismas. Zwischen den Grössen c und n besteht dann der Zusammenhang

$$c_0 n_0 = c_1 n_1 = c_2 n_2 \quad \text{.................................(1).}$$

Wir rechnen im Hauptschnitt des Prismas positiv den brechenden Winkel ω, wenn dessen Scheitel links von der Lichtstrahlenrichtung liegt, ferner den Einfallswinkel $\alpha_{\nu-1}$ und Brechungswinkel β_ν beim Uebergang in das νte Mittel, wenn die Drehung des Strahles in das Einfallslot im Sinne des Uhrzeigers erfolgt, und endlich den Ablenkungswinkel ϕ_ν, wenn bei der νten Brechung der Lichtstrahl in diesem Sinne gedreht wird. Dem Snellius-schen Brechungsgesetz zufolge erhalten wir dann das Gleichungssystem:

$$\frac{\sin\alpha_0}{\sin\beta_1} = \frac{c_0}{c_1} = \frac{n_1}{n_0}, \quad \alpha_1 - \beta_1 = -\omega, \quad \frac{\sin\alpha_1}{\sin\beta_2} = \frac{c_1}{c_2} = \frac{n_2}{n_1}, \quad \phi_1 = \alpha_0 - \beta_1, \quad \phi_2 = \alpha_1 - \beta_2 \ldots (2),$$

und hieraus für die Gesamtablenkung

$$\phi = \phi_1 + \phi_2 = \alpha_0 - \beta_2 - \omega \quad \text{..........................(3).}$$

9. Die Abbildung (Fig. 6) stellt einen solchen Strahlengang für den Fall dar, dass das Licht aus Luft in ein Glasprisma eintritt und aus diesem in Wasser austritt. Um die Figur möglichst übersichtlich zu gestalten, ist der Winkel ω so gross gewählt worden, dass α_1 und β_2 unserer obigen Festsetzung gemäss negativ werden. Ihre absoluten Werte seien durch $\bar{\alpha}_1$ und $\bar{\beta}_2$ angedeutet. Zur Ermittelung des gebrochenen Strahles aus dem einfallenden dient die bekannte Konstruktion von Reusch. Sie ist nicht an den beiden Uebergangsstellen O_1 und O_2 sondern, da parallelen Einfallsstrahlen auch parallele gebrochene Strahlen entsprechen, an der brechenden Kante O des Prismas ausgeführt worden. Wir geben zwei verschiedene Konstruktionen, deren Zusammenhang wichtig ist.

(1) Um O seien zunächst 3 konzentrische Kreise gezogen, deren Radien im Verhältnis $n_0 : n_1 : n_2$ stehen. Von dem Punkte N_0 auf Kreis n_0 des einfallenden Strahles ausgehend bestimmen die Einfallslote $N_1N_0 \| O_1L$ und $N_1N_2 \| LO_2$ die Richtungen des im Prisma verlaufenden Strahles $N_1O \| O_1O_2$ und des austretenden Strahles N_2O.

(2) Eine zweite ebenso einfache Konstruktion beginnt mit 3 um O konzentrischen Kreisen, deren Radien sich wie die Lichtgeschwindigkeiten $c_0 : c_1 : c_2$ verhalten. Von dem Punkte C_1' auf Kreis c_1 des einfallenden Strahles ausgehend,

bestimmt das Einfallslot $C_1'C_0 \parallel O_1L$ den im Prisma verlaufenden Strahl C_0C_2O und das durch C_2 gehende Lot $C_2C_1''' \parallel LO_2$ auf die Austrittsebene die Richtung $C_1'''O$ des austretenden Strahles.

In beiden eben besprochenen Konstruktionen ist das Brechungsgesetz gewahrt. Sie gelten für jeden, das Prisma durchsetzenden Strahlengang. Bei Vergleichung mehrerer Strahlengänge, wollen wir in der Figur C_2O festhalten, und alle übrigen Geraden, also auch die Einfallslote und Begrenzungsebenen des Prismas, die entsprechenden Drehungen ausführen lassen.

10. Mit Hilfe der Methode II *b* soll nun an der Hand der Figur eine Bedingung für das Minimum der Ablenkung gesucht werden. Wir nehmen an, der gezeichnete Strahlengang sei der des gesuchten Minimums. Eine ∞ kleine Aenderung $d\alpha_0$ des Einfallswinkels α_0 zieht dann ∞ kleine Aenderungen der übrigen Winkel nach sich. Nach $8_{(2)}$ ist hierbei $d\alpha_1 = d\beta_1$. Da wir vom Extremum des Ablenkungswinkels ϕ ausgehen, muss dessen ∞ kleine Aenderung $d\phi$ gegen die übrigen verschwinden. Es muss also $d\phi = d\phi_1 + d\phi_2 = 0$ sein. Hieraus folgt auf dem Kreise c_1 die Gleichheit der Bögen

$$c_1 d\phi_1 = -c_1 d\phi_2.$$

Wird auf $C_0C_1' = x_0$ in C_0 das Lot $C_0D' \parallel OO_1$, und ebenso auf $C_2C_1''' = x_2$ in C_2 das Lot $C_2D''' \parallel OO_2$ errichtet, so schneidet auf dem über $C_1D' = \dfrac{x_0}{\cos\alpha_0}$ als Durchmesser beschriebenen Kreise k_0 der ∞ kleine Winkel $d\beta_1$ einen Bogen $\dfrac{x_0}{\cos\alpha_0}d\beta_1$ ab und ebenso auf dem über $C_1'''D''' = \dfrac{x_2}{\cos\beta_2}$ als Durchmesser beschriebenen Kreise k_2 der ∞ kleine Winkel $d\bar{\alpha}_1$ einen Bogen $\dfrac{x_2}{\cos\beta_2}d\bar{\alpha}_1$ ab. Diese fallen aber mit den ∞ kleinen Bögen $c_1d\phi_1$ und $c_1d\phi_2$ zusammen, da die Kreise k_0 und k_2 sich mit dem Kreise c_1 in C_1' bezüglich C_1''' berühren. Wir erhalten also die Gleichungen

$$\frac{x_1}{\cos\alpha_0}d\beta_1 = c_1 d\phi_1 \quad \text{und} \quad \frac{x_2}{\cos\beta_2}d\bar{\alpha}_1 = c_1 d\phi_2,$$

aus denen nach oben

$$\frac{x_0}{\cos\alpha_0} = \frac{x_2}{\cos\beta_2},$$

d. h. die Gleichheit der Durchmesser der Kreise k_0 und k_2

$$D'C_1' = D'''C_1''' \text{ oder auch } D'O = D'''O \ldots\ldots\ldots\ldots\ldots\ldots (E_1)$$

folgt. Dies ist eine gesuchte geometrische Bedingung für das Extremum der Ablenkung. Für jeden anderen Strahlengang haben die Durchmesser der Kreise k_0 und k_2 verschiedene Länge.

Handelt es sich im Besonderen um ein Prisma, das an zwei gleiche optische Mittel angrenzt, ist also $c_2 = c_0$, so fallen C_2 und C_0 in einen Punkt zusammen und dieser ist dann ein Schnittpunkt der beiden gleichen Kreise k_0 und k_2. Da deren gemeinsame Sehne, hier C_0O, immer durch O geht, so ist, falls β_1 und $\alpha_1 = -\bar{\alpha}_1$ verschiedene Vorzeichen haben (wie in der Figur), diese die Symmetrieaxe der Figur. Es folgt also $\bar{\alpha}_1 = \beta_1$ und $\omega = 2\bar{\alpha}_1 = 2\beta_1$. Wenn dagegen β_1 und α_1 gleiche Vorzeichen haben, decken sich die beiden gleichen Kreise k_0 und k_2 und die Winkel α_1 und β_1, woraus folgt, dass $\omega = 0$ ist. Dieser Fall scheidet aus.

11. Haben wir jetzt für den allgemeinen Fall von 3 verschiedenen Mitteln eine geometrische Bedingung (E_1) für das Extremum gewonnen, so ist noch zu untersuchen, ob wir es mit einem Maximum oder Minimum zu tun haben. Da bei einem Prisma für kleine, aber endliche ω positive und negative Ablenkungen ϕ möglich sind (im besonderen Falle von nur 2 verschiedenen Mitteln dagegen nicht), so wollen wir von einem Maximum oder Minimum reden, je nachdem der absolute Wert $\bar{\phi}_m$ der extremen Ablenkung grösser oder kleiner ist, als die der benachbarten Strahlengänge. Wir führen diesen Nachweis ganz elementar nach Methode I.

Wir nehmen an, der durch obige geometrische Bedingung bestimmte Strahlengang liege gezeichnet vor. Für einen beliebigen anderen Strahlengang bleibe wieder C_0O fest, die Einfallslote und das Prisma sind dann um einen endlichen Winkel gedreht. Dabei schneiden sich die zwei neuen Einfallslote C_0A_0' und C_2A_2''' immer in einem Punkte A_ω auf einem festen durch C_0 und C_2 gehenden "*ω-Kreis.*" Auf den gleichen Kreisen k_0 und k_2 werden von ihnen die gleichen Bögen $C_1'A_0 = C_1'''A_2$ abgeschnitten. Wird nun durch Drehung um O der Kreis k_0 mit dem Kreis k_2 zur Deckung gebracht, wodurch alle mit bewegten Geraden um den Winkel ϕ_m gedreht werden, so deckt sich C_0A_0' mit $C_0'A_2'$. Es ist Winkel $A_0'OA_2' = C_1'OC_1'''$. Demnach ist der neue Ablenkungswinkel $A_0'OA_2''' \equiv \phi_a$ um $A_2'OA_2'''$ grösser als ϕ_m. Ebenso folgt, dass wenn die neuen Lote sich auf der anderen Seite von L_ω in B_ω schneiden, der neue Ablenkungswinkel $B_0'OB_2''' \equiv \phi_b$ um $B_0'OB_0'''$ grösser ist als ϕ_m. $\phi_a > \phi_m < \phi_b$. Es handelt sich also in unserer Figur um ein positives Minimum $\phi_m = \bar{\phi}_{min}$.

In gleicher Weise kann in jedem vorliegenden Falle Lage und Art der extremen Ablenkung bestimmt werden. Ehe wir jedoch die Ergebnisse für die verschiedenen Fälle zusammenstellen, sollen aus der oben gefundenen Bedingung einige andere abgeleitet, oder auch direkt gefunden werden, die zur Erledigung verschiedener Fragen besonders geeignet sind.

12. Bei jedem beliebigen Strahlengang gelten für die 4 durch die Punkte C_0 und N_1 bez. C_2 und N_1 auf die Begrenzungsebenen des Prismas gefällten Lote der beiden oben gegebenen Konstruktionen:

$$\frac{D'O}{C_1'O} = \frac{C_0L_0'}{C_1'L_0'} = \frac{N_1L_0}{N_0L_0} = \frac{N_1O}{M_0O}$$

und

$$\frac{D'''O}{C_1'''O} = \frac{C_2L_2'''}{C_1'''L_2'''} = \frac{N_1L_2}{N_2L_2} = \frac{N_1O}{M_2O}.$$

Für den extremen Strahlengang sind (E_1), d. h. unserer obigen Bedingung $D'O = D'''O$ gemäss alle Quotienten einander gleich, woraus $M_0O = M_2O \equiv MO$ folgt; d. h.:

(E_2) *Für den extremen Strahlengang schneiden sich die durch N_0 und N_2 zu den Begrenzungsebenen des Prismas gezogenen Parallelen in einem Punkte M auf N_1O, oder der durch $N_0N_1N_2$ bestimmte Kreis ω' berührt Kreis n_1.*

Dies lässt sich auch direkt ganz elementar nach Methode II *a* begründen. Entspricht das Viereck $N_0'N_1N_2'O$ einem beliebigen Strahlengange vom Mittelstrahl

N_1O, dessen Eintritts- und Austrittsstrahl den Winkel $N_0'ON_2' \equiv \phi$ einschliessen, so wird der ω'-Kreis vom Kreise n_1 nicht nur in N_1, sondern auch in einem anderen Punkte N_1' geschnitten. Das Viereck $N_0'N_1'N_2'O$ entspricht also ebenfalls einem Strahlengang vom Ablenkungswinkel ϕ. Für den extremen Strahlengang muss ϕ einen solchen Wert haben, dass die beiden Vierecke auch mit ihren Mittelstrahlen ($\equiv$ Diagonalen) sich decken, d. h. N_1 und N_1' müssen zusammenfallen, der ω'-Kreis also den Kreis n_1 berühren.

Aus (E_2) folgt ferner: (E_3) *Beim extremen Strahlendurchgang werden alle vier Einfallslote von den betreffenden Kreisen c oder n in demselben Verhältnis geschnitten.*

(E_4) Die Sehnenvierecke $N_0N_1N_2M$, $L_0N_1L_2O$, O_2OO_1L und $C_2O_\omega C_0L_\omega$ sind einander ähnlich.

13. Aus vorstehenden Sätzen kann nun leicht eine einfache

Konstruktion der Winkel α_1 und β_1 des extremen Strahlenganges

gefunden werden: Wir wählen das Sehnenviereck $L_0N_1L_2O$. Da $\beta_1 - \alpha_1 = \beta_1 + \bar{\alpha}_1 = \omega$, also bekannt ist, suchen wir $\cos\beta_1 : \cos\alpha_1 = N_1L_0 : N_1L_2$ zu bestimmen. Dies kann auf verschiedene Art elementar geschehen. Sehr einfach ist folgende reingeometrische Lösung: Nach (E_3) verhält sich

$$N_1L_0 : N_1L_2 = N_0L_0 : N_2L_2.$$

Durch entsprechende Addition und Subtraktion ergibt sich hieraus

$$(N_1L_0)^2 : (N_1L_2)^2 = (N_1L_0 + N_0L_0)(N_1L_0 - N_0L_0) : (N_1L_2 + N_2L_2)(N_1L_2 - N_2L_2).$$

Die beiden letzten Glieder sind nun Produkte von den Abschnitten zweier Sehnen im Kreise n_1, von denen die eine durch N_0, die andere durch N_2 geht. Nach dem Sehnensatze der Planimetrie sind diese Produkte aber gleich den Quadraten der entsprechenden halben kleinsten Sehnen im Kreise n_1, die durch N_0 und N_2 gehen. Es ist daher das gesuchte Verhältnis*

$$\cos\beta_1 : \cos\alpha_1 = N_1L_0 : N_2L_2 = \sqrt{n_1^2 - n_0^2} : \sqrt{n_1^2 - n_2^2}$$

oder: *Die von N_1 auf die Begrenzungsebenen des Prismas gefällten Lote verhalten sich wie die Sehnen im Kreise n_1, die die Kreise n_0 und n_2 berühren.* Dasselbe gilt für die Winkelschenkel von ω in allen Vierecken die in (E_4) aufgeführt worden sind. Eins dieser Vierecke kann nun leicht konstruiert werden.

14. Werden die in 12 an zweiter oder dritter Stelle übereinander stehenden Quotienten durch die Winkel α und β allein oder durch die Winkel und Radien c bez. n ausgedrückt, so erhält man für den extremen Strahlengang die Formeln

$$\frac{\operatorname{cotg}\beta_1}{\operatorname{cotg}\alpha_0} = \frac{\operatorname{cotg}\alpha_1}{\operatorname{cotg}\beta_2} \quad \text{und} \quad \frac{c_0\cos\beta_1}{c_1\cos\alpha_0} = \frac{c_2\cos\alpha_1}{c_1\cos\beta_2}$$

oder

$$\frac{\operatorname{tang}\alpha_0 \operatorname{tang}\alpha_1}{\operatorname{tang}\beta_1 \operatorname{tang}\beta_2} = 1 \quad \ldots\ldots\ldots\ldots (E_5),$$

und

$$\frac{\cos\alpha_0\cos\alpha_1}{\cos\beta_1\cos\beta_2} = \frac{c_0}{c_2} = \frac{n_2}{n_0} \quad \ldots\ldots\ldots\ldots (E_6).$$

* In der Diskussion, die dem Vortrag folgte, gab ich auf Anregung von Professor Schoute eine andre Ableitung mit Hilfe von Formel (E_{10}).

Von diesen beiden Bedingungsleichungen für den extremen Strahlengang ist besonders die erste wichtig, da sie nur die vier Winkelgrössen enthält und da aus ihr wichtige Folgerungen gezogen werden können.

(E_7) *In der Prismenfigur schneiden sich beim extremen Strahlengang der eintretende und austretende Strahl in einem Punkte S auf dem Durchmesser OL, wo L der Schnittpunkt der Einfallslote ist.* Zum Beweise fälle man von S Lote auf die Begrenzungsebenen und wende die Bedingung (E_5) an.

Da der Quotient der Tangenten zweier spitzer Winkel, ebenso wie der Quotient ihrer Sinus $\gtrless 1$ ist, je nachdem der Quotient der Winkel selbst es ist, so kann die Gleichung (E_5) nur dann bestehen, wenn ein Bruch > 1, der andere < 1 ist. Es muss daher

$$\text{entweder } \begin{matrix} c_0 > c_1 < c_2 \\ n_0 < n_1 > n_2 \end{matrix} \text{ oder } \begin{matrix} c_0 < c_1 > c_2 \\ n_0 > n_1 < n_2 \end{matrix} \text{ sein, d. h.:}$$

(E_8) *Beim Prisma ist ein extremer Durchgang nur dann möglich, wenn das Prisma dichter oder dünner als seine* **beiden** *umgebenden Mittel ist.*

15. Es fragt sich, wie weit, vom Extremum ausgehend, in der Figur die Einfallslote nach beiden Seiten gedreht werden können, damit noch ein Austreten des Strahls und nicht eine Totalreflexion an der zweiten Begrenzungsebene erfolge, wie weit also z. B. auf dem ω-Kreise A_ω oder B_ω wandern können. Ist $c_1 < c_0$, c_2, so bestimmt je eine von c_0 und von c_2 an den Kreis c_1 gelegte Tangente diese beiden Grenzlagen. Im allgemeinen ist keine ein Extremum, sondern nur in dem besonderen Falle, dass das oben betrachtete Extremum mit einer dieser Grenzlagen zusammenfällt. Wir gehen nicht näher darauf ein. Grenzlagen gibt es auch, wenn $c_0 < c_1 < c_2$ oder $c_0 > c_1 > c_2$ ist, also nach (E_8) **kein** Extremum vorhanden ist.

Während im Sonderfalle $c_2 = c_0$, wie einleitend in 8 bemerkt wurde, die Ablenkungswinkel ϕ und ihr Extremum ϕ_m entweder immer positiv oder immer negativ sind, so ist dies im allgemeinen Falle nicht so einfach (vergl. 11). Ausschlaggebend ist der ω-Kreis. Wenn dieser den Kreis c_1 berührt, ist

$$\sin \omega_0 = \frac{(c_0 - c_2)\, c_1}{c_0 c_2 - c_1^2} = \frac{(n_2 - n_0)\, n_1}{n_1^2 - n_0 n_2}.$$

Man erhält für die drei zu unterscheidenden Fälle $\omega \gtreqless \omega_0$ folgende

Uebersicht über die Vorzeichen der Ablenkungswinkel ϕ und die Art des Extremums ϕ_m,

je nachdem das Prisma

optisch dichter oder *optisch dünner,*

als seine Umgebung ist:

I, $\omega > \omega_0$.

Der ω-Kreis schneidet den Kreis c_1 nicht.

$\phi > 0,$	$\phi < 0,$
$\phi_m = + \bar{\phi}_{min},$	$\phi_m = - \bar{\phi}_{min}.$

II, $\omega = \omega_0$.

Der ω-Kreis berührt c_1 und fällt mit den beiden Kreisen k_0 und k_2 zusammen.

$$\phi \geqq 0, \qquad \phi \leqq 0.$$

$$\phi_m = 0, \qquad \phi_m = 0.$$

Das Prisma ist für den extremen Strahlendurchgang ein "Geradsichtsprisma."

III, $\omega < \omega_0$.

Der ω-Kreis schneidet den Kreis c_1.

$$\phi \gtreqless 0, \qquad \phi \lesseqgtr 0,$$

$$\phi_m = -\bar{\phi}_{max}, \qquad \phi_m = +\bar{\phi}_{max}.$$

Für zwei verschiedene Strahlendurchgänge, die den Schnittpunkten A_ω und B_ω des ω-Kreises mit c_1 entsprechen, wird $\phi = 0$, das Prisma also zum "Geradsichtsprisma." Diese beiden Strahlendurchgänge sind keine Extrema, sondern bilden nur die Uebergänge von positiver zu negativer Ablenkung.

16. (E_9) *Der Tangensbedingung* (E_5) *kann ein einfaches geometrisches Gewand in Gestalt eines geschlossenen Vierecks $H_0G_1H_1G_2$ gegeben werden, dessen Ecken abwechselnd auf zwei einander in O senkrecht schneidenden Geraden liegen* (Fig. 7). Sind für einen beliebigen Strahlengang die Winkel $H_0G_1O = \alpha_0$, $H_1G_1O = \beta_1$, $H_1G_2O = \alpha_1$, $H_2G_2O = \beta_2$, mithin $G_1H_1G_2 = \pi - \omega$, so sind H_2 und H_0 verschiedene Punkte, gehören die Winkel dagegen dem extremen Strahlengang an, so ist $H_2 \equiv H_0$. In letzterem Falle folgt, wenn $H_0O = h_0$, $H_1O = h_1$, $G_1O = g_1$, $G_2O = g_2$, aus der Identität

$$\frac{h_0}{g_1}\frac{h_1}{g_2} = \frac{h_1}{g_1}\frac{h_0}{g_2}$$

obige Tangensbedingung (E_5).

*Die Fusspunkte $A_0B_1A_1B_2$ der 4 von O auf die Seiten des Vierecks gefällten Lote a_0, b_1, a_1, b_2, liegen auf einem Kreise**, wie man leicht erkennt, wenn man die rechtwinkligen Dreiecke der Figur durch ihre umgeschriebenen Kreise ersetzt und die Figur von O aus invers transformiert. Diesem Satze J. Steiner's möchte ich hinzufügen, *dass A_0B_1 und A_1B_2 sich auf HO, A_0B_2 und B_1A_1, sich auf GO schneiden*, was ebenfalls durch Inversion leicht bewiesen werden kann.

Das Sehnenviereck $B_1OA_1H_1$ (Fig. 7) *ist den in* (E_4) *aufgeführten* (Fig. 6) *ähnlich*. Werden die reciproken Werte der Strecken g und h doppelt ausgedrückt, so folgt

$$\left.\begin{array}{l} \left\{\begin{array}{l} \dfrac{\sin\alpha_0}{a_0} = \dfrac{1}{g_1} = \dfrac{\sin\beta_1}{b_1} \\[2ex] \dfrac{\cos\alpha_0}{a_0} = \dfrac{1}{h_0} = \dfrac{\cos\beta_2}{b_2} \end{array}\right. \\[6ex] \left\{\begin{array}{l} \dfrac{\sin\alpha_1}{a_1} = \dfrac{1}{g_2} = \dfrac{\sin\beta_2}{b_2} \\[2ex] \dfrac{\cos\alpha_1}{a_1} = \dfrac{1}{h_1} = \dfrac{\cos\beta_1}{b_1} \end{array}\right. \end{array}\right\}.$$

Werden diese Gleichungen quadriert und die durch Klammern bezeichneten Paare

* J. Steiner, *Ges. W.*, II, S. 358, Lehrs. 2.

oder alle vier addiert, so folgen einfache Beziehungen zwischen den Strecken allein. In letzterem Falle ergibt sich

$$\frac{1}{a_0^2}+\frac{1}{a_1^2}=\frac{1}{b_1^2}+\frac{1}{b_2^2} \quad\ldots\ldots\ldots\ldots(E_{10}),$$

eine Beziehung, die für manche Berechnungen gute Dienste leistet.

17. Zum Schlusse noch einige Bemerkungen über den

Verlauf des Lichtstrahles im Hauptschnitte eines Prismensystems bei minimaler oder maximaler Ablenkung.

Abgesehen von einigen Sonderfällen ist es mir bis jetzt nicht gelungen, für dieses allgemeine Problem von beliebig vielen Brechungen synthetisch zu entsprechend einfachen elementargeometrischen Ergebnissen zu kommen, oder das Problem des Prismensystems auf das soeben behandelte von drei lichtbrechenden Medien zurückzuführen. Ich beschränke mich daher vorläufig darauf, mit Hilfe der Differentialrechnung die Herleitung der Bedingung (E) für den extremen Strahlengang kurz anzudeuten.

Von den drei verschiedenen Bedingungsgleichungen, die ich gebe, habe ich nur die Cosinusbedingung (E_I) in dem Werke: "*die Theorie der optischen Instrumente,* bearbeitet von wissenschaftlichen Mitarbeitern der optischen Werkstätte von Carl Zeiss," gefunden*. Die wichtigere Tangensbedingung (E_{II}) ist dem Bearbeiter des Kapitels über Prismensysteme entgangen, was um so auffälliger ist, als er diese für 2 Brechungen, allerdings nur in dem besonderen Falle $c_2=c_0$ des gewöhnlichen Prismas, angibt. Bei Aufstellung der Gleichungssysteme, deren Differentiation zur Bedingungsgleichung führt, bin ich möglichst dem oben angeführten Werke gefolgt.

Wir setzen k Begrenzungsebenen, $k-1$ Prismen von den brechenden Winkeln ω_ν und $k+1$ optische Mittel von den Lichtgeschwindigkeiten c_ν voraus. Ueber die Vorzeichen der Winkel α und β siehe die in 8 getroffenen Bestimmungen. Die einzelnen Ablenkungen seien ϕ_ν und die Gesamtablenkung ϕ. Dann gelten analog wie in 8 folgende Gleichungssysteme:

$$\frac{\sin\alpha_{\nu-1}}{\sin\beta_\nu}=\frac{c_{\nu-1}}{c_\nu}, \qquad (\nu=1,2,\ldots,k)\ldots\ldots\ldots(1),$$

$$\alpha_\nu-\beta_\nu=-\omega_\nu, \qquad (\nu=1,2,\ldots,k-1)\ldots\ldots\ldots(2),$$

$$\phi_\nu=\alpha_{\nu-1}-\beta_\nu, \qquad (\nu=1,2,\ldots,k),$$

$$\phi=\sum_1^k\phi_\nu=\alpha_0-\beta_k-\sum_1^{k-1}\omega_\nu \quad\ldots\ldots\ldots(3).$$

Für den extremen Wert erhalten wir durch Differentiation der Gleichungen (1) bis (3) und Elimination der c durch (1) die Gleichungssysteme:

$$\frac{d\alpha_{\nu-1}}{d\beta_\nu}=\frac{c_{\nu-1}}{c_\nu}\,\frac{\cos\beta_\nu}{\cos\alpha_{\nu-1}}=\frac{\operatorname{tang}\alpha_{\nu-1}}{\operatorname{tang}\beta_\nu}, \qquad (\nu=1,2,\ldots,k)\ldots\ldots(1'),$$

$$\frac{d\alpha_\nu}{d\beta_\nu}=1, \qquad (\nu=1,2,\ldots,k-1)\ \ldots(2'),$$

$$\frac{d\phi}{d\beta_k}=\frac{d\alpha_0}{d\beta_k}-1=0 \text{ oder } \frac{d\alpha_0}{d\beta_k}=1 \quad\ldots\ldots\ldots(3').$$

* a. a. O., I, Bd. *Die Bilderzeugung in optischen Instrumenten*; VIII Kap. F. Löwe (Czapski), *Prismen und Prismensysteme*, S. 422.

Die Multiplikation der Gleichungen (1′) ergibt nach (2′) und (3′) als Bedingung für den extremen Strahlengang,

$$\frac{\cos\alpha_0 \cos\alpha_1 \ldots \cos\alpha_{k-1}}{\cos\beta_1 \cos\beta_2 \ldots \cos\beta_k} = \frac{c_0}{c_k} = \frac{n_k}{n_0} \quad \ldots\ldots\ldots\ldots\ldots\ldots\ldots\ldots (\mathrm{E_I})$$

und

$$\frac{\operatorname{tang}\alpha_0 \operatorname{tang}\alpha_1 \ldots \operatorname{tang}\alpha_{k-1}}{\operatorname{tang}\beta_1 \operatorname{tang}\beta_2 \ldots \operatorname{tang}\beta_k} = 1 \quad \ldots\ldots\ldots\ldots\ldots\ldots (\mathrm{E_{II}}).$$

($\mathrm{E_{III}}$) Ganz analog wie in ($\mathrm{E_8}$) folgt hieraus, *dass ein Extremum ϕ_m der Ablenkung nur dann stattfinden kann, wenn weder alle Brechungen aus je einem dünneren in ein dichteres Medium noch umgekehrt erfolgen.*

($\mathrm{E_{IV}}$) Auch hier lässt sich die Tangensbedingung ($\mathrm{E_{II}}$) durch ein geschlossenes $2k$-Eck darstellen, dessen Seiten abwechselnd sich auf zwei zu einander senkrechten Geraden schneiden. Die vom Schnittpunkt beider auf die Vielecksseiten gefällten Lote genügen wie in ($\mathrm{E_{10}}$) der Gleichung:

$$\frac{1}{a_0^2} + \frac{1}{a_1^2} + \ldots + \frac{1}{a_{k-1}^2} = \frac{1}{b_1^2} + \frac{1}{b_2^2} + \ldots + \frac{1}{b_k^2} \quad \ldots\ldots\ldots\ldots (\mathrm{E_V}).$$

ON BINODES AND NODAL CURVES

BY MISS H. P. HUDSON.

The first part of this note shews that a binode can be studied by means of two auxiliary non-singular surfaces which approximate to the separate sheets of the given surface. The biplanes are the simplest of such surfaces; and if the binode is a B_i, i.e. if it reduces the class of the surface by i, then the approximation can be carried to the order $i-2$. This leads to a very convenient algebraic definition of a B_i. The second part investigates the reduction in class due to a general nodal curve; incidentally, the number of pinch-points is determined.

The method of § I. might be applied to unodes of all orders; then the two auxiliary surfaces have contact with one another, of a certain order which is the other characteristic of the singularity. The method of § II. might be applied to cuspidal, tacnodal, etc. curves, and also to curves of higher multiplicity, determining in each case the reduction in the class of the surface, and the number of its ordinary singularities.

REFERENCES.

SALMON, *Camb. and Dubl. Math. J.* II. p. 72, 1846.
CAYLEY, *Coll. Math. Papers*, VI. p. 123, 1868.
ROHN, *Math. Ann.* XXII. p. 124, 1883.
SEGRE, *Ann. di Mat.* (2) XXV. p. 1, 1896.
BASSET, *Geometry of Surfaces*, Cambridge, 1910.

I. (*Cartesian coordinates.*)

Theorem. If the origin O is an isolated B_i on a surface F, then

$$F = H \,.\, K + f_i \quad \ldots\ldots\ldots\ldots\ldots\ldots\ldots\ldots\ldots\ldots\text{(I)}$$

where f_i is small of order i near O, and H, K are surfaces each passing once only through O. The form of the equation shews that each of these surfaces has contact of order $i-2$ with one sheet of F; we may say that to this order of approximation, F is degenerate.

Let yz be the biplanes, and let F be expressed in the form

$$F = H \,.\, K + f_i,$$

where

$$H = y + \phi_2,$$

$$K = z + \psi_2,$$

and l has its greatest possible value, which implies that f_l contains a term in x^l. If $l=3$, the form assumed merely expresses that O is a binode. We have to prove $l=i$, the reduction in class.

Now the class of F is the number of points at which the tangent planes pass through an arbitrary straight line, i.e. the number of intersections other than O of F and the curve Γ of intersection of P_1, P_2, the first polars with regard to F of any two points $A_1(x_1y_1z_1)$, $A_2(x_2y_2z_2)$; and i is the number of intersections of F, Γ which coincide at O.

Then if $\Delta_1 \equiv x_1\frac{\partial}{\partial x}+y_1\frac{\partial}{\partial y}+z_1\frac{\partial}{\partial z}+w_1\frac{\partial}{\partial w}$, $(w=w_1=1)$ the equations of Γ are

$$\left.\begin{array}{l} P_1 \equiv H\Delta_1 K + K\Delta_1 H + \Delta_1 f_l = 0 \\ P_2 \equiv H\Delta_2 K + K\Delta_2 H + \Delta_2 f_l = 0 \end{array}\right\},$$

and Γ passes through O and touches the edge of the binode. At points of Γ near O we have: ΔH, ΔK finite; x small of order 1; y, z small of order 2; H, K, Δf_l small of order $l-1$; and since f_l contains a term in x^l, both f_l and F are small of order l, and exactly l intersections of F, Γ coincide at O.

Therefore $$l=i.$$

The converse is proved by reversing the argument, so that a B_i may be defined by means of the form (I), which is specially convenient for algebraic transformations.

If the B_i is decomposed by Segre's method into adjacent double points, we must suppose them arranged along the curve $H=K=0$, which is a generalization of the edge of the binode, just as H, K are generalizations of the biplanes. Note that H is not determinate, but may be replaced by any other surface having contact of order $i-2$ with H at O.

II. (*Homogeneous coordinates.*)

If the degree n of F is given, there is an upper limit to i, but its relation to n is unknown. If $n=3$, it is 6; if $n=4$, it is $\geqq 15$. If i exceeds this limit, then O ceases to be an isolated binode, and F acquires a nodal curve C through O. Then Γ degenerates into C and a residual Γ', and there are an infinite number of points near O common to F, P_1, P_2. The argument of § I. then shews that there exist non-singular surfaces having contact of any order with one sheet of F at O, but we cannot in this way determine the reduction in class.

Let C be of degree m and rank r, with no singularities except d nodes; then Γ' is of degree $(n-1)^2-m$, and the class of F is the number of intersections of F, Γ' not lying on C. Now Γ' meets C in

(i) the r points at which the tangent line to C meets the arbitrary straight line A_1A_2; these are points of contact of P_1, P_2, and at each of them Γ' meets F in two coincident points;

(ii) the q pinch-points of F, which are points of contact of P_1, P_2; these are ordinary singularities, and at each of them Γ' meets F in three points;

(iii) the d double points of C, which are tacnodes of F and points of osculation of P_1, P_2; through each there passes one branch of Γ', meeting F in four points;

(iv) the points of higher multiplicity of F on C; these are not ordinary singularities, and we only consider t triple points of F at simple points of C; each is a common node of P_1, P_2, and through it there pass three branches of Γ', each meeting F in three points.

Therefore the reduction in class is

$$mn + 2r + 3q + 4d + 9t$$

and it remains to determine q.

First let C be the straight line $z = w = 0$. Then the pair of tangent planes at any point $(xy00)$ of C is given by an expression of the form

$$T \equiv (U, V, W \between z, w)^2$$

where U, V, W are functions of x, y of degree $n-2$. The pinch-points are among the $2(n-2)$ intersections of C with the surface $V^2 - UW$; in general they are distinct, and there are no higher singularities. But each of the t triple points of F gives a squared linear factor in $V^2 - UW$, and in this case

$$q = 2(n - 2 - t).$$

If $V^2 - UW$ has another squared factor, then a pair of pinch-points coincide; through such a point there passes one branch of Γ', touching C and meeting F in six coincident points. If $V^2 - UW$ is a perfect square, all the pinch-points coincide in pairs. Now this is also the condition that T falls into two factors, each linear in z, w and rational in x, y; then there exist two auxiliary surfaces, given by these factors, each passing once only through C and touching one sheet of F. This last property defines an important class of double curves; it always implies that the pinch-points coincide in pairs, but if C is a general curve, the converse is not true.

Next let C be the total intersection of two surfaces Φ, Ψ of degrees k, l. Then

$$F \equiv (U, V, W \between \Phi, \Psi)^2$$

and the pinch-points are among the intersections of C with $V^2 - UW$, in number

$$2kl(n - k - l) = 2\{m(n-2) - r - 2d\}.$$

This number does not include the double points of C, for they do not in general lie on $V^2 - UW$, but it includes the triple points of F, each counted twice, and the number of pinch-points is

$$q = 2\{m(n-2) - r - 2d - t\} \quad \text{.........................(II).}$$

Finally, let C be a general curve, the partial intersection of Φ, Ψ, and let the residual C' (of degree m' etc.) meet C in I points, supposed ordinary points of C and of F. Take any two surfaces F_1, F_2, of degrees n_1, n_2, passing through C' and not through C, and having no multiple points on C'. Then the complex surface $F \cdot F_1 \cdot F_2$ has the complex curve of total intersection $C + C'$ as a nodal curve, and

$$F \cdot F_1 \cdot F_2 \equiv (U, V, W \between \Phi, \Psi)^2.$$

Now each of the I points is quadruple on $F.F_1.F_2$ and double on $V^2 - UW$, and absorbs four intersections of $V^2 - UW$, $C + C'$; there are $m'n - 2I$ other intersections of F, C' and $m(n_1 + n_2) - 2I$ of $F_1.F_2$, C, which are triple points on the complex surface; therefore the number of pinch-points of $F.F_1.F_2$ on $C + C'$ is

$$2[(m+m')(n+n_1+n_2-2)-(r+r')-2(d+d'+I)-\{t+m'n+m(n_1+n_2)-4I\}]-4I.$$

But the pinch-points on C' coincide in pairs at the $m'(n_1+n_2-2)-r'-2d'$ points of contact* of F_1, F_2 on C'. Therefore the number of pinch-points of F on C is as before

$$q = 2\{m(n-2) - r - 2d - t\}.$$

Substituting this value for q in II, we have the final formula for the reduction in the class of the surface due to a general nodal curve:

$$m(7n-12) - 4r - 8d + 3t \text{ ....(III).}$$

* Salmon, *Geom. of Three Dim.* 5th ed. p. 358.

ON THE CONFORMAL REPRESENTATION OF CONVEX DOMAINS

By E. Study.

We shall term *a monotonic graph* any continuous set of points (θ, Θ) in a Cartesian plane such that

1. If $\theta_2 - \theta_1 > 0$, then $\Theta_2 - \Theta_1 \geqq 0$.
2. If $\Theta_2 - \Theta_1 > 0$, then $\theta_2 - \theta_1 \geqq 0$.
3. The set is reproduced by the translation

$$\theta^* = \theta + 2\pi, \quad \Theta^* = \Theta + 2\pi.$$

By such a graph, which is a special kind of periodic curve, a sort of functional dependence $\Theta(\theta)$ is given. To a definite value of θ there will correspond either a definite value of Θ or a whole continuum of values, ranging between and including two extremes. These extreme values we may appropriately call $\Theta(\theta - 0)$ and $\Theta(\theta + 0)$, indicating thereby a process of passing to the limit by which they can be obtained. We shall speak in this case of a *vertical segment* ($\theta =$ const.) contained in the graph. Of course there may be *horizontal segments* ($\Theta =$ const.) as well. The sum of the lengths

$$\Theta(\theta + 0) - \Theta(\theta - 0)$$

of all vertical segments contained in any *primitive section* of the graph

$$\{\theta_0 \leqq \theta < \theta_0 + 2\pi, \quad \Theta_0 \leqq \Theta < \Theta_0 + 2\pi\}$$

cannot exceed 2π; therefore the values of θ belonging to this interval for which $\Theta(\theta - 0)$ and $\Theta(\theta + 0)$ do *not* coincide, will, at the utmost, form a denumerable set. *The primitive section may always be chosen so as not to divide any vertical or horizontal segment in two.* It then contains *one* representative of each set of segments that correspond in the group

$$\theta^* = \theta + 2\kappa\pi, \quad \Theta^* = \Theta + 2\kappa\pi \quad (\kappa = \pm 1, \pm 2, \ldots).$$

Finally it is clear that

$$H(\theta) = \frac{\Theta(\theta - 0) + \Theta(\theta + 0)}{2}$$

is a monotonic function, such that

$$H(\theta + 2\pi) = H(\theta) + 2\pi,$$

$$H(\theta) = \frac{H(\theta - 0) + H(\theta + 0)}{2},$$

determined by, and in its turn determining, the graph.

Graphs derived from one another by means of a translation

$$\{\theta_* = \theta + \text{const.}, \quad \Theta_* = \Theta + \text{const.}\},$$

and its inverse, we shall consider as not essentially different, or *equivalent*.

Monotonic graphs arise in connection with the conformal representation of convex domains on to one another.

Let us especially consider the (direct) mapping $w = w(z)$ of a circular domain, say $|z| < 1$, on to a second convex domain. The word *convex domain* we may take in a somewhat broader sense than is customary. We include under this denomination all domains that can be looked upon as limits of sequences of convex polygons, with the exception of the plane itself*. Thus a "convex domain" may extend to infinity; e.g. a parallel strip and a half-plane are convex domains.

Leaving aside, however, for the instant, infinite domains as well as all those with angular points or segments of straight lines on their boundaries, we readily see that from the mapping $w = w(z)$ a set of equivalent monotonic graphs can be derived: *θ and Θ are the angles between two (arbitrary) fixed straight lines and two corresponding tangents of the two boundaries.* It must be supposed, of course, that, starting from an initial couple of values (θ_0, Θ_0), θ and Θ change *continuously*, when the two tangents turn around the two domains.

Now this statement can be generalised so as to include all the limiting cases mentioned. Taking this for granted—the necessary definitions are too long to be reproduced here—we arrive at the conclusion:

I. *To every mapping $w = w(z)$ of the circular domain $|z| < 1$ on to another convex domain there corresponds a definite set of equivalent monotonic graphs. The same set corresponds, of course, to all the mappings given by the functions*

$$W(z) = Aw(z) + B \quad \{A, B = \text{const.}, A \neq 0\}.$$

The interest we attribute to this theorem consists in the possibility of its reversion:

II. *The set of mapping functions $W(z)$ is completely determined by the set of corresponding graphs, which, in its turn, can be prescribed arbitrarily (subject to conditions* 1. 2. 3.).

In order to get rid of the ambiguity contained in these assertions, we represent the points of the circle, once for all, by the formula

$$\zeta = e^{i\theta}.$$

On the other hand, we select from the set of functions $W(z)$ the one whose expression as a power series is

$$w(z) = z + \dots.$$

Then the functional dependence $\Theta(\theta)$ or the monotonic function $H(\theta)$ is determined apart from an additive constant, to which we may attribute an arbitrary value. If now we choose *out of a primitive section* of the graph two values θ_1, θ_2, such that

* Of course a more elementary (but less short) definition may be given.

$\theta_2 - \theta_1 > 0$, and that the corresponding values $\Theta_1 = \Theta(\theta_1)$ and $\Theta_2 = \Theta(\theta_2)$ are definite, we may look upon the difference

$$\Delta\Theta = \Theta_2 - \Theta_1 = H_2 - H_1$$

as a *mass*, positive or zero, situated in an arc of the circle, whose length is

$$\Delta\theta = \theta_2 - \theta_1.$$

It follows that the sum of all these masses is 2π, and that their distribution over the boundary $|\zeta| = 1$, determines the function $H(\theta)$ or the graph $\Theta(\theta)$, apart from an additive constant. Further it follows incidentally that

if $$\theta_1 < \theta < \theta_2, \quad \operatorname{Lim} \Delta\theta = 0,$$

then $$\operatorname{Lim} \Delta\Theta = \Theta(\theta + 0) - \Theta(\theta - 0).$$

Thus we see that the theorems I and II are equivalent to the following:

III. *If the function*

$$w(z) = z + \ldots$$

maps the circular domain $|z| < 1$ *on to another convex domain, it determines a distribution of masses, positive or zero, whose sum is* 2π, *over the boundary* $|\zeta| = 1$ *of the circle, and vice versa.*

Moreover it is possible to determine *explicitly* the function $w(z)$, if the distribution of masses (or one of the corresponding graphs) is given, and vice versa.

Starting from the graph as the given element, we form, by means of a so-called *Stieltjes-Integral*, the function

$$f(z) = -\frac{1}{\pi}\int d\Theta \,.\, \log\left(1 - \frac{z}{\zeta}\right) \quad \{|z| < 1\},$$

extending the integration, in the positive sense, over the boundary of the circle, viz., over a primitive interval of the graph. *Then*

$$w(z) = \int_0^z dz\, e^{f(z)}$$

is the function required.

If, conversely, $w(z)$ is known, we consider the function (any one branch of the function)

$$\phi(z) = \frac{1}{i}\log\left(z\frac{dw}{dz}\right) \quad \{|z| < 1\}.$$

Let the point z, *from the interior of the circle, approach the point* ζ *of the boundary on a straight line. Then the real part of* $\phi(z)$ *will tend towards a definite value, possibly depending upon the direction of approach. The range of values obtained thereby, including its lower and upper limits, is the range of values of* Θ *belonging to the value* θ *in one of the required graphs.*

The two limiting values of the range are

$$\Theta(\theta - 0) \text{ and } \Theta(\theta + 0);$$

approaching ζ on a radius, we obtain

$$H(\theta) = \frac{\Theta(\theta - 0) + \Theta(\theta + 0)}{2}.$$

Thus the study of the mapping of circular domains on to other convex domains is seen to be equivalent to the study of monotonic graphs.

To further elucidate this we add a few applications.

To any horizontal segment in the primitive interval of the graph there corresponds a segment of a straight line, forming part of the boundary of the map. If a part of the graph is analytic but not vertical, the corresponding part of the boundary is analytic too. Hence:

If the graph is nowhere analytic, except in its vertical segments, then the circle is a natural boundary to the mapping functions $W(z)$.

To any vertical segment of the graph there corresponds an angular point in the boundary of the map, whose internal angle is

$$\pi - \{\Theta(\theta+0) - \Theta(\theta-0)\},$$

provided this difference is positive.

If there is no vertical segment in the graph whose length reaches or exceeds π, *then the map remains entirely in the finite domain.*

If there is, in the primitive section of the graph, *one* vertical segment of length equal to π, the map extends to infinity as the interior of a parabola does, viz., in one direction only. If there are *two* such segments, the map is a parallel strip. If there is a vertical segment whose length exceeds π and is less than 2π, the map behaves like the interior of one branch of a hyperbola, extending to infinity in an infinity of directions. If there is a vertical segment of length 2π (and consequently one horizontal segment of the same length), the map is a half-plane.

Special interest belongs to the case in which the graph is entirely made up of horizontal and vertical segments, a finite (and necessarily even) number being contained in a primitive section of the graph. If we exclude the cases in which the length of a vertical segment reaches or exceeds π, the interior of the circle is mapped on to the interior of an ordinary (finite) *convex polygon*. The expression of $w(z)$ reduces to a well-known formula, due to H. A. Schwarz and Christoffel.

This formula was, in fact, the author's starting point. The process of passing from convex polygons to the other convex domains suggests also the methods of proof to be applied.

Of all the statements given the converse is true.

Finally it may be noticed that the "monotonic" graphs may be replaced by "graphs" of a more general description. Instead of our function $H(\theta)$ any such functions of limited variation (*fonctions à variation bornée*) as satisfy the conditions

$$H(\theta+2\pi) = H(\theta) + 2\pi,$$

$$H(\theta) = \frac{H(\theta-0) + H(\theta+0)}{2},$$

may be used. Substituting this in the formula for $w(z)$ we still obtain the mapping of the domain $|z| < 1$ on to another domain. But the author has not been successful in finding the characteristics of the map in this (much) more general case.

For proofs and further developments we refer to the booklet *Vorlesungen über ausgewählte Gegenstände der Geometrie*, II. (Leipzig, 1912).

ÜBER DIE BEGRIFFE "LINIE" UND "FLÄCHE"

Von Z. Janiszewski.

Mein Referat hat einen ganz negativen Charakter: ich will nur einige kritische Bemerkungen über den heutigen Stand einer Frage mitteilen und keine Lösung derselben. Ich stelle mich hier auf *rein geometrischen* (besser: *mengentheoretischen*) Boden und betrachte alle geometrischen Gebilde als Punktmengen. Der Einfachheit halber beschränke ich mich auf im dreidimensionalen euklidischen Raume nirgendsdichte Kontinua (d. h. Kontinua ohne innere Punkte).

Die geometrischen Gebilde, mit welchen wir gewöhnlich zu tun haben—die, so kompliziert sie uns auch erscheinen mögen, doch immer verhältnissmässig einfach sind—verteilen sich ganz natürlich in zwei grundverschiedene Arten: in Linien und Flächen. Es können zwar Gebilde auch aus Linien und Flächen zusammengesetzt sein—*wir beschränken uns aber auf solche, die in der Umgebung jedes Punktes dieselbe Dimensionsanzahl haben.*

Bei dieser Beschränkung scheint zwischen Linien und Flächen eine Kluft zu bestehen.

Die—bis heute offene—Frage, die ich hier an die Spitze stelle, lautet: ist es wirklich so, gilt dies auch von den allgemeinsten geometrischen Gebilden? Oder giebt es vielleicht solche Kontinua, die weder Linie, noch Fläche genannt werden können?

Für die Beantwortung dieser Hauptfrage bedarf man einer *ganz allgemeinen* Definition der Linie und der Fläche.

Für den zweidimensionalen Raum ist die Frage bekanntlich gelöst, oder sie ist vielmehr gegenstandslos, da dort jedes nirgendsdichte Kontinuum eine Linie ist (die Zoretti *Cantorsche Linie* nennt). Für den dreidimensionalen Raum aber besitzen wir keine solche allgemeingültige Definition, welche aus den nirgendsdichten Kontinuen die Linien aussonderte.

Viele beschränken sich von vornherein (offen oder stillschweigend) auf spezielle Gebilde (das gilt auch von den abstrakten, tiefen Untersuchungen von Enriques). Die am besten bekannte, Jordansche Definition steht gerade im Widerspruch zu den geometrischen Forderungen (wie die "Peanosche Kurve" zeigt) und interessirt uns also nicht*.

* Dabei ist es noch zu betonen, dass man überhaupt nicht die uns interessierende Frage als Untersuchung über den Funktionsbegriff auffassen darf. Der letztere scheint zu eng zu sein, es sei denn, dass man ganz unregelmässig mehrdeutige Funktionen betrachten will. Denn, wie will man z. B. die unten angegebene Kontinua als eindeutige Funktionen auffassen?

Eine wesentlich andere Untersuchung ist die von Fréchet, die von vornherein die Existenz verschiedener Dimensionstypen annimmt; Fréchet schreibt z. B. dem Kreise einen höheren Dimensionstypus zu, als der Strecke. Diese sehr interessante Idee aber schliesst unsere Frage nicht aus. Diese würde in Fréchet'scher Terminologie ungefähr so lauten: spaltet sich die Mannigfaltigkeit der den nirgendsdichten Kontinuen entsprechenden Dimensionstypen in zwei Klassen oder nicht?

Es könnte jedoch scheinen als liessen sich einige Aussagen als allgemeingültige Definitionen der Linie oder der Fläche aufstellen. Ich will nun eine Linie konstruiren die ein Gegenbeispiel zu einer solcher vermeinten Definition bildet und die als Vorbild oder Bestandsteil vieler anderer Gegenbeispiele zu anderen Versuchen der Definitionen dient.

Man könnte nämlich meinen dass ein Kontinuum eine *Fläche* ist, wenn es mit *jeder* es schneidenden Ebene, die einer gegebenen Ebene parallel ist, eine Linie gemein hat.

Nun, diese Eigenschaft besitzt folgendermassen konstruirte *Linie.*

Sei $\mathfrak{P}$ eine auf der Strecke PQ nirgends dichte, perfekte Punktmenge; wir verbinden jeden Punkt derselben mit einem ausserhalb der Geraden PQ liegenden Punkte A durch geraden Strecken. Die Gesammtheit dieser Strecken bildet ein Kontinuum $\mathfrak{C}$, das offenbar eine (Cantorsche) Linie ist.

In A richten wir eine Senkrechte AB zu der Ebene APQ. Jetzt ordnen wir den Punkten von AB die Punkte von $\mathfrak{P}$ (und also die Strecken des Kontinuums $\mathfrak{C}$) in der von Cantor angegebenen Weise zu: *jedem* Punkte der Strecke AB entspricht ein oder zwei Punkte der Menge $\mathfrak{P}$.

Wir betrachten nun $\mathfrak{C}$ als orthogonale Projektion der zu konstruirenden Linie.

In jedem Punkte der Strecke AB richten wir eine (ev. zwei) Senkrechte parallel und gleich der ihm entsprechenden Strecke des Kontinuums $\mathfrak{C}$; ihre Gesamtheit bildet das gesuchte Kontinuum $\mathfrak{K}$. Dass $\mathfrak{K}$ eine Linie ist können wir deshalb behaupten, weil es mit einer ebener Cantorschen Linie $\mathfrak{L}$ elementarverwandt (homeomorph) ist, die wir folgendermassen konstruiren: Wir verbinden durch gerade Strecken die sich entsprechenden Punkte einer Strecke AB und die Punkte einer perfekten Menge $\mathfrak{P}$, die auf einer zu AB parallelen Strecke PQ liegt und dort nirgends dicht ist.

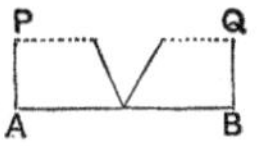

Es ist offenbar, dass dieses Kontinuum nirgends dicht ist; also ist es eine Linie. Ebenso ersichtlich ist seine Elementarverwandschaft mit dem vorher konstruirten Kontinuum.

$\mathfrak{K}$ ist also eine Linie und doch besitzt es die Eigenschaft mit jeder zu APQ parallelen und es schneidenden Ebene eine Linie (gerade Strecke) *gemein zu haben.*

Die Linien $\mathfrak{K}$ und $\mathfrak{L}$ sind noch dadurch merkwürdig, dass sie ein Kontinuum enthalten (die Strecke AB), von dem *jeder* Punkt ein Verzweigungspunkt ist. Es klingt in der Tat paradox dass die mehrfachen Punkte einer (Cantorschen, nicht Jordanschen) Linie ein Kontinuum bilden können*.

* Im Raume von unendlich vielen Dimensionen würde man mit Hülfe von $\mathfrak{L}$ eine Linie konstruiren können, von der *jeder* Punkt ein Verzweigungspunkt ist.

Ein ganz analoges Gegenbeispiel lässt sich für eine Definition der folgenden Art konstruiren: Die Umgebung eines Punktes A des gegebenen Kontinuum $\mathfrak{L}$ ist eine Fläche, wenn *jede* mit einem genügend kleinen Radius um ihn beschriebene Kugel mit $\mathfrak{L}$ eine Linie (z. B. einen Kreis) gemein hat. Dagegen wird diese Definition *wahrscheinlich* gut, wenn man dieselbe auch für eine unendliche Menge zu A benachbarter Punkte voraussetzt, welche von der Machtigkeit des Kontinuums ist. Eine ähnliche Bemerkung gilt von der zuerst versuchten Definition.

Ich gehe zu einem zweiten Beispiel über. Dieses beweist: *es gibt Kontinua, die keinen einfachen Bogen enthalten.* Es ist mir in der Tat gelungen ein Kontinuum zu konstruiren, derart, dass ein jedes Teilkontinuum ein Häufungskontinuum (continu de condensation) enthält*.

Man enthält sie als Grenze einer Folge von Linien, deren erste die Linie $y = \sin \frac{1}{x}$ ist, und die folgenden enthält man nach der Methode der Kondensation der Singularitäten indem man überall die einfachen Bogen durch die der Linie $y = \sin \frac{1}{x}$ elementarverwandte Linien ersetzt. Man muss nur dabei dafür sorgen, dass die Abänderungen die man macht, genügend schnell gegen Null konvergieren. Es scheint mir diese Linie insofern interessant zu sein, als alle bisherigen (mir bekannten) Beispiele der Topologie aus lauter geraden Strecken konstruirt waren oder werden konnten, und man meinen könnte, dies gelte allgemein.

Wir konstruiren jetzt auf dieser Linie als Directrix einen Zylinder. Die so erhaltene Fläche *enthält kein Abbild einer Kreisfläche.* Sie zeigt also, dass die Definitionen einer Fläche welche sich auf solche Abbildung stützen, nicht allgemein sind.

Im Vorhergehenden habe ich mich bei der Beurteilung der Definitionen auf gewisse Forderungen gestützt, die leicht zu formulieren sind; z. B., dass ein einer Cantorschen ebenen Linie elementarverwandtes Kontinuum selbst Linie heisse, und ebenso ein in eine endliche Anzahl von Linien zerlegbares Kontinuum; und dass ein Kontinuum, dass das Innere einer Kugel in Gebiete teilt, eine Fläche heisse. Sie sind notwendig und unmittelbar durch den intuitiven Sinn der Worte "Linie" und "Fläche" diktiert; sie scheinen aber nicht hinreichend zu sein. Die eventuellen weiteren Forderungen zu formulieren und ihre gegenseitige Unabhängigkeit zu beweisen scheint mir hier die nächste positive Aufgabe zu sein. Bevor dies getan wird kann man von keiner Definition behaupten, dass sie allgemeingültig ist.

Ich schliesse diese, nur negativen Entwickelungen mit der Bemerkung dass ich die so *ganz allgemeine* Frage nur zur Orientirung behandle. Ich meine dagegen, dass es bei dem jetzigem Stande der Wissenschaft, die Untersuchungen, die sich auf einfachere Gebilde beschränken, am ehesten hierauf ein Licht werfen können.

* Ich habe hier von Herrn Brouwer erfahren, dass ihm dasselbe Beispiel bekannt war.

ZUR ANALYSIS SITUS DER DOPPELMANNIGFALTIGKEITEN UND DER PROJEKTIVEN RÄUME

Von Dénes König.

In diesem Vortrage beabsichtigen wir die projektiven Räume einer beliebigen Dimensionszahl in bezug auf ihre Ein- und Zweiseitigkeit zu untersuchen. Und zwar so, dass die Resultate, so weit als möglich, mittels der kombinatorischen Methoden der Analysis situs abgeleitet werden. Dabei werden auch gewisse allgemeinere Resultate über sogen. Doppelmannigfaltigkeiten bewiesen.

§ 1. Die von Dehn und Heegaard begründete und von Steinitz weiter entwickelte kombinatorische Theorie* der Analysis situs stützt sich auf den Begriff der *Zellen* (Z^n) und *Sphären* (S^n) (der obere Index bezeichnet stets die Dimensionszahl). Als 0-dimensionale Zelle (Z^0) wird der Punkt, als 0-dimensionale Sphäre (S^0) das Punktepaar bezeichnet, als Z^1 eine von zwei Punkten (einer S^0) begrenzte "Strecke"; zwei Z^1 mit gemeinsamer Grenze bilden ein S^1, u. s. w. Im Allgemeinen: ein Z^n wird durch ein S^{n-1} begrenzt und zwei Z^n mit derselben S^{n-1}-Grenze bilden eine S^n.

Unter der *Indicatrix* einer $S^0=(Z_1^0, Z_2^0)$, die aus den Punkten Z_1^0 und Z_2^0 besteht, versteht man eine der zwei Reihenfolgen dieser zwei Punkte. Ebenso definirt man die zwei Indicatrizen einer Z^1, die durch diese S^0 begrenzt wird. Im Allgemeinen lässt sich die Indicatrix der n-dimensionalen Zellen und Sphären in rekursiver Weise folgendermassen definiren. Ist S^{n-1} die Grenze einer Z^n, so bestimmt die Indicatrix von S^{n-1} eine Indicatrix für Z^n, und umgekehrt. Die Indicatrix für $S^n=(Z_1^n, Z_2^n)$ wird angegeben, indem man für Z_1^n und Z_2^n je eine Indicatrix in der Weise bestimmt, dass die gemeinsame Grenze S^{n-1} von Z_1^n und Z_2^n hierdurch einmal die eine und einmal die andere Indicatrix erhält.—Die Indicatrix-Bestimmung der Zellen und Sphären läuft also im Grunde stets auf irgend eine Angabe über die Reihenfolge zweier Elemente hinaus.

Nun erweitert man die Bedeutung des Wortes Sphäre, indem man die aus S^n durch *interne Transformation* hervorgehende Mannigfaltigkeit ebenfalls als n-dimensionale Sphäre bezeichnet. (Durch S^n sollen stets nur die Sphären im engeren Sinne bezeichnet werden.) Die "interne Transformation" (für ihre präzise Erklärung verweisen wir auf die erwähnten Arbeiten†) besteht aus einer Reihe

* Dehn und Heegaard: "Analysis situs," *Encykl. d. Math.* B. III. p. 153 (1907); Steinitz: *Sitzungsber. d. Berliner Math. Ges.* B. 7, p. 29 (1908).—§ 1 unserer Note enthält im Wesentlichen nur Erklärungen dieser zwei Arbeiten in einer unseren Zwecken angepassten Darstellung.

† Dehn-Heegaard, p. 159; Steinitz, p. 32.

von Schritten, die—durch Einführung je eines neuen $Z^{\kappa-1}$—ein Z^κ durch zwei Z^κ ersetzen. Diese Schritte werden der Reihe nach für $\kappa = 0, 1, 2, \ldots, n$ ausgeführt. Nun kann also eine Sphäre durch eine beliebige Anzahl von Zellen gebildet werden. Auch diese allgemeinere Sphären n-ter Dimension können gewisse Z^{n+1} begrenzen.—Der Begriff der Indicatrix lässt sich auf diese allgemeinere Sphären und Zellen unmittelbar übertragen.

Eine endliche Menge gewisser Z^n bildet eine "*geschlossene Mannigfaltigkeit*" $\mathfrak{M}^n$, falls jede Z^{n-1} die zur Grenze einer dieser Z^n gehört, genau zweien dieser Z^n angehört. Für die allgemeine $\mathfrak{M}^n$ lässt sich der Begriff der internen Transformation ebenso erklären, wie für die S^n.

Wir wollen der Einfachheit halber den Begriff der geschlossenen Mannigfaltigkeit noch dadurch einschränken, dass wir annehmen, dass je zwei jener Z^n, welche die Mannigfaltigkeit bilden, höchstens eine gemeinsame Grenz-Z^{n-1} besitzen*.

Die Z^n, welche die Mannigfaltigkeit $\mathfrak{M}^n$ bilden, so wie die Z^{n-1}, welche die Grenze dieser Z^n bilden, u. s. w. bis zu den Z^0 (Punkten) herunter, bilden die *konstituirenden Elemente* von $\mathfrak{M}^n$. Zwei konstituirende Elemente Z^i und $Z^{i+\kappa}$ heissen *inzident*, wenn Z^i konstituirendes Element von $Z^{i+\kappa}$ ist.

Eine $\mathfrak{M}^n$ heisst bekanntlich zweiseitig, falls man ihren Z^n je eine Indicatrix in der Weise zuordnen kann, dass hierdurch jeder konstituirenden Z^{n-1} beidemal verschiedene Indicatrizen zugeordnet seien, d. h. wenn das verallgemeinerte Möbius'sche Gesetz† für die Z^n erfüllt werden kann. Ist dies möglich, so ist es —wenn wir noch $\mathfrak{M}^n$ als *zusammenhängend*‡ voraussetzen—auf zweierlei Weise möglich und durch diese zwei Möglichkeiten sind die zwei Indicatrizen einer $\mathfrak{M}^n$ definirt, die durch $\overrightarrow{\mathfrak{M}}^n$ und $\overleftarrow{\mathfrak{M}}^n$ bezeichnet werden können. Durch die Angabe einer Indicatrix für $\mathfrak{M}^n$ ist also jeder ihrer Z^n eine Indicatrix dem Möbius'schen Gesetze entsprechend zugeordnet. Wir werden von zwei mit Indicatrizen versehenen Z^n der $\mathfrak{M}^n$ sagen, dass sie "dieselbe" Indicatrix $\overrightarrow{Z}_1^n$ und $\overrightarrow{Z}_2^n$ erhalten haben (in Zeichen: $\overrightarrow{Z}_1^n = \overrightarrow{Z}_2^n$), wenn sie zur selben Indicatrix von $\mathfrak{M}^n$ gehören. (Dies hat natürlich nur für zweiseitige $\mathfrak{M}^n$ einen Sinn.)

§ 2. Seien

$$Z_1^0, Z_2^0, \ldots, Z_{i_0}^0, Z_1^1, Z_2^1, \ldots, Z_{i_1}^1, \ldots, Z_1^n, Z_2^n, Z_{i_n}^n$$

die konstituirenden Elemente der zusammenhängenden geschlossenen und zweiseitigen Mannigfaltigkeit $\mathfrak{M}^n$ und es sei eine Beziehung zwischen diesen Elementen in der Weise gegeben, dass jeder Z_ρ^κ in umkehrbar eindeutiger Weise ein Z_σ^κ von derselben Dimension entspricht, und zwar so, dass (1) stets $Z_\rho^\kappa \neq Z_\sigma^\kappa$ ist und (2) inzidenten Elementen wieder solche entsprechen. Eine solche Beziehung soll als eine R-

* Diese Eigenschaft bildet keine wesentliche Einschränkung, da sie sich durch interne Transformation stets erreichen lässt.

† Von Möbius für $n=2$ formulirt: *Werke*, II. p. 475.

‡ Wir wollen $\mathfrak{M}^n$ zusammenhängend nennen, wenn von jeder ihrer Z^n aus jede andere durch n-dimensionale Nachbarzellen sich erreichen lässt. Zwei Z^n heissen dabei Nachbarzellen, wenn ihre Grenzen eine gemeinsame Z^{n-1} besitzen.

Beziehung für $\mathfrak{M}^n$ bezeichnet werden. Z^κ_ρ und Z^κ_σ heissen "entsprechende" Elemente.

Ist für ein Z^κ_ρ eine Indicatrix $\overrightarrow{Z}^\kappa_\rho$ (durch das oben angegebene rekursive Verfahren) gegeben und ersetzt man in dieser Indicatrix-Bestimmung jede Zelle (von der Dimension 0, 1, 2...bis zu n) durch die entsprechende, so wird hierdurch für Z^n_σ eine bestimmte Indicatrix—sie sei $\overrightarrow{Z}^n_\sigma$—angegeben. Sie soll die der Indicatrix $\overrightarrow{Z}^n_\rho$ von Z^n_ρ entsprechende Indicatrix von Z^n_σ genannt werden.

(A) *Ist nun für die zweiseitige $\mathfrak{M}^n$ eine R-Beziehung fest gegeben, so ist für jedes ρ entweder stets $\overrightarrow{Z}^n_\rho = \overrightarrow{Z}^n_\sigma$ oder stets $\overrightarrow{Z}^n_\rho \neq \overrightarrow{Z}^n_\sigma$, also $\overrightarrow{Z}^n_\rho = \overleftarrow{Z}^n_\sigma$.*

Sind nämlich zwei Zellen Z^n_1 und Z^n_2 von $\mathfrak{M}^n$ so mit je einer Indicatrix versehen, dass ihre einzige gemeinsame Z^{n-1} beidemal verschiedene Indicatrizen hat, so gilt dasselbe für die entsprechenden Indicatrizen der entsprechenden zwei n-dimensionalen Zellen, d. h.: dann und nur dann haben Z^n_1 und Z^n_2 dieselbe Indicatrix, wenn diess für die entsprechenden Indicatrizen der entsprechenden Zellen der Fall ist. Da nun $\mathfrak{M}^n$ zusammenhängend ist, überträgt sich dieses Resultat unmittelbar auch auf den Fall, wo Z^n_1 und Z^n_2, beliebige, nicht speziell Nachbarzellen von $\mathfrak{M}^n$ sind.

Durch das jetzt bewiesene Resultat zerfallen die mit einer bestimmten R-Beziehung versehenen Mannigfaltigkeiten in *zwei Klassen I und II*, je nach dem stets $\overrightarrow{Z}^n_\rho = \overrightarrow{Z}^n_\sigma$ oder stets $\overrightarrow{Z}^n_\rho = \overleftarrow{Z}^n_\sigma$ ist.—Eine bestimmte Indicatrix von $\mathfrak{M}^n$ bestimmt für die Z^n von $\mathfrak{M}^n$ je eine Indicatrix. Ersetzt man diese Indicatrizen durch diejenigen, welche der (ursprünglichen) Indicatrix der entsprechenden Zellen entsprechen, so genügen, wie wir sahen, auch diese neuen Indicatrizen dem Möbius'schen Gesetz und bestimmen also eine Indicatrix für $\mathfrak{M}^n$. Diese soll die der ursprünglichen Indicatrix entsprechende Indicatrix von $\mathfrak{M}^n$ genannt werden. $\mathfrak{M}^n$ gehört also zur Klasse I oder II, je nach dem die entsprechende Indicatrix von $\mathfrak{M}^n$ mit der ursprünglichen übereinstimmt oder nicht.

Es soll nun speziell $\mathfrak{M}^n$ die Sphäre

$$S^n = (Z^n_1,\ Z^n_2)$$

bedeuten und die gemeinsame Grenze von Z^n_1 und Z^n_2 sei S^{n-1}. Da unter den konstituirenden Elementen

$$Z^0_1,\ Z^0_2,\ Z^1_1,\ Z^1_2,\ \ldots,\ Z^{n-1}_1,\ Z^{n-1}_2,\ Z^n_1,\ Z^n_2$$

der S^n nur zwei Zellen der selben Dimension vorkommen, so ist für S^n nur eine R-Beziehung möglich, durch welche die "diametralen" Zellen Z^κ_1 und Z^κ_2 als einander entsprechend gelten. In welche der zwei Klassen gehört nun S^n? Eine Indicatrix von S^n wird angegeben, wenn man zu S^{n-1} als Elemente von Z^n_1 und dann als dem Elemente von Z^n_2 die zwei verschiedenen Indicatrizen in der einen oder anderen Weise zuordnet. Aus einer Indicatrix von S^n ergibt sich die entsprechende—in Folge der rekursiven Indicatrix-Bestimmung—dadurch, dass man (1) die Indicatrix von S^{n-1} durch die entsprechende ersetzt und (2) Z_1^n und Z_2^n vertauscht. Ist also für S^{n-1} die entsprechende Indicatrix mit der ursprünglichen identisch, so ist (da dann nur (2) auszuführen ist) für S^n die entsprechende von der ursprünglichen verschieden, und umgekehrt. Also gehört S^n zur Klasse I oder II, je nach dem S^{n-1} zur Klasse II oder I gehört. Man sieht aber unmittelbar, dass $S^0=(Z_1^0, Z_2^0)$ zur Klasse II gehört, also gilt folgender Satz:

(B) *S^n gehört in die Klasse I oder II je nach dem n ungerade oder gerade ist.*

Mann kann leicht einsehn, dass die Zugehörigkeit zur Klasse I oder II (ebenso, wie die Ein- oder Zweiseitigkeit) eine in bezug auf interne Transformationen *invariante Eigenschaft* der Mannigfaltigkeiten bedeutet*. Diess bedeutet genauer Folgendes. Aus der mit einer R-Beziehung versehenen $\mathfrak{M}^n$ soll durch eine interne Transformation $\mathfrak{M}_1^n$ entstehen und es soll für $\mathfrak{M}_1^n$ eine R-Beziehung, R_1, von der Beschaffenheit existieren, dass zwei Zellen Z_1^n und Z_2^n von $\mathfrak{M}_1^n$ gemäss R_1 nur dann einander entsprechen, wenn dies gemäss R für diejenige zwei Zellen von $\mathfrak{M}^n$ gilt, aus denen Z_1^n resp. Z_2^n entstanden sind. In diesem Falle gehört $\mathfrak{M}_1^n$ in bezug auf R_1 in dieselbe Klasse I oder II, wie $\mathfrak{M}^n$ in bezug auf R.

§ 3. Sei wieder für die allgemeine zusammenhängende und zweiseitige Mannigfaltigkeit $\mathfrak{M}^n$ mit den konstituirenden Elementen

$$Z_1^0,\ Z_2^0,\ \ldots,\ Z_{i_0}^0,\ Z_1^1,\ Z_2^1,\ \ldots,\ Z_{i_1}^1,\ \ldots,\ Z_1^n,\ Z_2^n,\ \ldots,\ Z_{i_n}^n \quad \ldots\ldots\text{(I)}$$

eine R-Beziehung gegeben, welche die Zellen Z_ρ^κ und Z_σ^κ ($\kappa=0,\ 1,\ \ldots,\ n$; $\rho=1,\ 2,\ \ldots,\ i_\kappa$) als einander entsprechende Zellen charakterisirt. Die Zellen in (I) lassen sich also in Paare verschiedener Zellen derselben Dimension zusammenfassen. Sind $(Z_\rho^\kappa, Z_\sigma^\kappa)$ und (Z_ρ^l, Z_s^l) zwei solche Paare und sind Z_ρ^κ und Z_r^l zu einander inzident, so gilt das selbe für Z_σ^κ und Z_s^l, da ja eine R-Beziehung Inzidenzen niemals zerstört. In diesem Falle (d. h. wenn Z_ρ^κ mit Z_r^l oder mit Z_s^l inzident ist) werden wir diese zwei Paare selbst inzident nennen. Als Dimensionszahl des Paares $(Z_\rho^\kappa, Z_\sigma^\kappa)$ soll die Zahl κ bezeichnet werden. Es kann nun eventuell eine Mannigfaltigkeit $\overline{\mathfrak{M}}^n$ so angegeben werden, dass zwischen ihren konstituirenden Elementen und den oben genannten Paaren eine umkehrbar eindeutige Beziehung in der Weise besteht, dass jedem dieser Elemente ein Paar von derselben Dimension entspricht und inzidenten Elementen inzidente Paare entsprechen. In diesem Falle heisst $\mathfrak{M}^n$ eine *Doppelmannigfaltigkeit* von $\overline{\mathfrak{M}}^n$. Ist die R-Beziehung für $\mathfrak{M}^n$ gegeben, so ist $\overline{\mathfrak{M}}^n$ durch $\mathfrak{M}^n$ vollständig bestimmt, da es bei der Definition der Mannigfaltigkeiten nur auf die Inzidenzen ankommt.

Gehört nun $\mathfrak{M}^n$ zur Klasse I, und denkt man ihre Z^n dem Möbius'schen Gesetze entsprechend mit Indicatrizen versehn, so lassen sich diese Indicatrizen unmittelbar auf die Z^n von $\overline{\mathfrak{M}}^n$ übertragen, so dass also in diesem Falle auch $\overline{\mathfrak{M}}^n$ zweiseitig ist.

* Diess ist von Wichtigkeit, wenn man von der kombinatorischen Theorie zur analytisch-geometrischen Theorie übergeht, wo die Mannigfaltigkeiten als unendliche stetige Punktmengen betrachtet werden; da zwei $\mathfrak{M}^n$, von denen die eine aus der anderen durch interne Transformation hervorgeht, im Sinne der Analysis situs aequivalente Punktmengen bedeuten.

Ebenso sieht man, dass auch umgekehrt falls die Z^n von $\overline{\mathfrak{M}^n}$ dem Möbius'schen Gesetze entsprechend mit Indicatrizen versehen werden können, $\mathfrak{M}^n$ unbedingt zur Klasse I gehören muss. Man hat also den Satz:

(C) *$\overline{\mathfrak{M}^n}$ ist zweiseitig oder einseitig, je nach dem $\mathfrak{M}^n$ zur Klasse I oder II gehört.*

§ 4. Will man nun die hier gewonnenen Resultate auf den n-dimensionalen projektiven Raum, R^n, anwenden, so kann man natürlich analytische Hilfsmittel nicht vermeiden, da R^n, als die Gesamtheit der Verhältnisse

$$(x_1 : x_2 : \ldots : x_n : x_{n+1}),$$

selbst analytisch definirt wird.

Den bisher kombinatorisch definirten Begriffen entsprechen in bekannter Weise, was hier wohl nicht näher beschrieben werden muss, gewisse analytisch definirbare Gebiete des N-dimensionalen Raumes, d. h. der Gesamtheit der "Punkte" $(x_1, x_2, \ldots, x_N)$, wenn nur N genügend gross gewählt wird. Die in der kombinatorischen Theorie definirte Aussage "durch die R-Beziehung von $\mathfrak{M}^n$ erscheint $\mathfrak{M}^n$ als Doppelmannigfaltigkeit von $\overline{\mathfrak{M}^n}$" bedeutet in der analytisch-geometrischen Auffassung der Mannigfaltigkeiten Folgendes: "Man kann die Punkte von $\overline{\mathfrak{M}^n}$ und $\mathfrak{M}^n$ in vollkommen ein-zweideutiger und stetiger Weise auf einander so beziehen, dass zwei Gebiete (Zellen) von $\mathfrak{M}^n$, die durch die R-Beziehung einander entsprechen, demselben Gebiete von $\overline{\mathfrak{M}^n}$ entsprechen sollen." Diess lässt sich auch umkehren. Besteht nämlich eine ein-zweideutige und stetige punktweise Beziehung zwischen den Punktmengen $\overline{\mathfrak{M}^n}$ und $\mathfrak{M}^n$, so kann man diese zwei Mannigfaltigkeiten aus Zellen in der Weise aufbauen, dass $\mathfrak{M}^n$—im kombinatorischen Sinne—als Doppelmannigfaltigkeit von $\overline{\mathfrak{M}^n}$ erscheint. Dabei werden die entsprechenden Zellen der "Mannigfaltigkeit" $\mathfrak{M}^n$ aus entsprechenden Punkten der "Punktmenge" $\mathfrak{M}^n$ gebildet.

Die n-dimensionale Sphäre kann man als Punktmenge am einfachsten durch die Gesamtheit derjenigen Punkte $(x_1, x_2, \ldots, x_n, x_{n+1})$ des $n+1$-dimensionalen Raumes repräsentiren, für die

$$x_1^2 + x_2^2 + \ldots + x_{n+1}^2 = 1$$

ist. Lässt man den zwei diametralen Punkten

$$(\alpha_1, \alpha_2, \ldots, \alpha_{n+1}) \text{ und } (-\alpha_1, -\alpha_2, \ldots, -\alpha_{n+1})$$

dieser Sphäre den einzigen Punkt

$$(\alpha_1 : \alpha_2 : \ldots : \alpha_{n+1})$$

des n-dimensionalen projektiven Raumes, R^n, entsprechen, so ist hierdurch eine ein-zweideutige und stetige punktweise Beziehung für R^n und S^n gegeben.

Werden also die diametralen Elemente der Sphäre als einander entsprechend betrachtet, so ist S^n Doppelmannigfaltigkeit für R^n.

Unsere Sätze (B) und (C) ergeben also auch das Resultat*:

Der n-dimensionale projektive Raum ist einseitig für gerades n und zweiseitig für ungerades n.

* Diesen Satz habe ich schon im *Archiv der Math.* B. 19, p. 214 ausgesprochen. Dort wurde er ausführlich nur für $n=2$ und 3 bewiesen.

ON THE THEORY OF CONNEXES

By D. Sinzov.

I propose to give an account of the results I have obtained in the theory of connexes. Published only in Russian, they are not easily accessible to the mathematicians of other countries, and I hope it is not superfluous to present such an account.

A. Clebsch, in his remarkable paper, "Ueber eine Fundamentalaufgabe der Invariantentheorie," *Göttingen Abh.* XVII. 1872, had pointed out a new and vast domain of geometrical study—the investigation of configurations with composed elements (point, line) in the plane, (point, line), (point, plane), (line, plane), (point, line, plane) in the space. He himself published his investigations only on the first of these configurations—on the ternary point-line connex: "Ueber ein neues Grundgebilde der analytischen Geometrie der Ebene" (*Götting. Nachr.* 1872; *Math. Ann.* B. V. 1872). Important contributions to this theory were made by Mr Kyp. Stephanos ("On the conjugate connex," *Bull. Darboux,* (2) IV. 1880), MM. G. Darboux, L. Autonne and H. Poincaré on the algebraic integration of the differential equation of the first order and degree, connected with the connex $(m, 1)$, and by Mr G. Darboux on the theory of Singular Solutions. Further particular cases investigated are those of (1, 1)—bilinear connex, connected with collineation in the plane (A. Clebsch and P. Gordan and others*), of (2, 1) (Godt, Amodeo), and generally $(m, 1)$ (above mentioned, Müllendorf, Voss), (2, 2) (Armenante, Peano). There should be mentioned also the papers of G. Fouret on the systems of curves defined by an algebraical differential equation of the first order (*Bull. S. Math. France,* II. v, vi, xix).

At first inspired by the exposition of F. Lindemann (Clebsch-Lindemann, *Vorträge üb. analytische Geometrie,* B. I.) I tried eighteen years ago to extend the results above mentioned to the point-plane space connex ("The theory of connexes in space in connection with the theory of differential equations of the first order," *Kasan Univ.* 1894—1895, and separately 8vo. 254 pp.), a special case of which—namely (2, 1)—was studied by R. Krause (*M. An.* 14)†. I have given an account in *Bull. Darboux* (2) XXII. I cannot at the present time speak about my paper, the matter having been treated afterwards in a masterly manner by L. Autonne in his prize memoir (*Mém. cour. de Belgique,* 1902) and by M. Stuyvaert, "Recherches relatives

* Battaglini, Lazzeri, Voss; on collineation: Pasch, Muth, Rosanes, Kraus, Keller, St. Smith.

† A number of papers on the collineations in space and the bilinear space connex (Battaglini, Predella, Pittarelli, Loria, Segre, Richelot, St. Smith, Hirst, Lazzeri, Muth), and the papers of G. Fouret on the implexes are also to be mentioned.

aux connexes de l'espace," *Mém. Belg.* Coll. 8vo. t. 61, 1902. I would note only, that on the last pages I give a generalization to n dimensions, i.e. I mark shortly the character of the configurations with element: point, hyperplane in a space of n dimensions, and of the integration problem connected with it. The incompleteness of the theory of the plane (ternary) connex makes it necessary to return to this more simple case. In my further publications I had concentrated my efforts on the theory of conjugate connex and on the influence of certain singularities of the given connex on the class and order of its conjugate.

Instead of Clebsch's definition of a singular element, as an element to which corresponds more than one element in the conjugate connex, it seems to me to be more convenient to regard the singular elements as those whose point is a singular point of the corresponding connex-curve, or line is the singular line, or both together—or [which is quite the same] as those for which the equations of the *polar pair of the element* (X, U)

$$f(x, u) = 0, \quad \Sigma X f_x' = 0, \quad \Sigma U f_u' = 0 \quad \ldots\ldots\ldots\ldots\ldots\ldots (1)$$

are satisfied for every X, or for every U, which belong consequently to all the polar pairs of elements (X, U), whatever may be the point X, if the element is the *point-singular* one, i.e. if

$$f_{x_i}' = 0, \quad (i = 1, 2, 3) \ldots\ldots\ldots\ldots\ldots\ldots\ldots\ldots (a),$$

or whatever may be the line U, if the element (x, u) is the *line-singular* one, i.e. if

$$f_{u_i}' = 0, \quad (i = 1, 2, 3) \ldots\ldots\ldots\ldots\ldots\ldots\ldots\ldots (b),$$

or finally for every element (X, U), if both systems (a) and (b) are satisfied simultaneously—in which case the element (x, u) may be called a *properly-singular* element of the connex. By the convenient modification of the way of finding the order and class of the conjugate connex it is possible to appreciate the influence of certain singularities:

1. *A principal point (Hauptpunkt) diminishes the class of the conjugate connex by n, the order remaining unaltered. Reciprocally a principal line diminishes the order of the conjugate connex by m, the class remaining unaltered.*

2. *A pair of properly-singular elements* (μ, ν) (i.e. ∞^1 elements forming a pair of curves) *diminishes the order of the conjugate connex by μ, the class by ν.*

3. ∞^2 properly-singular elements, forming a coincidence, defined as the intersection of two connexes (μ, ν), (μ', ν') evoke diminutions:

$$\Delta m' = (3m-2)\nu\nu' + (3n-1)(\mu\nu' + \nu\mu') - [(\mu+\mu')\nu\nu' + (\nu+\nu')(\mu\nu' + \nu\mu')],$$
$$\Delta n' = (3m-1)(\mu\nu' + \nu\mu') + (3n-2)\mu\mu' - [(\mu+\mu')(\mu\nu' + \nu\mu') + (\nu+\nu')\mu\mu'].$$

The demonstration of the last theorem is based on the determination of the number of elements of intersection of the four connexes having a coincidence in common (Goettler).

It is perhaps noteworthy that in calculating examples to these theorems I have found an example of a connex

$$\kappa_1 x_2^2 x_3^2 u_1^3 + \kappa_2 x_3^2 x_1^2 u_2^3 + \kappa_3 x_1^2 x_2^2 u_3^3 = 0,$$

whose conjugate configuration is not a connex, but the coincidence

$$\frac{y_1^3}{\kappa_1 v_1^2} = \frac{y_2^3}{\kappa_2 v_2^2} = \frac{y_3^3}{\kappa_3 v_3^2}.$$

In following paper (1910) I have extended the results above indicated to the space point-plane connex: a principal point diminishes the class of the conjugate connex by n^2, a principal plane diminishes the order by m^2, a pair (curve in space, developable surface) (μ, ν) of properly-singular elements depresses the order of the conjugate connex by μ, the class by ν. Further formulae of reduction for the cases of ∞^2, ∞^3 and ∞^4 properly-singular elements defined as intersection of 4, 3 and 2 connexes, are too complicated to be given here.

I pass to another contribution to the theory of connexes,—the introduction of the osculating bilinear connex

$$\Sigma_{i,k} f_{ik}'' X_i U_k = 0 \quad (i, k = 1, 2, 3), \quad \left(f_{ik}'' = \frac{\partial^2 f}{\partial x_i \partial u_k}\right).$$

If we define the singular elements of a coincidence

$$f(x, u) = 0, \qquad \phi(x, u) = 0,$$

as those for which any one or both systems of equations are satisfied:

$$\frac{\frac{\partial f}{\partial x_1}}{\frac{\partial \phi}{\partial x_1}} = \frac{\frac{\partial f}{\partial x_2}}{\frac{\partial \phi}{\partial x_2}} = \frac{\frac{\partial f}{\partial x_3}}{\frac{\partial \phi}{\partial x_3}} \quad \ldots\ldots\ldots\ldots\ldots\ldots(a),$$

$$\frac{\frac{\partial f}{\partial u_1}}{\frac{\partial \phi}{\partial u_1}} = \frac{\frac{\partial f}{\partial u_2}}{\frac{\partial \phi}{\partial u_2}} = \frac{\frac{\partial f}{\partial u_3}}{\frac{\partial \phi}{\partial u_3}} \quad \ldots\ldots\ldots\ldots\ldots\ldots(b).$$

We can easily prove that *the osculation-element* (x, u) *is the properly-singular element of the coincidence-intersection of the connex with the osculating bilinear connex,* both systems (a) and (b) being satisfied.

The analogy with the theorem about the intersection of a surface with its tangential-plane gives an idea how to classify the elements of connex in accordance with the behaviour of the osculating bilinear connex, i.e. according to the properties of the collineation defined by it. The equation of the third degree

$$\begin{vmatrix} f_{11} - K & f_{21} & f_{31} \\ f_{12} & f_{22} - K & f_{32} \\ f_{13} & f_{23} & f_{33} - K \end{vmatrix} = 0,$$

or

$$K^3 - iK^2 + i'K - \Delta = 0$$

defines the values of K, corresponding to the invariant points and lines of the collineation

$$f_{1k} X_1 + f_{2k} X_2 + f_{3k} X_3 = K . X_k,$$

$$f_{j1} U_1 + f_{j2} U_2 + f_{j3} U_3 = K . U_j.$$

A. $\Delta \neq 0$.

I. 1. $K_1 \neq K_2 \neq K_3$ and all real. Hyperbolic element.

2. $K_1 \neq K_2 \neq K_3$ but two are complex. Elliptic element.

3. $K_1 = K_2 \neq K_3$. Parabolic element.

4. $K_1 = K_2 = K_3$. Only one invariant point, and one invariant line.

II. One of the invariant points coincides with the point of osculation-element $(X = x)$ [*point-singular element of the principal coincidence*]. The same cases are to be distinguished.

III. One of the invariant lines coincides with the line of the osculation-element $(U = u)$ [*line-singular element of the principal coincidence*].

IV. Both the point and line of osculation-element are invariant point and line [*properly-singular element of the principal coincidence*].

B. $\Delta = 0$. Collineation is a degenerate one.

I. Elements analogous to the inflexion-point of a curve.

II. $X = x$ for $K = 0$. Point-singular element of the connex.

III. $U = u$ for $K = 0$. Line-singular element of the connex.

IV. $X = x$, $U = u$ for $K = 0$. Properly-singular element of the connex.

Very interesting is the case of $n = 1$ i.e. of connex $(m, 1)$,

$$A_1 u_1 + A_2 u_2 + A_3 u_3 = 0.$$

Then the line-singular elements of the principal coincidence are determined by the equations

$$\frac{A_1(x)}{x_1} = \frac{A_2(x)}{x_2} = \frac{A_3(x)}{x_3},$$

i.e. their points are principal points of the connex. They are named in this case critical points of the equation, and the classification above given coincides with that given by H. Poincaré.

The same method is applicable to the classification of the elements of the space point-plane connex and of point-hyperplane connex in the space of n dimensions.

I adjourn to another occasion the account of my papers on the point-line-plane connexes.

ON PAIRS OF FRENETIAN TRIHEDRA

By Nicholas Hatzidakis.

A. *Historical notes.*

The first attempts at a generalization of the formulae of Frenet are those of Ph. Gilbert (cfr *Enz. d. m. W.* III. D. 3, p. 168), of Serret (*Calcul Diff. et Int.*, but only on normals of surfaces) and of Aoust in his various essays on *inclined curvature* (*courbure inclinée*) (*C.R.* v. 57th, pp. 217–9 (1863),—*Annali di Matematica*, Ser. I, v. 6th, pp. 65–87 (1864),—Ser. II, v. 2nd, pp. 39–64 (1868–9),—Ser. II, v. 3rd, pp. 55–69 (1869–70),—*Analyse infinitésimale des courbes tracées sur une surface quelconque*, Paris, 1869)*.

Prof. K. Żorawski, in recent times, has considered, in the Polish journal *Prace Matematyczno-fizyczne* (v. XVII, pp. 41–6: *Über Krümmungseigenschaften der Scharen von Linienelementen*), such a general trihedron of straight lines with applications to surfaces.

Dr Bruno Arndt's inaugural dissertation: *Über die Verallgemeinerung des Krümmungsbegriffes für Raumkurven* deals, also, with similar considerations.

Lastly, Prof. W. Fr. Meyer has extended the generalized formulae of Frenet to the n-dimensional space (*Jahresbericht der D. M.-V.*, v. XIX, 1910: *Ausdehnung der Frenet'schen Formeln und verwandter auf den R_n*) and, more recently, he has published a separate book: *Über die Theorie benachbarter Geraden und einen verallgemeinerten Krümmungsbegriff* (1911), which, however, mainly deals with the "shortest distances" of two infinitely near trihedra.

B. *The generalized formulae of Frenet.*

1. Let a *direction* TT' in space of three dimensions be given through its three cosines $\mathrm{A}, \mathrm{B}, \Gamma$, functions of an independent variable s. The direction HH' with the cosines $\Xi = \frac{\mathrm{A}'}{\sqrt{S\mathrm{A}'^2}}$, $\mathrm{H} = \frac{\mathrm{B}'}{\sqrt{S\mathrm{A}'^2}}$, $\mathrm{Z} = \frac{\Gamma'}{\sqrt{S\mathrm{A}'^2}}$†, evidently perpendicular to TT', and the direction BB', common perpendicular to TT' and HH', with the cosines

$$\Lambda = \epsilon(\mathrm{BZ} - \Gamma\mathrm{H}), \quad \mathrm{M} = \epsilon(\Gamma\Xi - \mathrm{AZ}), \quad \mathrm{N} = \epsilon(\mathrm{AH} - \mathrm{B}\Xi), \qquad (\epsilon^2 = 1),$$

* According to Prof. W. Fr. Meyer the generalization of the first three formulae of Frenet was known already by Jacobi.

† A' means $\frac{d\mathrm{A}}{ds}$ etc.

constitute, together with TT', *a generalized trihedron of Frenet.* TT' corresponds to the tangent in the case of a customary principal trihedron of a curve and may be called the *primitive* or *departure* direction; similarly, we give the names of *principal normal* direction to HH' and *binormal* direction to BB'. It is very easy to show how the *ordinary* formulae of Frenet can be demonstrated for this trihedron. Putting $\sqrt{S\mathrm{A}'^2} \equiv \frac{1}{P} (\gtrless 0)$, we get

$$\mathrm{A}' = \frac{\Xi}{P}, \quad \mathrm{B}' = \frac{\mathrm{H}}{P}, \quad \Gamma' = \frac{\mathrm{Z}}{P} \quad \ldots\ldots\ldots\ldots\ldots\ldots\ldots\ldots(1).$$

If we further put

$$d\Lambda = -\Xi_1 \sqrt{Sd\Lambda^2}, \quad d\mathrm{M} = -\mathrm{H}_1 \sqrt{Sd\Lambda^2}, \quad d\mathrm{N} = -\mathrm{Z}_1 \sqrt{Sd\Lambda^2},$$

the direction $(\Xi_1, \mathrm{H}_1, \mathrm{Z}_1)$ is the same as HH' $(\Xi, \mathrm{H}, \mathrm{Z})$; indeed, it is

$$S\Lambda^2 = 1, \quad S\Lambda d\Lambda = 0, \quad S\Lambda\Xi_1 = 0, \quad (\sqrt{Sd\Lambda^2} \gtrless 0),$$

and $\quad S\mathrm{A}\Lambda = 0, \quad S\mathrm{A}d\Lambda + S\Lambda d\mathrm{A} = 0, \quad \text{or} \quad -\sqrt{Sd\Lambda^2}\,.\,S\mathrm{A}\Xi_1 + \sqrt{Sd\mathrm{A}^2}\,.\,S\Lambda\Xi = 0\,;$

but $S\Lambda\Xi = 0$, consequently $S\mathrm{A}\Xi_1 = 0$; then, from the equations $S\Lambda\Xi_1 = 0$, $S\mathrm{A}\Xi_1 = 0$ we get $\Xi_1 = \Xi$, $\mathrm{H}_1 = \mathrm{H}$, $\mathrm{Z}_1 = \mathrm{Z}$, and therefore

$$\Lambda' = -\frac{\Xi}{R}, \quad \mathrm{M}' = -\frac{\mathrm{H}}{R}, \quad \mathrm{N}' = -\frac{\mathrm{Z}}{R}\left(\text{where } R \equiv \frac{1}{\sqrt{S\Lambda'^2}}\right) \quad \ldots\ldots(2).$$

We find, at last, solving the equations

$$S\Xi\Xi' = 0, \quad S\mathrm{A}\Xi' = -S\Xi\mathrm{A}' = -\frac{1}{P}, \quad S\Lambda\Xi' = -S\Xi\Lambda' = \frac{1}{R},$$

that

$$\Xi' = -\frac{\mathrm{A}}{P} + \frac{\Lambda}{R}, \quad \mathrm{H}' = -\frac{\mathrm{B}}{P} + \frac{\mathrm{M}}{R}, \quad \mathrm{Z}' = -\frac{\Gamma}{P} + \frac{\mathrm{N}}{R} \quad \ldots\ldots\ldots\ldots(3).$$

(1), (2) and (3) constitute the extended formulae of Frenet*. $\sqrt{Sd\mathrm{A}^2} \equiv d\Sigma$ is the elementary arc of the *first spherical indicatrix* of TT', or *angle of contingence*, $\frac{d\Sigma}{ds} = \frac{1}{P}$ its *curvature* and P its *first radius*. $\sqrt{Sd\Lambda^2} \equiv d\mathrm{T}$ is the *angle of torsion*, $\frac{d\mathrm{T}}{ds} = \frac{1}{R}$ the *torsion* and R the *second radius*. $\sqrt{Sd\Xi^2} = \sqrt{d\Sigma^2 + d\mathrm{T}^2} \equiv d\Pi$ is the *angle of total curvature*, $\frac{d\Pi}{ds} = \sqrt{\frac{1}{P^2} + \frac{1}{R^2}} \equiv \frac{1}{K}$ the *total curvature* and $K = \frac{PR}{\sqrt{P^2 + R^2}}$ the *radius of total curvature*. All these quantities depend, of course, on the choice of the variable s.

If, instead of three *directions*, we consider three determinate *straight lines* through the same point, this point describes a curve. $\frac{1}{P}, \frac{1}{R}, \frac{1}{K}$ are then, if we choose the arc of this curve as s, the curvatures *of the curve with respect to TT'*, as Prof. W. F. Meyer defines, or *the curvatures of TT' with respect to the curve.* The last designation is more general and more exact, the curvature belonging to TT', not

* Cfr Serret, *Calcul Diff. et Intégral.*

to the accidental presence of the curve. On the other hand, the first designation is more adapted to the customary conception of curvature, because, if TT' coincides with the tangent of the curve, HH', BB' become the two other principal directions, and $\frac{1}{P}$, $\frac{1}{R}$, $\frac{1}{K}$ the three customary curvatures.

C. *Relations between the curvatures of two Frenetian trihedra.*

2. Let us consider two trihedra of Frenet: (D) $[s, TT', HH', BB']$ and (D_1) $[s_1, T_1T_1', H_1H_1', B_1B_1']$; s_1 being given as function of s $(s_1 = \phi(s))$, we can express the cosines and the curvatures of (D_1) by means of the corresponding elements of (D) and the relative cosines of T_1T_1' with respect to (D).

Let, for this purpose, a, b, c be these relative cosines; we have then

$$\mathrm{A}_1 = \mathrm{A}a + \Xi b + \Lambda c, \quad \mathrm{B}_1 = \mathrm{B}a + \mathrm{H}b + \mathrm{M}c, \quad \Gamma_1 = \Gamma a + \mathrm{Z}b + \mathrm{N}c \quad \ldots(4),$$

and consequently, if we take (1), (2), (3) into consideration, we get

$$\left.\begin{aligned} \frac{d\mathrm{A}_1}{ds_1}\phi'(s) &= \mathrm{A}\left(a' - \frac{b}{P}\right) + \Xi\left(b' + \frac{a}{P} - \frac{c}{R}\right) + \Lambda\left(c' + \frac{b}{R}\right) \\ \frac{d\mathrm{B}_1}{ds_1}\phi'(s) &= \mathrm{B}\left(a' - \frac{b}{P}\right) + \mathrm{H}\left(b' + \frac{a}{P} - \frac{c}{R}\right) + \mathrm{M}\left(c' + \frac{b}{R}\right) \\ \frac{d\Gamma_1}{ds_1}\phi'(s) &= \Gamma\left(a' - \frac{b}{P}\right) + \mathrm{Z}\left(b' + \frac{a}{P} - \frac{c}{R}\right) + \mathrm{N}\left(c' + \frac{b}{R}\right) \end{aligned}\right\} \left(\text{where } a' = \frac{da}{ds} \text{ etc.}\right) \quad \ldots\ldots\ldots(5);$$

squaring and adding these relations, we get

$$\frac{\phi'(s)^2}{P_1^2} = \left(\frac{a}{P} - \frac{c}{R}\right)^2 + \frac{b^2}{K^2} + Sa'^2 + 2b\left(\frac{c'}{R} - \frac{a'}{P}\right) + 2b'\left(\frac{a}{P} - \frac{c}{R}\right) =$$

$$= \frac{a^2 + b^2}{P^2} + \frac{b^2 + c^2}{R^2} - \frac{2ac}{PR} + Sa'^2 + \frac{2}{P}\begin{vmatrix} a & b \\ a' & b' \end{vmatrix} + \frac{2}{R}\begin{vmatrix} b & c \\ b' & c' \end{vmatrix} \quad \ldots(6)^*;$$

multiplying (5) by A, B, Γ, or Ξ, H, Z, or Λ, M, N and adding, we get, since $\frac{d\mathrm{A}_1}{ds_1} = \frac{\Xi_1}{P_1}$ etc.,

$$\phi'(s) \cdot \frac{\mathfrak{x}}{P_1} = a' - \frac{b}{P}, \quad \phi'(s) \cdot \frac{\mathfrak{y}}{P_1} = b' + \frac{a}{P} - \frac{c}{R}, \quad \phi'(s) \cdot \frac{\mathfrak{z}}{P_1} = c' + \frac{b}{R},$$

or

$$\mathfrak{x} = \frac{U}{\sqrt{SU^2}}, \quad \mathfrak{y} = \frac{V}{\sqrt{SU^2}}, \quad \mathfrak{z} = \frac{W}{\sqrt{SU^2}} \quad \ldots\ldots\ldots\ldots\ldots\ldots(7),$$

$\mathfrak{x} = S\mathrm{A}\Xi_1$, $\mathfrak{y} = S\Xi\Xi_1$, $\mathfrak{z} = S\Lambda\Xi_1$ being the *relative* cosines of H_1H_1' with respect to (D) and U, V, W abbreviations for

$$a' - \frac{b}{P}, \quad b' + \frac{a}{P} - \frac{c}{R}, \quad c' + \frac{b}{R}.$$

* We have given this relation for the first time, in 1906. (Cfr Ἐπετηρὶς τοῦ ἐθνικοῦ Πανεπιστημίου, pp. 349—354: "Generalization of a formula from the theory of the surfaces.") It has been found independently by Dr Arndt in his dissertation (cfr *Historical Notes*).

The *relative* cosines of BB_1 are consequently

$$\left.\begin{aligned}
l &= \theta.(b\mathfrak{z} - c\mathfrak{y}) = \theta.\frac{bW - cV}{\sqrt{SU^2}} = \theta\frac{bc' - cb' + \dfrac{b^2+c^2}{R} - \dfrac{ac}{P}}{\sqrt{\left(a' - \dfrac{b}{P}\right)^2 + \left(b' + \dfrac{a}{P} - \dfrac{c}{R}\right)^2 + \left(c' + \dfrac{b}{R}\right)^2}}\\
m &= \theta.(c\mathfrak{x} - a\mathfrak{z}) = \theta.\frac{cU - aW}{\sqrt{SU^2}} = \theta\frac{ca' - ac' - b\left(\dfrac{c}{P} + \dfrac{a}{R}\right)}{\sqrt{\left(a' - \dfrac{b}{P}\right)^2 + \left(b' + \dfrac{a}{P} - \dfrac{c}{R}\right)^2 + \left(c' + \dfrac{b}{R}\right)^2}}\\
n &= \theta.(a\mathfrak{y} - b\mathfrak{x}) = \theta.\frac{aV - bU}{\sqrt{SU^2}} = \theta\frac{ab' - ba' + \dfrac{a^2+b^2}{P} - \dfrac{ca}{R}}{\sqrt{\left(a' - \dfrac{b}{P}\right)^2 + \left(b' + \dfrac{a}{P} - \dfrac{c}{R}\right)^2 + \left(c' + \dfrac{b}{R}\right)^2}}
\end{aligned}\right\}(\theta^2 = 1) \quad \ldots\ldots\ldots(8).$$

And the *absolute* cosines of (D_1) are the following:

$$\left.\begin{aligned}
&\mathrm{A}_1 = \mathrm{A}a + \Xi b + \Lambda c, \text{ etc.,} \quad \Xi_1 = \frac{\mathrm{A}U + \Xi V + \Lambda W}{\sqrt{SU^2}}, \text{ etc.}\\
&\Lambda_1 = \epsilon\frac{U(\Xi c - \Lambda b) + V(\Lambda a - \mathrm{A}c) + W(\mathrm{A}b - \Xi a)}{\sqrt{SU^2}}, \text{ etc.}
\end{aligned}\right\}(\epsilon^2 = 1)\ldots(9).$$

3. Let us now calculate the torsion of (D_1). Differentiating the formula

$$\Xi_1 = \frac{\mathrm{A}U + \Xi V + \Lambda W}{\sqrt{SU^2}}$$

we get $\dfrac{d\Xi_1}{ds_1}\phi'.\sqrt{SU^2} + \Xi_1(\sqrt{SU^2})' = \mathrm{A}U' + \Xi V' + \Lambda W' + U\mathrm{A}' + V\Xi' + W\Lambda'$,

or

$$\frac{d\Xi_1}{ds_1}\phi'.\sqrt{SU^2} + \Xi_1(\sqrt{SU^2})' = \mathrm{A}U' + \Xi V' + \Lambda W' + U\frac{\Xi}{P} + V\left(-\frac{\mathrm{A}}{P} + \frac{\Lambda}{R}\right) + W\left(-\frac{\Xi}{R}\right).$$

Squaring this equation and the similar ones for H_1, Z_1 and adding, we find

$$\frac{\phi'^2}{K_1^2}SU^2 + (\sqrt{SU^2}')^2 = SU'^2 + \left(\frac{U}{P} - \frac{W}{R}\right)^2 + \frac{V^2}{K^2} + \frac{2}{P}(UV' - U'V) + \frac{2}{R}(VW' - V'W)$$

or $$\frac{\phi'^2}{K_1^2}(SU^2)^2 = SU^2.SU'^2 - (SUU')^2 + SU^2\left[\left(\frac{U}{P} - \frac{W}{R}\right)^2 + \ldots + \frac{2}{R}(VW' - V'W)\right],$$

which can be written

$$\begin{aligned}\frac{\phi'^6}{P_1^4.K_1^2} &= S(UV' - VU')^2 +\\
&+ \frac{\phi'^2}{P_1^2}\left[\left(\frac{U}{P} - \frac{W}{R}\right)^2 + \frac{V^2}{K^2} + \frac{2}{P}(UV' - VU') + \frac{2}{R}(VW' - WV')\right] \quad \ldots(10).\end{aligned}$$

It remains to calculate the differences $UV' - VU'$, etc., and to replace $\dfrac{1}{K_1^2}$ by $\dfrac{1}{P_1^2} + \dfrac{1}{R_1^2}$ and $\dfrac{1}{P_1}$ by its value (6).

The formula thus found expresses the torsion of T_1T_1' by means of a, b, c, but, as this formula is very long, we omit it here.

Remark. We can, of course, find an expression for $\frac{1}{R_1}$, similar to the expression for $\frac{1}{P_1}$ (6), which contains l, m, n instead of a, b, c. But the notion of *torsion* requires (as well as for the customary torsion of a curve), that $\frac{1}{R_1}$ is expressed by means of the cosines of the *departure* direction.

An important special case is that of constant cosines a, b, c. The preceding formulae for cosines and curvatures become then shorter,

$$\frac{\phi'(s)^2}{P_1^2} = \left(\frac{a}{P} - \frac{c}{R}\right)^2 + \frac{b^2}{K^2} = \frac{a^2+b^2}{P^2} + \frac{b^2+c^2}{R^2} - \frac{2ac}{PR} \quad \ldots\ldots(6').$$

$$\mathfrak{x} = \frac{-\frac{b}{P}}{\sqrt{\left(\frac{a}{P}-\frac{c}{R}\right)^2 + \frac{b^2}{K^2}}}, \quad \mathfrak{y} = \frac{\frac{a}{P}-\frac{c}{R}}{\sqrt{\left(\frac{a}{P}-\frac{c}{R}\right)^2 + \frac{b^2}{K^2}}}, \quad \mathfrak{z} = \frac{\frac{b}{R}}{\sqrt{\left(\frac{a}{P}-\frac{c}{R}\right)^2 + \frac{b^2}{K^2}}} \quad \ldots\ldots(7').$$

$$l = \theta\frac{\frac{b^2+c^2}{R} - \frac{ac}{P}}{\sqrt{\left(\frac{a}{P}-\frac{c}{R}\right)^2 + \frac{b^2}{K^2}}}, \quad m = \theta\frac{-b\left(\frac{c}{P}+\frac{a}{R}\right)}{\sqrt{\left(\frac{a}{P}-\frac{c}{R}\right)^2 + \frac{b^2}{K^2}}}, \quad n = \theta\frac{\frac{a^2+b^2}{P} - \frac{ca}{R}}{\sqrt{\left(\frac{a}{P}-\frac{c}{R}\right)^2 + \frac{b^2}{K^2}}} \quad \ldots\ldots(8').$$

$$\frac{\phi'^6}{P_1^4 K_1^2} = b^2\left(\frac{1}{P}\left(\frac{1}{R}\right)' - \frac{1}{R}\left(\frac{1}{P}\right)'\right)^2 + \frac{\phi'^2}{P_1^2}\left[\left(\frac{U}{P} - \frac{W}{R}\right)^2 + \frac{V^2}{K^2} + 2b\left(\frac{c}{P}+\frac{a}{R}\right)\left(\frac{1}{P}\left(\frac{1}{R}\right)' - \frac{1}{R}\left(\frac{1}{P}\right)'\right)\right] \quad \ldots\ldots(10').$$

4. *Remark.* (D_1) is a Frenetian trihedron with respect to the fixed axes, but evidently not with respect to (D). We can, however, from T_1T_1' as departure direction, construct a third Frenetian trihedron (D_r): T_1T_1', H_rH_r', B_rB_r' with the cosines

$$a, b, c; \quad \mathfrak{x}_r = \frac{\frac{da}{ds_1}}{\sqrt{S\left(\frac{da}{ds_1}\right)^2}}, \text{ etc.}; \quad l_r = \theta . \frac{b\frac{dc}{ds_1} - c\frac{db}{ds_1}}{\sqrt{S\left(\frac{da}{ds_1}\right)^2}}; \quad (\theta^2 = 1), \text{ etc.}:$$

it is then
$$\frac{da}{ds_1} = \frac{\mathfrak{x}_r}{P_r}, \text{ etc.}; \quad \frac{d\mathfrak{x}_r}{ds_1} = -\frac{a}{P_r} + \frac{l_r}{R_r}, \text{ etc.}; \quad \frac{dl_r}{ds_1} = -\frac{\mathfrak{x}_r}{R_r},$$

where
$$\frac{1}{P_r} \equiv \sqrt{S\left(\frac{da}{ds_1}\right)^2} \text{ and } \frac{1}{R_r} \equiv \sqrt{S\left(\frac{dl_r}{ds_1}\right)^2}$$

are the *relative* curvatures of T_1T_1' with respect to (D).

(D_r) and (D_1) are in general different. They coincide only when

$$\frac{-\frac{b}{P}}{a'} = \frac{\frac{a}{P} - \frac{c}{R}}{b'} = \frac{\frac{b}{R}}{c'}.$$

These conditions are verified, (1) when $b = 0$, $\frac{a}{P} - \frac{c}{R} = 0$, equations which express that

T_1T_1' is the direction of the Mozzian axis of (D)*; (2) when $\frac{a'}{R}+\frac{c'}{P}=0$, because this condition is a consequence of

$$\frac{-\frac{b}{P}}{a'}=\frac{\frac{b}{R}}{c'},$$

when $b \gtrless 0$, and it verifies also the other equation

$$\frac{\frac{a}{P}-\frac{c}{R}}{b'}=\frac{\frac{b}{R}}{c'}\quad\left(\text{or}\quad\frac{-\frac{b}{P}}{a'}=\frac{\frac{a}{P}-\frac{c}{R}}{b'}\right).$$

The first case, $b=0$, $\frac{a}{P}=\frac{c}{R}$, is a special case of (2), for $b=0$; we have then $aa'+cc'=0$, or $\frac{a}{c'}=-\frac{c}{a'}$, and the condition

$$\frac{a'}{R}+\frac{c'}{P}=0 \quad \text{becomes} \quad \frac{a}{P}=\frac{c}{R}.$$

The condition (2) expresses that H_1H_1' *is perpendicular to the Mozzian axis of* (D).

D. *Relations between the normal and geodesic curvatures of two Frenetian trihedra.*

5. The mutual position of the two trihedra can also be defined by the following angles: the angle $a\equiv\Omega\equiv \widehat{TT', T_1T_1'}$, the angle $\varpi\equiv\widehat{HH', SS'}$ and the angle $\varpi_1\equiv\widehat{H_1H_1', SS'}$, SS' being the direction, which is the common perpendicular to TT' and T_1T_1'.

It is then very easy† to find three general formulae between the *normal* curvatures

$$\frac{\cos\varpi}{P}\equiv\frac{1}{P_n},\quad \frac{\cos\varpi_1}{P_1}\equiv\frac{1}{P_{1n}},$$

the *geodesic* curvatures

$$\frac{\sin\varpi}{P}\equiv\frac{1}{P_g},\quad \frac{\sin\varpi_1}{P_1}\equiv\frac{1}{P_{1g}},$$

and the *geodesic* torsions of the two trihedra

$$\frac{1}{R}-\frac{d\varpi}{ds}\equiv\frac{1}{R_g},\quad \frac{1}{R_1}-\frac{d\varpi_1}{ds_1}=\frac{1}{R_{1g}}.$$

These formulae are the following

$$\left.\begin{aligned}\frac{\phi'(s)}{P_{1n}}&=\frac{\cos\Omega}{P_n}-\frac{\sin\Omega}{R_g},\\ \frac{\phi'(s)}{P_{1g}}&=\frac{1}{P_g}+\frac{d\Omega}{ds},\\ \frac{\phi'(s)}{R_{1g}}&=\frac{\sin\Omega}{P_n}+\frac{\cos\Omega}{R_g}\end{aligned}\right\}\quad\ldots\ldots\ldots\ldots\ldots\ldots\ldots\ldots(11).$$

They can be shown by trigonometry in the same way as for the curves in my paper (*loc. cit.*).

* Viz. a direction parallel to the momentary axis of the *spherical image* of (D).

† Cfr *Nyt Tidsskrift for Matematik*, Copenhagen, v. XIV: N. Hatzidakis, "Om kurveteoretiske Invarianter."

ZUM 6-EBENENPROBLEM IM R_4

VON ROLAND WEITZENBÖCK.

Es gibt im R_4 "im allgemeinen" 5 Gerade, welche 6 gegebene Ebenen schneiden. Es wird geometrisch erläutert, wie man diese 5 Geraden erhalten kann, und es wird eine Gleichung fünften Grades aufgestellt, welche das Problem löst.

Der Gegenstand des Vortrages wird in einer ausführlichen Arbeit in den *Wiener Berichten* behandelt werden (voraussichtlich Januar 1913).

SUR LES ÉQUATIONS DIFFÉRENTIELLES DE LA GÉOMÉTRIE

PAR JULES DRACH.

La théorie d'intégration, que j'ai développée sous le nom de *théorie de la rationalité ou intégration logique* (et dont on trouvera les points essentiels dans ma Communication au Congrès, Section d'Analyse), donne pour les équations différentielles de la Géométrie *des méthodes régulières permettant de prévoir des réductions dans la difficulté de l'intégration*, c'est-à-dire *une réduction du groupe de rationalité.*

On ramène ainsi des propositions classiques, dues à l'ingéniosité des géomètres, à leur source analytique commune et l'on peut, sans difficulté, en indiquer de nouvelles.

J'appelle équations différentielles de la Géométrie, celles où interviennent ou des fonctions arbitraires ou des fonctions que l'on regarde comme *données* mais dont on ne précise pas la nature transcendante.

Le domaine des fonctions que l'on doit regarder comme données ou connues, se compose donc des fonctions *rationnelles* de certains éléments qui ont été représentés par un signe *explicite* (je dirai *explicitées*) et ce domaine comprendra toujours, avec une fonction $\phi(x, y)$ de deux variables par exemple, toutes ses dérivées $\frac{\partial \phi}{\partial x}, \frac{\partial \phi}{\partial y}, \ldots$ Ainsi il est bien entendu *qu'on n'explicite pas un signe fonctionnel d'opération*: $\log u$ ou $\sin u$ ne font pas nécessairement partie du domaine qui comprend u; les seules fonctions de u qui appartiennent toujours à ce domaine sont les fonctions rationnelles de u et de ses dérivées dont les coefficients sont rationnels par rapport aux autres éléments du domaine.

L'incertitude où l'on est de la nature des transcendantes *données* ne permet pas d'aboutir à des conclusions précises sur l'intégration, sauf dans le cas *général*; ce n'est que si les transcendantes sont définies par un système irréductible à partir des éléments rationnels absolus (variables et constantes arbitraires) que la théorie de *l'intégration logique* peut s'appliquer jusqu'au bout.

Il s'agit donc seulement ici d'une *application particulière et partielle de la théorie de la rationalité*: les propositions analytiques ne sont pas distinctes de celles qui interviennent dans cette théorie.

Équations du premier ordre.

1. Soit une équation du premier ordre

$$dv = m\,du \quad \ldots\ldots\ldots\ldots\ldots\ldots\ldots\ldots\ldots\ldots\ldots (1),$$

où m est une fonction *donnée* de u et v. *Supposons que l'on connaisse, pour les courbes*

$$\phi(u, v) = \text{const.}$$

qui satisfont à l'équation (1), *une propriété géométrique: de quelle utilité cette connaissance peut-elle être pour l'intégration de* (1)*?* Considérons d'abord le cas où les courbes étant tracées sur une surface (S), d'élément linéaire donné par

$$ds^2 = E du^2 + 2F du\,dv + G dv^2,$$

la propriété géométrique en question se conserve dans une déformation continue de (S); elle s'exprime alors par une relation,

$$R(\phi, I_1(\phi), I_2(\phi), \ldots, \Omega, I_1(\Omega), \ldots) = 0 \quad \ldots\ldots\ldots\ldots\ldots (2),$$

entre les *invariants de Beltrami* de la fonction ϕ (c'est-à-dire les paramètres différentiels $\Delta(\phi)$, $\Delta_2(\phi)$, $\Delta(\phi, \Delta\phi)$, ...) et *les invariants absolus* du ds^2 (ou de Minding) déduits de la courbure totale Ω par l'application à Ω des opérateurs $\Delta(\phi)$, $\Delta_2(\phi)$, ... Cette relation (2) peut être regardée comme une *forme réduite* de la relation

$$R(\Phi, I_1(\Phi), I_2(\Phi), \ldots, \Omega, I_1(\Omega), \ldots) = 0 \quad \ldots\ldots\ldots\ldots\ldots (3),$$

où Φ est une fonction arbitraire de ϕ.

Si la relation (3) ne renferme qu'en apparence la fonction Φ, c'est-à-dire si elle garde la même forme en $\phi, \frac{\partial\phi}{\partial u}, \frac{\partial\phi}{\partial v}, \ldots$ quel que soit Φ nous ne tirons aucun parti de la connaissance de (2): elle est simplement l'équation aux dérivées partielles qui définit le rapport $m = -\dfrac{\frac{\partial\phi}{\partial u}}{\frac{\partial\phi}{\partial v}}$. Elle ne renferme donc que le quotient $\dfrac{\frac{\partial\phi}{\partial u}}{\frac{\partial\phi}{\partial v}}$ et ses dérivées.

Exemples: La relation $\Delta\phi = 0$; la relation $\frac{1}{\rho_\phi} = F(\Omega, \Delta(\Omega), \ldots)$ où $\frac{1}{\rho_\phi}$ est la courbure géodésique de la courbe $\phi = \text{const.}$ et F une fonction quelconque des invariants absolus; la relation qui exprime que les lignes pour lesquelles $dv = m\,du$ sont des asymptotiques sur l'une des déformées de (S).

Si la relation (3) renferme $\Phi, \Phi', \Phi'', \ldots$ $\left(\text{où } \Phi' = \frac{d\Phi}{d\phi}, \ldots\right)$ lorsqu'on tient compte des formules de transformation:

$$\frac{\partial\Phi}{\partial u} = \Phi' \frac{\partial\phi}{\partial u}, \ldots, \frac{\partial^2\Phi}{\partial u^2} = \Phi' \frac{\partial^2\phi}{\partial u^2} + \Phi'' \left(\frac{\partial\phi}{\partial u}\right)^2, \ldots$$

et si *on l'identifie* à l'une de ses *réduites* (obtenue en particularisant Φ), on obtiendra un système différentiel qui définit, au moyen de ϕ, l'une des expressions suivantes:

$$\Phi,\ \Phi',\ \frac{\Phi''}{\Phi'},\ \frac{\Phi'''}{\Phi'} - \frac{3}{2}\left(\frac{\Phi''}{\Phi'}\right)^2.$$

Suivant les cas, on déduira de (3) par une *méthode régulière,* qui consiste essentiellement dans la formation des conditions d'intégrabilité de (1) et (3), c'est-à-dire, comme on l'a vu en Analyse, dans l'application de l'opération

$$X(f)=\frac{\partial f}{\partial u}+m\frac{\partial f}{\partial v},$$

qui donne par exemple:

$$X\left(\frac{\partial\phi}{\partial v}\right)+\frac{\partial m}{\partial v}\frac{\partial\phi}{\partial v}=0,$$

$$X\left(\frac{\partial^2\phi}{\partial v^2}\right)+2\frac{\partial m}{\partial v}\frac{\partial^2\phi}{\partial v^2}+\frac{\partial^2 m}{\partial v^2}\frac{\partial\phi}{\partial v}=0$$

....

une équation dont l'ordre différentiel ne dépasse pas trois et qui définit implicitement l'une des expressions:

$$\phi,\ \frac{\partial\phi}{\partial v},\ \frac{\partial^2\phi}{\partial v^2}:\frac{\partial\phi}{\partial v}\quad\text{ou}\quad\frac{\dfrac{\partial^3\phi}{\partial v^3}}{\dfrac{\partial\phi}{\partial v}}-\frac{3}{2}\left(\frac{\dfrac{\partial^2\phi}{\partial v^2}}{\dfrac{\partial\phi}{\partial v}}\right)^2.$$

Toutes les autres équations satisfaites par ϕ sont des conséquences de cette relation unique. On peut donc ajouter à l'équation:

$$X(\phi)=\frac{\partial\phi}{\partial u}+m\frac{\partial\phi}{\partial v}=0,$$

l'une des équations:

$$\phi=R,\quad\frac{\partial\phi}{\partial v}=K,\quad\frac{\partial^2\phi}{\partial v^2}=J\frac{\partial\phi}{\partial v},$$

$$\frac{\partial\phi}{\partial v}\frac{\partial^3\phi}{\partial v^3}-\frac{3}{2}\left(\frac{\partial^2\phi}{\partial v^2}\right)^2=I\left(\frac{\partial\phi}{\partial v}\right)^2,$$

où R, K, J, I sont connus en u, v *sans intégration.* Les équations qui définissent m sont alors, respectivement:

$$X(R)=0,\quad X(K)+K\frac{\partial m}{\partial v}=0,\quad X(J)+J\frac{\partial m}{\partial v}+\frac{\partial^2 m}{\partial v^2}=0,$$

$$X(I)+2I\frac{\partial m}{\partial v}+\frac{\partial^3 m}{\partial v^3}=0.$$

On observera que si les expressions de R, K, J, I sont, par rapport aux dérivées de m, d'ordre p, l'équation aux dérivées partielles que doit vérifier m est d'ordre au plus égal à $(p+3)$.

Si les éléments qui figurent dans (1) et dans (2) appartiennent à un domaine de rationalité $[\Delta]$ bien défini, on pourra préciser dans ce domaine la réduction ultérieure de l'intégration de (1), s'il en existe une qui ne soit pas uniquement celle signalée plus haut.

Remarque: Tout ce que nous venons de dire s'applique uniquement aux *solutions propres* de (2), c'est-à-dire à celles qui ne satisfont à aucune relation d'ordre inférieur par rapport aux dérivées de ϕ. L'équation (2) peut posséder des solutions particulières *impropres* qui pourront être traitées par la même méthode mais conduiront à des conclusions différentes. Ainsi $\Delta_2 \phi = 0$ qui définit sur une surface les lignes isothermes peut posséder des solutions de $\Delta\phi = F(\phi)$; cela arrive sur toutes les surfaces dont l'élément linéaire peut s'écrire $ds^2 = \frac{1}{F(\phi)}(d\phi^2 + d\psi^2)$, c'est-à-dire qui sont applicables sur des surfaces de révolution. Pour les solutions propres de $\Delta_2 \phi = 0$, la méthode conduit à la connaissance de J; pour les autres on a K.

Exemples: Si la relation (2) est:

$$\frac{1}{\rho_\phi} = \phi F(\Omega, \Delta(\Omega), \ldots)$$

où l'on désigne par ρ_ϕ le rayon de courbure géodésique des courbes $\phi(u, v) = \text{const.}$, elle donne ϕ sans intégration.

On a de même sans intégration, connaissant seulement leur équation différentielle, les lignes qui sur une déformée de (S) sont sur des sphères concentriques; pour celles qui sont dans des plans parallèles on connaîtra $\frac{\partial\phi}{\partial v} = K$ en même temps que $m(u, v)$, sauf si ces plans sont isotropes, auquel cas on ne connaîtra que

$$\frac{\partial^2\phi}{\partial v^2} : \frac{\partial\phi}{\partial v} = J.$$

Il peut exister, pour certaines formes du ds^2, des cas d'exception qui s'interprètent aisément.

La relation $$\Delta_2\phi = F\left(\Omega, \Delta(\Omega), \ldots, \frac{1}{\rho_\phi}, \ldots\right)$$

donnera $\frac{\partial\phi}{\partial v}$, c'est-à-dire un multiplicateur de l'équation (1) $dv - m\,du = 0$; les relations

$$\Delta(\phi, \Delta_2\phi) = 0 \quad \text{ou} \quad \Delta(\Delta\phi) = 0$$

donnent seulement $\frac{\partial^2\phi}{\partial v^2} : \frac{\partial\phi}{\partial v}$, c'est-à-dire la dérivée logarithmique d'un multiplicateur.

De même si l'on connaît l'équation différentielle

$$dv - m\,du = 0 \quad \ldots\ldots\ldots\ldots\ldots\ldots\ldots\ldots(1)$$

des lignes $\phi(u, v) = \text{const.}$ qui permettent de mettre un élément linéaire donné sous la forme

$$ds^2 = \alpha\,d\phi^2 + \frac{1}{\alpha}\,d\psi^2,$$

on connaît également $\frac{\partial^2\phi}{\partial v^2} : \frac{\partial\phi}{\partial v} = J$ sans intégration. Il en est de même pour les lignes $\phi(u, v) = \text{const.}$ telles que

$$ds^2 = \cos^2\alpha\,d\phi^2 + \sin^2\alpha\,d\psi^2.$$

Enfin si la relation (2) est

$$\Delta(\phi, \Delta_2\phi) - \frac{1}{2}(\Delta_2\phi)^2 - \frac{1}{4}\left[\frac{\Delta(\phi, \Delta\phi)}{\Delta\phi}\right]^2 = 0,$$

on peut obtenir sans intégration:

$$\frac{\frac{\partial^3\phi}{\partial v^3}}{\frac{\partial\phi}{\partial v}} - \frac{3}{2}\left(\frac{\frac{\partial^2\phi}{\partial v^2}}{\frac{\partial\phi}{\partial v}}\right)^2 = I,$$

et l'intégration de (1) se ramène, à deux quadratures près, à celle de deux équations de Riccati.

2. Il est clair que ce qui permet la théorie précédente c'est l'intervention du groupe ponctuel $\Phi = \Phi(\phi)$; on *peut donc appliquer la méthode à une propriété géométrique quelconque du faisceau de courbes* $\phi(u, v) = \text{const.}$ *dont on suppose connue l'équation différentielle*

$$dv = m\,du \quad \ldots\ldots\ldots\ldots\ldots\ldots\ldots\ldots(1),$$

cette propriété étant exprimée par une relation, de nature quelconque, entre les invariants géométriques, qui se ramène, en dernière analyse, à une relation entre $u, v, \phi, \frac{\partial\phi}{\partial v}, \frac{\partial^2\phi}{\partial v^2}, \frac{\partial^3\phi}{\partial v^3}$. Par exemple si l'on suppose que sur la surface

$$x = x(u, v), \quad y = y(u, v), \quad z = z(u, v) \quad \ldots\ldots\ldots\ldots(S),$$

les courbes définies par (1) sont planes, ou sphériques, ou tracées sur des surfaces algébriques de degré connu, on les obtient sans intégration.

Ainsi, soit

$$A(\phi) = \frac{\partial\phi}{\partial u} + m\frac{\partial\phi}{\partial v} = 0 \quad \ldots\ldots\ldots\ldots\ldots\ldots(a),$$

dans l'hypothèse où les courbes $\phi(u, v) = \text{const.}$ sont planes on peut, *si elles ne sont pas des droites,* définir l'intégrale de (a) par la relation: $z = \alpha x + \beta y + \gamma$ où α, β, γ sont des solutions de $A(\phi) = 0$. Deux d'entre elles sont des fonctions entièrement déterminées de la troisième.

En appliquant l'opération $A(f)$ on obtient les équations

$$A(z) = \alpha A(x) + \beta A(y),$$
$$A_2(z) = \alpha A_2(x) + \beta A_2(y), \quad \text{avec} \quad A_2(\omega) = A(A(\omega)),$$

et l'ensemble de ces trois relations détermine α, β, γ. Deux de ces expressions α et β seront d'ailleurs des fonctions de γ, si γ est variable.

L'équation aux dérivées partielles que doit vérifier m est simplement:

$$\begin{vmatrix} A(z) & A(x) & A(y) \\ A_2(z) & A_2(x) & A_2(y) \\ A_3(z) & A_3(x) & A_3(y) \end{vmatrix} = 0.$$

Le cas d'exception, où les lignes $\phi = \text{const.}$ sont des droites, se reconnaît à ce que:

$$\frac{A_2(z)}{A(z)} = \frac{A_2(x)}{A(x)} = \frac{A_2(y)}{A(y)};$$

on n'a que deux équations distinctes pour trouver α, β, γ et les droites sont les axes des faisceaux de plans ainsi obtenus. D'ailleurs elles sont à priori connues puisqu'en un point de (S) l'équation (1) donne la direction de la tangente à la courbe.

On connaît de même $\frac{\partial \phi}{\partial v}$ sur les lignes telles que la dérivée de ϕ suivant la direction conjuguée de la tangente à $\phi = \text{const.}$ est une fonction de ϕ (sauf si cette fonction est nulle): l'équation (2) est alors

$$\frac{\frac{\partial \phi}{\partial u}\left(D'\frac{\partial \phi}{\partial v} - D''\frac{\partial \phi}{\partial u}\right) + \frac{\partial \phi}{\partial v}\left(D'\frac{\partial \phi}{\partial u} - D\frac{\partial \phi}{\partial v}\right)}{\sqrt{E\left(D'\frac{\partial \phi}{\partial v} - D''\frac{\partial \phi}{\partial u}\right)^2 + 2F\left(D'\frac{\partial \phi}{\partial v} - D''\frac{\partial \phi}{\partial u}\right)\left(D'\frac{\partial \phi}{\partial u} - D\frac{\partial \phi}{\partial v}\right) + \dots}} = F(\phi).$$

On connaît simplement $\frac{\partial^2 \phi}{\partial v^2} : \frac{\partial \phi}{\partial v}$ pour les lignes qui satisfont à $\Delta_2(\phi) = 0$, Δ_2 étant un paramètre analogue à celui de Beltrami mais construit avec une forme différentielle quadratique quelconque, par exemple avec

$$D du^2 + 2D' du dv + D'' dv^2 \text{; etc.}$$

On forme sans difficulté des exemples où la propriété géométrique dont nous supposons l'existence n'entraîne que la connaissance de

$$\frac{\frac{\partial^3 \phi}{\partial v^3}}{\frac{\partial \phi}{\partial v}} - \frac{3}{2}\left(\frac{\frac{\partial^2 \phi}{\partial v^2}}{\frac{\partial \phi}{\partial v}}\right)^2 = I;$$

celui signalé plus haut s'applique encore lorsque $\Delta \phi$ et $\Delta_2 \phi$ sont construits avec une forme quadratique quelconque, définie sur (S), au lieu du carré de l'élément linéaire.

3. Les raisonnements précédents se transportent, sans y rien changer d'essentiel, à un système complet d'équations linéaires aux dérivées partielles qui n'admet qu'une solution.

On saura donc, par une méthode régulière, tirer parti de la connaissance d'une propriété géométrique de la famille de surfaces

$$\phi(x, y, z) = \text{const.}$$

pour la détermination de ϕ au moyen d'un système *complet* connu:

$$\frac{\frac{\partial \phi}{\partial x}}{X} = \frac{\frac{\partial \phi}{\partial y}}{Y} = \frac{\frac{\partial \phi}{\partial z}}{Z},$$

qu'on peut toujours regarder comme définissant les surfaces normales aux courbes de la congruence:

$$\frac{dx}{X} = \frac{dy}{Y} = \frac{dz}{Z}.$$

On se trouve ici dans le cas où, identiquement:

$$X\left(\frac{\partial Y}{\partial z} - \frac{\partial Z}{\partial y}\right) + Y\left(\frac{\partial Z}{\partial x} - \frac{\partial X}{\partial z}\right) + Z\left(\frac{\partial X}{\partial y} - \frac{\partial Y}{\partial x}\right) = 0.$$

Si l'on sait, par exemple, que les surfaces $\phi = \text{const.}$ forment une famille isotherme, on a:

$$\Delta_2\phi : \Delta\phi = F(\phi),$$

avec
$$\Delta\phi = \left(\frac{\partial\phi}{\partial x}\right)^2 + \left(\frac{\partial\phi}{\partial y}\right)^2 + \left(\frac{\partial\phi}{\partial z}\right)^2, \quad \Delta_2\phi = \frac{\partial^2\phi}{\partial x^2} + \frac{\partial^2\phi}{\partial y^2} + \frac{\partial^2\phi}{\partial z^2}.$$

On en déduit la connaissance de $\frac{\phi''}{\phi'}$, c'est-à-dire que ϕ sera défini à une transformation linéaire près, par des relations obtenues sans intégration. Si ces surfaces $\phi = \text{const.}$ sont des plans ou des sphères ou des surfaces algébriques de degré connu, ou même des surfaces transcendantes définies par une relation *de forme connue* en x, y, z:

$$\Omega(x, y, z) = 0,$$

on les obtient sans intégration.

Si les surfaces $\phi = \text{const.}$ sont orthogonales aux surfaces $\Delta\phi = \text{const.}$, auquel cas $\Delta(\phi, \Delta\phi) = 0$, on peut encore déduire de là $\frac{\phi''}{\phi'}$, c'est-à-dire que ϕ est défini aux transformations près du groupe linéaire. Enfin ϕ serait défini aux transformations près du groupe projectif, si l'on avait

$$\Delta(\phi, \Delta_2\phi) - \tfrac{1}{2}(\Delta_2\phi)^2 - \tfrac{1}{4}\left[\frac{\Delta(\phi, \Delta\phi)}{\Delta\phi}\right]^2 = \Delta\phi \,.\, F(x, y, z).$$

4. La méthode générale, employée pour édifier la théorie de la rationalité, donne aussi un procédé régulier pour *déduire de la connaissance d'une propriété géométrique du réseau des courbes*

$$\phi(u, v) = \text{const.}, \quad \psi(u, v) = \text{const.},$$

définies respectivement par les équations:

$$A(\phi) = \frac{\partial\phi}{\partial u} + m\frac{\partial\phi}{\partial v} = 0 \quad\ldots\ldots\ldots\ldots(1),$$

$$B(\psi) = \frac{\partial\psi}{\partial u} + n\frac{\partial\psi}{\partial v} = 0 \quad\ldots\ldots\ldots\ldots(2),$$

une réduction du groupe de rationalité de ces équations. Si l'on connaît une relation

$$R\left(u, v, \phi, \frac{\partial\phi}{\partial v}, \ldots, \psi, \frac{\partial\psi}{\partial v}, \ldots\right) = 0 \quad\ldots\ldots\ldots\ldots(3),$$

d'ordre quelconque par rapport aux dérivées de ϕ et de ψ, l'application de l'opérateur $A(\omega)$ qui donne

$$A\left(\frac{\partial\phi}{\partial v}\right) + \frac{\partial m}{\partial v}\frac{\partial\phi}{\partial v} = 0,$$

$$\ldots\ldots\ldots\ldots$$

et aussi
$$A(\psi) + (n - m)\frac{\partial\psi}{\partial v} = 0,$$

$$A\left(\frac{\partial\psi}{\partial v}\right) + (n - m)\frac{\partial^2\psi}{\partial v^2} + \frac{\partial n}{\partial v}\frac{\partial\psi}{\partial v} = 0,$$

$$\ldots\ldots\ldots\ldots,$$

et par suite n'introduit pas d'éléments $\phi, \frac{\partial\phi}{\partial v}, \ldots$ autres que ceux qui figurent dans R, conduit à une relation

$$F\left(\phi, \frac{\partial\phi}{\partial v}, \ldots\right) = G\left(u, v, \psi, \frac{\partial\psi}{\partial v}, \ldots, \frac{\partial^p\psi}{\partial v^p}\right) \quad \ldots\ldots\ldots\ldots\ldots\ldots(4),$$

dont (3) et toutes les autres sont des conséquences, et dont le premier membre est l'un des quatre invariants :

$$\phi, \quad \frac{\partial\phi}{\partial v} = K, \quad \frac{\partial^2\phi}{\partial v^2} : \frac{\partial\phi}{\partial v} = J, \quad \frac{\frac{\partial^3\phi}{\partial v^3}}{\frac{\partial\phi}{\partial v}} - \frac{3}{2}\left(\frac{\frac{\partial^2\phi}{\partial v^2}}{\frac{\partial\phi}{\partial v}}\right)^2 = I.$$

L'équation : $$A(F) = A(G) \quad \ldots\ldots\ldots\ldots\ldots\ldots\ldots\ldots\ldots\ldots(5),$$

peut alors, en tenant compte des équations résolvantes

$$A(\phi) = 0, \quad A(K) + K\frac{\partial m}{\partial v} = 0, \quad A(J) + J\frac{\partial m}{\partial v} + \frac{\partial^2 m}{\partial v^2} = 0, \quad A(I) + 2I\frac{\partial m}{\partial v} + \frac{\partial^2 m}{\partial v^2} = 0,$$

être débarrassée de ϕ et de ses dérivées. Elle est donc une relation *connue* entre

$$u, v, \psi, \frac{\partial\psi}{\partial v}, \ldots, \frac{\partial^{p+1}\psi}{\partial v^{p+1}}.$$

Mais nous savons qu'on en peut toujours déduire une relation analogue donnant explicitement en u, v l'un des quatre invariants :

$$\psi, \quad \frac{\partial\psi}{\partial v}, \quad \frac{\partial^2\psi}{\partial v^2} : \frac{\partial\psi}{\partial v}, \quad \frac{\frac{\partial^3\psi}{\partial v^3}}{\frac{\partial\psi}{\partial v}} - \frac{3}{2}\left(\frac{\frac{\partial^2\psi}{\partial v^2}}{\frac{\partial\psi}{\partial v}}\right)^2.$$

Il est donc loisible de supposer tout de suite que dans R (ou dans G) la fonction ψ ne figure qu'au second ordre.

Les types possibles de relations entre ϕ et ψ peuvent ainsi être établis a priori et la réduction de l'intégration de (1) *et de* (2) *précisée pour chacun d'eux.* On observera que l'équation (2) est nécessairement *spéciale*, et comme on aurait pu permuter le rôle de ϕ et ψ, il en est de même de (1). Toute propriété géométrique du réseau (ϕ, ψ) qui conduit à une relation (3) [c'est-à-dire qui ne s'exprime pas par une relation entre u, v, m, n et leurs dérivées] donnera donc, en dernière analyse, une des relations-types dont nous venons d'établir l'existence.

(*a*) Supposons, par exemple, qu'il existe une relation

$$\phi = F(u, v, \psi),$$

où ψ est connu, c'est-à-dire que l'intégration de (1) se ramène à celle de (2) et réciproquement (sans d'ailleurs qu'aucune d'elles s'intègre explicitement) ; on aura dans ce cas l'identité :

$$A(F) = \frac{\partial F}{\partial u} + m\frac{\partial F}{\partial v} + \frac{\partial F}{\partial\psi}(m-n)\frac{\partial\psi}{\partial v} = 0,$$

d'où résulte entre $\psi, \frac{\partial\psi}{\partial v}, u, v$ une relation du premier ordre, à laquelle nous pouvons supposer la forme :

$$\frac{\partial\psi}{\partial v} = K_1(u, v).$$

La fonction F (qui dépend de ψ) doit donc vérifier identiquement la relation

$$\frac{\partial F}{\partial u} + m\frac{\partial F}{\partial v} + \frac{\partial F}{\partial\psi}(m-n)K_1 = 0,$$

et cette condition est suffisante. On observe que F peut être supposé linéaire en ψ et l'on trouve immédiatement

$$F = \psi + \beta(u, v),$$

avec la condition : $$A(\beta) + (m-n)K_1 = 0.$$

En résumé *si l'équation* (2) :

$$B(\psi) = \frac{\partial\psi}{\partial u} + n\frac{\partial\psi}{\partial v} = 0,$$

possède le multiplicateur K_1, c'est-à-dire si

$$B(K_1) + K_1\frac{\partial n}{\partial v} = 0,$$

l'équation (1) $A(\phi) = \frac{\partial\phi}{\partial u} + m\frac{\partial\phi}{\partial v} = 0$, *où l'on a pris* m *de manière à vérifier la condition* :

$$\frac{\partial\beta}{\partial u} - nK_1 + m\left(\frac{\partial\beta}{\partial v} + K_1\right) = 0,$$

possède une solution ϕ, *liée à* ψ *par la relation* : $\phi = \psi + \beta$.

L'équation (1) possède évidemment le multiplicateur connu

$$\frac{\partial\phi}{\partial v} = K_1 + \frac{\partial\beta}{\partial v}.$$

(*b*) Si la relation *unique*, supposée connue, a la forme

$$\phi = F\left(u, v, \psi, \frac{\partial\psi}{\partial v}\right),$$

elle donnera de même

$$A(F) = \frac{\partial F}{\partial u} + m\frac{\partial F}{\partial v} + \frac{\partial F}{\partial\psi}(m-n)\frac{\partial\psi}{\partial v} + \frac{\partial F}{\partial\left(\frac{\partial\psi}{\partial v}\right)}\left\{(m-n)\frac{\partial^2\psi}{\partial v^2} - \frac{\partial n}{\partial v}\frac{\partial\psi}{\partial v}\right\} = 0,$$

d'où l'on conclut l'existence d'une relation connue :

$$\frac{\partial^2\psi}{\partial v^2} = J_1\frac{\partial\psi}{\partial v},$$

où J_1 ne dépend que de u, v et vérifie :

$$B(J_1) + J_1\frac{\partial n}{\partial v} + \frac{\partial^2 n}{\partial v^2} = 0.$$

L'identité

$$\frac{\partial F}{\partial u} + m\frac{\partial F}{\partial v} + (m-n)\frac{\partial\psi}{\partial v}\frac{\partial F}{\partial\psi} + \left\{(m-n)J_1 - \frac{\partial n}{\partial v}\right\}\frac{\partial\psi}{\partial v}\frac{\partial F}{\partial\left(\frac{\partial\psi}{\partial v}\right)} = 0,$$

conduit aisément à l'expression *unique*:

$$F = c\psi + \alpha \frac{\partial \psi}{\partial v},$$

acceptable sous la condition:

$$A(\alpha) + \alpha \left\{ (m-n) J_1 - \frac{\partial n}{\partial v} \right\} + c(m-n) = 0 \qquad (c \neq 0).$$

On peut encore prendre α arbitrairement et déterminer m par cette relation.

D'ailleurs l'équation:

$$\frac{\partial \phi}{\partial v} = \left(c + \frac{\partial \alpha}{\partial v}\right) \frac{\partial \psi}{\partial v} + \alpha J_1 \frac{\partial \psi}{\partial v},$$

jointe à celle qui donne ϕ, permet d'exprimer explicitement ψ et $\frac{\partial \psi}{\partial v}$ au moyen de ϕ et $\frac{\partial \phi}{\partial v}$ *lorsque* $c \neq 0$.

Si l'on a $c = 0$, auquel cas

$$\phi = \alpha \left(\frac{\partial \psi}{\partial v}\right),^{*} \quad \frac{\partial \phi}{\partial v} = \left(\frac{\partial \alpha}{\partial v} + \alpha J_1\right) \frac{\partial \psi}{\partial v},$$

log ϕ est donné par une quadrature; une autre quadrature est nécessaire pour obtenir ψ quand ϕ est supposé connu. La correspondance entre les solutions des équations (1) et (2) est évidente.

(*c*) Pour *épuiser* l'étude des cas où ϕ est explicitement connu quand on a intégré (2), il reste à examiner l'hypothèse:

$$\phi = F\left(u, v, \psi, \frac{\partial \psi}{\partial v}, \frac{\partial^2 \psi}{\partial v^2}\right).$$

La fonction F devra satisfaire identiquement à la relation:

$$\frac{\partial F}{\partial u} + m \frac{\partial F}{\partial v} + \frac{\partial F}{\partial \psi}(m-n)\frac{\partial \psi}{\partial v} + \frac{\partial F}{\partial \left(\frac{\partial \psi}{\partial v}\right)} \left\{ (m-n) \frac{\partial^2 \psi}{\partial v^2} - \frac{\partial n}{\partial v} \frac{\partial \psi}{\partial v} \right\}$$
$$+ \frac{\partial F}{\partial \left(\frac{\partial^2 \psi}{\partial v^2}\right)} \left\{ (m-n) \frac{\partial^3 \psi}{\partial v^3} - 2 \frac{\partial n}{\partial v} \frac{\partial^2 \psi}{\partial v^2} - \frac{\partial^2 n}{\partial v^2} \frac{\partial \psi}{\partial v} \right\} = 0,$$

où l'on suppose
$$\frac{\partial \psi}{\partial v} \frac{\partial^3 \psi}{\partial v^3} - \frac{3}{2} \left(\frac{\partial^2 \psi}{\partial v^2}\right)^2 = I_1 \left(\frac{\partial \psi}{\partial v}\right)^2,$$

I_1 étant connu en u, v et vérifiant:

$$B(I_1) + 2 I_1 \frac{\partial n}{\partial v} + \frac{\partial^3 n}{\partial v^3} = 0.$$

La réduction de F à sa forme la plus simple donnera, en observant que F est *déterminé* aux transformations près qui remplacent ψ par $f(\psi)$ où f est arbitraire:

$$F = \alpha \frac{\left(\frac{\partial^2 \psi}{\partial v^2}\right)^2}{\left(\frac{\partial \psi}{\partial v}\right)^3} + \beta \frac{\frac{\partial^2 \psi}{\partial v^2}}{\left(\frac{\partial \psi}{\partial v}\right)^2} + \frac{\gamma}{\left(\frac{\partial \psi}{\partial v}\right)},$$

* L'hypothèse $\phi = \alpha \left(\frac{\partial \phi}{\partial v}\right)^p$, où p est entier, se ramène à celle-là par l'extraction de la racine $p^{\text{ième}}$ de α.

où β, γ sont des fonctions de u, v seul qui s'expriment explicitement au moyen de $\alpha(u, v)$ et de ses dérivées. Cette dernière satisfait à une condition du troisième ordre facile à former, qui, jointe à la résolvante I_1, caractérise ce cas.

En général $\frac{\partial\psi}{\partial v}$ s'exprime rationnellement en ϕ, $\frac{\partial\phi}{\partial v}$, $\frac{\partial^2\phi}{\partial v^2}$, et il existe une relation donnant explicitement $\frac{\partial^2\phi}{\partial v^2}$ au moyen de ϕ, $\frac{\partial\phi}{\partial v}$, u, v.

Remarque. Il est clair que tout ce que nous venons de dire au sujet de l'intégration simultanée de deux équations du premier ordre, pourra se répéter, *avec des changements évidents,* pour trois ou un plus grand nombre d'équations.

Nous avons donc la possibilité *par une méthode régulière* d'utiliser la connaissance d'une propriété géométrique du système des courbes

$$\alpha(u, v) = \text{const.}, \quad \beta(u, v) = \text{const.}, \quad \gamma(u, v) = \text{const.}, \ldots$$

pour l'intégration des équations du premier ordre qui définissent séparément ces courbes.

Exemples géométriques.

Supposons qu'il s'agisse de mettre un élément linéaire connu sous la forme

$$ds^2 = a^2 d\phi^2 - 2d\phi d\psi + c^2 d\psi^2,$$

et que l'on connaisse les équations

$$A(\phi) = \frac{\partial\phi}{\partial u} + m\frac{\partial\phi}{\partial v} = 0, \qquad B(\psi) = \frac{\partial\psi}{\partial u} + n\frac{\partial\psi}{\partial v} = 0,$$

qui définissent ϕ et ψ. La relation

$$\frac{\Delta(\phi, \psi)}{\Delta\phi\Delta\psi - \Delta^2(\phi, \psi)} = 1$$

donnera explicitement
$$\frac{\partial\phi}{\partial v}\cdot\frac{\partial\psi}{\partial v} = \theta(u, v),$$

et l'on en déduira aisément les expressions de

$$\frac{\dfrac{\partial^2\phi}{\partial v^2}}{\dfrac{\partial\phi}{\partial v}} \quad \text{et de} \quad \frac{\dfrac{\partial^2\psi}{\partial v^2}}{\dfrac{\partial\psi}{\partial v}}$$

au moyen de u et v.

Supposons de même que les courbes $\phi = \text{const.}$, $\psi = \text{const.}$, définies par

$$A(\phi) = 0, \quad B(\psi) = 0,$$

satisfassent à la condition $\Delta(\phi, \Delta\psi) = \Delta\phi$; on en déduira une relation donnant $\frac{\partial\phi}{\partial v}$ au moyen de $\frac{\partial\psi}{\partial v}$, $\frac{\partial^2\psi}{\partial v^2}$, puis une relation en u, v, $\frac{\partial\psi}{\partial v}$, $\frac{\partial^2\psi}{\partial v^2}$, $\frac{\partial^3\psi}{\partial v^3}$, c'est-à-dire une expression explicite de I_1. On aurait de même $\frac{\partial^2\phi}{\partial v^2} : \frac{\partial\phi}{\partial v} = J(u, v)$, explicitement en u, v.

On retrouve aisément aussi les propositions classiques en Géométrie : s'il s'agit de mettre un ds^2 sous une forme connue

$$ds^2 = 2\Omega(\phi, \psi)\, d\phi d\psi,$$

on a, avec les équations différentielles des courbes $\phi = \text{const.}$, $\psi = \text{const.}$, la condition

$$\Delta(\phi, \psi) = \frac{1}{\Omega(\phi, \psi)},$$

d'où l'on déduira explicitement ϕ et ψ, sauf 1° si Ω a la forme $F(\phi - \psi)$ (F quelconque, surfaces de révolution) auquel cas on ne connaît que $\frac{\partial\phi}{\partial v}$ et $\frac{\partial\psi}{\partial v}$, 2° si Ω a la forme $\frac{1}{(\phi-\psi)^2}$ auquel cas on n'a que les invariants du groupe qui conserve $\frac{d\phi d\psi}{(\phi-\psi)^2}$, groupe étudié *dans ma communication* en Analyse : ϕ et ψ sont définis à une même transformation projective près.

Des remarques analogues peuvent se faire lorsque Ω n'est pas entièrement connu, mais satisfait à certaines équations aux dérivées partielles. Elles redonneraient, par exemple, la théorie de la réduction d'un ds^2 à la forme de Liouville, avec la discussion correspondante.

Équations du second ordre.

5. La méthode que nous venons d'exposer pour une ou plusieurs équations du premier ordre, s'applique quels que soient l'ordre et le nombre des équations différentielles ordinaires que l'on considère. En Géométrie, le cas des équations du second ordre (ou des équations linéaires aux dérivées partielles à trois variables x, y, z) est particulièrement important. Le groupe de transformations qui intervient est le groupe général à deux variables et le nombre des types de réductions possibles est voisin de soixante.

Si l'on définit la congruence des courbes Γ

$$\phi(x, y, z) = \text{const.}, \quad \psi(x, y, z) = \text{const.},$$

par les équations :

$$A(\phi) = A(\psi) = 0 \quad \ldots\ldots\ldots\ldots (1),$$

où

$$A(f) = X\frac{\partial f}{\partial x} + Y\frac{\partial f}{\partial y} + Z\frac{\partial f}{\partial z},$$

on saura *par une méthode régulière*, tirer parti, pour l'intégration des équations (1), de la connaissance d'une propriété géométrique quelconque des courbes Γ et l'on obtiendra toujours, en définitive, l'expression explicite, au moyen de x, y, z, des invariants différentiels en ϕ, ψ, $\frac{\partial\phi}{\partial y}$, $\frac{\partial\phi}{\partial z}$, ..., $\frac{\partial\psi}{\partial y}$, $\frac{\partial\psi}{\partial z}$, ... de l'un des *types* de groupes de transformations en ϕ, ψ.

Il en sera de même si l'on connaît pour un système de deux congruences de courbes $\phi = a$, $\psi = b$; $\phi_1 = a_1$, $\psi_1 = b_1$ une propriété géométrique et si l'on cherche à en tirer parti pour la détermination de l'une ou de l'autre congruence à partir de ses équations différentielles. Mais le nombre des types de relations possibles, entre ϕ, ψ et leurs dérivées d'une part, ϕ_1, ψ_1 et leurs dérivées d'autre part, types qu'on peut établir *a priori*, est extrêmement élevé.

Je me bornerai à indiquer des exemples tout à fait simples.

(*a*) Supposons que les courbes $\phi = \text{const.}$, $\psi = \text{const.}$, définies par le système

$$\frac{dx}{1} = \frac{dy}{m} = \frac{dz}{n} \qquad\qquad (1)$$

$\left(\text{ou les équations } A(\phi) = 0,\ A(\psi) = 0 \text{ avec } A(f) = \frac{\partial f}{\partial x} + m\frac{\partial f}{\partial y} + n\frac{\partial f}{\partial z}\right)$, soient *planes*.

On pourra poser: $z = \alpha x + \beta y + \gamma$, α, β, γ étant des fonctions à déterminer de ϕ et ψ.

Si deux de ces fonctions, α et β par exemple, sont *distinctes* en ϕ et ψ, on pourra prendre $\phi = \alpha$, $\psi = \beta$ et l'on aura en appliquant l'opération $A(f)$:

$$n = \alpha + \beta m,$$

$$A(n) = \quad \beta A(m),$$

qui donnent explicitement α et β en x, y, z. La condition qui exprime que les courbes sont planes est:

$$\frac{A_2(n)}{A(n)} = \frac{A_2(m)}{A(m)}.$$

Si α, β, γ sont fonctions d'un seul argument, et si α n'est pas constant, on pourra prendre $\phi = \alpha$; α et β seront donnés comme plus haut, mais β sera une fonction de α, d'ailleurs connue, et il en sera de même de γ. On connaîtra donc explicitement ϕ; la détermination de ψ dépendra d'une équation à deux variables, renfermant le paramètre ϕ.

En résumé si les plans qui renferment les courbes de la congruence dépendent de deux paramètres, chaque plan ne contient qu'un nombre limité de courbes qu'on obtient sans intégration; si ces plans ne dépendent que d'un paramètre, on a toujours ces plans sans intégration, mais les courbes qui sont dans chacun d'eux sont données par une équation différentielle du premier ordre qui peut être quelconque. Cela s'étend au cas où les courbes de la congruence sont sur des sphères, des surfaces algébriques de degré connu, ou même des surfaces transcendantes dont l'équation $\Omega(x, y, z) = 0$ a une forme connue en x, y, z.

(*b*) Supposons qu'il s'agisse de déterminer les trajectoires orthogonales d'une famille de surfaces

$$f(x, y, z) = \alpha,$$

qui soit une *famille de Lamé*. En posant:

$$\Delta(f, g) = \frac{\partial f}{\partial x}\frac{\partial g}{\partial x} + \frac{\partial f}{\partial y}\frac{\partial g}{\partial y} + \frac{\partial f}{\partial z}\frac{\partial g}{\partial z},$$

on a pour les inconnues ϕ, ψ les trois conditions:

$$\Delta(f, \phi) = 0, \quad \Delta(f, \psi) = 0, \quad \Delta(\phi, \psi) = 0.$$

La dernière, transformée par la méthode générale, s'écrit:

$$\frac{\partial\Phi}{\partial\phi}\frac{\partial\Psi}{\partial\phi}\Delta\phi + \left(\frac{\partial\Phi}{\partial\phi}\frac{\partial\Psi}{\partial\psi} + \frac{\partial\Phi}{\partial\psi}\frac{\partial\Psi}{\partial\phi}\right)\Delta(\phi, \psi) + \frac{\partial\Phi}{\partial\psi}\frac{\partial\Psi}{\partial\psi}\Delta\psi = 0;$$

elle ne garde sa forme, *si l'on ne peut avoir* $\Delta\phi = \Delta\psi = 0$, qu'en prenant

$$\frac{\partial \Phi}{\partial \psi} = 0, \quad \frac{\partial \Psi}{\partial \phi} = 0$$

(ou permutant ϕ et ψ). Les surfaces $\phi = \text{const.}$, $\psi = \text{const.}$ se déterminent donc séparément par des systèmes complets à une solution.

On sait bien en effet que ces surfaces définissent les deux systèmes de lignes de courbure des surfaces $f(x, y, z) = \alpha$.

Si l'on peut avoir $\Delta\phi = 0$, $\Delta\psi = 0$, les trajectoires sont intersections de développables isotropes: ceci ne se présente que si $f(x, y, z) = \alpha$ définit des plans ou des sphères (lignes de courbure indéterminées) et l'on retrouve alors les résultats classiques.

Remarque. Ces résultats, relatifs aux trajectoires orthogonales d'une famille de plans ou de sphères (et les résultats analogues relatifs aux trajectoires orthogonales d'une famille de droites ou de cercles dans le plan ou sur une surface connue), peuvent aussi s'obtenir par l'application directe de la théorie de la rationalité.

Si les équations différentielles à intégrer

$$\frac{dx}{1} = \frac{dy}{m} = \frac{dz}{n} \quad \text{.................................(1)},$$

sont données en coordonnées cartésiennes (x, y, z), on peut former explicitement au moyen de x, y, z, m, n et leurs dérivées les expressions des nouvelles variables qui ramènent le système (1) à sa forme la plus simple; on peut aussi donner explicitement au moyen des mêmes éléments les expressions des invariants du groupe Γ:

$$\Phi = f(\phi, \psi), \quad \Psi = g(\phi, \psi),$$

qui caractérise la réduction pour l'équation

$$A(F) = \frac{\partial F}{\partial x} + m\frac{\partial F}{\partial y} + n\frac{\partial F}{\partial z} = 0$$

dont ϕ et ψ sont deux solutions.

Par exemple, supposons que l'équation

$$\frac{dy}{dx} = m(x, y)$$

soit l'équation d'une famille de cercles ayant leurs centres sur l'axe ox et proposons nous de trouver leurs trajectoires orthogonales.

Si l'on pose: $$A(\phi) = \frac{\partial \phi}{\partial x} + m\frac{\partial \phi}{\partial y} = 0,$$

en écrivant l'équation des cercles:

$$x^2 + y^2 - 2\alpha x + \beta = 0,$$

on trouve $$x + my - \alpha = 0,$$

pour déterminer α et $\beta=\omega(\alpha)$, c'est-à-dire l'équation explicite des cercles. La condition que doit vérifier m [qui serait défini implicitement par les deux équations précédentes en y posant $\beta=\omega(\alpha)$, ω arbitraire] est simplement :

$$1+m^2+y\left(\frac{\partial m}{\partial x}+m\,\frac{\partial m}{\partial y}\right)=0.$$

L'équation qui définit les trajectoires orthogonales est :

$$B(\psi)=\frac{\partial\psi}{\partial x}-\frac{1}{m}\,\frac{\partial\psi}{\partial y}=0\,;$$

on en connaît *un multiplicateur* qui donne ψ par la quadrature :

$$d\psi=\frac{dx+m\,dy}{y\sqrt{1+m^2}}.$$

COMMUNICATION

SECTIONS III (*a*) AND III (*b*). JOINT MEETING

(ASTRONOMY AND STATISTICS)

ON THE LAW OF DISTRIBUTION OF ERRORS

By R. A. Sampson.

I wish to draw attention to certain points in the accepted Theory of Errors, in particular to the foundation of the law by which errors are distributed according to their magnitude. This is given by Gauss's Error Function $y = h\pi^{-\frac{1}{2}} \exp(-h^2x^2)$ and its study is a branch of the theory of probabilities, but following Chrystal's example (*Algebra*, Ch. XXXVIII.) I shall discard as far as possible the language of probabilities and speak of the conclusions as relating to averages only.

If there is but one quantity to determine and that is directly observed by observations of equal weight we accept as the value of the quantity the Arithmetic Mean of the observations, almost of necessity; but no rationally convincing reason has been given for doing so except in the case of no more than two observations, when the Arithmetic Mean does not favour one more than the other, either in method or result. The same cannot be said when the observations number three or more, for then the result favours the middle ones, and therefore the Arithmetic Mean is usually presented either as an axiom or as a practical necessity. But it does not end the difficulty to accept it, because we have to supersede it by a more general rule when the observations give the unknown affected by a variable coefficient, as $a_ix + n_i = 0$, $[i = 1, 2, \ldots s]$, where a_i varies from one observation to another. We cannot deduce any solution of these from the Arithmetic Mean, and therefore we take a rule which includes it as a particular case. Thus if we take a solution x, which, substituted in the members on the left hand of each observation-equation gives residuals Δ_i or $\nu_i \equiv a_ix + n_i$, we may adopt as determining x the rule

$$\Delta_1^2 + \Delta_2^2 + \ldots + \Delta_s^2 = minimum,$$

which furnishes us with a single equation for x, viz. $[aa]\,x + [an] = 0$, where as usual $[aa]$ stands for Σa_ia_i, $[i = 1, 2, \ldots s]$. The Arithmetic Mean can be deduced from this rule, known as the rule of Least Squares, by taking the case where

$$a_1 = a_2 = \ldots = a_s,$$

but the converse is not true, unless we supplement the Arithmetic Mean by postulating the existence of a law of distribution of errors by which the proportionate frequency of occurrence of any error between the magnitudes Δ and $\Delta + d\Delta$ may be denoted by $\phi(\Delta)\,d\Delta$.

Gauss who introduced this idea showed—provided at any rate $\phi(\Delta)$ may be differentiated—that if the rule of the Arithmetic Mean is not in any case to be contradicted we must have $\phi(\Delta)$ varying as $\exp(-h^2\Delta^2)$, or introducing a factor so as to make

$$\int_{-\infty}^{+\infty} \phi(\Delta)\, d\Delta = 1,$$

or $\phi(\Delta)\, d\Delta$ a measure of the proportionate occurrence of the individual error out of the whole,

$$\phi(\Delta) = h\pi^{-\frac{1}{2}} \exp(-h^2\Delta^2)$$

(*Theoria Motus*, § 177); and then we find that out of all possible values for the unknown x, leaving residuals $\Delta_1, \Delta_2, \ldots \Delta_s$ in the observations, that one is of most frequent occurrence, as measured by $\phi(\Delta_1)\phi(\Delta_2)\ldots\phi(\Delta_s)$ which makes $[\Delta\Delta]$ a minimum, and this is the rule of Least Squares.

It should not escape remark that this robs the simple Arithmetic Mean of its axiomatic character, since if an axiom requires to be generalised, it is the generalised form that is axiomatic; besides, neither the rule of Least Squares nor the existence of a differentiable function $\phi(\Delta)$ measuring the frequency of occurrence of each error Δ can be called an axiom. Indeed the Error Function introduces an idea which is entirely novel. In place of treating discrepancies between observation and adopted value as functions of the time or other circumstances of their origin, these features are ignored and they are rearranged merely in order of their magnitude; and the very sweeping conclusion is accepted that for every kind of observation and for every source of error where the word "error" is not even defined except as a residual, the errors arranged in a distribution in order of magnitude follow a law which is identically the same, depending only upon a single parameter controlling the scale. And I would here remark that discussions such as those with which Poincaré deals in his *Calcul des Probabilités* in which the law is made a function of the quantity measured as well as of the error, are not sufficient, since this quantity is only one of the circumstances which is left out of view. For example in the readings of a divided circle, the error may indeed be a function of the magnitude of the angle measured, but it may also be a function of the temperature, the hour of day, the season of the year, the attention of the observer and so forth, and all of these are equally ignored.

The first who put this striking theory to a practical test was Bessel. In the *Fundamenta Astronomiae* (1818), p. 20, he makes a comparison of the actual and the calculated number of errors within stated limits for 470 determinations of the right ascensions of *Sirius* and *Altair* by Bradley and gives the following table:

Between	Number of Errors by Theory	Number of Errors in fact
$0^s\cdot0$ and $0^s\cdot1$	95	94
·1 „ ·2	89	88
·2 „ ·3	78	78
·3 „ ·4	64	58
·4 „ ·5	50	51
·5 „ ·6	36	36
·6 „ ·7	24	26
·7 „ ·8	15	14
·8 „ ·9	9	10
·9 „ 1·0	5	7
over 1·0	5	8

The comparison has been made in numberless instances since and always shows an agreement of this kind, namely, one that is almost exact but shows a slight systematic preponderance of fact over theory in the occurrence of the largest errors, a feature which Bessel conjectures may be due to such causes as a slip of the pen or a disturbance of the instrument, or the like. The agreement is seldom so perfect as in these observations of Bradley's, but it is generally so good that Bauschinger in the *Encyklopädie der Math. Wiss.* (I. 2, p. 774) gives it in his opinion as the best demonstration of the law. But this can hardly be accepted as final. In the first place it is no explanation of the origin of the law to verify its truth, and besides though an error, in its statement, is treated merely as a residual, the law is not verified at all closely in the case of all residuals. The following will serve as an example. In the *Greenwich Observations*, 1909, there are 504 observations of the nadir point of the Transit Circle made by reflection of the wire in a mercury trough (besides stellar observations). If the residuals of these observations from their mean are taken and analysed according to their distribution in magnitude we get the numbers given below, while the error curve derived for comparison in the usual way indicates the numbers placed alongside them.

Greenwich Nadir Observations.

Limits of Error	Number of Observations	
	Actual	Calculated
0″·0 to 0″·2	79	95
0″·2 „ 0″·4	96	90
0″·4 „ 0″·6	75	80
0″·6 „ 0″·8	86	68
0″·8 „ 1″·0	55	54
1″·0 „ 1″·2	48	41
1″·2 „ 1″·4	20	29
1″·4 „ 1″·6	22	20
1″·6 „ 1″·8	16	12
1″·8 „ 2″·0	3	7
2″·0 „ 2″·2	3	4
2″·2 „ 2″·4	0	2
2″·4 „ 2″·6	1	1
2″·6 „ 2″·8	0	1
2″·8 „ 3″·0	0	0

The result is shown in Fig. 2 and may be compared with those from Bradley's transits in Fig. 1. It will be seen that the nadirs oscillate sharply to and fro across their error curve. The reason doubtless is that there is a strongly marked seasonal term in the position of the nadir, which is brought out when the observations are meaned by months, thus, recording only seconds.

January ...	55″·14	July	53″·68
February ...	5″·28	August ...	3″·18
March ...	5″·29	September ...	3″·77
April ...	4″·78	October ...	3″·86
May ...	4″·61	November ...	4″·46
June	4″·45	December ...	4″·56

I shall return to this question of oscillation, and now merely remark that its presence makes it an important question to define clearly what class of residuals

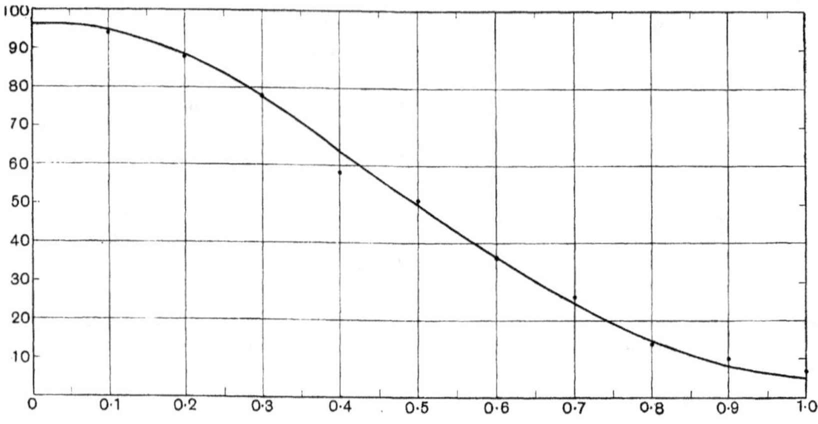

Fig. 1. Bradley's Transits and the Error Curve.

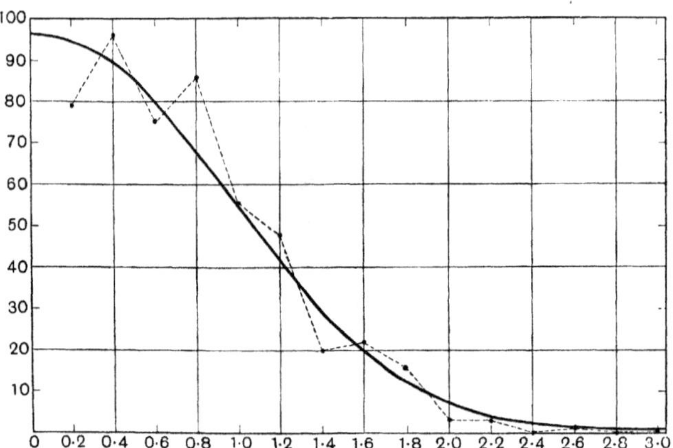

Fig. 2. Greenwich Nadirs, not cleared of Annual Fluctuation.

it is that really satisfies Gauss's law. Gauss does not occupy himself with this question. He simply introduces his function with the words "probabilitas...cuilibet errori Δ tribuenda exprimetur per functionem ipsius Δ quam per $\phi\Delta$ denotabimus," and argues certain properties of it, as that it may be discontinuous,—though he afterwards differentiates it. Laplace first offered a demonstration that the error function possessed a definite form and might be derived from the combination of an unlimited number of small errors individually following any arbitrary laws of distribution. A revised form of this is given by Poisson (*Connaissance des Temps*,

1827, p. 273) and this will be examined with interest because the theorem is remarkable, not to say improbable, and Poisson's proof has passed the scrutiny of such careful judges as Todhunter and Glaisher. The crucial point is the following. Taking the problem "If E is the sum of s errors, $\epsilon, \epsilon_1, \ldots \epsilon_{s-1}$ each multiplied by a coefficient, so that

$$E = \gamma\epsilon + \gamma_1\epsilon_1 + \ldots + \gamma_{s-1}\epsilon_{s-1},$$

and the law of frequency of occurrence of any value for the error ϵ_i is $f_i(\epsilon_i)$, to find the probability that E shall lie between $b - c$ and $b + c$, where the limits for the maximum error are $-a$ and $+a$," Poisson obtains the expression for the probability

$$p = \frac{2}{\pi}\int_0^\infty \rho\rho_1\rho_2 \ldots \rho_{s-1} \cos(\phi + \phi_1 + \ldots + \phi_{s-1} - b\alpha) \sin c\alpha \frac{d\alpha}{\alpha},$$

where

$$\rho_i \cos\phi_i = \int_{-a}^{+a} f_i(x) \cos\gamma_i \alpha x dx, \quad \rho_i \sin\phi_i = \int_{-a}^{+a} f_i(x) \sin\gamma_i \alpha x dx,$$

and it remains to be proved that p tends to a limit for $n \to \infty$, which is independent of the forms of all the functions f and of which the variable part is $\exp(-h^2x^2)$.

Poisson first proves that every ρ is less than unity by the following ingenious method:

$$\rho^2 = \left\{\int_{-a}^{+a} f(x) \cos\gamma\alpha x dx\right\}^2 + \left\{\int_{-a}^{+a} f(x) \sin\gamma\alpha x dx\right\}^2$$

$$= \int_{-a}^{+a} f(x) \cos\gamma\alpha x dx \int_{-a}^{+a} f(x') \cos\gamma\alpha x' dx' + \int_{-a}^{+a} f(x) \sin\gamma\alpha x dx \int_{-a}^{+a} f(x') \sin\gamma\alpha x' dx'$$

$$= \int_{-a}^{+a}\int_{-a}^{+a} f(x) f(x') [\cos\gamma\alpha x \cos\gamma\alpha x' + \sin\gamma\alpha x \sin\gamma\alpha x'] dx dx'$$

$$= \int_{-a}^{+a}\int_{-a}^{+a} f(x) f(x') \cos\gamma\alpha(x - x') dx dx'$$

$$< \int_{-a}^{+a}\int_{-a}^{+a} f(x) f(x') dx dx',$$

i.e. $< \left\{\int_{-a}^{+a} f(x) dx\right\}^2$, or < 1.

Further, if ρ be developed in ascending powers of α, so that $\rho = 1 - \gamma^2h^2\alpha^2 \ldots$ in order that the product $R = \rho\rho_1 \ldots \rho_{n-1}$ should converge to a limit other than zero it is necessary either that $\gamma^2h^2, \gamma_1^2h_1^2, \ldots$ should decrease regularly in a certain way, or else α^2 must be indefinitely small. Poisson gives an example of the former but treats it as exceptional; confining attention to the latter and keeping only the first two terms of each ρ, we have

$$\sum_i \log\rho \equiv \log R = -\alpha^2\Sigma\gamma_i^2h_i^2 - \tfrac{1}{2}\alpha^4\Sigma\gamma_i^4h_i^4 - \tfrac{1}{3}\alpha^6\Sigma\gamma_i^6h_i^6 \ldots,$$

and then if $\alpha \to 0$ but so that $\alpha^2\Sigma\gamma_i^2h_i^2$ is finite, it appears that $\alpha^4\Sigma\gamma_i^4h_i^4, \ldots$ all $\to 0$ and we have the form $R = \exp(-\alpha^2\Sigma\gamma_i^2h_i^2)$, from which the required result is shown to follow. But it will be remarked that this step draws a conclusion as to the coefficient of α^4 in the development of R, while the terms in α^4 in the separate factors ρ have been omitted. The conclusion is therefore unwarranted and there is no proof at all that peculiarities of the functions f efface themselves in the final result. I do not believe that the theorem is true. If the arbitrary distributions $f_i(x)$ have any zeros—and this is not excluded by the process of demonstration—I do not see how they can fail to reappear as zeros in the product R.

Poincaré remarks (*Calcul des Probabilités*, p. 149), "On a fait une hypothèse..., et cette hypothèse a été appelée loi des erreurs. Elle ne s'obtient pas par des déductions rigoureuses; plus d'une démonstration qu'on a voulu en donner est grossière, entre autres celle qui s'appuie sur l'affirmation que la probabilité des écarts est proportionnelle aux écarts*. Tout le monde y croit cependant, me disait un jour M. Lippmann, car les expérimentateurs s'imaginent que c'est un théorème de mathématiques, et les mathématiciens que c'est un fait expérimental." A large part of Poincaré's book is occupied with questions of the justification of this law, but I shall not refer to it further, for while it is very interesting it would lose by summarising and it does not touch upon the point of view which I give below.

I think it will be generally felt that the doctrine of the Error Function is so striking, so important, and so well supported by experience that it ought either to be proved or else reduced to a clear axiomatic position. I would therefore offer a few remarks towards its elucidation. These make no pretension to the generality aimed at by the earlier writers, but for that reason they are easier to bring under critical review.

The term "accidental error" has come to carry with it an undefined suggestion of a peculiar quality, but there seems no reason to treat an error otherwise than as a disregarded unknown, or more commonly, as a host of small disregarded unknowns. Let us include these unknowns as y, z, ... in the equations representing the observations so that we have

$$a_i x + b_i y + c_i z + \ldots + n_i = 0 \quad [i = 1, 2, \ldots s],$$

equations which are satisfied exactly. If we multiply by $a_i \ldots$ and add, the former solution $[a_i a_i]\,x + [a_i n_i] = 0$ gives the value of x correctly provided only

$$[a_i b_i] = [a_i c_i] = \ldots = 0,$$

but substituted in equations in the truncated form $a_i x + n_i$ it leaves residuals Δ_i or $v_i = -b_i y - c_i z - \ldots$. These are the quantities under discussion. We disregard any individuality they may have and schedule them according to their values and the number of times each value occurs.

It should be remarked that not until we reach very complicated cases will the graph of this schedule or distribution consist of anything but a few straight lines parallel to the axis of x. Thus for $y = \sin t$ and many others the distribution is represented by $\eta = 1$ for $-1 < \xi < 1$, and $\eta = 0$ beyond these limits (Fig. 3). Fig. 4 shows $y = \sin t + \sin 2t$ and its distribution. The distributions for $y = \log t$, $y = 1 + t$, $y = 1/(1 + t^3)$, ... are the same, namely a line parallel to the axis of ξ extending from $-\infty$ to $+\infty$. Those for $y = e^t$, $y = 1 + t^2$, $y = 1/(1 + t^2)$, ... consist of a line extending along $O\xi$ from $-\infty$ to 0, and parallel to $O\xi$ from 0 to $+\infty$. It will thus be seen that it is in making the distribution that the features of the functions which compose the

* Query, read *fonction des écarts*.

residual errors become effaced. It seems probable that Laplace and Poisson were on the wrong lines in attributing complete generality to the law of occurrence of their individual errors. In what follows I shall in all cases assume the graph of the

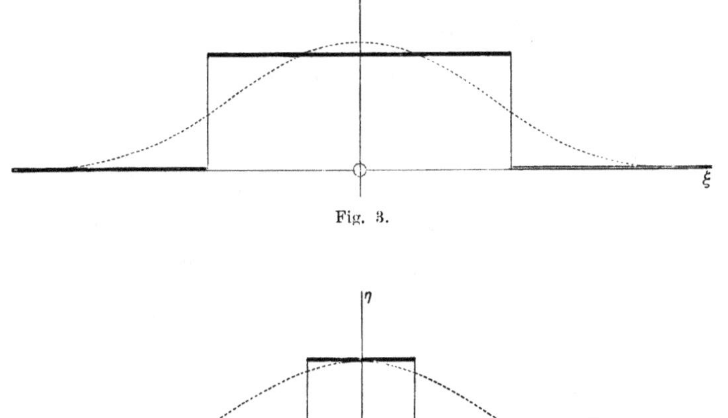

Fig. 3.

Fig. 4.

distribution shows no negative values (which would be meaningless) and converges to zero for $x = -\infty$ or $+\infty$ if not before.

Also, I shall deal only with complete distributions, in which every error is represented by its full proportionate number, giving a graph which is discontinuous only at a finite number of points, and shall postpone any remarks as to the degree to which this departs from actual experience.

I regard the superposition of two errors as the disturbance of one distribution by another, in the following way. The error Δ' is added to the error Δ. The latter occurs a proportionate number of times $\phi(\Delta)\,d\Delta$, and the former $\phi_1(\Delta')\,d\Delta'$, where

$$\int_{-\infty}^{+\infty} \phi(\Delta)\,d\Delta = 1, \qquad \int_{-\infty}^{+\infty} \phi_1(\Delta')\,d\Delta' = 1;$$

hence the combination $\Delta + \Delta'$ will occur the proportionate frequency

$$\phi(\Delta)\,\phi_1(\Delta')\,d\Delta d\Delta';$$

and if $\Theta \equiv \Delta + \Delta'$, the result of the disturbance of the first distribution by the second will be a distribution in which any error Θ occurs the proportionate number of times $d\Theta \int_{-\infty}^{+\infty} \phi(\Theta - \Delta')\,\phi_1(\Delta')\,d\Delta'$. I would now remark that if

$$\phi(\Delta) = h\pi^{-\frac{1}{2}} \exp(-h^2\Delta^2), \quad \phi_1(\Delta') = h'\pi^{-\frac{1}{2}} \exp(-h'^2\Delta'^2),$$

the step described reproduces for the distribution of Θ a function of the same type;

for

$$hh'\pi^{-1}\int_{-\infty}^{+\infty} d\Delta' \exp\{-h^2\Theta^2 + 2h^2\Theta\Delta' - H^2\Delta'^2\} \qquad [H^2 = h^2 + h'^2]$$

$$= hh'\pi^{-1}\exp\{-h^2\Theta^2 + h^4\Theta^2/H^2\}\int_{-\infty}^{+\infty} d\Delta'\{-(H\Delta' - h^2\Theta/H)^2\}$$

$$= hh'/H \,.\, \pi^{-\frac{1}{2}}\exp\{-(hh'\Theta/H)^2\},$$

and note that the "measure of precision," hh'/H follows the rule

$$\frac{H^2}{h^2h'^2} = \frac{1}{h^2} + \frac{1}{h'^2},$$

which is required for agreement with the ordinary theory.

We notice too that the step is commutative in order, that is, the result is the same whether Δ is disturbed by Δ', or Δ' by Δ.

It is clear that this reproduction of form is an important fact*; so far as I can determine it belongs with any generality only to the functions exp (x) and exp (x^2).

There are other cases where there is a partial reproduction of form. Thus if we take

$$\phi(\Delta) = \pi^{-1}/(1+\Delta^2), \quad \phi_1(\Delta') = \pi^{-1}/(1+\Delta'^2),$$

we have

$$\Phi(\Theta) = \pi^{-2}\int_{-\infty}^{+\infty}\frac{d\Delta'}{1+\Delta'^2}\cdot\frac{1}{1+(\Theta-\Delta')^2}$$

$$= \pi^{-2}\int_{-\infty}^{+\infty} d\Delta'\left\{\frac{\frac{1}{2}\Theta+\Delta'}{1+\Delta'^2}\cdot\frac{2}{\Theta(\Theta^2+4)} + \frac{\frac{1}{2}\Theta-(\Delta'-\Theta)}{1+(\Delta'-\Theta)^2}\cdot\frac{2}{\Theta(\Theta^2+4)}\right\}$$

$$= \frac{\pi^{-2}}{\Theta^2+4}\left[\{\tan^{-1}(\Delta'-\Theta)+\tan^{-1}\Delta'\} + \frac{1}{\Theta}\log\frac{1+\Delta'^2}{1+(\Delta'-\Theta)^2}\right]_{-\infty}^{+\infty}$$

$$= \frac{2\pi^{-1}}{\Theta^2+4},$$

which may be considered the same type; but if we take $\phi_1(\Delta') = a\pi^{-1}/(a^2+\Delta'^2)$ we find

$$\Phi(\Theta) = a\pi^{-2}\int_{-\infty}^{+\infty}\frac{d\Delta'}{a^2+\Delta'^2}\cdot\frac{1}{1+(\Theta-\Delta')^2}$$

$$= \frac{2a\pi^{-1}\Theta^2}{\Theta^4+2\Theta^2(a^2+1)+(a^2-1)^2},$$

which is a more complicated form. There is an obstacle to proving any definite result in the difficulty of saying what is to be understood by the same form. I must leave this question aside as it is not material to the present discussion to answer it. It will suffice to know that exp $(-h^2x^2)$, and of course also exp $(-h^2x^2)\cos(kx+\gamma)$ reproduce themselves. The latter form alone does not represent any real distribution because it contains negative values, but it is important when we have to consider the disturbance of one distribution by another, neither of which follows the law exp $(-h^2x^2)$. We may take as typical the form

$$\phi(\Delta) = h\pi^{-\frac{1}{2}}(1+A)^{-1}\exp(-h^2\Delta^2)[1+a\exp ik\Delta],$$

where $\exp ik\Delta$ is introduced in place of $\cos k\Delta$ to make the subsequent work more

* I learn from Father Willaert, S.J. that this is a known result and that he, and before him M. d'Ocagne, had published it. Cf. Willaert, *Ann. Soc. Scientifique de Bruxelles*, t. XXXI, April 1907. d'Ocagne, *Bull. Soc. Math. de France*, t. XXIII, 1895.

compact, and the phase-angle γ is included by reckoning a as complex. We must have $|a|$ in all cases less than 1, and in order to make $\int_{-\infty}^{+\infty} \phi(\Delta)\, d\Delta = 1$, we have

$$1 + A = 1 + a \exp(-k^2/4h^2).$$

The distribution then represents a simple case of fluctuation about the distribution

$$h\pi^{-\frac{1}{2}} \exp(-h^2\Delta^2).$$

Now let this distribution be disturbed by the distribution

$$\phi_1(\Delta') = h'\pi^{-\frac{1}{2}} (1 + A')^{-1} \exp(-h'^2\Delta'^2)\,[1 + a' \exp ik'\Delta'],$$

where $$|a'| < 1, \quad A' = a' \exp(-k'^2/4h'^2).$$

Consider in particular the term factored by aa',

$$\frac{hh'}{\pi(1 + A)(1 + A')} \exp(-h^2\Delta^2 - h'^2\Delta'^2)\,[\ldots aa' \exp i(k\Delta + k'\Delta')].$$

Write $\Delta = \Theta - \Delta'$;

$$\frac{hh'}{\pi(1 + A)(1 + A')} \cdot \exp\{-h^2\Theta^2 + 2h^2\Theta\Delta' - H^2\Delta'^2\}\,[\ldots aa' \exp i\{k\Theta + (k' - k)\Delta'\}]$$

$$= \frac{hh'}{\pi(1 + A)(1 + A')} \exp\left\{-\frac{h^2h'^2}{H^2}\Theta^2\right\} \exp\left\{-\left(H\Delta' - \frac{h^2\Theta}{H}\right)^2\right\}$$

$$\left[\ldots aa' \exp i\left\{\frac{h^2k' + h'^2k}{H^2}\Theta + \frac{k' - k}{H}\left(H\Delta' - \frac{h^2}{H}\Theta\right)\right\}\right].$$

Integrating with respect to Δ' from $-\infty$ to $+\infty$,

$$\frac{hh'/H}{\pi^{\frac{1}{2}}(1 + A)(1+A')} \cdot \exp\left\{-\frac{h^2h'^2}{H^2}\Theta^2\right\} \left[\ldots + aa' \exp\left\{-\frac{(k - k')^2}{4H^2}\right\} \exp\left\{i\,\frac{h^2k' + h'^2k}{H^2}\Theta\right\}\right].$$

Thus the expression for the resulting distribution with respect to Θ is

$$\Phi(\Theta) = \frac{hh'/H}{\pi^{\frac{1}{2}}(1 + A)(1 + A')} \exp\left\{-\frac{h^2h'^2}{H^2}\Theta^2\right\} \left[1 + a \exp\left\{-\frac{k^2}{4H^2}\right\} \exp\left\{i\,\frac{h'^2k}{H^2}\Theta\right\}\right.$$

$$\left. + a' \exp\left\{-\frac{k'^2}{4H^2}\right\} \exp\left\{i\,\frac{h^2k'}{H^2}\Theta\right\} + aa' \exp\left\{-\frac{(k - k')^2}{4H^2}\right\} \exp\left\{i\,\frac{h^2k' + h'^2k}{H^2}\Theta\right\}\right].$$

We notice that it is indifferent which of the two we take as the disturbing distribution.

Consider first the terms in a, a' compared with their primitive values. The effect of the disturbance is to make these fluctuating terms in every case recede in amplitude relatively to the non-fluctuating portion and at the same time to lengthen the span or period of the fluctuation. The more rapid the fluctuation originally, the more rapid is its decay. The fluctuating term will be smoothed out until its amplitude is so small and its period so long that it is indistinguishable from the non-fluctuating term.

We must also consider the term in aa', especially for the case $k = k'$; in this case the amplitude of this term is $|aa'|$ and the argument $k\Theta$ is reproduced, or, if the fluctuations of the two distributions are of the same period, this period is reproduced in their joint effect. The same would be true if subsequent disturbing distributions had this period, so that it would run unchanged through the whole and not become lengthened like the others. But it would remain a solitary term, and since

$|a|, |a'|, \ldots$ are all less than unity its amplitude would become less and less at every step.

Although the fluctuating terms decrease in amplitude they increase in number and it is desirable to examine the convergency of their sum if the steps are indefinitely multiplied.

It is evident that the expression will be always positive, since each step consists in an integration of elements each of which is positive. And the sum

$$\int_{-\infty}^{+\infty} \Phi(\Theta)\, d\Theta = 1,$$

since the terms within [ ] on p. 171 give respectively

$$\pi^{\frac{1}{2}} + a\pi^{\frac{1}{2}} \exp\left(-\frac{k^2}{4H^2}\right) \exp\left\{\frac{h'^2 k}{H^2} \cdot \frac{H}{2hh'}\right\}^2 + a'\pi^{\frac{1}{2}} \exp\left(-\frac{k'^2}{4H^2}\right) \exp\left\{\frac{h^2 k'}{H^2} \cdot \frac{H}{2hh'}\right\}^2$$

$$+ aa'\pi^{\frac{1}{2}} \exp\left\{-\frac{(k-k')^2}{4H^2}\right\} \exp\left\{\frac{h^2 k' + h'^2 k}{H^2} \cdot \frac{H}{2hh'}\right\}^2$$

$$= \pi^{\frac{1}{2}} \left[1 + a \exp\left(-\frac{k^2}{4h^2}\right) + a' \exp\left(-\frac{k'^2}{4h'^2}\right) + aa' \exp\left(-\frac{k^2}{4h^2} - \frac{k'^2}{4h'^2}\right)\right]$$

$$= \pi^{\frac{1}{2}} (1 + A)(1 + A').$$

With regard to individual values, it would seem that infinity is not excluded, since this may occur in the standard curve $y = h\pi^{-\frac{1}{2}} \exp(-h^2 x^2)$ for the special case of $h = \infty$.

It is of interest to examine the case when the disturbance is small but with a strongly marked oscillation; that is, when we have h' large compared to h and a' also large, subject to $|a'| < 1$.

We find the "measure of precision" becomes

$$\frac{hh'}{H} = h - \frac{1}{2}\frac{h^3}{h'^2},$$

and within [ ] the three variable terms are respectively

$$a\left(1 - \frac{k^2}{4h'^2}\right) \exp\left\{ik\left(1 - \frac{h^2}{h'^2}\right)\Theta\right\} + a'\left(1 - \frac{k'^2}{4h'^2}\right) \exp\left\{ik' \frac{h^2}{h'^2}\Theta\right\}$$

$$+ aa'\left(1 - \frac{(k-k')^2}{4h'^2}\right) \exp\left\{i\left[k\left(1 - \frac{h^2}{h'^2}\right) + k'\frac{h^2}{h'^2}\right]\Theta\right\}.$$

The most marked change is in the period of the oscillation factored by a'. The coefficient is not greatly altered, but the lengthening of period will have the effect of smoothing the oscillation away. In place of the oscillation with argument $k\Theta$, we now have two, with coefficients slightly different from and less than a, aa' respectively and arguments each slightly different from $k\Theta$.

The foregoing discussion is confined to the cases where the two distributions which disturb one another contain each only one or at any rate a finite number of fluctuating terms. To meet the cases of some of the distributions shown on p. 169 above, it would be necessary to consider an infinite number of terms, and it is certain that a number of points of difficulty would arise in the discussion, relating generally to questions of convergence. I shall not touch upon these, for as far as they are of interest they would be better dealt with by someone with more special knowledge of them than I; but if it is allowed to carry forward the conclusions of the previous pages to the general case the points are already made that seem necessary for understanding why in practice Gauss's error function always verifies itself. I would summarise these points as follows. We see that the graph of the error function is in its general run by no means remote from the graphs of such distributions as would most obviously occur; further if one of these distributions be treated as an error function qualified by fluctuations, and is then disturbed by the superposition of a second distribution of similar type, the superposition produces a general smoothing in every case, the non-fluctuating portion in the result being perpetuated and the fluctuating terms being relatively effaced both by spreading out their periods and by diminishing their intensities. If we assume that observed errors are due to a large number of small periodic errors this seems a sufficient explanation of the fact that observed errors do always present themselves as nearly pure illustrations of the graph of e^{-x^2} when distributed in order of their magnitude, whether we can or cannot detect any systematic feature in them by arranging them in order of their origin or otherwise. But if any error Δ occurs with a relative frequency

$$h\pi^{-\frac{1}{2}} \exp(-h^2\Delta^2)\, d\Delta,$$

and a number of observations are offered which condition but do not determine a set of errors $\Delta_1, \Delta_2, \dots$ among all possible solutions which may occur, as already remarked, that one will in practice occur most frequently which makes $\Delta_1^2 + \Delta_2^2 + \dots$ a minimum, which is a proof of the rule of Least Squares, which again contains as a particular case the Arithmetic Mean. It thus appears that there is no axiomatic element in the whole theory since the form of the error function follows from the definition of a distribution of superposed errors, and the other rules from the form of the error function. The fact that the Arithmetic Mean seems more primitive and axiomatic than the error function from which I derive it is no objection to this view, since results which are obvious may be deduced as particular cases from all correct theories.

It also appears exactly what is the degree of authority possessed by the rule of Least Squares. It must be regarded as one of an indefinite number of legitimate solutions,—or if we separate the solutions by finite steps,—one of a finite number, all of which will in practice regularly occur. Among these solutions the least square solution is the case that will actually occur more frequently than any other individual case. But there is nothing at all to show that it represents the fact or individual case that we want, nor even that it will occur more frequently than some one out of the *whole* of the discarded possible cases.

In the foregoing discussion the observations have been supposed so numerous as to give a continuous graph for the error function, and the successive superposed distributions of errors so numerous as to give a continuous disturbance of the first, and this to happen repeatedly, and we may ask what room there is for these infinitely numerous disturbances in practice where it always appears that a hundred or two observations will determine the error function quite closely. But there is no objection to regard the actual observations as a mere selection, taken at random or on any system not deliberately unrepresentative, from an infinite sequence which it is open to take, and so as generally yielding results that are representative of the whole.

COMMUNICATIONS

SECTION III (a)

(MECHANICS, PHYSICAL MATHEMATICS, ASTRONOMY)

ON DOUBLE LINES IN PERIODOGRAMS

By H. H. Turner.

We are indebted to Professor Schuster for the suggestion that for investigation of the vibrations or periodicities of a material system we should put aside preconceived ideas as to their probable special values and investigate *all* periods (within certain obvious limits) indifferently. We were not without hints in the past to this effect: thus Euler had specified the period of free oscillation of the earth (considered as a rotating rigid body) as about 10 months: when the coefficients of the terms of this period were found to be insensible it was concluded that there was no sensible free oscillation. Had the period not been assumed, the truth might have been discovered much earlier, viz. that owing to the earth's sensible departure from rigidity the period was much longer than this—about 14 months. But in spite of sporadic hints of this kind, it was reserved for Professor Schuster to formulate and emphasize the general necessity of approaching such problems without prejudices.

In the course of much work with his method of the "periodogram," a conviction has gradually been forced upon the writer that the number of "hidden periodicities" may be greater than has been supposed. Putting aside the problems of gravitational astronomy where the existence of numerous periodicities is already well recognized, it may fairly be said that the expectations of investigators have favoured a small number of real periodicities—such as an annual and perhaps semi-annual period in meteorological elements, an eleven year period in sunspots, and so on; perhaps one or two hitherto unknown outside these, but not many. Here again we have not been without hints to the contrary: when Professor Schuster investigated sunspots he found, not one period, but several, even many. But it may be questioned whether this very fact has not raised its own cloud of mistrust from the existing prejudices in favour of a single periodicity, or at least a strictly limited number.

It is unnecessary to debate at length whether periodicities are in general few or many: for patient investigation may be trusted to decide the matter. But it may be worth while to call attention to the doubt for the reason that the actual course of the investigation, at any rate in its early stages, may be modified by the attitude of the investigator. To shew this let us consider the extreme cases side by side. First suppose we have only a single noteworthy periodicity, such as the annual term in the rainfall. Calculating the coefficients of terms of all assumed periods, they will all be small except those near twelve months. They will not be zero, owing to the existence of accidental errors: and Professor Schuster has pointed out the analogy between the "accidental" part of the periodogram thus formed, and the faint continuous spectrum which in practice always accompanies a bright line spectrum. He has further used its average value as a standard of comparison for the reality of the "bright lines" themselves, tabulating the chances of "reality" when their intensities have assigned ratios to the average intensity of the accidental terms. And he remarks that we must, in determining this average value, exclude portions of the periodogram disturbed by "diffraction effects"; for example, near the annual term the coefficients will not fall abruptly to their accidental value, but will be enhanced by the neighbourhood of the annual term, in a manner strictly analogous to that in which bright lines are diffracted by a spectroscope of low resolving power. In the periodogram the equivalent of the "resolving power" is the total length of the series of observations, in units of the period under investigation: and even in our most extensive series this is in general small, so that the resolving power is low. Hence diffraction effects, i.e. spuriously enhanced coefficients, extend to some distance on each side of a definite periodicity.

Now consider the other extreme case—instead of a single well-marked periodicity suppose we have a periodogram full of periodicities, like a spectrum full of lines. Owing to the low resolving power each line will be accompanied by diffraction wings which may overlap those of the next line. The continuous or accidental spectrum will disappear, being covered up by the diffraction spectrum which will be much more intense: and no bright line will shew against the bright background with anything approaching the expected contrast. The investigator may be tempted to condemn the indications as too feeble to be worth notice, being actually misled by their very number and strength.

To avoid misunderstanding it may be well to repeat that this is only stated as a possibility and not as an ascertained fact. But there are hints that it may be well to keep the possibility in mind: and one practical consequence is that we must devise an alternative, or at any rate a supplement, to Professor Schuster's criterion for comparison with the continuous spectrum. We must be careful that we really get down to this continuous spectrum by removing the diffraction effects, and the quickest way to remove them is to remove their cause, i.e. to subtract any suspected periodicities from the material under investigation.

But the nature of the hints referred to, and of a suggestion for an improved criterion, will be best gathered from a couple of examples from recent experience.

(1) *Azimuths of Greenwich Transit Circle, Periods near* 14 *months.*

The Greenwich Transit Circle has been in position since 1851. The azimuth error shews slight variations, partly secular, partly annual, partly hitherto unexplained. Attention was recently called to the latter in connection with seismological questions; and for their more convenient investigation the secular and annual terms were first removed. The residual monthly means were then collected in periods of 14 months, and five consecutive periods grouped together to form 10 groups. The 14 means of each group were analyzed harmonically to obtain the constants of a term of the form

$$A \cos (\theta - \alpha),$$

A being expressed in units of $0''\cdot01$, and the unit for θ being such that $360°$ corresponds exactly to 14 months. The quantities found by Fourier analysis are of course

$$s = A \sin \alpha \text{ and } c = A \cos \alpha.$$

Now from the values of α given for each group as below, it is seen that α is certainly not constant, and hence there is no sensible term of exactly 14 months. But the values of α increase steadily by an average amount not far from 90°, as exemplified in the column β, deduced empirically. Had the differences $\alpha - \beta$ been irregular we might have set them down to accidental causes. But there is an

Table I. Greenwich Azimuths. 14 *months.*

Group	α	β	$\alpha - \beta$	A	$A \cos(\alpha - \beta)$	$A \sin(\alpha - \beta)$	$s - A \sin \beta_1$	$c - A \cos \beta_1$	α_2
I	154°	104°	+50°	33	+21	+25	− 2	−19	186°
II	231	194	+37	65	+52	+39	−39	−25	238
III	312	284	+28	23	+20	+11	− 1	+ 4	284
IV	348	14	−26	17	+15	− 8	−14	0	270
V	60	104	−44	32	+23	−22	+11	+27	22
VI	107	194	−87	21	+ 1	−21	+31	+13	68
VII	200	284	−84	12	+ 1	−12	+12	−22	152
VIII	63	14	+49	31	+20	+23	+16	− 2	107
IX	147	104	+43	19	+14	+13	− 6	− 5	230
X	223	194	+29	36	+31	+17	−13	−10	233
Mean					+20	+ 7	±15	±13	

obvious run about them to which we shall presently return. Meanwhile the consistently positive value of $A \cos(\alpha - \beta)$ shews that we have a periodicity of coefficient approximately 20 (i.e. $0''\cdot20$) and period such that the phase increases 90° in five periods of 14 months, i.e. the period is

$$14\left(1 + \frac{90°}{5 \times 360°}\right) = 14\cdot70 \text{ months} = 447\cdot1 \text{ days}.$$

The column $A \sin(\alpha - \beta)$ shews us whether we have hit off, by our preliminary discussion of α alone apart from A,

(*a*) Exactly the right *slope*: in the present case the earlier numbers are larger than the later. The value $\beta_1 = 89°$ would have been better than $\beta = 90°$. Trial and error is probably the quickest way of finding out the best value of β when both A and α are taken into account. For the present we shall accept the value 90° as good enough.

(*b*) Exactly the right *epoch*. We have adopted an epoch to make the mean value of $\alpha - \beta$ zero; but when we introduce the coefficients A we upset the compensation. Trial and error will often be the quickest way to put this right: in the present case it is found that we may profitably increase the values of β by 21°: and we will denote the corrected value of β by β_1.

We can now compare this calculated periodicity with the observations. The theoretical values of s and c are $A_1 \sin \beta_1$ and $A_1 \cos \beta_1$: and subtracting them from the observed values we get the 8th and 9th columns of Table I.

Now these columns may be utilized in two ways:—

Firstly, if there is nothing systematic about the residuals, they will give an indication of the accidental error. The mean numerical values are ± 15 and ± 13. The probable error of the mean of 10 may thus be put at ± 5: and the coefficient $+20$, being four times its probable error, has thus a strong claim to consideration. It is suggested that in this way these residuals provide the supplementary test of the reality of periodicities, the need for which was indicated above.

But secondly we may, before regarding them as purely accidental errors, discuss the columns for the existence of a possible neighbouring periodicity. In the present case this has already been suggested by the fourth column of Table I. We may proceed to investigate it by forming α_2 from the relation

$$\tan \alpha_2 = (s - A \sin \beta_1)/(c - A \cos \beta_1),$$

and from the values of α_2 given in the 10th column we deduce by a diagrammatic or other empirical method the values β_2 increasing uniformly by 47° as in Table II.

Table II. Greenwich Azimuths. 14 *months.*

Group	α_2	β_2	$\alpha_2 - \beta_2$	B	$B \cos(\alpha_2 - \beta_2)$	$B \sin(\alpha_2 - \beta_2)$	$s_2 - A_2 \sin \beta_2$	$c_2 - A_2 \cos \beta_2$	α_3	β_3
I	186	187	− 1	19	+19	− 0	0	+ 1	—	50°
II	238	234	+ 4	57	+57	+ 4	−23	−13	241	214
III	284	281	+ 3	4	+ 4	+ 0	+18	0	90	18
IV	270	328	−58	14	+ 7	−11	− 4	−17	193	182
V	22	15	+ 7	29	+29	+ 4	+ 6	+ 8	37	346
VI	68	62	+ 6	34	+34	+ 3	+14	+ 4	75	150
VII	152	109	+43	25	+18	+17	− 7	−15	192	314
VIII	107	156	−49	16	+10	−12	+ 8	+16	27	118
IX	230	203	+27	8	+ 7	+ 4	+ 2	+13	9	282
X	233	250	−17	17	+16	− 5	+ 6	− 3	116	86
Mean					+20		± 9	± 9		

It seems to me that the mere inspection of the differences $\alpha_2 - \beta_2$ is convincing that we are in the presence of a second periodicity. But we can proceed to test it just as before. Forming

$$B^2 = (s - A \sin \beta_1)^2 + (c - A \cos \beta_1)^2,$$

the values of B are given in the 5th column of Table II: and resolving the coefficient in the observed direction, we find the mean value (say A_2) of $B \cos(\alpha_2 - \beta_2)$ to be $A_2 = +20$ (i.e. $+0''\cdot20$): so that this line is equally intense with the former. Forming $B \sin(\alpha_2 - \beta_2)$ there is no suggestion of sensible error in slope or epoch. Forming $s_2 - A \sin \beta_2$ and $c_2 - A_1 \cos \beta_2$ we see that the mean numerical error has been reduced from ± 14 to ± 9; so that the probable error of a mean of 10 is now ± 3: and the coefficients are thus nearly seven times their probable errors. The residuals still shew traces of progression as exemplified by the columns α_3 and β_3 where β_3 steadily increases by 164°. But this corresponds to a period of

$$14\left(1 + \frac{164}{5 \times 360}\right) = 15\tfrac{1}{4} \text{ months approx.}$$

and is better dealt with in connection with means for 15 months. We have seen enough to realize the complex nature of the problem, and the manner in which it may be solved in steps.

Without repeating the detailed process we may notice a second example of an even more striking character. The azimuths of the Cape Transit Circle from 1856—1904 have been discussed in a manner similar to that above described for Greenwich, removing first the secular and annual terms and then dealing with the residuals. When these are grouped in periods of 12 months, so far from the removal of the annual term leaving residuals of an accidental nature, they shew a well-marked progression. On analysis this is found to be due to the presence of at least three periodicities near 12 months: and the Greenwich observations confirm them from quite independent material. The details will be gathered from Table III.

Table III. Azimuthal Terms near One Year
(after removing mean annual periodicity).

Period in Months	Coefficient		Phase		
	Greenwich	Cape	Gr.	Cape	G − C
11·76	0″·20	2″·6	294°	106°	188°
{11·60	0 ·19	1 ·0	315	145	170
{12·40	0 ·16	1 ·0	215	48	167

The 11·76 months period occurs in the Greenwich rainfall (over 80 years' observations) with a coefficient one quarter of the annual. The other pair of terms may be combined to give a variation in the annual term running through its cycle in 30 years; and this corresponds to what Mr Chandler has noticed in the variation of latitude. But the discussion of the origin of these terms cannot be undertaken here. The points to which the attention of mathematicians may appropriately be drawn are

Firstly, the evidence for groups of periodicities in certain cases: and their possible existence in others;

Secondly, the method of dealing with them above outlined.

RELATIONS AMONG FAMILIES OF PERIODIC ORBITS IN THE RESTRICTED PROBLEMS OF THREE BODIES

BY F. R. MOULTON.

The restricted problem of three bodies is that in which an infinitesimal body moves subject to the attractions of two finite masses revolving in circular orbits. In the present discussion only those orbits will be considered which lie in the plane of motion of the finite masses. The masses of the finite bodies will be denoted by $1-\mu$ and μ, and in the physical problem μ may have any value from zero to unity.

The differential equations which the motion of the infinitesimal body satisfies may be written in the form

$$\left.\begin{aligned} x'' &= F_1(\mu;\ x, x', y, y') \\ y'' &= F_2(\mu;\ x, x', y, y') \end{aligned}\right\} \qquad \text{.........................(1)},$$

where the accents denote derivatives. These equations admit the integral

$$x'^2 + y'^2 = F(\mu, x, y) - C \qquad \text{.........................(2)}.$$

Only orbits for which C is finite will be considered.

The functions F_1 and F_2 are regular except for such values of x and y that the infinitesimal body collides with one of the finite bodies, or when at least one of the equations $x' = \infty$, $y' = \infty$, $x = \infty$, or $y = \infty$ is satisfied. It follows from the integral (2) that $x' = \infty$ or $y' = \infty$ only in case of a collision or when at least one of the equations $x = \infty$, $y = \infty$ is satisfied.

The general solutions of equations (1) involve four arbitrary constants, one of which is additive to t because F_1 and F_2 do not involve t explicitly. One of the constants may be identified with C of the integral (2). The general solutions of (1) may be written in the form

$$\left.\begin{aligned} x &= f_1(t-t_0;\ \mu;\ c_1, c_2, C) \\ y &= f_2(t-t_0;\ \mu;\ c_1, c_2, C) \end{aligned}\right\} \qquad \text{.......................(3)},$$

where the constants of integration are t_0, c_1, c_2, and C.

When the initial conditions are such that the right members of equations (1) are regular, f_1 and f_2 are regular analytic functions of $t-t_0$, μ, c_1, c_2, and C in the neighbourhood of their initial values. When the initial conditions belong to a collision, x and y are regular functions of $(t-t_0)^{\frac{1}{3}}$, μ, c_1, c_2, and C in the neighbourhood of their initial values. There is an exception only when the collision is with a body of vanishing mass. If the collision is with μ, then f_1 and f_2 are functions of $\mu^{\frac{1}{3}}$ and $\mu = 0$ is an essential singular point.

In order that the solutions (3) shall be periodic with the period T three equations,

$$\left.\begin{array}{l}\phi_1(T;\ \mu;\ c_1, c_2, C)=0\\ \phi_2(T;\ \mu;\ c_1, c_2, C)=0\\ \phi_3(T;\ \mu;\ c_1, c_2, C)=0\end{array}\right\}\qquad\text{(4)},$$

must be satisfied. The properties of ϕ_1, ϕ_2, and ϕ_3 are similar to those of f_1 and f_2.

The method of discussion consists in establishing the existence of certain classes of periodic solutions for favorable values of the parameters; in determining the character of the possible changes they may undergo when the arbitrary parameters are varied; and in tracing out the connections among the various families of orbits from considerations of continuity and analysis situs, and from the results obtained by numerical calculations. The demonstrations of the existence and properties of classes of periodic solutions and the possible changes they may undergo can not be given in the limits of a single paper. The proofs of the theorems which follow will be found in a book on periodic orbits which is now being published by the Carnegie Institution of Washington. Moreover, the exigencies of time make it advisable to limit this report to the consideration of those orbits in which the motion of the infinitesimal body is retrograde, that is opposite to that of revolution of the finite bodies. The retrograde orbits are chosen because, in the first place, very little has been published on them, and because, in the second place, they are much simpler than those in which the motion is direct. The theory for the latter encounters the difficulties connected with the consideration of infinite periods.

Among the theorems upon which the discussion of the relations among classes of periodic orbits depends the following may be mentioned:

1. There are three families of small periodic orbits about each of the finite masses in which the motion is retrograde (also direct). When the orbits are sufficiently small the members of only one family are real, and they are symmetrical with respect to the line joining the finite bodies. These results are true for all values of μ between zero and unity.

2. There are six families of large periodic orbits about both of the finite masses in which the motion is retrograde. With respect to fixed axes the motion is retrograde in three families and direct in the other three. When the orbits are sufficiently large the members of only two families are real, and they are symmetrical with respect to the line joining the finite masses. These results are true for all values of μ between zero and unity.

3. There is a single family of small periodic orbits about each of the three straight-line Lagrangian equilibrium points. The motion is retrograde with respect to the equilibrium points, and the orbits are symmetrical with respect to the line joining the finite bodies. These results are true for all values of μ between zero and unity.

4. If $\mu < \frac{1}{2}(1-\sqrt{\frac{23}{27}})$, or if $1-\mu < \frac{1}{2}(1-\sqrt{\frac{23}{27}})$, there are two families of small periodic orbits about each of the equilateral triangular equilibrium points. The motion is retrograde with respect to the equilibrium points, and the orbits about one of the equilibrium points are, with respect to the line joining the finite bodies,

symmetrically opposite to those about the other equilibrium point. For μ greater than, but sufficiently near to, $\frac{1}{2}(1-\sqrt{\frac{23}{27}})$ there are two corresponding families about each of the equilateral triangular points, but they all have finite dimensions. They are the analytic continuation, with respect to μ, of the orbits for $\mu < \frac{1}{2}(1-\sqrt{\frac{23}{27}})$.

5. For μ sufficiently small there are closed ejectional orbits from $1-\mu$ of the form indicated in Fig. 1. For $1-\mu$ sufficiently small there are similar orbits of

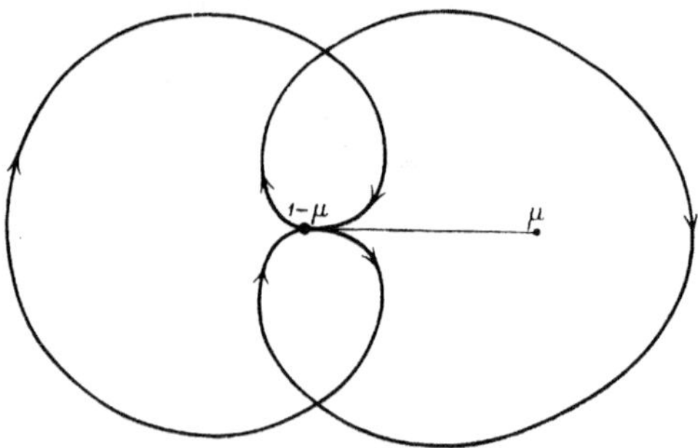

Fig. 1.

ejection from μ. In both cases there are also closed orbits of ejection making an arbitrary number of circuits around the finite masses. The closed orbits of ejection are not periodic, for their coordinates, considered as functions of t, have branch-points at which three branches permute when there is a collision.

6. There are two families of periodic orbits related to each closed orbit of ejection of the type considered in 5, and they have the closed orbits of ejection as limiting forms. A single closed orbit of ejection is given in heavy line in Fig. 2, and one member of each of the associated families of periodic orbits in lighter lines.

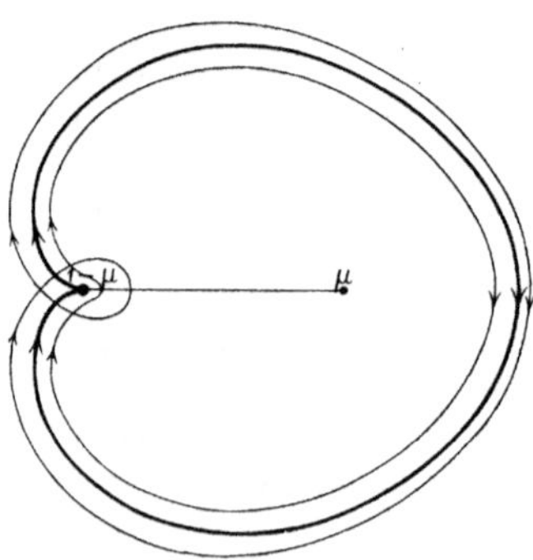

Fig. 2.

7. Real periodic orbits appear, or disappear, with the variation of C or μ only in united pairs. (Poincaré's Theorem.)

8. Let an orbit which, for a certain value of C, belongs to two series of orbits (a united pair) be called a *double orbit*. Such a double orbit is the beginning of Sir George Darwin's Satellites B and C, *Acta Mathematica*, Vol. XXI. Double orbits can appear, or disappear, with the variation of μ only in united pairs.

9. Orbits can unite, with variation of C or μ, only when they coincide throughout.

10. Orbits can acquire loops only by passing through an ejectional form, or by passing through a cusped form at a curve of zero relative velocity. Sir George Darwin's computations have proved that the periodic orbits can have cusps.

11. Only direct orbits, or orbits having direct loops, can have cusps. Cusps can appear or disappear, with the variation of μ, only in united pairs.

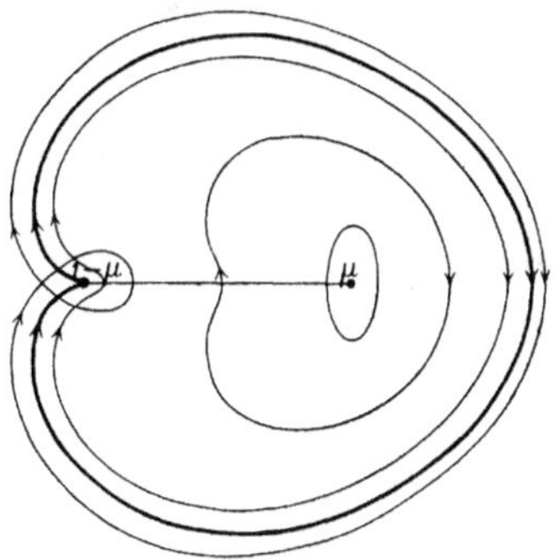

Fig. 3.

A number of illustrations of the evolution of certain families of retrograde orbits will now be given. It is to be understood that the series of changes indicated in the diagrams are inferred from the existence of the families of periodic orbits and the limiting cases mentioned in the foregoing theorems, and from the changes periodic orbits can undergo, and that the results are not fully established by direct analysis.

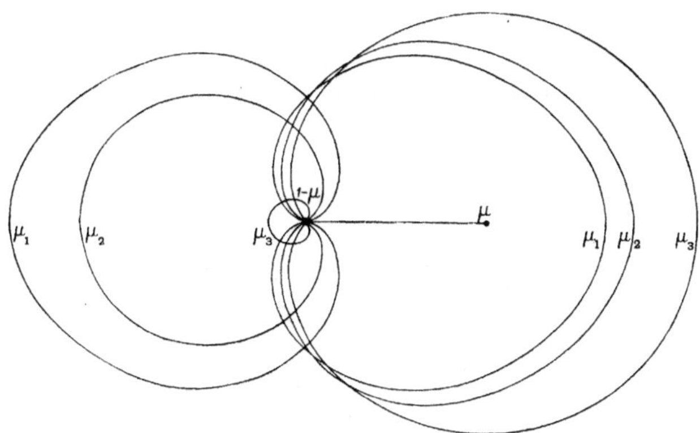

Fig. 4.

A. Evolution of retrograde orbit about μ showing the changes from a small approximately circular orbit (theorem 1) for large values of C to a closed orbit of ejection from $1-\mu$ (theorem 5) for a smaller value of C, and to an orbit of two loops, one enclosing $1-\mu$ and the other enclosing both finite bodies. The orbits are shown in Fig. 3.

B. Evolution of closed orbits of ejection from $1-\mu$ as μ varies from zero to unity. The notation is $0 \doteq \mu_1 < \mu_2 < \ldots < \mu_n \doteq 1$ and the corresponding orbits are shown in Fig. 4.

C. Evolution of the large retrograde periodic orbits enclosing both finite masses (theorem 2) for small C to an ejectional form (ejection from both finite masses simultaneously when $\mu = 1 - \mu = 0{\cdot}5$, the case shown in Fig. 5) for larger values of C, and finally to a direct orbit about both finite bodies consisting of a single loop. The evolution suggested here should be tested by a numerical calculation.

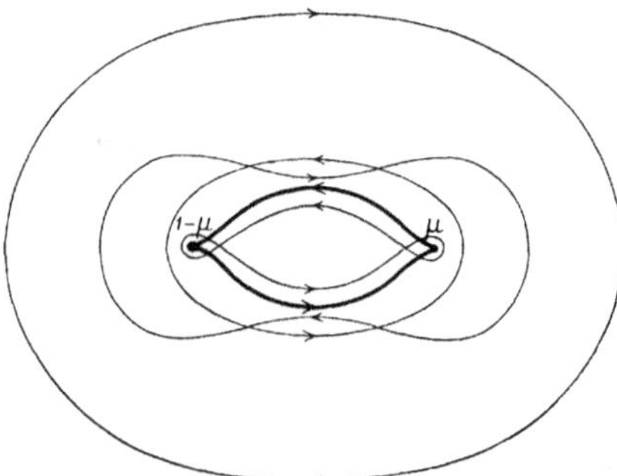

Fig. 5 $(1-\mu=\mu=\frac{1}{2})$.

D. Evolution of the oscillating satellites (theorem 3) about the straight line equilibrium points which are not between the finite masses, for decreasing values of C, to a closed orbit of ejection (theorem 5), and then to an orbit having a double loop about the finite body. The smaller loop expands and coincides with the larger when the orbit is a member of the retrograde series having a single loop (theorem 1). The orbits of two loops near this are Poincaré's orbits of the *deuxième genre*. The diagram, Fig. 6, gives only that series of orbits which passes μ.

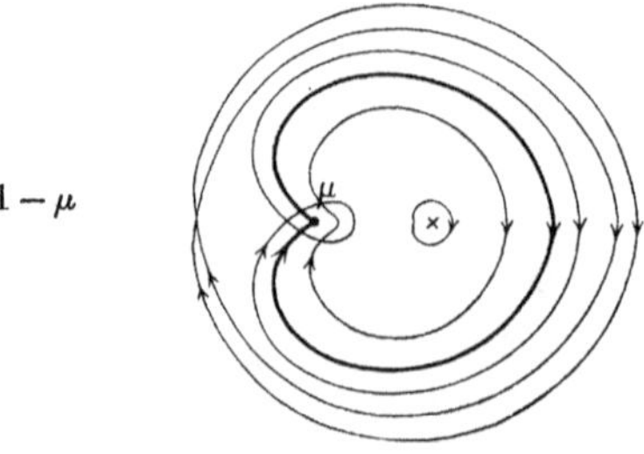

Fig. 6.

E. Evolution of the oscillating satellites (theorem 3) about the straight line equilibrium point which is between the finite bodies, for decreasing values of C, to orbits consisting of small loops about each of the finite bodies, and a large loop enclosing both finite bodies. The curves shown in Fig. 7 are for $\mu = 1 - \mu = 0{\cdot}5$, when the orbit becomes an orbit of ejection for both finite bodies simultaneously.

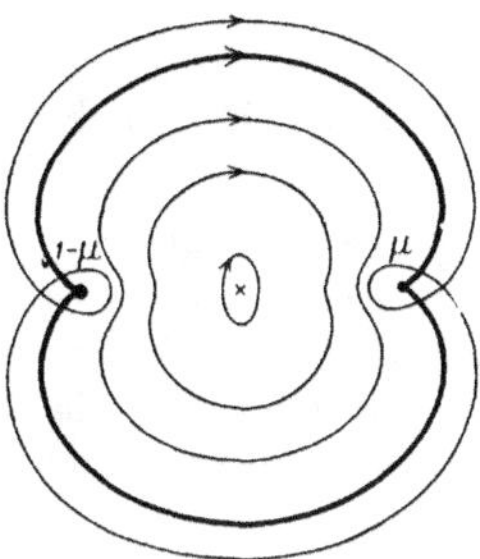

Fig. 7.

STABILE ANORDNUNGEN VON ELEKTRONEN IM ATOM

VON L. FÖPPL.

In meiner Göttinger Dissertation, die ich auf Anregung meines verehrten Lehrers Hilbert angefertigt habe, habe ich mir die Aufgabe gestellt, das in der Physik allgemein bekannte Thomson'sche Atommodell einer strengen mathematischen Untersuchung zu unterziehen.

J. J. Thomson denkt sich bekanntlich das Atom als eine Kugel, die räumlich homogen mit positiver Elektrizität geladen ist, während die negative Elektrizität in Form von Elektronen auftritt, die sich innerhalb der positiv geladenen Kugel reibungslos bewegen können. Diese Elektronen werden einerseits von der positiven Ladung der Kugel nach dem Kugelmittelpunkt mit der quasi-elastischen Kraft $K=\frac{\nu}{b^3}\cdot a$ angezogen, wobei ν die Anzahl der positiven Ladungseinheiten der Kugel, b der Radius der Kugel und a der Abstand des Elektrons vom Kugelmittelpunkt bedeuten; andererseits stossen sie sich umgekehrt proportional dem Quadrat ihres Abstandes von einander ab. Unter dem Einflusse dieser beiden Kräfte werden die Elektronen in der Kugel gewisse Gleichgewichtslagen aufsuchen. Die Aufgabe besteht nun darin, diese Gleichgewichtslagen auf Stabilität zu prüfen. Man kann von vorne herein sofort von jeder beliebigen Anzahl von Elektronen gewisse Gleichgewichtslagen angeben z.B. diejenigen, bei der alle n Elektronen in gleichen Abständen auf einem Kreis liegen, dessen Mittelpunkt mit dem Kugelmittelpunkt zusammenfällt. Jedoch wird man sofort zugeben, dass diese Gleichgewichtslagen für grössere Anzahlen n sicher labil sind, indem man das Gefühl hat, dass die Elektronen teilweise aus der Ebene heraustreten, um den Raum um den Kugelmittelpunkt herum möglichst gleichmässig auszufüllen. In der Tat habe ich auch bewiesen, dass unter allen möglichen Gleichgewichtslagen innerhalb der Kugel diejenige die stabile ist, bei der die Summe der Quadrate der Abstände von Kugelmittelpunkt möglichst klein ist.

J. J. Thomson hat in seiner Arbeit in dem *Phil. Mag.*, 1904 die Annahme gemacht, dass alle Elektronen mit gleicher Winkelgeschwindigkeit um eine in der Kugel feste durch den Kugelmittelpunkt gehende Achse rotieren. Mit Hilfe der durch die Rotation hervorgerufenen Zentrifugalkraft reduziert J. J. Thomson das ganze Problem auf ein ebenes. Er kommt zu dem interessanten Resultat, dass 1, 2, 3 Elektronen bei beliebiger Rotation, 4 und 5 Elektronen bei genügend starker Rotation in gleichen Abständen auf dem Umfang eines Kreises um den Kugelmittelpunkt stabile Anordnungen darstellen; dass dagegen 6 in gleicher Weise angeordnete Elektronen selbst unter Annahme noch so grosser Rotationsgeschwindigkeit labil sind und erst wieder durch ein siebentes Elektron in Mittelpunkt der Kugel stabilisiert werden. So weit führt J. J. Thomson seine Rechnungen streng durch. Bei grösseren Anzahl von Elektronen nimmt J. J. Thomson zur Vereinfachung der Rechnung an, dass alle Elektronen immer in einer Ebene bleiben. Offenbar ist diese Annahme für eine strenge Stabilitätsuntersuchung unzulässig, worauf J. J. Thomson auch hinweist.

Um diese Vernachlässigung zu vermeiden, habe ich von Rotation vorerst ganz abgesehen und einfach die stabilen Konfigurationen der Ruhe untersucht, wodurch sich die Aufgabe gegenüber der Thomson'schen zu einem dreidimensionalen Problem erweiterte. Im übrigen habe ich den von J. J. Thomson eingeschlagenen Weg der Stabilitätsuntersuchung benützt; nämlich die Methode der kleinen Schwingungen. Ich begann mit der allgemeinen Stabilitätsuntersuchung eines einzelnen Elektronenringes, d.h. alle n Elektronen sind in gleichen Abständen auf einem Kreis um den Kugelmittelpunkt angeordnet. Die Rechnung ergab die Thomson'schen Resultate, dass bloss für $n=1$, 2 und 3 Stabilität eintritt, während schon bei $n=4$ diejenige Schwingung labil ist, bei der zwei gegenüberliegende Elektronen des Ringes in dem einen Sinn, die beiden anderen im entgegengesetzten Sinn parallel der Symmetrieachse sich verschieben. Man überzeugt sich leicht, dass diese Verrückung im weiteren Verlauf der Bewegung auf die Tetraederanordnung der vier Elektronen führt. Diese ist aber, wie ich zeigen konnte, stabil.

Über die Hauptschwingungen eines Einzelringes in der Ringebene, die ich genau durchdiskutiert habe, liesse sich noch vieles Interessante sagen; doch verweise ich auf meine Dissertation, wo man sich an Hand von Figuren ein klares Bild machen kann.

Mathematisch ist vor allen Dingen interessant wie in der analytischen Behandlung der Stabilitätsfrage die geometrische Symmetrie der Anordnungen zum Ausdruck kommt. Man erhält nämlich bei Anwendung der Methode der kleinen Schwingungen für die Untersuchung der Stabilität von n Elektronen bekanntlich $3n$ Gleichungen und die charakteristische Determinante liefert für das Quadrat der Frequenzen eine Gleichung vom Grade $3n$, deren Lösung natürlich bei einigermassen grossem n unüberwindliche Schwierigkeiten machen würde, wenn nicht wegen des symmetrischen Baues der Determinante, welche in unserem Fall eine sogenannte Zirkulante ist, die eine Gleichung vom Grade $3n$ in n einzelne Gleichungen je vom Grade 3 zerfallen würde. Diese für die wirkliche zahlenmässige Berechnung natürlich fundamentale Tatsache tritt bei allen unseren Anordnungen wieder auf, sodass wir mit der Untersuchung von Gleichungen dritten Grades auskommen.

Doch kehre ich nun zur Beschreibung meiner Resultate zurück. Haben wir gesehen, dass 4 Elektronen nicht mehr in Ringanordnung stabil sind, sondern an den Ecken des regulären Tetraeders ihre stabile Lage besitzen, so werden wir bei 5 Elektronen wieder zur Ringanordnung zurückgeführt, aber mit dem Unterschiede gegen früher, dass nunmehr zu einem Elektronenring, dessen Mittelpunkt mit dem Kugelmittelpunkt zusammenfällt, noch beiderseits in gleichen Abständen von dem Elektronenring je ein weiteres Elektron hinzutritt—Polelektron, wie ich es nennen will. Die stabilen Anordnungen dieses Typs, den ich in Nachahmung der geometrischen Gestalt durch $n = 1 + m + 1$ bezeichnen will und bei dem die Rechnungen des isolierten Elektronenringes mit geringen Änderungen übernommen werden können, sind folgende:

$$5 = 1 + 3 + 1, \quad 6 = 1 + 4 + 1, \quad 7 = 1 + 5 + 1.$$

Dabei ist $6 = 1 + 4 + 1$ die Anordnung der Elektronen an den 6 Ecken des regulären Oktaeders. Dagegen sind die Anordnungen $8 = 1 + 6 + 1$ und $9 = 1 + 7 + 1$, wie ich gezeigt habe, labil.

Man wird vermuten, dass 8 Elektronen an den Ecken des Würfels stabil sind. Jedoch habe ich zeigen können, dass die Würfelanordnung labil ist und zwar sieht man, wenn man die labile Schwingungen des Würfels verfolgt, dass die 8 Elektronen eine Anordnung nach zwei parallelen Ringen von je 4 Elektronen liefern, die gegenseitig verschränkt liegen, so dass ihre Projektion auf eine Ebene senkrecht zur Symmetrieachse einen Achterring gibt. Auch diesen Anordnungstyp habe ich allgemein untersucht; auch schliesslich unter Hinzunahme von Polelektronen zu beiden Seiten des Doppelringes. Die langwierigen Zahlenrechnungen ergaben folgende stabilen Anordnungen:

$$8 = 4 + 4, \quad 10 = 1 + 4 + 4 + 1, \quad 12 = 1 + 5 + 5 + 1 \text{ (Ikosaeder)}, \quad 14 = 1 + 6 + 6 + 1.$$

Dagegen habe ich gefunden, dass die Anordnungen $16 = 1 + 7 + 7 + 1$ und $18 = 1 + 8 + 8 + 1$ labil sind. Eine labile Schwingung dieser letzteren Anordnung von 18 Elektronen ist diejenige, bei der die 4 Elektronen mit geraden Nummern jeder der beiden Achterringe in der einen Richtung, die 4 mit ungeraden Nummern in entgegengesetzter Richtung sich verschieben. Verfolgt man diese labile Verschiebung in ihrem weiteren Verlauf, so führt sie aus Symmetriegründen auf die folgende Anordnung der 18 Elektronen: $18 = 1 + 4 + 8 + 4 + 1$ d.h. abgesehen von den beiden Polelektronen treten drei Ringe auf: ein Achterring und zu beiden Seiten desselben je ein Viererring und zwar so, dass die Ringe gegenseitig verschränkt liegen. Diese Anordnung habe ich nicht mehr näher untergesucht. Die Stabilitätsuntersuchung macht hier grosse Schwierigkeiten; selbst die Bestimmung der Gleichgewichtslage ist hier keine leichte Aufgabe.

Ich habe nur noch den Fall von 20 Elektronen ins Auge gefasst, weil die 20 Elektronen ein besonderes Interesse beanspruchen, da sie möglicherweise an den 20 Ecken des Dodekaeders eine stabile Anordnung finden können; und zwar habe ich die Dodekaederanordnung der 20 Elektronen mit derjenigen Anordnung verglichen, die sich an die oben besprochenen Anordnungen von 12, 14, 18 Elektronen ungezwungen anreiht; nämlich $20 = 5 + 10 + 5$. Es stellte sich dabei heraus, dass die Dodekaederanordnung ein grösseres Gesamtpotential besitzt als diese letzere Anordnung und daher labil ist.

Es lassen sich daher von unseren Gesichtspunkt aus die 5 regulären Körper in zwei Gruppen einteilen: Tetraeder, Oktaeder und Ikosaeder liefern stabile, Würfel und Dodekaeder labile Anordnungen.

Abgesehen von den drei Anordnungen nach regulären Körpern habe ich—um nun einen zusammenfassenden Überblick zu geben—gefunden, dass bei allen stabilen Anordnungen eine Achse ausgezeichnet ist, indem sie Symmetrieachse ist. Ausserdem findet man bei allen Anordnungen mit mehr als einem Ring, dass die Ringe gegenseitig verschränkt oder wie man auch sagen kann verzahnt liegen. Nimmt man nun an, dass beide Eigenschaften allgemein für stabile Anordnungen Geltung besitzen, so kann man sich auch für grössere Anzahlen von Elektronen, deren strenge Stabilitätsuntersuchung wegen zu umständlicher Rechnungen nicht mehr gelingt, folgendes Bild von den stabilen Anordnungen entwerfen: es werden 1, 2 oder 3 Pole auftreten, nämlich entweder einer im Kugelmittelpunkt oder zwei zu beiden Seiten der Ringsysteme oder schliesslich gleichzeitig einer im Kugelmittelpunkt und je einer zu beiden Seiten der Elektronenringe. Abgesehen von diesen Polen werden sich die Elektronen in Form von Ringen um den Kugelmittelpunkt gruppieren. Dabei spielen natürlich die Teilbarkeitsverhältnisse der Zahlen eine grosse Rolle. Wegen dieser Eigenschaft lassen sich gewisse Zahlen aus der Zahlenreihe herausgreifen, bei denen immer wieder derselbe Anordnungstyp nach drei parallelen Ringen auftreten muss. Diese Zahlen wachsen abwechselnd um 24 und 48 Einheiten, so dass wir auf diese Weise auch formal das periodische System der Elemente nachahmen können.

Wegen näherer Ausführungen hierzu muss ich auf meine Dissertation verweisen. Was speziell die letzten Ausführungen betrifft, so möchte ich bemerken, dass ich den Boden der strengen Mathematik verlassen habe. Aber immerhin kommt diesen Darlegungen doch etwas mehr als nur einem Phantasiegebilde zu, da sie eine naheliegende Verallgemeinerung der mathematisch strengen Folgerungen sind, wie ich sie im Hauptteil meiner Arbeit für geringere Anzahlen von Elektronen abgeleitet habe.

ON THE PRACTICAL APPLICABILITY OF STOKES' LAW OF RESISTANCE, AND THE MODIFICATIONS OF IT REQUIRED IN CERTAIN CASES

By M. S. Smoluchowski.

§ 1. Stokes' law for the resistance of a sphere in a viscous liquid rests, as is well known, on the fundamental assumptions:

I. Slowness of motion, so that the inertia terms in the hydrodynamical equations may be neglected, in comparison with the effects of viscosity.

II. Complete adhesion without slip, of the liquid to the sphere, this being considered as a rigid body.

III. Unboundedness of the liquid and immobility at infinity.

In what follows I should like to contribute some remarks on this law with regard to certain cases of practical importance, where the underlying conditions are changed to some extent, which may be of some interest to those who are engaged with research work on subjects connected with Stokes' law.

First let us touch briefly the question of slipping, connected with the second of the above assumptions. Stokes' calculation can be generalised, by allowing the liquid to slip along the surface of the sphere, with a velocity proportional to the frictional force in a tangential direction [which in the case of a parallel laminar flow implies the surface condition $\beta u = \mu \frac{\partial u}{\partial y}$].

In this case, as Basset has shown, the simple law of Stokes has to be replaced by

$$F = 6\pi\mu Rc \frac{\beta R + 2\mu}{\beta R + 3\mu} \quad \text{..............................(1).}$$

Thus the minimal value of the resistance, for the case of infinite slip ($\beta = 0$), is two-thirds of the maximal value for no slip ($\beta = \infty$).

Now it is generally assumed, on account of the experimental researches of Poiseuille, Whetham, Couette, Ladenburg and others, that the slip of liquids along solid walls is negligibly small. Mr Arnold's* recent measurements prove, by their exact agreement with Stokes' law, that the coefficient of sliding friction β is certainly greater than 5,000 and probably greater than 50,000.

* H. D. Arnold, *Phil. Mag.* 22, p. 755 (1911).

§ 2. On the other side, his experiments, on bubbles of gas moving through liquid, gave the unexpected result that the slip at clean* surfaces between gas and liquid is infinite, as the velocity turned out too great by 50 per cent.

Now I think a different explanation of those experiments to be preferable, as in the case of gas bubbles or liquid drops also the interior liquid is subject to circulation. Some time ago I advised Mr Rybczynski in Lemberg to calculate the motion of a viscous sphere through viscous liquid. The calculation is quite easy and the result†, published January last year, and deduced also half a year later, quite independently of course, by M. Hadamard, is equally simple. It shows that for slow motion the inner liquid retains its spherical shape and that the resistance is

$$F = 6\pi\mu Rc\,\frac{3\mu' + 2\mu}{3\mu' + 3\mu} \quad\quad (2),$$

where μ' designates the viscosity of the liquid in the interior of the sphere.

Comparison with the above formula shows that the resistance experienced by a gas bubble or liquid drop without slip is the same as the resistance of a solid sphere with a coefficient of surface friction $\beta = \frac{3\mu'}{R}$; in fact the velocity and the stream lines of the outer liquid are identical in both cases. It would be interesting to verify the above formula by experiments on liquids with similar values of μ and μ'; in the case of Mr Arnold's experiments the viscosity in the interior was negligible in comparison with the viscosity of the outer medium, which had the same effect as if the surface slip were infinite. So far his results too are explained without the assumption of surface slip.

§ 3. However, there is a case where the existence of surface slip has been proved beyond doubt, namely in rarefied gases. As is well known, the magnitude of the coefficient of slipping $\gamma = \frac{\mu}{\beta}$ is, according to the kinetic theory and also to the old experiments of Kundt and Warburg, roughly equal to the mean length of the free path of the gas molecules; therefore the phenomenon plays an important part even at ordinary pressures in the motion of very minute droplets, as in Millikan's experiments.

Now unfortunately one cannot use formula (1) for this case, with substitution of the empirical value for β, except for the case of comparatively small slip. For if the mean length λ is comparable with the dimensions of the moving sphere, the ordinary hydrodynamical equations cease to be valid altogether, since the implicit assumption underlying them, that the state of the gas is varying little for distances comparable with λ, is impaired.

Therefore also the interesting deduction of a corrected formula by Prof. E. Cunningham‡ is not to be considered as a demonstration and Messrs Knudsen and

* I.e. provided the surface be not contaminated with solid films.

† W. Rybczynski, *Bull. Acad. d. Sciences Cracovie*, 1911, p. 40; J. Hadamard, *Comptes Rendus*, 152, p. 1735 (1911); 154, p. 109 (1912).

‡ E. Cunningham, *Proc. Roy. Soc.* 83, p. 357 (1910).

S. Weber may be right in trying to get closer approximation by other, purely empirical formulas*. At any rate the formula proposed by Cunningham

$$F = 6\pi\mu Rc\left[1 + A\frac{\lambda}{R}\right]^{-1}$$

serves remarkably well for interpolation, considering the experiments of those authors and those of Mr McKeehan†. It is preferable to write it in the form

$$F = 6\pi\mu Rc\left[1 + \frac{B}{R\rho}\right]^{-1},$$

where ρ is the density of the gas, as mistakes are easily involved by using the mean length of free path λ, which is a very indefinite term and really has no precise meaning.

For great rarefaction the resistance is proportional to the cross-section of the sphere, and for this case the calculation can be carried out exactly, if the way is known, how the interaction between the surface of the sphere and the gas molecules takes place. If they rebound like elastic bodies, we get in accordance with Cunningham

$$F = \frac{4}{3}\sqrt{\frac{8}{3\pi}}\,R^2\pi\rho c V,$$

where V is the square root of the mean square of molecular velocity. The numerical coefficient, as calculated from the experiments mentioned above, is considerably larger, it amounts to 1·65 (Knudsen and Weber) or 1·84 (McKeehan). McKeehan concludes that molecules are reflected from the surface of the sphere only in a normal direction; I think, however, that his theoretical formula is not quite exact and at any rate his conclusion seems to me at variance with fundamental principles of the kinetic theory of gases. I think that the experimental results are explained best by the view, supported also by other researches of this kind, especially those of Knudsen, that a solid surface acts in scattering the impinging molecules irregularly in all directions whether with or without change of mean kinetic energy. We shall not go into these questions now, however, as they belong to the kinetic theory of gases, not to hydrodynamics.

§ 4. Now let us consider what modifications are required in Stokes' law, if the third of the fundamental assumptions is impaired, the liquid being limited by solid walls, or a greater number of similar spherical bodies being contained in it.

In this case the linear form of the hydrodynamical equations makes it possible to attain their solution by a method of successive approximations, analogous to the method of images used in the theory of electrostatic potential. It consists in the successive superposition of solutions formed as if the fluid would extend to infinity, but so chosen as to annul the residual motion at the boundaries, with increasing approximation.

This method was used first by H. Lorentz in order to determine the influence of an infinite plane wall on the progressive movement of a sphere, and we shall refer

* M. Knudsen and S. Weber, *Ann. d. Phys.* 36, p. 981 (1911).

† McKeehan, *Physik. Zeitsch.* 12, p. 707 (1911).

to his formulas later on*. He found that the resistance of the sphere is increased by a fraction amounting to $\frac{9}{8}\frac{R}{a}$ for normal motion, $\frac{9}{16}\frac{R}{a}$ for parallel motion, if a denotes the distance from the wall. Mr Stock in Lemberg has extended the calculation for the second case to the fourth order of approximation, including terms with $\left(\frac{R}{a}\right)^4$†.

In a somewhat similar way Ladenburg‡ calculated the resistance experienced by a sphere, when moving along the axis of an unlimited cylindrical tube, and his result, indicating an increase in comparison with the usual formula of Stokes in the proportion of $1 : 1 + 2{\cdot}4\frac{R}{\rho}$ (where ρ = radius of the tube), has been verified with very satisfactory approximation by his own experiments and by those of Mr Arnold.

§ 5. Now let us apply this method to the case where a greater number of similar spheres are in motion, and extend a little further now an investigation which I had begun in a paper published last year§. Imagine a sphere of radius R, moving with the velocity c along the X-axis, its centre being situated at the distance x from the origin. It would produce at the point P (with coordinates ξ, η, ζ) certain current velocities u_0, v_0, w_0, of order $\frac{Rc}{r}$, defined by Stokes' equations, if the fluid be unlimited. But if we assume this point P to be the centre of a solid sphere of radius R, we have to superpose a fluid motion $u_1 v_1 w_1$, chosen so as to annul the velocities of the primary motion at the points of this sphere and satisfying the conditions of rest for infinity.

This motion may be called the "reflected" motion; it can be found with any degree of approximation, by making use of the solution of the hydrodynamical equations given by Lamb, in form of a development in spherical harmonics. But as it is of order $\frac{Rc}{r}$ at the surface of the second sphere, which is its origin, it seems probable, *a priori*, that its magnitude at the first sphere will be of order $c\left(\frac{R}{r}\right)^2$, and I have verified this as well as the following results by explicit calculation. Thus if we confine ourselves to terms of order $c\left(\frac{R}{r}\right)^2$, we can apply a simplified method of evaluating the mutual influence of such spheres, by neglecting the difference between the velocity at the centre of the second sphere and at its surface.

That is to say, the sphere P, being at rest, is subjected to frictional forces

$$X = 6\pi\mu R u_0,$$
$$Y = 6\pi\mu R v_0,$$
$$Z = 6\pi\mu R w_0$$

* H. A. Lorentz, *Abhandlungen ü. th. Physik*, I. p. 23 (1906). In Millikan's determinations of the ionic charge the increase of resistance due to the presence of the condenser plates may produce an increase of the order of one-thousandth.

† J. Stock, *Bull. Acad. d. Sciences Cracovie*, 1911, p. 18.

‡ R. Ladenburg, *Ann. d. Phys.* 23, p. 447 (1907).

§ M. S. Smoluchowski, *Bull. Acad. d. Sciences Cracovie*, 1911, p. 28.

on account of the motion of the first sphere; on the other side, the moving sphere experiences a reaction by virtue of the presence of the sphere P, such as if this would execute simultaneously the three motions $-u_0, -v_0, -w_0$; the three current systems resulting therefrom, according to the usual formulas of Stokes, produce at the centre of the first sphere nine current components, giving rise to nine components of frictional force, to be calculated each according to Stokes' law of resistance.

If both spheres are in simultaneous motion, the mechanical effects are found by superposition of the forces corresponding to the two cases where one of them is moving and the other one at rest.

In this way an interesting conclusion is obtained for the case where both spheres are moving in parallel directions with equal velocity: then both are subjected to equal additional forces in the same direction, one component in the direction of motion, tending to diminish the resistance by the amount $\frac{9}{2}\frac{R^2\pi\mu c}{r}\left[1-\frac{3}{4}\frac{R}{r}\right]$, the other component along the line joining the centres, towards the sphere which is going ahead, of amount $\frac{9}{2}\frac{R^2\pi\mu c\cos\theta}{r}\left[1-\frac{9}{4}\frac{R}{r}\right]$ [where θ is the angle between the line of centres and the direction of motion].

Thus two heavy spheres of this kind would sink faster than Stokes' law is indicating and, besides, their path must be deflected from the vertical towards the line of centres by an angle ϵ defined by

$$\sin\epsilon=\frac{3}{4}\frac{R}{r}\left[1-\frac{3}{2}\frac{R}{r}\right]\sin\theta\cos\theta.$$

§ 6. Analogous methods are applicable to a greater assemblage of spheres. The motion results from superposition of simpler solutions, where one sphere is supposed moving and all the other ones resting. Each of the component solutions comprises the direct action, and for higher approximation also its "reflections."

Now if the parallel motion of a cloud of n similar spheres is considered, the resistance of each of them will be diminished by an expression proceeding after powers of R, the first term of which will be of the order of magnitude $\mu cR^2\Sigma\frac{1}{r}$. We see that these developments would be divergent for an infinite number of spheres. It is evident that for instance an infinite row of spherical particles, arranged at equal distances, would acquire infinite velocity, by virtue of their gravity, as also an infinite cylinder would behave in the same way. This applies *a fortiori* to two-dimensional infinite assemblages.

Stokes' law of resistance will not be true even approximately, and the development will cease to be convergent in general, unless $\frac{nR}{S}$ is small, where S denotes a kind of mean distance, comparable with the linear dimensions of the cloud.

§ 7. The same result follows from the following simple reasoning. Imagine a spherical cloud of radius S, containing n spherical particles, each of radius R and density σ, suspended in a medium of viscosity μ, of negligible density, for example a cloud of minute drops of water in air. Then currents will take place in the spherical cloud and it will attain a certain velocity as a whole, which may be calculated after the formula (2), just as if the cloud would form a homogeneous medium of density $n\left(\frac{R}{S}\right)^3$ and of the same viscosity as the outer medium. The mass velocity resulting therefrom, of amount $\frac{4}{15}\frac{nR^3g\sigma}{S\mu}$, is superposed upon the displacement of the particles, relative to the moving cloud, taking place with velocity $\frac{2}{9}\frac{R^2g\sigma}{\mu}$. Thus evidently the downward velocity will be much increased, and Stokes' law cannot be true even approximately, unless $\frac{nR}{S}$ is small in comparison to unity. This condition shows that Stokes' law can be applied only to particles constituting clouds of exceedingly scarce crowding, and it is easily seen that it would be quite erroneous to apply it to actual fogs or actual clouds in the atmosphere, with diminished transparency [as in this case the aggregate cross-section of the particles $nR^2\pi$ is comparable with the cross-section of the cloud $S^2\pi$]. As an illustration how cautious we must be in this respect, I may mention that the ratio $\frac{nR}{S}$ amounts to 10 and even to 100 for a cubic centimetre cloud as produced by Sir J. J. Thomson and H. A. Wilson, in their experiments on the determination of the ionic charge.

§ 8. What has been said applies of course only to clouds moving in an otherwise unlimited medium. The conditions of motion are quite different for a cloud contained in a closed vessel, as in the experiments just referred to. Prof. E. Cunningham has attempted to evaluate the order of magnitude of the correction to be applied to Stokes' law in this case. His estimate is founded on the supposition that each particle moves approximately in such a way, as if it were contained in a rigid spherical envelope, of radius comparable with half the distance to its next neighbours. Now this supposition does not seem quite evident, although we shall see that it leads to a result of the right order.

We can calculate the resultant motion in quite an exact way, if we consider a homogeneous assemblage of equal spherical particles, moving all of them with the same velocity c in the direction of negative X, towards an infinite rigid wall, which we assume to be the plane YZ. In this case we see, by making use of H. A. Lorentz's calculation before alluded to, that a moving sphere x, y, z produces at a point ξ, situated on the axis of X, a velocity component

$$u = -\frac{3}{4}\frac{Rc}{r}\left[1+\left(\frac{\xi - x}{r}\right)^2\right] + \frac{3}{4}\frac{Rc}{\rho}\left[1+\frac{x^2+\xi^2}{\rho^2}+\frac{6x\xi(x+\xi)^2}{\rho^4}\right] \ldots\ldots (3).$$

The first part of this expression, containing $r=\sqrt{(x-\xi)^2+y^2+z^2}$, is the component of direct motion, according to Stokes; the second part is the component caused by "reflection" at the plane YZ; it contains the distance between the point ξ and the reflected source $\rho=\sqrt{(x+\xi)^2+y^2+z^2}$.

The terms with higher powers of $\frac{R}{r}$ have been neglected, as we confine ourselves to the first approximation. The total current produced in the point ξ by the motion of all the particles is equal to $U = \Sigma u$, where the summation is to be extended over all their values of x, y, z. Now we might consider it right to replace the summation by an integration, since one particle corresponds to a space A^3, if A denotes a sort of mean distance between the particles. In this case the result would be very simple, for we should have

$$U = \frac{1}{A^3}\iiint u\,dx\,dy\,dz.$$

The integrals of the separate terms constituting u can be evaluated explicitly if we extend them to a cylinder with YZ as basis, of height h and of radius G. Then we can use the well-known expression for the potential of a disk in points of its axis, and expressions derivable from it by differentiation with respect to ξ, and by these means we find the unexpected result that the integral current U is zero, if we extend the summation to an infinite value of G.

But in reality U is not defined by integration but by summation. Evidently both operations lead to the same result for distant parts of the space, but not for those parts whose distance from the point ξ is comparable with the distances A between two particles. Therefore the resultant current U in points at a great distance (in comparison with A) from the wall will be given by

$$\left.\begin{aligned} U &= \frac{3}{4}\frac{Rc}{A}\beta, \\ \text{where}\quad \beta &= \frac{1}{A^2}\iiint \frac{1}{r}\left(1+\frac{x^2}{r^2}\right)dx\,dy\,dz - \Sigma\frac{A}{r}\left(1+\frac{x^2}{r^2}\right) \end{aligned}\right\}\quad\text{...............(4)},$$

to be extended over a space great in comparison with A, is a purely numerical coefficient.

In order to evaluate β we must know how the particles are arranged. If we suppose an arrangement in rectangular order, we can get easily an approximate value by explicit calculation and by integrating over a cube of height H, constructed around the point ξ, which gives

$$\iiint \frac{1}{r}\left(1+\frac{x^2}{r^2}\right)dx\,dy\,dz = 8H^2\left[\log(1+\sqrt{3}) - \tfrac{1}{2}\log 2 - \frac{\pi}{12}\right].$$

It is sufficient to take H equal to a small uneven multiple of $\frac{A}{2}$, as the expression for β is rapidly converging with extension of the limits of integration. In this way I have found the approximate value $\beta = 3{\cdot}09$, and therefore the resistance for one particle will be

$$F = 6\pi\mu Rc\left[1+\frac{3}{4}\frac{R\beta}{A}\right] = 6\pi\mu Rc\left[1+2{\cdot}32\frac{R}{A}\right]\quad\text{...........(5)}.$$

This formula would apply, of course, also if the particles were arranged in a different way, but then the numerical value of β would be different. Our result agrees to the order of magnitude with Prof. Cunningham's estimate, which led him for the case of an equilateral arrangement to a similar formula, with a *coefficient of* $\frac{R}{A}$ *included* within the limits 3·67 and 4·5.

§ 9. However, the practical application of this formula is rather questionable, as it applies only to a regular arrangement of particles. If they were arranged in clusters, the correction might even become negative. It is interesting to note that the average value of β, for a particle whose position relatively to the other ones is defined by pure accident, would be zero, and that seems quite natural, as the average current of liquid U in the cross-section must be zero. Thus it follows, what we should not have expected at first sight, that Stokes' law applies for the particles of an actual cloud, on an average with no correction whatever, of this order of magnitude.

The evaluation of the quadratic terms would be much more complicated of course, as then all possible kinds of single reflections caused by any one sphere have to be taken into account.

The general result of our calculation shows at any rate that Stokes' law is undergoing but small corrections if applied to the particles of a uniform cloud filling a closed vessel. But it is important to note that things will change entirely if the cloud is not of quite uniform density or if it does not fill the whole empty space between the walls. Then as a rule convective currents will arise, which in certain cases may be of preponderant influence. Their velocity may be calculated approximately by considering the medium as a homogeneous liquid subjected to certain forces, the intensity of which per unit volume corresponds to the aggregate force acting on the particles contained in it.

Consider for instance an electrolyte in an electric field. If it is conducting in accordance with Ohm's law, the average electric density is zero and no currents will take place. But in bad liquid conductors, with deviations from Boyle's law, convective currents may arise, which may influence also materially the apparent value of the conductivity. They have been observed long ago, for instance by Warburg*.

Similar movements may be produced in ionised gases, and I think more attention ought to be paid to them than usually is done. In experiments where the saturation current of strong radio-active material is observed between condenser plates wide apart†, these phenomena may be of importance as producing an apparently greater mobility of the ions than under normal conditions.

§ 10. There is another application of the theoretical methods exposed above which may be mentioned. Imagine a two-dimensional infinite assemblage of equal spherical particles, distributed uniformly over the plane $x = l$, whilst the plane YZ again may be supposed to be a rigid wall. Now let all these particles be moving along the plane in direction Y with equal velocity c; what motion will be produced in the surrounding liquid, and what will be the resistance experienced by every particle?

According to Lorentz again the motion produced by a single sphere moving parallel to a fixed wall is, when higher powers of the ratio $\frac{R}{l}$, which we suppose to be a small quantity, are neglected:

$$v = \frac{3}{4}\frac{Rc}{r}\left[1 + \left(\frac{y}{r}\right)^2\right] - \frac{3}{4}\frac{Rc}{\rho}\left[1 + \left(\frac{y}{\rho}\right)^2\right] - \frac{3}{2}\frac{Rcx(x+\xi)}{\rho^3} + \frac{9}{2}\frac{Rcxy^2(x+\xi)}{\rho^5},$$

where the first term is the direct current according to Stokes, while the remaining terms represent the current reflected by the wall, just as in the former example.

* E. Warburg, *Wied. Ann. d. Phys.* 54, p. 396 (1895).

† Cf. Rutherford, *Radio-activity*, pp. 35, 84.

We might also in this case calculate the resultant current by forming Σv over all values of y and z, and derive therefrom the resistance of a single particle. But we shall confine ourselves to the following remarks.

In the extreme case where the particles are so crowded, as nearly to touch one another, a lamellar flow will take place in the liquid between the fixed wall and the plane $x=l$ with a velocity $v=\frac{cx}{l}$, while on the other side of the plane $x=l$ the liquid will be dragged along by the sheet of moving particles with the constant velocity c. The frictional force per unit of surface of the plane $x=l$ is evidently equal to $\frac{\mu c}{l}$, therefore the resistance experienced by each particle is

$$F=\frac{\mu c A^2}{l},$$

which is much smaller than Stokes' law would indicate, as A is of the order of R but the distance l is supposed to be of higher order.

Now consider the other extreme case, where the distances A between the particles are so great that Stokes' law is approximately valid, which requires A to be of order l. Let us calculate the resultant motion of the liquid for points at infinite distance from the wall ($\xi=\infty$). For such points the summation mentioned above can be replaced by integration; besides we can put $\frac{1}{r}-\frac{1}{\rho}=\frac{2l\xi}{r^3}$, $\frac{1}{r^3}-\frac{1}{\rho^3}=\frac{6l\xi}{r^5}$; and thus we get

$$V_\infty=\Sigma v=\frac{9Rcl\xi}{A^2}\iint\frac{y^2dydz}{(\xi^2+y^2+z^2)^{\frac{5}{2}}}.$$

This integral can be transformed by putting $y=s\sin\phi$, $z=s\cos\phi$, $dydz=sdsd\phi$, and we get finally

$$V_\infty=\frac{6Rl\pi c}{A^2}.$$

By comparing this with Stokes' law for the resistance F we have

$$V_\infty=\frac{F}{A^2}\frac{l}{\mu};$$

that means that in both cases the liquid at a great distance from the wall will be dragged along, in a parallel direction to it, with such a velocity as if the force corresponding to unit surface $\frac{F}{A^2}$ were distributed uniformly over the liquid, in a plane at a distance l from the fixed wall. This result, which can be generalised for a greater number of similar layers, seems natural enough if the distances between the particles are small in comparison with their distance from the wall, so that the assemblage can be considered as if forming a homogeneous medium, but we see it remains true for particles widely apart. Without going into further details, I may only mention that this result has an important bearing on the theory of electric endosmose, which will be explained elsewhere with full details.

§ 11. I may conclude with a brief remark about the influence of the inertia terms in the hydrodynamical equations (assumption I), which have been neglected as well in Stokes' original calculations as in the above reasonings. It is well known that this neglection is justified only if the ratio $\frac{Rc\sigma}{\mu}$ is small in comparison to unity. But it has been proved by Oseen* in an important paper, commented upon in a very interesting way by H. Lamb, that the solution given by Stokes is defective even if this criterion is fulfilled; for at distances r where $\frac{rc\sigma}{\mu}$ is large, the inertia terms must be of prevalent influence over viscosity. Oseen himself has given a solution which is different from Stokes' equations for those distant parts of the space and gives better approximation there. However, the resistance of the sphere depends only on the state of movement in its immediate neighbourhood, therefore the resistance law of Stokes is not impaired by those results. The condition of its validity may be defined more exactly by means of the recent experiments of Mr Arnold, which have shown that it holds with very good accuracy (one half per cent.) for spheres moving under influence of gravity, provided their radius is smaller than $0{\cdot}6\,\bar{r}$, where the critical radius $\bar{r}$ is defined by the relation $\frac{\bar{r}\bar{v}\sigma}{\mu}=1$. This means that the ratio $\frac{Rc\sigma}{\mu}$ must be smaller than $(0{\cdot}6)^3=0{\cdot}22$.

§ 12. The inertia terms are of greater importance, in the case before alluded to, where the motion of a greater number of similar spheres is considered. For it is legitimate to calculate the forces of reaction between such spheres by using Stokes' equations for slow motion only if they are lying within the space where viscosity is predominant over inertia. Mr Oseen has generalised recently† the calculation of the interaction of two spheres given by me by introducing in it his solution of Stokes' problem. The forces exerted on the two spheres come out unequal in this case and are given by much more complicated expressions. They become identical with the first approximation given by me if the distance r between the two spheres satisfies the condition that $\frac{rc\sigma}{2\mu}$ is small. Mr Oseen thinks this to be a great restriction on the validity of those formulas for experimental purposes, but he omits the factor σ in the above expression. We satisfy ourselves easily that, for instance, in the case of water-drops in air, as in Sir J. J. Thomson's and H. A. Wilson's condensation experiments, the limit of validity for r is of the order of several centimetres; in Perrin's experiments on the applicability of Stokes' law to the particles of emulsions it would amount to hundreds of metres. It is also sufficiently great for direct experiments, when highly viscous liquids are used, as Ladenburg did in his elaborate research. Ordinary hydraulic experiments, with water and spheres of a size to be handled conveniently, are excluded of course when Stokes' law or any of those modifications are in question.

One might try to apply Oseen's method of approximate correction for inertia also to the other cases treated above, but it will imply rather cumbersome calculations and, besides, for movements in closed vessels it will be generally of lesser importance than in a liquid extending to infinity.

* Oseen, *Arkiv f. mat. astr. fysik*, 6 (1911); H. Lamb, *Phil. Mag.* 21, p. 112 (1911).

† F. Oseen, *Arkiv f. mat. astr. fysik*, 7 (1912).

THE APPLICATION OF THE METHOD OF W. RITZ TO THE THEORY OF THE TIDES

By A. E. H. Love.

Four years ago W. Ritz published a memoir* in which he propounded a new method of solving problems of mathematical physics; and in this and a subsequent memoir† he gave examples of the application of the method to obtain approximate numerical solutions of problems which had long defied the efforts of analysts. The salient features of the method may be described as follows:—First the problem is formulated as a problem of Calculus of Variations, that is to say the differential equations of the problem are recognised as conditions required to be satisfied in order to minimize, or render stationary under suitable restrictions, an integral, which may, for example, be the expression for the potential energy of a vibrating system. In the next place the dependent variable, which may, for example, denote a component of displacement in a vibrating system, is taken to be expanded in an infinite series of functions. These functions may be polynomials, or the series may be a Fourier series, or the functions in question may be normal functions of some quite different vibrating system. It matters little what functions are employed provided that an arbitrary function, arbitrary in the sense in which functions that occur in mathematical physics are arbitrary, can be expanded in terms of them. The only necessary restriction affecting the choice of the functions in a series of which the dependent variable is expanded is that, if the variation of the integral is to be subject to special conditions which hold at a boundary, the functions in question must satisfy similar conditions. If, for example, the dependent variable is a displacement, and the boundary is so fixed that the displacement there vanishes, the dependent variable must be expanded in a series of functions which vanish at the boundary. In the practical application of the method the choice of a set of normal functions is conditioned by the circumstance that, unless a suitable set of normal functions is chosen, the integral to be rendered stationary may be intractable. After the formulation of the problem as a problem of Calculus of Variations, and choice of a suitable set of normal functions in a series of which the dependent variable is to be expanded,

* W. Ritz, "Ueber eine neue Methode zur Lösung gewisser Variationsprobleme der mathematischen Physik," *J. f. Math.* (*Crelle*), Bd 135 (1908).

† W. Ritz, "Theorie der Transversalschwingungen einer quadratischen Platte mit freien Rändern," *Ann. d. Phys.* (4te *Folge*), Bd 28 (1909). The two memoirs are reprinted in "Gesammelte Werke Walther Ritz, Œuvres publiées par la Société Suisse de Physique," Paris, 1911.

the next step is to assume the dependent variable to be expressed approximately by a *terminated* series of such functions fitted with arbitrary coefficients. This series is then substituted in the integral, and the integral is thus expressed as a rational integral function, frequently a homogeneous quadratic function, of the coefficients in the terminated series. The coefficients are then adjusted to make this rational integral function stationary subject to the satisfaction of some specified conditions. The approximation to the correct value for the dependent variable improves as the number of terms in the assumed series increases, but it is often quite good when the number of terms is small. The approximate expressions which are thus obtained for the dependent variable, or variables, do not in general satisfy the differential equations of the problem*. They satisfy boundary conditions of the type above described, and the functions to which they are approximations satisfy the differential equations; but the series, when differentiated term by term, do not in general satisfy these differential equations even formally. The security for the series being approximate expressions for functions which do satisfy the differential equations is furnished by the satisfaction of the conditions that arise from the formulation of the problem as a problem in the Calculus of Variations.

The suggestion that this method might be applied to the dynamical theory of the tides was made by Poincaré†. Taking as the dependent variable the vertical displacement of the surface of the sea at any point of specified latitude and longitude, he determined the form of the integral which must be stationary when the differential equations are satisfied, and pointed out some difficulties that beset the application of the method. One of these difficulties is the possible existence of critical latitudes at which some terms of the integral would be infinite or indeterminate. This particular difficulty may be turned by choosing a different dependent variable. I propose to explain a method of procedure, differing slightly from Poincaré's, and shall restrict the investigation to the problem of the free oscillations of a sheet of water of uniform depth covering a rotating globe. The water will be treated as frictionless, and the effects of the self-attraction will be neglected. The problem in the more extended form in which account is taken of the self-attraction having been solved very completely by S. S. Hough‡, the comparison of the results to be obtained by the present method with those obtained by him makes it possible to proceed with great confidence to problems relating to sheets of water of varying depths enclosed by rigid boundaries. I have gone some way with such applications but the work is as yet unfinished.

Let a denote the radius of the rotating globe, h the depth of the water, g the value of gravity at the surface. Let ω denote the angular velocity of the rotation. For the sake of brevity the globe will be compared with the earth, but it is to be understood that no forces except an attraction towards the centre, specified by g, are

* The idea of leaving the differential equations of the problem out of account when once the integral to be rendered stationary has been found is perhaps the most novel feature of Ritz's method. The method has much in common with approximate methods previously used by Lord Rayleigh. In regard to this matter reference may be made to a paper by Lord Rayleigh in *Phil. Mag.* (ser. 6), vol. 22 (1911), pp. 225—229.

† H. Poincaré, *Leçons de Mécanique Céleste*, t. 3 (1910), pp. 297—303.

‡ S. S. Hough, "The application of harmonic analysis to the dynamical theory of the tides," Parts I and II, *London, Phil. Trans. R. Soc.*, vols. 189 (1897) and 191 (1898).

supposed to act upon the water. The position of a point on the surface is specified by co-latitude θ, measured from the North Pole, and longitude ϕ, measured eastwards from a fixed meridian. Let ζ denote the vertical elevation, or radial displacement, at any point at time t; u, v the southward and eastward components of velocity at the same point and the same time. Then the equations of motion are known* to be

$$\frac{\partial u}{\partial t} - 2\omega v \cos\theta = -\frac{g}{a}\frac{\partial \zeta}{\partial \theta} \qquad\text{(1)},$$

$$\frac{\partial v}{\partial t} + 2\omega u \cos\theta = -\frac{g}{a\sin\theta}\frac{\partial \zeta}{\partial \phi} \qquad\text{(2)},$$

while the equation of continuity takes the form

$$\frac{\partial \zeta}{\partial t} = -\frac{h}{a\sin\theta}\left\{\frac{\partial}{\partial \theta}(u\sin\theta) + \frac{\partial v}{\partial \phi}\right\} \qquad\text{(3)}.$$

These equations will determine an oscillatory motion if u, v, ζ are all proportional to simple harmonic functions of the time; but, since the oscillation is about a state of steady motion, not a state of equilibrium, the phases of the different simple harmonic functions are not the same, and are not related to one another in a very simple way.

With a view to the formulation of the problem as a problem in the Calculus of Variations we introduce the notation

$$u\sin\theta = \frac{\partial \xi}{\partial t}, \quad v = \sin\theta\frac{\partial \eta}{\partial t} \qquad\text{(4)}.$$

Then $\xi \operatorname{cosec}\theta$ and $\eta\sin\theta$ are, with sufficient approximation, expressions for the southward and eastward components of the displacement of the water. We introduce also the notation

$$\cos\theta = \mu \qquad\text{(5)},$$

which is usual in problems involving spherical harmonic analysis. It is evident that ξ must vanish when $\mu = \pm 1$. Then equation (3) becomes

$$\zeta = \frac{h}{a}\left(\frac{\partial \xi}{\partial \mu} - \frac{\partial \eta}{\partial \phi}\right) \qquad\text{(6)}.$$

When u, v, ζ are eliminated from equations (1) and (2) by means of equations (4)—(6) there result the two equations

$$\frac{1}{\sin\theta}\frac{\partial^2 \xi}{\partial t^2} - 2\omega\sin\theta\cos\theta\frac{\partial \eta}{\partial t} = -\frac{gh}{a^2}\frac{\partial}{\partial \theta}\left(\frac{\partial \xi}{\partial \mu} - \frac{\partial \eta}{\partial \phi}\right) \qquad\text{(7)},$$

$$\sin\theta\frac{\partial^2 \eta}{\partial t^2} + 2\omega\cot\theta\frac{\partial \xi}{\partial t} = -\frac{gh}{a^2\sin\theta}\frac{\partial}{\partial \phi}\left(\frac{\partial \xi}{\partial \mu} - \frac{\partial \eta}{\partial \phi}\right) \qquad\text{(8)}.$$

Now, if we suppose that ξ and η are proportional to simple harmonic functions of t of period $2\pi/\sigma$, and write

$$\left.\begin{aligned}\xi &= \xi_1\cos\sigma t + \xi_2\sin\sigma t\\ \eta &= \eta_1\cos\sigma t + \eta_2\sin\sigma t\end{aligned}\right\} \qquad\text{(9)},$$

* H. Lamb, *Hydrodynamics*, 3rd edn. (Cambridge, 1906), p. 313.

equations (7) and (8) give rise to the equations

$$\left.\begin{aligned}
&\frac{\sigma^2}{1-\mu^2}\xi_1 + 2\omega\sigma\mu\eta_2 + \frac{gh}{a^2}\frac{\partial}{\partial\mu}\left(\frac{\partial\xi_1}{\partial\mu} - \frac{\partial\eta_1}{\partial\phi}\right)\\
&\frac{\sigma^2}{1-\mu^2}\xi_2 - 2\omega\sigma\mu\eta_1 + \frac{gh}{a^2}\frac{\partial}{\partial\mu}\left(\frac{\partial\xi_2}{\partial\mu} - \frac{\partial\eta_2}{\partial\phi}\right)\\
&(1-\mu^2)\,\sigma^2\eta_1 - 2\omega\sigma\mu\xi_2 - \frac{gh}{a^2}\frac{\partial}{\partial\phi}\left(\frac{\partial\xi_1}{\partial\mu} - \frac{\partial\eta_1}{\partial\phi}\right)\\
&(1-\mu^2)\,\sigma^2\eta_2 + 2\omega\sigma\mu\xi_1 - \frac{gh}{a^2}\frac{\partial}{\partial\phi}\left(\frac{\partial\xi_2}{\partial\mu} - \frac{\partial\eta_2}{\partial\phi}\right)
\end{aligned}\right\} \quad \text{..............(10)},$$

and these equations are the principal equations of that Calculus of Variations problem in which the integral J, where

$$J = \int_{-1}^{1} d\mu \int_0^{2\pi} d\phi \left[\sigma^2\left\{\frac{\xi_1^2+\xi_2^2}{1-\mu^2} + (1-\mu^2)(\eta_1^2+\eta_2^2)\right\} + 4\omega\sigma\mu(\xi_1\eta_2 - \xi_2\eta_1) - \frac{gh}{a^2}\left\{\left(\frac{\partial\xi_1}{\partial\mu} - \frac{\partial\eta_1}{\partial\phi}\right)^2 + \left(\frac{\partial\xi_2}{\partial\mu} - \frac{\partial\eta_2}{\partial\phi}\right)^2\right\}\right] \quad \text{...(11)},$$

is made stationary. The variations must, of course, be subject to the conditions that ξ_1 and ξ_2 vanish at $\mu = \pm 1$, while all the quantities ξ_1, ξ_2, η_1, η_2, considered as functions of ϕ, are periodic with period 2π. These conditions imply that the two variations $\delta\xi_1$, $\delta\xi_2$ vanish at $\mu = \pm 1$, and the two variations $\delta\eta_1$, $\delta\eta_2$ have the same values at $\phi = 2\pi$ as at $\phi = 0$. We write now

$$\frac{\sigma}{2\omega} = f, \quad \frac{4\omega^2 a^2}{gh} = \beta \quad \text{..............................(12)},$$

and

$$J = 4\omega^2 I \quad \text{..................................(13)},$$

then the integral I to be made stationary is given by the formula

$$I = \int_{-1}^{1} d\mu \int_0^{2\pi} d\phi \left[f^2\left\{\frac{\xi_1^2+\xi_2^2}{1-\mu^2} + (1-\mu^2)(\eta_1^2+\eta_2^2)\right\} + 2f\mu(\xi_1\eta_2 - \xi_2\eta_1) - \frac{1}{\beta}\left\{\left(\frac{\partial\xi_1}{\partial\mu} - \frac{\partial\eta_1}{\partial\phi}\right)^2 + \left(\frac{\partial\xi_2}{\partial\mu} - \frac{\partial\eta_2}{\partial\phi}\right)^2\right\}\right] \quad \text{...(14)},$$

and we have

$$\begin{aligned}
\delta I = &-\frac{2}{\beta}\int_0^{2\pi}\left[\left(\frac{\partial\xi_1}{\partial\mu} - \frac{\partial\eta_1}{\partial\phi}\right)\delta\xi_1\right]_{-1}^{1} d\phi - \frac{2}{\beta}\int_0^{2\pi}\left[\left(\frac{\partial\xi_2}{\partial\mu} - \frac{\partial\eta_2}{\partial\phi}\right)\delta\xi_2\right]_{-1}^{1} d\phi\\
&+\frac{2}{\beta}\int_{-1}^{1}\left[\left(\frac{\partial\xi_1}{\partial\mu} - \frac{\partial\eta_1}{\partial\phi}\right)\delta\eta_1\right]_0^{2\pi} d\mu + \frac{2}{\beta}\int_{-1}^{1}\left[\left(\frac{\partial\xi_2}{\partial\mu} - \frac{\partial\eta_2}{\partial\phi}\right)\delta\eta_2\right]_0^{2\pi} d\mu\\
&+ \text{terms which vanish by (10)} \quad \text{......................................(15)},
\end{aligned}$$

so that δI vanishes in virtue of equations (10), the conditions above laid down in regard to the periodicity of the functions, and the special conditions which hold at the limits of integration.

The only difference that would be made in the above work if it were applied to the problem of free oscillations of an ocean of uniform depth bounded by vertical cliffs would be that the limits of integration in respect to μ and ϕ would have to be assigned in a suitable way, and that the variation would have to be conducted in accordance with the condition that the horizontal component of displacement along the normal to any boundary vanishes at the boundary. In the application to an ocean covering the globe we have merely to bear in mind that ξ_1 and ξ_2 must vanish at $\mu = \pm 1$. For this application we suppose in general that

$$\left.\begin{array}{ll} \xi_1 = \xi_0 \cos s\phi, & \xi_2 = -\xi_0 \sin s\phi \\ \eta_1 = \eta_0 \sin s\phi, & \eta_2 = \eta_0 \cos s\phi \end{array}\right\} \quad\text{.....................(16)},$$

where s may be zero or any positive integer, and ξ_0, η_0 are functions of μ only.

When $s = 0$, that is to say when the displacement is independent of longitude, it is convenient to eliminate η_0 by means of the equation

$$(1 - \mu^2) f\eta_0 + \mu\xi_0 = 0 \quad\text{..........................(17)},$$

which follows at once from (10). Then the integral I to be made stationary is given by the equation

$$I = \int_{-1}^{1} \left\{\frac{f^2 - \mu^2}{1 - \mu^2} \xi_0^2 - \frac{1}{\beta}\left(\frac{d\xi_0}{d\mu}\right)^2\right\} d\mu \quad\text{.....................(18)}.$$

It may be noted here that, when ζ is used as dependent variable instead of ξ_0, the differential equation satisfied by ζ is the principal equation of that Calculus of Variations problem in which the integral

$$\int_{-1}^{1} \left\{\frac{1 - \mu^2}{f^2 - \mu^2}\left(\frac{d\zeta}{d\mu}\right)^2 - \beta\zeta^2\right\} d\mu \quad\text{........................(19)}$$

is made stationary; and the difficulty noted by Poincaré, as mentioned above, in regard to the critical latitudes arises through the presence of $f^2 - \mu^2$ in the denominator. The critical latitudes are those at which $\mu = \pm f$, and they are real if $\sigma < 2\omega$, or the period is longer than half a day.

We have now, in accordance with the general method, to substitute for ξ_0 in (18) a series of the form $\sum_{\kappa=1}^{n} A_\kappa u_\kappa(\mu)$, where the u_κ are functions in terms of which an arbitrary function can be expanded, and every one of the set of functions u_κ vanishes in a sufficiently high order at $\mu = \pm 1$. The functions of the type

$$(1 - \mu^2)\frac{dP_\kappa}{d\mu},$$

where P_κ, or $P_\kappa(\mu)$, denotes Legendre's κth coefficient, possess the required properties. In fact the expression

$$\left\{(1 - \mu^2)\sum_{\kappa=1}^{n}\left(A_\kappa \frac{dP_\kappa}{d\mu}\right)\right\}$$

vanishes at $\mu = \pm 1$ for all values of the coefficients A_κ, and it has been shown by R. A. Sampson* that an arbitrary function of μ may be expanded, within the interval $1 > \mu > -1$, in a series of the form

$$\sum_{\kappa=1}^{\infty} A_\kappa (1 - \mu^2)\frac{dP_\kappa}{d\mu}.$$

We assume then for ξ_0

$$\xi_0 = \sum_{\kappa=1}^{n} A_\kappa (1 - \mu^2)\frac{dP_\kappa}{d\mu} \quad\text{..........................(20)},$$

and observe that (18) has the form

$$I = \int_{-1}^{1} \left\{\xi_0^2 - \frac{1 - f^2}{1 - \mu^2}\xi_0^2 - \frac{1}{\beta}\left(\frac{d\xi_0}{d\mu}\right)^2\right\} d\mu \quad\text{.................(21)}.$$

* R. A. Sampson, "On Stokes's current function," *London, Phil. Trans. R. Soc.*, vol. 182 (1891).

Now it follows from well-known formulae that

$$\int_{-1}^{1}\left(\frac{d\xi_0}{d\mu}\right)^2 d\mu=\sum_{\kappa=1}^{n}\frac{2\kappa^2(\kappa+1)^2}{2\kappa+1}A_\kappa^2 \quad\text{.....................(22)},$$

$$\int_{-1}^{1}\frac{\xi_0^2}{1-\mu^2}d\mu=\sum_{\kappa=1}^{n}\frac{2\kappa(\kappa+1)}{2\kappa+1}A_\kappa^2 \quad\text{.....................(23)},$$

and we have also

$$\int_{-1}^{1}\xi_0^2 d\mu=\int_{-1}^{1}\left[\sum_{\kappa=1}^{n}\left\{(1-\mu^2)A_\kappa\frac{dP_\kappa}{d\mu}\right\}\right]^2 d\mu=\int_{-1}^{1}\left[\sum_{\kappa=1}^{n}\left\{A_\kappa\frac{\kappa(\kappa+1)}{2\kappa+1}(P_{\kappa-1}-P_{\kappa+1})\right\}\right]^2 d\mu$$

or

$$\int_{-1}^{1}\xi_0^2 d\mu=\sum_{\kappa=1}^{n}\left\{A_\kappa^2\frac{\kappa^2(\kappa+1)^2}{(2\kappa+1)^2}\left(\frac{2}{2\kappa-1}+\frac{2}{2\kappa+3}\right)\right.$$

$$\left.+2A_\kappa A_{\kappa+2}\frac{\kappa(\kappa+1)(\kappa+2)(\kappa+3)}{(2\kappa+1)(2\kappa+5)}\frac{2}{2\kappa+3}\right\}\text{...(24)},$$

in which A_{n+2} is to be replaced by zero.

On substituting in (21) and putting $\partial I/\partial A_\kappa=0$ for all values of κ from 1 to n we obtain a series of equations to determine the ratios of the coefficients A_κ, and, on elimination of these coefficients, there results an equation connecting f and β which is an approximation to the period equation.

Now equations (6), (9), (16), (20) give, when $s=0$,

$$\zeta=-\frac{h}{a}\cos\sigma t\sum_{\kappa=0}^{n}A_\kappa\kappa(\kappa+1)P_\kappa \quad\text{.....................(25)},$$

and thus, if we put

$$\zeta=\cos\sigma t\times\operatorname*{Lt}_{n\to\infty}\sum_{\kappa=1}^{n}C_\kappa P_\kappa \quad\text{........................(26)},$$

we shall have a series of equations of the type

$$\frac{C_{\kappa-2}}{(2\kappa-3)(2\kappa-1)}+\frac{C_{\kappa+2}}{(2\kappa+3)(2\kappa+5)}$$

$$+C_\kappa\left\{\frac{1}{\beta}+\frac{1-f^2}{\kappa(\kappa+1)}-\frac{2}{(2\kappa-1)(2\kappa+3)}\right\}=0\text{ ...(27)},$$

which agrees with the sequence equation for the C's obtained by Hough (Part I. p. 221) when that equation is simplified by neglecting the effect of the self-attraction.

In the more general case where $s\neq 0$, we substitute from (16) in (14) and find

$$I=\int_{-1}^{1}\left[f^2\left\{\frac{\xi_0^2}{1-\mu^2}+(1-\mu^2)\eta_0^2\right\}+2f\mu\xi_0\eta_0-\frac{1}{\beta}\left(\frac{d\xi_0}{d\mu}-s\eta_0\right)^2\right]d\mu\text{...(28)}.$$

It is now convenient to write

$$\frac{d\xi_0}{d\mu}-s\eta_0=\zeta_0 \quad\text{................................(29)},$$

so that

$$\zeta=\frac{h}{a}\cos(s\phi+\sigma t)\zeta_0\text{.............................(30)}.$$

Then (28) becomes

$$I=\int_{-1}^{1}\left[f^2\frac{\xi_0^2}{1-\mu^2}+2\frac{f}{s}\mu\xi_0\frac{d\xi_0}{d\mu}+\frac{f^2}{s^2}(1-\mu^2)\left(\frac{d\xi_0}{d\mu}\right)^2\right.$$

$$\left.-2\frac{f^2}{s^2}(1-\mu^2)\zeta_0\frac{d\xi_0}{d\mu}-2\frac{f}{s}\mu\zeta_0\xi_0+\frac{f^2}{s^2}(1-\mu^2)\zeta_0^2-\frac{1}{\beta}\zeta_0^2\right]d\mu\text{ ...(31)}.$$

We now assume for ξ_0, ζ_0 series of the form

$$\xi_0=\sum_{\kappa=1}^{n}A_\kappa^{(s)}P_\kappa^{(s)}(\mu),\quad \zeta_0=\sum_{\kappa=1}^{n}C_\kappa^{(s)}P_\kappa^{(s)}(\mu)\text{ ....(32)},$$

where $$P_\kappa^{(s)}(\mu) = (1-\mu^2)^{s/2}\frac{d^s P_\kappa(\mu)}{d\mu^s} \qquad\qquad (33).$$

We transform the integrals

$$\int_{-1}^{1}(1-\mu^2)\left(\frac{d\xi_0}{d\mu}\right)^2 d\mu \text{ and } \int_{-1}^{1}\mu\xi_0\frac{d\xi_0}{d\mu}d\mu$$

by integration by parts into

$$-\int_{-1}^{1}\xi_0\frac{d}{d\mu}\left\{(1-\mu^2)\frac{d\xi_0}{d\mu}\right\}d\mu \text{ and } -\tfrac{1}{2}\int_{-1}^{1}\xi_0^2\,d\mu,$$

and use known formulae* in regard to the "associated functions" $P_\kappa^{(s)}(\mu)$. Then we find after some reductions

$$\begin{aligned}I = {}&\frac{f^2}{s^2}\Sigma\frac{2}{2\kappa+1}\kappa(\kappa+1)\frac{(\kappa+s)!}{(\kappa-s)!}\{A_\kappa^{(s)}\}^2 - \frac{f}{s}\Sigma\frac{2}{2\kappa+1}\frac{(\kappa+s)!}{(\kappa-s)!}\{A_\kappa^{(s)}\}^2\\
&-\frac{2f^2}{s^2}\Sigma C_\kappa^{(s)}A_{\kappa+1}^{(s)}\frac{(\kappa+2)(\kappa+1+s)}{2\kappa+3}\frac{2}{2\kappa+1}\frac{(\kappa+s)!}{(\kappa-s)!}\\
&\qquad+\frac{2f^2}{s^2}\Sigma C_\kappa^{(s)}A_{\kappa-1}^{(s)}\frac{(\kappa-1)(\kappa-s)}{2\kappa-1}\frac{2}{2\kappa+1}\frac{(\kappa+s)!}{(\kappa-s)!}\\
&-\frac{2f}{s}\Sigma C_\kappa^{(s)}A_{\kappa+1}^{(s)}\frac{\kappa+1+s}{2\kappa+3}\frac{2}{2\kappa+1}\frac{(\kappa+s)!}{(\kappa-s)!} - \frac{2f}{s}\Sigma C_\kappa^{(s)}A_{\kappa-1}^{(s)}\frac{\kappa-s}{2\kappa-1}\frac{2}{2\kappa+1}\frac{(\kappa+s)!}{(\kappa-s)!}\\
&+\Sigma\left[\frac{f^2}{s^2}\left\{1-\frac{(\kappa+1)^2-s^2}{(2\kappa+1)(2\kappa+3)}-\frac{\kappa^2-s^2}{(2\kappa-1)(2\kappa+1)}\right\}-\frac{1}{\beta}\right]\frac{2}{2\kappa+1}\frac{(\kappa+s)!}{(\kappa-s)!}\{C_\kappa^{(s)}\}^2\\
&-2\frac{f^2}{s^2}\Sigma C_\kappa^{(s)}C_{\kappa+2}^{(s)}\frac{(\kappa+s+2)(\kappa+s+1)}{(2\kappa+3)(2\kappa+5)}\frac{2}{2\kappa+1}\frac{(\kappa+s)!}{(\kappa-s)!} \qquad\qquad (34).\end{aligned}$$

The equations of the type $\partial I/\partial A_\kappa^{(s)} = 0$ enable us to express any of the coefficients $A_\kappa^{(s)}$ by the formula

$$\left\{\kappa(\kappa+1)-\frac{s}{f}\right\}A_\kappa^{(s)} = \frac{\kappa-s}{2\kappa-1}\left(\kappa+1+\frac{s}{f}\right)C_{\kappa-1}^{(s)} - \frac{\kappa+s+1}{2\kappa+3}\left(\kappa-\frac{s}{f}\right)C_{\kappa+1}^{(s)} \ldots (35),$$

while the equations of the type $\partial I/\partial C_\kappa^{(s)} = 0$ are found to be

$$\begin{aligned}&-\left(\kappa+2+\frac{s}{f}\right)\frac{\kappa+s+1}{2\kappa+3}A_{\kappa+1}^{(s)} + \left(\kappa-1-\frac{s}{f}\right)\frac{\kappa-s}{2\kappa-1}A_{\kappa-1}^{(s)}\\
&-\frac{(\kappa-s)(\kappa-s-1)}{(2\kappa-1)(2\kappa-3)}C_{\kappa-2}^{(s)} - \frac{(\kappa+s+2)(\kappa+s+1)}{(2\kappa+3)(2\kappa+5)}C_{\kappa+2}^{(s)}\\
&+\left\{1-\frac{(\kappa+1)^2-s^2}{(2\kappa+1)(2\kappa+3)}-\frac{\kappa^2-s^2}{(2\kappa-1)(2\kappa+1)}-\frac{s^2}{\beta f^2}\right\}C_\kappa^{(s)} = 0 \ldots (36).\end{aligned}$$

On elimination of the coefficients $A_\kappa^{(s)}$ there results a sequence equation for the coefficients $C_\kappa^{(s)}$, and I have verified that this equation becomes identical with the sequence equation obtained by Hough (Part II. p. 153) for the same coefficients when his equation is simplified by neglecting the effect of the self-attraction.

It will thus be seen that in the problem under discussion the method of Ritz yields very promptly the results already known. It is further noteworthy that the series assumed in this method to represent the components of displacement of the water do formally satisfy the differential equations of the problem, although this is not necessary to the success of the method, and need not be expected to hold in other applications.

* The necessary formulae are all given by Hough in the second of the memoirs cited above.

ON THE LAPLACEAN ORBIT METHODS

By A. O. Leuschner.

The object of this paper is to set forth briefly to what extent the principles proposed by Laplace in his *Mécanique Céleste,* tome I, livre 2, may be made available in practice for the derivation of preliminary orbits of comets, minor planets, and satellites. Owing to its theoretical elegance, Laplace's method has engaged the attention of investigators at various times, but with indifferent success. In attempting to apply his principles to the formulation of practical orbit methods, mathematicians and astronomers have encountered difficulties which have led to the abandonment of Laplace's in favour of the more customary methods.

It is not necessary to recite here the familiar history of Laplace's method, especially in view of the full and accurate historical account included by Charlier in his recent remarkable papers in the *Meddelande* of the Lund Observatory, Nos. 45—47. It is my purpose, however, to show that the difficulties referred to are either not real or that they may be overcome without difficulty, and that the Laplacean principles admit of being cast into orbit methods of greater practical efficiency than the methods ordinarily in use.

If this statement appears to be in direct contradiction to the opinion held by many eminent theoretical astronomers, the contradiction does not lie so much in an actual difference of mathematical results—for no mathematical problem, if correctly solved, admits of contradictory results—but in the difference of interpretation of the requirements of an orbit method. These requirements may be formulated either from an essentially theoretical or from an essentially practical point of view.

The theoretical point of view demands methods which are applicable to the solution of an orbit from the necessary and sufficient conditions under any and all circumstances, as for instance for intervals of any length and in any ratio between the dates of three complete observations. The practical point of view demands methods which admit of as rapid and as accurate a solution of an orbit as the observational material immediately available in the case of a newly discovered object permits. In either case an orbit method is the more efficient the less numerical work is required in producing a result which represents the observational data. But since, in the present state of astronomical science, the observational conditions necessary and sufficient for the derivation of a preliminary orbit are available within a short time after discovery, the theoretical point of view must be of more consequence hereafter to the pure mathematician than to the theoretical astronomer. In other words, astronomical science no longer requires the solution of the purely theoretical problem. What astronomical science does demand is a direct orbit method which admits of the derivation of the best possible preliminary orbit within a short time after discovery and of a method which admits of the convenient correction of the preliminary orbit on the basis of such additional observational material as may subsequently become available. To meet this demand of astronomical science has been the aim of my development of Laplace's principles into direct methods which I have termed "Short Methods."

For several years these methods have been in process of printing in volume VII. of the Publications of the Lick Observatory, but their published appearance has been delayed owing to unavoidable obstacles in the State Printing Office of California. In a very condensed form they have been included in the third edition of Klinkerfues, *Theoretische Astronomie*, by H. Buchholz.

Before proceeding to a discussion of the difficulties referred to above as having been encountered by various investigators in attempting to formulate practical orbit methods on the basis of Laplace's principles, and before demonstrating the advantages of the methods which I have termed "Short Methods," it is necessary to state that in emphasizing the practical value of the Laplacean principles I do not intend to detract in the least from their great but astronomically less important theoretical value. I shall therefore first consider in brief the general theoretical value of the Laplacean principles by means of a summary of their essential features.

Laplace starts with the three linear differential equations of motion of the second order for the two-body problem, under the Newtonian law of attraction as applied to the motion of a material point (the object) about the Sun. He then expresses the heliocentric rectangular co-ordinates of the object in terms of its geocentric co-ordinates and the heliocentric co-ordinates of the Earth. The rectangular co-ordinates are then replaced in terms of polar co-ordinates, and thereby three equations are derived which give the geocentric distance ρ, its velocity ρ', and its acceleration ρ'' at the epoch in terms of the observed co-ordinates (for which we may choose α and δ), their velocities (α', δ'), and their accelerations (α'', δ''), and also in terms of the unknown heliocentric distance r of the object and known quantities depending upon the motion of the Earth about the Sun.

If, therefore, for the present, we assume the co-ordinates α, δ, their velocities α', δ', and their accelerations α'', δ'' to be known, we have three fundamental equations for the solution of the four unknowns ρ, ρ', ρ'', and r. The fourth equation is derived from the triangle Sun-Earth-object, and involves ρ, r, and known quantities. By elimination the problem reduces to an equation of the seventh degree in ρ with not more than two positive real roots, which may be interpolated to six decimals from a table which I have prepared for this purpose, so that the solution may be accomplished without the hitherto necessary laborious numerical approximations.

The direct solution which has just been outlined corresponds to the so-called first hypothesis of other methods. It is evident that the accuracy of Laplace's direct solution depends upon the accuracy of the fundamental observational data for which we have chosen α, δ; α', δ'; α'', δ''. If the epoch is chosen to coincide with the date of one of the observations, then α, δ are fixed numbers, and the accuracy of the Laplacean solution depends upon the accuracy of the adopted values of their velocities and accelerations or, which is an equivalent statement, upon the accuracy of their first and second differential coefficients. In practically all other methods the accuracy of the solution depends upon the accuracy of the adopted values of the ratios of the triangles.

Perhaps the first shot was fired at Laplace's method by Lagrange in a letter (*Oeuvres*, t. XIV., p. 106), in which he says that while analytically Laplace's method constitutes the simplest solution of the problem, in practice it does not afford corresponding advantages because the differential coefficients could not be determined with the necessary accuracy. This far-reaching statement Lagrange intended as a mere opinion, which he proposed to verify later by mathematical demonstration. It is a remarkable fact that Lagrange's opinion, although never verified by himself, has been the chief cause of retarding the further development of the Laplacean method until recent times. It is quoted by Bauschinger, again without demonstration, in *Die Bahnbestimmung der Himmelskörper*, Leipzig, 1906, p. 393, and apparently confirmed by Bauschinger's statement (*loc. cit.*) that every application of methods based on Laplace's furnishes the proof of the laboriousness of the calculation and of the slight accuracy of the result. Harzer, to whom I am indebted for directing my attention to the Laplacean method by his valuable paper in *A. N.*, No. 3371, has abandoned the researches which he commenced with such signal success, and has sought other lines of attack of the orbit problem, as indicated by his recent memoirs in the Publications of the Kiel Observatory. Bruns, who has made a most important contribution to the subject in *A. N.*, No. 2824, particularly with reference to the barycentric parallax, clearly states the theoretical degree of approximation of the Laplacean methods, but renders a final verdict to the effect that the Laplacean methods will remain impracticable, because at least two direct solutions would be necessary to get rid of the parallax.

To cite further authorities in explanation of why astronomers have been loth to adopt the Laplacean methods would seem superfluous.

In analyzing the contention that the Laplacean methods are inferior theoretically, it is necessary to draw a distinction between the approximation of the so-called first hypothesis, direct solution, and that of subsequent hypotheses.

The degree of accuracy of the Laplacean methods has been discussed by Poincaré in a masterly manner in *Bulletin Astronomique*, t. XXIII., and also by myself in Publ. of the L. O., vol. VII. part 7, pp. 267–278. In agreement with Bruns' statement, it is found that for three observations the direct solution by Laplace's method is superior to the first hypothesis of Gauss, inasmuch as for unequal intervals Laplace's error in the geocentric distance is of the second order with respect to the intervals as compared to Gauss's error of the first order, provided the epoch be chosen to coincide with the mean date of the observations. For equal intervals the accuracy is the same, the error being of the second order. For unequal intervals, if the epoch in the Laplacean methods be chosen to coincide with the date of one of the observations (generally the middle), the error is also the same, being of the first order.

It is only recently that Bauschinger himself has published a discussion of the Laplacean methods in comparison to that of Gauss (*Festschrift Heinrich Weber*, Verlag von B. G. Teubner, 1912). Contrary to his previous contention (in his *Bahnbestimmung*), Bauschinger agrees in his *Festschrift* with the foregoing conclusions and independently demonstrates the same.

It would seem, therefore, that the theoretical accuracy of the *direct* solution is no longer doubted, and this question requires no further comment.

Nevertheless Bauschinger still considers it to be a fact, repeatedly confirmed by experience, that the methods based on Laplace do not accomplish in practice what they promise theoretically and supports this contention by the statement that although he has attempted many orbit solutions by Laplace's method, he has rarely succeeded in obtaining a satisfactory result, and that he has learned of similar failures on the part of other computers. In view of the conceded theoretical value of the direct solution, these failures on the part of Bauschinger are the more remarkable, as my own "Short Methods" in their present form have not failed in a single instance upon which they were tried. In parts 8 and 10 of the forthcoming volume VII. of the Publications of the L. O. no less than fifteen complete solutions by the "Short Methods" are reproduced in detail by Professor Crawford. Many of these were selected from the available computational material as severe test cases. For an additional number the accuracy of the results is carefully compared with solutions by other methods.

Contrary to the experience of Bauschinger, in practically every one of these cases experience has demonstrated the superior accuracy and greater rapidity of the "Short Methods."

It is very much to be deplored that the author of the *Festschrift* has refrained from publishing the details of the computations which he has marked as failures, so that the "Short Methods" in their present form might be tested on those examples. Such tests by direct solution, although very desirable, are, however, perhaps unnecessary in view of the following admission by the author (*Festschrift*, p. 10):

"If it can be foreseen that one will succeed in the first approximation (for minor planets this will be the case for an interval up to 30 days between the extreme observations; for comets, for an interval of 6 to 8 days on the average), then neither of the two methods (Laplace's and Gauss's) has an advantage over the other; in particular no cause exists to designate Laplace's method as especially short (Leuschner's "Short Method"), nor Gauss's method as especially accurate."

It is clear, therefore, that the failures did not apply to the intervals referred to above. Furthermore my own experience with the "Short Method" in its present form and with Gauss's method leaves no doubt that the solution by the former would have been not only more accurate, but also shorter.

Now since it is conceded that for the intervals referred to the difficulties apparently encountered with the Laplacean method are not real, so that the "Short Method" is not inferior to that of Gauss, we may proceed to the examination of the concluding statement of the *Festschrift*:

"But if a computation by approximations is in prospect, then it is not economical to work by Laplace's method, whether one carries out the calculation of hypotheses or desists from the solution proper (by successive hypothesis) and undertakes an orbit correction; for then Gauss's method maintains the advantage theoretically as well as in practice."

Can this contention, which, aside from Bruns's verdict regarding the parallax, represents the only remaining criticism of the Laplacean method, be substantiated? The answer is *yes*, if applied to Bauschinger's own definition or interpretation of what ought to constitute "the process of the second hypothesis" in the Laplacean methods. The answer is emphatically *no*, if applied to the "Short Methods" in their present form.

Bauschinger (*Festschrift*, p. 2) recognizes as Laplace's method that indicated by Laplace and later further developed by Bruns (and Schwarzschild), but not the transformations which are due to Harzer (and by his example to myself). He then proceeds to a discussion of the Laplacean method as thus defined and disposes of Harzer's and my own methods by declaring that they do not represent real orbit solutions in the sense of Gauss. Although the "Short Methods" are thus distinguished from the Laplacean methods, and although their convergence is not further discussed, they are included in his final judgment of the Laplacean methods, as quoted above.

It is not my purpose to dwell on this inconsistency, but rather to emphasize that Bauschinger's difficulty arises from his own interpretation of what constitutes an orbit solution. He adheres to the purely theoretical point of view referred to above. But there can be no doubt that in astronomical science that method must be considered as the most satisfactory which solves the problem with the greatest accuracy and the least numerical effort, whether such method be developed in the sense of Gauss or any other sense.

The following free quotation from the *Festschrift* (p. 2) will make Bauschinger's position perfectly clear:

"Laplace and his successors accomplish the approximations following the first in a logical manner by improving the differential coefficients of the observed co-ordinates before obtaining the definitive Elements, and offer therefore a real orbit method in the sense of Gauss. Harzer and Leuschner, on the contrary, recognizing the slight convergence of this procedure, have adopted another course, in that they compare the observations with the results of the first approximation and correct the elements by means of the residuals; that is, then, naturally not an orbit *determination* in the strict sense, but an orbit *correction* and the problem proper is not solved"!

Before proceeding, might it not be well to ask whether a real solution has not been accomplished and whether the problem proper is not solved, no matter how the orbit has been produced, so long as the result is satisfactory and conveniently obtained? Would it not be rather illogical to adhere to Laplace's original process, after its slow convergence has been universally recognized?

On p. 9 of the *Festschrift* we read:

"In Laplace's method the first and second derivatives of the observed co-ordinates are improved by calculating step by step the third, fourth, etc. derivatives...," and also:

"I am not aware that this process has ever been applied in practice; suitable formulae for computation are not to be found even in the more elaborate reproductions of Laplace's method."

Bauschinger's failures with Laplace's method in the second hypothesis cannot, therefore, refer to his own interpretation of Laplace's method, but must have been experienced with Harzer's or my own methods. The limitations of Harzer's method were fully recognized by myself (see Publ. of the L. O., vol. VII., part 1), and gave the impetus to my "Short Methods."

If, then, finally any of Bauschinger's failures are due to the "Short Methods," the computations must have been performed by the "Short Method" in its original form (Publ. of the L. O., vol. VII., part 1) and not by the "Short Methods" in their present form, in which they have appeared only recently for the first time in Buchholz's third edition of *Klinkerfues.*

It is not difficult to state why the *original* "Short Method" may have failed in certain cases in the second approximation. The differential correction is based on series which in certain cases do not converge satisfactorily, as for instance for fast moving comets near perihelion and for long arcs. But this difficulty is now fully overcome in the present form of the "Short Methods" by the introduction of *closed* expressions.

I must therefore contend that Bauschinger's final conclusion, quoted above, that it is not economical to work in the second approximation by Laplace's methods, does not refer to the "Short Methods" in their present form, as these were not known to him when he wrote the *Festschrift.* I must also contend (and the proofs of this contention are contained in Publ. of the L. O., vol. VII., parts 7 to 10) that in their present form the "Short Methods" answer all present requirements of astronomical science, in that they produce results of the greatest accuracy, as far as the available data will permit by *any* method, and in the most convenient form.

The author of the *Festschrift* contends (p. 2) that to raise Laplace's solution to the level of Gauss's it would be necessary to invent devices, such as Gauss gained in so incomparable a manner by the introduction of the ratio of sector to triangle, upon which ultimately the whole brevity and the whole success of his method depends.

It is not necessary to review here the familiar details of preparing for successive hypothesis by the computation of the ratios of sector to triangle in the method of Gauss; let it be sufficient to recall that each ratio must be computed by successive approximation, that their object is to express accurately the ratios of the triangles on the basis of the previous hypothesis, and that each hypothesis is worked out with the ratios of the triangles belonging to the previous hypothesis. This process, upon which Bauschinger lays great stress, is wholly indirect and is entirely dispensed with in the direct differential correction introduced by Harzer and myself. In this connection I may quote from my Introduction to part 7, vol. VII., of the Publ. of the L. O., as follows:

"Instead of forming a second hypothesis in the usual sense by correcting the geocentric velocities and accelerations on the basis of the first solution, it is here proposed, as in part 1, to determine at once from differential relations such corrections to the geocentric distance and to the velocities in the heliocentric rectangular co-ordinates—which, together with the geocentric co-ordinates at the middle date, practically, are the elements of the orbit—as will remove whatever residuals the first solution may leave in the first and third geocentric places. This mode of approximation is not comparable to the method of forming successive hypotheses, but obviously leads directly to the required result. Hitherto it has been applied to the solution of more or less definitive orbits from normal places, with the aid of the differential coefficients of the geocentric co-ordinates with respect to the geometrical elements.

"The advantages of the method of differential correction, over the method of successive hypotheses in improving an orbit, are exactly the same as the advantages of the method of differential correction over the ordinary trials in the successive approximations for the distances in Olbers's and Gauss's methods, etc. In the course of the ordinary trials, the final values of the distances of one trial form the initial values in the next trial, etc. In the method of differential correction, such corrections to the initial values of one trial are derived differentially from the differences between the initial and final values in the same trial, as will produce an agreement of the initial and final values in the next trial. But the number of approximations required by the ordinary trials is, in general, far in excess of that required by the method of differential correction.

"Another simplification involved in the direct solution and the differential correction proposed here is that the consideration of various topics essential to the older methods becomes unnecessary, as, for instance, the consideration of the ratio sector to triangle, etc.

"If it had been deemed desirable or essential, formulae on the plan of successive hypotheses might have been set up without difficulty. For this purpose it is not necessary to determine the corrections to the initial values of the velocities and accelerations from the results of the first solution. Instead, an expression of $\sigma = \rho \cos \delta$ may be derived in which account is taken of the third and higher derivatives expressed in terms of the co-ordinates and their lower derivatives. The actual derivation of these expressions of the higher derivatives is somewhat complicated, although theoretically quite simple, but the resulting formula for κ admits of a second hypothesis as soon as r and $r' = \frac{1}{k}\frac{dr}{dt}$ become known in the first hypothesis."

Preparatory to the differential correction in the "Short Methods" the residuals are not computed, as Bauschinger implies, after the elements have been derived, but they are obtained in the course of the direct solution from the geocentric distance ρ and the heliocentric velocities x', y', z'. The elements are computed after the residuals are found satisfactory, which is generally the case in the first hypothesis (direct solution) for moderate intervals.

Nor should any emphasis be laid upon the fact (*Festschrift*, p. 8) that Laplace's direct solution may lead to considerable errors for fast moving comets, even for short intervals, and to wholly illusory results for long intervals. Comet 1910 a was the fastest moving comet of recent years at the time of discovery. The orbit solution was performed without the least difficulty from one-day intervals by the "Short Method."

The question of a direct solution from very long intervals does not arise in practice, as previous approximations are invariably available in such cases. Long intervals are apt to lead to inaccurate results in any method, if applied to a first approximation instead of being used as a basis of orbit improvement. In computing from a long arc, every computer, no matter what method he may favour, will derive a first approximation of his fundamental data, be it the ratios of triangles, with or without the aid of the ratio of sector to triangle, or be it my starting values ρ, x', y', z', from existing preliminary orbits.

As repeatedly referred to above, the solution of a first approximation from a long arc is of purely theoretical interest, and does not fall within the requirements of astronomical science. Nor does there appear to be any reason why for a long arc the velocities and accelerations of the observed co-ordinates should not be based upon more than three observations if four may become necessary in Gauss's method.

The verdict of Bruns regarding the parallax, referred to above, pertains to a difficulty which I have readily overcome by a process of eliminating the parallax in the direct solution. A striking example of the effect of parallax was furnished by minor planet 1911 MT, where the direct solution (by Messrs Haynes and Pitman, first approximation), with the elimination of parallax, yielded the desired result. In this connection I ought to state that the computers find that they unfortunately committed a numerical error in the representation of their direct solution and then removed residuals which did not exist, so that the differential correction published in L. O. Bulletin 210 was wholly superfluous.

The foregoing discussion inevitably leads to the conclusion that the difficulties that appear to have been encountered in the application of Laplace's method are either not real or that they have finally been overcome in the present form of the "Short Methods." It is also evident that whatever difference of opinion may still exist regarding the value of my own "Short Methods" resolves itself into a difference of interpretation of the requirements of an orbit method.

At the outset of this paper I stated it to be my object to set forth to what extent the principles proposed by Laplace may be made available for derivation of practical orbit methods.

Having discussed and disposed of the arguments against its practical value, I desire to direct attention to the fact that there have existed other and far more serious causes of failure not referred to by any one in his published criticisms, but fortunately overcome in my forthcoming papers in volume VII. of the Publications of the Lick Observatory.

In this connection I may mention that hitherto comet orbits were computed by the general method of Laplace, without hypothesis regarding the eccentricity, instead of by a direct parabolic method; that closed expressions were not available for the second approximation so that it could not be applied to any and all conditions, particularly to arcs of any length; that the solution of the distances was performed by laborious approximations; that the accuracy attainable in each case could not be ascertained in advance; that numerical criteria did not exist to distinguish the physical from the mathematical solutions in the case of three roots in the parabolic method; that, as already referred to, no method existed for completely eliminating the parallax, if necessary; that no criteria were available to distinguish between the feasibility of solutions with or without assumption regarding the eccentricity; that no provision had been made for passing from one class of orbit to another in the course of computation without repeating the solution; that no provision had been made for taking immediate account of the perturbations, and so forth.

For the details of these and many other improvements I must refer to my forthcoming papers and to the examples discussed by Professor Crawford in volume VII. of the Publications of the Lick Observatory.

I trust that I may not appear to have been over-enthusiastic in proclaiming the possibilities of the orbit method initiated by Laplace, and too severe in refuting the objections that have been raised against it. I hope that its latest formulations, whether they still are to be counted among the Laplacean methods or not, will be given frequent and exhaustive trial and that these trials may lead to further improvements of importance.

THE BALANCING OF THE FOUR-CRANK ENGINE

By G. T. Bennett.

§ 1. The problem of balancing the primary and secondary hammering and the primary tilting of a four-cylinder engine with equal cranks has been treated by several writers. Special reference may be made to Schubert ("Zur Theorie des Schlickschen Problems," *Mathematische Gesellschaft*, Hamburg, 1898), Lorenz (*Dynamik der Kurbelgetriebe*, Leipzig, 1901), and papers by Mallock, Macfarlane Gray, Dalby, Schlick, Inglis and others in the *Transactions of the Institution of Naval Architects* (1886–1911).

The present paper is in the main a geometrical study, having the particular requirements of the practical mechanical problem as its basis. The method differs considerably from those hitherto used; and leads not only to some graphical constructions which may assist the draughtsman, but also to some fresh formulae which may facilitate calculations. The results as a whole will, it is hoped, be found to supplement and complete those which are at present available.

The material may best be presented in a sequence determined rather by convenience of theoretical development. At the end of the paper the chief results are summarized and rearranged in a form which may appeal more directly to the engineer engaged in designing a balanced engine.

§ 2. The problem consists virtually in placing four masses M_1, M_2, M_3, M_4 at the ends of equal cranks N_1Q_1, N_2Q_2, N_3Q_3, N_4Q_4, set out perpendicular to a shaft $N_1N_2N_3N_4$, so as to satisfy the six equations

$$\left.\begin{array}{ll}\Sigma M_1 x_1 = 0 & \ldots\ldots(1)\\ \Sigma M_1 y_1 = 0 & \ldots\ldots(2)\\ \Sigma M_1 (x_1^2 - y_1^2) = 0 & \ldots\ldots(3)\\ \Sigma M_1 x_1 y_1 = 0 & \ldots\ldots(4)\\ \Sigma M_1 x_1 z_1 = 0 & \ldots\ldots(5)\\ \Sigma M_1 y_1 z_1 = 0 & \ldots\ldots(6)\end{array}\right\} \text{ or } \left\{\begin{array}{ll}\Sigma M_1 \cos\theta_1 = 0 & \ldots\ldots(7)\\ \Sigma M_1 \sin\theta_1 = 0 & \ldots\ldots(8)\\ \Sigma M_1 \cos 2\theta_1 = 0 & \ldots\ldots(9)\\ \Sigma M_1 \sin 2\theta_1 = 0 & \ldots\ldots(10)\\ \Sigma M_1 z_1 \cos\theta_1 = 0 & \ldots\ldots(11)\\ \Sigma M_1 z_1 \sin\theta_1 = 0 & \ldots\ldots(12)\end{array}\right.$$

where M_1 is the mass at $Q_1(x_1y_1z_1)$ on the circular cylinder $x^2 + y^2 = a^2$; with $x_1 = a\cos\theta_1$, $y_1 = a\sin\theta_1$ for the equivalent forms of equation in cylindrical coordinates; and similarly for the other three masses.

Of the first six equations, or their equivalents, (1) and (2) secure the balance of the primary forces, (3) and (4) secure the balance of the secondary forces, and (5) and (6) secure the balance of the primary couples.

It is convenient to take the origin at the centroid and to associate with the above equations the additional three

$$\Sigma M_1 = M \quad \text{.........(13)},$$

$$\Sigma M_1 z_1 = 0 \quad \text{.........(14)},$$

$$\Sigma M_1 z_1^2 = Mc^2 \quad \text{.........(15)}.$$

The set of six equations (7—12) involve only eight essentially independent quantities; namely the ratios of four masses, the differences of four angles, and the ratios of the differences of four lengths. Any two of the eight, as data, should lead to the values of the rest, and any three should be related. In particular the four crank-angles are necessarily related (20—31) and the four masses are related (95—97). The distances between the cranks are independent, but the distances of the cranks from the centroid are also four related quantities (62—63).

§ 3. We may first consider (1), (2), (3), (4) alone. Then if P_1, P_2, P_3, P_4 are the projections of Q_1, Q_2, Q_3, Q_4 on the plane $z = 0$ the equations may be interpreted, two-dimensionally, as meaning that masses M_1, M_2, M_3, M_4 at P_1, P_2, P_3, P_4 are "dynamically circular." The centroid is at the centre of the circle

$$K \equiv x^2 + y^2 - a^2 = 0 \quad \text{.........(16)}$$

and the moment of inertia about any diameter is the same and equal to

$$\Sigma M_1 x_1^2 = \Sigma M_1 y_1^2 = \tfrac{1}{2} Ma^2 \text{.........(17)}.$$

The product of inertia about any two perpendicular diameters is zero: but the general case of any two lines whatever giving zero product of inertia may be considered: say

$$lx + my + 1 = 0$$

and

$$l'x + m'y + 1 = 0$$

give zero product of inertia. So

$$\Sigma M_1 (lx_1 + my_1 + 1)(l'x_1 + m'y_1 + 1) = 0,$$

and hence

$$\tfrac{1}{2} Ma^2 (ll' + mm') + M = 0;$$

that is

$$ll' + mm' + 2/a^2 = 0 \text{.........(18)}.$$

Hence the two lines are conjugate lines with regard to the imaginary polarizing circle

$$I \equiv x^2 + y^2 + a^2/2 = 0 \quad \text{.........(19)};$$

each line, that is, passes through the pole of the other with regard to the circle I. Pole and polar are on opposite sides of the centre O at distances having a product equal to $a^2/2$. (In particular any equilateral triangle inscribed in K is self-polar with regard to I.)

Take now specially the pair of chords P_1P_2 and P_3P_4, for which the product of inertia is zero. It follows that these lines are conjugate. The pair of chords P_2P_3 and P_1P_4 are similarly conjugate, and also P_3P_1 and P_2P_4. If the points P_1, P_2, P_3

are placed arbitrarily on the circle K, the point P_4 may be found thus:—*Construct the pole F of P_1P_2 with regard to the circle I; then P_3F meets the circle again in P_4* (Fig. 1). Permutations of P_1, P_2, P_3 give three similar constructions all leading to the

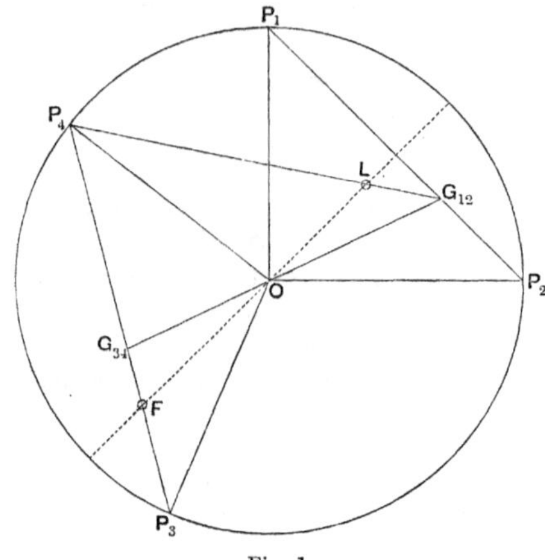

Fig. 1.

same point P_4. This construction gives a graphical form to the relation among the four crank-angles θ_1, θ_2, θ_3, θ_4. It may be remarked that the pole of any chord of K is on the diameter perpendicular to the chord, at the point where it is cut by the side of an equilateral triangle inscribed in K, with either end of the chord as the opposite vertex.

§ 4. With any four crank-angles θ_1, θ_2, θ_3, θ_4 are associated masses M_1, M_2, M_3, M_4 with ratios determinable from any three of the equations (7), (8), (9), (10). These mass-ratios may, however, be found graphically by further use of the method of products of inertia.

Lines through P_3 and P_4 conjugate with regard to I give a zero product of inertia for the four masses, and therefore also for M_1 and M_2 alone. Hence the products of inertia for M_1 and M_2 are equal and opposite. If one of the two lines is drawn parallel to P_1P_2 the other then cuts P_1P_2 in G_{12}, the centroid of M_1 at P_1 and M_2 at P_2. Hence the construction:—*Draw through P_3 a line parallel to P_1P_2 and find L its pole with regard to I. Then P_4L meets P_1P_2 in G_{12}, the centroid of M_1 and M_2 at P_1 and P_2* (Fig. 1). The line $G_{12}O$ will meet P_3P_4 in G_{34}, and the ratios of the four masses may all be got from the segments of the lines $P_1G_{12}P_2$, $P_3G_{34}P_4$ and $G_{12}OG_{34}$: or the construction may be repeated for each pair of masses separately.

§ 5. The relation connecting the differences $\theta_{12} \equiv \theta_1 - \theta_2$, etc. of the crank-angles may be put in various forms, each being an equivalent of the determinant eliminant of equations (7), (8), (9), (10). They are necessary for use later in connection with the formulae involving the masses and the crank-intervals.

The chords P_1P_2 and P_3P_4 of circle K are

$$x\cos\tfrac{1}{2}(\theta_1+\theta_2)+y\sin\tfrac{1}{2}(\theta_1+\theta_2)-a\cos\tfrac{1}{2}\theta_{12}=0,$$

$$x\cos\tfrac{1}{2}(\theta_3+\theta_4)+y\sin\tfrac{1}{2}(\theta_3+\theta_4)-a\cos\tfrac{1}{2}\theta_{34}=0,$$

and these are conjugate with regard to circle I (19) if

$$\cos\tfrac{1}{2}(\theta_1+\theta_2-\theta_3-\theta_4)+2\cos\tfrac{1}{2}\theta_{12}\cos\tfrac{1}{2}\theta_{34}=0 \quad\ldots\ldots\ldots\ldots(20),$$

two other equivalent equations having the same form*.

The symmetric equations

$$\Sigma\cos\tfrac{1}{2}(\theta_1+\theta_2-\theta_3-\theta_4)=0 \quad\ldots\ldots\ldots\ldots\ldots\ldots(21)$$

and

$$\Sigma\cos\tfrac{1}{2}\theta_{12}\,.\,\cos\tfrac{1}{2}\theta_{34}=0 \quad\ldots\ldots\ldots\ldots\ldots\ldots(22)$$

are equivalents of (20).

Multiplied by $2\sin\sigma$ and $2\cos\sigma$, where $2\sigma=\Sigma\theta_1$, (21) gives

$$\Sigma\sin(\theta_1+\theta_2)=0 \quad\text{or}\quad \Sigma x_1y_2=0 \quad\ldots\ldots\ldots\ldots\ldots(23),$$

and

$$\Sigma\cos(\theta_1+\theta_2)=0 \quad\text{or}\quad \Sigma x_1x_2=\Sigma y_1y_2\ldots\ldots\ldots\ldots\ldots(24).$$

From (21) or (22), by expanding,

$$3(1+\tan\tfrac{1}{2}\theta_1\tan\tfrac{1}{2}\theta_2\tan\tfrac{1}{2}\theta_3\tan\tfrac{1}{2}\theta_4)+\Sigma\tan\tfrac{1}{2}\theta_1\tan\tfrac{1}{2}\theta_2=0\ldots\ldots(25),$$

from which is obtainable, without ambiguity of value, the fourth angle associated with any three. This last is a function of the differences of the angles, though not so in appearance, and may be written in each of four forms such as

$$3+\tan\tfrac{1}{2}\theta_{12}\tan\tfrac{1}{2}\theta_{13}+\tan\tfrac{1}{2}\theta_{13}\tan\tfrac{1}{2}\theta_{14}+\tan\tfrac{1}{2}\theta_{14}\tan\tfrac{1}{2}\theta_{12}=0\ldots\ldots(26).$$

This equation gives immediately any one of the angles, θ_1 excepted, from known values of the other three.

The equation (20) multiplied by $2\sin\frac{1}{2}\theta_{12}$ gives

$$\sin\tfrac{1}{2}(\theta_{13}+\theta_{14})-\sin\tfrac{1}{2}(\theta_{23}+\theta_{24})+\sin\tfrac{1}{2}(2\theta_{12}+\theta_{34})+\sin\tfrac{1}{2}(2\theta_{12}-\theta_{34})=0,$$

and so, pairing inside and outside terms,

$$\sin\tfrac{1}{2}(\theta_{12}+\theta_{13})\cos\tfrac{1}{2}\theta_{24}-\sin\tfrac{1}{2}(\theta_{21}+\theta_{23})\cos\tfrac{1}{2}\theta_{14}=0.$$

Hence six formulae of the type

$$\frac{\sin\frac{1}{2}(\theta_{12}+\theta_{13})}{\sin\frac{1}{2}(\theta_{21}+\theta_{23})}=\frac{\cos\frac{1}{2}\theta_{14}}{\cos\frac{1}{2}\theta_{24}} \quad\ldots\ldots\ldots\ldots\ldots\ldots(27).$$

By multiplication of three of these last equations

$$\sin\tfrac{1}{2}(\theta_{13}+\theta_{14})\sin\tfrac{1}{2}(\theta_{21}+\theta_{24})\sin\tfrac{1}{2}(\theta_{32}+\theta_{34})$$

$$=\sin\tfrac{1}{2}(\theta_{23}+\theta_{24})\sin\tfrac{1}{2}(\theta_{31}+\theta_{34})\sin\tfrac{1}{2}(\theta_{12}+\theta_{14}) \quad\ldots\ldots\ldots\ldots(28),$$

of which type there are four formulae.

From (20)

$$\cos(\theta_1+\theta_2-\theta_3-\theta_4)=2(1+\cos\theta_{12})(1+\cos\theta_{34})-1$$

$$=(1+2\cos\theta_{12})(1+2\cos\theta_{34})-2\cos\theta_{12}\cos\theta_{34}.$$

So

$$(1+2\cos\theta_{12})(1+2\cos\theta_{34})=\Sigma\cos(\theta_1+\theta_2-\theta_3-\theta_4)\ldots\ldots\ldots(29),$$

* In any formula which is unsymmetric in the suffixes, they may be permuted, and give rise to a group of equations all of the same type. Any equation quoted by its reference number will be understood to belong to the type so named, but not necessarily to be the particular specimen given.

and hence from the identity $1+2\cos\theta \equiv \sin\frac{3}{2}\theta/\sin\frac{1}{2}\theta$

$$\frac{\sin\frac{3}{2}\theta_{12}\,.\,\sin\frac{3}{2}\theta_{34}}{\sin\frac{1}{2}\theta_{12}\,.\,\sin\frac{1}{2}\theta_{34}} = \text{cet. sim.} \quad \ldots\ldots\ldots\ldots\ldots\ldots(30).$$

From (29)

$$\frac{1+2\cos\theta_{14}}{1+2\cos\theta_{24}} = \frac{1+2\cos\theta_{13}}{1+2\cos\theta_{23}} = \frac{\cos\theta_{14}-\cos\theta_{13}}{\cos\theta_{24}-\cos\theta_{23}} = \frac{\sin\frac{1}{2}(\theta_{13}+\theta_{14})}{\sin\frac{1}{2}(\theta_{23}+\theta_{24})}.$$

Hence six formulae of the type

$$\frac{\sin\frac{1}{2}(\theta_{13}+\theta_{14})}{\sin\frac{1}{2}(\theta_{23}+\theta_{24})} = \frac{\sin\frac{3}{2}\theta_{14}\sin\frac{1}{2}\theta_{24}}{\sin\frac{3}{2}\theta_{24}\sin\frac{1}{2}\theta_{14}} \quad \ldots\ldots\ldots\ldots\ldots\ldots(31).$$

Further, an equation in θ may be put down of which $\theta_1, \theta_2, \theta_3, \theta_4$ are the four solutions. With $2\sigma = \Sigma\theta_1$ the equation

$$\cos(2\theta-\sigma) - \Sigma\cos(\theta+\theta_1-\sigma) + \Sigma\cos\tfrac{1}{2}(\theta_1+\theta_2-\theta_3-\theta_4) = 0 \quad \ldots(32)$$

is identically satisfied by $\theta_1, \theta_2, \theta_3, \theta_4$. Hence, by (21), they satisfy

$$\cos(2\theta-\sigma) = \Sigma\cos(\theta+\theta_1-\sigma) \quad \ldots\ldots\ldots\ldots\ldots\ldots(33).$$

Putting

$$\Sigma\sin\theta_1 = \lambda\sin\epsilon \quad \ldots\ldots\ldots\ldots\ldots\ldots(34),$$

$$\Sigma\cos\theta_1 = \lambda\cos\epsilon \quad \ldots\ldots\ldots\ldots\ldots\ldots(35),$$

so that

$$\lambda^2 = 4 + 2\Sigma\cos\theta_{12} \quad \ldots\ldots\ldots\ldots\ldots\ldots(36)$$

and

$$\tan\epsilon = \frac{\Sigma\sin\theta_1}{\Sigma\cos\theta_1} \quad \text{or} \quad \Sigma\sin(\theta_1-\epsilon) = 0 \quad \ldots\ldots\ldots\ldots\ldots\ldots(37),$$

(33) may be written

$$\cos(2\theta-\sigma) = \lambda\cos(\theta+\epsilon-\sigma) \quad \ldots\ldots\ldots\ldots\ldots\ldots(38).$$

And further it may be noticed that

$$\lambda^2\sin 2\epsilon = \Sigma\sin 2\theta_1 + 2\Sigma\sin(\theta_1+\theta_2) = \Sigma\sin 2\theta_1 \quad \text{from (23)}\ldots\ldots(39),$$

$$\lambda^2\cos 2\epsilon = \Sigma\cos 2\theta_1 + 2\Sigma\cos(\theta_1+\theta_2) = \Sigma\cos 2\theta_1 \quad \text{from (24)}\ldots\ldots(40),$$

hence

$$\lambda^4 = 4 + 2\Sigma\cos 2\theta_{12} \quad \ldots\ldots\ldots\ldots\ldots\ldots(41)$$

and

$$\tan 2\epsilon = \frac{\Sigma\sin 2\theta_1}{\Sigma\cos 2\theta_1} \quad \text{or} \quad \Sigma\sin 2(\theta_1-\epsilon) = 0 \quad \ldots\ldots\ldots\ldots(42).$$

§ 6. Some of the equations of § 5, besides the parent form (20), admit of a geometric interpretation.

The formula (30) shows the angles $\theta_1, \theta_2, \theta_3, \theta_4$ to be such that a pencil of directions given by $\frac{1}{2}\theta_1, \frac{1}{2}\theta_2, \frac{1}{2}\theta_3, \frac{1}{2}\theta_4$ is homographic with a pencil given by $\frac{3}{2}\theta_1, \frac{3}{2}\theta_2, \frac{3}{2}\theta_3, \frac{3}{2}\theta_4$. The same result appears also from (38), which may be rewritten as a homographic relation between $\tan\frac{1}{2}\theta$ and $\tan\frac{3}{2}\theta$.

Formula (28) shows an involution pencil to be given by directions $\theta_1, \theta_2, \theta_3$ paired with $\frac{1}{2}(\theta_1+\theta_4), \frac{1}{2}(\theta_2+\theta_4), \frac{1}{2}(\theta_3+\theta_4)$: and this result may be obtained directly from the figure and construction of § 3. For a quadrilateral may be taken whose sides are the three sides of the triangle $P_1P_2P_3$ and the polar of P_4 with regard to I; the three pairs of vertices being P_1, P_2, P_3 paired with the poles of P_1P_4, P_2P_4, P_3P_4. Projection of these vertices from O gives an involution pencil having the directions of its rays given by the angles in question.

A companion involution may be associated with the last. If angles θ_1, θ_2, θ_3 paired with ϕ_1, ϕ_2, ϕ_3 give an involution, then ϕ_2, ϕ_3 may be replaced by $\theta_1+\phi_1-\phi_3$ and $\theta_1+\phi_1-\phi_2$ respectively and a new involution is obtained. On the present occasion the new involution is given by angles θ_1, θ_2, θ_3 paired with $\frac{1}{2}(\theta_1+\theta_4)$, $\frac{1}{2}(3\theta_1-\theta_3)$, $\frac{1}{2}(3\theta_1-\theta_2)$. This is also shown independently by equation (31). The last two directions are obtainable by taking the images in OP_1 of the perpendiculars from O on P_1P_3 and P_1P_2. This affords a fresh construction, alternative to the polar method of § 3, deriving P_4 from P_1, P_2, P_3; namely:—*Let the images in OP_1 of the perpendiculars from O upon P_1P_3 and P_1P_2 meet these chords P_1P_3 and P_1P_2 in U and V, and let UV meet P_2P_3 in W. Then P_1P_4 is perpendicular to OW.*

Another construction for this involution may be made independently of the circle as follows. Draw a triangle (Fig. 2) with sides α_1, α_2, α_3 representing the given directions $\theta_1, \theta_2, \theta_3$. Take the line which bisects the angle between α_1 and α_2 (observing the proper senses of α_1 and α_2) and reflect it in α_1, giving a line β_3. Similarly, reflect the angle bisector of α_1 and α_3 in α_1 and so get β_2. Draw β_1 joining the point $\beta_2\beta_3$ to the vertex $\alpha_2\alpha_3$. Then reflecting α_1 (with sense included) in β_1 gives the direction θ_4.

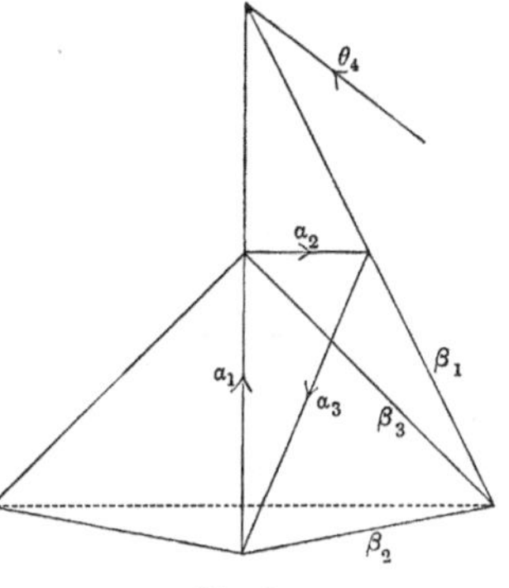

Fig. 2.

A slightly condensed variant of this method may be used. *Take the point of intersection of the angle-bisectors for the vertices on α_1 (being the centre of one of the circles touching the sides of the triangle $\alpha_1\alpha_2\alpha_3$) and take its image in the side α_1. Joining this to the opposite vertex gives β_1, and the reflection of α_1 in β_1 gives the direction θ_4.*

Formulae (34), (35), (39), (40) may be given a graphical interpretation analogous to the mass-quadrilaterals derivable from the equations (7), (8), (9), (10). From (7) and (8) it follows that a closed quadrilateral may be drawn whose sides have the directions θ_1, θ_2, θ_3, θ_4 and lengths proportional to M_1, M_2, M_3, M_4: and (9), (10) show that another quadrilateral may be drawn with the sides unaltered and the angles doubled. Similarly (34), (35) show that four successive unit vectors in directions θ_1, θ_2, θ_3, θ_4 give a resultant vector λ in direction ϵ: and (39), (40) show that four unit vectors in directions $2\theta_1$, $2\theta_2$, $2\theta_3$, $2\theta_4$ give a resultant λ^2 in direction 2ϵ. This affords a test for the proper relation among the directions of the cranks. It has the appearance of a twofold condition; but if, for the doubled angles, the resultant comes squared, then (in general) its angle is doubled, and conversely.

A similar interpretation may be given to (23), (24): namely that a closed equilateral hexagon can be drawn with its sides, say 12, etc., having directions given by $\theta_1+\theta_2$, etc. If the sides are drawn (Fig. 3) in the order 12, 34, 23, 14, 31, 24 the alternate vertices (beginning with the first) lie on a line whose direction is represented by σ, and the zero sum of projections of the sides on this line interprets equation (21).

A variant of this last arises on taking the resultants of vectors 12, 23, 31 and 14, 24, 34 as equal and opposite. Rotating both through the angle $-(\theta_1+\theta_2+\theta_3)$ it follows easily that *equal vectors in directions* θ_1, θ_2, θ_3 *give a resultant whose direction is* $\sigma-\theta_4+\pi/2$: deriving θ_4 again from θ_1, θ_2, θ_3.

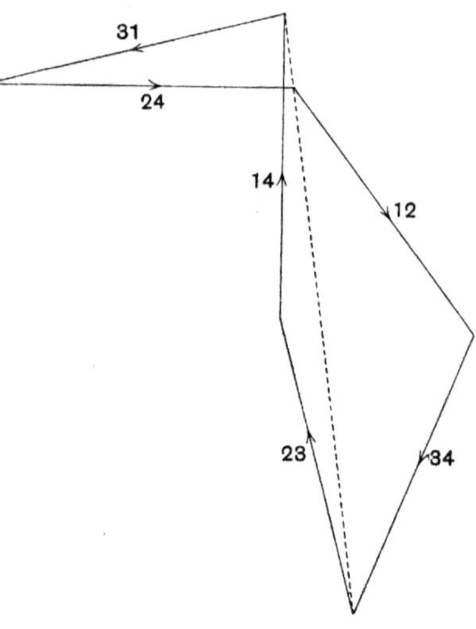

Fig. 3.

§ 7. The zero determinant obtained by eliminating the ratios of the masses from equations (1), (2), (3), (4) may be interpreted as meaning that a rectangular hyperbola (Fig. 4) passes through the points P_1, P_2, P_3, P_4 and O. Or, as equivalent, taking any conic $F(x, y)=0$ through the four points P_1, P_2, P_3, P_4, then $\Sigma M_1 F(x_1, y_1)=0$: and hence, using (1), (2), (3), (4), it follows that if the conic passes through O it is a rectangular hyperbola, and conversely. The polar equations of the hyperbola and the circle K give by elimination of the radius vector the equation (38) for the vectorial angles of the common points. The polar equation of the hyperbola is therefore

$$r\cos(2\theta-\sigma)=\lambda a\cos(\theta+\epsilon-\sigma) \qquad (43).$$

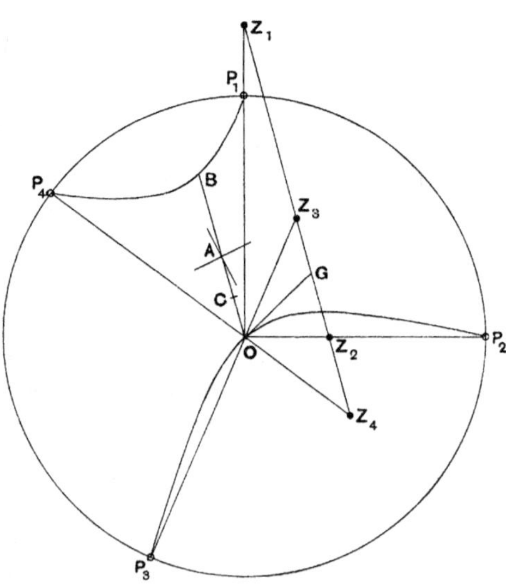

Fig. 4.

The directions of the asymptotes are given by

$$\theta \equiv \tfrac{1}{2}(\pi/2+\sigma) \quad (\text{mod. } \pi/2) \qquad (44),$$

and hence the directions of the axes by

$$\theta \equiv \sigma/2 \quad (\text{mod. } \pi/2) \qquad (45),$$

being the mean values of θ_1, θ_2, θ_3, θ_4 (mod. 2π). This agrees with the equal inclination of any pair of sides of the quadrangle $P_1P_2P_3P_4$ to the axes of the hyperbola. The tangent at O is

$$\theta = \pi/2 + \sigma - \epsilon \qquad (46),$$

and the diameter OAB is $\theta = \epsilon$, with $(\lambda a, \epsilon)$ for coordinates of B: the diameter and tangent being equally inclined to the asymptotes.

It will be found useful later to be able to construct the line OAB without making use of the hyperbola itself. From the resultant of unit vectors in directions θ_1, θ_2, θ_3, θ_4 given in § 6 it follows here that the resultant of vectors OP_1, OP_2, OP_3, OP_4 is OB. This is in agreement with a theorem, known of any rectangular hyperbola cut by a circle, that the centre of mean position C of the four common points is midway between the centres, A and O. As alternative constructions, or as checks on the drawing, it may be remarked that the centre A lies on the nine-point circles of the four triangles formed by the points P_1, P_2, P_3, P_4. Further, also, each of four such pairs of lines as P_1O and P_1B are equally inclined to the known directions of the asymptotes.

§ 8. It may now be shown that any transversal drawn parallel to the line $OCAB$ (Fig. 4) cuts the radii OP_1, OP_2, OP_3, OP_4 in points Z_1, Z_2, Z_3, Z_4 giving a range similar to that of the ends of the cranks N_1, N_2, N_3, N_4 on the axis of the shaft. For this purpose the line $OCAB$ may conveniently be taken, by rotation ϵ of the former axes, as the line $y=0$. The equations of the circle and the hyperbola are then

$$K \equiv x^2 + y^2 - a^2 = 0 \qquad (47),$$

$$H \equiv -x^2 + y^2 + 2\mu xy + \lambda a(x - \mu y) = 0 \qquad (48),$$

where

$$\mu = \tan(2\epsilon - \sigma) \qquad (49).$$

From these equations are derivable $\Sigma x_1 = \lambda a$, $\Sigma y_1 = 0$, $\Sigma x_1 y_1 = 0$; confirming the mean centre C and the equation (42).

Let the cubic curve

$$J \equiv (x - \mu y)K + (x + \mu y + a/\lambda)H = 0 \qquad (50)$$

now be taken, passing through the four points P_1, P_2, P_3, P_4. It reduces to the form

$$J\lambda/a \equiv 2\lambda(1+\mu^2)xy^2/a + (\lambda^2-1)x^2 + 2\mu xy + (1-\lambda^2\mu^2)y^2 = 0 \qquad (51)$$

(in which only the terms, and not the precise coefficients, are here essential).

Taking $\Sigma M_1 J_1/y_1 = 0$

and quoting (1), (2), (4) it follows that

$$\Sigma M_1 x_1^2/y_1 = 0 \qquad (52),$$

or, as equivalent, since $x_1^2 + y_1^2 = a^2$,

$$\Sigma M_1/y_1 = 0 \qquad (53).$$

The equation (48) then gives

$$\Sigma M_1 H_1/y_1 = 0,$$

and so $$\Sigma M_1 x_1/y_1 = \mu M \quad\text{.................................(54).}$$

Take now a transversal $y = h$ cutting OP_1, OP_2, OP_3, OP_4 in points Z_1, Z_2, Z_3, Z_4; and let the tangent $x - \mu y = 0$ cut $y = h$ in $G(\mu h, h)$. Let the lengths GZ_1, GZ_2, GZ_3, GZ_4 be, by temporary anticipation, z_1, z_2, z_3, z_4. So

$$\mu h + z_1 = hx_1/y_1,$$

that is $$z_1 = h(x_1/y_1 - \mu) \quad\text{..............................(55).}$$

Then there follow as consequences

$$\Sigma M_1 z_1 = 0 \quad \text{from (54)} \quad\text{..........................(56),}$$

$$\Sigma M_1 x_1 z_1 = \Sigma M_1 h(x_1^2/y_1 - \mu x_1) = 0 \quad \text{from (52)} \quad\text{...........(57),}$$

and $$\Sigma M_1 y_1 z_1 = \Sigma M_1 h(x_1 - \mu y_1) = 0 \quad\text{........................(58).}$$

Hence lengths z_1, z_2, z_3, z_4 above constructed satisfy (56), (57), (58), and so have ratios which satisfy the equations (5), (6) and (14). *A transversal, therefore, drawn parallel to the line of centres OA is cut by the crank-radii in points Z_1, Z_2, Z_3, Z_4 giving the spacing of the cylinders; and is cut by the tangent at O in the point G which represents the centroid of the four masses M_1, M_2, M_3, M_4 at Z_1, Z_2, Z_3, Z_4.*

§ 9. In terms of three dimensions it now appears, from (55), that the points Q_1, Q_2, Q_3, Q_4 lie on the rectangular hyperbolic paraboloid

$$yz = h(x - \mu y) \quad\text{................................(59)}$$

as well as upon the circular and hyperbolic cylinders

$$K \equiv x^2 + y^2 - a^2 = 0$$

and $$H \equiv -x^2 + y^2 + 2\mu xy + \lambda a(x - \mu y) = 0.$$

The two cylinders have four common generators each cutting the paraboloid in one finite point, and so giving the four points Q_1, Q_2, Q_3, Q_4.

Eliminating x and y from the equations of the three quadrics there results a quartic equation in z, with roots z_1, z_2, z_3, z_4, which may be written

$$\lambda^2 z^2 (z^2 - 2z_0 z + \rho^2) = (z^2 - \rho^2)^2 \quad\text{........................(60),}$$

where $z_0 = -\mu h$; so that G is $(-z_0, h)$ and $\rho = OG$. This quartic is of the form

$$f(z) \equiv z^4 - pz^3 + qz^2 + s = 0 \quad\text{........................(61),}$$

where $p = \Sigma z_1$, $q = \Sigma z_1 z_2$, $s = z_1 z_2 z_3 z_4$; and shows

$$\Sigma z_1 z_2 z_3 = 0 \quad\text{..................................(62),}$$

or, as equivalent, $$\Sigma 1/z_1 = 0 \quad\text{..................................(63).}$$

This is the relation referred to at the outset (§ 2) as connecting the distances of the cranks from the centroid.

If the four points Z_1, Z_2, Z_3, Z_4 are taken as data, and z_1, z_2, z_3, z_4, measured from an arbitrary origin, fail to satisfy the equation (63), then to correct the values and determine G as new origin the values z_1, z_2, z_3, z_4 must all be reduced by a value z determined by

$$\Sigma 1/(z_1 - z) = 0 \quad\text{................................(64),}$$

that is $$\frac{d}{dz}[(z-z_1)(z-z_2)(z-z_3)(z-z_4)]=0 \qquad (65).$$

This cubic in z has three real roots separating the four real roots of the quartic. The root to be chosen (see later, § 14) is the central one of the three. Its value may be calculated by any of the usual methods, Horner's or otherwise. The trigonometric method will be found to apply somewhat favourably; giving

$$z=\Sigma z_1/4+k\cos\phi \qquad (66),$$

where $$12k^2=\Sigma(z_1-z_2)^2 \qquad (67)$$

and $$8k^3\cos 3\phi=\Sigma(z_1-z_2)(z_1-z_3)(z_1-z_4) \qquad (68).$$

Comparison of the equivalent quartics (60), (61) gives

$$\lambda^2-1=\frac{2\lambda^2 z_0}{p}=\frac{(\lambda^2+2)\rho^2}{q}=\frac{-\rho^4}{s} \qquad (69),$$

and hence $$\rho^4-q\rho^2-3s=0 \qquad (70),$$

a quadratic for ρ^2, and $$\frac{2z_0}{p}=\frac{3\rho^2}{2\rho^2+q} \qquad (71),$$

giving z_0. The equations (70), (71) serve to determine the position of O relatively to four given positions of Z_1, Z_2, Z_3, Z_4, and so determine the crank-lines as rays of the pencil projecting Z_1, Z_2, Z_3, Z_4 from O.

§ 10. Equation (60) gives, for $z=z_1$,

$$\lambda OZ_1=(z_1^2-\rho^2)/z_1 \qquad (72),$$

and if $GY_1=GZ_1$ with Y_1 on OZ_1, then $OY_1 . OZ_1=\rho^2-z_1^2$, and so $OY_1=\lambda GY_1$ (Fig. 5).

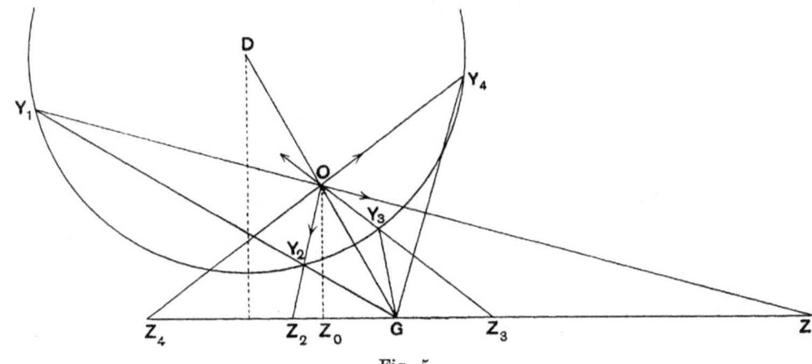

Fig. 5.

The sides and angles of the triangle OGY_1 give a geometric interpretation to the formula (38). Further it follows that the four points Y_1, Y_2, Y_3, Y_4 lie on a circle

$$Y\equiv\lambda^2[(x-\mu h)^2+(y-h)^2]-(x^2+y^2)=0 \qquad (73)$$

with O and G for inverse points. The centre D on OG is $(-p/2, -p/2\mu)$ by use of (69); also $GD=s/\rho^3$; $DO/DG=\lambda^2$; and the square of the tangent from G to the circle Y is s/ρ^2.

It may be noticed that, along with the pencil of lines OY_1, OY_2, OY_3, OY_4 making angles θ_1, θ_2, θ_3, θ_4 with the original initial line, may be associated the pencil GY_1, GY_2, GY_3, GY_4 making angles $2\theta_1-\epsilon$, $2\theta_2-\epsilon$, $2\theta_3-\epsilon$, $2\theta_4-\epsilon$, and the pencil DY_1, DY_2, DY_3, DY_4 making with the same line angles $3\theta_1-\sigma-\pi/2$, etc.

§ 11. If the equations (1), (2), (5), (6) are considered alone they represent the equilibrium of four centrifugal force vectors due to revolving masses M_1, M_2, M_3, M_4, namely $M_1 . Q_1N_1\omega^2$ on Q_1N_1, etc., the lines of action being all normal to the axis of the shaft. Four further conditions of equilibrium are necessary, of which one is geometrical. Any line meeting three of the forces must, for zero moments, meet the other also: and hence the four lines Q_1N_1, etc. are generators of a rectangular hyperbolic paraboloid, as has appeared analytically in § 9 (59). The four generators Q_1N_1, etc. give the same cross-ratio for the pencil $O(P_1P_2P_3P_4)$, representing their directions, as for the range $N_1N_2N_3N_4$ in which the axis z meets them. This is accountable for the possibility (§ 8) of cutting the pencil $O(P_1P_2P_3P_4)$ by a transversal $Z_1Z_2Z_3Z_4$ giving a range similar to $N_1N_2N_3N_4$.

The three further conditions give the ratios necessary for the forces, and hence for the masses. Taking moments about the line through N_3 parallel to Q_4N_4, so that two of the four forces have no moment,

$$M_1z_{13}\sin\theta_{14}+M_2z_{23}\sin\theta_{24}=0,$$

where $z_{13}=z_1-z_3$, etc.; hence

$$\frac{M_1}{M_2}=-\frac{z_{23}\sin\theta_{24}}{z_{13}\sin\theta_{14}} \quad\text{.............................(74)},$$

and similarly

$$\frac{M_1}{M_2}=-\frac{z_{24}\sin\theta_{23}}{z_{14}\sin\theta_{13}},$$

so

$$\frac{z_{13}.z_{24}}{z_{14}.z_{23}}=\frac{\sin\theta_{13}.\sin\theta_{24}}{\sin\theta_{14}.\sin\theta_{23}} \quad\text{...........................(75)},$$

showing again the equal cross-ratios already referred to.

A geometrical equivalent of (74) is

$$\frac{M_1}{M_2}=-\frac{Z_2Z_3.Z_2Z_4.OZ_1}{Z_1Z_3.Z_1Z_4.OZ_2},$$

and hence the ratios of the masses are symmetrically given by

$$M_1.Z_1Z_2.Z_1Z_3.Z_1Z_4/OZ_1=\text{cet. sim.} \quad\text{..................(76)},$$

where OZ_1, as in (72), has its sign determined by comparison of its sense with that of OP_1. From (72) and (76)

$$M_1z_1z_{12}z_{13}z_{14}/(z_1^2-\rho^2)=\text{cet. sim.} \quad\text{.....................(77)}.$$

Now from (61) may be derived the identity

$$zf'(z)-3f(z)\equiv z^4-qz^2-3s$$

$$\equiv(z^2-\rho^2)(z^2+\rho^2-q)+(\rho^4-q\rho^2-3s).$$

Hence, with (70), $\quad z_1f'(z_1)=(z_1^2-\rho^2)(z_1^2+\rho^2-q),$

and so $\quad z_1z_{12}z_{13}z_{14}/(z_1^2-\rho^2)=z_1^2+\rho^2-q \quad\text{....................(78)}.$

Hence $\quad M_1(z_1^2+\rho^2-q)=\text{cet. sim.} \quad\text{........................(79)},$

a formula deriving the ratios of the masses directly from the spacing of the cranks.

§ 12. The equality of the cross-ratios of the range $N_1N_2N_3N_4$ and the pencil of rays $O(P_1P_2P_3P_4)$ permits not only a direction for a transversal cutting the rays of the pencil in a similar range (§ 11), but also, by exchange of two pairs of elements, allows a direction for a transversal cutting the rays, in order, in $Z_2Z_1Z_4Z_3$: and similarly two other directions giving $Z_3Z_4Z_1Z_2$ and $Z_4Z_3Z_2Z_1$. These three directions, known as readily derivable from a mass-quadrilateral by taking the direction of a diagonal, suffice for their purpose but lack the direct correspondence given by the transversal of § 8.

In the figure and construction of § 4 the line $G_{12}OG_{34}$ is the line of the resultant of the vectors $M_1 . OP_1$ and $M_2 . OP_2$, and also of $M_3 . OP_3$ and $M_4 . OP_4$; hence it is parallel to a diagonal of a mass-quadrilateral for which M_1 and M_2 are adjacent sides. The figure of § 4 thus gives at once the three directions of transversals which cut the pencil $O(P_1P_2P_3P_4)$ in a range of points giving the crank-spacings, but with double exchange of two pairs of elements. The transversal of §§ 7—8, drawn parallel to the resultant of four equal vectors parallel to the rays, gives the spacing also, with the advantage of direct correspondence of angle and position.

It may be useful to add here a statement of the relation of the four vectors and the four transversal directions for the general case of the quadrilateral construction used for balancing primary forces and primary couples only. Let p_1, p_2, p_3, p_4, rays of a pencil, give the directions of the four vectors; let t_{14}, t_{24}, t_{34} give the directions of the resultants of pairs of the vectors (being the same lines as t_{23}, t_{31}, t_{12}) and let t have the direction of a fourth transversal giving direct correspondence. Then the four pairs of directions p_1 and p_2, p_3 and p_4, t_{14} and t_{24}, t_{34} and t are in involution. There are three such involutions. These two tetrads of concurrent rays if cut by any conic, conveniently a circle, passing through their common point, give rise to two quadrangles of points $P_1P_2P_3P_4$ and $T_{14}T_{24}T_{34}T$ on the conic, and these two quadrangles have the same diagonal triangle. This (or the usual ruler construction for the sixth element of an involution) gives a means of finding the transversal giving direct correspondence.

In the case of Fig. 4 of § 7 the hyperbola H is itself a conic through O; and the points P_1, P_2, P_3, P_4 give the first quadrangle. The second quadrangle consists of the three points in which OG_{14}, OG_{24}, OG_{34} meet H again, together with the point B, the end of the diameter OAB. This quadrangle has the same diagonal triangle as $P_1P_2P_3P_4$.

§ 13. The product of inertia method of § 3 may now be extended to three dimensions and applied to the system of equations (1—6), (13—15), (17), which have the effect of making the disposition of the four masses at Q_1, Q_2, Q_3, Q_4 "dynamically cylindrical."

Consider any two planes

$$lx + my + nz + 1 = 0,$$
$$l'x + m'y + n'z + 1 = 0,$$

and suppose the product of inertia of the four masses in relation to them to be zero, so that

$$\Sigma M_1 (lx_1 + my_1 + nz_1 + 1)(l'x_1 + m'y_1 + n'z_1 + 1) = 0,$$

whence, from the equations referred to

$$\tfrac{1}{2}a^2(ll'+mm')+c^2nn'+1=0 \quad\text{.......................(80)},$$

showing that the two planes are conjugate with regard to the imaginary spheroid of revolution

$$S\equiv 2(x^2+y^2)/a^2+z^2/c^2+1=0 \quad\text{....................(81)}.$$

The product of inertia is zero if, in particular, one of the planes contains three of the four points Q_1, Q_2, Q_3, Q_4 and the second plane passes through the remaining point. Hence each point is the pole, with regard to S, of the plane of the other three points, and the four points form a self-polar tetrahedron. The same consequence appears otherwise on taking the quadric whose envelope equation is

$$\Sigma M_1(lx_1+my_1+nz_1+1)^2=0 \quad\text{.......................(82)},$$

for its form shows the tetrahedron $Q_1Q_2Q_3Q_4$ to be self-polar, and the equation reduces to the form associated with (80).

The conjugacy of Q_1 and Q_2 with regard to S gives

$$2(x_1x_2+y_1y_2)/a^2+z_1z_2/c^2+1=0,$$

or in cylindrical coordinates

$$2\cos(\theta_1-\theta_2)+z_1z_2/c^2+1=0.$$

Hence $$-z_1z_2/c^2=1+2\cos\theta_{12} \quad\text{..........................(83)},$$

or, as equivalent, $$-z_1z_2/c^2=\sin\tfrac{3}{2}\theta_{12}/\sin\tfrac{1}{2}\theta_{12} \quad\text{.......................(84)}.$$

From (83) $$z_{12}z_3=4c^2\sin\tfrac{1}{2}\theta_{12}\,.\,\sin\tfrac{1}{2}(\theta_{13}+\theta_{23}) \quad\text{.................(85)},$$

and hence from (20) $$z_{12}z_{34}=4c^2\sin\theta_{12}\,.\,\sin\theta_{34} \quad\text{............................(86)},$$

verifying (75) again.

From (83) and (31) is also derivable

$$\frac{z_1}{z_2}=\frac{\sin\tfrac{1}{2}(\theta_{13}+\theta_{14})}{\sin\tfrac{1}{2}(\theta_{23}+\theta_{24})} \quad\text{..........................(87)},$$

and from (85)

$$\frac{z_{12}}{z_{13}}=\frac{\sin\tfrac{1}{2}\theta_{12}\sin\tfrac{1}{2}(\theta_{41}+\theta_{42})}{\sin\tfrac{1}{2}\theta_{13}\sin\tfrac{1}{2}(\theta_{41}+\theta_{43})} \quad\text{.......................(88)}.$$

The equation (74) with (27—28) hence gives the symmetrical formula

$$\frac{M_1}{M_2}=-\frac{\sin\tfrac{1}{2}\theta_{23}\,.\,\sin\tfrac{1}{2}\theta_{24}\,.\,\sin\tfrac{1}{2}(\theta_{23}+\theta_{24})}{\sin\tfrac{1}{2}\theta_{13}\,.\,\sin\tfrac{1}{2}\theta_{14}\,.\,\sin\tfrac{1}{2}(\theta_{13}+\theta_{14})} \quad\text{.................(89)}$$

for the mass-ratios in terms of the angles.

§ 14. Equations (7), (8), multiplied by $\cos\theta_1$ and $\sin\theta_1$ and added, give

$$M_1+M_2\cos\theta_{12}+M_3\cos\theta_{13}+M_4\cos\theta_{14}=0,$$

which is obtainable also directly by resolving the four vectors parallel to the first. Using (83)

$$-2M_1c^2+M_2(z_1z_2+c^2)+M_3(z_1z_3+c^2)+M_4(z_1z_4+c^2)=0,$$

and hence, from (13), (14),

$$M_1(z_1^2+3c^2)=Mc^2 \quad\text{..........................(90)},$$

and $$z_1^2/c^2=(M-3M_1)/M_1 \quad\text{..........................(91)}.$$

Equation (90) confirms the form of (79) and shows further, with (70), that

$$c^2 = (\rho^2 - q)/3 = s/\rho^2 \quad \text{.........................(92)},$$

so that c^2 is the positive root of the quadratic

$$3c^4 + qc^2 - s = 0 \quad \text{............................(93)}.$$

The product $z_1 z_2 z_3 z_4$ ($= s$) is necessarily positive, from (92), and the choice of G (§ 9) is restricted so as to have two of the points Z_1, Z_2, Z_3, Z_4 on each side of it.

Divided by z_1 (90) gives

$$\Sigma M_1 z_1 + 3c^2 \Sigma M_1/z_1 = Mc^2 \Sigma 1/z_1,$$

and hence, from (14) and (63),

$$\Sigma M_1/z_1 = 0 \quad \text{................................(94)}.$$

From (14) and (91) it follows that

$$\Sigma \left[M_1 (M - 3M_1)\right]^{\frac{1}{2}} = 0 \quad \text{........................(95)},$$

the signs of the radicals matching those of the corresponding z's. This is the relation among the masses referred to in § 2.

Again (91) and (63) give

$$\Sigma \left[M_1/(M - 3M_1)\right]^{\frac{1}{2}} = 0 \quad \text{..........................(96)}.$$

These two equations, (95) and (96), must necessarily be only different forms of the same relation. And it may, in fact, be shown that the rationalized form of (95) is a factor of that of (96): so that (95) may be used as involving (96) as well. The rationalized form of (95) is found to be

$$\mu_1 (\mu_1^3 - 4\mu_1\mu_2 + 9\mu_3)^2 = 4\mu_4 (7\mu_1^3 - 27\mu_1\mu_2 + 54\mu_3) \quad \text{........(97)},$$

μ_1 ($= M$), μ_2, μ_3, μ_4 being the sums of the products of M_1, M_2, M_3, M_4 taken one, two, three and four at a time.

[The algebraic theorem which occurs here—of a form somewhat peculiar—is that if $x = (a + b + c + d)/3$ gives a root of the equation

$$\sqrt{a(x-a)} + \sqrt{b(x-b)} + \sqrt{c(x-c)} + \sqrt{d(x-d)} = 0,$$

it is necessarily a double root.]

§ 15. One or two extra formulae may finally be noted as close associates of some of the foregoing:

$$\Sigma M_1/z_1^3 = \tfrac{1}{3} M \Sigma 1/z_1^3 \quad \text{..........................(98)},$$

of the same type as (14), (15) and (94);

$$z_1/h = \cot(\theta_1 - \epsilon) + \tan(\sigma - 2\epsilon) \quad \text{.......................(99)},$$

giving, with (37) and (42), z's in terms of θ's, an equivalent of (55);

$$\tan(\sigma - \theta_4 + \pi/2) = \frac{\sin\theta_1 + \sin\theta_2 + \sin\theta_3}{\cos\theta_1 + \cos\theta_2 + \cos\theta_3} \quad \text{...........(100)};$$

an immediate and unsymmetrical equivalent of (21), confirming the last construction of § 6; and

$$(\rho + z_1)(\rho + z_2)(\rho + z_3)(\rho + z_4)/(\rho + z_0)$$
$$= (\rho - z_1)(\rho - z_2)(\rho - z_3)(\rho - z_4)/(\rho - z_0) \quad \text{..............(101)},$$

verifiable from (69), (70). This last presents itself if the ratios of the M's are taken from (77) and are substituted in (3) and (4) by way of verification.

§ 16. The unbalanced secondary couple remains to be noticed. The vector polygon representing it has sides M_1z_1, etc. in directions $2\theta_1$, etc., and the direction and magnitude of their resultant may be found. Equation (38) gives

$$\Sigma M_1 z_1 \cos(2\theta_1 - \sigma) = \lambda \Sigma M_1 z_1 \cos(\theta_1 + \epsilon - \sigma),$$

and hence, from (7) and (8),

$$\cos\sigma \,.\, \Sigma M_1 z_1 \cos 2\theta_1 + \sin\sigma \,.\, \Sigma M_1 z_1 \sin 2\theta_1 = 0 \ldots\ldots\ldots\ldots(102),$$

showing the resultant to have direction $\sigma + \pi/2$. As the shaft revolves, the secondary couple takes alternately a zero or maximum value as an axis or an asymptote of the rectangular hyperbola H becomes parallel to the plane of the axes of the cylinders.

The square of the resultant is

$$(\Sigma M_1 z_1 \cos 2\theta_1)^2 + (\Sigma M_1 z_1 \sin 2\theta_1)^2 = \Sigma M_1^2 z_1^2 + 2\Sigma M_1 M_2 z_1 z_2 \cos 2\theta_{12}.$$

This may be put in terms of M's and z's by use of (83) and may be then reduced by use of (91) and (92) to $(M\rho)^2$. The resultant side closing the secondary couple polygon is thus $M\rho$ in direction $\sigma + \pi/2$.

§ 17. Some modes of application of the various results to the practical problem of design may now, in conclusion, be presented. Four chief cases may be sufficient to consider, differing according as the data consist of crank-angles or cylinder-spacings, and differing according as the work is to be analytical or mainly graphical. For each case a brief itinerary is here sketched, with references to the fuller details of the earlier paragraphs, and with indications of alternative routes.

I. *Graphical method: the directions of three cranks as data.*

(i) To find the direction of the fourth crank. Use the polarizing construction of § 3. Radii OP_1, OP_2, OP_3 of circle radius a represent given cranks: and P_3P_4 must pass through the pole of P_1P_2 with regard to circle I, centre O and radius-squared equal to $-a^2/2$. (Fig. 1.)

Three alternative constructions are given in § 6.

(ii) To find the mass-ratios. Join P_4 to the pole of the line through P_3 parallel to P_1P_2. This line cuts P_1P_2 in G_{12}, the centroid of M_1 at P_1 and M_2 at P_2. (Fig. 1, § 4.)

(iii) To find the cylinder-spacing. Find the resultant OB of equal vectors OP_1, OP_2, OP_3, OP_4 and draw a parallel to OB cutting the vectors in Z_1, Z_2, Z_3, Z_4. (§ 8.) For other constructions for the line OB see § 7. Also alternatively take OG_{12} and draw a parallel cutting the vectors in Z_2, Z_1, Z_4, Z_3. (§ 12.)

As a check on (i), if the angles of the equal vectors of (iii) are doubled the angle of the resultant should come doubled and its magnitude OB^2/OP_1. (§ 6.)

II. *Analytical method: the angles θ_1, θ_2, θ_3 for three cranks being given.*

(i) For the fourth angle θ_4 use (25) or (26) or (100).

(ii) For the mass-ratios use (89).

(iii) For the spacings use (87) or (88): or as alternatives (37) and (99).

III. *Graphical method: collinear points Z_1, Z_2, Z_3, Z_4 giving the spacing.* (Fig. 5.)

(i) Find G so that $\Sigma 1/z_1 = 0$ (two terms positive and two negative), where $z_1 = GZ_1$, etc. (63, § 9.)

(ii) With $p = \Sigma z_1$, $q = \Sigma z_1 z_2$, $s = z_1 z_2 z_3 z_4$, solve the quadratic $\rho^4 - q\rho^2 - 3s = 0$. Draw circle centre G and radius ρ. (70.)

(iii) Take Z_0 so that $GZ_0 = \frac{3}{2}p\rho^2/(2\rho^2 + q)$ and draw the line perpendicular to $Z_1Z_2Z_3Z_4$ through Z_0.

(iv) From O, either common point of circle (ii) and line (iii), project Z_1, Z_2, Z_3, Z_4. These rays give the crank-directions. To give them the proper senses take those of GZ_1, GZ_2, GZ_3, GZ_4 in the order of occurrence of the Z's, and alternately repeat and reverse them.

(v) The mass-ratios are given by

$$M_1 . Z_1Z_2 . Z_1Z_3 . Z_1Z_4/OZ_1 = \text{cet. sim.} \quad (76.)$$

Or as alternative draw GX perpendicular to $Z_1Z_2Z_3Z_4$ and make $GX^2 = 3s/\rho^2$ $(= 3c^2)$ and then (90)

$$M_1 . XZ_1^2 = M_2 . XZ_2^2 = M_3 . XZ_3^2 = M_4 . XZ_4^2.$$

The properties of the circle Y (§ 10) may be used as a check on the working of (ii) and (iii). Its centre is at D on GO where $GD = s/\rho^3$: and the foot of the perpendicular from D on $Z_1Z_2Z_3Z_4$ is distant $-\frac{1}{2}\Sigma GZ_1$ from Z_0. The tangents from G to the circle Y have length c, the inertia constant, and the angle between them is $2 \sin^{-1} \lambda$.

IV. *Analytical method: the differences of z_1, z_2, z_3, z_4 as data.*

(i) Determine the zero so that $\Sigma 1/z_1 = 0$. (66—68.)

(ii) Find the positive root of the quadratic in c^2

$$3c^4 + qc^2 - s = 0. \quad (93.)$$

(iii) To find the angles use (83). The angles, as determined by their cosines, are ambiguous only in sign; and are to be taken so that $\theta_{12} + \theta_{23} + \theta_{31} = 0$, etc.

(iv) To find the mass-ratios use (90).

The selections above made are framed as being the most apt, possibly, for handling the practical problem of construction: but, as may be seen, they are not entirely exhaustive of the copious supply of simple material which is available for the purpose.

§ 18. It may be useful to collect here also the various vector polygons which occur:

(i) Four vectors $(1, \theta_1)$, etc. give resultant (λ, ϵ). The angle ϵ gives the direction of the shaft-transversal.

(ii) Four vectors $(1, 2\theta_1)$, etc. give resultant $(\lambda^2, 2\epsilon)$.

(iii) Three vectors $(1, \theta_1)$, $(1, \theta_2)$, $(1, \theta_3)$ give resultant in direction $\sigma - \theta_4 + \pi/2$ giving θ_4.

(iv) Six vectors $(1, \theta_1 + \theta_2)$, etc. give a closed hexagon.

(v) Four vectors (M_1, θ_1), etc. give a closed quadrilateral for the primary forces.

(vi) Four vectors $(M_1, 2\theta_1)$, etc. give a closed quadrilateral for the secondary forces.

(vii) Four vectors $(M_1 z_1, \theta_1)$, etc. give a closed quadrilateral for the primary couples.

(viii) Four vectors $(M_1 z_1, 2\theta_1)$, etc. give a resultant $(M\rho, \sigma + \pi/2)$ representing the unbalanced secondary couple.

The illustrative figures, though differing in scale, have all been drawn for the case of the exact values

$$\tan \tfrac{1}{2}\theta_1 = 1, \quad z_1 = 715, \quad M_1 = 81,$$
$$\tan \tfrac{1}{2}\theta_2 = 0, \quad z_2 = -195, \quad M_2 = 165,$$
$$\tan \tfrac{1}{2}\theta_3 = -\tfrac{3}{2}, \quad z_3 = 165, \quad M_3 = 169,$$
$$\tan \tfrac{1}{2}\theta_4 = 3, \quad z_4 = -429, \quad M_4 = 125,$$

with $z_0 = -132$, $h = 231$, $\rho = 33\sqrt{65}$, $c = 65\sqrt{33}$, $\mu = \tfrac{4}{7}$, $\lambda^2 = \tfrac{32}{65}$, $\tan \epsilon = -\tfrac{11}{3}$, $\tan \sigma = \tfrac{7}{4}$.

Note. In discussion of the above paper Professor Morley remarked that the directions of the sides of any triangle and of its Euler line (joining the circumcentre to the centroid) are symmetrically related (*Mathematische Annalen*, 1899, pp. 411—412. "Some polar constructions," F. Morley); and that the relation is of the type given by the equations of § 5. From this follows a further construction for the direction of the fourth crank: namely, if a triangle is drawn with sides given by angles $\tfrac{1}{2}\theta_1$, $\tfrac{1}{2}\theta_2$, $\tfrac{1}{2}\theta_3$, the direction of the Euler line gives the angle $\tfrac{1}{2}\theta_4$.

The construction may be immediately associated with Fig. 1 and the unit polygon of § 6 and § 18 (iii). The Euler line of the triangle $P_1P_2P_3$ joins O to the centroid and so has direction $\sigma - \theta_4 + \pi/2$: and if the sides of the triangle and its Euler line are reflected in a line having direction $\tfrac{1}{4}(\theta_1 + \theta_2 + \theta_3 + \pi)$ the directions obtained are $\tfrac{1}{2}\theta_1$, $\tfrac{1}{2}\theta_2$, $\tfrac{1}{2}\theta_3$, $\tfrac{1}{2}\theta_4$.

SOME THEOREMS RELATING TO THE RESISTANCE OF COMPOUND CONDUCTORS

BY T. J. I'A. BROMWICH.

In what follows some of the familiar theorems for a system of linear conductors are extended to include cases of solid conductors, and also cases when both solid and linear conductors are present in the system.

The only paper dealing with this topic with which I am acquainted is one published last year by Dr G. F. C. Searle*, in which the method of attacking the problem is quite different. In modern practical work with low standard resistances, the use of four-terminal standards has proved important, and it seems worth while, therefore, to sketch two simple applications of the method to such standards.

I. *The variational equation for the system of currents flowing in a solid conductor, or a system of conductors.*

The currents being supposed *steady*, the electric force (X, Y, Z) at any point of the conductor is derived from a potential V; thus if (u, v, w) denote the actual components of current and (u', v', w') any geometrically possible components of current, we find

$$Xu + Yv + Zw = -\frac{\partial}{\partial x}(uV) - \frac{\partial}{\partial y}(vV) - \frac{\partial}{\partial z}(wV),$$

and
$$Xu' + Yv' + Zw' = -\frac{\partial}{\partial x}(u'V) - \frac{\partial}{\partial y}(v'V) - \frac{\partial}{\partial z}(w'V),$$

because both (u, v, w) and (u', v', w') satisfy the equation of continuity

$$\frac{\partial u}{\partial x} + \frac{\partial v}{\partial y} + \frac{\partial w}{\partial z} = 0.$$

Hence on applying Green's lemma we find the equation

$$\int (Xu' + Yv' + Zw')\, d\tau = -\int (lu' + mv' + nw')\, V dS \text{ ....(1)},$$

with a similar equation containing u, v, w.

In equation (1) the volume-integral extends throughout the interior of any continuous conductor, and the surface-integral to the boundary of the conductor, l, m, n being the direction-cosines of the outward normal.

* *Electrician*, March 31—April 21, 1911.

But we can now extend (1) to cover the whole system of conductors; for at any interface between two conductors the potential V and the normal component of current $(lu' + mv' + nw')$ will be continuous. Thus the contributions to the right-hand side of (1) from the interfaces of adjoining conductors will cancel out, and *we are left with a surface-integral extending only to the places at which current is led into or out of the system.*

In most cases of practical interest the leads may be regarded as confined to a number of isolated points; and if these points are denoted by 1, 2, ..., n, the corresponding currents (which are *supplied*) being called ξ_1', ξ_2', ..., ξ_n' (in our hypothetical system), we can write

$$-\int (lu' + mv' + nw')\, dS_1 = \xi_1', \text{ etc.,}$$

so that equation (1) takes the modified form

$$\int (Xu' + Yv' + Zw')\, d\tau = \xi_1' V_1 + \xi_2' V_2 + \dots + \xi_n' V_n \dots\dots\dots\dots (2),$$

where
$$0 = \xi_1' + \xi_2' + \dots + \xi_n'.$$

Since (as we have already remarked) this equation remains true when the *actual* current system replaces the hypothetical, we see that

$$\int (X\,\delta u + Y\,\delta v + Z\,\delta w)\, d\tau = V_1 \delta\xi_1 + V_2 \delta\xi_2 + \dots + V_n \delta\xi_n \dots\dots\dots (3),$$

where
$$0 = \delta\xi_1 + \delta\xi_2 + \dots + \delta\xi_n,$$

and
$$\delta u = u' - u, \quad \delta v = v' - v, \quad \delta w = w' - w.$$

This equation (3) is analogous to the equation of virtual work in mechanical systems; here it is simply an expression of the fact that X, Y, Z is derived from a potential-function.

If we suppose the conductor to be isotropic (but not necessarily uniform), the relation between X, Y, Z and u, v, w is

$$X = \rho u, \quad Y = \rho v, \quad Z = \rho w,$$

where ρ is the specific resistance at the point considered. Thus if u', v', w' represents another current system in the same conductor, we have

$$Xu' + Yv' + Zw' = \rho(uu' + vv' + ww') = X'u + Y'v + Z'w \dots\dots\dots (4).$$

More generally, if the conductor is crystalline, all experimental evidence shews that the matrix of resistance coefficients is symmetrical, so that we can write

$$X = au + hv + gw, \quad Y = hu + bv + fw, \quad Z = gu + fv + cw,$$

and equation (4) remains true.

Combining (2) and (4) we then see that

$$\xi_1' V_1 + \xi_2' V_2 + \dots + \xi_n' V_n = \xi_1 V_1' + \xi_2 V_2' + \dots + \xi_n V_n',$$

which at once yields the reciprocal relation:

If unit current is led in at 1 and out at 2, the potential difference between 3 and 4 $(V_3 - V_4)$ *is the same as that between 1 and 2* $(V_1' - V_2')$ *when unit current is led in at 3 and out at 4.*

This theorem has long been known for the case of *linear* conductors; it was first proved in its general form in the paper by Dr Searle already quoted (§§ 3, 7).

Returning now to the variational form of equation we have the results

$$\int \rho\,(u\delta u + v\delta v + w\delta w)\,d\tau = \Sigma V_r\delta\xi_r,$$

or

$$\int \{au\delta u + h\,(u\delta v + v\delta u) + \ldots\}\,d\tau = \Sigma V_r\delta\xi_r,$$

the former referring to an isotropic system, the latter to a crystalline system.

Thus, writing

$$H = \tfrac{1}{2}\int \rho\,(u^2 + v^2 + w^2)\,d\tau \text{ or } \tfrac{1}{2}\int (au^2 + 2huv + \ldots)\,d\tau \quad \ldots\ldots\ldots(5),$$

as the case may be, we see that when the variations are *small* we have the variational equation

$$\delta H = \Sigma V_r\delta\xi_r, \text{ where } 0 = \Sigma\delta\xi_r \quad \ldots\ldots\ldots\ldots\ldots\ldots\ldots\ldots(6),$$

which is correct to the first order.

And when the *exact* form of δH is needed we can write (for an isotropic system)

$$\delta H = \int \rho\,(u\delta u + v\delta v + w\delta w)\,d\tau + \tfrac{1}{2}\int \rho\,\{(\delta u)^2 + (\delta v)^2 + (\delta w)^2\}\,d\tau = \Sigma V_r\delta\xi_r + \delta_2 H, \text{ say} \quad \ldots\ldots\ldots(7).$$

The necessary modification in (7) for a crystalline system is obvious.

It is easy to modify these results to include the case when some (or all) of the conductors are treated as linear wires, some of which may contain batteries. For instance suppose that 1, 2 are points joined by a wire of resistance R containing a battery E, which carries a current I: then $\xi_1 = -I$, $\xi_2 = +I$, and

$$V_1 - V_2 + E = RI.$$

Thus (6) takes the form

$$\delta H_1 = E\delta I - RI\delta I + V_3\delta\xi_3 + \ldots + V_n\delta\xi_n,$$

where H_1 refers only to the solids. Hence if

$$H = H_1 + \tfrac{1}{2}RI^2,$$

we find

$$\delta H = E\delta I + V_3\delta\xi_3 + \ldots + V_n\delta\xi_n.$$

It is easy to generalize the last equation to include any number of wires and then we obtain the result

$$\delta H = \Sigma E_s\delta I_s + \Sigma V_r\delta\xi_r \quad \ldots\ldots\ldots\ldots\ldots\ldots\ldots\ldots(6\,a),$$

where

$$H = H_1 + \tfrac{1}{2}\Sigma R_s I_s^2, \text{ and } 0 = \Sigma\delta\xi_r.$$

Similarly in (7) it is only necessary to add the term $\Sigma E_s\delta I_s$ to the right-hand side, in order to include the effect of batteries.

II. *Deductions from the variational equation.*

A. If the system is free from internal batteries, and if the currents supplied from external sources are *prescribed*, we have in (6 *a*)

$$E_s = 0, \quad \delta\xi_r = 0,$$

and so to the first order

$$\delta H = 0.$$

Thus H satisfies the condition of being *stationary*: and using (7) we see that accurately

$$\delta H = \delta_2 H,$$

and $\delta_2 H$ is essentially positive. Consequently *with prescribed currents at the points of supply, the value of H is a minimum in the actual current system.* This is the analogue of Kelvin's theorem for a dynamical system; it is of course a familiar result for a system of linear conductors.

B. When batteries are present, it is easy to extend (A) to shew that $(H - \Sigma E_s I_s)$ is now a minimum, by using the extended form of (7).

C. It follows from (A), that when zero currents are supplied and no batteries are present, no currents will flow in the system; for the case of zero current makes H zero, and (since H is in general *positive*) this current system makes H a *minimum*, and is therefore the actual system.

Suppose now that v_r denotes the potential when unit current is supplied at r and taken out at n; then the function

$$U = V - (\xi_1 v_1 + \xi_2 v_2 + \ldots + \xi_{n-1} v_{n-1})$$

will be a potential-function which corresponds to the case in which zero current is supplied at each terminal. Consequently U corresponds to the case of zero current; or *U is a constant.*

Hence in general we can write

$$V = (\xi_1 v_1 + \xi_2 v_2 + \ldots + \xi_{n-1} v_{n-1}) + \text{const.}$$

when the currents $\xi_1, \xi_2, \ldots, \xi_{n-1}$ are supplied at $1, 2, \ldots, (n-1)$, and the whole of them are taken out at n; this of course is equivalent to supposing $\xi_1, \xi_2, \ldots, \xi_n$ supplied at $1, 2, \ldots, n$, with the condition

$$\xi_1 + \xi_2 + \ldots + \xi_n = 0.$$

D. It is clear on substituting the value of V found in (C) that H becomes a quadratic function of $\xi_1, \xi_2, \ldots, \xi_{n-1}$; and so (6 a) leads to the result

$$\delta H = V_1 \delta\xi_1 + V_2 \delta\xi_2 + \ldots + V_{n-1} \delta\xi_{n-1} - V_n (\delta\xi_1 + \delta\xi_2 + \ldots + \delta\xi_{n-1}),$$

provided that $\delta I_s = 0$. Consequently

$$\frac{\partial H}{\partial \xi_1} = V_1 - V_n, \quad \frac{\partial H}{\partial \xi_2} = V_2 - V_n, \quad \ldots, \quad \frac{\partial H}{\partial \xi_{n-1}} = V_{n-1} - V_n \quad \ldots\ldots\ldots(8)$$

are the general equations for our system.

III. *Special consideration of four-terminal systems.*

In modern accurate work on standards of low resistance it is usual to employ conducting strips with four terminals: two of these terminals may be used as current leads, and the other two will then be used for determination of potential-difference.

The so-called resistance of the standard will then be the potential difference corresponding to unit current supplied.

In such cases we have four points 1, 2, 3, 4 and

$$\xi_1 + \xi_2 + \xi_3 + \xi_4 = 0.$$

Thus H can be regarded as a quadratic function of ξ_1, ξ_2, ξ_3; it will therefore contain *six* resistance coefficients and so *the* resistance of the standard depends on the arrangement of the terminals. Of course in practical cases two definite terminals (and only these) are used as leads, and the other two for potential readings only: thus, for instance, let the potential readings be taken at 2, 3, the leads being at 1, 4. Then we can write

$$H = \tfrac{1}{2}(a\xi_1^2 + b\xi_2^2 + c\xi_3^2 + 2f\xi_2\xi_3 + 2g\xi_3\xi_1 + 2h\xi_1\xi_2) \quad \text{...........(9)}$$

in general; and so from (8) and (9) we have

$$\left.\begin{aligned} V_1 - V_4 &= a\xi_1 + h\xi_2 + g\xi_3 \\ V_2 - V_4 &= h\xi_1 + b\xi_2 + f\xi_3 \\ V_3 - V_4 &= g\xi_1 + f\xi_2 + c\xi_3 \end{aligned}\right\} \quad \text{........................(10).}$$

Thus when unit current enters at 1 and leaves at 4, so that $\xi_1 = 1$, $\xi_2 = 0$, $\xi_3 = 0$, we find

$$V_2 - V_3 = h - g.$$

Hence *in this arrangement of leads, the resistance of the conductor is* $(h - g)$.

An interchange of the leads with the potential-terminals will not alter the result: this is evident from the reciprocal relation above (p. 236), or directly from (10) by taking the case $\xi_1 = 0$, $\xi_2 = 1$, $\xi_3 = -1$, which gives $V_1 - V_4 = h - g$, or the same potential-difference as in the first arrangement.

In some of the earlier experimental work it was supposed that any four-terminal system could be effectively represented by a model of five linear resistances with four terminals as indicated roughly in the figure.

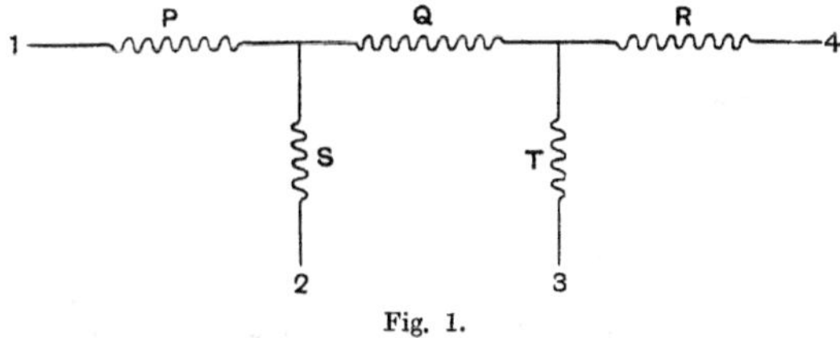

Fig. 1.

It is evident that this system cannot be equivalent to the general conductor, which contains *six* coefficients (instead of *five*): in this special case the form for H is

$$\tfrac{1}{2}\{P\xi_1^2 + S\xi_2^2 + T\xi_3^2 + Q(\xi_1 + \xi_2)^2 + R(\xi_1 + \xi_2 + \xi_3)^2\}.$$

Hence in this model $f = R = g$; which is not usually the case. In fact if the points 1, 2 were used as leads and 3, 4 for potential readings, it is evident that the apparent resistance (being $g - f$ in general) would be zero in the model; a fact which may be seen from the figure by inspection.

To conclude we shall give the theory (from our present standpoint) of one of the laboratory methods for finding the resistance of a four-terminal standard.

A bridge is constructed as in the rough sketch; a battery is placed between LM, and the resistances R, S are adjusted until a balance is obtained when a galvanometer joins 2 to K. Then the galvanometer is placed between 3 and K, and the resistances R, S are adjusted to R', S' so as to again produce a balance.

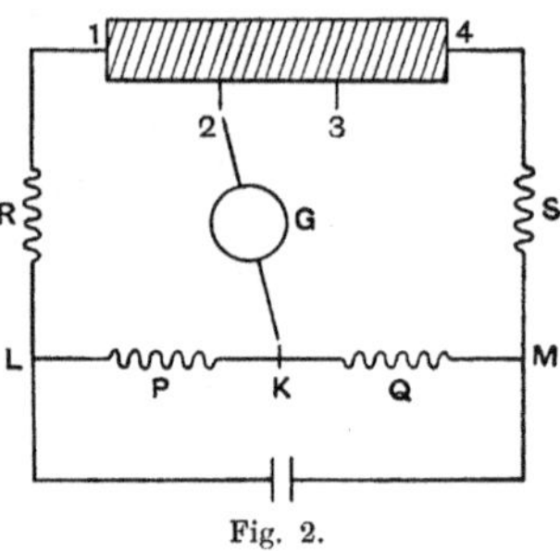

Fig. 2.

In the first arrangement (which is that sketched) $\xi_3 = 0$, and so equations (10) give

$$V_1 - V_4 = a\xi_1 + h\xi_2, \quad V_2 - V_4 = h\xi_1 + b\xi_2.$$

Thus when the balance has been obtained (so that $\xi_2 = 0$),

$$V_1 - V_2 = (a - h)\,\xi_1, \quad V_2 - V_4 = h\xi_1.$$

Hence $$V_L - V_2 = (R + a - h)\,\xi_1, \quad V_2 - V_M = (S + h)\,\xi_1,$$

and so as in the usual Wheatstone Bridge method we have

$$(R + a - h)/P = (S + h)/Q \quad \text{........................(11)},$$

because $V_2 = V_K$.

In the second arrangement we get similarly

$$V_1 - V_3 = (a - g)\,\xi_1, \quad V_3 - V_4 = g\xi_1,$$

leading to $$(R' + a - g)/P = (S' + g)/Q \quad \text{........................(12)}.$$

Hence on subtracting (11) from (12) we eliminate a, and find

$$\{(R' - R) + (h - g)\}/P = \{(S' - S) - (h - g)\}/Q,$$

which gives *the* resistance in the form

$$h - g = \{P\,(S' - S) - Q\,(R' - R)\}/(P + Q).$$

In the special case when $P = Q$, the formula reduces to

$$h - g = \tfrac{1}{2}\{(S' - S) - (R' - R)\},$$

which is very convenient for practical work.

It is noteworthy that only the *changes* $R' - R$, $S' - S$ in the resistances employed need to be observed (as well, of course, as P and Q).

This experimental arrangement and the final result are given in § 26 of Dr Searle's paper: those interested will find other examples (derived from experimental arrangements used in various standard Laboratories) in the concluding sections of his paper, and to any of these examples the analysis described above can be easily applied.

DISPERSION AND DOUBLE-REFRACTION OF ELECTRONS IN RECTANGULAR GROUPING (CRYSTALS)

By P. P. Ewald.

There are two ways of reducing the aeolotropic properties of a crystal to the properties of the molecules: first you may suppose every molecule to be aeolotropic having the same symmetry as the large crystal. For the theory of double-refraction this would mean that there are, in each molecule, three different periods of free vibrations in three principal directions. These aeolotropic molecules have to face the same direction in the crystal—this being the only restriction as to their arrangement. That these assumptions really lead to a definite value for the double-refraction has recently been shown by Langevin.

The other possible explanation of crystalline properties has been given in the theory of crystal-structure by Bravais, Sohnke and Schönfliess. In this theory the ultimate parts constituting the crystal are assumed isotropic and the aeolotropy is produced solely by the arrangement of these particles in a regular space-grouping. This theory accounts for all classes of symmetry possible in crystals, and all qualitative conclusions from this theory seem to show the truth of the fundamental idea. I may add, that recent experiments by M. Laue on the passage of Roentgen-rays through crystal slabs furnish the most direct evidence of the reality of the space-grouping.

It seems therefore to be of interest to know to what extent double-refraction may be considered to result from the arrangement of isotropic molecules, and this is the question I have studied.

It so chances that the preciseness of the problem leads to a stricter treatment than has hitherto been given, of questions arising in the general theory of refraction in transparent bodies. These questions, concerning mainly the formation of the refracted wave in the bordering layers of a body of finite extent, will be mentioned at the end of this lecture.

Consider the simplest grouping which has the symmetry of the rhombical system. The planes of the grouping intersect at right angles and have the distances (a, b, c) in the (x, y, z) directions. (Fig. 1.) In each of the points of this grouping let there be situated a molecule with an electron capable of being excited by a periodic electromagnetic field. Let there be further a quasi-elastic force between the molecule and the electron and let this be equal for displacements of the electron in all directions. It is easy to extend the theory to cases where the latter assumption is not true, but we will here only consider isotropic molecules.

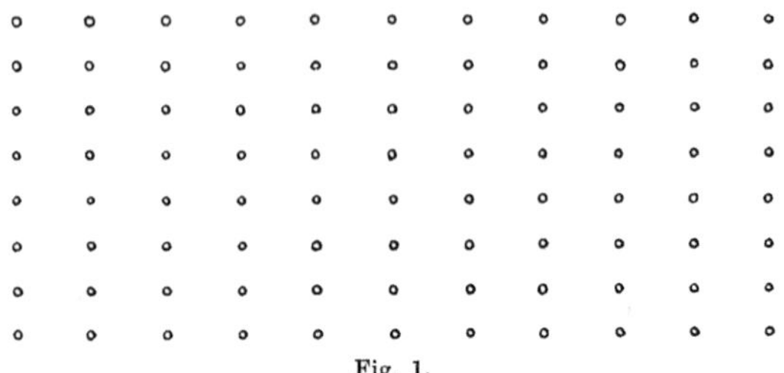

Fig. 1.

When an electron in the grouping is displaced from its position of equilibrium there is not only the quasi-elastic force acting on it, but also a force arising from the action of the neighbouring molecules. This force will not be the same for displacements in the different directions x and owing to the difference of intervals a and b. This is how the interaction of the neighbouring molecules gives rise to double-refraction.

We have now to put these ideas in a mathematical form. For this purpose we assume that the oscillations of the electrons take place as if they were induced by a plane wave sweeping over the system with a certain (unknown) velocity q. This means that all electrons in a certain plane passing through the origin of coordinates have the same phase and that the phase of any other electron is given by its distance S from this plane. So that we may describe the field of any of the electrons by the well-known Hertzian vector-potential

$$P = a\frac{e^{-in\left(t-\frac{R}{c}+\frac{S}{q}\right)}}{R},$$

where a is the maximum electric momentum of the electron, R the distance between electron and point of observation, n the frequency, $c = 3{\cdot}10^{10}\,\frac{\text{cm}}{\text{sec}}$; and where the term S/q in the exponent produces the phase difference demanded. The determination of the velocity q will presently furnish the refractive index $\nu = c/q$.

Before determining ν we have to consider the field produced by the assumed oscillations. The potential of this field is ΣP, the sum extended over all electrons of the grouping. This sum can be formed by means of complex integrals and is found to be the product of two terms; one of these represents a wave of the same character as the wave we used for describing the phase of the electrons. The other term is periodically reproduced round each molecule and only comes in by its mean value, as the distance of the molecules is small compared to the wave-length. So that we find the optical field to consist of a plane wave of the velocity q.

It is startling at first sight to find by this result that we may not introduce into our system the incident wave at all. This wave of velocity c (and perhaps other direction) would not combine with the field of the electrons to form a plane wave in the interior of the crystal. The reason why the incident wave has to be left away will appear later on considering the influence of bordering layers.

We now proceed to find the velocity q. For this we demand that the electrons be in dynamical equilibrium with the periodic field. This means that each oscillation must be produced only by the "exciting field," i.e. the field radiated to the electron under consideration by all other electrons of the grouping. The potential of this "exciting field" is $\Sigma' P$, where the accent denotes the process of leaving out one of the terms of the sum. We now have the differential equation for the oscillations:

$$m\ddot{y} + g\dot{y} + mn_0^2 y = e \times \text{exciting field},$$

where $y = \frac{a}{e} \cdot e^{-in\left(t + \frac{S}{q}\right)}$ is the vectorial elongation, e the charge of the electron and n_0 its period of free vibration. The exciting field is found from the potential by twice applying the operation of curl, and as both sides of the equation depend on time in the same periodic way we get the equation

$$a \cdot \frac{m}{e^2}\left(n_0^2 - n^2 - in\frac{g}{m}\right) = \text{curl curl } \Sigma' P$$

$$= \text{curl curl } \Sigma' a \frac{e^{-in\left(t - \frac{R}{c} + \frac{S}{q}\right)}}{R}.$$

This is a linear homogeneous vector-equation for a and it is equivalent to three simultaneous scalar-equations of the same kind for the components of the amplitude a. These are only consistent if the determinant of the coefficients of a_x, a_y, a_z vanishes. These coefficients are formed of the second derivatives of the sum Σ' and depend therefore on the value of q. The only values of q and of the refractive index $\nu = c/q$ which are in accordance with the condition of dynamical equilibrium of the system can be found from this determinant.

As the results of this theory I may state: we find a dispersion-formula

$$\nu^2 = 1 + \frac{1}{\frac{m}{Ne^2}(n_0^2 - n^2) + \psi}$$

which resembles closely the formula found by Lorentz and Planck. The only difference is found in the term ψ; this is a number depending on the ratio $a:b:c$ of the intervals of the grouping and on the direction of propagation and polarisation of the optical wave. The numerical value of ψ can be calculated with $a:b:c$ given. In the case of a cubical arrangement, $\psi = -\frac{1}{3}$ and we obtain the exact formula of Planck-Lorentz.

The invariant relation due to T. H. Havelock,

$$D_{xy} = \frac{1}{\nu_x^2 - 1} - \frac{1}{\nu_y^2 - 1} = \psi_x - \psi_y,$$

between the refractive indices in the x and y directions follows from the formula given. D is independent of the number and nature of the electrons and of the frequency of the wave; by it we obtain the rational measure of Double-refraction caused by structure.

It is of interest to find out whether the influence of structure produces the same order of magnitude of double-refraction as has been found in crystals. By calculating $\psi_x - \psi_y$ for a grouping of which the intervals have the same ratios as the crystallographic axes of Anhydrite this is found to be the case. Even if in many crystals we shall not feel entitled to attribute the whole of double-refraction to the arrangement of isotropic particles, we see from this example that the structural effect plays a great part in the production of optical aeolotropy.

It can further be shown that in the grouping the refractive index varies with the direction of the optical wave in the same way as in crystals, the surface of wave-normals being the same in both cases.

Let me add some remarks of interest for the general conception of refraction. We have found that electrons oscillating with the assumed connection of phases give rise to a plane wave of velocity q and that therefore the incident wave may not be superposed on the field due to the electrons. We conclude that in a slab of a crystal of finite thickness the incident light wave is annihilated by the electrons on the border of the slab. In the interior of this crystal there exists, therefore, only the refracted wave. In letting the thickness of the slab grow indefinitely we remove the border; but its influence is felt in the fact that the incident wave is not detected in the interior.

This action of the bordering layer is perhaps formulated here for the first time. It seems of highest theoretical and practical interest to follow it in a more precise manner, because it shows us the exact process by which an incident wave of one direction is transformed into the refracted wave of other direction and velocity. By considering the bordering layers of a body we study this fundamental change, whereas in the usual theory of dispersion we find the conditions under which the changed state, if inaugurated, may continue to exist. The dynamical treatment of a system with a border is more difficult than that of an unlimited system, but I hope soon to present some studies of this—it seems to me, important—point.

THE GRAPHICAL RECORDING OF SOUND WAVES; EFFECT OF FREE PERIODS OF THE RECORDING APPARATUS

By Dayton C. Miller.

I have devised an instrument which may be called the "Phonodeik" for graphically recording sound waves. In the phonodeik the sound is in effect concentrated by a conical horn upon a diaphragm and the movements of the latter are recorded photographically or are projected upon a screen.

It is well known that when waves of various frequencies are impressed upon a diaphragm the response of the latter is not proportional to the amplitude of the wave; the diaphragm having its own natural period, its response to impressed waves of frequencies near its own is exaggerated in degrees depending upon the damping. The theory for simple cases is complete, but what actually happens in a given practical case is much complicated by indeterminate conditions and cannot be predicted.

It is also well known that the resonating horn modifies the waves which may be said to enter it. Therefore it follows that the resultant motion of the diaphragm is quite different from that of the original wave in air.

These same difficulties are present in the reproductions of the talking machine and the telephone, perhaps in a doubled degree since they may occur once in making the record and again in reproducing it. It is argued that the photographic records of the phonodeik are more truthful than the best talking machine records because the movements of the first diaphragm which is nearly free are directly recorded without further mechanical disturbance. The justification of the phonodeik for research upon complex sound waves requires that it be shown that the diaphragm responds to all the frequencies being investigated; that it responds properly to any combination of simple waves; that it does not introduce fictitious frequencies; it is necessary to determine the relation between the response to a wave of any frequency and the relative intensity of that wave; it is necessary to show that the recording attachment faithfully transmits the motions of the diaphragm.

The actual response of the phonodeik has been investigated by reference to an arbitrary standard scale of organ pipes. Two sets of pipes, one *open diapason* set of metal and one *stopped diapason* set of wood, were especially voiced and made of uniform loudness according to the judgment of a skilled organ voicer. The uniformity of these pipes has been improved and verified by two experimental methods. Each set contains eighty pipes, giving each musical semitone from a frequency of 129 to that of 12400. The sounds from the various pipes of the two sets have been photographed and analysed; the open diapason pipes have a strong octave accompanying the fundamental; the lowest octave of stopped pipes have a weak third partial tone accompanying the fundamental and no other partials, while all the other pipes of this set give practically simple tones.

With these pipes the response of the phonodeik under varying conditions has been determined by sounding the pipes in succession and making a photographic record of the amplitude of vibration. A curve is then plotted which shows the response to the successive tones. Such a curve is shown by the irregular line in Fig. 1; in this diagram the abscissae are proportional to the logarithms of the frequencies, and the ordinates represent the amplitudes of vibration of the diaphragm corresponding to these frequencies. The smooth curve represents an ideal response, the ordinates of this curve are the undistorted amplitudes of vibration for the various frequencies on the supposition that all the tones of the scale are of the same intensity. Constant intensity as here implied is defined by the condition that the square of the

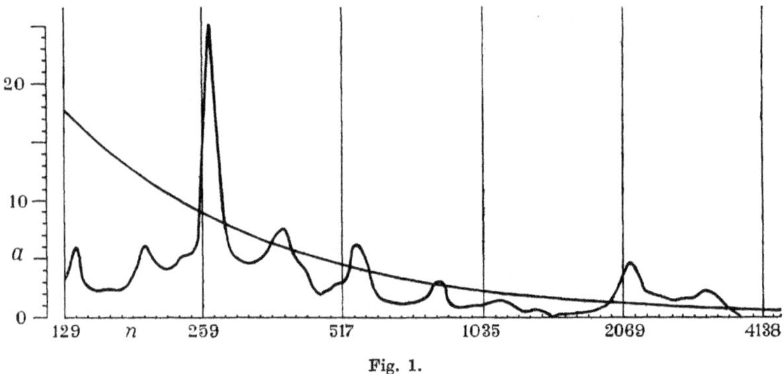

Fig. 1.

amplitude times the square of the frequency is constant; that is the energy of vibration is constant.

The departure of the actual response from the ideal is very marked; five causes were suspected and each was investigated. These distortions might be produced by unequal loudness of the pipes, by the attached vibrating mirror, by the mounting and housing of the diaphragm, by the free periods of the diaphragm, and by the free periods of the resonating horn. The effects due to the first three causes were found to be practically negligible in the analysis of the actual photographic curves; the effects due to the diaphragm and the horn are important.

The response of the diaphragm, with no horn and with no cover on either the front or back side, was determined with the sets of standard pipes. Fig. 2 shows the results plotted in the manner described for Fig. 1. The curve shows three distinct peaks, and indicates that the response is very small for low frequencies.

In order to learn the condition of vibration of the diaphragm, Chladni's method of sand figures was used; the figures were drawn and photographed for the eighty tones of the standard scale for frequencies from 129 to 12400. The first peak of Fig. 2 corresponds to the real free period of the diaphragm, when it vibrates as a whole most vigorously. The diaphragm then forms various irregular nodal patterns giving small response as the frequency is increased, till it forms one concentric circular nodal line, corresponding to the first over-tone, whose frequency is 3·28 times that of the fundamental; this corresponds to the middle peak of Fig. 2. With increasing frequency, smaller irregular nodal figures occur passing into two concentric circles, giving the second over-tone represented by the third peak in Fig. 2. The frequency of this tone is 6·72 times that of the fundamental.

Nodal patterns of three, four, and five perfect concentric circles were obtained, corresponding to frequencies beyond the limits of the diagrams shown here.

The presence of the three natural periods of the diaphragm was proven, also, by making a direct photographic record in the phonodeik of the vibrations resulting from a single impulse produced by a hand-spat in front of the diaphragm.

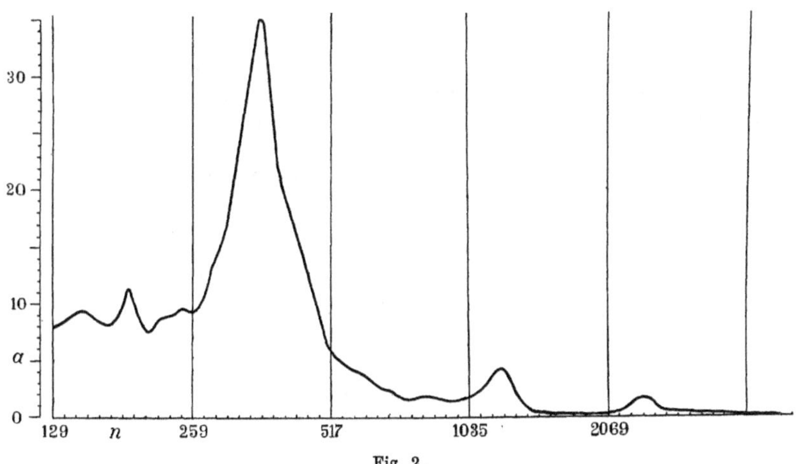

Fig. 2.

An elaborate study was made of the effects of the concentrating horn, and it is shown that it acts as a resonator, as does the body of a violin or the sound-board of a piano. But the horn adds its own free period effects; it has all the tones of the bugle, shown by at least ten peaks in the curve of Fig. 1; this makes it a very annoying adjunct to the phonodeik, but as it increases the response of the diaphragm several thousand-fold, it has seemed desirable to retain it.

Experiments have been made with horns of a great variety of materials and of varying sizes and shapes. The results of these investigations must be given elsewhere. A long slender cone seems to be the best shape of horn for the phonodeik.

The effect of such a horn on the response curve is shown in Figs. 3 and 4; the dotted line, *a*, shows the response of the diaphragm alone, as already described; the difference between this and the actual response, *b*, is mainly due to the horn; line *c* represents the ideal response. If the ordinates of the curve *b* are multiplied by such factors as would cause curve *a* to coincide with curve *c*, the curve *d* is obtained; the difference between curves *d* and *c* shows clearly the peaks due to the complete series of partial tones of the horn; ten or more of these partials may be identified.

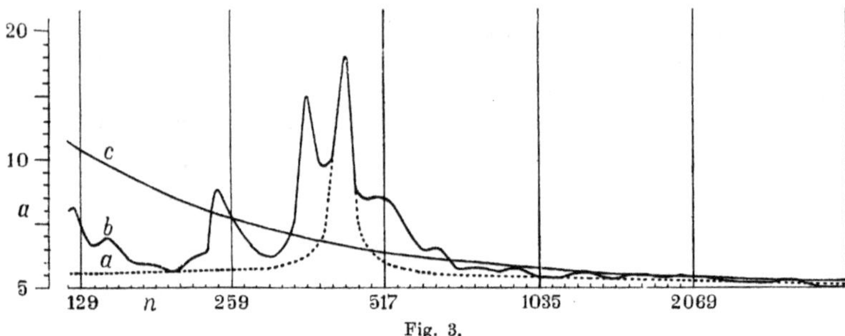

Fig. 3.

The practical application of these results in the quantitative analysis of sound is as follows. Photographs are made of the waves from the unknown sound; at the same time a response curve is made with the set of standard pipes, as described. The wave is analysed, and each separate component is corrected by being multiplied by a factor depending upon the frequency of the component, and determined by reference to the response curve. The factor is such a number, greater or less than

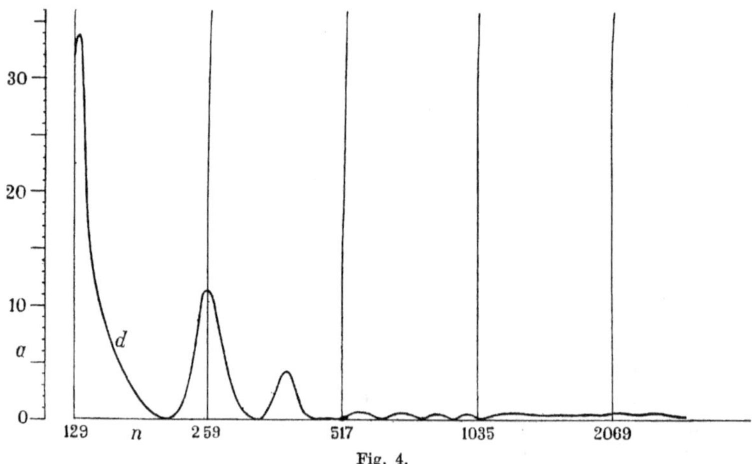

Fig. 4.

unity, that when multiplied into the ordinate of the response curve for the specified frequency will make the product equal to the corresponding ordinate of the ideal curve, as shown in Fig. 1. The corrected amplitudes of the various components are those which would have been obtained if the response of the phonodeik had been equal to the ideal specified, and had been free from the effects of all free period and other distorting effects.

Practical justification of the method has been obtained by photographing the same tone from a musical instrument with widely differing conditions of the phonodeik, as with diaphragms and horns of different free periods. The photographs so obtained are quite unlike in appearance; but after the components of each have been properly corrected new curves synthesized from these corrected components are similar.

Detailed descriptions of the instruments and methods employed in these experiments, together with results of the analytical study of various musical and vocal tones, are to be given in papers which will be published in various journals devoted to physics.

SUR LE MOUVEMENT D'UN FIL

PAR E. TERRADAS.

1. Je vais m'occuper du mouvement d'un fil dans le cas où tous ses points décrivent la même trajectoire relativement à des axes animés d'un mouvement de rotation*. On fera usage des notations suivantes: δ masse par unité de longueur; σ arc de trajectoire par rapport à un point fixe; s arc du fil; T tension; v vitesse commune à tous les points, fonction seulement du temps t; r distance d'un élément à l'axe z de rotation; ρ rayon de première courbure; w vitesse angulaire constante des axes; λ, μ, ν cosinus directeurs de la binormale; $\mathfrak{T}$, $\mathfrak{N}$, $\mathfrak{B}$ projections sur la tangente, normale et binormale des forces extérieures par unité de masse. Les équations du mouvement relatif donnent

$$\left.\begin{aligned}
\delta\left[\frac{dv}{dt}-\mathfrak{T}\right] &= \frac{\partial T}{\partial \sigma}+\frac{\delta w^2}{2}\frac{dr^2}{d\sigma}\\
\delta\left[v^2-\rho\mathfrak{N}\right] &= T+\frac{\delta}{2}\left[w^2\rho^2\left[\frac{d^2r^2}{d\sigma^2}-2\right]-2wv\rho\right]\\
w^2\left[\lambda x+\mu y\right] &= 2wv\left[\frac{dx}{d\sigma}\mu-\frac{dy}{d\sigma}\lambda\right]+\mathfrak{B}
\end{aligned}\right\} \quad\ldots\ldots\ldots\ldots(1)$$

$$\sigma=\int_0^t vdt+s;\quad d\sigma^2=dx^2+dy^2+dz^2.$$

Au moyen de ces équations on doit pouvoir exprimer x, y, z en fonction de σ, v en fonction de t, et il faut encore $T>0$. Dans le cas $v=\text{const.}$ ce sont les équations d'équilibre d'un fil dont la tension serait $T-v^2$. Quand $\mathfrak{T}=\mathfrak{N}=\mathfrak{B}=v=0$, ce sont les équations de la corde à sauter.

2. Supposons d'abord $\mathfrak{B}$ fonction de v ou de t; la dernière équation montre que, ou v est constant, ou, v étant variable et $\mathfrak{B}$ fonction linéaire de v, la projection de r sur la binormale doit être constante, et l'angle de la binormale avec l'axe doit être aussi constant. Dans le cas où $\mathfrak{B}=0$, v peut ne pas être constante quand les angles précedents sont égaux à 90°, c'est-à-dire que, si v n'est pas constante, la courbe doit être une hélice ou une courbe plane. Si c'est une hélice, $\mathfrak{N}$ étant convenable, δ peut suivre une loi quelconque, mais, en général, il faudra que δ suive une loi exponentielle ou soit constante. Si $\mathfrak{B}$ est fonction seulement de σ, le mouvement est, en général, impossible, à moins de satisfaire $\mathfrak{B}$ à certaines conditions. Si $\mathfrak{B}$ étant 0, $\mathfrak{T}$ et $\mathfrak{N}$ sont fonctions seulement de σ, elles ne peuvent pas être quelconques. Si elles satisfont à la relation $\mathfrak{T}-\rho\frac{\partial\mathfrak{N}}{\partial\sigma}=\kappa$, κ étant constante, le mouvement est donné par $\frac{dv}{dt}=\kappa$.

* V. aussi Arnoult, *Thèse*, Nancy, 1911.

3. Dans le cas d'une courbe plane, supposons, en premier lieu, $\mathfrak{T}$ et $\mathfrak{N}$ fonctions de v et $\delta = 1$. L'élimination de T des deux premières équations (1) conduit à une équation de la forme $T_1 - T_2 S_1 - S_2 = 0$, T_1 et T_2 étant des fonctions de t, S_1 et S_2 de σ seulement. En suivant la méthode qui a été indiquée par M. Appell dans son mémoire des Acta*, on en déduit que, ou la vitesse est constante dans la trajectoire $\frac{K_2 - S_2}{K_1 - S_1} = K_3$, les K étant des constantes, ou la trajectoire est un cercle dont le centre est à l'origine, la loi du mouvement étant $\frac{dv}{dt} - \mathfrak{T} = 0$. Toutefois, si $\mathfrak{N}$ fût tel que $2wv - \mathfrak{N} = K_2$, K_2 étant une constante, le mouvement selon la loi $\frac{dv}{dt} - \mathfrak{T} = K_1$ est possible dans la trajectoire $\frac{K_2 - S_2}{K_1 - S_1} = K_3$.

Si $\mathfrak{T}$ et $\mathfrak{N}$ sont fonctions seulement de σ, il en résulte que, ou v est constante, ou si $\mathfrak{T}$ et $\mathfrak{N}$ satisfont à certaines conditions, la trajectoire peut être une spirale logarithmique ou un cercle parcouru avec vitesse variable.

Si $\mathfrak{T}$ et $\mathfrak{N}$ sont la somme d'une fonction de v et d'une fonction de σ, on arrive à des resultats semblables; si $\mathfrak{T}$ et $\mathfrak{N}$ sont le produit d'une fonction de v et d'une fonction de σ, on peut considérer divers cas selon le nombre des liaisons linéaires parmi les fonctions de t ou de σ de l'équation qu'on obtient en éliminant la tension. S'il n'y a qu'une relation, ou si leur nombre est 4, la vitesse est constante.

4. Nous allons nous occuper d'un cas particulièrement intéressant où l'on peut réaliser facilement la courbe. Il suffit de prendre une chaîne sans fin, et de la faire tourner en la pendant de l'avant-bras. La chaîne a une figure permanente par rapport à des axes entrainés suivant la rotation. Il est admis qu'il n'y a pas de glissement entre la chaîne et son support.

La vitesse constante des éléments, ou vitesse du fil, est $v = wr_1$, r_1 étant le rayon du cylindre qui sert d'appui à la chaîne.

On obtiendra les équations qui conviennent à ce cas en posant

$$\mathfrak{T} = \mathfrak{N} = 0\,; \quad v = wr_1.$$

Eliminant T entre les deux premières équations (1) on obtient après intégration, et puisque

$$\rho = \frac{2p}{2 - \frac{d^2 r^2}{d\sigma^2}},$$

(p distance de l'origine à la tangente)

$$r^2 + 2p\rho + 4r_1\rho = h_1,$$

h_1 étant constante. Mais comme $\rho = r\frac{dr}{dp}$ on aura, éliminant ρ, intégrant et en égard à ce que pour $r = r_1$, $p = -r_1$,

$$(p + 2r_1)(h_1 - r^2) = r_1(h_1 - r_1^2) \quad \ldots\ldots\ldots\ldots\ldots\ldots\ldots\ldots (2).$$

* V. Appell, *Acta mathematica*, tome x.

On peut remarquer que w n'intervient pas; c'est-à-dire, que la courbe trajectoire du fil est indépendante de la vitesse de rotation. En augmentant w on augmente T selon la relation suivante où $z = r^2$,

$$\frac{T}{w^2} = h_1 + 2z_1 - z.$$

On se rend compte facilement de ce que la forme de la courbe soit indépendante de w. En effet: les forces extérieures sont uniquement la force centrifuge ordinaire et celle de Coriolis; la première est proportionelle à w^2, la deuxième aussi en vertu de $v = wr_1$. Donc, la direction de la resultante est indépendante de w.

De l'équation (2) on en déduit, θ étant l'angle polaire compté à partir du minimum r_m de r,

$$\pm \frac{\theta}{r_1} = \int_{z_m}^{z} \frac{2z - z_1 - h_1}{2z\sqrt{R}} dz,$$

$$R = (z - z_1)[z^2 - (2h_1 + 3z_1) z + (h_1 + z_1)^2] = (z - z_1)(z - z_a)(z - z_6).$$

La condition $R > 0$ jointe à $T > 0$ donne les conditions limitatives de r. On peut distinguer deux cas:

(1) $h_1 > z_1$. Les trois racines z_6, z_a, z_1 sont réelles, supposons $z_6 > z_a > z_1$. La courbe est comprise dans la couronne $r = r_1$, $r = r_a$. Elle est formée d'une infinité de branches simétriques ou congruentes à la branche qui va de $r = r_1$ à $r = r_a$. Dans les points correspondants, elle touche les cercles limites de la couronne. Il y a un point où p s'annule, dont le r_p est donné par $2r_p^2 = h_1 + z_1$.

Le rayon de courbure donné par

$$2r_1\rho = \frac{(h_1 - z)^2}{h_1 - z_1}$$

est maximum pour $r = r_1$, se réduit à la quatrième partie pour $r = r_p$ et devient minimum pour $r = r_a$. La courbe a la forme de la figure (1). On peut l'appeler courbe extérieure.

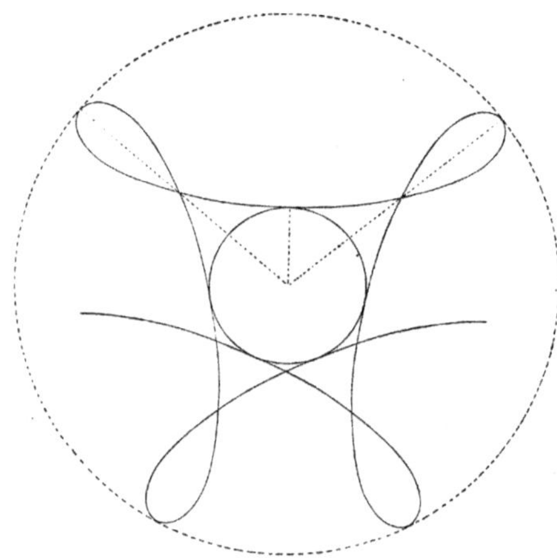

Fig. 1.

(2) $h_1 < z_1$. (a) $-z_1 < h_1 < z_1$. Dans ce cas, $z_\beta > z_1 > z_a$. La courbe est entièrement comprise dans la couronne $r = r_1$, $r = r_a$. Elle passe aussi de concave à convexe par rapport à l'origine, le rayon de courbure est aussi maximum pour $r = r_1$ et minimum pour $r = r_a$. Comme dans le cas antérieur il ne s'annule jamais. La courbe a la forme de la figure 2. On peut l'appeler courbe intérieure.

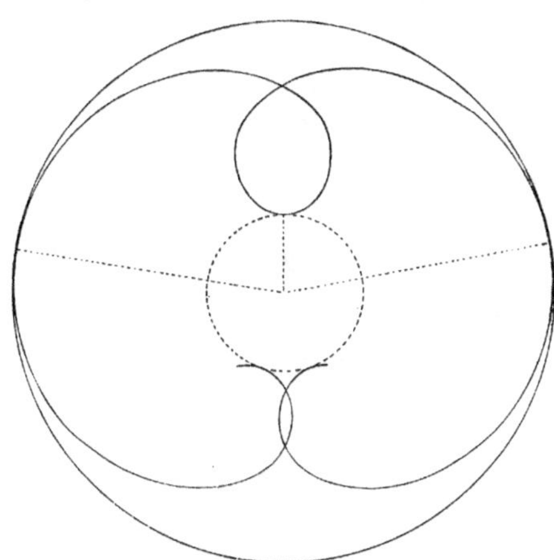

Fig. 2.

(b) $h_1 = -z_1$. C'est aussi une courbe intérieure dont l'équation est

$$\pm \theta = l \, . \frac{r_1 - r}{r_1 + r}.$$

Elle tend asimptotiquement à la circonference $r = r_1$.

Il n'y a pas d'autres cas possibles. Cependant il est à considérer à part le cas $h_1 = z_1$, qui correspond à la circonference $r = r_1$, ce qui se déduit de la valeur de p.

5. Considérons la courbe extérieure. Avec $z = \sqrt[3]{4}x + j$, $j = \frac{2h_1 + 4z_1}{3}$, on passe à la forme de Weierstrass. En introduisant l'intégrale

$$I = \int_{e_3}^{x} \frac{dx}{\sqrt{R}}$$

on trouve
$$r^2 = \sqrt[3]{4} p (I + w_2) + j \quad \ldots\ldots\ldots\ldots\ldots\ldots\ldots (3).$$

Si l'on fait $-j = \sqrt[3]{4} p(v)$, on a $p'(v) = -ir_1(h_1 + z_1)$, et

$$\theta = [\sqrt[3]{4} r_1 - i\zeta(v)] I - \frac{i}{2} l \frac{\sigma(I + w_2 - v)\sigma(w_2 + v)}{\sigma(I + w_2 + v)\sigma(w_2 - v)}.$$

Si l'on veut profiter des tables, on pourra se servir des valeurs suivantes :

$$\text{sen}^2\alpha = \kappa^2 = \frac{e_2 - e_3}{e_1 - e_2}, \quad \text{sen}^2\phi = \text{sn}^2 [I\sqrt{e_1 - e_3}] = \frac{r^2 - r_1^2}{r_a^2 - r_1^2}, \quad w_1 = \frac{K}{\sqrt{e_1 - e_3}};$$

$$\text{sen}^2\alpha' = 1 - \kappa^2, \quad \text{sen}^2\phi' = \text{sn}^2 \frac{v\sqrt{e_1 - e_3}}{i} = \frac{r_b^2 - r_1^2}{r_b^2}.$$

Avec lesquels on aura w_1, v et diverses valeurs de I pour les valeurs correspondantes de r. La valeur de θ peut être mise dans la forme:

$$\theta = (\sqrt[3]{4}\, r_1 - \pi V)\, I + \tfrac{1}{2} J \quad \ldots\ldots\ldots\ldots (4),$$

avec
$$\pi V = i\left[\frac{v}{w_1}\zeta(w_1) - \zeta(v)\right] = \frac{\pi}{2w_1}\,\frac{\cosh.\dfrac{\pi v}{2iw_1} - 3q^2\cosh.\dfrac{3\pi v}{2iw_1}\cdots}{\operatorname{senh}.\dfrac{\pi v}{2iw_1} - q^2\operatorname{senh}.\dfrac{3\pi v}{2iw_1}\cdots},$$

$$tg J = \frac{2AB}{A^2 - B^2}, \quad q = e^{-\pi\frac{K'}{K}},$$

$$A = 1 - 2q\cosh.\frac{\pi v}{iw_1}\cos\frac{\pi}{w_1} I + 2q^4\cosh.\frac{2\pi v}{iw_1}\cos\frac{2\pi}{w_1} I \ldots$$

$$B = -2q\operatorname{senh}.\frac{\pi v}{iw_1}\operatorname{sen}\frac{\pi}{w_1} I + 2q^4\operatorname{senh}.\frac{2\pi v}{iw_1}\operatorname{sen}\frac{2\pi}{w_1} I \ldots.$$

C'est avec ces formules que la courbe fig. (1) a été calculée.

La valeur de l'angle θ_m que font les rayons $r = r_1$ et $r = r_a$ resulte de

$$\theta_m = (\sqrt[3]{4}\, r_1 - \pi V)\, w_1.$$

La longueur est donnée par

$$\frac{2\sigma}{\sqrt[3]{4}} = \frac{1}{\sqrt[3]{4}}\int_{z_1}^{z}\frac{h_1 - z}{\sqrt{R}}\,dz = (h_1 - j)\, I - \frac{1}{2}\sqrt{\frac{(r_a^2 - r^2)(r_b^2 - r^2)}{r^2 - r_1^2}} + \sqrt[3]{4}\,\zeta(I).$$

La demi-longueur l d'une branche est:

$$l = \frac{\sqrt[3]{4}}{2}\left[h_1 - r_1^2 - \sqrt[3]{4}\,\frac{2\pi^2}{w_1^2}\,\Sigma\,\frac{q^n}{(1-q^n)^2}\right] w_1.$$

S'il y a n branches, la longueur totale de la courbe sera

$$L = 2n\,[l + \pi r_1 - \theta_m].$$

6. Ces courbes dépendent de deux paramètres, par exemple, z_1 et h_1. Toutes les courbes pour lesquelles $\dfrac{z_1}{h_1} = X$ a la même valeur, sont homotétiques. On obtient la totalité des courbes extérieures, en faisant varier X de 1 à 0, et les courbes intérieures en faisant varier X de 1 à -1, et en y ajoutant toutes les homotétiques. A mesure que h_1 augmente, r_1 étant invariable, θ_m augmente, jusqu'à $\dfrac{\pi}{2}$ dans les extérieures. Lorsque h_1 diminue dans les intérieures, θ_m augmente sans limite.

A toute courbe extérieure definie par les valeurs de r_1 et r_a, on peut faire correspondre une courbe intérieure, qu'on pourrait appeler conjuguée et pour laquelle $r_1' = r_a$ et $r_a' = r_1$, les accents se rapportant à la courbe intérieure. Une fois construite l'une de ces courbes, p. e. l'extérieure, on obtient la conjuguée par les formules

$$\theta' = \theta + \sqrt[3]{4}\,(r_a - r_1)\, I, \quad r' = r, \quad l' = l + \frac{\sqrt[3]{4}}{2}\, w_1\,[z_a - z_1 - (h_1 + h_1')].$$

Effectivement: des formules qui donnent z_a et z_b on en déduit en éliminant h_1

$$\frac{z_b}{2} = \frac{z_1 + z_a}{2} + \sqrt{z_1 z_a}.$$

Les trois racines de R sont communes aux courbes conjuguées. Par conséquent, j, v, w_1, I, J sont les mêmes aussi. Les formules antérieures s'obtiennent immédiatement en restant les valeurs de θ et l correspondantes. La figure 2 est la courbe conjuguée de la figure 1.

On déduit encore

$$h_1 = z_a - z_1 + \sqrt{z_1 z_a}\,; \quad h_1' = z_1 - z_a - \sqrt{z_1 z_a}.$$

7. On peut exprimer $x + iy$ en fonction uniforme de I. On a en effet

$$x + iy = re^{i\theta} = i\sqrt[3]{2}\, e^{v\zeta(w_2) + i[\sqrt[3]{4}r_1 - i\zeta(v)]I}\, \frac{\sigma(I + w_2 - v)}{\sigma(I + w_2)\,\sigma(v)}\,;$$

$x + iy$ est, par conséquent, fonction doublement périodique de seconde espèce à multiplicateurs non spéciaux.

Avec cette valeur de $x + iy$ on peut analiser s'il n'y aurait pas des courbes algébriques qui correspondraient à des valeurs de v qui seraient des fractions de la période w_2. On sait que si le coefficient de I dans l'exposant est $v\,\dfrac{\zeta(w_2)}{w_2}$ et v est égal à $\dfrac{2nw_2}{m}$ on peut mettre $(x + iy)^m$ en fonction algébrique de $\wp\,(I + w_2)$, c'est-à-dire de $x^2 + y^2$. Mais il faut pour cela en employant les notations courantes*

$$\psi_m(v) = 0, \quad i[\sqrt[3]{4}r_1 - i\zeta(v)] = \frac{2n}{m}\,\zeta(w_2) \quad \ldots\ldots\ldots\ldots\ldots\ldots(5).$$

De cette dernière on en déduit que, s'il existe une courbe algébrique, elle doit être fermée et telle que $2\theta_m$ soit les $\dfrac{n}{m}$ de la circonference. En effet : la condition de fermeture est selon (4)

$$2\theta_m = 2\pi\,\frac{n'}{m'} = 2w_1\left[\sqrt[3]{4}r_1 + i\left(\frac{v}{w_1}\,\zeta(w_1) - \zeta(v)\right)\right]$$

ce qui, avec la dernière condition, se reduit à $\dfrac{n'}{m'} = \dfrac{n}{m}$.

Toutefois, s'il est bien évident qu'il existe des courbes fermées, ces courbes ne correspondent pas à $v = \dfrac{2n}{m}\,w_2$, du moins en général, car les deux conditions (5) doivent être satisfaites par une valeur de X, les n et m étant entiers. Elles sont donc transcendentes et fermées. Le cas $n = 1$, $m = 4$ (particulièrement simple, car

$$\wp(v) = \wp\left(\frac{w_2}{2}\right)$$

est facile à calculer) conduit à $z_1(z_a + z_b) = z_a z_b$, et d'autre part, le deuxième (5) conduit pour la courbe extérieure à $h_1 = 3z_1$, deux conditions incompatibles. On arrive au même résultat pour la courbe intérieure.

* Halphen, *Fonctions elliptiques*, tome II, p. 180.

DAS GRAVITATIONSFELD

Von Max Abraham.

Nachdem es Maxwell gelungen war, die Fernkräfte aus der Elektrodynamik zu verbannen, entstand die Aufgabe, auch die Gravitation auf Nahewirkungen zurückzuführen. Eine solche Nahewirkungstheorie muss Differentialgleichungen des Schwerkraftfeldes angeben, durch deren Integration das Newtonsche Gesetz, wenigstens mit der durch die astronomischen Erfahrungen gewährleisteten Annäherung, folgt. Sie muss ferner die vom Gravitationsfelde an die Materie abgegebene Energie und Bewegungsgrösse aus dem Energiestrom und den fiktiven Spannungen, sowie aus Dichte der Energie und Bewegungsgrösse des Feldes ableiten.

Der erste Ansatz für eine Theorie des statischen Gravitationsfeldes rührt von Maxwell selbst her*. Er ist der Elektrostatik nachgebildet; entsprechend dem verschiedenen Vorzeichen der Kraft zwischen gleichnamigen Massen wird indessen die Energiedichte des Schwerkraftfeldes negativ, wenn man sie bei verschwindendem Felde gleich null setzt. An dieser Schwierigkeit scheitert jeder Versuch, die Theorie des Schwerkraftfeldes nach dem Vorbilde des elektromagnetischen Feldes zu konstruieren, d. h. sie auf die Wechselwirkungen zweier Vektoren zurückzuführen, welche durch Differentialgleichungen von der Art der Maxwellschen miteinander verknüpft sind.

Eine glückliche Idee von A. Einstein† bot die Aussicht dar, von einer anderen Seite aus dem Problem der Schwerkraft beizukommen. Er nahm an, dass eine Funktionsbeziehung zwischen dem Gravitationspotential und der Lichtgeschwindigkeit bestehe. Hieran anknüpfend, habe ich Ausdrücke für die fiktiven Spannungen, die Energiedichte, den Energiestrom ($\mathfrak{S}$) und die Impulsdichte ($\mathfrak{g}$) des Schwerkraftfeldes angegeben, welche diese durch die ersten Ableitungen des Gravitationspotentials nach den Koordinaten und nach der Zeit ausdrücken‡; diese Ausdrücke umfassen auch die Theorie des zeitlich veränderlichen Feldes. Dabei wird zwischen Impulsdichte und Energiestrom die in der Elektrodynamik gültige Beziehung ($\mathfrak{g} = \mathfrak{S}/c^2$) angenommen. In der Sprache der vierdimensionalen Vektoranalysis kann man sagen: Die zehn Grössen, nämlich die sechs Spannungskomponenten, die drei Komponenten des Vektors $c\mathfrak{g} = \mathfrak{S}/c$, und die Energiedichte ϵ bilden die Komponenten

* J. Cl. Maxwell, *Scientific papers* I, S. 570.

† A. Einstein, *Ann. d. Phys.* 35, 898, 1911.

‡ M. Abraham, *Physik. Zeitschr.* 13, 1, 1912.

eines vierdimensionalen Tensors. Dabei hängen die Komponenten des Gravitationstensors von denjenigen eines Vierervektors ab, nämlich des Gradienten des Gravitationspotentials, während diejenigen des elektromagnetischen Zehnertensors durch die Komponenten eines Sechservektors bestimmt sind.

Ich habe mich soeben der Sprache der Relativitätstheorie bedient. Doch wird sich zeigen, dass diese Theorie mit den hier vorgetragenen Ansichten über die Schwerkraft nicht zu vereinbaren ist, schon darum nicht, weil das Axiom von der Konstanz der Lichtgeschwindigkeit aufgegeben wird. Ich habe in meinen früheren Arbeiten über die Gravitation versucht, wenigstens im unendlich kleinen, die Invarianz gegenüber den Lorentz-Transformationen zu bewahren. Doch habe ich mich davon überzeugt, dass meine Bewegungsgleichungen des materiellen Punktes sich nicht mit den Prinzipien der analytischen Mechanik vereinbaren lassen. Andrerseits hat sich die von Einstein zugrunde gelegte "Äquivalenzhypothese" ebenfalls als unhaltbar erwiesen. Ich möchte es daher hier vorziehen, die neue Gravitationstheorie zu entwickeln, ohne auf das Raum-Zeit-Problem einzugehen. Ich gehe aus von den geeignet verallgemeinerten Ausdrücken der zehn Komponenten des Gravitationstensors und von den Lagrangeschen Bewegungsgleichungen, und suche im übrigen die Voraussetzungen möglichst wenig einzuschränken.

Es sei

$$w = w(c) \quad \ldots\ldots (1)$$

eine zunächst beliebige Funktion der Lichtgeschwindigkeit c, α eine universelle, auch von c unabhängige Konstante. Unsere Ausdrücke für die fiktiven Spannungen lauten

$$\left.\begin{aligned} \alpha X_x &= -\frac{1}{2}\left(\frac{\partial w}{\partial x}\right)^2 + \frac{1}{2}\left(\frac{\partial w}{\partial y}\right)^2 + \frac{1}{2}\left(\frac{\partial w}{\partial z}\right)^2 - \frac{1}{2c^2}\left(\frac{\partial w}{\partial t}\right)^2 \\ \alpha Y_y &= \frac{1}{2}\left(\frac{\partial w}{\partial x}\right)^2 - \frac{1}{2}\left(\frac{\partial w}{\partial y}\right)^2 + \frac{1}{2}\left(\frac{\partial w}{\partial z}\right)^2 - \frac{1}{2c^2}\left(\frac{\partial w}{\partial t}\right)^2 \\ \alpha Z_z &= \frac{1}{2}\left(\frac{\partial w}{\partial x}\right)^2 + \frac{1}{2}\left(\frac{\partial w}{\partial y}\right)^2 - \frac{1}{2}\left(\frac{\partial w}{\partial z}\right)^2 - \frac{1}{2c^2}\left(\frac{\partial w}{\partial t}\right)^2 \end{aligned}\right\} \quad \ldots\ldots (1a),$$

$$\left.\begin{aligned} \alpha X_y &= \alpha Y_x = -\frac{\partial w}{\partial x}\frac{\partial w}{\partial y} \\ \alpha Y_z &= \alpha Z_y = -\frac{\partial w}{\partial y}\frac{\partial w}{\partial z} \\ \alpha Z_x &= \alpha X_z = -\frac{\partial w}{\partial z}\frac{\partial w}{\partial x} \end{aligned}\right\} \quad \ldots\ldots (1b).$$

Die Vektoren, Energiestrom und Impulsdichte seien bestimmt durch

$$\alpha\mathfrak{S} = \alpha c^2 \mathfrak{g} = -\frac{\partial w}{\partial t}\,\mathrm{grad}\, w \quad \ldots\ldots (1c),$$

und die Energiedichte ϵ sei

$$\alpha\epsilon = \frac{1}{2}(\mathrm{grad}\, w)^2 + \frac{1}{2c^2}\left(\frac{\partial w}{\partial t}\right)^2 \quad \ldots\ldots (1d).$$

Es ist bemerkenswert, dass diese auch im zeitlich veränderlichen Felde stets positiv ausfällt.

Zur Berechnung der Schwerkraft $\mathfrak{f}$, die auf die Volumeinheit der Materie wirkt, bedienen wir uns des Impulssatzes, indem wir setzen

$$\left.\begin{aligned}\mathfrak{f}_x&=\frac{\partial X_x}{\partial x}+\frac{\partial X_y}{\partial y}+\frac{\partial X_z}{\partial z}-\frac{\partial \mathfrak{g}_x}{\partial t}\\ \mathfrak{f}_y&=\frac{\partial Y_x}{\partial x}+\frac{\partial Y_y}{\partial y}+\frac{\partial Y_z}{\partial z}-\frac{\partial \mathfrak{g}_y}{\partial t}\\ \mathfrak{f}_z&=\frac{\partial Z_x}{\partial x}+\frac{\partial Z_y}{\partial y}+\frac{\partial Z_z}{\partial z}-\frac{\partial \mathfrak{g}_z}{\partial t}\end{aligned}\right\} \qquad (2).$$

Entsprechend berechnet sich aus dem Energiesatze die vom Gravitationsfelde an die Volumeinheit der Materie abgegebene Energie

$$k_t=-\operatorname{div}\mathfrak{S}-\frac{\partial \epsilon}{\partial t} \qquad (2\,a).$$

Setzt man in (2) und (2 a) die Ausdrücke (1 a—d) ein, so erhält man

$$\alpha\mathfrak{f}_x=-\frac{\partial w}{\partial x}\left(\frac{\partial^2 w}{\partial x^2}+\frac{\partial^2 w}{\partial y^2}+\frac{\partial^2 w}{\partial z^2}\right)$$
$$-\frac{1}{c}\frac{\partial w}{\partial t}\cdot\frac{\partial}{\partial x}\left(\frac{1}{c}\frac{\partial w}{\partial t}\right)+\frac{\partial}{\partial t}\left(\frac{1}{c}\frac{\partial w}{\partial t}\cdot\frac{1}{c}\frac{\partial w}{\partial x}\right),$$

oder, wenn man abkürzungshalber schreibt,

$$\Box w=\frac{\partial^2 w}{\partial x^2}+\frac{\partial^2 w}{\partial y^2}+\frac{\partial^2 w}{\partial z^2}-\frac{1}{c}\frac{\partial}{\partial t}\left(\frac{1}{c}\frac{\partial w}{\partial t}\right) \qquad (3),$$

$$\alpha\mathfrak{f}_x=-\frac{\partial w}{\partial x}\Box w+\frac{1}{c}\frac{\partial w}{\partial t}\left\{\frac{\partial}{\partial t}\left(\frac{1}{c}\frac{\partial w}{\partial x}\right)-\frac{\partial}{\partial x}\left(\frac{1}{c}\frac{\partial w}{\partial t}\right)\right\} \qquad (3\,a).$$

Da nun, vermöge Gl. (1), w nur von c, d. h. auch von $1/c$, abhängt, so gilt

$$\frac{\partial \frac{1}{c}}{\partial t}\frac{\partial w}{\partial x}-\frac{\partial \frac{1}{c}}{\partial x}\frac{\partial w}{\partial t}=0 \qquad (3\,b),$$

und es ergibt sich

$$\alpha\mathfrak{f}_x=-\Box w\,.\frac{\partial w}{\partial x}.$$

Es folgt also aus dem Impulssatz (2) die Schwerkraft pro Volumeinheit

$$\mathfrak{f}=-\frac{1}{\alpha}\Box w\,.\operatorname{grad} w \qquad (4),$$

andererseits berechnet sich aus dem Energiesatz (2a) die pro Raum- und Zeiteinheit an die Materie abgegebene Energie*

$$k_t=\frac{1}{\alpha}\Box w\,.\frac{\partial w}{\partial t} \qquad (4\,a).$$

Noch haben wir keine Annahme über die Funktion $w(c)$ gemacht. Wir wollen annehmen, es sei w gleich c^λ. Wir sagen dann, w ist "vom Grade λ." Es mag hinsichtlich der Längenmessung die folgende Festsetzung getroffen werden: dieselbe hat so zu geschehen, dass die Abmessungen eines hinreichend kleinen Körpers die

* Obwohl die Impulsgleichungen (2) und die Energiegleichung (2 a) nicht die geeignete Form haben, um allgemein aus einem Zehnertensor einen Vierervektor abzuleiten, so haben doch in diesem besonderen Falle infolge der Relation (3 b) die Ausdrücke (4), (4 a) für die Schwerkraft die Form eines Vierervektors.

gleichen bleiben, wenn er an verschiedene Orte des Schwerkraftfeldes gebracht wird. Dann ersieht man aus (1*d*), dass die Dichte der Energie, und folglich auch die Energie selbst, vom Grade 2λ sind. Dasselbe wird auch von der Energie der Materie gelten müssen.

Es sei nun

$$E = M \, . \, \Phi(c) \quad \ldots\ldots\ldots\ldots\ldots\ldots\ldots\ldots(5)$$

die Energie eines ruhenden materiellen Punktes im statischen Schwerkraftfelde; M sei eine, dem materiellen Punkte individuelle, von c unabhängige Konstante (Massenkonstante), von zunächst unbekannter Dimension.

Dann ist, wie E, auch Φ, das Gravitationspotential, vom Grade 2λ.

Aus der Arbeit bei Verschiebung des Punktes im Felde $-dE = -Md\Phi$ folgt für die Schwerkraft der Wert

$$\mathfrak{K} = -M \operatorname{grad} \Phi \quad \ldots\ldots\ldots\ldots\ldots\ldots\ldots\ldots(5a)$$

oder nach (5)

$$\mathfrak{K} = -\frac{E}{\Phi} \, . \operatorname{grad} \Phi \ldots\ldots\ldots\ldots\ldots\ldots\ldots\ldots(5b).$$

Wir wollen nun annehmen, dass dieser letztere Ausdruck für die Schwerkraft auch im Falle der Bewegung gelte, sowohl für einen materiellen Punkt als auch für ein beliebiges physikalisches System von hinreichend geringer Ausdehnung. Diese Annahme besagt: Es soll allgemein die Schwere eines Systems proportional seinem Energieinhalt sein. Die Gesetze der Erhaltung der Energie und der Erhaltung der Schwere werden dann identisch. Doch ist natürlich das Verhältnis von Energie und schwerer Masse von c, d. h. vom Orte im Gravitationsfelde, abhängig.

Wir wenden nun, um über den Grad (2λ) von Φ, und damit auch über den Grad (λ) von w, Aufschluss zu erhalten, die Gleichungen der analytischen Mechanik an. Es sei die Lagrangesche Funktion des materiellen Punktes

$$L = L(v, \Phi) \quad \ldots\ldots\ldots\ldots\ldots\ldots\ldots\ldots(6).$$

Dann sind Betrag des Impulses (G) und Energie

$$G = \frac{\partial L}{\partial v}, \quad E = v\frac{\partial L}{\partial v} - L \ldots\ldots\ldots\ldots\ldots\ldots\ldots\ldots(6a).$$

Da man hat

$$\frac{\partial L}{\partial x} = \frac{\partial L}{\partial \Phi}\frac{\partial \Phi}{\partial x}, \quad \frac{\partial L}{\partial \dot{x}} = \frac{\partial L}{\partial v} \, . \, \frac{\dot{x}}{v} = \mathfrak{G}_x,$$

so ergeben die Lagrangeschen Gleichungen

$$-\frac{d}{dt}\left(\frac{\partial L}{\partial \dot{x}}\right) + \frac{\partial L}{\partial x} = 0 \text{ usw.} \ldots\ldots\ldots\ldots\ldots\ldots\ldots\ldots(6b)$$

die Vektorgleichung

$$-\frac{d\mathfrak{G}}{dt} + \frac{\partial L}{\partial \Phi} \operatorname{grad} \Phi = 0 \quad \ldots\ldots\ldots\ldots\ldots\ldots\ldots\ldots(7),$$

wobei das erste Glied die Trägheitskraft, das zweite die Schwerkraft darstellt.

Durch Vergleichung mit (5 *b*) folgt

$$E = -\Phi \frac{\partial L}{\partial \Phi} \quad \text{.............................(7 } a\text{)}.$$

Hieraus und aus (6 *a*) erhalten wir

$$L = v \frac{\partial L}{\partial v} + \Phi \frac{\partial L}{\partial \Phi} \quad \text{.............................(8)}.$$

Es muss also, nach einem Satze von Euler, die **Lagrangesche Funktion eine homogene Funktion ersten Grades von *v* und Φ** sein. Wir können diese, ohne Einschränkung der Allgemeinheit, schreiben

$$L = -M\Phi f\left(\frac{v}{\Phi}\right) \text{.................................(9)},$$

wo $f(0) = 1$.

Da das Vorzeichen von *v* nicht von Einfluss sein kann, wird die Reihenentwicklung von *f* nach Potenzen des Quadrats $(v/\Phi)^2$ fortschreiten

$$f\left(\frac{v}{\Phi}\right) = 1 - \frac{a}{2}\left(\frac{v}{\Phi}\right)^2 + \dots \quad \text{.........................(9 } a\text{)}$$

Die Koeffizienten dieser Reihenentwicklung müssen universelle Konstanten sein; denn die Massenkonstante ist die einzige dem materiellen Punkte individuelle Konstante.

Sei nun $$\Phi = c^{2\lambda} \quad \text{...................................(10)};$$

alsdann ist der Koeffizient *a* im allgemeinen keine dimensionslose Grösse, sondern die $(4\lambda - 2)$te Potenz einer Geschwindigkeit. Man wird also dazu genötigt, eine universelle Konstante c_0 von der Dimension einer Geschwindigkeit einzuführen und zu setzen

$$a = \zeta \,.\, c_0^{4\lambda-2} \quad \text{................................(10 } a\text{)}$$

(ζ eine reine Zahl).

Diese neue Konstante c_0 geht in den Ausdruck der trägen Masse des materiellen Punktes ein, deren Grenzwert für kleine Geschwindigkeit (Ruhmasse) definiert ist durch

$$m_0 = \lim_{v=0} (G/v) = \lim_{v=0} \left(\frac{1}{v}\frac{\partial L}{\partial v}\right);$$

aus (9), (9 *a*), (10) und (10 *a*) folgt nämlich

$$m_0 = M \,.\, \frac{a}{\Phi} = \zeta M \,.\, \frac{c_0^{4\lambda-2}}{c^{2\lambda}} \quad \text{........................(10 } b\text{)}.$$

Wenn man die Einführung einer universellen Geschwindigkeit von zweifelhafter physikalischer Bedeutung in den Ausdruck der Ruhmasse vermeiden will und diese als lediglich durch die Massenkonstante *M* und durch die Lichtgeschwindigkeit in dem betreffenden Punkte bestimmt ansieht, so bleibt nichts anderes übrig, als zu setzen

$$2\lambda = 1, \quad \lambda = \tfrac{1}{2}.$$

Dann wird $$\Phi = c \quad \text{......................................(11)}$$

und $$w = \sqrt{c} \quad \text{..................................(11 } a\text{)}.$$

Damit ist die Funktion $w(c)$ bestimmt, von der die Komponenten ($1\,a$—d) des Gravitationstensors und diejenigen der Schwerkraft (4, $4a$) abhängen*. Die Ruhmasse ist, nach ($10\,b$),

$$m_0 = \zeta \cdot \frac{M}{c} \quad \text{.................................(11 } b\text{).}$$

Die träge Ruhmasse ist also vom Grade -1 in c, entsprechend meiner früheren Behauptung†.

Kehren wir nun zurück zu den Ausdrücken (4), ($4\,a$), welche die pro Raum- und Zeiteinheit vom Schwerkraftfelde an die Materie abgegebenen Beträge von Impuls und Energie bestimmen. Für den von Materie und Energie leeren Raum folgt aus ihnen

$$\Box w = 0 \quad \text{....................................(12),}$$

mit $(w = \sqrt{c})$

als Differentialgleichung des Gravitationsfeldes (verallgemeinerte Laplacesche Gleichung). Wo aber Materie mit der Energiedichte η den Raum erfüllt, müssen wir setzen

$$w \Box w = 2\alpha\eta \quad \text{................................(12 } a\text{)}$$

(verallgemeinerte Poissonsche Gleichung)‡.

Es folgt nämlich dann aus (4) für die Schwerkraft pro Volumeinheit

$$\mathfrak{f} = -\frac{2\eta}{w}\operatorname{grad} w = -\frac{\eta}{c}\operatorname{grad} c \quad \text{.....................(13),}$$

und durch Integration über das Volumen ergibt sich die Gesamtkraft

$$\mathfrak{K} = -\frac{E}{c}\operatorname{grad} c \quad \text{...........................(13 } a\text{),}$$

in Übereinstimmung mit ($5\,b$) und (11).

Entsprechend folgt aus ($4\,a$)

$$k_t = \frac{2\eta}{w}\frac{\partial w}{\partial t} = \frac{\eta}{c}\frac{\partial c}{\partial t},$$

und durch Volumintegration

$$K_t = \frac{E}{c}\frac{\partial c}{\partial t} \quad \text{.................................(13 } b\text{)}$$

als Wert der in der Sekunde vom veränderlichen Schwerefelde abgegebenen Energie.

* Die Ausdrücke ($1\,a$—d) mit $w = \sqrt{c}$ gehen aus denen meiner ersten Note hervor, indem man setzt

$$\Phi = \tfrac{1}{2}c^2 = \tfrac{1}{2}w^4, \quad \pi\gamma = ac^2 = aw^6,$$

dabei bedeutet γ die Gravitationskonstante im gewöhnlichen Sinne, die somit vom Grade 3 in c ist.

† M. Abraham, *Physik. Zeitschr.* 13, 314, 1912. Allerdings sehe ich die dort angegebene Ableitung nicht mehr als stichhaltig an.

‡ In ($12\,a$) ist die linke Seite invariant gegenüber Lorentz-Transformationen, die rechte Seite aber nicht, weil die Energiedichte keine Invariante im Sinne der vierdimensionalen Vektoranalysis ist. Es ist demnach gerade die physikalisch so plausibele Hypothese der Proportionalität von Energie und Schwere, welche dazu zwingt, der Differentialgleichung des Schwerefeldes eine der Relativitätstheorie nicht entsprechende Gestalt zu geben.

Es lauten somit Impuls- und Energiesatz für einen materiellen Punkt (und für ein System, das einem solchen äquivalent ist)

$$\frac{d\mathfrak{G}}{dt} = -\frac{E}{c}\operatorname{grad} c \qquad \text{(14)},$$

$$\frac{dE}{dt} = \frac{E}{c}\frac{\partial c}{\partial t} \qquad \text{(14 a)}.$$

Es bedingt also ein Gefälle von c ein Anwachsen des Impulsvektors, eine lokale zeitliche Zunahme von c ein Anwachsen der Energie des Punktes; im statischen Felde bleibt die Energie des bewegten materiellen Punktes konstant, wenn ausser der Schwerkraft keine andere Kraft wirkt.

Die Feldgleichungen (12), (12 a) gehen für den speziellen Fall des statischen Feldes in die von A. Einstein angegebenen* über. Doch erscheinen dessen Entwicklungen als einigermassen willkürlich und nicht widerspruchsfrei, insofern, als sie auf der Äquivalenzhypothese beruhen, die sich gerade mit diesen Feldgleichungen nicht vereinbaren lässt. Auf die Bewegungsgleichungen (14), (14 a) kommen wir am Schlusse zurück.

Wir betrachten zwei ruhende materielle Punkte, P und P', von den Massenkonstanten M und M'. Auf P wirkt, nach (5 a) und (11), die Kraft

$$\mathfrak{K} = -M\operatorname{grad} c = -2Mw\operatorname{grad} w \qquad \text{(15)},$$

in dem von P' erregten Felde; dieses Feld bestimmt sich gemäss (12 a) durch Integration von

$$\frac{\partial^2 w}{\partial x^2} + \frac{\partial^2 w}{\partial y^2} + \frac{\partial^2 w}{\partial z^2} = \frac{2\alpha\eta}{w}$$

für den Grenzfall einer in P' konzentrierten Energie $E' = M'c'$. Es folgt

$$w = w_\infty - \frac{\alpha}{2\pi}\frac{M'w'}{r} \qquad \text{(15 a)},$$

wo w_∞ den Wert von $\sqrt{c}$ für $r = \infty$ angibt. Aus (15) erhält man für den Betrag der Kraft, mit der P' den Punkt P anzieht,

$$K = 2Mw\frac{\partial w}{\partial r} = \frac{\alpha}{\pi}\frac{Mw \cdot M'w'}{r^2} \qquad \text{(15 b)}.$$

Da nun aber, nach (15 a), w selbst von r abhängt, so ist die Anziehungskraft nicht streng proportional zu r^{-2}. Man hat vielmehr†

$$K = \frac{\alpha}{\pi} \cdot \frac{MM'}{r^2} \cdot w_\infty w' \left(1 - \frac{\alpha}{2\pi} \cdot \frac{w'}{w_\infty} \cdot \frac{M'}{r}\right) \qquad \text{(15 c)}.$$

Nach der hier entwickelten Theorie gilt also das Newtonsche Gesetz nicht streng; es ist durch ein Gesetz von der Form

$$K = \frac{A}{r^2} - \frac{B}{r^3} \qquad \text{(16)}$$

zu ersetzen.

* A. Einstein, *Ann. d. Phys.* 38, 443, 1912.

† Streng genommen hängt auch w' von r ab; jedoch in dem Falle, wo M klein gegen M' ist (z. B. wenn P den Planeten, P' die Sonne bedeutet), ist dieser Umstand ohne Einfluss.

Für die numerische Berechnung des Quotienten B/A genügt es, zu setzen: $\alpha = \pi\gamma/c^3$ (γ Gravitationskonstante), $M' = m'c$, wo m' die Masse des anziehenden Körpers bedeutet,

$$w'/w_\infty = 1.$$

Man erhält dann für die Sonne, wenn r in astronomischen Einheiten gemessen wird,

$$B/A = \gamma m'/2c^2 = 10^{-8}.$$

Die in (16) eingeführte Korrektion des Newtonschen Gesetzes dürfte mithin zu klein sein, um die Planetenbewegung merklich zu beeinflussen.

Ich möchte betonen, dass die hier vorgetragene Theorie von besonderen Annahmen über die Lagrangesche Funktion unabhängig ist. Letztere ist, wie wir gesehen haben, eine homogene Funktion von v und $\Phi = c$, die wir schreiben können (vgl. 9, 9 *a*)

$$L = -Mcf(\beta), \quad \beta = \frac{v}{c} \quad \text{.........................(17)},$$

mit

$$f(\beta) = 1 - \frac{\zeta}{2}.\beta^2 + \dots \quad \text{.......................(17 } a\text{)}.$$

Setzt man speziell*, indem man für konstantes c, d. h. ausserhalb des Schwerefeldes, die Relativtheorie als gültig ansieht,

$$f(\beta) = \sqrt{1 - \beta^2} = 1 - \tfrac{1}{2}\beta^2 + \dots \quad \text{....................(18)},$$

so wird

$$L = -M.\sqrt{c^2 - v^2} \quad \text{..........................(18 } a\text{)},$$

und die Bewegungsgleichung (14) nimmt dann die von A. Einstein angegebene Form an. Da hier $\zeta = 1$ ist, so folgt aus (11 *b*) für die Ruhmasse der Wert

$$m_0 = \frac{M}{c} = \frac{E_0}{c^2} \quad (E_0 \text{ Ruhenergie}) \text{....................(18 } b\text{)}.$$

Doch könnte sehr wohl $\zeta = 1$ sein, ohne dass die Koeffizienten der höheren Glieder in der Reihenentwicklung der Funktion $f(\beta)$ mit den von der Relativtheorie geforderten übereinstimmen.

In meiner Dynamik des Elektrons ist†

$$f(\beta) = \frac{1 - \beta^2}{2\beta}.\ln\left(\frac{1 + \beta}{1 - \beta}\right) = 1 - \frac{2}{3}\beta^2 + \dots \quad \text{.................(19)},$$

man hat also $\zeta = \frac{4}{3}$, und daher

$$m_0 = \frac{4}{3}\frac{M}{c} = \frac{4}{3}\frac{E_0}{c^2} \quad \text{.........................(19 } a\text{)}.$$

Von den schwebenden Fragen der Dynamik des Elektrons ist die oben entwickelte Theorie des Gravitationsfeldes ganz unabhängig. Ihre Voraussetzungen sind, ausser der Grundannahme, dass das Schwerkraftfeld durch die Lichtgeschwindigkeit bestimmt sei: Die Gültigkeit der Ausdrücke (1 *a*—*d*) für den Gravitationstensor, die Prinzipien der Mechanik (d. h. die Impuls- und Energiesätze und die Lagrangeschen Gleichungen) und endlich die Hypothese der Proportionalität von Schwere und Energie.

* M. Planck, *Verh. d. D. Phys. Ges.* 8, 140, 1906.

† M. Abraham, *Theorie der Elektrizität* II, 2. Aufl. S. 167, 1908.

AETHER, MATTER AND GRAVITY

By S. B. McLaren.

I denote the coordinates of a point in space of three dimensions by x, y, z, in four by x, y, z, s. The four-dimensional space is "Euclidean," the distance from the origin being

$$(x^2 + y^2 + z^2 + s^2)^{\frac{1}{2}}.$$

The vector notation is reserved for three dimensions. τ or t is the time. The velocity of a point is denoted by

$$\mathbf{u} = \frac{d\mathbf{r}}{dt},$$

or in four dimensions it has the components

$$\frac{d\mathbf{r}}{d\tau} \text{ with } \frac{ds}{d\tau}.$$

dv and $dvds$ are elements of volume. I write the suffix a to denote reference to a particle of aether.

$\frac{d}{dt}$ or $\frac{d}{d\tau}$, denotes the time rate of change at a moving point.

§ 1. *Time and Space.*

For mathematical physics time is merely the independent variable in a scheme of differential equations where the measurable variables are positions. So long as the form of these equations is unaltered so are all the results of measurement.

With Minkowski the physical universe is a steady state in four dimensions. The time variable is the distance of any of a series of parallel sections from one of them. For him all the physical characteristics of any section of the universe are fixed by the value of this independent variable. For us too the time t, as it appears in physics, is still the independently variable distance s of the sections. If we write

$$s = ct \qquad \text{(1)},$$

that merely indicates some scale of measurement of velocities.

But though I suppose the state of the universe to be steady, it is a state of steady motion. Aether and matter are both incompressible fluids moving in unchanging streams. To fix the coordinates of any particle of aether or matter the absolute time τ is required. In any stream line therefore x, y, z, s are functions of τ. Thus I have one more disposable variable than Minkowski, but otherwise I have fewer physically ultimate conceptions. Electric charge is explained as a flux of matter, electric current also.

The time of physics is still s, for this is still the independent variable whose value gives the "present" physical state, though it does not give the actual coordinates of any particle of matter or aether.

The equation of continuity is

$$\operatorname{Div}\left(m\frac{d\mathbf{r}}{d\tau}\right)+\frac{d}{ds}\left(m\frac{ds}{d\tau}\right)=0 \quad \ldots\ldots\ldots\ldots(2).$$

Along any stream line $\mathbf{r}$ is a function of s.

Let

$$m\frac{ds}{d\tau}=\rho a \quad \ldots\ldots\ldots\ldots(3),$$

where a is a constant. (2) becomes

$$\operatorname{Div}\left(\rho\frac{d\mathbf{r}}{ds}\right)+\frac{d\rho}{ds}=0 \quad \ldots\ldots\ldots\ldots(4).$$

By using (1) the equation of continuity reappears in the ordinary form. Thus the density in three dimensions is proportional to the flux along the fourth dimension.

ρ I identify with the density of electric fluid; m is proportional to the absolute weight per unit volume.

§ 2. *The equations of Motion.*

Maxwell's electromagnetic scheme is extended from three to four dimensions, and all the formulae are deduced from the principle of least action. To begin with three dimensions,

$$\delta\iiiint L\,dv\,dt=0 \quad \ldots\ldots\ldots\ldots(5)$$

$$L=(8\pi)^{-1}\left(\frac{d\mathbf{F}}{c\,dt}+\nabla\phi\right)^2-(8\pi)^{-1}(\operatorname{Curl}\mathbf{F})^2+\rho c^{-1}(\mathbf{uF}-c\phi)+\tfrac{1}{2}\epsilon\rho^2(\mathbf{u}^2-c^2) \quad \ldots(6).$$

The last term assigns intrinsic energy and momentum to electricity and makes internal equilibrium of the electron possible.

(5) requires $\qquad \delta(\rho\,dv)=0.$

$\mathbf{E}$ and $\mathbf{H}$ the electric and magnetic intensities are given by

$$\mathbf{E}=-\frac{d\mathbf{F}}{c\,dt}-\nabla\phi,\quad \mathbf{H}=\operatorname{Curl}\mathbf{F} \quad \ldots\ldots\ldots\ldots(7).$$

(6) is invariant under the Lorentz-Einstein substitution

$$\left.\begin{array}{ll} ct_1=\beta(ct-\alpha x), & \beta^2(1-\alpha^2)=1 \\ x_1=\beta(x-\alpha ct), & \alpha<1 \\ y_1=y, & \\ z_1=z, & \end{array}\right\} \quad \ldots\ldots\ldots\ldots(8).$$

(5) and (6) give Maxwell's electromagnetic equations together with

$$\frac{d}{dt'}(\epsilon\rho^2\mathbf{u}\,dv)+\nabla\left(\tfrac{1}{2}\epsilon\rho^2c^2-\tfrac{1}{2}\epsilon\rho^2\mathbf{u}^2\right)dv=\mathbf{E}'\rho\,dv \quad \ldots\ldots\ldots\ldots(9),$$

$$\mathbf{E}'=\mathbf{E}+c^{-1}[\mathbf{uH}] \quad \ldots\ldots\ldots\ldots(10).$$

The electric fluid has per unit volume the momentum $\epsilon\rho^2\mathbf{u}$ and the energy $\frac{1}{2}\epsilon\rho^2(c^2+\mathbf{u}^2)$ with the apparent pressure

$$\tfrac{1}{2}\epsilon\rho^2(c^2-\mathbf{u}^2) \qquad \text{.....(11).}$$

The integral expression

$$\int(\epsilon\rho\,\mathbf{u}+c^{-1}\mathbf{F})\,d\mathbf{r}$$

round any closed curve moving with the fluid is constant. Permanent "irrotational" motion is possible such that

$$\epsilon\rho\,\mathbf{u}+c^{-1}\mathbf{F}=\nabla\chi, \quad \epsilon\rho c+c^{-1}\phi=-\frac{d\chi}{cdt} \qquad \text{.....(12).}$$

With (12) Maxwell's equations give

$$\left.\begin{aligned}\left(\nabla^2-\frac{1}{c^2}\frac{d^2}{dt^2}-\frac{4\pi}{\epsilon c^2}\right)\rho\mathbf{u}=0\\ \left(\nabla^2-\frac{1}{c^2}\frac{d^2}{dt^2}-\frac{4\pi}{\epsilon c^2}\right)\rho=0\end{aligned}\right\} \qquad \text{.....(13).}$$

(5) and (6) can be extended by analogy to four dimensions

$$\delta\iiiiint L_1\,dv_1\,ds_1\,d\tau_1=0 \qquad \text{.....(5)}_1,$$

$$\begin{aligned}L_1=(8\pi)^{-1}\left(\frac{d\mathbf{F}_1}{ad\tau_1}+\nabla J_1\right)^2+(8\pi)^{-1}\left(\frac{d\phi_1}{ad\tau_1}+\frac{dJ_1}{ds_1}\right)^2\\ -(8\pi)^{-1}\left(\frac{d\mathbf{F}_1}{ds_1}-\nabla\phi_1\right)^2-(8\pi)^{-1}(\text{Curl}\,\mathbf{F}_1)^2\\ +m_1a^{-1}\left(\frac{d\mathbf{r}_1}{d\tau_1}\mathbf{F}_1+\frac{ds_1}{d\tau_1}\phi_1-aJ_1\right)\\ +\tfrac{1}{2}k^2m_1^2\left\{\left(\frac{d\mathbf{r}_1}{d\tau_1}\right)^2+\left(\frac{ds_1}{d\tau_1}\right)^2-a^2\right\}\end{aligned} \qquad \text{.....(6)}_1,$$

$$\frac{dm_1}{d\tau_1}+\text{Div}\left(m_1\frac{d\mathbf{r}_1}{d\tau_1}\right)+\frac{d}{ds_1}\left(m_1\frac{ds_1}{d\tau_1}\right)=0 \qquad \text{.....(2)}_1.$$

Transform by the substitution

$$\begin{aligned}a\tau&=\beta_1(a\tau_1-\alpha_1s_1), &\quad \beta_1^2(\alpha_1^2-1)&=1,\\ s&=\beta_1(s_1-\alpha_1a\tau_1), &\quad \alpha_1&>1,\\ \mathbf{r}&=\mathbf{r}_1.\end{aligned}$$

This involves a reference to axes moving with a velocity $a\alpha_1$ greater than a.

It is referred to these axes that I suppose the state a steady one. There is translation as a whole with a velocity greater than that quantity a which corresponds to the velocity c of light in (6). Then $(2)_1$ becomes transformed into (2) and I now postulate that $(5)_1$ is to hold under the restriction that m is a constant, or that the liquid is incompressible; this condition introduces a liquid pressure p_m in matter.

$(5)_1$ therefore becomes for steady states

$$\delta \iiiiint L\, dv\, ds\, d\tau \quad \text{......(14)},$$

$$L = -(8\pi)^{-1}(\nabla J)^2 + (8\pi)^{-1}\left(\frac{dJ}{ds}\right)^2 + mJ$$

$$+ (8\pi)^{-1}\left(\frac{d\mathbf{F}}{ds} + \nabla\phi\right)^2 - (8\pi)^{-1}(\text{Curl}\,\mathbf{F})^2$$

$$+ ma^{-1}\left(\frac{d\mathbf{r}}{d\tau}\mathbf{F} - \frac{ds}{d\tau}\phi\right) + \tfrac{1}{2}k^2 m^2\left\{\left(\frac{d\mathbf{r}}{d\tau}\right)^2 - \left(\frac{ds}{d\tau}\right)^2\right\} \quad \text{......(15)},$$

with

$$\delta\,(dv\,ds) = 0 \quad \text{......(16)}.$$

Variation of $\mathbf{F}$, ϕ and J gives Maxwell's electromagnetic equations together with

$$\left(\nabla^2 - \frac{1}{c^2}\frac{d^2}{dt^2}\right)J + 4\pi m = 0 \quad \text{......(17)}.$$

Here (1) and (3) are used.

Variation of the distribution of matter gives four equations of motion. These involve the pressure of matter p_m whose amount can be found by an integration as in Hydrodynamics

$$p_m = -\,\mathrm{a} + \tfrac{1}{2}\epsilon\rho^2(c^2 - \mathbf{u}^2) + mJ \quad \text{......(18)},$$

where

$$k^2 a^2 = \epsilon c^2,$$

and a is a constant of integration.

The condition at the free surface of matter is

$$p_m = 0$$

or

$$\tfrac{1}{2}\epsilon\rho^2(c^2 - \mathbf{u}^2) = \mathrm{a} - mJ \quad \text{......(19)}.$$

Three equations of motion are

$$\frac{d}{dt'}(\epsilon\rho^2\mathbf{u}\,dv) + \nabla p_m dv = \mathbf{E}'\rho\,dv + m\nabla J\,dv \quad \text{......(20)}.$$

By (18) these are reduced to (9). Regarding the left of (19) as the pressure, there appear to be the externally applied pressure a, and tension mJ at the external surface of matter.

§ 3. *Aether a polarizable fluid; matter an aether sink.*

If in (6) all terms were omitted except the last, then (12) would become

$$\epsilon\rho\mathbf{u} = \nabla\chi, \quad \epsilon\rho c = -\frac{d\chi}{c\,dt}.$$

It is now obvious how the terms involving differential coefficients of J in (15) are to be interpreted.

Write

$$\left.\begin{aligned} m_a\frac{d\mathbf{r}_a}{d\tau} &= \sigma\mathbf{u}_a = (4\pi)^{-\frac{1}{2}}\nabla J \\ m_a\frac{ds_a}{d\tau} &= \sigma c = -(4\pi)^{-\frac{1}{2}}\frac{dJ}{ds} \end{aligned}\right\} \quad \text{......(21)}.$$

In (21) I understand m_a to be the absolute density of aether, σ the density in three dimensions. (17) is now

$$\text{Div}(\sigma\mathbf{u}_a) + \frac{d\sigma}{dt} = -(4\pi)^{\frac{1}{2}} m \quad \text{.........................(22)},$$

or

$$\text{Div}\left(m_a \frac{d\mathbf{r}_a}{d\tau}\right) + \frac{d}{ds}\left(m_a \frac{ds_a}{d\tau}\right) = -(4\pi)^{\frac{1}{2}} m \quad \text{..............(23)}.$$

Thus by (22) the quantity $(4\pi)^{\frac{1}{2}} m$ of aether disappears per second and per unit volume of space. Or by (23) the absolute volume $dv_a ds_a$ shrinks at a rate given by

$$m_a \frac{d}{d\tau'}(dv_a ds_a) = -(4\pi)^{\frac{1}{2}} m \, dv_a ds_a \quad \text{....................(24)}.$$

The aether thus absorbed is conceived of as annihilated, its momentum being given to matter. It may be shown that it is possible for the aether to move "irrotationally" even when matter is present.

The equations of motion for aether can be deduced by putting for its Lagrangian function L_a,

$$L_a = -\tfrac{1}{2} m_a^2 \left(\frac{d\mathbf{r}_a}{d\tau}\right)^2 + \tfrac{1}{2} m_a^2 \left(\frac{ds_a}{d\tau}\right)^2 + (8\pi)^{-1}(\mathbf{E}^2 - \mathbf{H}^2) \quad \text{........(25)},$$

and introducing what appears as the external friction on unit quantity of the aether

$$(4\pi)^{\frac{1}{2}} m\mathbf{u}_a.$$

This force is really the rate of transfer of momentum due to the destruction of aether by matter.

The pressure p_a is found to be

$$p_a = b + \tfrac{1}{2}\sigma^2 (c^2 - \mathbf{u}_a^2) - (8\pi)^{-1}(\mathbf{E}^2 - \mathbf{H}^2) \quad \text{..............(26)}.$$

On restoring the Newtonian Potential J the principles of least action and conservation of energy are saved by a mathematical fiction which attributes energy to matter of density $-mJ$ and gives (14) and (15) once more.

§ 4. *Optical Phenomena not affected by the motion of the aether.*

The gravitational energy is wholly kinetic, it is per unit volume

$$\tfrac{1}{2} m_a^2 \left(\frac{d\mathbf{r}_a}{d\tau}\right)^2 + \tfrac{1}{2} m_a^2 \left(\frac{ds_a}{d\tau}\right)^2.$$

The electromagnetic energy is wholly potential

$$(8\pi)^{-1}\left(\frac{d\mathbf{F}}{ds} + \nabla\phi\right)^2 + (8\pi)^{-1}(\text{Curl}\,\mathbf{F})^2.$$

These two forms of energy may be put down as due to the motion of translation of the atoms of a fluid and their polarization. If the polarizing apparatus be regarded as centred about each atom on a universal joint, then there is a statical balance of the electromagnetic forces in spite of the aether's motion. They produce also a resultant force on aether of intensity

$$-(8\pi)^{-1}\nabla(\mathbf{E}^2 - \mathbf{H}^2)$$

in unit volume.

I would also urge for those whom this does not satisfy that the absolute velocity of the aether can be made as small as we please by increasing m_a. The force of gravity is the only phenomenon imposing any restriction. It is

$$m_a \frac{d\mathbf{r}_a}{d\tau} \text{ or } (4\pi)^{-\frac{1}{2}} \nabla J,$$

and we therefore know only the product of mass and velocity.

§ 5. *Cavities in the Aether, Electron and Magneton, Positive and Negative charge.*

It may be that cavities exist in the aether. The surface of any such cavity must also be a boundary for matter. Conditions sufficient to satisfy the principles of conservation both of energy and momentum are

$$\mathbf{E}_t + c^{-1} [\mathbf{u}_n \mathbf{H}] = 0, \quad \mathbf{H}_n = 0 \quad \text{.......................(27)},$$

$$\frac{dJ}{dn} + \mathbf{u}_n \frac{dJ}{c^2 dt} = 0 \quad \text{..................................(28)},$$

$$p_a + p_m = 0 \quad \text{..................................(29)}.$$

The suffix n denotes a normal and t a tangential component, $\mathbf{u}_n$ the velocity of the surface normal to itself.

By (27) the surface is a perfect reflector; by (28) its normal velocity is equal to the normal velocity of the aether itself and by (29) the total pressure of aether and matter on the surface of the cavity vanishes. Such a surface may carry what is effectively an electric surface charge. For by Maxwell's equations the total normal electric induction is invariant. If it is multiply connected it may serve as a permanent magnet, for the total magnetic flux across any aperture is invariant.

Also there is here a possible distinction between two kinds of electricity. The one a volume distribution, the other an integral induction over the surface of the cavity.

(14) and (15) lead to the same expressions for the energy and momentum of the Gravitational field as have already been given by Max Abraham. It is therefore now unnecessary for me to do more than refer the reader to his paper in this volume for these results which I had myself obtained independently.

SOPRA UN CRITERIO DI CLASSIFICAZIONE DEI MASSIMI E DEI MINIMI DELLE FUNZIONI DI PIU VARIABILI

DI C. SOMIGLIANA.

Se si osserva che in una regione montuosa le linee, che dividono i diversi bacini idrografici, ossia le creste, vanno a congiungersi nelle vette della regione, nasce spontanea l'idea di classificare le vette mediante il numero delle creste che in esse concorrano. Anche le linee di valle, cosi chiamando le linee percorse dalle correnti d'acqua, hanno in generale origine dalle vette e sono per ogni vetta in numero uguale a quello dalle creste. Si hanno cosi vette di 1°, 2°, 3°, ... ordine secondo che da esse partono *una* linea di cresta ed *una* di valle, oppure *due* linee di cresta e *due* di valle, oppure *tre* linee di cresta e *tre* di valle e cosi via.

Portando questo criterio nel campo delle funzioni di due variabili per ottenere una classificazione dei massimi e dei minimi, occorre anzitutto definire in modo preciso analiticamente le linee di cresta e le linee di valle. A ciò si arriva considerando—sopra ogni linea di livello—i massimi ed i minimi dell' angolo che la verticale fa colla normale alla superficie. I luoghi di questi massimi e di questi minimi costituiscono rispettivamente sulla superficie le linee di cresta e le linee di valle.

L'equazione complessiva di tutte queste linee sia di cresta che di valle è la segmente

$$\frac{\partial \phi}{\partial x}\frac{\partial \Delta_1 \phi}{\partial y} - \frac{\partial \phi}{\partial y}\frac{\partial \Delta_1 \phi}{\partial x} = 0 \quad \dots\dots\dots\dots(1),$$

dove $\phi(x, y)$ è la funzione che si considera, e

$$\Delta_1 \phi = \left(\frac{\partial \phi}{\partial x}\right)^2 + \left(\frac{\partial \phi}{\partial y}\right)^2.$$

La determinazione di queste linee non richiede perciò alcuna integrazione.

Nelle vicinanze di un massimo o di un minimo la equazione (1) si può porre in generale sotto la forma:

$$u_1(x, y)\, v_1(x, y)\, u_2(x, y)\, v_2(x, y) \dots u_n(x, y)\, v_n(x, y) = 0.$$

Il numero n rappresenta allora *l'ordine* del massimo o del minimo considerato.

Non è difficile estendere la definizione di linee di cresta e di valle alle funzioni di un numero qualunque di variabili.

Queste linee hanno interesse anche sotto un altro punto di vista. Esse possono essere considerate come le linee secondo le quali, a partire da un massimo, o da un minimo, la funzione decresce, o cresce, colla maggiore o colla minore rapidità. Esse portano quindi in certo modo ad una estensione del concetto di massimo e di minimo per le funzioni che dipendono da più di una variabile, in quanto rappresentano degli enti, non più a dimensione nulla, come sono i massimi e minimi ordinari, ma a dimensione finita (uguale ad uno) caratterizzati da una certa prevalenza, o deficienza, in essi dei valori della funzione.

Sotto questo punto di vista la considerazione delle linee di cresta e di valle può tornare assai utile nello studiare l'andamento di una funzione di due o più variabili nel campo in cui essa è definita.

ON A LAW OF CONNEXION BETWEEN TWO PHENOMENA WHICH INFLUENCE EACH OTHER

By W. Esson.

1. It is proposed to measure the influence of one phenomenon upon another by the ratio of the percentage increase of one to the percentage increase of the other.

2. One advantage of this method is that the measure is a pure number independent of the units in terms of which the phenomena are expressed. Another advantage is that the mode in which one phenomenon influences the other need not be known.

3. If A and B are the number of units of the two phenomena present at any time and a small increase dA determines a small increase dB, the measure here proposed is the ratio $A^{-1}dA/B^{-1}dB = m$.

4. In the case in which the changes of A and B do not influence either the nature of the phenomena or the mode in which they react upon each other it is reasonable to suppose that m is constant. In this case the relation between the phenomena is expressed by $A/A_0 = (B/B_0)^m$.

5. In other cases in which the conditions affecting the phenomena are not so simple m will be a function of A/A_0 and B/B_0.

6. This method of computing the mutual influence of two phenomena occurred to the author forty years ago during the course of a research conducted by Harcourt and himself upon the connexion between the conditions of a chemical change and its amount. The results of this research were published in the Bakerian Lecture of the Royal Society in 1895.

7. In the early part of this year Harcourt made a further research on the variation with temperature of the rate of a chemical change and requested the author to discuss his experiments as well as the experiments of other chemists on the same subject. This discussion has been published in the *Transactions of the Royal Society*, Series A, Vol. CCXII. pp. 187–204.

8. In the Bakerian Lecture it was shown that the rate of chemical action k is connected with the temperature t by the relation $k/k_0 = \{(c+t)/(c+t_0)\}^m$ and further that the value of c is 272·6, so that the relation may be written $k/k_0 = (T/T_0)^m$, T being the absolute temperature. Thus it was shown that the zero of chemical energy is the same as the zero of the molecular energy of temperature.

m is a constant depending on the kind of medium in which the chemical action takes place and is the measure $k^{-1}dk/T^{-1}dT$ of the effect of thermal energy upon chemical energy. If u is the mean velocity of the atoms or electrons the energy of which causes the chemical action and v is the mean velocity of the molecules, the measure of the effect may be expressed as $m = v^{-1}dv/u^{-1}du$.

This relation is satisfied by the results of experiments by other chemists in which due precaution has been taken to keep constant the nature of the chemical action and of the medium.

It is hoped that the relation may be of use in the measurement of the mutual influence of other than chemical phenomena.

SELF-CONTAINED ELECTROMAGNETIC VIBRATIONS OF A SPHERE AS A POSSIBLE MODEL OF THE ATOMIC STORE OF LATENT ENERGY

By L. Silberstein.

This communication will be published in the *Philosophical Magazine* in the course of 1913.

MULTIPLY-CHARGED ATOMS

BY SIR J. J. THOMSON.

In the photographs of the positive rays (see, for example, those given in the *Phil. Mag.* Aug. 1912) the mercury line is remarkable for the exceptionally small displacement of the head of its parabola. Even when the electric and magnetic fields are strong enough to produce deflexions of several millimetres in the heads of the parabolas corresponding to the other elements, the head of the mercury parabola is so little deflected that at first sight it seems to coincide with the origin. When, however, the electric field used to deflect the particles is made very large, in our experiments from 5000 to 10,000 volts per centimetre, the head of this parabola is distinctly displaced, and on measuring the electrostatic displacement it is found to be 1/8 of the normal displacement of the heads of the parabolas corresponding to the other elements.

The displacement due to the electric field is inversely proportional to the kinetic energy of the particle displaced, so that the atoms which produce the head of the mercury parabola must have eight times the maximum amount of energy possessed by the normal atoms. This could be accounted for if some of the mercury atoms in the discharge-tube had lost 8 corpuscles, for then the energy communicated to the atom by the electric field would be eight times the energy communicated to an atom with the normal charge. Eight is a very large number of corpuscles to lose, much larger than the number lost by the other elements which get multiple charges. I had previously to these experiments never found a case in which this number exceeded three; so that in my paper in the *Phil. Mag.* for August 1912, I suggested another explanation of the behaviour of the mercury line. A study of the plates taken with large electrostatic deflexions has revealed the existence of 7 parabolas due to mercury, corresponding to the mercury atom with 1, 2, 3, 4, 5, 6, 7 charges respectively, the parabola corresponding to 8 charges has not been detected, but as the parabolas get in general fainter for each additional charge on the atom, it is probably there but too faint to be detected. Fig. 1. taken from a photograph when the gas in the tube was the residual gas left after exhaustion by the Gaede pump, shows these lines very well. The measurements of m/e for these parabolas gave the following values:

m/e	
200	
102	200/2
66·3	200/3
50·4	200/4
44	this not a mercury line but is due to CO_2.
39·8	200/5
33·7	200/6
28·6	200/7

It will be noticed that the heads of the parabolas corresponding to 1, 2, 3 charges respectively lie on a straight line passing through the origin, indicating that the velocities of the particles producing the heads of these parabolas are all equal, and therefore, since each particle is an atom of mercury, that the kinetic energy of the particles at the heads of the parabolas is constant. This is what we should have expected, for the heads of all the parabolas are due to particles which had originally lost 8 corpuscles, the particle at the head of the parabola corresponding to one charge has regained 7 of these after passing through the cathode, that at the head of the parabola corresponding to two charges, six, and so on; these particles when in the discharge-tube were all in the same condition, and so acquired the same amount of kinetic energy.

The question now arises as to how the mercury atom acquires these very various charges. When a mercury atom is ionized, can it lose any number of corpuscles from one to eight? Taking for example a mercury atom with 5 positive charges, has it got into this condition by losing 5 corpuscles when it was ionized, or did it originally lose the maximum number 8 and regain 3 subsequently? The photographs prove, I think, that the second supposition is the correct one, and that in the discharge-tube there are two, and only two, kinds of ionization; in one of these kinds the mercury atom loses 1 corpuscle, while in the other kind it loses 8, and that there are no indications of ionization of such a character as to deprive the mercury atom of 7, 6, 5, 4, 3, or 2 corpuscles.

The evidence for this is as follows: let us suppose for a moment that atoms with any charge from 1 to 8 were produced by the ionization of the mercury atoms in the discharge-tube. Consider now the parabola due to the mercury atom which, when it passed through the electric and magnetic fields, had one positive charge. This parabola would result from atoms of the following kinds:

1. Atoms which had lost 8 corpuscles on ionization and regained 7;
2. Atoms which had lost 7 and regained 6;
3. Atoms which had lost 6 and regained 5,

and so on, the eighth and last members of the series being atoms which had lost one corpuscle on ionization and had not regained it.

The parabola on the photographic plate would be due to the superposition of the 8 parabolas due to these types of atoms.

If $8d$ is the horizontal distance from the vertical axis of the head of the parabola due to the atom which lost one corpuscle on ionization, the horizontal distances of the heads of the parabolas due to atoms of the 1st, 2nd, 3rd types will be respectively $8d/8$, $8d/7$, $8d/6$, $8d/5$, $8d/4$, $8d/3$, $8d/2$, $8d$. Thus up to the horizontal distance $8d/7$ there would be only one parabola, at $8d/7$ another parabola would be added, this would produce an abrupt increase in the intensity of the photograph, there would be another abrupt increase at $8d/6$, another at $8d/5$, and so on. Thus the intensity of the parabola on the photograph would not be continuous, there would be places where the intensity was suddenly increased giving a beaded appearance to the photograph. The abrupt increase at $8d$ is very marked on this parabola, the others are not visible; but as the intensity of this parabola is very great it might be thought that they escaped detection owing to the breadth of the parabola. Let us therefore consider one of the finer parabolas, say, that due to the atom with four charges, to which this objection does not apply.

This parabola might arise from atoms which had lost 8 charges on ionization and regained 4, from those which had lost 7 and regained 3, and so on, the last being atoms which had lost 4 and not regained any. Then if d has the same meaning as before, the horizontal distances from the vertical axis of the heads of the parabolas would be $4 \times 8d/8$, $4 \times 8d/7$, $4 \times 8d/6$, $4 \times 8d/5$, $4 \times 8d/4$; there would, therefore, be abrupt increases in intensity at $32d/7$, $32d/6$, $32d/5$, $32d/4$. The photographs show, however, that the intensity is perfectly continuous, and not one of these abrupt increases is to be seen. We conclude, therefore, that there are no atoms which begin with 7, 6, 5, 4, 3, 2 charges, and that in this case there are two and only two types of ionization, in the one type an atom loses a single corpuscle, in the other it loses 8.

This result suggests that ionization takes place in the discharge-tube in two ways. In the first method the ionizing agents are the rapidly moving corpuscles which constitute the cathode rays, these very small particles penetrate into the atom and come into collision with the corpuscles inside it individually, the collision in favourable cases causing the corpuscle struck to escape from the atom; this type of ionization results in the atom losing a single charge. In the other type of ionization we suppose that the mercury atom is struck by a rapidly moving atom and not by a corpuscle; after the collision the mercury atom starts off with a very considerable velocity, which at first is not shared by the corpuscles inside it. The tendency of the corpuscles to leave the atom depends only upon the *relative* velocity of the atom and the corpuscles inside it, so that the ionizing effect produced by the collision is the same as if the atom were at rest, and *all* the corpuscles were moving with the velocity acquired by the atom in the collision. Thus if there were eight corpuscles in the mercury atom connected with about the same firmness to the atom, the result of the atom acquiring a high velocity in a collision might be the detachment of the set of eight leaving the atom with a charge of 8 units of positive electricity. We see in this way how the cathode particles might produce one type of ionization resulting in singly charged atoms, while the atoms forming the positive rays might produce another type of ionization resulting in multiply-charged atoms.

All the elements I have examined give multiply-charged positive atoms with the exception of hydrogen, on which I have never observed more than one charge; in no other case, however, have I observed charges approaching that possessed by mercury. The majority of the elements seem to acquire only two charges; this is the number acquired by helium, and this case is interesting since in the vacuum-tube the helium atom occurs with both single and double charges, whilst as an α particle it always seems to have two charges, suggesting that the process by which the α particle acquires its charge is analogous to the process by which multiply-charged atoms are produced in the discharge-tube.

I have observed nitrogen atoms with three charges, but the parabola due to the triply-charged atom is exceedingly faint. Argon shows triply-charged atoms very distinctly as can be seen from fig. 2, where the parabolas I, II, III due to Arg_{+}, Arg_{++}, Arg_{+++} are all very distinct; the Arg_{++} parabola has probably a parabola due to neon superposed on it.

This plate shows the helium line, and thus incidentally gives us an estimate of the sensitiveness of this method of detecting small quantities of a gas. The volume of the discharge-tube was about two litres, the pressure 1/300 of a mm. of mercury: there was thus in the discharge-tube about 1/100 c.c. of argon at atmospheric pressure. This is about the amount in 1 c.c. of air at this pressure, and as the helium line was visible along with the argon, we see that this method can detect the amount of helium in 1 c.c. of air, which, according to Sir William Ramsay, is about 4×10^{-6} c.c., even though this is mixed with an enormous excess of argon. The tube used for this photograph was not at all well adapted for detecting small quantities of an impurity, as the cathode had been in use for several weeks and the tube through it was almost silted up by the sand-blast action of the positive rays, and was just about to be replaced by a new tube.

An interesting result, which is now being investigated, is that when very pure nitrogen is in the discharge-tube the mercury line corresponding to the atom with five charges becomes abnormally bright, brighter than those for the atom with four or even three charges, though in other gases the greater the charge on the atom the fainter the line.

I have much pleasure in thanking Mr F. W. Aston, B.A., Trinity College, for the great assistance he has given me in these investigations.

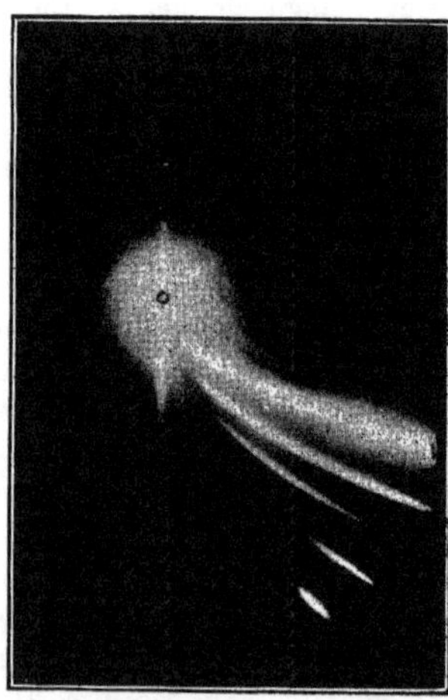

Fig. 1.

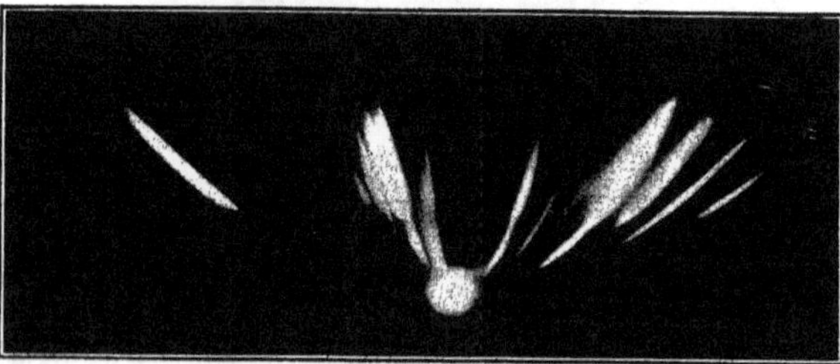

Fig. 2.

ON WAVE-TRAINS DUE TO A SINGLE IMPULSE

By Horace Lamb.

The object of this note is to call attention to some graphical methods by which the general character of the wave-train generated by a single impulse in a dispersive medium of any given constitution can be ascertained. For simplicity, cases of propagation in one dimension are alone considered.

The properties of the medium are sufficiently determined, for the present purpose, by the relation between the constants σ and k in the formula

$$u = \cos kx \cos \sigma t \quad \text{.................................(1)},$$

which represents an indefinitely extended system of standing waves of simple-harmonic type. If τ denote the period, and λ the wave-length, we have of course

$$\tau = 2\pi/\sigma, \quad \lambda = 2\pi/k \quad \text{..............................(2)}.$$

The velocity of propagation of a progressive train of simple-harmonic waves

$$u = \cos k\,(Vt \pm x) \quad \text{.................................(3)}$$

is

$$V = \lambda/\tau = \sigma/k \quad \text{.................................(4)},$$

and the corresponding value of the group-velocity is

$$U = \frac{d\sigma}{dk} = V - \lambda \frac{dV}{d\lambda} \quad \text{..............................(5)}.$$

By superposition of wave-systems of the type (1) we can obtain the result corresponding to an initial impulse (at time $t = 0$) whose space-distribution is given by an arbitrary relation

$$u = \phi\,(x) \quad \text{...................................(6)},$$

viz. we have

$$u = \frac{1}{\pi}\int_0^\infty \cos \sigma t \cos kx\, f(k)\, dk + \frac{1}{\pi}\int_0^\infty \cos \sigma t \sin kx\, F(k)\, dk \text{.........(7)},$$

where

$$f(k) = \int_{-\infty}^{\infty} \phi\,(\alpha) \cos k\alpha\, d\alpha, \quad F(k) = \int_{-\infty}^{\infty} \phi\,(\alpha) \sin k\alpha\, d\alpha \quad \text{.........(8)}.$$

If the initial distribution be symmetrical with respect to the origin O, we have $F(k) = 0$; and if we further assume it to be concentrated in the immediate neighbourhood of O, in such a way that

$$\int_{-\epsilon}^{\epsilon} \phi\,(\alpha)\, d\alpha = 1 \quad \text{................................(9)},$$

where ϵ is infinitesimal, we have finally

$$u = \frac{1}{\pi}\int_0^\infty \cos\sigma t \cos kx\,dk \qquad (10),$$

which is the expression to be elucidated. It may be noted that the integral will not converge at the upper limit unless the wave-velocity (σ/k) is infinite for infinitely short waves $(k=\infty)$. The difficulty, when it arises, may sometimes be met by assuming, in (6),

$$\phi(x) = \frac{1}{\pi}\cdot\frac{\beta}{x^2+\beta^2} \qquad (11),$$

which makes $$f(k) = e^{-k\beta}, \quad F(k) = 0 \qquad (12),$$

and examining the form which (7) assumes as β is diminished. In some cases, e.g. that of sound-waves, we cannot expect a determinate result from an absolutely concentrated impulse.

The integral in (10), even when convergent, cannot be evaluated except in a few special cases, the most important of which is that of deep-water waves*; but Lord Kelvin has shewn† how to obtain an approximation, under certain conditions, in the general case. The disturbance is, in the first place, analysed into two progressive systems of simple-harmonic waves, travelling to the right and left, respectively; thus

$$u = \frac{1}{2\pi}\int_0^\infty \cos(\sigma t - kx)\,dk + \frac{1}{2\pi}\int_0^\infty \cos(\sigma t + kx)\,dk \qquad (13).$$

In strictness both these integrals have to be taken into account in calculating the disturbance at any given place and time; but usually the first term is alone important at points lying to the right of the origin. It may be remarked, however, that if the group-velocity is negative, as it may be in certain imaginable cases‡, it is the *second* integral which is predominant at such points. Taking, however, the more usual case, and ignoring the second term, we have, for $x > 0$,

$$u = \frac{1}{2\pi}\int_0^\infty \cos(\sigma t - kx)\,dk \qquad (14).$$

The disturbance thus represented is made up of an aggregate of simple-harmonic trains, of wave-lengths varying from ∞ to 0, each travelling with its proper velocity. Owing to differences of phase between the various trains, we have "interference," in the optical sense, and the main effect, at time t and place x, will be due to those trains whose wave-length is such that the phase is stationary, or nearly so. The equation

$$t\frac{d\sigma}{dk} - x = 0, \quad \text{or} \quad x = Ut \qquad (15),$$

therefore determines the wave-lengths which are predominant at the time and place considered. Lord Kelvin proceeds to shew, by a process which need not be

* Discussed by Poisson and Cauchy; see *Proc. Lond. Math. Soc.* (2) vol. 2, p. 371.

† *Proc. Roy. Soc.* vol. 42, p. 80 (1887); *Math. and Phys. Papers*, vol. 4, p. 303.

‡ See *Proc. Lond. Math. Soc.* (2) vol. 1, p. 473 (1904).

reproduced*, that corresponding to every value of k which satisfies (15) we have a term

$$\frac{1}{\sqrt{\{2\pi t\,|\,d^2\sigma/dk^2\,|\}}}\,.\cos(\sigma t - kx - \tfrac{1}{4}\pi) \quad\text{..................(16)}$$

in the value of u.

The process and the result may be illustrated by various graphical constructions. The simplest, in some respects, is based on a slight modification of a diagram already employed by the writer†. We construct a curve with λ as abscissa and Vt as ordinate, where t denotes of course the time that has elapsed from the beginning of the disturbance. To ascertain the nature of the wave-system in the neighbourhood of any point x, we measure off a length OQ equal to x, along the axis of ordinates. If PN be the ordinate corresponding to any given abscissa λ, the phase of the disturbance at the point x, due to the elementary wave-train whose wave-length is λ, will be given by the gradient of the line QP; for if we draw QR parallel to ON we have

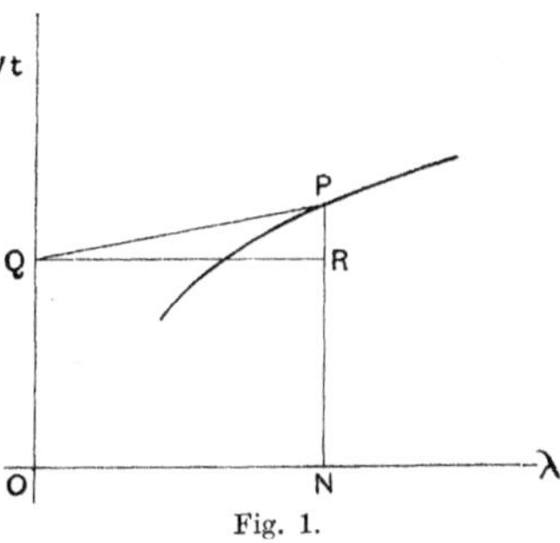

Fig. 1.

$$\frac{PR}{QR} = \frac{PN - OQ}{ON} = \frac{Vt - x}{\lambda} = \frac{\sigma t - kx}{2\pi} \quad\text{..................(17).}$$

Hence the phase will be stationary if QP be a tangent to the curve; and the predominant wave-lengths at the point x are accordingly given by the abscissae of the points of contact of the several tangents which can be drawn from Q. These are characterised by the property that the group-velocity has a given value x/t.

If we imagine the point Q to travel along the straight line on which it lies, we get an indication of the distribution of wave-lengths in the medium, at the instant t for which the diagram has been constructed. If we wish to follow the changes which take place in time at a given point x, we may either imagine the ordinates to be altered in the ratio of the respective times, or we may imagine the point Q to approach O in such a way that OQ varies inversely as t.

We may consider a few particular cases. In the case of deep-water waves, where

$$\sigma^2 = gk, \quad V^2t^2 = gt^2\,.\,\lambda/2\pi \quad\text{........................(18),}$$

the curve is a parabola (Fig. 2), and one tangent can be drawn from any position of Q. It appears at once that in the procession of waves at any instant the wave-length diminishes continually from front to rear; and that the waves which pass any assigned point will have their wave-lengths continually diminishing. It will be found that these statements are reversed in the case of waves due to capillarity only where

$$\sigma^2 = Tk^3, \quad V^2t^2 = 2\pi Tt^2/\lambda \quad\text{........................(19).}$$

* See also Rayleigh, *Phil. Mag.*, vol. 21, p. 181 (1911).

† *Manch. Memoirs*, vol. 44 (1900); *Hydrodynamics*, 3rd ed., pp. 362, 439.

When gravity and capillarity are both operative, we have

$$\sigma^2 = gk + Tk^3, \quad V^2 = g\lambda/2\pi + 2\pi T/\lambda \quad \text{.................(20).}$$

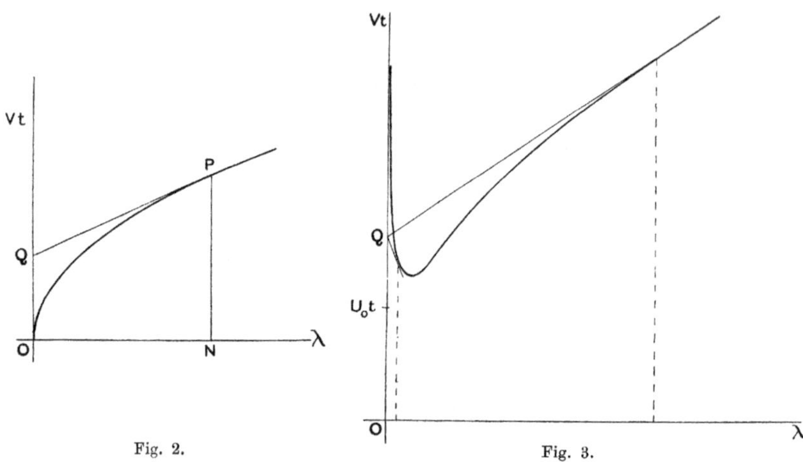

Fig. 2. Fig. 3.

The curve is shewn in Fig. 3. It appears that two tangents can be drawn from a point Q on the axis of ordinates provided

$$OQ < U_0 t \quad \text{...................................(21),}$$

where U_0 is the minimum group-velocity, corresponding to the point of inflexion on the curve*. We have, then, *two* predominant wave-lengths at any point of the train, which is moreover limited in the rear.

Finally, we may notice the case where there is a limiting value to the wave-velocity, as in the case of gravity waves on water of uniform depth. The relation is now

$$\sigma^2 = gk \tanh kh \quad \text{............................(22);}$$

and the curve is shewn in Fig. 4. The condition of stationary phase is now fulfilled approximately for *all* very long waves, and we infer that the main† disturbance begins with a solitary wave about the time $x/(gh)^{\frac{1}{2}}$ followed by a train of oscillatory waves whose general character towards the rear approximates to that which is characteristic of deep-water waves.

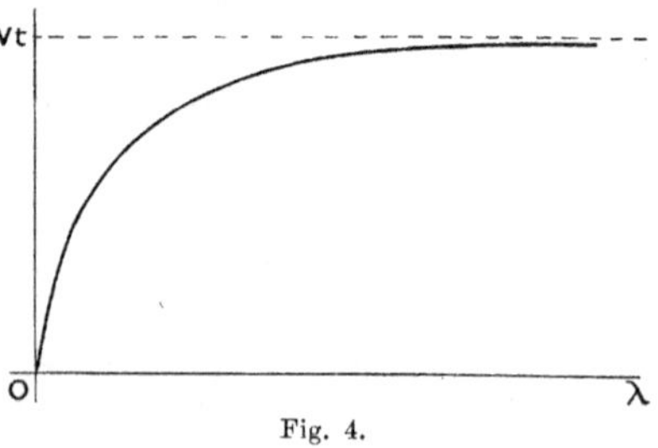

Fig. 4.

The construction above explained has the defect that it gives no indication of the relative amplitudes in different parts of the wave-train. For this purpose

* *Manch. Memoirs*, *l.c.* See also Rayleigh, *l.c.*

† This is preceded by a minor inequality, whose initial stages have been investigated by Rayleigh; see *Phil. Mag.*, vol. 18 (1909), and *Sc. Papers*, vol. 5, p. 514.

we may construct the curve which gives the relation between σt as ordinate and k as abscissa. If we draw a line through the origin whose gradient is x, the phase due to a particular wave-train, viz. $\sigma t - kx$, will be represented by the difference of the ordinates of the curve and the straight line. This difference will be stationary when the tangent to the curve is parallel to the straight line, i.e. when

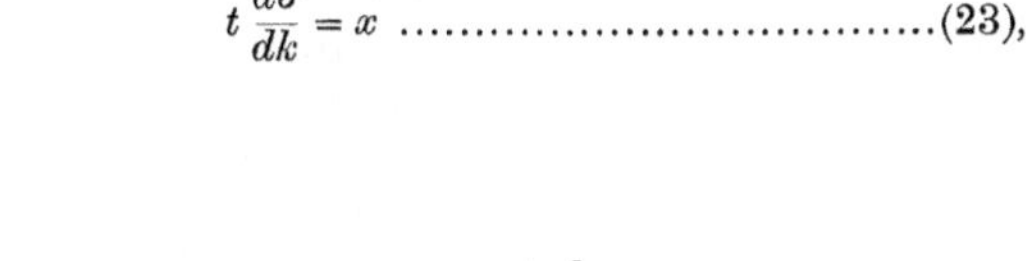

$$t\frac{d\sigma}{dk} = x \qquad \text{(23)},$$

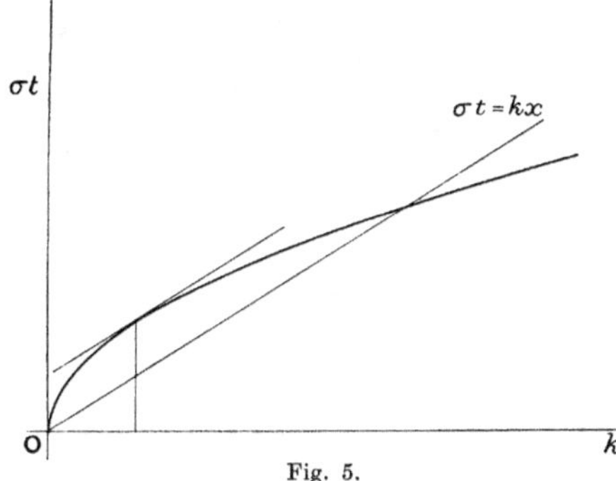

Fig. 5.

as already found by the previous method. It is further evident that the phase-difference, for elementary trains of slightly different wave-lengths, will vary ultimately as the square of the increment of k. Also that the range of values of k for which the phase is sensibly the same will be greater, and consequently the resulting disturbance will be more intense, the greater the vertical chord of curvature of the curve. This explains the occurrence of the quantity $td^2\sigma/dk^2$ in the denominator of Kelvin's formula (16).

The annexed Fig. 6, which corresponds to the case of water waves under gravity

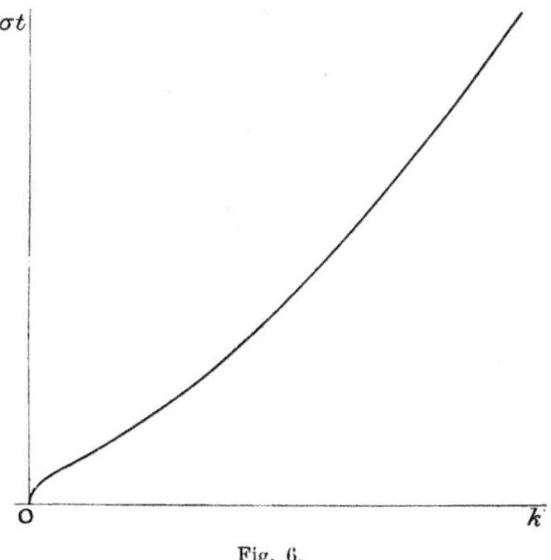

Fig. 6.

and capillarity combined, indicates that, of the two systems of waves in the more advanced portion of the procession, one consists of short and high waves, the other of long and low ones*.

It may naturally be asked whether the above methods are capable of throwing any light on the theory of earthquake waves, or on the interpretation of seismometric records. All that can be said, at present, is that they do afford a means of testing, to some extent, particular theories which attribute the observed effects to dispersion. In particular it may be noted that the period of the waves which pass any particular point will become shorter or longer as time goes on, according as the curve which shews the relation between V and λ is convex or concave upwards. The same conclusion may be obtained analytically as follows. If we regard λ, and therefore k, as a function of x and t, we have, by differentiation of (15),

$$\frac{d\sigma}{dk} + t\frac{d^2\sigma}{dk^2}\frac{\partial k}{\partial t} = 0 \quad \text{.............................(24).}$$

Hence

$$\frac{\partial\sigma}{\partial t} = \frac{d\sigma}{dk}\frac{\partial k}{\partial t} = -\left(\frac{d\sigma}{dk}\right)^2 \div t\frac{d^2\sigma}{dk^2} \quad \text{........................(25).}$$

But from (5) we find

$$\frac{d^2\sigma}{dk^2} = \lambda^3\frac{d^2V}{d\lambda^2} \quad \text{.................................(26).}$$

Hence $\partial\sigma/\partial t$ is positive, i.e. the frequency at the point x increases, if $d^2V/d\lambda^2$ is negative.

It may be noted that if an analytical representation of the wave-system, in any case of dispersion, could be obtained, the relation between σ and k could be deduced as follows. We should have

$$\sigma t - kx = f(x, t) \quad \text{.............................(27),}$$

a known function of x and t, with the condition

$$t\frac{d\sigma}{dk} = x \quad \text{....................................(28).}$$

Hence

$$\sigma - k\frac{d\sigma}{dk} = \frac{1}{t}f\left(t\frac{d\sigma}{dk}, t\right) \quad \text{..........................(29).}$$

Since the right hand side must be a function of k only, the function $f(x, t)$ must be homogeneous in respect to x and t, of degree one. Hence, putting

$$f(x, t) = tF\left(\frac{x}{t}\right) \quad \text{..............................(30),}$$

we have

$$\sigma = k\frac{d\sigma}{dk} + F\left(\frac{d\sigma}{dk}\right) \quad \text{..........................(31),}$$

which is of Clairaut's form.

This is illustrated in the case of gravity waves, or capillary waves. A formula which includes several interesting cases is

$$F\left(\frac{x}{t}\right) = m\left\{1 - \left(\frac{x}{ct}\right)^{\frac{1}{n}}\right\}^n \quad \text{..........................(32).}$$

* It indicates also that the shorter waves have the greater amplitude, but this circumstance is greatly modified (by interference) in the case of an initial disturbance which is diffused over a finite range of x instead of being concentrated at a point.

This leads to
$$\sigma = \frac{mkc}{\{m^{\frac{1}{n-1}} + (kc)^{\frac{1}{n-1}}\}^{n-1}} \qquad \text{.....(33).}$$

Thus if $n = 2$, we have
$$\sigma = \frac{kc}{1 + kb} \qquad \text{.....(34);}$$

if $n = \frac{1}{2}$,
$$\sigma^2 = k^2c^2 + m^2 \qquad \text{.....(35),}$$

which is the case of waves propagated vertically in an isothermal atmosphere; if $n = \frac{3}{2}$,
$$\sigma^2 = \frac{k^2c^2}{1 + k^2b^2} \qquad \text{.....(36),}$$

which is the case of a tense chain destitute of stiffness, but possessing rotatory inertia.

HOW ATWOOD'S MACHINE SHOWS THE ROTATION OF THE EARTH EVEN QUANTITATIVELY

By John G. Hagen.

The *Treatise* of G. Atwood in which the machine, known by his name, is described and mathematically explained, was published in Cambridge as early as 1784, or six years before Guglielmini's experiments (1790—92) in Bologna. It is a curious fact that Benzenberg in his book, *Die Versuche über die Umdrehung der Erde* (1804), mentions Atwood's machine, illustrates it by a figure (1, Plate VII) and (p. 94) praises its advantage of reducing the velocity of falling bodies to any desired minor quantity. And yet he, who had a double experience in Hamburg (1802) and Schlebusch (1804), with velocities from heights of 76 and 85 metres, never thought of profiting from this advantage. On the contrary, he proposed both the Pantheon and St Peter's for a repetition of his experiments. Perhaps contrary to expectation his double wish was fulfilled a century later, at Paris in 1903 and at Rome in 1912, but not with the heights that he had in view. In the Pantheon only 68 m. were utilized, and in Rome only 23 m., in a building close to St Peter's Basilica. Even that height was unnecessarily large. The free fall would have given us an easterly deviation of 1·8 mm.; this was reduced to one half, or 0·9 mm., by the machine. The experiments proved that less than one half of this quantity can be observed quite exactly with the new method.

I. *Description of the Apparatus and the Experiments.*

1. The pulley was mounted in an upper room of what is called the *Nicchione*. The floor of the room consists of a stone vault which covers a triangular staircase, 23 m. high. Within the staircase a shaft was built of wood and cloth to protect the descending weight and thread from air-currents. The shaft communicates with the room above by a circular opening in the floor.

2. The pulley consists of a thin, solid disc of aluminium, with steel-points in the axis, and was made new for the purpose. The rest of the apparatus is part of an old seismograph: a brass tube, which holds the pulley excentrically so that the weight hangs inside the tube, and is thus protected from lateral currents. A bell-jar, put over the whole apparatus, excludes vertical air-currents. The tube has three legs and is screwed to a wooden structure which is firmly imbedded in wall and floor. The old weight was an empty cylinder, with a hole in the centre of the cover for the

thread. It was filled with shot until it weighed about 50 gr. The counterweight was made three-fourths as heavy; it is, at its upper part, provided with an arrow-shaped spring, and at its lower end with a hook.

The spring just mentioned is part of the damping-apparatus. If the counterweight were arrested directly and suddenly, the thread would be apt to break or, in consequence of its expansion and following contraction, jump off the pulley. The damper consists of a brass plate which is lifted by the counterweight to a height of about 70 cm., as soon as it enters the upper room. The thread passes through a central hole in the plate and pulls the arrow-shaped spring through the same, but not the weight. The plate is thus firmly caught between the spring and the weight. When the plate is lifted, it glides between two guiding bars, which also prevent it from falling back. For this purpose another pair of springs is needed. In our case the springs were attached to the guiding bars, under the supposition that the plate would always reach the same height. This is not the case on account of its varying friction against the bars. The springs should be attached to the plate and the bars provided with rack-incisions.

3. The shaft reaches to the ground-floor, where it is connected, air-tight, with a wooden box. Inside, at the bottom, are the binding posts of an electric current, which hold the fusible wire on which the counterweight is hooked.

Higher up, the box is provided with two windows, in the shape of biplanar glass-plates. They are on a level with a theodolite which is mounted outside against the wall and independent of the floor.

Through the theodolite the thread of the descending weight is observed from two directions. Viewed from the north, the easterly deviations are seen, and viewed from the east, the deviations within the meridian. The narrow locality necessitated the use of a mirror, which was mounted on the wall, north or east of the thread, opposite either of the windows in the shaft. The mirror can be turned on vertical hinges, like a door, by means of a rod from the eye end of the theodolite, until a certain mark, inside the shaft, throws its image on a certain line of the micrometer scale. In the shaft there are two marks, in the shape of black threads suspended against the white wall: one in the meridian and the other in the prime vertical of the descending thread.

The micrometer scale is divided on glass and visible, with the descending thread, through the strong illumination inside the shaft.

4. The experiments proceeded in the following way. The observer closed the circuit of the fusible wire by a switch close to the theodolite. About ten seconds later he saw the thread appear in the field, but never the weight. During the first trials the entire attention was directed to the oscillations of the pendulum after the counterweight was arrested.

One fact revealed by these trials was that, viewed from the north, the first oscillation always went from east to west, and never once from west to east. This fact alone constituted a qualitative proof of the terrestrial rotation as it had never been reached by observing the free fall of bodies.

It was found out, however, that the oscillations of the pendulum did not represent the easterly deviation quantitatively. The amplitudes grew alternately smaller and larger; and the same thing was observed from the east side. The weight, after its descent, evidently described spherical ellipses with revolving axes. The cause of this was the expansion of the thread, to the extent of one-third of a metre, and the bouncing back of the weight. At the same time the thread began to spin and to revolve the weight about an axis which may not have coincided with its geometrical axis. The thread would sometimes show vibrations like a musical chord. It became evident that oscillations after the arresting of the counterweight would not yield any quantitative measures, except the direction of the vertical. Five elongations on either side were deemed sufficient to fix the position of the plumb-line on the micrometer scale.

Fortunately, after some practice, it was found that the fall of the thread could be observed with great exactness, before the counterweight struck the damping apparatus.

5. For an instant the thread would appear on the micrometer scale as a perfectly steady fine line, whose position could be estimated to tenths of a scale division, and even to half tenths. The value of one division was determined by placing a millimetre scale in contact with the thread and perpendicular to the line of sight. The number of millimetres covered by the micrometer gave the following results:

View from E., distance 6·24 m.; one division = 1·697 mm.,
" " N., " 4·30 m.; " " = 1·170 mm.

6. The crucial test for all the methods of measuring the easterly deviation of falling bodies is the southern deviation. Both Gauss and Laplace have proved, from theory, that the southerly deviation of falling bodies is inappreciable. Beginning was therefore made with this, and special care was taken to avoid windy days, and also hours of the day when the Museum, close by, was frequented by visitors. After the method had stood the test of 22 experiments, with no deviation to the south, a larger number of experiments was made for the easterly deviation, and less care was taken in the selection of quiet days and hours.

In computing the final results no experiment, which had succeeded mechanically, was discarded. They are as follows:

Deviation S.; 22 experiments: $+0·010 \pm 0·027$ (1 exp. $\pm 0·128$),
" E.; 66 " : $+0·899 \pm 0·027$ (1 " $\pm 0·217$).

All measures are expressed in millimetres. The stricter choice of the circumstances is evidenced in the smaller P.E. of one experiment for southerly deviations.

II. *The Theory of the Experiments.*

As the theory of these experiments will be published in detail in a forthcoming volume of the Vatican Observatory, it need only be outlined here. The coordinate system is connected with the place of the experiments: the origin is the centre of gravity of the descending weight in its original position, and the axes have the directions:

$$+x = \text{south}, \quad +y = \text{east}, \quad +z = \text{nadir}.$$

1. The following notations are used:

ω = angular velocity of earth . . . . = 0·000073,

ϕ = latitude of the place of experiments = 41° 54′,

g = acceleration of free fall . . . . = 9·80 m.,

γ = „ „ retarded fall . . . = $\alpha + \beta z$.

The fundamental equations for the relative motion of a body suspended from a string are, as deduced from Gauss and Binet:

$$\left.\begin{aligned} \ddot{x} &= -(g-\gamma)\,.\,x/z \\ \ddot{y} &= +2\omega\cos\phi\,.\,\dot{z} - (g-\gamma)\,.\,y/z \\ \ddot{z} &= +\gamma \end{aligned}\right\} \quad \ldots\ldots\ldots\ldots\ldots\ldots\ldots(1).$$

The equations are only approximate: firstly, all terms with powers ω^2 are omitted; secondly, in the fractions x/z and y/z the denominator z should strictly be replaced by $z+d$, where d is the distance of the weight from the point where the string leaves the pulley. In the experiment the distance d should be kept as small as possible. A third approximation will be the assumption of a constant value for γ. It was well known to Atwood that γ increases constantly on account of the thread adding continually to the descending weight*. In his *Treatise* (1784), 110, he gives γ the linear form:

$$\gamma = \alpha + \beta z \ldots (\alpha \text{ and } \beta \text{ positive constants}) \quad \ldots\ldots\ldots\ldots(2).$$

There is no difficulty in integrating the third equation (1), as was known to Atwood from the writings of Euler. Yet the substitution of the exponential functions in the first equations of (1) would make the latter unmanageable.

The assumption of some equivalent constant value γ_0 for $\ddot{z}$ yields the following integrals:

$$\left.\begin{aligned} x &= 0 \\ y &= \frac{\gamma_0^2}{2\gamma_0 + g}\,\omega\cos\phi\,.\,t^3 \\ z &= \tfrac{1}{2}\gamma_0 t^2 \end{aligned}\right\} \quad \ldots\ldots\ldots\ldots\ldots\ldots\ldots\ldots(3).$$

Evidently, when γ_0 approaches g, the equations will represent the free fall of bodies.

2. The hypothetical constant value γ_0, which will approximately satisfy the system (1), has to be determined from experiments by means of the last equation (3). *À priori* it would be impossible to say, for what value of z the experiment should be made. The safest way seemed to be, to deduce γ_0 for a number of equidistant heights

* In the forthcoming volume it will be shown how γ can be kept constant during the fall.

and to take their mean. Accordingly the height of 22·96 m. was divided into 3 nearly equal parts, and the duration t of each fall observed. These are the results:

z	t	γ_0
7·47 m.	s. 6·52	0·352 m.
19·94	8·71	394
22·96	10·48	418
	Mean =	0·388 m.

Adopting $\gamma_0 = 0{\cdot}388$ as the nearest approximation possible to an equivalent constant value of the acceleration, we can now compute the easterly deviation y from the second equation (3), as follows:

$$\frac{\gamma_0^2}{2\gamma_0 + g} = 0{\cdot}014229 \text{ m.}, \quad \omega \cos \phi = 0{\cdot}00005426,$$

$$t = 10{\cdot}48, \quad t^3 = 1151{\cdot}0184,$$

$$y = +\,0{\cdot}889 \text{ mm.}$$

Incidentally, it may be observed that the three determinations of γ_0 serve also to compute the constants α and β. After the exact equation

$$\ddot{z} = \alpha + \beta z$$

is integrated, the three pairs (z, t) give three conditions, which yield, as the most probable solution:

$$\gamma = 0{\cdot}315 + 0{\cdot}036\,z.$$

The variable γ reaches the value $\gamma_0 = 0{\cdot}388$ m. after a distance of only 2 metres.

The main results may be tabulated in the following manner:

Deviation	South, 22 Exp.	East, 66 Exp.
By theory	$x = 0$	$y = +0{\cdot}889$ mm.
,, experiment	$x = +0{\cdot}010 \pm 0{\cdot}027$ mm.	$y = +0{\cdot}899 \pm 0{\cdot}027$ mm.
Exp.—Theory	+0·010 mm.	+0·010 mm.

Although the close agreement between experiment and theory seems to be partly accidental, yet it certainly is greater than it ever was in experiments with freely falling bodies. It is quite likely, therefore, that, after Atwood's machine has proved its adaptability to the experiment proposed by Newton, trials with freely falling bodies will not be resumed again.

REZIPROKE KRÄFTEPLÄNE ZU DEN SPANNUNGEN IN EINER BIEGSAMEN HAUT

Von Wilhelm Blaschke.

Nach einigen älteren Versuchen, die bis auf Johann Bernoulli zurückgehen, ist vor nunmehr etwa dreissig Jahren von Lecornu und Beltrami die Statik der biegsamen und undehnbaren Häute entwickelt worden*. Ich möchte hier zeigen' dass man dieser nicht bloss für den Geometer sondern vielleicht auch für den Techniker interessanten Theorie eine besonders durchsichtige Fassung geben kann, wenn man bemerkt, dass die in die graphische Statik von J. C. Maxwell eingeführten reziproken Kräftepläne sich auch zur Bestimmung der Spannungen in einer biegsamen Haut anwenden lassen. Dabei kommt man auch von einer neuen Seite aus auf die von L. Bianchi zum Studium der Flächendeformation benutzten "assoziierten" Flächen†.

Denken wir uns ein Stück eines Polyeders etwa von lauter Dreiecksflächen begrenzt. Sind seine Kanten aus einem starren Material ausgeführt und in ihren Endpunkten gelenkig verbunden, so haben wir ein besonderes räumliches Fachwerk vor uns, das A. Föppl als ein *Flechtwerk* bezeichnet. Wirken auf dieses Flechtwerk nur in den Eckpunkten seines Randes äussere Kräfte ein, so kann man für den Fall des Gleichgewichts zu den Spannungen in seinen Kanten einen reziproken Kräfteplan entwerfen, in dem die Spannungen durch Strecken dargestellt sind, die zu den entsprechenden Kanten parallel laufen.

Das Polyeder unterwerfen wir nun einem Grenzübergang, bei dem es in eine krumme Fläche übergeht. Sehen wir das Polyeder als Fachwerk an, so müssen wir in der Grenze der Fläche notwendig folgende mechanische Eigenschaften zuschreiben: die völlige Biegsamkeit und die Undehnbarkeit. Führt man diesen Grenzprozess in geeigneter Weise durch, so geht auch das Polyeder des zugehörigen Kräfteplans in ein Flächenstück über, das man als Kräfteplan zu den Spannungen in der ersten Fläche ansehen kann. Die Verhältnisse liegen nämlich, wie man das nach diesen Grenzbetrachtungen schon vorhersieht, so.

* L. Lecornu, "Sur l'équilibre des surfaces flexibles et inextensibles," *Journal de l'École Polytechnique*, cahier 48 (1880), S. 1—109.

E. Beltrami, "Sull' equilibrio delle superficie flessibili ed inestendibili," *Opere matematiche*, 3. Bd, S. 420—464 oder *Rendiconti del Istituto Lombardo*, Serie 2, Bd 15 (1882), S. 217—265.

Man vergleiche dazu auch L. Bianchi, *Lezioni di Geometria differenziale* (Pisa 1903), Bd 2, S. 31 u. ff.

† Man siehe etwa Bianchi (Lukat), *Vorlesungen über Differentialgeometrie* (Leipzig 1910), Kapitel 11.

Es sei eine biegsame und undehnbare Haut, nennen wir sie H, unter dem Einfluss äusserer Kräfte, die nur an ihrem Rande angreifen, im Gleichgewicht. Dann kann man zur Fläche H eine zweite Fläche K mit folgenden Eigenschaften auffinden. Wir bilden H derart punktweise auf K ab, dass in entsprechenden Punkten die Flächennormalen parallel ausfallen. Will man nun zu einem Linienelement ds der Haut H die hier wirksame Spannung angeben, so braucht man nur das entsprechende Linienelement $d\bar{s}$ von K aufzusuchen: die Richtung von $d\bar{s}$ gibt bei zweckmässiger Vorzeichenwahl die *Richtung* und das Verhältnis $d\bar{s} : ds$ gibt die *Grösse* der auf die Längeneinheit bezogenen Spannung. So können wir also mit Recht K als reziproken Kräfteplan der Spannungen in H bezeichnen. "Reziprok" deshalb, weil die Beziehung zwischen den beiden Flächen H und K wechselseitig ist, man kann nämlich in derselben Weise auch H als Kräfteplan zu K ansehen.

Nimmt man H als gegeben, K als gesucht an, so ergibt sich für K eine partielle Differentialgleichung zweiter Ordnung, die auch in Hilberts Untersuchungen über Integralgleichungen eine Rolle spielt. Die auf den Rand von H wirkenden äusseren Kräfte liefern die zugehörigen Randbedingungen. Aus dieser partiellen Differentialgleichung ersieht man aber, dass K nichts anderes ist als die assoziierte Fläche einer geeigneten "infinitesimalen Verbiegung" von H.

Es sei darauf hingewiesen, wie man diese assoziierten Flächen Bianchis, für die wir eben eine *statische* Deutung gefunden haben, auch in einfacher Weise *kinematisch* erklären kann.

Denken wir uns die Fläche H stetig ohne Dehnung und ohne Knick verbogen und greifen wir einen Augenblick dieses Biegungsvorgangs heraus. Ist mit einem Flächenelement von H ein starrer Körper verbunden, so führt er in dem Augenblick eine Schraubung aus (die "tangierende" Schraubung). Die mit dieser Schraubung verbundene Drehung stellen wir in üblicher Weise durch einen Vektor dar, der zur Drehachse parallel läuft. Zu jedem Punkte von H gehört in dieser Weise ein Vektor. Alle diese Vektoren tragen wir von einem festen Punkt aus an, dann erfüllen die Endpunkte im Allgemeinen eine Fläche und die ist die assoziierte Fläche K der Biegung von H.

Ist H eine geschlossene konvexe Fläche, so reduziert sich, wie man unschwer nachweisen kann, die zugehörige Fläche K notwendig auf einen Punkt. Deutet man diese Beziehung (zum Beispiel) kinematisch, so findet man den Satz von Jellett und Liebmann über die *Unverbiegbarkeit einer geschlossenen konvexen Fläche.*

Einige Formeln werden die vorgetragenen geometrischen und mechanischen Überlegungen deutlicher machen. Nehmen wir an, unser reelles Flächenstück H hänge einfach zusammen und werde durch parallele Normalen eineindeutig auf die Kugel abgebildet. Nennen wir die Richtungskosinus der Normalen X, Y, Z und die Entfernung der Tangentialebene vom Koordinatenanfang p, so können wir die Fläche durch eine solche Gleichung darstellen

$$p = H(X, Y, Z).$$

Dabei bedeutet es keine Einschränkung, wenn wir von der Funktion H annehmen, es sei

$$H(cX, cY, cZ) = cH(X, Y, Z)$$

für alle positiven c. Ein Punkt von H hat dann die Koordinaten

$$x = H_x, \quad y = H_y, \quad z = H_z,$$

wenn die Marken partielle Differentiationen andeuten.

Wir setzen das Flächenstück H als biegsam voraus—die Undehnbarkeit ist unwesentlich—und nehmen an, H sei im Gleichgewicht unter der Einwirkung äusserer Kräfte, die tangentiell am Rande von H angreifen.

Es sei nun S die Spannung mit den Komponenten $d\tilde{x}/ds$, $d\tilde{y}/ds$, $d\tilde{z}/ds$, die auf das Linienelement ds von H wirkt. Soll Gleichgewicht herrschen, so müssen längs jeder geschlossenen Kurve auf H die drei Integrale

$$\tilde{x} = \int d\tilde{x}, \quad \tilde{y} = \int d\tilde{y}, \quad \tilde{z} = \int d\tilde{z}$$

Null ergeben. Erstreckt man daher diese Integrale von einem festen Punkt P_0 zu einem veränderlichen Punkt P der Fläche H, so sind die Integrale unabhängig vom Weg und nur abhängig von der Lage des Punktes P. Beschreibt P die Fläche H, so durchläuft $\tilde{P}\{\tilde{x}, \tilde{y}, \tilde{z}\}$ eine Fläche $\tilde{H}$. Da die Richtung $\{d\tilde{x} : d\tilde{y} : d\tilde{z}\}$ der Spannung S in der Tangentialebene an H liegt—gerade dadurch drückt sich die Biegsamkeit der Fläche aus—so sind die Flächen H und $\tilde{H}$ so auf einander bezogen, dass in entsprechenden Punkten P und $\tilde{P}$ die Tangentialebenen parallel sind. Deshalb kann man $\tilde{H}$ analog wie H so darstellen

$$\tilde{p} = \tilde{H}(X, Y, Z).$$

Die partielle Differentialgleichung für die Fläche $\tilde{H}$, die nichts anderes ist als der Kräfteplan K zur Haut H, gewinnt man durch folgende Betrachtung. Längs jeder geschlossenen Kurve auf H müssen die Integrale über die Momente der Spannung um die Koordinatenachsen

$$x^* = \int y\, d\tilde{z} - z\, d\tilde{y}, \quad y^* = \int z\, d\tilde{x} - x\, d\tilde{z}, \quad z^* = \int x\, d\tilde{y} - y\, d\tilde{x}$$

verschwinden oder, was dasselbe bedeutet, diese Integrale sind erstreckt zwischen den Punkten P_0 und P von H unabhängig vom Integrationsweg. x^*, y^*, z^* sind also Funktionen der Richtungskosinus X, Y, Z der Flächennormalen in P.

Die drei Integranden oder die zweireihigen Determinanten der Matrix

$$\left\| \begin{matrix} H_x & \tilde{H}_{xx}dX + \tilde{H}_{xy}dY + \tilde{H}_{xz}dZ \\ H_y & \tilde{H}_{yx}dX + \tilde{H}_{yy}dY + \tilde{H}_{yz}dZ \\ H_z & \tilde{H}_{zx}dX + \tilde{H}_{zy}dY + \tilde{H}_{zz}dZ \end{matrix} \right\|$$

müssen vollständige Differentiale sein. Aus der Homogeneität von H und $\tilde{H}$ folgen die identischen Beziehungen

$$\frac{H_{yy}\tilde{H}_{zz} - 2H_{yz}\tilde{H}_{yz} + H_{zz}\tilde{H}_{yy}}{XX} = \frac{H_{xy}\tilde{H}_{zx} + H_{zx}\tilde{H}_{xy} - H_{xx}\tilde{H}_{yz} - H_{yz}\tilde{H}_{xx}}{YZ}$$

$$= \frac{H_{zz}\tilde{H}_{xx} - 2H_{zx}\tilde{H}_{zx} + H_{xx}\tilde{H}_{zz}}{YY} = \frac{H_{yz}\tilde{H}_{xy} + H_{xy}\tilde{H}_{yz} - H_{yy}\tilde{H}_{zx} - H_{zx}\tilde{H}_{yy}}{ZX}$$

$$= \frac{H_{xx}\tilde{H}_{yy} - 2H_{xy}\tilde{H}_{xy} + H_{yy}\tilde{H}_{xx}}{ZZ} = \frac{H_{zx}\tilde{H}_{yz} + H_{yz}\tilde{H}_{zx} - H_{zz}\tilde{H}_{xy} - H_{xy}\tilde{H}_{zz}}{XY}.$$

Führt man nach Hilbert für diese sechs gleichwertigen Ausdrücke die Abkürzung $(H, \tilde{H})$ ein, so reduzieren sich die Integrabilitätsbedingungen auf die einzige Gleichung

$$(H, \tilde{H}) = 0.$$

Derselben partiellen Differentialgleichung für H genügen aber, wie ich gezeigt habe*, die den Biegungen der Fläche H assoziierten Flächen.

Zum Schluss sei noch erwähnt, dass man auch die vom Punkte $P^*\{x^*, y^*, z^*\}$ beschriebene Fläche, die auf $\tilde{H}$ unter Orthogonalität entsprechender Linienelemente bezogen ist, in gewissem Sinne als Kräfteplan für die Spannungen der Haut H verwenden kann. Derartige Kräfteplane sind für Fachwerke kürzlich von L. Henneberg betrachtet worden†.

* Ein Beweis für die Unverbiegbarkeit geschlossener konvexer Flächen, *Göttinger Nachrichten* 1912.

† Über das Gleichgewicht an Seilnetzen..., *H. Weber-Festschrift* (Leipzig 1912), S. 111—129.

EXTENSION DE LA NOTION DES VALEURS CRITIQUES AUX ÉQUATIONS À QUATRE VARIABLES D'ORDRE NOMOGRAPHIQUE SUPÉRIEUR

PAR FARID BOULAD.

Dans notre mémoire intitulé: "Application de la notion des valeurs critiques à la disjonction des variables dans les équations d'ordre nomographique supérieur" (*Bull. de la Soc. Math. de France* t. XXXIX. 1911, p. 105), nous avons exposé une méthode permettant, au moyen de cette notion, d'effectuer la disjonction des trois variables de l'équation $F_{123} = 0$* en vue de la construction d'un nomogramme à points alignés, dans le cas où cette équation se présente sous la forme pratique générale†

$$F_1 . F_{23} + G_1 . G_{23} + H_1 . H_{23} = 0.$$

Nous nous proposons ici d'étendre cette notion très féconde, due à M. le Professeur d'Ocagne, à la résolution du problème de la disjonction des 4 variables z_1, z_2, z_3, z_4 de l'équation $F_{1234} = 0$ en vue de sa représentation par un nomogramme à double alignement, dans le cas fréquent dans la pratique, où cette équation est mise sous les deux formes suivantes nomographiquement rationnelles‡.

$$\left.\begin{aligned} F_{1234} = F_1 . F_{234} + G_1 . G_{234} + H_1 . H_{234} = 0 \\ F_{1234} = F_4 . F_{123} + G_4 . G_{123} + H_4 . H_{123} = 0 \end{aligned}\right\} \quad \ldots\ldots\ldots\ldots\ldots\ldots(1).$$

D'après le principe du double alignement établi par M. d'Ocagne§, pour que le problème ci-dessus soit résolu, il suffit de déterminer deux systèmes de trois fonctions (F_2, G_2, H_2) et (F_3, G_3, H_3) des deux variables z_2 et z_3 et un système de trois fonctions (F_0, G_0, H_0) d'une seule variable auxiliaire z_0, tels que l'équation proposée (1) puisse être obtenue par l'élimination de cette variable z_0 entre les deux déterminants

$$\begin{vmatrix} F_1 & G_1 & H_1 \\ F_2 & G_2 & H_2 \\ F_0 & G_0 & H_0 \end{vmatrix} = 0 \ldots\ldots(2), \qquad \begin{vmatrix} F_4 & G_4 & H_4 \\ F_3 & G_3 & H_3 \\ F_0 & G_0 & H_0 \end{vmatrix} = 0 \ldots\ldots\ldots(3).$$

* Nous employons ici la notation de M. d'Ocagne qui consiste, lorsqu'on affecte un indice à chacune des variables z_1, z_2, z_3, ..., à placer à la suite de chaque signe fonctionnel les indices de toutes les variables sur lesquelles porte ce signe.

† Cette méthode a fait précédemment l'objet d'une communication par nous à l'Acad. des Sc. de Paris (*Comptes rendus du* 14 *Février* 1910, p. 379).

‡ Si, toutefois, l'équation $F_{1234}=0$ se présente seulement sous l'une des deux formes ci-dessus (1), la notion des valeurs critiques peut être utilisée aussi d'une façon plus générale pour effectuer la disjonction des variables de cette équation, ainsi que nous l'avons indiqué dans notre récent mémoire "Sur les équations à 4 variables d'ordre nomographique supérieur" (*Bull. de la Soc. Math. de France* t. XL. 1912).

§ *Traité de Nomographie* par M. d'Ocagne, p. 213, et son *Cours de Calcul Graphique et Nomographique*, p. 305.

Or, remarquons que si nous considérons les deux formes (1) comme étant les deux développements respectifs des deux déterminants (2) and (3) par rapport aux éléments de leurs premières lignes, nous pourrons admettre que chacun des éléments F_0, G_0, H_0 figurant dans le déterminant (2) est une fonction des deux variables z_3 et z_4 et que chacun de ces mêmes éléments figurant dans le déterminant (3) est une fonction des deux variables z_1 et z_2; avec la condition que, si l'on écrit, d'après cette remarque, les deux déterminants ci-dessus comme suit:

$$F_{1234} = \begin{vmatrix} F_1 & G_1 & H_1 \\ F_2 & G_2 & H_2 \\ F_{34} & G_{34} & H_{34} \end{vmatrix} = 0 \ldots\ldots (2'), \qquad F_{1234} = \begin{vmatrix} F_4 & G_4 & H_4 \\ F_3 & G_3 & H_3 \\ F_{12} & G_{12} & H_{12} \end{vmatrix} = 0 \ldots\ldots (3'),$$

il faut que l'on ait l'une ou l'autre des conditions suivantes:

1°—ou bien

$$\frac{F_{34}}{F_{12}} = \frac{G_{34}}{G_{12}} = \frac{H_{34}}{H_{12}} \quad \ldots\ldots\ldots\ldots\ldots\ldots\ldots\ldots\ldots\ldots (4),$$

2°—ou bien, en rapportant à un même système d'axes (Ox, Oy) les deux nomogrammes à simple alignement représentatifs des deux équations (2) et (3), il faut que les équations des supports des deux échelles (z_0) de ces deux nomogrammes soit identiques afin que l'on puisse superposer ces deux nomogrammes partiels en faisant coïncider les supports de ces deux échelles sans se préoccuper de leur graduation. De cette façon on obtiendra un nomogramme à double alignement représentatif de l'équation (1) ayant comme *charnière* le support commun des deux échelles (z_0).

L'équation du support de l'échelle (z_0) du nomogramme $(z_1 z_2 z_0)$ s'obtient en éliminant (z_3) et (z_4) entre les équations suivantes:

$$x = F_{34}, \quad y = G_{34}, \quad z = H_{34} \quad \ldots\ldots\ldots\ldots\ldots\ldots\ldots\ldots (5),$$

définissant cette échelle en coordonnées cartésiennes et homogènes. De même l'équation du support de l'échelle (z_0) correspondante au second nomogramme $(z_0 z_3 z_4)$ s'obtient en éliminant z_1, z_2 entre les équations suivantes de cette dernière échelle,

$$x = F_{12}, \quad y = G_{12}, \quad z = H_{12} \quad \ldots\ldots\ldots\ldots\ldots\ldots\ldots\ldots (6).$$

Il est intéressant de remarquer que chacune des deux conditions (4) n'est autre chose que l'équation (1) mise sous la forme $\Phi_{12} = \Phi_{34}$.

Cela posé, les deux systèmes des fonctions inconnues (F_2, G_2, H_2) et (F_{34}, G_{34}, H_{34}) s'obtiennent en résolvant les deux identités suivantes:

$$F_2 . F_{234} + G_2 . G_{234} + H_2 . H_{234} \equiv 0 \text{ (quels que soient } z_3 \text{ et } z_4) \quad \ldots\ldots (7),$$

$$F_{34} . F_{234} + G_{34} . G_{234} + H_{34} . H_{234} \equiv 0 \text{ (quel que soit } z_2) \quad \ldots\ldots\ldots\ldots\ldots (8),$$

lesquelles avec l'équation (1) donnent par élimination le premier déterminant cherché (2').

De même, les deux systèmes des fonctions inconnues (F_3, G_3, H_3) et (F_{12}, G_{12}, H_{12}) sont fournies par les deux identités suivantes:

$$F_3 . F_{123} + G_3 . G_{123} + H_3 . H_{123} \equiv 0 \text{ (quels que soient } z_1 \text{ et } z_2) \ldots\ldots\ldots (9),$$

$$F_{12} . F_{123} + G_{12} . G_{123} + H_{12} . H_{123} \equiv 0 \text{ (quel que soit } z_3) \quad \ldots\ldots\ldots\ldots\ldots (10),$$

lesquelles avec l'équation (1) donnent également par élimination le second déterminant (3′).

La résolution des identités fonctionnelles ci-dessus s'effectue aisément par la méthode de la détermination des valeurs critiques exposée dans nos deux mémoires précités.

Remarque importante. Il arrive dans la pratique que les équations des supports des deux échelles (z_0) des nomogrammes $(z_1 z_2 z_0)$ et $(z_0 z_3 z_4)$ rapportés à un même système d'axes sont réelles et homogènes en x, y, z, mais non identiques. Dans ce cas, si ces équations représentent des droites ou des coniques, ces deux nomogrammes pourront être toujours superposés d'une infinité de manières en ayant recours au principe de l'homographie.

Pour cela, il suffira de faire subir respectivement aux deux déterminants générateurs (2′) et (3′) les deux transformations homographiques les plus générales obtenues par la multiplication de ces deux déterminants respectivement par les deux déterminants transformateurs suivants:

$$\Delta_1 = |\alpha \beta' \gamma''| \neq 0, \qquad \Delta_2 = |\lambda \mu' \nu''| \neq 0.$$

En identifiant les deux équations obtenues pour les deux échelles transformées (z_0), on aura les valeurs des paramètres $\alpha, \beta, \ldots \mu'', \nu''$ qui réalisent la coïncidence de ces deux échelles et par suite la représentation par double alignement de l'équation donnée.

Application.

Equation à 4 variables d'ordre nomographique total 6° la plus générale.

Cette équation s'écrit sous la forme suivante:

$$f_1(f_4 A_{23} + g_4 B_{23} + h_4 C_{23}) + g_1(f_4 A_{23}' + g_4 B_{23}' + h_4 C_{23}') + h_1(f_4 A_{23}'' + g_4 B_{23}'' + h_4 C_{23}'') = 0 \quad (11)$$

d'ordre 2 par rapport à chacune des deux variables z_1 et z_4 et d'ordre 1 par rapport à chacune des deux autres variables z_2 et z_3, et dans laquelle on a d'une manière générale

$$\begin{aligned} A_{23} &= a_1 f_2 f_3 + a_2 f_2 + a_3 f_3 + a_4, \\ B_{23} &= b_1 f_2 f_3 + b_2 f_2 + b_3 f_3 + b_4, \\ C_{23} &= c_1 f_2 f_3 + c_2 f_2 + c_3 f_3 + c_4, \\ &\cdots\cdots\cdots\cdots\cdots\cdots \\ C_{23}'' &= c_1'' f_2 f_3 + c_2'' f_2 + c_3'' f_3 + c_4'', \end{aligned}$$

f_2 et f_3 désignant des fonctions quelconques des deux variables z_2 et z_3 et $a_1, a_2, \ldots c_3'', c_4''$ des constantes quelconques.

Cette même équation peut s'écrire aussi sous la 2° forme,

$$f_4(f_1 A_{23} + g_1 A_{23}' + h_1 A_{23}'') + g_4(f_1 B_{23} + g_1 B_{23}' + h_1 B_{23}'') + h_4(f_1 C_{23} + g_1 C_{23}' + h_1 C_{23}'') = 0 \quad (12).$$

Cherchons par la méthode ci-dessus à mettre les deux formes (11) et (12) de cette équation respectivement sous les deux formes des deux déterminants (2′) et (3′).

Pour cela prenons les éléments:

$$F_1 = f_1, \quad G_1 = g_1, \quad H_1 = h_1,$$
$$F_4 = f_4, \quad G_4 = g_4, \quad H_4 = h_4.$$

Pour avoir les éléments (F_2, G_2, H_2) du déterminant (2′), substituons ces lettres respectivement à f_1, g_1, h_1 dans la forme (11). Ensuite ordonnons nomographiquement par rapport aux 6 quantités ou groupe de quantités $f_3f_4, f_3g_4, f_3, f_4, g_4, h_4$. En exprimant que l'identité obtenue a lieu quels que soient z_2 et z_3 nous aurons un système de 6 équations linéaires et homogènes en F_2, G_2, H_2 que nous appellerons (Σ).

Cela posé, pour déterminer aussi les éléments F_{34}, G_{34}, H_{34} substituons ces derniers respectivement à f_1, g_1, h_1 dans l'équation (11). Puis rendons indéterminée la variable z_2 dans l'identité obtenue, nous aurons un système des deux équations linéaires et homogènes en F_{34}, G_{34}, H_{34} que nous appellerons (Φ). A présent, posons d'une manière générale

$$A_i = a_ix + a_i'y + a_i'',$$

$$B_i = b_ix + b_i'y + b_i'',$$

$$C_i = c_ix + c_i'y + c_i''.$$

Je dis que, *si les rapports*

$$\frac{B_1}{B_3} \equiv \frac{C_1}{C_3} \equiv \frac{A_2}{A_4} \equiv \frac{B_2}{B_4} \equiv \frac{C_2}{C_4} \quad \text{..........................(13)}$$

sont identiquement égaux, quels que soient x et y, les équations du système (Σ) *sont compatibles et les échelles* (z_2) *et* (z_0) *du nomogramme* ($z_1 z_2 z_0$) *correspondant à la forme* (11) *sont situées sur une même conique définie par l'équation*

$$\frac{A_1}{A_3} = \frac{B_1}{B_3} \quad \text{...................................(14).}$$

En effet, substituons, dans le système d'équations (Σ), x, y, l respectivement à F_2, G_2, H_2. Puis éliminons f_2 entre la première équation de ce système et chacune des 5 autres, nous aurons l'égalité des 6 rapports

$$\frac{A_1}{A_3} = \frac{B_1}{B_3} = \frac{C_1}{C_3} = \frac{A_2}{A_4} = \frac{B_2}{B_4} = \frac{C_2}{C_4}.$$

Or, en remarquant que chacune des cinq relations qu'on obtient en égalant l'un quelconque de ces 6 rapports, le premier par exemple avec les 5 autres, doit représenter l'équation du support de l'échelle (z_2), il résulte que pour rendre compatible le système d'équations (Σ), il faut et il suffit que l'on rende identiques ces cinq relations. En effectuant ainsi cette identification, on obtient précisément les conditions (13).

D'autre part, substituons x, y, 1, respectivement, à F_{34}, G_{34}, H_{34} dans le système d'équations (Φ). Puis ordonnons par rapport aux 6 quantités ci-dessus

$$f_3f_4,\ f_3g_4,\ f_3,\ f_4,\ g_4,\ h_4.$$

En ayant égard aux conditions (13) on trouve immédiatement pour le support de l'échelle (z_0) la même équation (14) que pour le support de l'échelle (z_2). Donc ces deux échelles (z_0) et (z_2) sont situées sur une même conique.

A présent, si l'on cherche les éléments F_3, G_3, H_3 et F_{12}, G_{12}, H_{12} du déterminant (3′) correspondant à la forme (12) et si l'on pose d'une manière générale,

$$\Delta_i = a_ix + b_iy + c_i,$$

$$\Delta_i' = a_i'x + b_i'y + c_i',$$

$$\Delta_i'' = a_i''x + b_i''y + c_i'',$$

on démontre, par une marche analogue à la précédente, que si les conditions

$$\frac{\Delta_1'}{\Delta_2'} \equiv \frac{\Delta_1''}{\Delta_2''} \equiv \frac{\Delta_3}{\Delta_4} \equiv \frac{\Delta_3'}{\Delta_4'} \equiv \frac{\Delta_3''}{\Delta_4''} \quad \text{........................(15)}$$

sont remplies, l'équation (12) peut être mise sous la forme du déterminant (3') et que les deux échelles (z_0) et (z_3) du nomogramme $(z_0\, z_3\, z_4)$ correspondant à ce déterminant sont situées sur une même conique définie par l'équation

$$\frac{\Delta_1}{\Delta_2} = \frac{\Delta_1'}{\Delta_2'}.$$

Comme les deux échelles (z_0) correspondantes des deux nomogrammes partiels $(z_1\, z_2\, z_0)$ et $(z_0\, z_3\, z_4)$ ont leurs supports coniques il résulte, en vertu de la *Remarque* ci-dessus, que ces deux nomogrammes pourront être superposés par l'homographie de façon à obtenir un nomogramme à double alignement représentatif de l'équation (11).

De là, le théorème suivant :

Si les conditions (13) *et* (15) *sont satisfaites, l'équation* (11) *à* 4 *variables d'ordre total* 6° *la plus générale est représentable par un nomogramme à double alignement sur lequel la charnière est une conique servant de support commun aux deux échelles* (z_2) *et* (z_3).

Exemple pratique : Comme cas particulier de ce nomogramme signalons celui construit par M. Wolff pour la représentation de l'équation suivante :

$$\frac{V^{\frac{3}{2}}}{40^{\frac{3}{2}} s^3} = \frac{bd + d^2}{b + \sqrt{2}d}$$

(*Cours de Calcul Graphique et Nomographique,* par M. d'Ocagne, p. 332), qui donne la vitesse d'écoulement V d'un canal à section trapézoïdale avec talus à 1/1 en fonction de la largeur au plafond b, de la hauteur d du plan d'eau et de la pente longitudinale s.

INTEGRALS IN DYNAMICS AND THE PROBLEM OF THREE BODIES

By Selig Brodetsky.

§ 1. The general dynamical problem is defined in the following manner. We have n "coordinates" $q_1, q_2, \ldots, q_n$; n "momenta" $p_1, p_2, \ldots, p_n$; and the independent variable t. Between these variables we have the Hamiltonian system of equations:

$$\frac{dq_r}{dt} = \frac{\partial H}{\partial p_r}; \quad \frac{dp_r}{dt} = -\frac{\partial H}{\partial q_r}; \quad (r = 1, 2, \ldots, n) \quad \ldots\ldots\ldots\ldots\ldots(1);$$

H is a given function of the $2n+1$ variables. The problem is to find a function $W(q_1, q_2, \ldots, q_n, p_1, p_2, \ldots, p_n, t)$ such that $\frac{dW}{dt} = 0$ in virtue of the equations (1). Thus if $W = \text{const.}$ is an integral we have

$$\frac{\partial W}{\partial t} + \sum_{r=1}^{n} \left(\frac{\partial W}{\partial q_r} \frac{dq_r}{dt} + \frac{\partial W}{\partial p_r} \frac{dp_r}{dt} \right) = 0;$$

i.e.

$$\frac{\partial W}{\partial t} + \sum_{r=1}^{n} \left(\frac{\partial W}{\partial q_r} \frac{\partial H}{\partial p_r} - \frac{\partial W}{\partial p_r} \frac{\partial H}{\partial q_r} \right) = 0;$$

i.e.

$$\frac{\partial W}{\partial t} + (W, H) = 0 \quad \ldots\ldots\ldots\ldots\ldots\ldots\ldots\ldots\ldots\ldots\ldots\ldots(2),$$

using the Poisson bracket expression.

This gives

$$\frac{dH}{dt} \equiv \frac{\partial H}{\partial t} + (H, H) \equiv \frac{\partial H}{\partial t}.$$

Thus if H does not involve t explicitly, $\frac{dH}{dt} = 0$, and $H = \text{const.}$ is an integral, the energy integral.

§ 2. H can involve the momenta in any manner we choose. One case is however of great importance in natural problems, namely when H is a quadratic function of the p's, involving only squares; i.e.

$$H \equiv \tfrac{1}{2}P_1 p_1^2 + \tfrac{1}{2}P_2 p_2^2 + \ldots + \tfrac{1}{2}P_n p_n^2 + P \ldots\ldots\ldots\ldots\ldots\ldots(3),$$

where $P_1, P_2, \ldots, P_n, P$ are functions of $q_1, q_2, \ldots, q_n, t$.

When H has the form given by equation (3), we can prove an elementary but fundamental property of the integrals. Let $W = \text{const.}$ be an integral not involving one of the momenta, say p_1. $\dfrac{dW}{dt} = 0$ gives

$$\frac{\partial W}{\partial t} + \frac{\partial W}{\partial q_1}\frac{\partial H}{\partial p_1} + \sum_{r=2}^{n}\left(\frac{\partial W}{\partial q_r}\frac{\partial H}{\partial p_r} - \frac{\partial W}{\partial p_r}\frac{\partial H}{\partial q_r}\right) = 0;$$

i.e. $$\frac{\partial W}{\partial t} + P_1 p_1 \frac{\partial W}{\partial q_1} + \sum_{r=2}^{n}\left(P_r p_r \frac{\partial W}{\partial q_r} - \left[\frac{1}{2}\frac{\partial P_1}{\partial q_r} p_1^2 + \ldots + \frac{1}{2}\frac{\partial P_n}{\partial q_r} p_n^2 + \frac{\partial P}{\partial q_r}\right]\frac{\partial W}{\partial p_r}\right) = 0.$$

Now p_1 occurs to the first power only in the term $P_1 p_1 \dfrac{\partial W}{\partial q_1}$. Otherwise it occurs either in the form p_1^2 or not at all. Thus we must have

$$P_1 \frac{\partial W}{\partial q_1} = 0;$$

i.e. $\dfrac{\partial W}{\partial q_1} = 0$, and W does not contain q_1.

In other words if an integral of a system (3) can be found which does not contain a certain one of the momenta, then it does not contain the corresponding coordinate. The converse is of course not necessarily true.

§ 3. This result is true of the problem of three bodies reduced by means of the integrals of linear momentum. We here have $n = 6$; and

$$H = T - U,$$

where $$T = \frac{1}{2\mu}(p_1^2 + p_2^2 + p_3^2) + \frac{1}{2\mu'}(p_4^2 + p_5^2 + p_6^2);$$

$$U = m_1 m_2/r_{12} + m_2 m_3/r_{13} + m_3 m_1/r_{23};$$

and $r_{12} = (q_1^2 + q_2^2 + q_3^2)^{\frac{1}{2}}$;

$$r_{13} = \left\{(q_4^2 + q_5^2 + q_6^2) + \frac{2m_2}{m_1 + m_2}(q_1 q_4 + q_2 q_5 + q_3 q_6) + \left(\frac{m_2}{m_1 + m_2}\right)^2 (q_1^2 + q_2^2 + q_3^2)\right\}^{\frac{1}{2}};$$

$$r_{23} = \left\{(q_4^2 + q_5^2 + q_6^2) - \frac{2m_1}{m_1 + m_2}(q_1 q_4 + q_2 q_5 + q_3 q_6) + \left(\frac{m_1}{m_1 + m_2}\right)^2 (q_1^2 + q_2^2 + q_3^2)\right\}^{\frac{1}{2}};$$

$$\mu = m_1 m_2/(m_1 + m_2); \quad \mu' = m_3(m_1 + m_2)/(m_1 + m_2 + m_3);$$

m_1, m_2, m_3 are the masses of the bodies.

H is of the form given by (3). The result of § 2 holds, and if $W = \text{const.}$ is an integral not containing p_1, say, then it does not involve q_1. The differential equation for W is now

$$\frac{\partial W}{\partial t} + \frac{p_2}{\mu}\frac{\partial W}{\partial q_2} + \frac{p_3}{\mu}\frac{\partial W}{\partial q_3} + \frac{p_4}{\mu'}\frac{\partial W}{\partial q_4} + \frac{p_5}{\mu'}\frac{\partial W}{\partial q_5} + \frac{p_6}{\mu'}\frac{\partial W}{\partial q_6}$$
$$- \frac{\partial W}{\partial p_2}\left[\frac{m_1 m_2 q_2}{r_{12}^3} + \frac{m_1 m_3}{r_{13}^3}\left\{\left(\frac{m_2}{m_1 + m_2}\right)^2 q_2 + \left(\frac{m_2}{m_1 + m_2}\right) q_5\right\}\right.$$
$$\left. + \frac{m_2 m_3}{r_{23}^3}\left\{\left(\frac{m_1}{m_1 + m_2}\right)^2 q_2 - \left(\frac{m_1}{m_1 + m_2}\right) q_5\right\}\right]$$

$$-\frac{\partial W}{\partial p_3}\left[\frac{m_1 m_2 q_3}{r_{12}^3}+\frac{m_1 m_3}{r_{13}^3}\left\{\left(\frac{m_2}{m_1+m_2}\right)^2 q_3+\left(\frac{m_2}{m_1+m_2}\right) q_6\right\}\right.$$
$$\left.+\frac{m_2 m_3}{r_{23}^3}\left\{\left(\frac{m_1}{m_1+m_2}\right)^2 q_3-\left(\frac{m_1}{m_1+m_2}\right) q_6\right\}\right]$$
$$-\frac{\partial W}{\partial p_4}\left[\frac{m_1 m_3}{r_{13}^3}\left\{\left(\frac{m_2}{m_1+m_2}\right) q_1+q_4\right\}+\frac{m_2 m_3}{r_{23}^3}\left\{-\left(\frac{m_1}{m_1+m_2}\right) q_1+q_4\right\}\right]$$
$$-\frac{\partial W}{\partial p_5}\left[\frac{m_1 m_3}{r_{13}^3}\left\{\left(\frac{m_2}{m_1+m_2}\right) q_2+q_5\right\}+\frac{m_2 m_3}{r_{23}^3}\left\{-\left(\frac{m_1}{m_1+m_2}\right) q_2+q_5\right\}\right]$$
$$-\frac{\partial W}{\partial p_6}\left[\frac{m_1 m_3}{r_{13}^3}\left\{\left(\frac{m_2}{m_1+m_2}\right) q_3+q_6\right\}+\frac{m_2 m_3}{r_{23}^3}\left\{-\left(\frac{m_1}{m_1+m_2}\right) q_3+q_6\right\}\right]=0 \qquad \text{.........(4).}$$

q_1 is present only in the square brackets and in terms of

$$\frac{1}{r_{12}^3};\ \frac{q_1}{r_{13}^3};\ \frac{1}{r_{13}^3};\ \frac{q_1}{r_{23}^3};\ \frac{1}{r_{23}^3}.$$

r_{12}, r_{23}, r_{31} cannot make one another disappear in any manner; they are also all irrational to q_1. Thus the five forms in which q_1 appears in (4) must separately vanish. Hence $\frac{\partial W}{\partial p_4}=0$, which also gives $\frac{\partial W}{\partial q_4}=0$. Further we get

$$q_2\frac{\partial W}{\partial p_2}+q_3\frac{\partial W}{\partial p_3}=0 \qquad \text{..............................(5).}$$

Also $$\left(\frac{m_2}{m_1+m_2}\right)\left(q_5\frac{\partial W}{\partial p_2}+q_2\frac{\partial W}{\partial p_5}+q_6\frac{\partial W}{\partial p_3}+q_3\frac{\partial W}{\partial p_6}\right)+q_5\frac{\partial W}{\partial p_5}+q_6\frac{\partial W}{\partial p_6}=0;$$

and $$-\left(\frac{m_1}{m_1+m_2}\right)\left(q_5\frac{\partial W}{\partial p_2}+q_2\frac{\partial W}{\partial p_5}+q_6\frac{\partial W}{\partial p_3}+q_3\frac{\partial W}{\partial p_6}\right)+q_5\frac{\partial W}{\partial p_5}+q_6\frac{\partial W}{\partial p_6}=0.$$

These two equations are equivalent to

$$q_5\frac{\partial W}{\partial p_5}+q_6\frac{\partial W}{\partial p_6}=0 \qquad \text{..............................(6);}$$

and $$q_5\frac{\partial W}{\partial p_2}+q_2\frac{\partial W}{\partial p_5}+q_6\frac{\partial W}{\partial p_3}+q_3\frac{\partial W}{\partial p_6}=0 \qquad \text{.....................(7).}$$

Finally we have the equation

$$\frac{\partial W}{\partial t}+\frac{p_2}{\mu}\frac{\partial W}{\partial q_2}+\frac{p_3}{\mu}\frac{\partial W}{\partial q_3}+\frac{p_5}{\mu'}\frac{\partial W}{\partial q_5}+\frac{p_6}{\mu'}\frac{\partial W}{\partial q_6}=0 \qquad \text{.................(8).}$$

Equations (5) and (6) give us

$$W=W'(q_2p_3-q_3p_2,\quad q_5p_6-q_6p_5,\quad q_2, q_3, q_5, q_6, t).$$

Substitute in (7). We get

$$\frac{\partial W'}{\partial(q_2p_3-q_3p_2)}-\frac{\partial W'}{\partial(q_5p_6-q_6p_5)}=0;$$

hence $$W=W''(q_2p_3-q_3p_2+q_5p_6-q_6p_5,\quad q_2, q_3, q_5, q_6, t)$$
$$=W''(L, q_2, q_3, q_5, q_6, t),$$

where $L = \text{const.}$ is that integral of angular momentum which does not contain p_1. Substituting in (8), we get

$$\frac{\partial W''}{\partial t} + \frac{p_2}{\mu}\left(\frac{\partial W''}{\partial q_2} + p_3\frac{\partial W''}{\partial L}\right) + \frac{p_3}{\mu}\left(\frac{\partial W''}{\partial q_3} - p_2\frac{\partial W''}{\partial L}\right)$$
$$+ \frac{p_5}{\mu'}\left(\frac{\partial W''}{\partial q_5} + p_6\frac{\partial W''}{\partial L}\right) + \frac{p_6}{\mu'}\left(\frac{\partial W''}{\partial q_6} - p_5\frac{\partial W''}{\partial L}\right) = 0;$$

i.e.
$$\frac{\partial W''}{\partial t} + \frac{p_2}{\mu}\frac{\partial W''}{\partial q_2} + \frac{p_3}{\mu}\frac{\partial W''}{\partial q_3} + \frac{p_5}{\mu'}\frac{\partial W''}{\partial q_5} + \frac{p_6}{\mu'}\frac{\partial W''}{\partial q_6} = 0,$$

in which we look upon L as a constant. This equation is possible only if

$$\frac{\partial W''}{\partial t} = \frac{\partial W''}{\partial q_2} = \frac{\partial W''}{\partial q_3} = \frac{\partial W''}{\partial q_5} = \frac{\partial W''}{\partial q_6} = 0.$$

Hence
$$W = W''(L).$$

Thus an integral of the problem of three bodies in the reduced form given above, from which p_1 is absent, is merely a function of that integral of angular momentum from which p_1 is absent. The same is true for any one of the momenta. It is seen that the momenta are associated in couples. When p_1 is absent, then p_4, as well as q_1, q_4, is absent, and the integral is a function of L. When p_2 is absent, then p_5, as well as q_2, q_5, is absent, and the integral is a function of M, where $M \equiv q_3p_1 - q_1p_3 + q_6p_4 - q_4p_6$. When p_3 is absent, then p_6, as well as q_3, q_6, is absent, and the integral is a function of N, where $N \equiv q_1p_2 - q_2p_1 + q_4p_5 - q_5p_4$.

There are no integrals in which momenta from more than one set are absent, e.g. p_1 and p_2, or p_3 and p_5, etc.

It follows that integrals independent of the integrals of angular momentum must involve all the momenta.

The argument here given applies equally well if we are given that $W = \text{const.}$ is an integral from which one of the coordinates, say q_1, is absent. We get q_1 present in terms of the form

$$\frac{q_1}{r_{12}^3};\ \frac{1}{r_{12}^3};\ \frac{q_1}{r_{13}^3};\ \frac{1}{r_{13}^3};\ \frac{q_1}{r_{23}^3};\ \frac{1}{r_{23}^3}.$$

To make the term in $\frac{q_1}{r_{12}^3}$ go out, we must make $\frac{\partial W}{\partial p_1} = 0$, and the same results follow. Thus if q_1 or q_4 is absent from an integral, then the integral is a function of L; etc.

It follows that integrals independent of the integrals of angular momentum must involve all the coordinates.

Hence we have the important result:

In the problem of three bodies, integrals independent of the integrals of angular momentum must involve all the coordinates and all the momenta.

This is true of n bodies, $n \geqslant 3$; also for all laws of force which give U in odd powers of r_{12}, r_{13}, r_{23}; i.e. for which the law of attraction is according to a positive or negative even power of the mutual distance.

§ 4. The energy integral is an integral independent of L, M, N. It is seen that it involves all the coordinates and all the momenta. Do other such integrals exist?

Bruns has shewn that when the coordinates, the momenta and the time are all supposed to occur algebraically, then no integrals exist independent of the energy integral and L, M, N.

Painlevé has extended this to include integrals in which only the momenta are supposed to occur algebraically, the coordinates and the time occurring anyhow.

We shall here consider the more general type of integrals in which only one of the momenta is supposed to occur algebraically, the other momenta and the coordinates and the time occurring anyhow; we shall however restrict the occurrence of the single momentum thus considered to even powers only.

Moreover we shall consider not only the problem of three bodies and the allied problems mentioned at the end of the last article, but also all problems defined by

$$H = \tfrac{1}{2}P_1 p_1^2 + \tfrac{1}{2}P_2 p_2^2 + \dots + \tfrac{1}{2}P_n p_n^2 + P \quad \dots\dots\dots\dots\dots\dots(9),$$

where $P_1, P_2, \dots, P_n$ *are constants,* and P does not involve the time. There exists then the energy integral $H = \text{const.}$

As in Bruns' theorem we need only consider integrals rational and algebraic in p_1, the momentum supposed to occur algebraically. We write then

$$W \equiv W_1/W_2 = \text{const.}$$

is an integral, where W_1, W_2 are rational and algebraic in p_1^2 containing only zero and positive powers. We say p_1^2 because we suppose p_1 to occur only in even powers.

§ 5. We suppose then that

$$W \equiv W_1/W_2 = \text{const.}$$

is an integral of a system given by (9). We can suppose that W contains no factor H in the numerator or the denominator, for if H does occur as a factor we may put the energy constant for it and the result equated to a constant must still be an integral.

Now we can write

$$W_1 \equiv HW_1' + W_1''; \qquad W_2 \equiv HW_2' + W_2'';$$

where W_1', W_2' are all of the same form as W_1, W_2, but containing p_1 two degrees lower respectively, and where W_1'', W_2'' do not contain p_1 at all. We have

$$\frac{dW}{dt} = 0;$$

i.e.
$$\frac{d}{dt}[(HW_1' + W_1'')/(HW_2' + W_2'')] = 0;$$

i.e.
$$(HW_2' + W_2'')\frac{d}{dt}(HW_1' + W_1'') - (HW_1' + W_1'')\frac{d}{dt}(HW_2' + W_2'') = 0;$$

i.e.
$$W_2''\frac{dW_1''}{dt} - W_1''\frac{dW_2''}{dt} = H

multiplied by a certain function, since

$$\frac{dH}{dt}=0.$$

Now, in (9), $$\frac{\partial H}{\partial q_1}=\frac{\partial P}{\partial q_1}; \quad \frac{\partial H}{\partial q_2}=\frac{\partial P}{\partial q_2}; \text{ etc.}$$

Hence $$\frac{dw}{dt}=\frac{\partial w}{\partial t}+\Sigma\left(P_r p_r\frac{\partial w}{\partial q_r}-\frac{\partial w}{\partial p_r}\frac{\partial P}{\partial q_r}\right),$$

so that if w does not contain p_1, $\frac{dw}{dt}$ is linear in p_1. Further W_1'', W_2'' do not contain p_1. Hence $W_2''\frac{dW_1''}{dt}-W_1''\frac{dW_2''}{dt}$ is at most linear in p_1. It cannot then be a multiple of H unless it vanishes. Thus we must have

$$\frac{d}{dt}(W_1''/W_2'')=0.$$

Hence $W_1''/W_2''=\text{const.}$ is an integral. But it does not contain p_1. Thus W_1''/W_2'' is a function of those independent integrals which do not contain p_1. Call this function L_1.

We get $$\begin{aligned}W&=(HW_1'+L_1W_2'')/(HW_2'+W_2'')\\&=L_1+H(W_1'-L_1W_2')/(HW_2'+W_2'').\end{aligned}$$

$W=\text{const.}$, $L_1=\text{const.}$ are integrals. Hence $H(W_1'-L_1W_2')/(HW_2'+W_2'')=\text{const.}$ is an integral, i.e. $(W_1'-L_1W_2')/(HW_1'+W_2'')=\text{const.}$ is an integral. Call it $W'=\text{const.}$

Suppose W_1 is of order $2n$ in p_1, W_2 of order $2m$. We may suppose $2n\geqslant 2m$, since if this is not so we can use the inverted form of W for our integral. W' now has p_1^{2n-2} at most in the numerator, and p_1^{2m} in the denominator. If p_1 still occurs to a higher degree in the numerator than in the denominator, we repeat the above process for W'. We get, say,

$$W'=L_2+HW'',$$

where L_2 is a function of integrals independent of p_1, and $W''=\text{const.}$ is an integral with p_1^{2n-4} at most in the numerator, and p_1^{2m} in the denominator. We repeat the process till we have reduced W to the form

$$W=L_1+H(L_2+H(L_3+H(\ldots+H(L_\lambda+HW^{(\lambda)})\ldots))),$$

where $W^{(\lambda)}=\text{constant}$ is an integral with p_1^{2m} in the denominator and lower powers of p_1, say up to $p_1^{2n'}$ in the numerator. We invert $W^{(\lambda)}$, and we reduce it by the same process to

$$1/W^{(\lambda)}=L_{\lambda+1}+H(L_{\lambda+2}+\ldots+H(L_\mu+HW^{(\mu)})\ldots),$$

where $W^{(\mu)}=\text{const.}$ is an integral with $p_1^{2n'}$ in the denominator, and lower powers of p_1, say up to $p_1^{2n''}$ in the numerator. We invert $W^{(\mu)}$ and repeat the process. Proceeding in this manner we must ultimately get W in the form

$$W=L_1+H\left(L_2+\ldots+H\left(L_\lambda+\cfrac{H}{L_{\lambda+1}+H\left(L_{\lambda+2}+\ldots+H\left(L_\mu+\cfrac{H}{L_{\mu+1}+\ldots}\right)\ldots\right)}\right)\ldots\right)$$
$$+HW^{(\nu)},$$

where $W^{(\nu)} = \text{const.}$ is an integral not containing p_1 at all, i.e. $W^{(\nu)}$ is a function of the independent integrals independent of p_1, say $L_{\nu+1}$. We get then W in terms of H and the functions $L_1, L_2, \ldots, L_\lambda, L_{\lambda+1}, \ldots, L_\mu, L_{\mu+1}, \ldots, L_\nu, L_{\nu+1}$; i.e. W is a function of the energy integral and those independent integrals which do not contain p_1.

This is true for any momentum.

If two momenta occur in the manner indicated, then W is a function of the energy integral and the independent integrals independent of these two momenta. Similarly for any number of momenta occurring in this manner.

If all the momenta are involved algebraically and only in even powers, W is a function of H only, because integrals independent of all the momenta reduce to mere constants.

In these results if a momentum does not occur at all, it is equivalent to its occurring rationally algebraically in even powers.

Thus in problems defined by (9), after we have found the independent integrals in which one or more of the momenta is absent, then there are no new integrals of the type considered except the energy integral $H = \text{const.}$

§ 6. We apply this to the problem of three or more bodies with the proper laws of force, the Newtonian being one. Here the only integrals in which one or more of the momenta is absent are

$$L = \text{const.}, \quad M = \text{const.}, \quad N = \text{const.}$$

Thus if p_1 occurs in the manner indicated in § 5, W is a function of H and L; the same holds for p_4, and in fact the property is reciprocal—if p_1 occurs in the prescribed form then p_4 occurs in this form, and vice versa. Similarly for the couples p_2, p_5 to which corresponds M, and for p_3, p_6 to which corresponds N.

If two momenta out of different sets, say p_1 and p_2, occur in the prescribed manner, then the integral is merely a function of H, and then all the momenta occur in this manner.

The conclusions here arrived at are of great generality, and should throw light upon the unknown integrals of the problem of three or more bodies.

§ 7. We consider once more the problems given by (9). Suppose that the power series in p_1

$$W \equiv A_n p_1^n + A_{n-1} p_1^{n-1} + \ldots + A_1 p_1 + A_0 = \text{const.} \quad \ldots\ldots\ldots(10)$$

is an integral. We have

$$\frac{dW}{dt} \equiv p_1^n \frac{dA_n}{dt} + p_1^{n-1}\frac{dA_{n-1}}{dt} + \ldots + p_1 \frac{dA_1}{dt} + \frac{dA_0}{dt}$$
$$- np_1^{n-1} A_n \frac{\partial P}{\partial q_1} - (n-1)\, p_1^{n-2} A_{n-1} \frac{\partial P}{\partial q_1} - \ldots - A_1 \frac{\partial P}{\partial q_1}.$$

Now $A_n, A_{n-1}, \ldots, A_1, A_0$ contain no p_1. Hence $\frac{dA_n}{dt}, \frac{dA_{n-1}}{dt}, \ldots, \frac{dA_0}{dt}$, are, using (9), linear in p_1. We can write

$$\frac{dA_n}{dt} = P_1 p_1 \frac{\partial A_n}{\partial q_1} + \left\{\frac{dA_n}{dt} - P_1 p_1 \frac{\partial A_n}{\partial q_1}\right\}; \text{ etc.}$$

We get

$$\frac{dW}{dt} \equiv p_1^n \left[P_1 p_1 \frac{\partial A_n}{\partial q_1} + \left\{\frac{dA_n}{dt} - P_1 p_1 \frac{\partial A_n}{\partial q_1}\right\}\right] + p_1^{n-1}\left[P_1 p_1 \frac{\partial A_{n-1}}{\partial q_1} + \left\{\frac{dA_{n-1}}{dt} - P_1 p_1 \frac{\partial A_{n-1}}{\partial q_1}\right\}\right]$$
$$+ \ldots + \left[P_1 p_1 \frac{\partial A_0}{\partial q_1} + \left\{\frac{dA_0}{dt} - P_1 p_1 \frac{\partial A_0}{\partial q_1}\right\}\right] - [n p_1^{n-1} A_n + (n-1) p_1^{n-2} A_{n-1} + \ldots + A_1] \frac{\partial P}{\partial q_1}.$$

The expressions in the curly brackets contain no p_1. We now equate $\frac{dW}{dt}$ to zero. Equating coefficients of powers of p_1 we get the results:

$$\left.\begin{aligned}
[a_{n+2}] &\equiv P_1 p_1 \frac{\partial A_n}{\partial q_1} = 0;\\
[a_{n+1}] &\equiv \left\{\frac{dA_n}{dt} - P_1 p_1 \frac{\partial A_n}{\partial q_1}\right\} + P_1 p_1 \frac{\partial A_{n-1}}{\partial q_1} = 0;\\
[a_n] &\equiv \left\{\frac{dA_{n-1}}{dt} - P_1 p_1 \frac{\partial A_{n-1}}{\partial q_1}\right\} + P_1 p_1 \frac{\partial A_{n-2}}{\partial q_1} - n\, A_n \frac{\partial P}{\partial q_1} = 0; \text{ etc.};\\
[a_m] &\equiv \left\{\frac{dA_{m-1}}{dt} - P_1 p_1 \frac{\partial A_{m-1}}{\partial q_1}\right\} + P_1 p_1 \frac{\partial A_{m-2}}{\partial q_1} - m\, A_m \frac{\partial P}{\partial q_1} = 0; \text{ etc.};\\
[a_2] &\equiv \left\{\frac{dA_1}{dt} - P_1 p_1 \frac{\partial A_1}{\partial q_1}\right\} + P_1 p_1 \frac{\partial A_0}{\partial q_1} - 2A_2 \frac{\partial P}{\partial q_1} = 0;\\
[a_1] &\equiv \left\{\frac{dA_0}{dt} - P_1 p_1 \frac{\partial A_0}{\partial q_1}\right\} - A_1 \frac{\partial P}{\partial q_1} = 0;
\end{aligned}\right\} \quad \ldots\ldots(11).$$

We get $[a_1]$ as a case of $[a_m]$ by using the value $A_{-1} = 0$. In the same way $[a_{n+1}]$ and $[a_{n+2}]$ are cases of $[a_m]$ in which A_{n+1} and A_{n+2} are taken to be zero.

$[a_{n+2}] = 0$ gives us A_n independent of q_1. $[a_{n+1}] = 0$ gives us by a partial integration, A_{n-1} in terms of A_n. $[a_n] = 0$ gives us A_{n-2} in terms of A_n and A_{n-1}, i.e. in terms of A_n. Thus working backwards till $[a_2] = 0$ we get $A_{n-1}, A_{n-2}, \ldots, A_1, A_0$ all in terms of A_n. The last equation gives another condition which A_n must satisfy besides being independent of q_1.

We can use the equations (11) upwards. $[a_1] = 0$ gives us

$$A_1 = \left\{\frac{dA_0}{dt} - P_1 p_1 \frac{\partial A_0}{\partial q_1}\right\} \Big/ \frac{\partial P}{\partial q_1};$$

$[a_2] = 0$ gives $$A_2 = \left[\left\{\frac{dA_1}{dt} - P_1 p_1 \frac{\partial A_1}{\partial q_1}\right\} + P_1 p_1 \frac{\partial A_0}{\partial q_1}\right] \Big/ 2 \frac{\partial P}{\partial q_1};$$

i.e. we have A_2 in terms of differentials of A_0. Similarly we get $A_3, A_4, \ldots, A_n$ in terms of differentials of A_0, given by the equations $[a_3] = [a_4] = \ldots = [a_n] = 0$. The two equations $[a_{n+1}] = [a_{n+2}] = 0$ give us two conditions which A_0 must satisfy in order that $W = \text{const.}$ may be an integral.

The conditions $[a_{n+1}] = [a_{n+2}] = 0$ are really equivalent to getting $A_{n+1} = A_{n+2} = 0$ from the general equation

$$A_m = \left[\left\{\frac{dA_{m-1}}{dt} - P_1 p_1 \frac{\partial A_{m-1}}{\partial q_1}\right\} + P_1 p_1 \frac{\partial A_{m-2}}{\partial q_1}\right] \Big/ m \frac{\partial P}{\partial q_1} \quad \ldots\ldots\ldots(12).$$

Thus if A_0 is chosen such that on obtaining A_1, A_2, etc. from equations of the type (12) we get two consecutive A's, say A_{n+1}, A_{n+2}, zero, then A_0 will "generate" a finite power series which is an integral of the system. We may call A_0 a "finite generating function."

If A_0 is such that we never get two consecutive A's vanishing then we do not get a finite power series in p_1 as an integral. But assuming convergency, we can obtain an infinite power series which equated to zero will give us an integral. Any value of A_0 will lead to such an infinite power series, and may be called an "infinite generating function."

But A_0 must not itself be an integral. For if $A_0 = \text{const.}$ is an integral, then as it contains no p_1, it also contains no q_1. Thus $\frac{dA_0}{dt} = \frac{\partial A_0}{\partial q_1} = 0$. We get $A_1 = A_2 = A_3 =$ etc. all zero. Thus we do not get a power series at all. In particular A_0 must not be a mere constant. Thus we cannot have

$$A_1 p_1 + A_2 p_1^2 + \ldots = \text{const.}$$

as an integral, whether as a finite or as an infinite power series.

It is obvious that the algebraic sum of any number of generating functions will be the generating function of the algebraic sum of their integrals, whether finite or infinite. Further, assuming convergency, the product or quotient of two generating functions will be the generating function of the product or quotient of their integrals. This can be extended to any number of operations. Thus we get the general result that any algebraic (or other well-behaved) function of generating functions is the generating function of this function of their integrals. We assume the necessary convergency.

§ 8. It seems that this result applied to the problem of three or more bodies with the appropriate laws of force, should enable us to extend Bruns' and Painlevé's results mentioned above (§ 4) to shew that integrals independent of the classical integrals must contain all the coordinates and all the momenta, not one of the latter occurring algebraically. I have not yet, however, been able to construct a perfectly rigorous demonstration.

§ 9. We now consider the general theory of integrals in dynamics. If the system is defined by

$$\frac{dq_r}{dt} = \frac{\partial H}{\partial p_r}; \quad \frac{dp_r}{dt} = -\frac{\partial H}{\partial q_r}; \quad (r = 1, 2, \ldots, n);$$

then $W = \text{const.}$ is an integral if

$$\frac{\partial W}{\partial t} + \sum_{r=1}^{n} \left(\frac{\partial W}{\partial q_r} \frac{\partial H}{\partial p_r} - \frac{\partial W}{\partial p_r} \frac{\partial H}{\partial q_r} \right) = 0;$$

i.e. if $\frac{\partial W}{\partial t} + (W, H) = 0$, using Poisson's bracket expression.

Now we may discuss a dynamical problem from two points of view. One way is the consideration of integrals already exemplified. This gives us general solutions independent of the conditions special to the problem, such as initial positions and motions. But in any given problem we may also take into account given conditions and solutions already known. Thus it is not necessary to have $\frac{\partial W}{\partial t}+(W,H)\equiv 0$. If $\frac{\partial W}{\partial t}+(W,H)\equiv W_1$, where W_1 is known to vanish for our problem, then $W=$const. is still an integral. When $\frac{\partial W}{\partial t}+(W,H)\equiv 0$, we may call $W=$const. a primary integral.

If $\frac{\partial W}{\partial t}+(W,H)\equiv MW$, there are two cases

(1) $M\equiv\frac{1}{L}\frac{dL}{dt}$: then $\frac{d}{dt}(LW)=0$, and $LW=$const. is a primary integral.

We may call W a sub-integral (Forsyth).

(2) If we cannot get $M\equiv\frac{1}{L}\frac{dL}{dt}$, then $W=0$ may be an integral if W vanishes for a given moment of time. For then $\frac{dW}{dt}$ will vanish at this moment. Also

$$\frac{d}{dt}\frac{dW}{dt}=\frac{d}{dt}(MW)=M\frac{dW}{dt}+W\frac{dM}{dt}=0$$

at this moment. And so on we may shew that all the derivatives of W with respect to the time (totally) vanish, so that $W=0$ is true of the problem. Painlevé calls such an integral particularised; we may add the epithet primary, since it is possible to have

$$\frac{\partial W}{\partial t}+(W,H)\equiv MW+W_1,$$

where W_1 vanishes in virtue of the conditions or known solutions. In this case too we get sub and particularised integrals. But a distinction has to be made. This distinction we obtain as follows.

If $\frac{\partial W}{\partial t}+(W,H)\equiv W_1$, where W_1 vanishes because of conditions or integrals, but W_1 is independent of W, then $W=$const. is an integral of our problem, which we shall call secondary (Forsyth calls it composite). If then $\frac{\partial W}{\partial t}+(W,H)\equiv MW+W_1$, then $W=$const. may be a secondary particularised integral, or it may be possible to get a function L such that $M\equiv\frac{1}{L}\frac{dL}{dt}$, so that $LW=$const. is a secondary integral. In this case W is a secondary sub-integral. These definitions may be extended.

§ 10. Cases of primary integrals are of course very well known. The energy integral where it exists is one. Secondary integrals are e.g. Liouville's integrals for the form: $T=\frac{1}{2}M(Q_1\dot{q}_1^2+\ldots)$; $M(V-C)=F_1(q_1)+F(q_2,\ldots)$. In fact if $W_1=C_1$ is the Liouville integral, we get $\frac{dW_1}{dt}\equiv\dot{q}_1\frac{\partial M}{\partial q_1}(T+V-C)$; hence if $T+V-C=0$ is the energy integral, Liouville's integral exists, but only in virtue of the energy integral. It is therefore secondary.

Again, if T is homogeneous in p's of order 2, and in q's of order m; and V is homogeneous in q's of order $-m-2$; then Jacobi has shewn that

$$q_1 \frac{\partial T}{\partial \dot{q}_1} + \ldots + q_n \frac{\partial T}{\partial \dot{q}_n} - (m+2)\, Ct - A = 0$$

is true of the problem. This also depends on the energy integral and is in general secondary.

Examples of particularised and other integrals will also readily occur to one.

§ 11. For the further discussion we suppose the constant of integration if it exists to be absorbed in W. Now let $\frac{\partial W}{\partial t} + (W, H) \equiv W_1$. We use the operator $(H)\, W \equiv \frac{\partial W}{\partial t} + (W, H)$. Then $(H)\, W \equiv W_1$. If now $W = 0$ is true of the problem, so is $W_1 = 0$. Again if $(H)^2 W \equiv (H)\, W_1 \equiv W_2$, then $W_2 = 0$ must be true. Proceeding in this way, it follows that if $(H)^n W \equiv (H)^{n-1} W_1 \equiv \ldots \equiv (H)\, W_{n-1} \equiv W_n$, then $W_n = 0$ is also true of the problem.

This argument also works backwards. Thus if $W_1 = 0$ is true, then $(H)\, W = 0$, so that $W = 0$ is true (with constant of integration absorbed). If $(H)^{-1} W \equiv W_{-1}$, then $W_{-1} = 0$ is true. In general if $(H)^{-n} W \equiv W_{-n}$, then $W_{-n} = 0$ is true of the problem.

Now positive powers of the operators give us nothing new. In fact the relations $(H)\, W, (H)^2 W, \ldots, (H)^n W, \ldots = 0$ are merely conditions that $W = 0$ may be true. But negative powers of (H) can give us new results, as is seen in the case of Liouville's and Jacobi's integrals already exemplified.

Thus in a dynamical problem, if $W = 0$ is an integral, we can take FW, where F is any function and find $(H)^{-1} FW, (H)^{-2} FW$, etc. These functions equated to zero will give us new integrals of our problem. In general we have the energy integral which may be used for this purpose. But when no energy integral exists, we can use another integral for the purpose. It is easy to shew that integrals given by ignorable coordinates are of no use in this connection.

§ 12. When no energy integral exists and no other integral is known, we may construct an artificial energy integral by increasing the order of the system by one. Thus Whittaker has shewn that when

$$H(q_1, q_2, \ldots, q_n, p_1, p_2, \ldots, p_n) + h = 0$$

is the energy integral of a problem, then if we solve for p_1 in the form

$$p_1 + K(q_1, q_2, \ldots, q_n, p_1, p_2, \ldots, p_n, h) = 0,$$

the problem can be represented by a Hamiltonian system with variables $q_2, \ldots, q_n$, $p_2, \ldots, p_n$, and independent variable q_1, in the form

$$\frac{dq_r}{dq_1} = \frac{\partial K}{\partial p_r}; \quad \frac{dp_r}{dq_1} = -\frac{\partial K}{\partial q_r}; \quad (r = 2, 3, \ldots, n).$$

If now t occurs in H in a dynamical problem, we reverse Whittaker's process. Let us then have

$$\frac{dq_r}{dt}=\frac{\partial H}{\partial p_r};\quad \frac{dp_r}{dt}=-\frac{\partial H}{\partial q_r};\quad (r=1, 2, \ldots, n);$$

H involves t explicitly. Take a new independent variable τ, and a new momentum T. Let

$$H'+h'\equiv(T+H)F,$$

where F is any function of the q's, p's, T and t. We get a new system

$$\left.\begin{aligned}\frac{dq_r}{d\tau}&=\frac{\partial H'}{\partial p_r};\quad \frac{dp_r}{d\tau}=-\frac{\partial H'}{\partial q_r};\quad (r=1, 2, \ldots, n)\\ \frac{dt}{d\tau}&=\frac{\partial H'}{\partial T};\quad \frac{dT}{d\tau}=-\frac{\partial H'}{\partial t}\end{aligned}\right\}\quad \ldots\ldots\ldots\ldots(13).$$

H' does not involve τ. We have then an energy integral $H'+h'=0$. We can now use $H'+h'$ for deriving secondary integrals. When obtained these may be reduced back to the problem as originally stated with independent variable t and variables $q_1, q_2, \ldots, q_n, p_1, p_2, \ldots, p_n$.

The method of obtaining secondary integrals here sketched is useful in obtaining particular integrals. Thus if $(H)\,W\equiv W_1$ and W_1 can be made to vanish by means of some particular condition, we have an integral of a particular case.

§ 13. The notion of secondary integrals can be used to obtain integrals involving the energy where an energy integral does not really exist. Let H involve t explicitly. Then as seen in § 1, $\frac{dH}{dt}\equiv\frac{\partial H}{\partial t}$. If now the conditions of the problem make $\frac{\partial H}{\partial t}$ zero, $H=h$ is an integral in spite of t being present in H. We may call such an integral a "quasi energy integral." If H can be split up into a kinetic energy, which is (as is usually the case) independent of the time explicitly, and a potential energy V, then the condition is that $\frac{\partial V}{\partial t}=0$ should be true of the problem. We get a particular solution.

Thus let a particle move in a plane under a central force due to a potential which involves the time. Using polar coordinates,

$$H=\tfrac{1}{2}(\dot{r}^2+r^2\dot{\theta}^2)+V(r, t).$$

There is no energy integral. But if $\frac{\partial V}{\partial t}=0$ is true of the problem, then

$$H=h$$

is a quasi energy integral, if $\frac{\partial V}{\partial t}=0$ and $H=h$ are compatible. $r^2\dot{\theta}=\text{const.}$ is true of the system. Let the constant be K. Put

$$V'=V+K^2/2r^2.$$

The conditions governing V' are found to be

$$\left.\begin{aligned} 2(h - V') &= V_{tt}'^2 / V_{rt}'^2; \\ \text{and} \quad V_r' V_{rt}'^3 &= V_{rrt}' V_{tt}'^2 - 2V_{rtt}' V_{tt}' V_{rt}' + V_{ttt}' V_{rt}'^2 \end{aligned}\right\} \quad \text{............(14);}$$

where suffixes denote partial differentiations. It is easy to find V' from these differential equations. The result is of no particular interest except as an illustration of the existence of a quasi energy integral.

§ 14. Again we may have a "quasi ignorable coordinate." Thus if

$$W \equiv p_1 - C_1 = 0$$

is to be an integral, then we must have $\frac{\partial H}{\partial q_1} = 0$. If then $\frac{\partial H}{\partial q_1}$ contains as a factor an expression which is known to vanish for the problem, we get the quasi ignorable coordinate q_1, and $p_1 - C_1 = 0$ is an integral.

A simple illustration is the following. A particle moves in a plane under a central force varying inversely as the cube of the distance. With polar coordinates

$$H = \tfrac{1}{2}(\dot{r}^2 + r^2\dot{\theta}^2) + \frac{L}{r^2}.$$

We convert into the Hamiltonian form, by putting $r = q_1$, $\theta = q_2$, $\dot{r} = p_1$, $r^2\dot{\theta} = p_2$,

$$H = \frac{1}{2}\left(p_1^2 + \frac{p_2^2}{q_1^2}\right) + \frac{L}{q_1^2}.$$

q_2 is an ignorable coordinate. Let $p_2 - C_2 = 0$ be the integral; then

$$\frac{d}{dt}(p_1 - C_1) \equiv -\frac{\partial H}{\partial q_1} \equiv \frac{p_2^2 + 2L}{q_1^3} = \frac{C_2^2 + 2L}{q_1^3}$$

for our problem. Thus if $2L = -C_2^2$, we get the quasi ignorable coordinate q_1, and $p_1 - C_1 = 0$ is true of our problem, as a particular solution.

We get $\dot{r} = C_1$; $r^2\dot{\theta} = C_2$. Hence $r = C_1 t + C_1'$; $\dot{\theta} = C_2/(C_1 t + C_1')^2$; so that

$$\theta = -\frac{C_2}{C_1}\Big/(C_1 t + C_1') + C_2' = -C_2/C_1 r + C_2'.$$

Thus we get the reciprocal spiral $\theta r = \text{constant}$ by a proper choice of the zero for θ.

Several problems can be solved in this manner, and integrals can be obtained replacing the energy integral when it does not exist, as exemplified in § 13. But we shall conclude with an extension of the well known Liouville's integrals.

§ 15. Let $\quad H = \tfrac{1}{2}P_1 p_1^2 + \tfrac{1}{2}P_2 p_2^2 + \ldots + \tfrac{1}{2}P_n p_n^2 + P,$

where $P_1, P_2, \ldots, P_n, P$ are functions of $(q_1, q_2, \ldots, q_n)$. $H = h$ is an integral. We shall examine when we can have an integral of the form

$$W \equiv L + \tfrac{1}{2}L_1 p_1^2 + \tfrac{1}{2}L_2 p_2^2 + \ldots + \tfrac{1}{2}L_n p_n^2 = \text{const.}$$

depending on the energy integral; $L_1, L_2, \ldots, L_n, L$ are also functions of $(q_1, q_2, \ldots, q_n)$.

We put $\frac{dW}{dt} \equiv (M + M_1 p_1 + \ldots + M_n p_n)(H - h)$, where $M_1, M_2, \ldots, M_n, M$ are functions of $(q_1, q_2, \ldots, q_n)$. We get

$$\sum_{r=1}^{n} P_r p_r \left(\frac{\partial L}{\partial q_r} + \frac{1}{2}\frac{\partial L_1}{\partial q_r} p_1^2 + \ldots + \frac{1}{2}\frac{\partial L_n}{\partial q_r} p_n^2\right) - \sum_{r=1}^{n} L_r p_r \left(\frac{\partial P}{\partial q_r} + \frac{1}{2}\frac{\partial P_1}{\partial q_r} p_1^2 + \ldots + \frac{1}{2}\frac{\partial P_n}{\partial q_r} p_n^2\right)$$
$$\equiv (M + M_1 p_1 + \ldots + M_n p_n)(P - h + \tfrac{1}{2}P_1 p_1^2 + \ldots + \tfrac{1}{2}P_n p_n^2).$$

Equating coefficients of powers and products of the p's, we get at once $M = 0$. We also get

$$P_r \frac{\partial L}{\partial q_r} - L_r \frac{\partial P}{\partial q_r} = M_r (P - h); \qquad (r = 1, 2, \ldots, n);$$

and
$$P_r \frac{\partial L_s}{\partial q_r} - L_r \frac{\partial P_s}{\partial q_r} = M_r P_s; \qquad (r, s = 1, 2, \ldots, n).$$

These give us
$$\frac{\partial}{\partial q_r}\left\{\frac{L_r}{P_r}(P - h)\right\} = \frac{\partial L}{\partial q_r} \qquad (r = 1, 2, \ldots, n).$$

We put
$$\frac{L_r}{P_r}(P - h) = L + F_r(q_1, q_2, \ldots, q_{r-1}, q_{r+1}, \ldots, q_n); \qquad (r = 1, 2, \ldots, n).$$

F_r does not contain q_r. We then obtain after a little reduction

$$\frac{\partial}{\partial q_r}\left\{\frac{P_s}{P - h}(F_r - F_s)\right\} = 0 \qquad (r \neq s, r, s = 1, 2, \ldots, n).$$

We put

$$\frac{P_s}{P - h}(F_r - F_s) = F_{rs}(q_1, q_2, \ldots, q_{r-1}, q_{r+1}, \ldots, q_n); \qquad (r = 1, 2, \ldots, s-1, s+1, \ldots, n).$$

F_{rs} is again independent of r.

Thus if we can find quantities F_r, F_{rs}, $r = 1, 2, \ldots, n$; $s = 1, 2, \ldots, n$; $r \neq s$, such that F_r, F_{rs} do not contain r, and

$$P_s/(P - h) = F_{rs}/(F_r - F_s) \quad \ldots\ldots\ldots\ldots\ldots\ldots(15),$$

then we have an integral

$$W \equiv L + \tfrac{1}{2}L_1 p_1^2 + \ldots + \tfrac{1}{2}L_n p_n^2 = \text{const.},$$

where
$$L_r = (L + F_r) P_r/(P - h) \quad \ldots\ldots\ldots\ldots\ldots\ldots(16).$$

Equation (15) gives us $(n-1)$ values for $P_s/(P-h)$. There must then be $(n-2)$ relations among the quantities

$$F_1, F_2, \ldots, F_{s-1}, F_s, F_{s+1}, \ldots, F_n, F_{1,s}, F_{2,s}, \ldots, F_{s-1,s}, F_{s+1,s}, \ldots, F_{ns}.$$

The integral is

$$L + \tfrac{1}{2}\sum_{r=1}^{n}(L + F_r) P_r/(P - h) \,.\, p_r^2 = \text{const.},$$

i.e. $L/(P - h) \,.\, \{(P - h) + \tfrac{1}{2}\Sigma P_r p_r^2\} + \tfrac{1}{2}\sum_{r=1}^{n} P_r F_r/(P - h) \,.\, p_r^2 = \text{const.} = K$, say.

Using the energy integral $H - h = 0$, we get

$$\sum_{r=1}^{n} P_r F_r p_r^2 = 2K(P - h).$$

Equation (15) defines a generalised form of Liouville's system.

For $n=1$, we get the energy integral.

For $n=2$, we get a Liouville's system as usually defined.

For $n=3$, we get the first extension of a Liouville's system.

Equation (15) gives $P_1/(P-h)=F_{21}/(F_2-F_1)=F_{31}/(F_3-F_1)$ for $s=1$. Hence $F_{21}/(F_2-F_1)=F_{31}/(F_3-F_1)$. Similarly we get

$$F_{12}/(F_1-F_2)=F_{32}/(F_3-F_2); \quad F_{13}/(F_1-F_3)=F_{23}/(F_2-F_3).$$

Thus we get

$$F_{12}\,.\,F_{23}\,.\,F_{31}+F_{21}\,.\,F_{32}\,.\,F_{13}=0 \quad \text{.....................(17)}.$$

It is to be remembered that each F is independent of that coordinate which has for suffix the first number in the suffix of F. If there is only one number in the suffix of F, then we take this number as the suffix of the absent coordinate.

CONTRIBUTION À LA THÉORIE DE LA CHUTE DES CORPS, EN AYANT ÉGARD À LA ROTATION DE LA TERRE

PAR ALFRED DENIZOT.

C'est Newton qui a découvert la déviation vers l'est d'un corps tombant sur la terre. Dans la théorie du mouvement relatif on ramène ce phénomène à l'effet d'une force fictive, à laquelle on a donné le nom de la force centrifuge composée. Le calcul donne encore une déviation du corps vers le sud qui se présente comme l'effet de la même force fictive, mais qui est trop petite pour être confirmée par l'expérience.

À d'autres occasions, spécialement dans une séance de la "Berliner mathematische Gesellschaft*," j'ai montré qu'on a négligé dans les équations générales quelques termes contenants le carré de la vitesse angulaire de la terre. Ces termes sont l'expression d'une autre force fictive que j'ai nommée la force centrifuge instantanée. Surtout j'ai appelé l'attention à ce que la force centrifuge composée ne peut pas expliquer la rotation apparente du plan d'oscillation d'un pendule, tandis que la force centrifuge instantanée peut servir pour donner une telle explication.

Aussi dans le problème de la chute libre d'un corps les termes, négligés totalement jusqu'à présent, conduisent à des résultats rémarquables. En conservant ces termes dans les équations différentielles du mouvement, leur intégration s'effectue d'une manière très simple.

Prenons d'abord les équations qui se rapportent au système suivant des coordinées rectangulaires: L'origine O soit un point de la verticale du lieu sous la latitude ϕ, l'axe $O\Xi$ perpendiculaire à l'axe de rotation de la terre, pris positif dans la direction opposée au centre du parallèle, l'axe OZ parallèle à l'axe de la terre dans un sens convenablement choisi. Les deux axes sont alors situés dans le plan méridien; l'axe OH sera dirigé vers l'est, dans la direction normale à ce plan.

En signifiant par g la pesanteur et par ω la vitesse angulaire de la terre, on reçoit par rapport au système mentionné les équations suivantes

$$\left.\begin{aligned}\ddot{\xi} &= -g\cos\phi + \omega^2\xi + 2\omega\dot{\eta}\\ \ddot{\eta} &= \qquad\qquad\quad \omega^2\eta - 2\omega\dot{\xi}\\ \ddot{\zeta} &= -g\sin\phi\end{aligned}\right\} \quad \ldots\ldots\ldots\ldots\ldots\ldots\ldots(1).$$

* *Sitzungsberichte*, 1906, p. 78 (aussi v. *Bulletin de l'Acad. d. Sciences de Cracovie*, 1904, p. 449; *Annalen der Physik*, 1905, p. 299; *Physikalische Zeitschrift*, VI. 1905, p. 342, VII. 1906, p. 507).

Les intégrales générales de ces équations différentielles donnent les expressions

$$\left.\begin{aligned}\xi &= (At + C)\sin \omega t - (Bt + D)\cos \omega t + \frac{g}{\omega^2}\cos \phi \\ \eta &= (At + C)\cos \omega t + (Bt + D)\sin \omega t \\ \zeta &= \tfrac{1}{2}gt^2 \sin \phi + Et + F\end{aligned}\right\} \quad \ldots\ldots\ldots\ldots(2).$$

L'état du corps au commencement du mouvement détermine les constantes A, B, C, D, E, F. Dans notre cas, au moment $t = 0$, le point matériel doit coïncider avec l'origine O des coordinées, c'est-à-dire $\xi_0 = \eta_0 = \zeta_0 = 0$, et les vitesses relatives doivent être $\dot{\xi}_0 = \dot{\eta}_0 = \dot{\zeta}_0 = 0$. Ces six conditions donnent les constantes. On trouve

$$\left.\begin{aligned}\xi &= \frac{g}{\omega^2}\cos \phi\,(1 - \cos \omega t - \omega t \sin \omega t) \\ \eta &= \frac{g}{\omega^2}\cos \phi\,(\sin \omega t - \omega t \cos \omega t) \\ \zeta &= -\tfrac{1}{2}gt^2 \sin \phi\end{aligned}\right\} \quad \ldots\ldots\ldots\ldots\ldots\ldots(3).$$

Si nous posons $\xi - \frac{g}{\omega^2}\cos \phi = \xi^1$, nous aurons au lieu de ξ

$$\xi^1 = \frac{g}{\omega^2}\cos \phi\,(\cos \omega t + \omega t \sin \omega t),$$

et nous voyons que la trajectoire du point dans le plan ΞOH est la développante du cercle, dont le rayon est égal à $\frac{g}{\omega^2}\cos \phi$.

En se servant des séries pour $\sin \omega t$ et $\cos \omega t$, on trouve ensuite

$$\left.\begin{aligned}\xi &= -\tfrac{1}{2}gt^2 \cos \phi - \tfrac{1}{8}g\omega^2 t^4 \cos \phi \\ \eta &= \tfrac{1}{3}g\omega t^3 \cos \phi \\ \zeta &= -\tfrac{1}{2}gt^2 \sin \phi\end{aligned}\right\} \quad \ldots\ldots\ldots\ldots\ldots\ldots\ldots\ldots(4).$$

Ordinairement on rapporte les équations différentielles à un système des coordonnées rectangulaires $O(X, Y, Z)$, dont deux axes sont situés dans le plan horizontal et le troisième axe coïncide avec la verticale du lieu. Nous choisissons le système suivant: L'origine des coordonnées coïncide avec l'origine précédente, de même l'axe OY avec l'axe OH. Les axes OX et OZ se trouvent donc dans le plan méridien: l'axe des x est la trace du plan horizontal mené par l'origine O et dirigé vers le nord, l'axe des z coïncide avec la verticale.

Les nouveaux axes sont liés aux précédents par les relations suivantes

$$\left.\begin{aligned}\xi &= -x \sin \phi - z \cos \phi \\ \eta &= y \\ \zeta &= x \cos \phi - z \sin \phi\end{aligned}\right\} \quad \ldots\ldots\ldots\ldots\ldots\ldots\ldots\ldots\ldots(5).$$

À l'aide de ces relations nous transformons les équations (1) et nous obtenons

$$\left.\begin{aligned}\ddot{x} &= \omega^2(x \sin \phi + z \cos \phi)\sin \phi - 2\omega\dot{y}\sin \phi \\ \ddot{y} &= \omega^2 y + 2\omega(\dot{x}\sin \phi + \dot{z}\cos \phi) \\ \ddot{z} &= g + \omega^2(x \sin \phi + z \cos \phi)\cos \phi - 2\omega\dot{y}\cos \phi\end{aligned}\right\} \quad \ldots\ldots(6).$$

Si nous voulons déterminer les intégrales de ces équations, il nous faut à l'aide de (5) transformer les valeurs (4) de ξ, η, ζ en x, y, z. De cette manière nous aurons

$$\left.\begin{aligned} x &= -\tfrac{1}{8}g\omega^2t^4 \sin\phi\cos\phi \\ y &= \tfrac{1}{3}\, g\omega t^3 \cos\phi \\ z &= \tfrac{1}{2}gt^2 - \tfrac{1}{8}g\omega^2t^4\cos^2\phi \end{aligned}\right\} \quad \ldots\ldots\ldots\ldots\ldots\ldots\ldots\ldots (7).$$

Comme nous voyons il y a une déviation (y) vers l'est et une autre (x) vers le sud. La valeur pour la déviation vers l'est est la même que donne la théorie adoptée jusqu'à présent; mais cette dernière donne pour la déviation vers le sud la valeur $\frac{1}{6}g\omega^2t^4 \sin\phi\cos\phi$, donc la valeur trouvée auparavant est plus petite de

$$\tfrac{1}{24}g\omega^2t^4 \sin\phi\cos\phi.$$

Cette différence, il est vrai, ne joue aucun rôle quant à l'expérience, parce que les erreurs des observations la surpassent, néanmoins elle présente quelque intérêt quant à la question, dans quel dégré les deux forces fictives contribuent au phénomène.

Supposons d'abord que le temps de l'observation soit si court que l'influence de ces forces fictives reste insensible, nous aurons les équations du mouvement

$$\ddot{x} = 0,\ \ddot{y} = 0,\ \ddot{z} = g,$$

dont les intégrales seront

$$x = 0,\ y = 0,\ z = \tfrac{1}{2}gt^2 \ \ldots\ldots\ldots\ldots\ldots\ldots\ldots\ldots\ldots (a).$$

Si la chute du corps dure plus longtemps, nous aurons des influences des deux forces fictives, mais nous voulons d'abord supposer que ce ne soit que la force centrifuge composée qui se développe et que la force centrifuge instantanée reste encore sans effet. En négligeant alors cette dernière, nous nous servons d'un procédé qu'on a employé jusqu'à présent.

Donc nous aurons les équations

$$\begin{aligned} \ddot{x} &= -\ \ 2\omega\dot{y}\sin\phi \\ \ddot{y} &= \ \ 2\omega(\dot{x}\sin\phi + \dot{z}\cos\phi) \\ \ddot{z} &= g + 2\omega\dot{y}\cos\phi. \end{aligned}$$

On reçoit par l'intégration les expressions

$$\left.\begin{aligned} x &= -\tfrac{1}{6}g\omega^2t^4 \sin\phi\cos\phi \\ y &= \ \ \tfrac{1}{3}g\omega t^3\cos\phi \\ z &= \ \ \tfrac{1}{2}gt^2 - \tfrac{1}{6}g\omega^2t^4\cos^2\phi \end{aligned}\right\} \quad \ldots\ldots\ldots\ldots\ldots\ldots\ldots\ldots (b).$$

C'est le même résultat, auquel Gauss est parvenu le premier. Pour la déviation vers l'est nous avons obtenu la valeur déjà indiquée sous (7); donc on peut considérer la déviation vers l'est comme l'effet produit seulement par la force centrifuge composée.

Enfin nous voulons admettre que dans les équations générales (6) on puisse supprimer la force centrifuge composée et ne conserver que des termes représentants la force centrifuge instantanée. Dans ce cas on a les équations suivantes

$$\begin{aligned} \ddot{x} &= \ \ \omega^2(x\sin\phi + z\cos\phi)\sin\phi \\ \ddot{y} &= \ \ \omega^2 y \\ \ddot{z} &= g + \omega^2(x\sin\phi + z\cos\phi)\cos\phi. \end{aligned}$$

Nous trouvons

$$\left.\begin{array}{l} x = \frac{1}{24} g\omega^2 t^4 \sin\phi \cos\phi \\ y = 0 \\ z = \frac{1}{2} g t^2 + \frac{1}{24} g\omega^2 t^4 \cos^2\phi \end{array}\right\} \ldots\ldots\ldots\ldots\ldots\ldots\ldots\ldots (c).$$

Maintenant nous avons reçu une déviation du corps vers le nord, mais aucune vers l'est.

En faisant alors la somme des x, y, sous (a), (b), (c), nous obtenons les expressions déjà indiquées sous (7), reçues par nous comme résultat de l'intégration des équations générales (6).

En étudiant le phénomène de Newton nous sommes parvenus au résultat suivant:

La déviation vers l'est est l'effet de la force centrifuge composée, la déviation vers le sud est une superposition des effets de la force centrifuge composée et de la force centrifuge instantanée.

ÜBER ASYMPTOTISCHE INTEGRATION VON DIFFERENTIAL-GLEICHUNGEN MIT ANWENDUNG AUF DIE BERECHNUNG VON SPANNUNGEN IN KUGELSCHALEN

VON OTTO BLUMENTHAL.

Das mechanische Problem, das ich behandeln will, ist die Bestimmung des elastischen Spannungszustandes einer Kugelkalotte unter der Navierschen Annahme, dass bei der elastischen Verzerrung Normalen zur neutralen Faser geradlinig und normal bleiben. Der Nachweis der Zulässigkeit dieser Annahme bei kleinen Wandstärken und der Ansatz des Problems ist in den allgemeinen Formeln enthalten, die Love für die elastischen Deformationen dünner Schalen gegeben hat*. Reissner hat kürzlich gezeigt†, dass die vollständige Lösung dieses Problems des Gleichgewichtes dünner Kugelschalen im Falle rotationssymmetrischer Belastung auf die Integration einer gewöhnlichen linearen Differentialgleichung 4. Ordnung führt. Ich werde mich hier mit der Behandlung dieser Gleichung 4. Ordnung beschäftigen, denn sie ist nach den gewöhnlichen Integrationsmethoden durch Reihenentwicklung praktisch nicht durchführbar; man muss vielmehr andere Methoden heranziehen, die ich als *asymptotische Integration* bezeichnet habe.

1. Es handelt sich um folgende Aufgabe und folgende zu berechnende Grössen. Eine Kugelkalotte von dem Radius a und der Dicke $2h$ mit dem Öffnungswinkel θ_0 erfahre keine äusseren Kräfte, nur an dem freien Rande θ_0 mögen über den ganzen Rand gleichmässig verteilte Spannungen und Biegungsmomente angreifen, unter deren Einfluss die Schale im Gleichgewicht ist. Es soll der Spannungs- und Dehnungszustand in der ganzen Kalotte berechnet werden. Die in Betracht kommenden Komponenten der Spannungen und Verrückungen sind folgende:

Normalspannungen T_1, T_2; eine Schubspannung N; Biegungsmomente G_1, G_2; Verschiebungen u, w. (Siehe Figur auf der folgenden Seite.)

Die Lösung des Problems wird zurückgeführt auf die Integration einer Differentialgleichung für N:

$$\left.\begin{aligned} &\sin^4\theta\, N'''' + 2\sin^3\theta\cos\theta\, N''' - \sin^2\theta\,(3-\sin^2\theta)\, N'' \\ &\qquad + \sin\theta\cos\theta\,(3+2\sin^2\theta)\, N' + (b^4\sin^4\theta - 3)\, N = 0 \\ &b^4 = (1-\sigma^2)\left(1+\frac{3a^2}{h^2}\right), \qquad \sigma = \text{Poissonsches Verhältnis} \end{aligned}\right\} \quad (A).$$

* Love, *Treatise on Elasticity*, 2nd ed., p. 488—510.

† *Festschrift Müller-Breslau* (Leipzig, Kröner, 1912), S. 181—193.

Aus N berechnen sich dann die übrigen Grössen wie folgt:

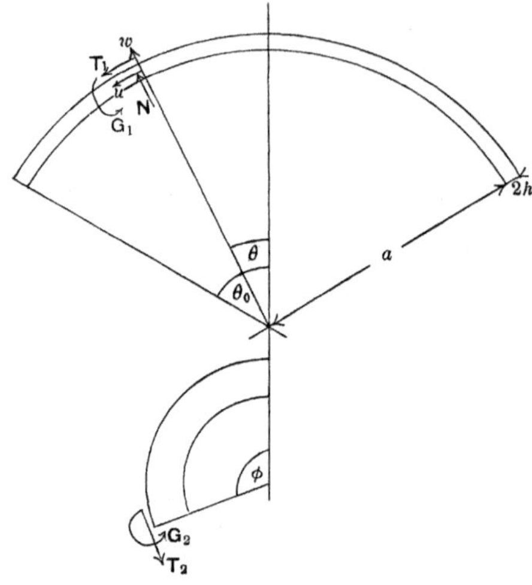

$$T_1 = N \cot\theta, \qquad T_2 = N',$$

$$G_1 = -\frac{h^2}{3a(1-\sigma)}\left[\frac{N'''}{1+\sigma} + \cot\theta\, N'' - \left(\frac{2-\sigma}{1+\sigma}\cot^2\theta + \frac{1-\sigma}{1+\sigma}\right)N' + \left(\sigma + \frac{2-\sigma}{1+\sigma}\frac{1}{\sin^2\theta}\right)\cot\theta\, N\right],$$

$$G_2 = -\frac{h^2}{3a(1-\sigma)}\left[\sigma\frac{N'''}{1+\sigma} + \cot\theta\, N'' + \left(\frac{1-2\sigma}{1+\sigma}\cot^2\theta - \sigma\frac{1-\sigma}{1+\sigma}\right)N' + \left(1 - \frac{1-2\sigma}{1+\sigma}\frac{1}{\sin^2\theta}\right)\cot\theta\, N\right],$$

$$u = \frac{1}{2}\frac{a}{h}\frac{1+\sigma}{E}\sin\theta\,(B+\mathfrak{J}),$$

$$w = -\frac{1}{2}\frac{a}{h}\frac{1}{E}[\sigma\cot\theta\, N - N' + (1+\sigma)\cos\theta\,(B+\mathfrak{J})], \qquad \text{.........(A')}$$

B = willkürliche Konstante,

E = Elastizitätsmodul,

$$\mathfrak{J} = \int\frac{1}{\sin\theta}(N\cot\theta - N')\,d\theta.$$

Die Abzählung der Integrationskonstanten ergiebt in einfacher Weise die geometrischen und dynamischen Randbedingungen, unter denen Gleichgewicht möglich ist. Durch die Festsetzung, dass Spannungen und Biegungsmomente im Punkte $\theta = 0$ der Kugel endlich bleiben sollen, werden 2 Integrationskonstanten absorbiert, wie die Reihenentwicklung der Integrale der Gleichung (A) in der Umgebung von $\theta = 0$ zeigt. Es sind dann noch 3 Konstanten übrig, von denen eine nur in die Verschiebungen, nicht in die Kräfte eingeht. Dies entspricht der Tatsache, dass der ganzen Schale noch eine Translation erteilt werden kann. Abgesehen von dieser Willkürlichkeit sind noch zwei Bedingungen am Rande vorgebbar, die sich entweder auf die angreifenden Kräfte und Momente, oder auf die geometrische Gestalt des Randes beziehen können. Ein einfaches Bedingungssystem ist z. B., dass der Rand etwas auseinander gedehnt ist, und kein Moment an ihm herrschen soll. Es soll aber hier auf spezielle Probleme nicht eingegangen werden.

2. Die Schwierigkeit besteht nun darin, dass die Gleichung (A) für N einen nach Voraussetzung sehr grossen Parameter b^4 enthält. Da nämlich bei der ganzen Theorie die Kugelschale als sehr dünn vorausgesetzt werden muss, ist $\frac{a}{h}$ notwendig eine grosse Zahl und demgemäss b^4 gross. Dies hat zur Folge, dass die üblichen Reihenentwicklungen nach Potenzen von θ (oder $\sin\theta$) nur in der nächsten Umgebung des Nullpunktes mit praktisch brauchbarer Schnelligkeit konvergieren werden, während in einigermassen grösserer Entfernung $\left(\text{z. B. } \theta = \frac{1}{\sqrt{b}}\right)$ infolge des grossen Koeffizienten b^4 die Glieder der Reihe zunächst sogar stark anwachsen werden, um erst nach einer mit wachsendem b gegen Unendlich gehenden Zahl von Gliedern abzunehmen. Diese Reihen sind also zur Übersicht über den Verlauf der Erscheinung gänzlich ungeeignet. *Hier greift die asymptotische Integration ein.*

Die asymptotische Integration setzt sich zur Aufgabe, gerade die grossen Werte des Parameters zu berücksichtigen, indem sie die Integrale nach absteigenden Potenzen des Parameters entwickelt. Das Verfahren wird wohl am übersichtlichsten und für die Rechnung am bequemsten so dargestellt*.

Durch die Substitution $N = \frac{M}{\sqrt{\sin\theta}}$ wird zuerst das Glied mit dem 3. Differentialquotienten weggeschafft und für M folgende Gleichung erhalten:

$$\left.\begin{array}{l} M'''' + a_2 M'' + a_1 M' + (b^4 + a_0) M = 0, \\ a_2 = -\frac{3}{2}\frac{1}{\sin^2\theta} + \frac{5}{2}, \quad a_1 = 3\frac{\cos\theta}{\sin^3\theta}, \quad a_0 = -\frac{63}{16}\frac{1}{\sin^4\theta} + \frac{9}{8}\frac{1}{\sin^2\theta} + \frac{9}{16} \end{array}\right\} \ldots\text{(B)}.$$

Wächst nun bei festem $\theta > 0$ der Parameter b beliebig an, so werden die Funktionen a_0, a_1, a_2 immer mehr gegen b^4 verschwinden. Man kann daraus schliessen, dass die Integrale M näherungsweise übereinstimmen mit den Lösungen der vereinfachten Gleichung

$$M'''' + b^4 M = 0 \ldots\ldots\ldots\ldots\text{(B')}.$$

Die Theorie der asymptotischen Integration hat die Aufgabe, diesen Schluss genau zu begründen und den Fehler der Näherung abzuschätzen.

3. Ich will die Fehlerabschätzung ausführlich geben, weil ich sie in hohem Masse werde benutzen müssen†. Seien

$$\lambda_1 = \frac{b}{\sqrt{2}}(1+i), \quad \lambda_2 = \frac{b}{\sqrt{2}}(1-i), \quad \lambda_3 = \frac{b}{\sqrt{2}}(-1+i), \quad \lambda_4 = \frac{b}{\sqrt{2}}(-1-i)$$

* Siehe auch meine Darstellung des Verfahrens in *Arch. Math. Phys.* (3) 19 (1912), S. 137 f.

† Ich habe diese Methode der Restabschätzung in *Arch. Math. Phys.* (3) 19 (1912), S. 141 ff., veröffentlicht und damals für neu gehalten. Seitdem bin ich durch Herrn Bôcher in freundlicher Weise darauf aufmerksam gemacht worden, dass dieselbe Methode bereits 1908 von Birkhoff, *Trans. Am. Math. Soc.* 9, angegeben worden ist.

die 4 Wurzeln der Gleichung $\lambda^4 + b^4 = 0$, und M_1 das Integral von (B), das durch $e^{\lambda_1\theta}$ asymptotisch dargestellt wird, schliesslich

$$\eta_1 = M_1 - e^{\lambda_1\theta} \quad \ldots\ldots (1).$$

Das Fehlerglied η_1 genügt der Beziehung

$$\eta_1'''' + b^4\eta_1 = -(a_2 M_1'' + a_1 M_1' + a_0 M_1),$$

und ein diese Gleichung befriedigender Ausdruck ist

$$\eta_1 = \frac{1}{4b^4}\left[\sum_1^4 \lambda_i e^{\lambda_i\theta}\int_{\theta_1}^{\theta} e^{-\lambda_i\theta}(a_2 M_1'' + a_1 M_1' + a_0 M_1)\, d\theta\right] \quad \ldots\ldots (2),$$

wo mit θ_1 die untere Grenze des Integrationsintervalls bezeichnet sei, das sich von dort nach θ_0 erstreckt. Mit Einsatz dieses Wertes in (1) findet sich

$$a_2 M_1'' + a_1 M_1' + a_0 M_1 = (\lambda_1^2 a_1 + \lambda_1 a_2 + a_0)\, e^{\lambda_1\theta}$$
$$+ \frac{1}{4b^4}\sum_1^4 (\lambda_i^3 a_2 + \lambda_i^2 a_1 + \lambda_i a_0)\, e^{\lambda_i\theta}\int_{\theta_1}^{\theta} e^{-\lambda_i\theta}(a_2 M_1'' + a_1 M_1' + a_0 M_1)\, d\theta \quad \ldots (2').$$

Aus dieser Gleichung finden wir nach einem auf Liouville zurückgehenden Verfahren zunächst eine obere Grenze für $|a_2 M_1'' + a_1 M_1' + a_0 M_1|\, e^{-\lambda_1\theta}$. Dazu gebrauchen wir eine Abschätzung der Integrale der rechten Seite, die ich sofort in einer für die praktischen Zwecke besonders geeigneten allgemeinen Form geben will. Sei $\phi(\theta)$ eine im Intervall (θ_1, θ_0) nirgends verschwindende, positive, stetige Funktion. Dann ist

$$\int_{\theta_1}^{\theta} e^{-\lambda_i\theta}(a_2 M_1'' + a_1 M_1' + a_0 M_1)\, d\theta = \int_{\theta_1}^{\theta} e^{(\lambda_1-\lambda_i)\theta}\left(\frac{a_2 M_1'' + a_1 M_1' + a_0 M_1}{\phi(\theta)}\, e^{-\lambda_1\theta}\right)\phi(\theta)\, d\theta,$$

und daher, da $e^{(\lambda_1-\lambda_i)\theta}$ nirgends in dem Intervall grösseren absoluten Betrag hat als am Punkte θ,

$$\left|\int_{\theta_1}^{\theta} e^{-\lambda_i\theta}(a_2 M_1'' + a_1 M_1' + a_0 M_1)\, d\theta\right|$$
$$< \left|e^{(\lambda_1-\lambda_i)\theta}\right| \max_{(\theta_1, \theta)}\left|\frac{a_2 M_1'' + a_1 M_1' + a_0 M_1}{\phi(\theta)}\, e^{-\lambda_1\theta}\right|\int_{\theta_1}^{\theta}\phi(\theta)\, d\theta.$$

Daraus folgt aber, bei Einsatz in (2'),

$$\max_{(\theta_1, \theta_0)}\left|\frac{a_2 M_1'' + a_1 M_1' + a_0 M_1}{\phi(\theta)}\, e^{-\lambda_1\theta}\right| < \frac{\max\limits_{(\theta_1, \theta_0)}\left|\dfrac{\lambda_1^2 a_2 + \lambda_1 a_1 + a_0}{\phi(\theta)}\right|}{1 - \dfrac{1}{4b^4}\max\limits_{(\theta_1, \theta_0)}\sum_1^4\left|\dfrac{\lambda_i(\lambda_i^2 a_2 + \lambda_i a_1 + a_0)}{\phi(\theta)}\right|\displaystyle\int_{\theta_1}^{\theta_0}\phi(\theta)\, d\theta}$$

$$< \frac{\max\limits_{(\theta_1, \theta_0)}\dfrac{\frac{b^2}{2}|a_2| + \frac{b}{\sqrt{2}}|a_1| + |a_0|}{\phi(\theta)}}{1 - \dfrac{1}{\sqrt{2}b^3}\max\limits_{(\theta_1, \theta_0)}\dfrac{\frac{b^2}{2}|a_2| + \frac{b}{\sqrt{2}}|a_1| + |a_0|}{\phi(\theta)}\displaystyle\int_{\theta_1}^{\theta_0}\phi(\theta)\, d\theta},$$

und dieselbe Ungleichung gilt auch für M_2, M_3, M_4. Setzen wir die gefundene obere Grenze nochmals in (2) ein, so erhalten wir für die Reste η folgende Abschätzung:

$$\eta = e^{\lambda\theta}\frac{\Theta}{b^3},\quad \Theta < \frac{1}{\sqrt{2}}\int_{\theta_1}^{\theta_0}\phi(\theta)\,d\theta\,\frac{\max\limits_{(\theta_1,\,\theta_0)}\dfrac{\frac{b^2}{2}|a_2|+\frac{b}{\sqrt{2}}|a_1|+|a_0|}{\phi(\theta)}}{1-\dfrac{1}{\sqrt{2}b^3}\max\limits_{(\theta_1,\,\theta_0)}\dfrac{\frac{b^2}{2}|a_2|+\frac{b}{\sqrt{2}}|a_1|+|a_0|}{\phi(\theta)}\displaystyle\int_{\theta_1}^{\theta_0}\phi(\theta)\,d\theta} \quad\text{.........(3).}$$

Im Falle der Gleichung (2′) erhält man eine günstige Abschätzung von η, indem man $\phi(\theta) = \frac{1}{\sin^2\theta}$ setzt. Es ergiebt sich so mit unwesentlichen Vernachlässigungen:

$$\Theta < \frac{1}{\sqrt{2}}\cot\theta_1\frac{\dfrac{3b^2}{4}+\dfrac{3b}{\sqrt{2}}\cot\theta_1+\dfrac{63}{16}\dfrac{1}{\sin^2\theta_1}}{1-\dfrac{1}{\sqrt{2}b^3}\cot\theta_1\left(\dfrac{3b^2}{4}+\dfrac{3b}{\sqrt{2}}\cot\theta_1+\dfrac{63}{16}\dfrac{1}{\sin^2\theta_1}\right)} \quad\text{.........(3′).}$$

Die Ungleichung (3) besagt, dass η von der Ordnung $\left|\frac{e^{\lambda\theta}}{b}\right|$ ist. Aus der Ungleichung (3′) schliesst man spezieller, dass der Rest mit wachsendem b gegen das Hauptglied verschwindet, wenn man innerhalb solcher Intervalle (θ_1, θ_0) bleibt, für die $b\sin\theta$ mit b unendlich wird. Dasselbe gilt für die Differentialquotienten aller Ordnungen.

4. Es lassen sich also mit relativ verschwindendem Fehler 4 Integrale der Gleichung (A) in folgender reeller Form ansetzen:

$$\left.\begin{aligned} N_1 &= \frac{1}{\sqrt{\sin\theta}}e^{\frac{b}{\sqrt{2}}\theta}\cos\frac{b}{\sqrt{2}}\theta, & N_2 &= \frac{1}{\sqrt{\sin\theta}}e^{\frac{b}{\sqrt{2}}\theta}\sin\frac{b}{\sqrt{2}}\theta,\\ N_3 &= \frac{1}{\sqrt{\sin\theta}}e^{-\frac{b}{\sqrt{2}}\theta}\cos\frac{b}{\sqrt{2}}\theta, & N_4 &= \frac{1}{\sqrt{\sin\theta}}e^{-\frac{b}{\sqrt{2}}\theta}\sin\frac{b}{\sqrt{2}}\theta \end{aligned}\right\}\quad\text{...(4).}$$

Jetzt aber kommt der schwierigste Punkt. Es handelt sich darum, wie sich das den Endlichkeitsbedingungen für $\theta = 0$ und Randbedingungen für $\theta = \theta_0$ genügende Integral von (A) aus diesen partikulären Integralen zusammensetzt. Die Randbedingungen machen keine Schwierigkeit, da ja für $\theta = \theta_0$ die asymptotische Darstellung gültig ist. Dagegen ist es unmöglich, auf direktem Wege zu entscheiden, welche der Integrale (4) die *Endlichkeitsbedingungen* am Punkte $\theta = 0$ befriedigen: denn an diesem Punkte versagt die asymptotische Darstellung augenscheinlich, denn der Rest (3) wird dort grösser als das Hauptglied. Man kann aber die Entscheidung zunächst auf unstrengem Wege aus der physikalischen Anschauung ableiten, indem man als Tatsache annimmt, dass die Spannungen N von dem Rande nach dem Pole $\theta = 0$ abklingen. Die einzigen Funktionen (4), die diese Eigenschaft haben*, sind die Funktionen N_1 und N_2, während N_3 und N_4 mit wachsender Entfernung von dem

* Wir betrachten dabei nur solche Intervalle (θ, θ_0), die in genügender Entfernung von $\theta = 0$ bleiben, sodass das langsame Abnehmen des Nenners $\sqrt{\sin\theta}$ nicht in Betracht kommt.

freien Rande immer grösser werden. Unser Integral der Gleichung (A) wird also asymptotisch als

$$N = A_1 N_1 + A_2 N_2 \ldots\ldots\ldots\ldots\ldots\ldots\ldots\ldots\ldots\ldots(5)$$

anzusetzen sein, und die Konstanten A_1 und A_2 werden dann zur Erfüllung zweier Bedingungen am freien Rande gerade ausreichen.

Die Kräfte, und ebenso, nach den Gleichungen (A′), die Biegungsmomente und Verrückungen, stellen sich daher als sehr schnelle Schwingungen dar, die von dem freien Rande nach dem Pol exponentiell sehr rasch abklingen. Das logarithmische Dekrement auf die Bogeneinheit ist nämlich $\frac{b}{\sqrt{2}}$. Dieses übersichtliche Resultat hätte mit Hülfe der gewöhnlichen Integrationsmethoden durch Potenzreihen nicht gewonnen werden können.

5. Es bleibt übrig, streng nachzuweisen, dass der Ansatz (5) richtig ist, d.h. dass die beiden Integrale N_3 und N_4 am Punkte $\theta = 0$ die Endlichkeitsbedingungen nicht befriedigen. Eine Aufgabe dieser Art findet sich wohl in den meisten praktisch wichtigen Problemen asymptotischer Integration. Sie lässt sich allgemein so formulieren: ein exaktes Integral einer Differentialgleichung ist festgelegt durch Bedingungen an einem Punkte, an dem die asymptotische Darstellung versagt; trotzdem sollen auf einem Umweg die Integrationskonstanten der asymptotischen Darstellung bestimmt werden. Allgemein zu reden, ist das Problem höchst schwierig, und ich sehe keine allgemeine Methode, es zu behandeln. Ein in manchen praktischen Fällen anwendbares Verfahren habe ich an dem Beispiel der Legendreschen Polynome in meiner citierten Arbeit durchgeführt*. Dieses Verfahren ist aber bei unserem jetzigen Beispiele nicht anwendbar. Wohl aber kann ich die Frage auch hier exakt entscheiden, wenn auch auf einem nicht eleganten Wege.

Ich gehe dazu aus von der Darstellung (3′) des Restes. Man sieht leicht, dass für $\theta_1 \geqq \frac{3}{b}$ der Rest kleiner als die Hälfte des Hauptgliedes ist. Andererseits lassen sich die am Punkte $\theta = 0$ den Endlichkeitsbedingungen genügenden Integrale für solche Werte von θ, die etwa kleiner sind als $\frac{12}{b}$, ohne Schwierigkeiten numerisch berechnen. *Ich werde zeigen, dass man durch numerische Berechnung der Werte eines den Endlichkeitsbedingungen genügenden Integrals für* 4 *Punkte des Bereiches*

$$\frac{3}{b} < \theta < \frac{12}{b}$$

beweisen kann, dass die den Endlichkeitsbedingungen genügenden Integrale notwendig die Form (5) *haben müssen.*

In der Tat haben *alle Integrale* (5) *und nur diese* folgende charakteristische Eigenschaft.

Ersetzen wir die asymptotische Darstellung (5) durch Zufügung der Reste durch eine exakte, so lautet diese, wie leichte Rechnung zeigt, für die Funktion $M = \sqrt{\sin\theta}\, N$

$$M = e^{\frac{1}{\sqrt{2}}b\theta}\left[\left(A_1 \cos\frac{1}{\sqrt{2}}b\theta + A_2 \sin\frac{1}{\sqrt{2}}b\theta\right) + \left(a_1(\theta)\cos\frac{1}{\sqrt{2}}b\theta + a_2(\theta)\sin\frac{1}{\sqrt{2}}b\theta\right)\right],$$

* l.c. S. 150—163.

wobei, mit Einführung der in (3) definierten Restgrösse $\frac{|\Theta|}{b^3} = \tau(\theta)$

$$a_1(\theta)^2 + a_2(\theta)^2 = \tau(\theta)^2 (A_1^2 + A_2^2)$$

ist. Wir betrachten die zwei Werte $\theta_1' = \pi \frac{\sqrt{2}}{b}$ und $\theta_1'' = \frac{3\pi}{2}\frac{\sqrt{2}}{b}$. Dann ist

$$\begin{aligned} M(\theta_1')^2 + e^{-\pi} M(\theta_1'')^2 &= [(A_1 + a_1(\theta_1'))^2 + (A_2 + a_2(\theta_1''))^2]\, e^{\sqrt{2}b\theta_1'} \\ &= (A_1^2 + A_2^2)\left(1 - \epsilon\sqrt{\frac{a_1(\theta_1')^2 + a_2(\theta_1'')^2}{A_1^2 + A_2^2}}\right)^2 e^{\sqrt{2}b\theta_1'} \\ &= (A_1^2 + A_2^2)(1 - \sqrt{2}\tau_1)^2 e^{\sqrt{2}b\theta_1'}, \end{aligned}$$

wo ϵ eine Zahl vom absoluten Betrage < 1 bezeichnet und τ_1 eine Grösse, deren absoluter Betrag den grösseren der beiden Werte $\tau(\theta_1')$, $\tau(\theta_1'')$ nicht übersteigt. Machen wir dieselbe Rechnung für $\theta_2' = 2\pi\frac{\sqrt{2}}{b}$, $\theta_2'' = \frac{5\pi}{2}\frac{\sqrt{2}}{b}$ und dividieren, so folgt

$$\frac{M(\theta_2')^2 + e^{-\pi} M(\theta_2'')^2}{M(\theta_1')^2 + e^{-\pi} M(\theta_1'')^2} = e^{\sqrt{2}\pi}\frac{(1 - \sqrt{2}\tau_2)^2}{(1 - \sqrt{2}\tau_1)^2} > \left(e^{\frac{\pi}{\sqrt{2}}}\frac{1 - \sqrt{2}\tau}{1 + \sqrt{2}\tau}\right)^2 \quad \text{.........(6)},$$

wo τ eine obere Grenze für die beiden Zahlen τ_1 und τ_2 bezeichnet. Alle 4 Punkte θ_1', θ_1'', θ_2', θ_2'' liegen in dem Intervall $\left(\frac{3}{b}, \frac{12}{b}\right)$. *Demnach ist aber $\tau < \frac{1}{2}$ und daher der betrachtete Quotient grösser als* 1. Dies lässt sich rechnerisch für die beiden den Endlichkeitsbedingungen genügenden Integrale N verifizieren, und damit beweisen, dass diese in der Form (5) darstellbar sind.

Die genauere Ausführung ist folgende: Vor allem ist zu bemerken, dass der Nachweis, dass die den Endlichkeitsbedingungen genügenden Integrale die Ungleichung (6) befriedigen und daher nur aus N_1 und N_2 zusammengesetzt sind, aber N_3 und N_4 nicht enthalten, *nur für genügend grosse Werte von b* geführt werden muss. Aus den Entwicklungen des § 3 meiner citierten Arbeit folgt nämlich, dass die Koeffizienten der Darstellung dieser Integrale durch die asymptotischen Fundamentalintegrale *analytische Funktionen* von b sind. Sind daher die Koeffizienten von N_3 und N_4 für genügend grosse Werte von b Null, so sind sie es für alle Werte.

Nun schreiten die einfachsten Entwicklungen für die den Endlichkeitsbedingungen genügenden Integrale N—aus denen sich ja die entsprechenden M sofort ergeben—nach ungeraden Potenzen von $\sin\theta$ fort. Die Rekursion für die Koeffizienten schreibt sich am einfachsten in folgender Form. Ich setze zuerst $N = \Sigma c_n \sin^n\theta$, dann $c_n \sin^n\theta = d_n$, sodass $N = \Sigma d_n$ wird, schliesslich $\sin\theta = a/b$. Dann findet sich

$$d_n = 2\frac{(n-2)(n-3)\,a^2}{(n+1)(n-1)\,b^2}d_{n-2} - \left[\frac{(n-2)(n-4)(n-5)\,a^4}{(n+1)(n-1)^2\,b^4} + \frac{a^4}{(n+1)(n-1)^2(n-3)}\right]d_{n-4}.$$

Nun brauchen wir die Werte der Integrale N nur mit sehr geringer Genauigkeit (vielleicht 10°/$_\circ$ Fehler). Daher können wir zunächst bei genügend grossem b den $\sin\theta$ durch θ ersetzen und haben also N für die vier Werte $a = \pi\sqrt{2}$, $3\pi/\sqrt{2}$, $2\pi\sqrt{2}$, $5\pi/\sqrt{2}$ zu berechnen, die alle zwischen 3 und 12 liegen. Aus den Rekursionsformeln lassen sich dann mit Leichtigkeit für genügend grosses b Näherungswerte für die ersten d_n ableiten. Die rechte Seite dieser Formeln reduziert sich nämlich praktisch auf das zweite Glied der Klammer. Man findet, dass die d_n im allgemeinen infolge von $a > 3$ zuerst zunehmen, dann aber rasch abnehmen. Die Summe N berechnet sich leicht mit genügender Genauigkeit aus wenigen (höchstens 7) Gliedern.

6. Die asymptotische Darstellung ist nicht, wie wir bisher getan haben, auf das erste Glied beschränkt, sondern lässt sich zu beliebiger Gliederzahl fortsetzen. Wir kommen allgemein für die Integrale M der Gleichung (B) zu folgender asymptotischen Näherung:

$$M_i = e^{\lambda_i \theta}\left(1 + \frac{1}{b} f_{i1}(\theta) + \frac{1}{b^2} f_{i2}(\theta) + \ldots + \frac{1}{b^n} f_{in}(\theta)\right) \quad \ldots\ldots\ldots\ldots(7).$$

Die Funktionen f werden gefunden, indem man den Ausdruck (7) in (B) einsetzt und die Koeffizienten aller Potenzen von b bis einschliesslich $\frac{1}{b^{n-3}}$ Null setzt. Dies ergiebt die folgenden n Gleichungen, wobei zur Abkürzung $\epsilon_i = \lambda_i / b$ eingeführt ist:

$$\begin{aligned} 4\epsilon_i^3 f'_{i,\,\kappa+3} + \epsilon_i^2 (6f''_{i,\,\kappa+2} + a_2 f_{i,\,\kappa+2}) + \epsilon_i (4f'''_{i,\,\kappa+1} + 2a_2 f'_{i,\,\kappa+1} + a_1 f_{i,\,\kappa+1}) \\ + (f''''_{i,\,\kappa} + a_2 f''_{i,\,\kappa} + a_1 f'_{i,\,\kappa} + a_0 f_{i,\,\kappa}) = 0 \quad (\kappa = 1, 2, \ldots, n-3) \ \ldots\ldots(8). \end{aligned}$$

Aus diesen berechnen sich die Funktionen f der Reihe nach durch Quadraturen Für unsere Gleichung ergiebt sich z. B.

$$f_{i1} = -\frac{1}{4\epsilon_i}\left(\frac{3}{2}\cot\theta + \frac{5}{2}\theta\right),$$

und auch alle weiteren Funktionen f_{in} lassen sich in geschlossener Form anschreiben.

Die Reste η_{in} der Ausdrücke (7) befriedigen die inhomogene Differentialgleichung

$$\eta''''_{in} + a_2 \eta''_{in} + a_1 \eta'_{in} + (b^4 + a_0)\,\eta_{in} = -\frac{e^{\lambda_i \theta}}{b^{n-2}} F_{in},$$

$$\begin{aligned} F_{in} = \{&(f''''_{i,\,n-2} + a_2 f''_{i,\,n-2} + a_1 f'_{i,\,n-2} + a_0 f_{i,\,n-2}) \\ &+ \epsilon_i (4f'''_{i,\,n-1} + 2a_2 f'_{i,\,n-1} + a_1 f_{i,\,n-1}) + \epsilon_i^2 (6f''_{i,\,n} + a_2 f_{i,\,n})\} \\ &+ \frac{1}{b}\{(f''''_{i,\,n-1} + a_2 f''_{i,\,n-1} + a_1 f'_{i,\,n-1} + a_0 f_{i,\,n-1}) + \epsilon_i (4f'''_{i,\,n} + 2a_2 f'_{i,\,n} + a_1 f_{i,\,n})\} \\ &+ \frac{1}{b^2}\{f''''_{i,\,n} + a_2 f''_{i,\,n} + a_1 f'_{i,\,n} + a_0 f_{i,\,n}\}. \end{aligned}$$

Nach der Methode der Variation der Konstanten folgt hieraus z. B. für η_{1n} die Darstellung

$$\eta_{1n} = -\frac{1}{b^{n-2}} \sum_1^4 M_i \int_{\theta_1}^{\theta} \mu_i e^{\lambda_1 \theta} F_{1n} d\theta \ \ldots\ldots\ldots\ldots\ldots\ldots\ldots(9),$$

wo μ_i bekannte Determinantenquotienten sind, in die die M_i nebst ihren drei ersten Ableitungen eingehen. Die Gleichung (3) oder die äquivalente

$$M_i = e^{\lambda_i \theta}\left(1 + \frac{\Theta_i}{b^3}\right) \ \ldots\ldots\ldots\ldots\ldots\ldots\ldots\ldots(3a)$$

zeigt, dass die μ_i von der Form $-\frac{\lambda_i}{4b^4} e^{-\lambda_i \theta}\left(1 + \frac{\sigma_i}{b}\right)$ sein müssen. Zu ihrer wirklichen

Berechnung—mit genauer Abschätzung der Fehlerglieder σ—aber sind die Determinantenausdrücke wenig geeignet. Man geht dazu vielmehr besser davon aus, dass die μ_i Integrale der adjungierten Differentialgleichung

$$\mu'''' + \frac{d^2}{d\theta^2}(a_2\mu) - \frac{d}{d\theta}(a_1\mu) + (b^4 + a_0)\,\mu = 0$$

sind*. Diese ist von derselben Form wie (B), und man erhält daher asymptotische Darstellung und Rest ihrer Integrale nach Formel (3). In unserem Falle ergiebt sich das besonders günstige Resultat, dass (B) mit ihrer adjungierten zusammenfällt. Daher ist bis auf konstante und für unsere Zwecke belanglose Faktoren der Form $1 + \frac{\tau_i}{b}$ einfach $\mu_i = -\frac{\lambda_i}{4b^4} M_{5-i}$. Damit aber wird augenscheinlich, nach den in Nr. 3 gebrauchten Methoden der Integralabschätzung,

$$|\eta_{in}| < \frac{1}{\sqrt{2}} \frac{|e^{\lambda_i\theta}|}{b^{n+1}} \left(1 + \frac{\Theta_{\max}}{b^3}\right)^2 \int_{\theta_1}^{\theta_0} |F_{in}|\, d\theta \;\ldots\ldots\ldots\ldots\ldots\ldots\ldots(10),$$

wo $\Theta_{\max}$ die in (3′) angegebene obere Grenze der $|\Theta_i|$ bedeutet.

Um möglichst gute asymptotische Näherung zu erhalten, hat man für jeden Wert von b und θ_1 dasjenige n zu bestimmen, das die kleinsten oberen Grenzen für die Reste η_{in} liefert.

* Schlesinger, *Handbuch der Differentialgleichungen* I, S. 52—76.

THE METHOD OF PERMANENT AND TRANSITORY MODES OF EQUILIBRIUM IN THE THEORY OF THIN ELASTIC BODIES

By J. Dougall.

1. The Mathematical Theory of Elastic Solids contains no chapters more fascinating or with greater claims to the attention of the physicist than those which deal with the approximate laws of strain in thin bodies subjected to given forces. The results are simple and should, one cannot but think, be susceptible of correspondingly simple rigorous proof, that is to say, formal deduction from the fundamental differential equations. But as a matter of history the processes by which they have been obtained are to a marked degree tentative and incomplete, and it has only been by the gradual elimination of errors and inconsistencies that they have come to take the forms in which they stand at present. Even now it may not be too much to say that one's belief in the broad correctness of these Approximate Theories rests not on any confidence in the reasoning by which they are attached to the general equations, but rather on the evidence of their agreement with direct experiment and with various exact solutions that have been given for special cases.

With a view to the improvement and development of the existing evidence of the latter sort I have worked out exact solutions of the general problems of equilibrium for certain bodies in analytical forms which make it possible to apply them immediately to the derivation of approximate results, and I propose to give here a brief statement of the nature of the method, illustrated by an example of its application to a simple problem.

The bodies for which I have worked out, or am at present engaged in working out, the solutions are the plate, circular cylinder, spherical shell and circular cone. In the present paper I refer only to the isotropic plate and circular cylinder*.

For the necessary leisure to prosecute the investigations, which are more tedious than might be expected, I have to express my great indebtedness to the Carnegie Research Scheme in connection with the Universities of Scotland.

2. The essential feature of the method is the determination of solutions for any point *source of strain* (i.e. for a force applied at a single point either in the interior or on the surface of a body), in terms of *fundamental solutions*, each of which is a solution of the body and surface equations of equilibrium under no forces.

* The following paper has already been published: "An Analytical Theory of the Equilibrium of an Isotropic Elastic Plate," *Edinburgh Royal Soc. Trans.* Vol. XLI, Part I (1904), pp. 129—228.

The existence of such solutions for the physical problem before us—they are common enough in other departments of physical mathematics, the theory of vibrations for example—seems at first sight to conflict with the well-known result that the only solutions of the elastic equations of equilibrium under no forces are the rigid body displacements. The exemption of the fundamental solutions spoken of here from this theorem has its origin in the fact that they possess *singularities*, at infinity in the case of the cylinder, at infinity or at some right line in the case of the plate.

In each case the fundamental solutions fall into two classes, consisting of what may be called *permanent* and *transitory* modes of equilibrium respectively. From one point of view the former are degenerate forms of the latter, but they are distinguished from these by the possession of certain important properties.

In a *circular cylinder*, for example, the *transitory modes* are defined in terms of harmonic functions of the form

$$e^{\alpha z} J_m(\alpha\rho) \cos m(\omega - \omega_0),$$

where the fundamental numbers α are real or complex, being the roots of a certain transcendental equation whose form depends on the value of the integer m.

The *permanent modes* correspond in a certain way to $\alpha = 0$, and involve only rational integral functions of the rectangular coordinates of a point. They are in fact the six modes of equilibrium discovered by Saint Venant, along with the six rigid body displacements.

A solution for any given source can be found which is expressed in two distinct analytical forms on the two sides of the source, so that only those transitory modes occur on either side which vanish at infinity on that side. Unless some condition at infinity is laid down, the coefficients of the permanent modes are to a certain extent arbitrary, but the *difference* of the coefficients with which each of the twelve permanent modes occurs on the two sides of the source is determinate.

In a *plate*, the *transitory modes* are defined in terms of harmonic functions of the forms

$$(e^{\kappa z}, e^{-\kappa z}) f(x, y),$$

where

$$\left(\frac{\partial^2}{\partial x^2} + \frac{\partial^2}{\partial y^2} + \kappa^2\right) f = 0,$$

and the fundamental numbers κ are roots of the equations

$$\sinh 2\kappa h \pm 2\kappa h = 0, \text{ and } \sinh 2\kappa h = 0,$$

where $2h$ is the thickness of the plate.

The *permanent modes* are rational integral functions of z, and involve functions of x, y which are solutions of the equation

$$\left(\frac{\partial^2}{\partial x^2} + \frac{\partial^2}{\partial y^2}\right)^2 F = 0.$$

For a point source in the plate, the function f of the transitory modes is

$$G_0(\kappa R),$$

where $G_0(\alpha)$ is that solution of Bessel's equation which tends to zero when the imaginary part of α tends to $+\infty i$, and the function F of the permanent modes is

$$\tfrac{1}{4}R^2 \log \frac{R}{2h} - \tfrac{1}{4}R^2,$$

where $$R^2 = (x-x')^2 + (y-y')^2.$$

3. It is usually easy to determine an infinite number of modes of equilibrium under no forces, and when this is so, if we *assume* that any deformation within a part of the body (bounded by two transverse sections) where no force is applied is expressible in terms of these modes, the well-known reciprocal property known as Betti's Theorem leads to a simple method of defining the coefficients under which they occur in terms of the given forces applied at other parts of the body. It might be possible, say by an application of the Theory of Integral Equations, to complete this process by justifying the assumption referred to, and this would be useful, for the permanent modes can easily be found for many forms of cylindrical boundary for which it is not at all likely that the transitory modes can be determined in manageable forms, and an approximate theory can be rigorously deduced from the knowledge of the permanent modes and of the expressibility of any source solution in terms of these and transitory modes.

The method I am to describe, however, does not depend on the above or any similar assumption.

4. The first step is to find a solution for given surface tractions, and especially for those tractions which arise from the known solutions for a single force at a point of an infinite solid, in terms of particular solutions having no singularity anywhere in the solid. These particular solutions of which this, the classical form of solution, is composed, involve functions of the type

$$e^{i\kappa z},\ \cos m(\omega-\omega_0), \text{ in the circular cylinder,}$$

$$P_n^m(\cos\theta),\ \cos m(\phi-\phi_0), \text{ in the spherical shell,}$$

where κ and n are *real*.

Solutions of this class have been given by previous writers, notably for the spherical shell by Lamé and by Lord Kelvin. The solutions for the plate by Lamé and Clapeyron, and for the circular cylinder by Pochhammer* belong to this class, but they are not perfectly general. From this classical form of solution the solution in terms of permanent and transitory modes is deduced by an application of Cauchy's Theory of Residues.

5. The permanent and transitory modes are distinguished from each other most strikingly by their respective characters at a distance from the source, for the disturbance due to a transitory mode diminishes with great rapidity as the distance from the source increases, in virtue of the presence of a factor which, either exactly or approximately, is an exponential function of a negative multiple of the ratio of this distance to the thickness of the rod or plate. But even close to the source, although the transitory modes have an important local effect, yet in general their

* "Beitrag zur Theorie der Biegung des Kreiscylinders," *J. f. Math.* (Crelle), Bd 81 (1876).

contributions to the strain are of a higher order in the thickness than those from the permanent modes due to the distant sources.

As a consequence of this it can be shown that even if we neglect the transitory modes altogether and in the solution for a continuous distribution of applied force, which is obtained by integration from the source solutions, retain the permanent terms alone, the values of the displacements, and of the most important stress components are found correctly to a first approximation.

6. The solutions for an infinite cylinder or plate give of course *particular* solutions for the body force and the traction at the surface of a finite cylinder, or at the faces of a finite plate. If we retain the terms derived by integration both from the permanent and the transitory modes of the source solutions, the particular solution is exact, but by means of integration by parts in the case of the cylinder, and an analogous transformation in the case of the plate, this exact solution can be transformed into a solution in the form of series of terms of ascending order in the thickness, with a remainder. These series may or may not be convergent according to the character of the distribution of force, but they can be used to determine the displacement and stress to any assigned degree of approximation.

At a cross-section at which the applied force or any of its space derivatives becomes discontinuous, certain terms of the transitory type are introduced in the course of the integration by parts referred to a few lines above. These transitory terms, however, do not interfere with the first approximation to the displacement and stress.

7. Lastly, it can be shown by an application of Betti's Theorem similar to a well-known application of Green's Theorem in the Theory of the Potential that the deformation in a finite cylinder or plate due to tractions at the ends or edges alone is expressible in terms of the permanent and transitory modes found for an infinite cylinder or plate. And the conditions at the ends or edges which define the permanent modes (which alone are of importance except close to the ends or edges), can be deduced, either exactly as in the case of the cylinder, or with sufficient approximation in the case of the plate. Thus, for example, by the use of the Green's function method, I have found* to a second approximation the edge conditions for the permanent flexural mode in a plate, originally given by Kirchhoff to a first approximation in correction of Poisson.

I have prepared and hope to have published soon a somewhat elaborate account of the method as applied to the circular cylinder.

The illustrative example given below is only meant to give a general idea of the nature of the analysis.

8. *Infinite circular cylinder. Solutions in terms of harmonic functions.*

If (x, y, z) or (ρ, ω, z) be the rectangular or cylindrical coordinates of a point the equations of equilibrium under no body forces are

$$\mu\nabla^2(u_x, u_y, u_z) + (\lambda+\mu)\left(\frac{\partial}{\partial x}, \frac{\partial}{\partial y}, \frac{\partial}{\partial z}\right)\Delta = 0 \quad \text{..............(1).}$$

* *Edinburgh Roy. Soc. Trans.* l.c. Article 46.

The following three general solutions are easily verified:

$$u_x = x\frac{\partial^2\phi}{\partial z^2},\quad u_y = y\frac{\partial^2\phi}{\partial z^2},\quad u_z = -x\frac{\partial^2\phi}{\partial x\partial z} - y\frac{\partial^2\phi}{\partial y\partial z} - \frac{2(\lambda+2\mu)}{\lambda+\mu}\frac{\partial\phi}{\partial z}\ \ldots(2);$$

$$u_x = \frac{\partial\theta}{\partial x},\quad u_y = \frac{\partial\theta}{\partial y},\quad u_z = \frac{\partial\theta}{\partial z}\ \ldots\ldots(3);$$

$$u_x = \frac{\partial\psi}{\partial y},\quad u_y = -\frac{\partial\psi}{\partial x},\quad u_z = 0\ \ldots\ldots(4);$$

where ϕ, θ, ψ are harmonic functions.

We shall use the symbol ν for

$$\frac{2(\lambda+2\mu)}{\lambda+\mu} = 4(1-\sigma)\ \ldots\ldots(5).$$

In (2)
$$\Delta = \frac{\partial u_x}{\partial x} + \frac{\partial u_y}{\partial y} + \frac{\partial u_z}{\partial z} = (2-\nu)\frac{\partial^2\phi}{\partial z^2};$$

in (3) and (4)
$$\Delta = 0\ \ldots\ldots(6).$$

The displacements in cylindrical coordinates can be written down at a glance from (2), (3) and (4). They are

$$\left.\begin{aligned} u_\rho &= \rho\frac{\partial^2\phi}{\partial z^2} + \frac{\partial\theta}{\partial\rho} + \frac{1}{\rho}\frac{\partial\psi}{\partial\omega} \\ u_\omega &= \frac{1}{\rho}\frac{\partial\theta}{\partial\omega} - \frac{\partial\psi}{\partial\rho} \\ u_z &= -\rho\frac{\partial^2\phi}{\partial\rho\partial z} - \nu\frac{\partial\phi}{\partial z} + \frac{\partial\theta}{\partial z} \end{aligned}\right\}\ \ldots\ldots(7).$$

Since

$$\widehat{\rho\rho} = \lambda\Delta + 2\mu\frac{\partial u_\rho}{\partial\rho},\quad \widehat{\rho\omega} = \mu\left(\frac{\partial u_\omega}{\partial\rho} - \frac{u_\omega}{\rho} + \frac{1}{\rho}\frac{\partial u_\rho}{\partial\omega}\right),\quad \widehat{\rho z} = \mu\left(\frac{\partial u_\rho}{\partial z} + \frac{\partial u_z}{\partial\rho}\right)\quad (8),$$

we obtain from (6) and (7)

$$\left.\begin{aligned} \frac{\widehat{\rho\rho}}{\mu} &= (\nu-2)\frac{\partial^2\phi}{\partial z^2} + 2\rho\frac{\partial^3\phi}{\partial\rho\partial z^2} + 2\frac{\partial^2\theta}{\partial\rho^2} + \frac{2}{\rho}\frac{\partial^2\psi}{\partial\rho\partial\omega} - \frac{2}{\rho^2}\frac{\partial\psi}{\partial\omega} \\ \frac{\widehat{\rho\omega}}{\mu} &= \frac{\partial^3\phi}{\partial\omega\partial z^2} + \frac{2}{\rho}\frac{\partial^2\theta}{\partial\rho\partial\omega} - \frac{2}{\rho^2}\frac{\partial\theta}{\partial\omega} - \frac{\partial^2\psi}{\partial\rho^2} + \frac{1}{\rho}\frac{\partial\psi}{\partial\rho} + \frac{1}{\rho^2}\frac{\partial^2\psi}{\partial\omega^2} \\ \frac{\widehat{\rho z}}{\mu} &= 2\rho\frac{\partial^3\phi}{\partial z^3} - \nu\frac{\partial^2\phi}{\partial\rho\partial z} + \frac{1}{\rho}\frac{\partial^3\phi}{\partial\omega^2\partial z} + 2\frac{\partial^2\theta}{\partial\rho\partial z} + \frac{1}{\rho}\frac{\partial^2\psi}{\partial\omega\partial z} \end{aligned}\right\}\ (9).$$

9. *Solution for normal traction* $N(z)\cos(\omega-\omega')$.

We shall suppose that the function $N(z)$ vanishes unless z lies between z_1 and z_2, where $z_2 > z_1$. Then by Fourier's Theorem

$$N(z) = \frac{1}{\pi}\int_0^\infty d\kappa\int_{z_1}^{z_2} N(z')\cos\kappa(z-z')\,dz'\ \ldots\ldots(10).$$

In view of this theorem, we begin with the simplified problem with surface conditions

$$\widehat{\rho\rho} = \cos\kappa(z-z')\cos(\omega-\omega'),\quad \widehat{\rho\omega} = 0,\quad \widehat{\rho z} = 0\ \ldots\ldots(11).$$

Assume
$$\begin{pmatrix}\phi,\ \theta \\ \psi\end{pmatrix} = \begin{pmatrix}A,\ B \\ C\end{pmatrix} J_1(i\kappa\rho)\cos\kappa(z-z')\,{\cos \atop \sin}(\omega-\omega')\ \ldots\ldots(12).$$

These values of ϕ, θ, ψ give at $\rho = a$, from (9),

$$\left.\begin{aligned}
\widehat{\rho\rho}\frac{a^2}{\mu} &= \{A\,(\overline{2-\nu}\kappa^2a^2J - 2i\kappa^3a^3J') + B\,(-2\kappa^2a^2J'') \quad + C\,(-2)\,(J - i\kappa aJ')\} \\
&\qquad\qquad \cos\kappa\,(z-z')\cos(\omega-\omega') \\
\widehat{\rho\varpi}\frac{a^2}{\mu} &= \{A\,\kappa^2a^2J \qquad\qquad + B\,.\,2\,(J - i\kappa aJ') + C\,(\kappa^2a^2J'' + i\kappa aJ' - J)\} \\
&\qquad\qquad \cos\kappa\,(z-z')\sin(\omega-\omega') \\
\widehat{\rho z}\frac{a}{\mu\kappa} &= \{A\,(2\kappa^2a^2J + \nu i\kappa aJ' + J) \;+ B\,(-2i\kappa aJ') \quad + C\,(-J)\} \\
&\qquad\qquad \sin\kappa\,(z-z')\cos(\omega-\omega')
\end{aligned}\right\}\ (13)$$

where the argument of the J functions and derivatives is $i\kappa a$.

In (13) we shall write the coefficients of A, B, C for brevity in the forms

$$\left.\begin{aligned}
&Aa_1 + Bb_1 + Cc_1 \\
&Aa_2 + Bb_2 + Cc_2 \\
&Aa_3 + Bb_3 + Cc_3
\end{aligned}\right\} \quad \ldots\ldots\ldots\ldots\ldots\ldots\ldots(14).$$

A solution of the problem with surface conditions (11) is therefore (12), with

$$A = \frac{a^2}{\mu}\frac{A_1}{D}, \quad B = \frac{a^2}{\mu}\frac{B_1}{D}, \quad C = \frac{a^2}{\mu}\frac{C_1}{D} \quad \ldots\ldots\ldots\ldots\ldots(15),$$

where

$$D = \begin{vmatrix} a_1, & b_1, & c_1 \\ a_2, & b_2, & c_2 \\ a_3, & b_3, & c_3 \end{vmatrix} \quad \ldots\ldots\ldots\ldots\ldots\ldots\ldots(16),$$

and A_1, B_1, C_1 are the first minors multiplying a_1, b_1, c_1 in D.

In order to obtain a solution of the problem of normal traction $N(z)\cos(\omega-\omega')$ it is now natural to try the result of multiplying the values in (12) by $N(z')/\pi$ and integrating as to z' from z_1 to z_2, then as to κ from 0 to ∞. It will be found, however, that unless the form of the function $N(z)$ is considerably restricted, the κ integrations cannot be carried out, on account of the form of the functions ϕ, θ, ψ defined by (12) and (15) near $\kappa = 0$*. But the difficulty is easily surmounted. For the terms of negative degree in the ϕ, θ, ψ of (12) and (15) can contribute nothing to the surface tractions at $\rho = a$, simply because the values of $\widehat{\rho\rho}$, $\widehat{\rho\varpi}$ and $\widehat{\rho z}$ in (11) contain no terms of negative degree in their κ power expansions. Moreover these terms of negative degree in ϕ, θ, ψ of (12) and (15) are of the form $H_4/\kappa^4 + H_2/\kappa^2$, so that all we have to do is to *subtract* them.

Write the terms to be subtracted as $(\phi_0, \theta_0, \psi_0)\dfrac{a^2}{\mu}$.

We thus obtain for normal traction $N(z)\cos(\omega-\omega')$ the tentative solution

$$\begin{pmatrix}\phi, \theta \\ \psi\end{pmatrix} = \frac{a^2}{\pi\mu}\int_0^\infty d\kappa\int_{z_1}^{z_2}\left[\frac{1}{D}\left\{\begin{pmatrix}A_1, B_1 \\ C_1\end{pmatrix} J_1(i\kappa\rho)\cos\kappa\,(z-z')\,{\cos \atop \sin}(\omega-\omega')\right\} - \begin{pmatrix}\phi_0, \theta_0 \\ \psi_0\end{pmatrix}\right] N(z')\,dz' \quad \ldots\ldots(17).$$

It is not difficult to show, by calculating $\widehat{\rho\rho}$, $\widehat{\rho\varpi}$, $\widehat{\rho z}$ for any *internal* point from (17), and proving that these pass continuously into $N(z)\cos(\omega-\omega')$, 0, 0, respectively as ρ approaches a, that (17) is actually a solution of the problem proposed.

* In Pochhammer's solution l.c. κ is an integer, and he avoids a similar difficulty to that which arises here by restricting the form of $N(z)$.

The order of integration in (17) can be changed, and we can put

$$\begin{pmatrix}\phi, \theta \\ \psi\end{pmatrix} = \int_{z_1}^{z_2} \begin{pmatrix}\phi_1, \theta_1 \\ \psi_1\end{pmatrix} N(z')\,dz' \quad \text{.......................(18)},$$

where

$$\begin{pmatrix}\phi_1, \theta_1 \\ \psi_1\end{pmatrix} = \frac{a^2}{\pi\mu}\int_0^\infty \left[\frac{1}{D}\left\{\begin{pmatrix}A_1, B_1 \\ C_1\end{pmatrix} J_1(i\kappa\rho)\cos\kappa(z-z')\,{\cos \atop \sin}(\omega-\omega')\right\} - \begin{pmatrix}\phi_0, \theta_0 \\ \psi_0\end{pmatrix}\right] d\kappa \quad \text{...(19)}.$$

Equations (18) and (19) define a solution of the problem proposed at the head of this article, of the type referred to in Art. 4.

10. *Expression of the definite integrals as complex integrals.*

In each of the integrals of (19) the integrand is a uniform even function of κ of the form

$$E(\kappa)\cos\kappa(z-z') - E_0(\kappa),$$

where $E_0(\kappa)$ stands for the terms of negative degree in κ in the power expansion of $E(\kappa)\cos\kappa(z-z')$.

Now

$$\int_0^\infty \{E(\kappa)\cos\kappa(z-z') - E_0(\kappa)\}\,d\kappa = \tfrac{1}{2}\int_{-\infty}^{\infty} \{E(\kappa)\cos\kappa(z-z') - E_0(\kappa)\}\,d\kappa \quad \text{...(20)}.$$

The latter integral can be regarded as a complex integral, and its path of integration can be deformed.

We define a *first* and a *second* path of integration as passing from $-\infty$ to ∞ along the real axis, except near the origin, where they follow a small semicircle above and below 0 respectively; also a *third* and a *fourth* path as passing along the imaginary axis from $-\infty i$ to ∞i except near the origin where they follow a small semicircle to the left and right of 0 respectively.

Then the right-hand member of (20)

$$= \tfrac{1}{2}\int \{E(\kappa)\cos\kappa(z-z') - E_0(\kappa)\}\,d\kappa \quad \text{(first path)}$$

$$= \tfrac{1}{2}\int E\kappa\cos\kappa(z-z')\,d\kappa \quad \text{(first path)} \quad \text{..........................(21)}$$

$\left(\text{for evidently } \int \frac{d\kappa}{\kappa^2} \text{ or } \int \frac{d\kappa}{\kappa^4} \text{ over this path vanishes}\right)$.

This again

$$= \tfrac{1}{4}\int E(\kappa)\,e^{-i\kappa(z-z')}\,d\kappa \text{ (first path)} + \tfrac{1}{4}\int E(\kappa)\,e^{i\kappa(z-z')}\,d\kappa \text{ (first path)} \quad \text{......(22)}$$

$$= \tfrac{1}{4}\int E(\kappa)\,e^{-i\kappa(z-z')}\,d\kappa \text{ (first path)} + \tfrac{1}{4}\int E(\kappa)\,e^{-i\kappa(z-z')}\,d\kappa \text{ (second path)} \quad \text{...(23)},$$

as we see by changing κ into $-\kappa$ in the second integral of (22).

The result may be written

$$\int_0^\infty \{E(\kappa)\cos\kappa(z-z') - E_0(\kappa)\}\,d\kappa = \tfrac{1}{4}\int E(\kappa)\,e^{-i\kappa(z-z')}\,d\kappa \text{ (first and second paths)} \quad \text{...(24)}.$$

In the integrals on the right of (24) change the variable of integration from κ to β where

$$\beta = i\kappa a \quad \text{..................................(25)}.$$

If $E(\kappa)$ thus becomes E_β, we obtain

$$\int_0^\infty \{E(\kappa)\cos\kappa(z-z') - E_0(\kappa)\}\,d\kappa = \frac{1}{4ia}\int E_\beta e^{-\frac{\beta(z-z')}{a}}\,d\beta \text{ (third and fourth paths) (26).}$$

11. *Transformation of the complex integrals into series.*

Applying the transformation (26) to the integrals of (19), we find

$$\begin{pmatrix}\phi_1, \theta_1\\ \psi_1\end{pmatrix} = \frac{a}{4\pi i\mu}\int \frac{1}{D}\begin{pmatrix}A_1, B_1\\ C_1\end{pmatrix} J_1\frac{\beta\rho}{a} e^{-\frac{\beta(z-z')}{a}} \frac{\cos}{\sin}(\omega-\omega')\,d\beta \text{ (third and fourth paths) (27),}$$

where the functions D, A_1, B_1, C_1 defined at (16) are to be expressed in terms of β by means of (25).

The integrand in (27) vanishes at infinity in such a way that, by Cauchy's Theorem, each integral can be replaced by $(-2\pi i)$ (sum of the residues of the integrand at the poles of D to the right or left of the path according as $z-z'$ is positive or negative).

Hence, if $z > z'$,

$$\left.\begin{aligned}\begin{pmatrix}\phi_1, \theta_1\\ \psi_1\end{pmatrix} &= -\frac{a}{\mu}\sum_\beta \frac{1}{dD/d\beta}\begin{pmatrix}A_1, B_1\\ C_1\end{pmatrix} J_1\frac{\beta\rho}{a} e^{-\frac{\beta(z-z')}{a}} \frac{\cos}{\sin}(\omega-\omega')\\ &\quad -\frac{a}{2\mu}\cdot\text{coefficient of }\frac{1}{\beta}\text{ in }\frac{1}{D}\begin{pmatrix}A_1, B_1\\ C_1\end{pmatrix} J_1\frac{\beta\rho}{a} e^{-\frac{\beta(z-z')}{a}} \frac{\cos}{\sin}(\omega-\omega')\end{aligned}\right\}\ldots(28),$$

where $\sum_\beta$ denotes summation over the zeroes of D whose real part is positive; it can be proved that these are all simple.

If $z < z'$, the corresponding results are found most simply by interchanging z and z' in (28), retaining the meaning of $\sum_\beta$ unchanged.

12. *Permanent terms of the solution* (28).

The principal terms in the expansions of A_1, B_1, C_1 and $1/D$ near $\beta = 0$ are

$$\left.\begin{aligned}A_1 = \tfrac{1}{8}\beta^4 - \tfrac{7}{64}\beta^6,\quad A_2 = -\tfrac{1}{8}\beta^4 - \tfrac{1}{64}\beta^6,\quad A_3 = -\tfrac{1}{8}\beta^6 + \tfrac{3}{64}\beta^8\\ \frac{1}{D} = \frac{64}{(8-\nu)^2}\left\{(8-\nu)\beta^{-9} + \frac{64-7\nu}{24}\beta^{-7}\right\}\end{aligned}\right\}\ldots(29).$$

With the help of these results, the partial values of ϕ_1, θ_1, ψ_1 in the second line of (28) are easily calculated, and lead by (7) to the displacements $(z > z')$,

$$\left.\begin{aligned}u_\rho &= \left\{\tfrac{1}{3}(z-z')^3 + \frac{4-\nu}{4}(z-z')\rho^2\right\}\cos(\omega-\omega')\\ u_\omega &= \left\{-\tfrac{1}{3}(z-z')^3 + \frac{4-\nu}{4}(z-z')\rho^2\right\}\sin(\omega-\omega')\\ u_z &= \left\{-(z-z')^2\rho + \tfrac{1}{2}\rho^3 + \frac{\nu-10}{4}a^2\rho\right\}\cos(\omega-\omega')\end{aligned}\right\}\cdot\frac{2}{\pi\mu(8-\nu)a^4}\ldots\ldots(30)$$

with the rigid body rotation

$$\left.\begin{aligned}u_\rho &= (z-z')\cos(\omega-\omega')\\ u_\omega &= -(z-z')\sin(\omega-\omega')\\ u_z &= -\rho\cos(\omega-\omega')\end{aligned}\right\}\cdot\left\{\frac{4-\nu}{4}\frac{2}{\pi\mu(8-\nu)} - \frac{\nu^2-20\nu+92}{3\pi\mu(8-\nu)^2}\right\}\frac{1}{a^2}\ldots\ldots(31).$$

For $z < z'$ the signs of all these displacements have to be changed.

The functions of ν which occur here are easily expressed in terms of the familiar Poisson's ratio σ and Young's modulus E by the relations

$$\left.\begin{aligned} \frac{4-\nu}{4} &= \sigma \\ \mu(8-\nu) &= 2E \\ \frac{\nu^2 - 20\nu + 92}{(8-\nu)^2} &= 1 + \frac{2\mu}{E} - \frac{\mu^2}{E^2} \end{aligned}\right\} \qquad \text{(32).}$$

13. *Transitory terms of the solution* (28).

The function D possesses an infinite number both of real and of complex zeroes.

For the large real roots

$$\beta = N\pi + \frac{\pi}{4} - \frac{1}{N\pi}\left(\frac{19}{8}\right)$$

and for the large complex roots

$$\beta = N\pi + \frac{\pi}{2} - \frac{\log 4N\pi}{4N\pi} \pm i\left(\tfrac{1}{2}\log 4N\pi + \frac{1}{4N}\right) \qquad \text{(33),}$$

approximately, where N is a positive integer.

The asymptotic forms of the high transitory modes in the first line of (28) are different in the two cases of real and complex values of β.

Real roots of D.

$$\left.\begin{aligned} A_1 = -\frac{4(\nu+2)}{N\pi^2}, \quad B_1 = -\frac{2(\nu+2)^2}{N\pi^2}, \quad C_1 = 4(\nu+2)N \\ \frac{dD}{d\beta} = 8\sqrt{2}\pi^{-\frac{3}{2}}\cos N\pi\,(N\pi)^{\frac{9}{2}} \end{aligned}\right\} \qquad \text{(34),}$$

$$\left.\begin{array}{lll} u_\rho = & \left(\frac{a}{\rho} - \frac{\rho}{a}\right)\left(\frac{\rho}{a}\right)^{-\frac{1}{2}} & \sin\left(\frac{N\pi\rho}{a} + \frac{\pi}{4}\frac{\rho - a}{a}\right)\cos(\omega - \omega') \\ u_\omega = -N\pi & \quad \text{,,} & \cos(\quad \text{,,} \quad)\sin(\text{ ,, }) \\ u_z = -\frac{\rho}{a} & \quad \text{,,} & \cos(\quad \text{,,} \quad)\cos(\text{ ,, }) \\ \frac{\widehat{\rho\rho}}{\mu} = 2N\pi\left(\frac{1}{\rho} - \frac{\rho}{a^2}\right) & \quad \text{,,} & \cos(\quad \text{,,} \quad)\cos(\text{ ,, }) \\ \frac{\widehat{\rho\omega}}{\mu} = \frac{N^2\pi^2}{a} & \quad \text{,,} & \sin(\quad \text{,,} \quad)\sin(\text{ ,, }) \\ \frac{\widehat{\rho z}}{\mu} = N\pi\left(\frac{2\rho}{a^2} - \frac{1}{\rho}\right) & \quad \text{,,} & \sin(\quad \text{,,} \quad)\cos(\text{ ,, }) \end{array}\right\} (-1)^{N-1}\frac{\nu+2}{2\mu}\frac{1}{N^4\pi^4}\, e^{-\left(N\pi + \frac{\pi}{4}\right)\frac{z-z'}{a}} \qquad \text{(35),}$$

where $z > z'$.

Complex roots of D.

$$\left.\begin{aligned} A_1 = 4N^3\pi^2, \quad B_1 = -4iN^4\pi^3, \quad C_1 = -4N^3\pi^2 \\ \frac{dD}{d\beta} = -16\sqrt{2}\pi^{-\frac{3}{2}}\cos N\pi\, e^{-\frac{i\pi}{4}}(N\pi)^5 \end{aligned}\right\} \qquad \text{(36),}$$

$$\begin{pmatrix} u_\rho, u_z \\ u_\omega \end{pmatrix} = \begin{pmatrix} \frac{\rho}{a} - 1, \; i\left(1 - \frac{\rho}{a}\right) \\ \frac{i}{N\pi}\left(\frac{a}{\rho} - 1\right) \end{pmatrix} \left(\frac{\rho}{a}\right)^{-\frac{1}{2}} e^{-\frac{i\beta\rho}{a}} e^{-\frac{\beta(z-z')}{a}} \begin{matrix}\cos \\ \sin\end{matrix}(\omega - \omega') \cdot \frac{1}{8\mu}(-1)^{N-1}(N\pi)^{-\frac{1}{2}} \quad \text{......(37)},$$

$$\begin{pmatrix} \widehat{\rho\rho}, \widehat{\rho z} \\ \widehat{\rho\omega} \end{pmatrix} = \frac{2\mu N\pi}{a} \begin{pmatrix} i\left(1 - \frac{\rho}{a}\right), \; 1 - \frac{\rho}{a} \\ \frac{1}{N\pi}\left(\frac{a}{\rho} - 1\right) \end{pmatrix} \;\; ,, \;\; ,, \;\; ,, \;\; ,, \;\; ,, \;\; ,, \;\; ,, \;\; ,, \quad \text{......(38)},$$

where $z > z'$.

14. *Solution for distribution of normal traction.*

The solution for normal traction $N(z)\cos(\omega - \omega')$ from $z = z_1$ to $z = z_2$ is given by (18), in which we can take for ϕ_1, θ_1, ψ_1 their values as in (27) or in (28).

Let any displacement u be calculated from (27) and (7), and let the result be

$$u = -\frac{1}{4\pi i}\int \frac{U}{D} e^{-\frac{\beta(z-z')}{a}} d\beta \text{ (third and fourth paths) ....(39).}$$

Here U is a uniform function of β, even for u_ρ and u_ω, odd for u_z.

For definiteness suppose now that u is u_ρ.

The transformation by Cauchy's Theorem gives, for $z > z'$,

$$u_\rho = \tfrac{1}{2} \text{ coeff. of } \frac{1}{\beta} \text{ in } \frac{U}{D} e^{-\frac{\beta(z-z')}{a}} + \sum_\beta \frac{U}{dD/d\beta} e^{-\frac{\beta(z-z')}{a}} \quad \text{.........(40)},$$

for $z < z'$,

$$u_\rho = \tfrac{1}{2} \text{ coeff. of } \frac{1}{\beta} \text{ in } \frac{U}{D} e^{\frac{\beta(z-z')}{a}} + \sum_\beta \frac{U}{dD/d\beta} e^{\frac{\beta(z-z')}{a}} \quad \text{............(41).}$$

The value of u_ρ corresponding to (18) is now (when $z_1 < z < z_2$),

$$u_\rho = \int_{z_1}^{z} N(z')\{u_\rho \text{ of } (40)\}\, dz' + \int_{z}^{z_2} N(z')\{u_\rho \text{ of } (41)\}\, dz' \quad \text{......(42).}$$

From the asymptotic forms of u_ρ in Art. 13 it will be found that the integrations as to z' here can be performed term by term.

Thus, for $z_1 < z < z_2$,

$$\left.\begin{aligned} u_\rho = &\int_{z_1}^{z} \left\{\tfrac{1}{2} \text{ coeff. of } \frac{1}{\beta} \text{ in } \frac{U}{D} e^{-\frac{\beta(z-z')}{a}}\right\} N(z')\, dz' + \int_{z}^{z_2} \left\{\tfrac{1}{2} \text{ coeff. of } \frac{1}{\beta} \text{ in } \frac{U}{D} e^{\frac{\beta(z-z')}{a}} N(z')\, dz'\right\} \\ &+ \sum_\beta \left[\frac{U}{dD/d\beta}\left\{\int_{z_1}^{z} e^{-\frac{\beta(z-z')}{a}} N(z')\, dz' + \int_{z}^{z_2} e^{\frac{\beta(z-z')}{a}} N(z')\, dz'\right\}\right] \end{aligned}\right\} \quad \text{......(43)},$$

for $z > z_2$,

$$u_\rho = \int_{z_1}^{z_2} \left\{\tfrac{1}{2} \text{ coeff. of } \frac{1}{\beta} \text{ in } \frac{U}{D} e^{-\frac{\beta(z-z')}{a}}\right\} N(z')\, dz' + \sum_\beta \left\{\frac{U}{dD/d\beta}\int_{z_1}^{z_2} e^{-\frac{\beta(z-z')}{a}} N(z')\, dz'\right\} \text{...(44)},$$

and, for $z < z_1$,

$$u_\rho = \int_{z_1}^{z_2} \left\{\tfrac{1}{2} \text{ coeff. of } \frac{1}{\beta} \text{ in } \frac{U}{D} e^{\frac{\beta(z-z')}{a}}\right\} N(z')\, dz' + \sum_\beta \left\{\frac{U}{dD/d\beta}\int_{z_1}^{z_2} e^{\frac{\beta(z-z')}{a}} N(z')\, dz'\right\} \text{......(45).}$$

The values of u_ρ in (40) and (41) are simply what we should get by calculating u_ρ from (28); and similarly with u_ω and u_z.

The nature of the solution for $z > z_2$ or $z < z_1$ is obvious. It is compounded of permanent and transitory modes, being obtained by the integration of such with respect to a variable and between limits which do not depend on the current coordinates.

For $z_1 < z < z_2$ the case is different.

15. *Nature of the solution within the region where the force is applied.*

Consider the second line of (43), say

$$I = \sum_\beta \left[\frac{U}{dD/d\beta}\left\{\int_{z_1}^{z} e^{-\frac{\beta(z-z')}{a}} N(z')\,dz' + \int_z^{z_2} e^{\frac{\beta(z-z')}{a}} N(z')\,dz'\right\}\right] \quad \ldots(46).$$

Repeated integration by parts gives

$$\left.\begin{aligned} I = &\sum_\beta \left[\frac{U}{dD/d\beta}\left\{\frac{2a}{\beta} N(z) + \frac{2a^3}{\beta^3}\frac{d^2}{dz^2} N(z) \ldots \text{ to the term in } a^{n-1} \text{ or } a^n\right\}\right] \\ &- \sum_\beta \left[\frac{U}{dD/d\beta} e^{-\frac{\beta(z-z_1)}{a}}\left\{\frac{a}{\beta} N(z_1) - \frac{a^2}{\beta^2}\frac{d}{dz_1} N(z_1) + \ldots + (-1)^{n-1}\frac{a^n}{\beta^n}\frac{d^{n-1}}{dz_1^{n-1}} N(z_1)\right\}\right] \\ &- \sum_\beta \left[\frac{U}{dD/d\beta} e^{-\frac{\beta(z_2-z)}{a}}\left\{\frac{a}{\beta} N(z_2) + \frac{a^2}{\beta^2}\frac{d}{dz_2} N(z_2) + \ldots + \frac{a^n}{\beta^n}\frac{d^{n-1}}{dz_2^{n-1}} N(z_2)\right\}\right] \\ &+ \sum_\beta \left[\frac{U}{dD/d\beta}\left\{(-1)^n \frac{a^n}{\beta^n}\int_{z_1}^{z} e^{-\frac{\beta(z-z')}{a}}\frac{d^n}{dz'^n} N(z')\,dz' + \frac{a^n}{\beta^n}\int_z^{z_2} e^{\frac{\beta(z-z')}{a}}\frac{d^n}{dz'^n} N(z')\,dz'\right\}\right] \end{aligned}\right\}$$

$$\ldots\ldots(47).$$

Consider separately the four series in the four lines on the right of (47).

In the *first line*, the sum of the series can be found in finite terms, for it is

$$\sum_\beta \left[\text{Residue at } \beta \text{ of } \frac{U}{D}\left\{\frac{2a}{\beta} N(z) + \frac{2a^3}{\beta^3}\frac{d^2}{dz^2} N(z) + \ldots \text{ to the term in } a^{n-1} \text{ or } a^n\right\}\right] \quad (48).$$

The function of which the residue is taken here is an odd uniform function of β which vanishes at infinity in such a way that the sum of all its residues vanishes.

But the residues at $\pm\beta$ are equal, so that

$\sum_\beta [\;]$ in (48) $= -\frac{1}{2}$ residue at 0 of the function of β in $[\;]$ in (48)

$$= -\tfrac{1}{2} \text{ coeff. of } \frac{1}{\beta} \text{ in } \frac{U}{D}\left\{\frac{2a}{\beta} N(z) + \frac{2a^3}{\beta^3}\frac{d^2}{dz^2} N(z) + \ldots \text{ to term in } a^{n-1} \text{ or } a^n\right\}$$

$$\ldots\ldots(49).$$

Now the ascending power expansion of U/D has the form

$$\frac{U}{D} = U^{(-4)}\beta^{-4} + U^{(-2)}\beta^{-2} + U^{(0)}\beta^0 + U^{(2)}\beta^2 + \ldots \quad \ldots\ldots\ldots\ldots(50).$$

Hence the *first line* of I in (47)

$$= -aU^{(0)} N(z) - a^3 U^{(2)}\frac{d^2}{dz^2} N(z) - \ldots \text{ to the term in } a^{n-1} \text{ or } a^n \quad \ldots(51).$$

This will be considered in next article in combination with the first line of (43).

In the *second line* of (47) the general term is proportional to the displacement

$$u_\rho = Ue^{-\frac{\beta(z-z_1)}{a}}$$

belonging to one of the transitory modes.

When a is small the corresponding deformation is insensible except very close to the section $z = z_1$.

Similarly the *third* line of (47) represents a *local perturbation* at the end $z = z_2$.

Lastly the *fourth* line of (47) contains a series of the same general character as the series (46); but this series is at once of a higher order in a on account of the factor a^n and more highly convergent on account of the factor β^{-n} in the general term.

16. *The asymptotic particular solution within the region of applied force.*

We consider now the first line of (43) along with (51); let the part of u_ρ arising from the sum of these be denoted by $u_\rho^{(0)}$.

By (50)
$$\frac{U}{D} = U^{(-4)}\beta^{-4} + U^{(-2)}\beta^{-2} + U^{(0)}\beta^0 + U^{(2)}\beta^2 + \dots.$$

Also
$$\left.\begin{aligned} e^{-\frac{\beta(z-z')}{a}} &= 1 - \frac{\beta(z-z')}{a} + \frac{\beta^2(z-z')^2}{2a^2} - \frac{\beta^3(z-z')^3}{6a^3} + \dots \\ e^{\frac{\beta(z-z')}{a}} &= 1 + \frac{\beta(z-z')}{a} + \frac{\beta^2(z-z')^2}{2a^2} + \frac{\beta^3(z-z')^3}{6a^3} + \dots \end{aligned}\right\} \quad \text{......(52).}$$

Thus the first line of (43) is

$$\left.\begin{aligned} &\int_{z_1}^{z} \{-\tfrac{1}{12}a^{-3}U^{(-4)}(z-z')^3 - \tfrac{1}{2}a^{-1}U^{(-2)}(z-z')\}\, N(z')\,dz' \\ +&\int_{z}^{z_2} \{\tfrac{1}{12}a^{-3}U^{(-4)}(z-z')^3 + \tfrac{1}{2}a^{-1}U^{(-2)}(z-z')\}\, N(z')\,dz' \end{aligned}\right\} \quad \text{......(53).}$$

It is convenient to introduce a function F, defined as follows:

$$\left.\begin{aligned} &\text{when } z < z_1, && F = -\int_{z_1}^{z_2} \tfrac{1}{12}(z-z')^3 N(z')\,dz' \\ &\text{when } z_1 < z < z_2, && F = \int_{z_1}^{z} \tfrac{1}{12}(z-z')^3 N(z')\,dz' - \int_{z}^{z_2} \tfrac{1}{12}(z-z')^3 N(z')\,dz' \\ &\text{when } z > z_2, && F = \int_{z_1}^{z_2} \tfrac{1}{12}(z-z')^3 N(z')\,dz' \end{aligned}\right\} \quad \text{...(54).}$$

It can be seen at once that F and its first three derivatives are continuous functions of z, and that

$$\frac{d^4F}{dz^4} = N(z).$$

We shall write, as equivalent to the definition (54),

$$F = D_z^{-4} N(z) \quad \text{..................................(55).}$$

Also we shall write $D_z^{-2} N(z)$ for $\dfrac{d^2F}{dz^2}$.

Thus (53) becomes

$$-a^{-3}U^{(-4)}D_z^{-4}N(z) - a^{-1}U^{(-2)}D_z^{-2}N(z) \quad \text{..............(56).}$$

Hence the sum of the first line of (43) along with (51) is

$$u_\rho^{(0)} = -a^{-3}U^{(-4)}D_z^{-4}N(z) - a^{-1}U^{(-2)}D_z^{-2}N(z) - aU^{(0)}N(z) - a^3U^{(2)}\frac{d^2}{dz^2}N(z) - \dots$$

to the term in a^{n-1} or a^n ......(57).

The functions $U^{(-4)}$, $U^{(-2)}$, $U^{(0)}$, etc., are all of order 0 in a.

The value of $u_\rho^{(0)}$ in (57) along with the fourth line of (47) gives the value of u_ρ in an exact particular solution.

In (57) the terms of negative order in a, and involving D_z^{-4} and $D_z^{-2}N(z)$, are just those which come from the permanent terms in the source solution, i.e. the terms in (30) and (31).

Hence if these permanent terms only are used, the displacements neglected are of an order in a higher by two units than any of those retained.

It is not difficult to show that this must be so, independently, by consideration solely of the dimensions of the terms in the unit of length.

If $N(z)$ is a rational integral function of z the series (57) continued indefinitely terminates, and we obtain a particular solution in finite terms.

In the general case when $N(z)$ has an unlimited number of successive derivatives it may happen that the remainder (in the fourth line of (47)) tends to zero with increasing n.

The condition for this can be shown to be that the power series in ζ

$$\sum_n a^n \frac{d^n}{dz^n} N(z)\, \zeta^n$$

should have a radius of convergence greater than $1/|\beta_0|$, where β_0 is the zero of D with smallest modulus.

In general, however, the series (57) is only asymptotic.

COMMUNICATIONS

SECTION III (*b*)

(ECONOMICS, ACTUARIAL SCIENCE, STATISTICS)

EQUILIBRIUM AND DISTURBANCE IN THE DISTRIBUTION OF WEALTH.

BY R. A. LEHFELDT.

It has been said that the element of time is the centre of difficulty in economic problems. The following is an attempt at a method of expressing these time relations in mathematical form, so that they may be apprehended with the clarity that always attaches to mathematical language. The paper arose from Prof. Edgeworth's recent articles in the *Economic Journal* (Sept. and Dec. 1911).

I.

In a single business, the entrepreneur must be regarded as a constant factor. If by association with quantities $x, y, \ldots$ of other factors he generates an amount of product $f(x, y, \ldots)$; if ϕ be the price of the product, and $\xi, \eta, \ldots$ the prices of the factors, then the profit is

$$v = \phi f - \xi x - \eta y - \ldots.$$

First assume that all the prices are constant; then to make v a maximum, we must have

$$\phi \frac{df}{dx} - \xi = 0, \qquad \phi \frac{df}{dy} - \eta = 0, \quad \text{etc.,}$$

the solution of which is, say,

$$x = x_0, \qquad y = y_0, \quad \text{etc.}$$

The business should therefore grow to and stay at this size, and the quantity $v_{\text{max.}}$ will be the return to the entrepreneur. The question has been raised whether this quantity tends to a normal level: this is considered below, but while dealing with a single business, it is best to put the entrepreneur in a different category from the other factors.

The assumption of constant prices depends on the insignificance of the business, compared with the industry or the nation of which it forms part. There are, of course, limits: a very great increase in a single manufacturer's product might lower its price, or a very great increase in his demand for skilled labour result in higher wages. Usually the assumption is practically true in those respects, but fails sooner in respect of capital. A very great increase in the manufacturer's demand for capital would probably make the lenders (the banks) hesitate, and raise their price for it. Otherwise $v_{\text{max.}}$ might conceivably become indefinitely great.

If all the conditions are constant, there is no problem. The real question is, What are the consequences of something new happening? There may be a gradual or a sudden change in conditions: we take the latter first. Let x be altered abruptly to x_1. In general, its price will alter also, say to ξ'. Then the equation of profit becomes

$$v' = \phi f - \xi' x - \eta y \ldots,$$

and there is a new state of equilibrium characterised by

$$\frac{dv'}{dx} = 0 = \phi \frac{df}{dx} - \xi', \qquad \frac{dv'}{dy} = 0 = \phi \frac{df}{dy} - \eta, \text{ etc.,}$$

whose solution is

$$x = x_0', \; y = y_0', \text{ etc.}$$

If $x_1 \gtrless x_0$ then $\xi' \lesseqgtr \xi$ and x_0' will usually lie between x_0 and x_1. (I omit complications, such as functions with more than one maximum.) y_0' may be $>$ $=$ or $<$ y_0 according to the nature of the function. The business will then move towards the new maximum.

Consider particular cases, in order to bring out the meaning of these equations. The abrupt change may occur within or without the business. Changes within the business do not, however, usually alter prices, and hence do not cause a change in the function v; this is nearly always true as concerns labour; changes of capital require more consideration.

(*a*) Some of the capital may be abruptly destroyed, as by a fire (not covered by insurance). Then the entrepreneur will probably have no difficulty, if the business was making normal profits, in persuading his banker that the damaged buildings or machinery should be renewed, and the banker will lend the money at normal interest.

(*b*) A sudden influx of capital, by floating a debenture issue. If the business were really in equilibrium before, this money could only be used to fund floating liabilities, and any balance over would have to be lent outside. No appreciable change would occur in the price of capital used, and the profits function would tend towards the same maximum as before.

Next take changes outside the business. Capital might become suddenly scarce by destruction on a large scale, say by a war or an earthquake: or plentiful by a boom on the Stock Exchange turning investors' attention to the industry. (Similar, but smaller changes are constantly happening and showing themselves in fluctuations of the bank rate.) Then the price of capital will be changed and the business will tend towards the new maximum.

Labour, also, may change in price, owing to emigration, immigration, sudden demand for some other industry, famine, epidemic and so on: or temporarily through a strike.

Another cause leading to disturbance and a new position of equilibrium is invention. The adoption of a new process may be expressed conveniently by introducing a new term in the equation, say $-\zeta z$ (z having previously been zero); and this is usually characterised by the fact that $\frac{dv'}{dz}$ is exceptionally large, so that as soon as that direction is rendered available by the invention, there is a strong tendency for the business to move along it (with the corresponding readjustments in $x, y, \ldots$)*.

The business then will, after the disturbance, tend to move towards the position of equilibrium (the new one if a price has been altered, the original position if not). To take the simple case of two variable factors, the tendency will be to move in the direction which makes with the axis of x an angle whose tangent is

$$\left(\frac{dv'}{dy} \div \frac{dv'}{dx}\right)_{(x=x_1,\ y=y_0)}.$$

The tendency, or driving force is in this direction, but the resistance to changes along x and along y will not in general be the same (these resistances include economic friction of various sorts). Hence the movement of the business will not be in the direction above indicated.

Call the driving force $\frac{dv'}{dx} = X$. Then we may write as a first approximation

$$X = h \frac{dx}{dt}$$

as an equation of motion ($t =$ time, $h =$ constant)†.

X depends for its amount on the departure from equilibrium $x - x_0'$. If we took it to be $\psi(y)(x - x_0')$ the solution would be in the well-known exponential form, but as near states of equilibrium, the surface is curved about a maximum, it would be better to take $X = \psi(y)(x - x_0')^2$, which yields the solution

$$-(x - x_0') = \frac{h}{\psi(y)\, t} = \frac{q}{t} \text{ say.}$$

Here q represents the slowness (or inertia) of recovery from a disturbance of the given type.

The constant q in the case of disturbances within the business, would express the delay in reconstructing machinery after a fire, or in taking on new hands after a strike, and so on. In the case of outside disturbances, the much longer delay in saving new capital, or in growing and educating new skilled workmen. By means of it, greater precision can be given to the ideas associated with such terms as "normal price," "in the long run," etc.

In any case the direction in which a business will move after a disturbance depends on the q's as well as on dv'/dx, dv'/dy. If, as is often the case, one q is much greater than another the business will move, first, approximately in the direction of the smallest q, then in that of the next larger, and so on.

Gradual changes in the conditions will differ from abrupt changes, in that $x - x_0$ the departure from equilibrium will never be large. The business continually adjusts

* $\frac{dv}{dx}$ may be called the "profit gradient" in the direction of x.

† x as well as y, f, etc. are really fluxes or velocities; hence $\frac{dx}{dt}$ corresponds to acceleration in dynamics, and h is in some aspects a measure of inertia rather than of friction.

itself towards the equilibrium state, i.e. x tends towards x_0, but as the latter is changing continuously, x never reaches it. The departure from (or lag behind) equilibrium, in any direction, will depend on the value of q in that direction.

II.

From the point of view of an entire industry or of the community at large, there being a number of entrepreneurs, it is convenient to regard their services as paid for at normal rates, according to ability, and introduce a term vu where u represents the quantity and v the price of this factor of production, giving

$$v = \phi f - vu - \xi x - \eta y - \ldots.$$

Although the entrepreneur gets paid in quite a different way from the capitalist or the workman, there can, I think, be no doubt that in the long run his receipts usually tend to an average depending on his productivity, so that the symbol v has a definite meaning. If this is so, then on the average and in the long run v in the last equation will be zero (provided the number of separate businesses is enough to justify statistical treatment). At any time when v is not zero, it constitutes a surplus or deficiency in profits belonging to all the entrepreneurs.

More generally, if a change in conditions alters ξ to ξ', then the whole surplus accruing, as a consequence, to a unit of x is

$$\int (\xi' - \xi)\, dt,$$

which could be found if the relation of ξ to x (i.e. the elasticity of demand) and q be known.

I CARATTERI MATEMATICI DELLA SCIENZA ECONOMICA

DI LUIGI AMOROSO.

(ABSTRACT.)

Proposing to apply to economics methods which have succeeded in mathematical physics the author obtains a partial differential equation relating to the distribution of incomes, from the following postulates:

I. There exists a continuous function $z = F(x, t)$ with continuous derivatives such that $F(x, t)\,dx\,dt$ represents for every value of x and t the number of individuals who during the interval of time between t and $t + dt$ have an income comprised between x and $x + dx$.

II. If x_0 and x_1 are two very closely neighbouring values of x, the number of persons who at the moment t pass from income x_0 to income x_1 is, other things being equal, proportioned to the difference between the number of persons who at the moment under consideration have the income x_1 and the number of persons who at the same moment have the income x_0.

From these postulates the author deduces partial differential equations, such as

$$\frac{dz}{dt} = \rho \frac{d^2z}{dx^2} \quad (\rho \text{ a constant}),$$

for which it is claimed that like the partial differential equations in the physical sciences—that of heat for example—they may lead to important practical conclusions, such as the distribution of incomes at any moment t if three data are given:

1. The initial distribution at moment t_0.
2. The increment of population in the same interval from t_0 to t.
3. The constant ρ.

It is supposed that in the interval t_0, t no radical perturbation of incomes happens for any reason whatever. These perturbations would render the integral not regular.

REDUCTION OF ERRORS BY MEANS OF NEGLIGIBLE DIFFERENCES

By W. F. Sheppard.

Abstract.

The so-called "graduation" practised by actuaries consists in replacing each term of a sequence by the weighted mean of a number of terms, so as to give an opportunity for neighbouring errors of opposite signs to balance one another. This process is not legitimate unless the difference between the original and the substituted term is a difference which would be negligible if the original terms did not contain errors. The underlying assumption therefore is that differences of the true values, beyond those of a certain order, are negligible; and the substitution is only legitimate if the mistake thus introduced is expressible in terms of these negligible differences.

In the same way the existence of alternative formulae for quadrature, etc., implies that the difference between two formulae represents negligible differences of the terms of the true sequence.

Conversely, any formula expressing a quantity by a linear function of a set of terms of a sequence may, within certain limits, be modified by the addition of terms representing the negligible differences.

The added portion will involve arbitrary constants, and we can choose these so as to give the best result for the new formula.

For comparison of goodness, some criterion is required. We take the mean square of error as the criterion; the problem therefore is to find the values of the coefficients when the mean square of error of the total expression is a minimum.

It is not possible to give more than a formal solution for the general case; but, for the standard class of cases in which the errors of the terms of the sequence are independent and are all equally variable, the solution is obtained explicitly in terms of central differences.

The method differs in essentials from the method of least squares and the method of moments, but it is found to give, for certain problems, the same results as would be given by these methods. The formulae obtained may therefore be regarded as giving (for the standard class of cases) a general solution, in terms of central differences, of the equations which arise in these methods.

The method is extended to particular classes of cases where the errors are not independent or are not all equally variable.

Contents.

I. Introductory.

1. The preliminary property, on which the methods considered in this paper are based, may be seen by differencing an ordinary mathematical table. The following is an example, the position of the decimal point being, as usual, ignored in writing down the differences.

x	$\log_{10} x$	1st diff.	2nd diff.	3rd diff.	4th diff.	5th diff.	6th diff.	7th diff.	8th diff.
5·4	·7324		−1		−2		+8		−32
		+80		−1		+4		−16	
5·5	·7404		−2		+2		−8		+33
		+78		+1		−4		+17	
5·6	·7482		−1		−2		+9		−38
		+77		−1		+5		−21	
5·7	·7559		−2		+3		−12		+46
		+75		+2		−7		+25	
5·8	·7634		0		−4		+13		−44

Here the successive differences at first are regular and decrease (numerically), and then become irregular and increase. The reason of this is that each term in the column $\log_{10} x$ consists of two portions, viz. the true value and an "error" which has to be added (algebraically) in order to obtain the corrected final figure. The effect of this error may be seen by writing down the true value to 7 places of decimals, and the error, for each term.

x	$10^7 \log_{10} x$	1st diff.	2nd diff.	3rd diff.
5·4	7323938+062		−1490+0490	
		+79689+311		+54−1054
5·5	7403627+373		−1436−0564	
		+78253−253		+52+0948
5·6	7481880+120		−1384+0384	
		+76869+131		+46−1046
5·7	7558749+251		−1338−0662	
		+75531−531		+47+1953
5·8	7634280−280		−1291+1291	

It will be seen that the differences of the true values decrease numerically (the apparent irregularities in the 3rd differences being due to our having only taken the true values to 7 places), while the differences of the errors increase.

This property, which is the preliminary condition for the application of the methods of the paper, may be stated as follows.

(i) We are dealing with a sequence

$$\ldots u_{-1},\ u_0,\ u_1,\ u_2,\ \ldots.$$

(ii) The terms of this sequence correspond to those of another sequence $\ldots U_{-1}$, U_0, U_1, U_2, ..., but differ from them by errors ... e_{-1}, e_0, e_1, e_2, ..., so that, if u_r is any term of the former sequence,

$$u_r = U_r + e_r.$$

(iii) The terms of the second sequence, i.e., the true values of the u's, are so related that their differences beyond those of a certain order, which we shall throughout denote by j, are negligible. We can define more precisely later on (see § 3) the meaning of the word "negligible."

The sequences which possess this property are of many different kinds. We have taken a mathematical table as an example; but this is rather because our knowledge of the U's enables us to illustrate and test our methods than because of any actual utility of the methods in reference to such tables. The cases to which our methods have to be applied cover, on the one hand, cases in which the U's are actual magnitudes, such as lengths, pressures, velocities, etc. measured more or less imperfectly at a series of points of space or of time, and, on the other hand, cases in which the u's are statistical data which are taken as varying from hypothetical U's by errors due to the paucity of the observations.

2. The problem which we have to consider is a generalisation of, and includes, the problem of what is—not very happily—described by actuaries as "graduation"; and we can consider the nature of this particular case before proceeding to the general problem.

The conditions are as follows. The U's are certain statistical ratios, whose values, as found from the limited data available, are the u's. The U's, if we had them, would proceed regularly, in the sense that condition (iii) of § 1 would be satisfied. But the u's do not proceed regularly, since their errors produce an irregularity, analogous to the irregularity discussed in § 1. "Graduation" consists in replacing the sequence ... u_{-1}, u_0, u_1, ... by another sequence v_{-1}, v_0, v_1, ..., such that each v is a linear compound* of the corresponding u and n others on each side of it, i.e., that

$$v_0 = p_n u_n + p_{n-1} u_{n-1} + \dots + p_{-n} u_{-n},$$

where p_n, p_{n-1}, ... p_{-n} are numerical coefficients; but the process may be regarded in two different ways.

Usually the object seems to be to obtain a sequence which is more regular than that of the u's; i.e., the process is merely one of "smoothing." But this is not a scientific method, unless some criterion is adopted for deciding which of the possible new sequences is the best; and it is difficult to apply such a criterion except to the sequence as a whole, in which case the process becomes one of "fitting."

The better view, which enables us to treat each term separately, is that we wish to apply a correction to the term in order to make its error as small as possible. If we can do this, the new sequence will be smoother than the old one, not merely because the new errors will be smaller, but also because the errors of adjoining terms will be more closely correlated. But this smoothness should be regarded rather as an incident than as an object.

In any case, certain conditions must be satisfied in order that it may be legitimate to replace u_0 by v_0. We have therefore to see (1) what these conditions

* I use this term in preference to the ordinary term "linear function," since we are not concerned, for the most part, with functionality; what we have usually to consider is the proportion in which the various u's, or the differences of various orders, have to be compounded in order to produce the required result.

are, and (2) what criterion we are to adopt for determining whether one set of p's which satisfies these conditions is better than another.

3. Since $$u_r = U_r + e_r \quad \text{.....................(1)},$$
we shall have $$v_0 = V_0 + f_0 \quad \text{.....................(2)},$$
where $$V_0 \equiv p_n U_n + p_{n-1} U_{n-1} + \dots + p_{-n} U_{-n} \quad \text{.....................(3)},$$
$$f_0 \equiv p_n e_n + p_{n-1} e_{n-1} + \dots + p_{-n} e_{-n} \quad \text{.....................(4)},$$
so that in substituting v_0 for u_0 we really substitute V_0 for U_0 and f_0 for e_0. Assuming (as a condition of our method) that the mean value of each e is zero, it follows that the substitution of v_0 for u_0 is not legitimate unless it would also be legitimate to substitute V_0 for U_0, i.e., unless the difference between V_0 and U_0 is negligible. This difference can be expressed in terms of U_0 and its central differences up to that of order $2n$ inclusive. But, by hypothesis, differences of orders 1 to j are not negligible, while those of higher orders are negligible. Hence, for practical purposes, the necessary condition for replacing u_0 by v_0 is that $V_0 - U_0$ should only involve differences of U_0 of orders $j+1$ and upwards; i.e., that $v_0 - u_0$ should only involve differences of u_0 of orders $j+1$ and upwards. We may therefore say that v_0 is formed from u_0 by adding terms involving these differences.

Whether, even then, the substitution is legitimate in the particular case, will depend on the actual magnitude of these differences of U_0 and on the magnitude of their coefficients in the formula. We must therefore, in describing the differences as negligible, remember that this description is a relative one, and that the differences may cease to be negligible if the numerical coefficients by which they are multiplied become too great.

4. The condition stated in § 3 implies $j+1$ relations between the coefficients $p_{-n}, p_{-n+1}, \dots p_n$ in the expression for v_0; and, subject to these relations, we want to choose the p's so as to make f_0, the error of v_0, as small as possible. This we cannot do. But it is necessary to adopt some criterion to determine whether one formula for v_0 is better or worse than another. For reasons familiar in the theory of error, we take the mean square of error as a criterion, and say that one formula is better or worse than another according as it gives a smaller or a greater mean square of error of v_0. Hence the problem of "graduation," in the more correct sense explained in § 2, is to find the p's so as to make the mean square of

$$f_0 \equiv p_n e_n + p_{n-1} e_{n-1} + \dots + p_{-n} e_{-n}$$

as small as possible, subject to the conditions stated above.

(The adoption of the above criterion is, of course, only a delimitation of the cases we have to consider. We must adopt some criterion, in order to apply mathematical methods; if in any particular case the criterion is unsuitable, this only means that our results do not apply to that particular case.)

5. This problem leads to its own extension, if we consider what we are to do at the extremities of the range. Suppose, for instance, that $j = 3$, $m = 9$; i.e., that differences beyond the 3rd are negligible, and that we are ordinarily using 9 consecutive u's to improve the value of the middle one. Let the values at the beginning of the sequence be denoted by $u_1, u_2, u_3, \dots$. Then the above statement of

the problem applies to the improvement of u_5, u_6, u_7, ...; but what are we to do with u_4, u_3, u_2, and u_1?

(i) We can leave them as they are. This means that the improvement of the sequence is incomplete.

(ii) We can improve u_4 and u_3 by using 7 and 5 u's respectively. But this does not enable us to do anything to u_2 (since the first 3 terms only give 1st differences and a 2nd difference) or to u_1.

(iii) The difficulty is met by extending our method so as to enable us to use the 9 terms u_1, u_2, ... u_9 for the improvement not only of u_5 but also of u_4, u_3, u_2, and u_1. Taking any term, such as u_2, we compound it with differences of orders 4 and upwards; and we then choose the coefficients in the resulting expression so as to make its mean square of error a minimum.

6. This leads us to the general problem, which may be stated as follows. Our data are m consecutive u's, corresponding to m consecutive U's. There is a quantity W which either (a) is one of the U's, or (b) is a linear compound of two or more of them, or (c) can be approximately (i.e., within the limits of accuracy within which we are working) represented by a linear compound of this kind. Let w be the first approximation to W, obtained by replacing the U's in W by the corresponding u's. Then the problem is to determine the form of a quantity z such that (i) z is a linear compound of the u's, (ii) z differs from w by differences of u of order $j+1$ and upwards, and (iii) the mean square of error of z is as small as possible.

For any particular set of data, the particular problems included in the above fall into two classes. The first class comprises those in which the total number of u's is greater than m, so that we apply one process to a number of different sets of u's, dropping one u at each stage and taking up another. This class includes the problems of (1) "graduation," (2) calculation of ordinates when the data represent areas, or of areas of consecutive strips when the data represent ordinates, and (3) determination of direction of tangents, for the purpose, e.g., of determining modal values (values corresponding to maximum ordinates). The second class comprises the cases in which we have only the m u's and are dealing with them all together. Here the specially important problems are those of determining the best formulae for quadrature and for calculation of moments. Interpolation is a matter which requires to be dealt with specially.

7. In the cases coming under (c) of § 6 there will usually be several alternative formulae for W, i.e., there will be several linear compounds of the U's, any one of which would, if the U's were known, give W with sufficient accuracy. But this will not affect our results, since the differences between the alternative expressions will themselves only involve differences of U of orders exceeding j. The following cases will illustrate this; j being taken $= 2$.

(i) Suppose that the U's correspond to values ... x_{-1}, x_0, x_1, ... of a known quantity x, these values differing by a constant increment h, so that $x_r = x_0 + rh$; and suppose that there are intermediate values of U, which can be represented by a polynomial in x of degree not exceeding $m-1$. Then, if U_θ is the value of U corresponding to $x = x_0 + \theta h$, where $0 < \theta < 1$, we have as possible formulae for U_θ

$$U_\theta = U_0 + \theta\Delta U_0 + \tfrac{1}{2}\theta(\theta-1)\Delta^2 U_0,$$

$$U_\theta = U_0 + \theta\Delta U_0 + \tfrac{1}{2}\theta(\theta-1)\Delta^2 U_{-1}.$$

But these differ by $\frac{1}{2}\theta(\theta-1)\Delta^3 U_{-1}$.

(ii) On the same suppositions, let $m=5$, and let the required quantity be the area of the figure bounded by the ordinates U_1 and U_5. Then the values of $U_1' \equiv (dU/dx)_1$ and of $U_5' \equiv (dU/dx)_5$ will be given by

$$hU_1' = (U_2 - U_1) - \tfrac{1}{2}(U_3 - 2U_2 + U_1) = -\tfrac{1}{2}(U_3 - 4U_2 + 3U_1),$$

$$hU_5' = (U_5 - U_4) + \tfrac{1}{2}(U_5 - 2U_4 + U_3) = \tfrac{1}{2}(3U_5 - 4U_4 + U_3);$$

and therefore, by well-known formulae, we have as alternative expressions for the area

$$W = \tfrac{1}{3}.h(U_1 + 4U_2 + 2U_3 + 4U_4 + U_5),$$

$$W = \tfrac{1}{2}.h(U_1 + 2U_2 + 2U_3 + 2U_4 + U_5) - \tfrac{1}{24}.h(3U_1 - 4U_2 + U_3) - \tfrac{1}{24}.h(U_3 - 4U_4 + 3U_5)$$

$$= \tfrac{1}{24}.h(9U_1 + 28U_2 + 22U_3 + 28U_4 + 9U_5).$$

It will be found that these differ by $\frac{1}{24}(U_5 - 4U_4 + 6U_3 - 4U_2 + U_1) = \frac{1}{24}.\Delta^4 U_1$.

8. *Notation.* (i) Our formulae will be expressed in terms of central differences; and therefore, supposing the data are m u's, the formulae will be of different kinds according as m is odd or even.

(a) Let $m = 2n+1$. Then we take the u's to be $u_{-n}, u_{-n+1}, \ldots u_n$, so that the central u is u_0. The central differences of u_0 are $\mu\delta u_0, \delta^2 u_0, \mu\delta^3 u_0, \ldots$, where

$$\mu\delta u_0 \equiv \tfrac{1}{2}\{(u_1 - u_0) + (u_0 - u_{-1})\}, \quad \delta^2 u_0 \equiv u_1 - 2u_0 + u_{-1},$$

$$\mu\delta^{2r+1}u_0 \equiv \mu\delta.\delta^{2r}u_0, \quad \delta^{2r+2}u_0 \equiv \delta^2.\delta^{2r}u_0.$$

We can then express $z \equiv p_n u_n + p_{n-1}u_{n-1} + \ldots + p_{-n}u_{-n}$ in the form

$$z = q_0 u_0 + q_1\mu\delta u_0 + q_2\delta^2 u_0 + \ldots + q_{2n}\delta^{2n}u_0.$$

(b) Let $m = 2n$. Then we take the u's to be $u_{-n+1}, u_{-n+2}, \ldots u_n$, so that the two u's in the middle are u_0 and u_1. The central-difference formulae then involve $\mu u_{\frac{1}{2}}$ and its central differences $\delta u_{\frac{1}{2}}, \mu\delta^2 u_{\frac{1}{2}}, \delta^3 u_{\frac{1}{2}}, \ldots$, where

$$\mu u_{\frac{1}{2}} \equiv \tfrac{1}{2}(u_1 + u_0), \quad \delta u_{\frac{1}{2}} \equiv u_1 - u_0,$$

$$\mu\delta^{2r}u_{\frac{1}{2}} \equiv \mu.\delta^{2r}u_{\frac{1}{2}}, \quad \delta^{2r+1}u_{\frac{1}{2}} \equiv \delta.\delta^{2r}u_{\frac{1}{2}};$$

and $z \equiv p_n u_n + p_{n-1}u_{n-1} + \ldots + p_{-n+1}u_{-n+1}$ can be expressed in the form

$$z = q_0\mu u_{\frac{1}{2}} + q_1\delta u_{\frac{1}{2}} + q_2\mu\delta^2 u_{\frac{1}{2}} + \ldots + q_{2n-1}\delta^{2n-1}u_{\frac{1}{2}}.$$

(ii) We can deal with the powers of μ and δ as operators obeying the laws of arithmetic. If, e.g., we take $\ldots\ v_{-1} \equiv \delta^2 u_{-1},\ v_0 \equiv \delta^2 u_0,\ v_1 \equiv \delta^2 u_1, \ldots$, and then take $\xi_{\frac{1}{2}} \equiv \mu\delta^2 v_{\frac{1}{2}}$, we shall have $\xi_{\frac{1}{2}} = \mu\delta^4 u_{\frac{1}{2}}$. We can therefore detach the operators, and can, e.g., write $u_0 + \delta^2 u_0 + \frac{1}{2}\delta^4 u_0$ in the form $(1 + \delta^2 + \frac{1}{2}\delta^4)u_0$.

(iii) For stating the relations between the p's and the q's in (a) or (b) of (i), and for our purposes generally, it is convenient to express the binomial coefficients, and certain coefficients derived from them, in a notation suitable to central differences. If we write*

$$(n,\ r) \equiv n(n-1)\dots(n-r+1)/r!, \quad [n,\ r] \equiv n(n+1)\dots(n+r-1)/r!,$$

we can also write

$$(n,\ 2s+1] \equiv (n-s)(n-s+1)\dots(n+s)/(2s+1)!$$
$$= n(n^2-1^2)(n^2-2^2)\dots(n^2-s^2)/(2s+1)!,$$
$$(n+\tfrac{1}{2},\ 2s] \equiv (n-s+1)(n-s+2)\dots(n+s)/(2s)!;$$

this latter being equivalent to

$$(n,\ 2s] \equiv \{n^2-(\tfrac{1}{2})^2\}\{n^2-(\tfrac{3}{2})^2\}\dots\{n^2-(s-\tfrac{1}{2})^2\}/(2s)!.$$

These definitions hold whether n or $n+\frac{1}{2}$ is an integer or not. If in $(n,\ r]$ we give n a series of values with constant increment 1, we obtain a sequence whose 1st differences are the values of $(n,\ r-1]$ for intermediate values of n; i.e.,

$$(n+1,\ r]-(n,\ r]=(n+\tfrac{1}{2},\ r-1].$$

If now we construct central differences of these coefficients in the usual way, we obtain a new set of coefficients which may be written

$$[n,\ 2s) \equiv \tfrac{1}{2}\{(n+\tfrac{1}{2},\ 2s]+(n-\tfrac{1}{2},\ 2s]\}$$
$$= n^2(n^2-1^2)(n^2-2^2)\dots\{n^2-(s-1)^2\}/(2s)!,$$
$$[n+\tfrac{1}{2},\ 2s+1) \equiv \tfrac{1}{2}\{(n+1,\ 2s+1]+(n,\ 2s+1]\}$$
$$= (n+\tfrac{1}{2}).(n-s+1)(n-s+2)\dots(n+s)/(2s+1)!;$$

this latter being equivalent to

$$[n,\ 2s+1) = n\{n^2-(\tfrac{1}{2})^2\}\{n^2-(\tfrac{3}{2})^2\}\dots\{n^2-(s-\tfrac{1}{2})^2\}/(2s+1)!.$$

(iv) We then have the following relations:

(*a*) For (*a*) of (i), r being a positive integer,

$$\delta^{2r}u_0 = u_r-(2r,\ 1)\,u_{r-1}+(2r,\ 2)\,u_{r-2}\dots+(-)^r(2r,\ r)\,u_0+\dots+u_{-r}\dots(5),$$
$$2r.\mu\delta^{2r-1}u_0 = ru_r-(r-1)(2r,\ 1)\,u_{r-1}+(r-2)(2r,\ 2)\,u_{r-2}\dots-ru_{-r}\ (6),$$
$$u_r = u_0+(r,\ 1]\,\mu\delta u_0+[r,\ 2)\,\delta^2u_0+(r,\ 3]\,\mu\delta^3u_0+\dots+[r,\ 2r)\,\delta^{2r}u_0\ \dots(7),$$
$$u_{-r} = u_0-(r,\ 1]\,\mu\delta u_0+[r,\ 2)\,\delta^2u_0-(r,\ 3]\,\mu\delta^3u_0+\dots+[r,\ 2r)\,\delta^{2r}u_0\dots(8).$$

Formulae (5) and (6) give the p's in terms of the q's, and (7) and (8) give

$$\left.\begin{aligned} q_0 &= p_0+(p_1+p_{-1})+(p_2+p_{-2})+\dots+(p_n+p_{-n}) \\ q_{2s} &= [0,\ 2s)\,p_0+[1,\ 2s)(p_1+p_{-1})+[2,\ 2s)(p_2+p_{-2})+\dots+[n,\ 2s)(p_n+p_{-n}) \end{aligned}\right\}\ (9),$$

$$q_{2s-1} = (1,\ 2s-1](p_1-p_{-1})+(2,\ 2s-1](p_2-p_{-2})+\dots+(n,\ 2s-1](p_n-p_{-n})\ (10).$$

(*b*) For (*b*) of (i)

$$\delta^{2r-1}u_{\frac{1}{2}} = u_r-(2r-1,\ 1)\,u_{r-1}+(2r-1,\ 2)\,u_{r-2}-\dots-u_{-r+1}\dots\dots(11),$$
$$(2r+1)\,\mu\delta^{2r}u_{\frac{1}{2}} = (r+\tfrac{1}{2})\,u_{r+1}-(r-\tfrac{1}{2})(2r+1,\ 1)\,u_r$$
$$+(r-\tfrac{3}{2})(2r+1,\ 2)\,u_{r-1}-\dots-(r+\tfrac{1}{2})\,u_{-r}\dots(12),$$
$$u_{r+1} = \mu u_{\frac{1}{2}}+[r+\tfrac{1}{2},\ 1)\,\delta u_{\frac{1}{2}}+(r+\tfrac{1}{2},\ 2]\,\mu\delta^2u_{\frac{1}{2}}+\dots+[r+\tfrac{1}{2},\ 2r+1)\,\delta^{2r+1}u_{\frac{1}{2}}\ (13),$$
$$u_{-r} = \mu u_{\frac{1}{2}}-[r+\tfrac{1}{2},\ 1)\,\delta u_{\frac{1}{2}}+(r+\tfrac{1}{2},\ 2]\,\mu\delta^2u_{\frac{1}{2}}-\dots-[r+\tfrac{1}{2},\ 2r+1)\,\delta^{2r+1}u_{\frac{1}{2}}\ (14).$$

* The alternative notation for $(n,\ r)$, $[n,\ r]$, $(n,\ r]$, $[n,\ r)$, is $n_{(r)}$, $n_{[r]}$, $n_{(r]}$, $n_{[r)}$. But the former are more convenient for printing.

Formulae (11) and (12) give the p's in terms of the q's, and (13) and (14) give

$$\left.\begin{aligned} q_0 &= (p_1+p_0)+(p_2+p_{-1})+(p_3+p_{-2})+\ldots+(p_n+p_{-n+1}) \\ q_{2s} &= (\tfrac{1}{2},\,2s]\,(p_1+p_0)+(\tfrac{3}{2},\,2s]\,(p_2+p_{-1})+\ldots+(n-\tfrac{1}{2},\,2s]\,(p_n+p_{-n+1}) \end{aligned}\right\}\ldots(15),$$

$$q_{2s-1}=[\tfrac{1}{2},\,2s-1)\,(p_1-p_0)+[\tfrac{3}{2},\,2s-1)\,(p_2-p_{-1})+\ldots$$
$$\ldots+[n-\tfrac{1}{2},\,2s-1)\,(p_n-p_{-n+1})\ldots(16).$$

II. Formal Solution of General Problem.

9. We take first the case in which there are $m\equiv 2n+1$ u's, which we denote by $u_{-n}, u_{-n+1}, \ldots u_n$. Since, by hypothesis, W is a linear compound of the U's, we can write it in the form

$$W=c_0U_0+c_1\mu\delta U_0+c_2\delta^2U_0+\ldots \quad\ldots\ldots\ldots\ldots\ldots\ldots\ldots(17).$$

Then our first approximation to W is

$$w=c_0u_0+c_1\mu\delta u_0+c_2\delta^2u_0+\ldots \quad\ldots\ldots\ldots\ldots\ldots\ldots\ldots(18).$$

Now suppose we take

$$z\equiv p_nu_n+p_{n-1}u_{n-1}+\ldots+p_{-n}u_{-n} \quad\ldots\ldots\ldots\ldots\ldots\ldots\ldots(19)$$

$$\equiv q_0u_0+q_1\mu\delta u_0+q_2\delta^2u_0+\ldots+q_{2n}\delta^{2n}u_0 \quad\ldots\ldots\ldots\ldots(19\text{ A}).$$

Then, in order that w and z may differ by differences of the u's of orders exceeding j, we must have

$$q_0=c_0,\quad q_1=c_1,\quad q_2=c_2,\quad \ldots\quad q_j=c_j \quad\ldots\ldots\ldots\ldots\ldots\ldots(20).$$

The error of z is

$$f\equiv p_ne_n+p_{n-1}e_{n-1}+\ldots+p_{-n}e_{-n}\quad\ldots\ldots\ldots\ldots\ldots\ldots\ldots(21).$$

Let the mean square of error of u_r be denoted by $\pi_{r,r}a^2$, and the mean product of errors of u_r and u_s by $\pi_{r,s}a^2$, where a is a quantity of the same kind as the e's, so that the π's are numerical. Then, if the mean square of error of z is R^2a^2, we have

$$R^2=\sum_r\sum_s \pi_{r,s}p_rp_s \quad\ldots\ldots\ldots\ldots\ldots\ldots\ldots\ldots\ldots(22),$$

where $\sum_r$ and $\sum_s$ denote summations for all integral values of r and of s from $-n$ to n, so that the term involving $\pi_{r,r}$ is $\pi_{r,r}p_r^2$, and the term involving $\pi_{r,s}$ is $2\pi_{r,s}p_rp_s$. The problem is then to determine the values of the p's which will make R^2 a minimum, subject to the conditions (20).

10. If in (20) we express the q's in terms of the p's, in accordance with (19) and (19 A), we shall have $j+1$ conditions

$$\left.\begin{array}{lll} p_n+p_{n-1}+\ldots+p_1+p_0+p_{-1}+\ldots+p_{-n} & & =c_0 \\ (n,1]\,p_n+(n-1,1]\,p_{n-1}+\ldots+(1,1]\,p_1 & -(1,1]\,p_{-1}-\ldots-(n,1]\,p_{-n} & =c_1 \\ [n,2)\,p_n+[n-1,2)\,p_{n-1}+\ldots+[1,2)\,p_1 & +[1,2)\,p_{-1}+\ldots+[n,2)\,p_{-n} & =c_2 \\ (n,3]\,p_n+(n-1,3]\,p_{n-1}+\ldots+(1,3]\,p_1 & -(1,3]\,p_{-1}-\ldots-(n,3]\,p_{-n} & =c_3 \\ & \text{etc.} & \end{array}\right\}(23),$$

the last being

$$(n,j]\,p_n+(n-1,j]\,p_{n-1}+\ldots+(-n,j]\,p_{-n}=c_j$$

or

$$[n,j)\,p_n+[n-1,j)\,p_{n-1}+\ldots+[-n,j)\,p_{-n}=c_j,$$

according as j is odd or even. We shall, for shortness, assume j to be odd; the general results will be independent of this assumption. To make R^2 a minimum, subject to the above conditions, we proceed in the usual way, by taking differentials and equating them to zero. From (22) we obtain

$$\sum_r \{\sum_s \pi_{r,s} p_s\} dp_r = 0;$$

and hence, by (23), we find that we must have, if j is odd,

$$\left.\begin{aligned} \pi_{n,n}p_n + \pi_{n,n-1}p_{n-1} + \ldots + \pi_{n,-n}p_{-n} &= \lambda_0 + (n, 1]\lambda_1 + [n, 2)\lambda_2 + \ldots + (n, j]\lambda_j \\ \pi_{n-1,n}p_n + \pi_{n-1,n-1}p_{n-1} + \ldots + \pi_{n-1,-n}p_{-n} &= \lambda_0 + (n-1, 1]\lambda_1 + [n-1, 2)\lambda_2 + \ldots \\ &\qquad + (n-1, j]\lambda_j \\ &\;\;\vdots \\ \pi_{-n,n}p_n + \pi_{-n,n-1}p_{n-1} + \ldots + \pi_{-n,-n}p_{-n} &= \lambda_0 + (-n, 1]\lambda_1 + [-n, 2)\lambda_2 + \ldots \\ &\qquad + (-n, j]\lambda_j \end{aligned}\right\} \quad \ldots\ldots(24),$$

where the λ's are unknown multipliers. These are $2n+1$ equations by means of which we can formally (using determinants) express $p_n, p_{n-1}, \ldots p_{-n}$ in terms of $\lambda_0, \lambda_1, \ldots \lambda_j$. Substituting in (23), we have $j+1$ equations to determine $\lambda_0, \lambda_1, \ldots \lambda_j$ in terms of $c_0, c_1, \ldots c_j$; and these give $p_n, p_{n-1}, \ldots p_{-n}$, so that the solution is theoretically complete.

11. The result of these operations will be to give the p's, and therefore the q's, as linear compounds of the c's. This leads to the following conclusions.

(i) Expressing the q's in terms of the c's, let z, as given by (19 A), become

$$z = c_0t_0 + c_1t_1 + c_2t_2 + \ldots + c_jt_j \quad \ldots\ldots\ldots\ldots\ldots\ldots\ldots(25).$$

Then the forms of the t's are independent of the values of the c's, and therefore, if g is any one of the numbers 0, 1, 2, ... j, we should obtain $z = t_g$ by putting $c_g = 1$ and each of the other c's $= 0$. Hence $t_0, t_1, t_2, \ldots t_j$ are the values of z obtained by making R^2 a minimum, subject to the respective sets of conditions

$$\left.\begin{array}{l} q_0 = 1,\; q_1 = 0,\; q_2 = 0,\; q_3 = 0, \ldots q_j = 0 \\ q_0 = 0,\; q_1 = 1,\; q_2 = 0,\; q_3 = 0, \ldots q_j = 0 \\ q_0 = 0,\; q_1 = 0,\; q_2 = 1,\; q_3 = 0, \ldots q_j = 0 \\ \quad\vdots \\ q_0 = 0,\; q_1 = 0,\; \ldots\ldots\ldots\; q_{j-1} = 0,\; q_j = 1 \end{array}\right\} \quad \ldots\ldots\ldots\ldots(26).$$

(ii) Let W be U_r, the true value of one of the given u's. Then, by (7),

$$w = u_r = u_0 + (r, 1]\mu\delta u_0 + [r, 2)\delta^2 u_0 + (r, 3]\mu\delta^3 u_0 + \ldots \quad \ldots\ldots(27).$$

Hence, if we denote the new value of u_r by v_r, we shall have

$$v_r = t_0 + (r, 1]t_1 + [r, 2)t_2 + (r, 3]t_3 + \ldots \quad \ldots\ldots\ldots\ldots\ldots(28),$$

the series being continued up to the term in t_j. But this will be a polynomial in r of degree not exceeding j. The differences of the v's of order exceeding j will therefore be zero, so that we shall have

$$v_r = v_0 + (r, 1]\mu\delta v_0 + [r, 2)\delta^2 v_0 + \ldots \quad \ldots\ldots\ldots\ldots\ldots\ldots(29),$$

the series being continued up to the difference of order j. But, since this is true for all the v's, we see by comparing (29) with (28) that we must have

$$t_0 = v_0, \quad t_1 = \mu\delta v_0, \quad t_2 = \delta^2 v_0, \ldots.$$

(iii) Hence we get the following result. Let us, as in (19 A), write

$$z \equiv q_0 u_0 + q_1 \mu\delta u_0 + q_2 \delta^2 u_0 + \ldots + q_{2n}\delta^{2n} u_0.$$

Taking each set of conditions in (26) separately, find the values of q which, for each set, make R^2 a minimum. Let the resulting values of z be denoted by v_0, $\mu\delta v_0$, $\delta^2 v_0$, ..., up to $\delta^{2k} v_0$ or $\mu\delta^{2k+1} v_0$ according as $j = 2k$ or $= 2k+1$. Regarding $\mu\delta v_0$, $\delta^2 v_0$, ... as the central differences of v_0 derived from a sequence $v_{-n}, v_{-n+1}, \ldots v_n$, whose differences of order $j+1$ are zero, construct this sequence, or suppose it to be constructed. Then, whatever compound W may be of the U's or of U_0, $\mu\delta U_0$, $\delta^2 U_0$, ..., z will be the same compound of the v's or of v_0, $\mu\delta v_0$, $\delta^2 v_0$, ... There may, as mentioned in § 7, be alternative formulae for W, but this will not affect our results, provided that the alternative formulae differ by differences of the U's of order exceeding j; for, since the corresponding differences of the v's will be zero, the formulae will all give the same z.

12. To find the mean square of error of z, we have, by (22), (23), and (24),

$$\begin{aligned} R^2 &= \sum_r \{\sum_s \pi_{r,s} p_s\} p_r \\ &= \sum_r \{\lambda_0 + (r, 1]\lambda_1 + [r, 2)\lambda_2 + \ldots + (r, j]\lambda_j\} p_r \\ &= \lambda_0 \sum_r p_r + \lambda_1 \sum_r (r, 1] p_r + \ldots + \lambda_j \sum_r (r, j] p_r \\ &= \lambda_0 c_0 + \lambda_1 c_1 + \lambda_2 c_2 + \lambda_3 c_3 + \ldots + \lambda_j c_j. \end{aligned}$$

Let the value of λ_f obtained from the $(g+1)$th set of conditions in (26) be $\lambda_{f,g}$. Then the total value of λ_f is $\lambda_{f,0} c_0 + \lambda_{f,1} c_1 + \ldots + \lambda_{f,j} c_j$, and therefore

$$R^2 = \sum_f \sum_g c_f c_g \lambda_{f,g} \quad \ldots\ldots\ldots\ldots\ldots\ldots\ldots\ldots\ldots (30).$$

13. We shall obtain similar results for the case in which there are $m \equiv 2n$ u's. We denote these by $u_{-n+1}, u_{-n+2}, \ldots u_n$. Suppose that

$$w = c_0 \mu u_{\frac{1}{2}} + c_1 \delta u_{\frac{1}{2}} + c_2 \mu\delta^2 u_{\frac{1}{2}} + \ldots,$$

and that we replace this by

$$z \equiv q_0 \mu u_{\frac{1}{2}} + q_1 \delta u_{\frac{1}{2}} + q_2 \mu\delta^2 u_{\frac{1}{2}} + \ldots + q_{2n-1}\delta^{2n-1} u_{\frac{1}{2}}.$$

Then the mean square of error of z will be found to be a minimum when

$$z = c_0 \mu v_{\frac{1}{2}} + c_1 \delta v_{\frac{1}{2}} + c_2 \mu\delta^2 v_{\frac{1}{2}} + \ldots,$$

the last term being $c_{2k}\mu\delta^{2k} v_{\frac{1}{2}}$ or $c_{2k+1}\delta^{2k+1} v_{\frac{1}{2}}$ according as $j = 2k$ or $= 2k+1$; the quantities $\mu v_{\frac{1}{2}}$, $\delta v_{\frac{1}{2}}$, $\mu\delta^2 v_{\frac{1}{2}}$, ... being the values of z obtained by making its mean square of error a minimum, subject to the sets of conditions (26). The equations involved will be (23) and (24), with the omission of the terms involving p_{-n} and the last equation in (24).

III. Case of Errors Independent and Equally Variable.

(A.) *General Solution.*

14. The simplest case is that in which the errors of the u's are independent and have all the same mean square, which will be denoted by a^2. We shall first suppose that $m = 2n + 1$.

We can write z in the form

$$z \equiv p_n u_n + p_{n-1} u_{n-1} + \dots + p_{-n} u_{-n}$$
$$= p_0 u_0 + \tfrac{1}{2}(p_1 + p_{-1})(u_1 + u_{-1}) + \tfrac{1}{2}(p_2 + p_{-2})(u_2 + u_{-2}) + \dots + \tfrac{1}{2}(p_n + p_{-n})(u_n + u_{-n})$$
$$+ \tfrac{1}{2}(p_1 - p_{-1})(u_1 - u_{-1}) + \tfrac{1}{2}(p_2 - p_{-2})(u_2 - u_{-2}) + \dots + \tfrac{1}{2}(p_n - p_{-n})(u_n - u_{-n}) \quad \dots\dots\dots(31).$$

Also, if the mean square of error of z is $R^2 a^2$,

$$R^2 = p_n^2 + p_{n-1}^2 + \dots + p_{-n}^2$$
$$= p_0^2 + \tfrac{1}{2}(p_1 + p_{-1})^2 + \tfrac{1}{2}(p_2 + p_{-2})^2 + \dots + \tfrac{1}{2}(p_n + p_{-n})^2$$
$$+ \tfrac{1}{2}(p_1 - p_{-1})^2 + \tfrac{1}{2}(p_2 - p_{-2})^2 + \dots + \tfrac{1}{2}(p_n - p_{-n})^2 \dots\dots\dots(32).$$

From (31) we see that z may be regarded as based on two separate sequences, in one of which, if r has any of the values 0, 1, 2, ... n, the coefficient of u_r and of u_{-r} is $\tfrac{1}{2}(p_r + p_{-r})$, while in the other the coefficients of u_r and of u_{-r} are respectively $\tfrac{1}{2}(p_r - p_{-r})$ and $\tfrac{1}{2}(p_{-r} - p_r)$. And from (32) we see that the total value of R^2 is the sum of the values for these two sequences. The first sequence gives rise to the portion $q_0 u_0 + q_2 \delta^2 u_0 + \dots$ of z, and the second to the portion $q_1 \mu\delta u_0 + q_3 \mu\delta^3 u_0 + \dots$. We can therefore treat these portions independently, and make their mean squares of error a minimum separately, the respective conditions being

$$q_0 = c_0, \quad q_2 = c_2, \quad q_4 = c_4, \dots,$$

and
$$q_1 = c_1, \quad q_3 = c_3, \quad q_5 = c_5, \dots.$$

The effect is the same as if we first took

$$W = c_0 U_0 + c_2 \delta^2 U_0 + \dots,$$
$$z = q_0 u_0 + q_2 \delta^2 u_0 + \dots + q_{2n} \delta^{2n} u_0,$$

and then took
$$W = c_1 \mu\delta U_0 + c_3 \mu\delta^3 U_0 + \dots,$$
$$z = q_1 \mu\delta u_0 + q_3 \mu\delta^3 u_0 + \dots + q_{2n-1} \mu\delta^{2n-1} u_0,$$

and then added the results obtained by making the mean square of error of each z a minimum, subject to the above conditions.

15. Taking first
$$W = c_0 U_0 + c_2 \delta^2 U_0 + \dots \quad \dots\dots\dots(33),$$
let us write

$$z = q_0 u_0 + q_2 \delta^2 u_0 + \dots + q_{2n} \delta^{2n} u_0 \quad \dots\dots\dots(34)$$
$$= p_0 u_0 + p_1(u_1 + u_{-1}) + p_2(u_2 + u_{-2}) + \dots + p_n(u_n + u_{-n}) \quad \dots(34\text{A}),$$

where the new p_r is the old $\tfrac{1}{2}(p_r + p_{-r})$.

(i) The mean square of error of z is

$$R^2 = p_0^2 + 2p_1^2 + 2p_2^2 + \dots + 2p_n^2 \quad \dots\dots\dots(35),$$

and the conditions are, if $j = 2k$ or $2k+1$,

$$\left.\begin{array}{r} p_0 + 2p_1 + 2p_2 + \dots + 2p_n = c_0 \\ [1, 2)p_1 + [2, 2)p_2 + \dots + [n, 2)p_n = \tfrac{1}{2}c_2 \\ \vdots \\ [1, 2k)p_1 + [2, 2k)p_2 + \dots + [n, 2k)p_n = \tfrac{1}{2}c_{2k} \end{array}\right\} \dots\dots\dots(36).$$

Making R^2 a minimum, subject to these conditions, we find that

$$\left.\begin{aligned} p_0 &= \lambda_0 \\ p_1 &= \lambda_0 + [1,\ 2)\lambda_2 + [1,\ 4)\lambda_4 + \ldots + [1,\ 2k)\lambda_{2k} \\ p_2 &= \lambda_0 + [2,\ 2)\lambda_2 + [2,\ 4)\lambda_4 + \ldots + [2,\ 2k)\lambda_{2k} \\ &\vdots \\ p_n &= \lambda_0 + [n,\ 2)\lambda_2 + [n,\ 4)\lambda_4 + \ldots + [n,\ 2k)\lambda_{2k} \end{aligned}\right\} \quad \ldots\ldots\ldots\ldots(37).$$

For any given value of k, we could find the p's from these equations in the manner explained in § 10, and thence we could find the q's. But this requires each formula to be worked out separately. The following is the general solution.

(ii) The value of z is, as in § 11,

$$z = c_0 v_0 + c_2 \delta^2 v_0 + \ldots + c_{2k} \delta^{2k} v_0 \quad \ldots\ldots\ldots\ldots\ldots\ldots(38),$$

where v_0, $\delta^2 v_0$, ... $\delta^{2k} v_0$ are the values of z obtained by making its mean square of error a minimum, subject to the conditions

$$\begin{gathered} q_0 = 1,\ q_2 = 0,\ q_4 = 0,\ \ldots\ q_{2k-2} = 0,\ q_{2k} = 0, \\ q_0 = 0,\ q_2 = 1,\ q_4 = 0,\ \ldots\ q_{2k-2} = 0,\ q_{2k} = 0, \\ \vdots \\ q_0 = 0,\ q_2 = 0,\ q_4 = 0,\ \ldots\ q_{2k-2} = 0,\ q_{2k} = 1. \end{gathered}$$

(iii) The conditions for $\delta^{2t} v_0$ are

$$q_0 = q_2 = \ldots = q_{2t-2} = 0,\ q_{2t} = 1,\ q_{2t+2} = \ldots = q_{2k} = 0 \quad \ldots\ldots\ldots\ldots(39).$$

Making R^2 a minimum, subject to these conditions, we obtain equations of the form (37). These equations show that p_r is a polynomial in r^2 of degree k; and it may therefore be written in the form

$$p_r = \alpha_0 + \alpha_2 (r,\ 2] + \alpha_4 (r,\ 4] + \ldots + \alpha_{2k} (r,\ 2k] \quad \ldots\ldots\ldots\ldots(40).$$

Hence, since

$$z = p_0 u_0 + p_1 (u_1 + u_{-1}) + p_2 (u_2 + u_{-2}) + \ldots + p_n (u_n + u_{-n}),$$

it follows that z is of the form

$$z = \alpha_0 \Sigma u_r + \alpha_2 \Sigma (r,\ 2]\, u_r + \alpha_4 \Sigma (r,\ 4]\, u_r + \ldots + \alpha_{2k} \Sigma (r,\ 2k]\, u_r \quad \ldots(41),$$

the summation being in each case from $r = -n$ to $r = n$.

But it may be shown that

$$\Sigma (r,\ 2g]\, u_r = (4g + 2)(n + \tfrac{1}{2},\ 2g + 1] \sum_{s=0}^{s=n} (n + \tfrac{1}{2},\ 2s]/(2s + 2g + 1) \,.\, \delta^{2s} u_0 \ \ldots(42);$$

and therefore, by substitution in (41), we see that

$$z = \sum_{s=0}^{s=n} \phi(s)/\{(2s+1)(2s+3)\ldots(2s+2k+1)\} \,.\, (n + \tfrac{1}{2},\ 2s]\, \delta^{2s} u_0,$$

where $\phi(s)$ is some polynomial in s of degree k. In other words

$$q_{2s} = \phi(s)/\{(2s+1)(2s+3)\ldots(2s+2k+1)\} \,.\, (n + \tfrac{1}{2},\ 2s] \quad \ldots\ldots(43)$$

for all values of s from 0 to n. Now introduce the conditions (39). Then these, omitting $q_{2t} = 1$, show that $\phi(s)$ contains $s,\ s-1, \ldots\ s-t+1,\ s-t-1, \ldots\ s-k$ as factors; and the condition $q_{2t}=1$ determines the numerical coefficient. Thus q_{2s} is completely determined, and we find that

$$\delta^{2t}v_0=(-)^{k-t}\frac{[t+\tfrac{1}{2},k+1]}{t!(k-t)!(n+\tfrac{1}{2},2t]}\sum_{s=0}^{s=n}\frac{s(s-1)\ldots(s-k)}{(s-t)[s+\tfrac{1}{2},k+1]}(n+\tfrac{1}{2},2s]\,\delta^{2s}u_0 \quad \ldots(44).$$

(iv) Hence we find that, in the complete expression for z, s having any value from 0 to n,

$$q_{2s}=\frac{s(s-1)\ldots(s-k)}{k!}\frac{(n+\tfrac{1}{2},2s]}{[s+\tfrac{1}{2},k+1]}\sum_{t=0}^{t=k}(-)^{k-t}(k,t)\frac{[t+\tfrac{1}{2},k+1]}{(n+\tfrac{1}{2},2t]}\frac{c_{2t}}{s-t} \quad \ldots(45).$$

We might have obtained this without using the method of § 11. We should find, as in (iii) that q_{2s} is of the form given by (43); then the conditions

$$q_0=c_0,\; q_2=c_2,\;\ldots\; q_{2k}=c_{2k}$$

would give the values of $\phi(0)$, $\phi(1)$, $\phi(2),\ldots\phi(k)$; and thence, by Lagrange's interpolation-formula, we should obtain the value of $\phi(s)$, which would give (45).

(v) We see from (44) that the process of calculating the values of v_0, $\delta^2 v_0$, ..., for any value or set of values of n, is very simple. We can start with the formula for v_0 for the case of $k=0$ ($j=0$ or 1), viz.:—

$$v_0=\sum_{s=0}^{s=n}(n+\tfrac{1}{2},2s]/(2s+1)\,.\,\delta^{2s}u_0 \quad \ldots\ldots\ldots\ldots(46)$$

$$=1/(n+\tfrac{1}{2})\,.\sum_{s=0}^{s=n}[n+\tfrac{1}{2},2s+1)\,\delta^{2s}u_0 \quad \ldots\ldots\ldots\ldots(47),$$

which is, of course, equivalent to

$$v_0=(u_n+u_{n-1}+\ldots+u_{-n})/(2n+1).$$

To obtain the formula for $\delta^{2t}v_0$, for any other value of k, we divide the series in (47) by the coefficient of $\delta^{2t}u_0$, and then multiply the terms by a set of multipliers, which depend on k and t but are the same for all values of n. These multipliers, written for each set as a series, are as follows, for values of j from 0 to 5.

(a) For v_0.

$$(j=2 \text{ or } 3)\,[-3(s-1)/(2s+3)=]\,1+0-\frac{3}{7}-\frac{2}{3}-\frac{9}{11}-\frac{12}{13}-1-\frac{18}{17}-\frac{21}{19}-\frac{8}{7}-\frac{27}{23}\ldots,$$

$$(j=4 \text{ or } 5)\,[15(s-1)(s-2)/\{2(2s+3)(2s+5)\}=]\,1+0-0+\frac{5}{33}+\frac{45}{143}+\frac{6}{13}$$
$$+\frac{10}{17}+\frac{225}{323}+\frac{15}{19}+\frac{20}{23}+\frac{108}{115}+\ldots.$$

(b) For $\delta^2 v_0$.

$$(j=2 \text{ or } 3)\,[5s/(2s+3)=]\,0+1+\frac{10}{7}+\frac{5}{3}+\frac{20}{11}+\frac{25}{13}+2+\frac{35}{17}+\frac{40}{19}+\frac{15}{7}+\frac{50}{23}+\ldots,$$

$$(j=4 \text{ or } 5)\,[-35s(s-2)/\{(2s+3)(2s+5)\}=]\,0+1+0-\frac{35}{33}-\frac{280}{143}-\frac{35}{13}-\frac{56}{17}$$
$$-\frac{1225}{323}-\frac{80}{19}-\frac{105}{23}-\frac{112}{23}-\ldots.$$

(c) For $\delta^4 v_0$.

$$(j=4 \text{ or } 5)\ [63s(s-1)/\{2(2s+3)(2s+5)\} =]\ 0+0+1+\frac{21}{11}+\frac{378}{143}+\frac{42}{13}+\frac{63}{17}+\frac{1323}{323}+\frac{84}{19}+\frac{108}{23}+\frac{567}{115}+\ldots.$$

The coefficients in the resulting formulae, for values of n up to 9 ($m=19$), are given in Table I. A (i).

The expressions so found are connected by certain linear relations. If

$$P_{2k} \equiv (n+\tfrac{1}{2},\, 2k]/(2k+1)\,.\,\delta^{2k} v_0 \quad \ldots\ldots\ldots\ldots(48),$$

so that $P_0 = v_0$ as given by (47), and P_2, P_4, ... are found by multiplying the coefficients in P_0 by the values of $5s/(2s+3)$, $63s(s-1)/\{2(2s+3)(2s+5)\}$, ..., as given above, then the values of v_0, $\delta^2 v_0$, $\delta^4 v_0$, ... for $k=1, 2, \ldots$ are given by

$$(j=2 \text{ or } 3)\left\{\begin{aligned} v_0 &= P_0 - P_2 \\ (n+\tfrac{1}{2},\, 2]/3\,.\,\delta^2 v_0 &= \phantom{P_0 -{}} P_2 \end{aligned}\right\} \quad \ldots\ldots\ldots\ldots(49),$$

$$(j=4 \text{ or } 5)\left\{\begin{aligned} v_0 &= P_0 - P_2 + \tfrac{3}{7}P_4 \\ (n+\tfrac{1}{2},\, 2]/3\,.\,\delta^2 v_0 &= \phantom{P_0 -{}} P_2 - \tfrac{10}{7}P_4 \\ (n+\tfrac{1}{2},\, 4]/5\,.\,\delta^4 v_0 &= \phantom{P_0 - P_2 + \tfrac{3}{7}{}} P_4 \end{aligned}\right\} \quad \ldots\ldots\ldots\ldots(50),$$

etc.,

so that it would be sufficient to tabulate P_0, P_2, P_4, ... But it is simpler to calculate the terms of v_0, $\delta^2 v_0$, $\delta^4 v_0$, ... for each value of k by multiplications by the appropriate factors, either independently or successively.

16. Next taking $$W = c_1 \mu\delta U_0 + c_3 \mu\delta^3 U_0 + \ldots \quad \ldots\ldots\ldots\ldots(51),$$
we find in the same way, using the formula

$$\Sigma\,[r,\, 2g+1)\,u_r = 2\,(n+\tfrac{1}{2},\, 2g+1]\sum_{s=1}^{s=n} 2s\,(n+\tfrac{1}{2},\, 2s]/(2s+2g+1)\,.\,\mu\delta^{2s-1}u_0 \ldots(52),$$

that
$$z = c_1\mu\delta v_0 + c_3\mu\delta^3 v_0 + \ldots + c_{2k-1}\mu\delta^{2k-1}v_0 \quad \ldots\ldots\ldots\ldots(53)$$
$$= q_1\mu\delta u_0 + q_3\mu\delta^3 u_0 + \ldots + q_{2n-1}\mu\delta^{2n-1}u_0 \quad \ldots\ldots\ldots\ldots(54),$$

where

$$\mu\delta^{2t-1}v_0 = (-)^{k-t}\frac{[t+\tfrac{1}{2},\, k]}{t!\,(k-t)!\,(n+\tfrac{1}{2},\, 2t]}\sum_{s=1}^{s=n}\frac{s(s-1)\ldots(s-k)}{(s-t)\,[s+\tfrac{1}{2},\, k]}(n+\tfrac{1}{2},\, 2s]\,\mu\delta^{2s-1}u_0 \ldots(55),$$

and that the coefficient of $\mu\delta^{2s-1}u_0$ in the complete expression for z is

$$q_{2s-1} = \frac{s(s-1)\ldots(s-k)}{k!}\,\frac{(n+\tfrac{1}{2},\, 2s]}{[s+\tfrac{1}{2},\, k]}\sum_{t=1}^{t=k}(-)^{k-t}(k,\, t)\,\frac{[t+\tfrac{1}{2},\, k]}{(n+\tfrac{1}{2},\, 2t]}\,\frac{c_{2t-1}}{s-t} \ldots\ldots(56).$$

The formula for $\mu\delta v_0$ for the case of $j=1$ or 2 is therefore found from the formula for v_0 in (47) by omitting the 1st term, altering $\delta^2 u_0$, $\delta^4 u_0$, ... to $\mu\delta u_0$, $\mu\delta^3 u_0$, ..., dividing this new series by the coefficient of $\mu\delta u_0$, and then multiplying the terms by 1, 2, 3, ... The formula for $\mu\delta^{2t-1}v_0$, for any other value of j, is obtained by dividing this series by the coefficient of $\mu\delta^{2t-1}u_0$ and then multiplying the terms by a set of multipliers independent of n. These multipliers, for values of j up to 6, are as follows.

(a) For $\mu\delta v_0$.

$$(j=3 \text{ or } 4)\,[-5(s-2)/(2s+3)=]\,1+0-\frac{5}{9}-\frac{10}{11}-\frac{15}{13}-\frac{4}{3}-\frac{25}{17}-\frac{30}{19}-\frac{5}{3}-\frac{40}{23}-\ldots,$$

$$(j=5 \text{ or } 6)\,[35(s-2)(s-3)/\{2(2s+3)(2s+5)\}=]\,1+0-0+\frac{35}{143}+\frac{7}{13}+\frac{14}{17}+\frac{350}{323}+\frac{25}{19}+\frac{35}{23}+\frac{196}{115}+\ldots.$$

(b) For $\mu\delta^3 v_0$.

$$(j=3 \text{ or } 4)\,[7(s-1)/(2s+3)=]\,0+1+\frac{14}{9}+\frac{21}{11}+\frac{28}{13}+\frac{7}{3}+\frac{42}{17}+\frac{49}{19}+\frac{8}{3}+\frac{63}{23}+\ldots,$$

$$(j=5 \text{ or } 6)\,[-63(s-1)(s-3)/\{(2s+3)(2s+5)\}=]\,0+1+0-\frac{189}{143}-\frac{168}{65}-\frac{63}{17}-\frac{1512}{323}-\frac{105}{19}-\frac{144}{23}-\frac{3969}{575}-\ldots.$$

(c) For $\mu\delta^5 v_0$.

$$(j=5 \text{ or } 6)\,[99(s-1)(s-2)/\{2(2s+3)(2s+5)\}=]\,0+0+1+\frac{297}{143}+\frac{198}{65}+\frac{66}{17}+\frac{1485}{323}+\frac{99}{19}+\frac{132}{23}+\frac{3564}{575}+\ldots.$$

The coefficients in the resulting formulae, for values of n up to 9, are given in Table I. A (ii).

17. For $m=2n$, suppose that

$$W=c_0\mu U_{\frac{1}{2}}+c_1\delta U_{\frac{1}{2}}+c_2\mu\delta^2 U_{\frac{1}{2}}+\ldots \quad\ldots\ldots(57).$$

Then it will be found (cf. §13) that

$$z=c_0\mu v_{\frac{1}{2}}+c_1\delta v_{\frac{1}{2}}+c_2\mu\delta^2 v_{\frac{1}{2}}+\ldots \quad\ldots\ldots(58)$$

$$=q_0\mu u_{\frac{1}{2}}+q_1\delta u_{\frac{1}{2}}+q_2\mu\delta^2 u_{\frac{1}{2}}+\ldots+q_{2n-1}\delta^{2n-1}u_{\frac{1}{2}} \quad\ldots\ldots(59),$$

where, for $j=2k$ or $2k+1$,

$$q_{2s}=\frac{s(s-1)\ldots(s-k)}{k!}\frac{(n,\,2s+1]}{[s+\frac{3}{2},\,k]}\sum_{t=0}^{t=k}(-)^{k-t}(k,\,t)\frac{[t+\frac{3}{2},\,k]}{(n,\,2t+1]}\frac{c_{2t}}{s-t} \quad\ldots(60),$$

and, for $j=2k+1$ or $2k+2$,

$$q_{2s+1}=\frac{s(s-1)\ldots(s-k)}{k!}\frac{(n,\,2s+1]}{[s+\frac{3}{2},\,k+1]}\sum_{t=0}^{t=k}(-)^{k-t}(k,\,t)\frac{[t+\frac{3}{2},\,k+1]}{(n,\,2t+1]}\frac{c_{2t+1}}{s-t} \quad\ldots(61),$$

so that $\mu v_{\frac{1}{2}}$, $\delta v_{\frac{1}{2}}$, $\mu\delta^2 v_{\frac{1}{2}}$, ... are obtained by a method similar to that of §§ 15 and 16.

(i) For $\mu v_{\frac{1}{2}}$, $\mu\delta^2 v_{\frac{1}{2}}$, ... the standard series is

$$(j=0 \text{ or } 1)\;\mu v_{\frac{1}{2}}=1/n\,.\sum_{s=0}^{s=n-1}(n,\,2s+1]\,\mu\delta^{2s}u_{\frac{1}{2}} \quad\ldots\ldots(62);$$

and other expressions are obtained by a process similar to that of §15 (v), the multipliers (after dividing by a coefficient) being as follows.

(a) For $\mu v_{\frac{1}{2}}$.

$$(j=2 \text{ or } 3)\,[-3\,(s-1)/(2s+3)=]\,1+0-\frac{3}{7}-\frac{2}{3}-\frac{9}{11}-\frac{12}{13}-1-\frac{18}{17}-\frac{21}{19}-\frac{8}{7}-\ldots,$$

$$(j=4 \text{ or } 5)\,[15\,(s-1)\,(s-2)/\{2\,(2s+3)\,(2s+5)\}=]\,1+0-0+\frac{5}{33}+\frac{45}{143}$$
$$+\frac{6}{13}+\frac{10}{17}+\frac{225}{323}+\frac{15}{19}+\frac{20}{23}+\ldots.$$

(b) For $\mu\delta^2 v_{\frac{1}{2}}$.

$$(j=2 \text{ or } 3)\,[5s/(2s+3)=]\,0+1+\frac{10}{7}+\frac{5}{3}+\frac{20}{11}+\frac{25}{13}+2+\frac{35}{17}+\frac{40}{19}+\frac{15}{7}+\ldots,$$

$$(j=4 \text{ or } 5)\,[-35s\,(s-2)/\{(2s+3)\,(2s+5)\}=]\,0+1+0-\frac{35}{33}-\frac{280}{143}-\frac{35}{13}$$
$$-\frac{56}{17}-\frac{1225}{323}-\frac{80}{19}-\frac{105}{23}-\ldots.$$

(c) For $\mu\delta^4 v_{\frac{1}{2}}$.

$$(j=4 \text{ or } 5)\,[63s\,(s-1)/\{2\,(2s+3)\,(2s+5)\}=]\,0+0+1+\frac{21}{11}+\frac{378}{143}+\frac{42}{13}$$
$$+\frac{63}{17}+\frac{1323}{323}+\frac{84}{19}+\frac{108}{23}+\ldots.$$

(ii) For $\delta v_{\frac{1}{2}}, \delta^3 v_{\frac{1}{2}}, \ldots$ the standard series is

$$(j=1 \text{ or } 2)\;\delta v_{\frac{1}{2}}=1/n\,.\sum_{s=0}^{s=n-1} 3\,(n,\,2s+1]/(2s+3)\,.\,\delta^{2s+1}u_{\frac{1}{2}} \quad \ldots\ldots\ldots(63),$$

the coefficients in which are obtained from those in the standard series of (i) by multiplying by the values of $3/(2s+3)$, i.e. by $1, \frac{3}{5}, \frac{3}{7}, \frac{1}{3}, \frac{3}{11}, \ldots$; and the multipliers for other expressions (after dividing by a coefficient) are as follows.

(a) For $\delta v_{\frac{1}{2}}$.

$$(j=3 \text{ or } 4)\,[-5\,(s-1)/(2s+5)=]\,1+0-\frac{5}{9}-\frac{10}{11}-\frac{15}{13}-\frac{4}{3}-\frac{25}{17}-\frac{30}{19}-\frac{5}{3}-\frac{40}{23}-\ldots,$$

$$(j=5 \text{ or } 6)\,[35\,(s-1)\,(s-2)/\{2\,(2s+5)\,(2s+7)\}=]\,1+0-0+\frac{35}{143}+\frac{7}{13}$$
$$+\frac{14}{17}+\frac{350}{323}+\frac{25}{19}+\frac{35}{23}+\frac{196}{115}+\ldots.$$

(b) For $\delta^3 v_{\frac{1}{2}}$.

$$(j=3 \text{ or } 4)\,[7s/(2s+5)=]\,0+1+\frac{14}{9}+\frac{21}{11}+\frac{28}{13}+\frac{7}{3}+\frac{42}{17}+\frac{49}{19}+\frac{8}{3}+\frac{63}{23}+\ldots,$$

$$(j=5 \text{ or } 6)\,[-63s\,(s-2)/\{(2s+5)\,(2s+7)\}=]\,0+1+0-\frac{189}{143}-\frac{168}{65}-\frac{63}{17}$$
$$-\frac{1512}{323}-\frac{105}{19}-\frac{144}{23}-\frac{3969}{575}-\ldots.$$

(c) For $\delta^5 v_{\frac{1}{2}}$.

$$(j=5 \text{ or } 6)\,[99s(s-1)/\{2(2s+5)(2s+7)\}=]\,0+0+1+\frac{27}{13}+\frac{198}{65}+\frac{66}{17}$$

$$+\frac{1485}{323}+\frac{99}{19}+\frac{132}{23}+\frac{3564}{575}+\ldots.$$

The coefficients in the resulting formulae, for values of n up to 10 ($m=20$), are given in Table I. B (i) and (ii).

(B.) *Applications.*

18. "*Graduation.*" Here $W=U_0$, so that z is v_0 as given in § 15; i.e. by (44),

$$v_0=u_0+(-)^k\frac{1.3.5\ldots(2k+1)}{1.2\ldots k}\sum_{s=k+1}^{s=n}\frac{(s-1)(s-2)\ldots(s-k)}{(2s+1)(2s+3)\ldots(2s+2k+1)}(n+\tfrac{1}{2},2s]\,\delta^{2s}u_0 \quad \ldots\ldots(64).$$

(i) The ratio of the new mean square of error to the old is (cf. § 12), by (37),

$$\begin{aligned}
R^2 &= p_0^2+2p_1^2+2p_2^2+\ldots+2p_n^2\\
&= p_0\lambda_0+2p_1\{\lambda_0+[1,2)\lambda_2+\ldots+[1,2k)\lambda_{2k}\}+\ldots\\
&\qquad\ldots+2p_n\{\lambda_0+[n,2)\lambda_2+\ldots+[n,2k)\lambda_{2k}\}\\
&= \lambda_0\{p_0+2p_1+2p_2+\ldots+2p_n\}\\
&\qquad+2\lambda_2\{[1,2)p_1+[2,2)p_2+\ldots+[n,2)p_n\}\\
&\qquad+\ldots\\
&\qquad+2\lambda_{2k}\{[1,2k)p_1+[2,2k)p_2+\ldots+[n,2k)p_n\}\\
&= \lambda_0=p_0 \quad\ldots\ldots(65),
\end{aligned}$$

by (36) and (37), since $c_0=1$, $c_2=c_4=\ldots=c_{2k}=0$.

But p_0 is the coefficient of u_0 when $\delta^{2k+2}u_0$, $\delta^{2k+4}u_0$, ... $\delta^{2n}u_0$ are expressed in terms of the u's. Hence we find that

$$R^2=1+\sum_{s=k+1}^{s=n}(-)^{k+s}\frac{1.3.5\ldots(2k+1)}{1.2\ldots k}\,\frac{(n-s+1)(n-s+2)\ldots(n+s)}{s!(s-k-1)!\,s(2s+1)(2s+3)\ldots(2s+2k+1)} \quad \ldots\ldots\ldots(66).$$

Denoting this by R_k^2, we shall have, in the usual notation of hypergeometric series,

$$\begin{aligned}
R_k^2-R_{k-1}^2 &= (4k+1)\frac{1.3.5\ldots(2k-1)}{1.2\ldots k}\\
&\qquad\times\sum_{s=k}^{s=n}(-)^{k+s}\frac{(n-s+1)(n-s+2)\ldots(n+s)}{s!(s-k)!(2s+1)(2s+3)\ldots(2s+2k+1)}\\
&= (4k+1)\frac{1.3.5\ldots(2k-1)}{1.2.3\ldots k}\,\frac{(n-k+1)(n-k+2)\ldots(n+k)}{k!(2k+1)(2k+3)\ldots(4k+1)}\\
&\qquad\times F\{-n+k,\,n+k+1,\,k+\tfrac{1}{2};\;1,\,k+1,\,2k+\tfrac{3}{2};\;1\}\ldots(67).
\end{aligned}$$

The value of $F\{\ \}$ is given by the formula*

$$F\{-n,\,2\theta+n-1,\,\phi-\theta;\;1,\,\theta,\,\phi;\;1\}=[2\theta-\phi,n]/[\phi,n]\ldots\ldots(68),$$

* See *Proceedings of the London Mathematical Society*, series 2, vol. 10, p. 475.

so that

$$R_k^2 - R_{k-1}^2 = (4k+1)\left\{\frac{1.3.5\ldots(2k-1)}{1.2.3\ldots k}\right\}^2 \frac{(n-k+1)(n-k+2)\ldots(n+k)}{(2n-2k+1)(2n-2k+3)\ldots(2n+2k+1)}$$

$$= (4k+1)\left\{\frac{1.3.5\ldots(2k-1)}{2.4.6\ldots(2k)}\right\}^2 \frac{(m^2-1^2)(m^2-3^2)\ldots\{m^2-(2k-1)^2\}}{m(m^2-2^2)(m^2-4^2)\ldots\{m^2-(2k)^2\}} \quad \ldots\ldots\ldots(69),$$

and thence

$$R_k^2 = \frac{1}{m} + 5\left(\frac{1}{2}\right)^2 \frac{m^2-1^2}{m(m^2-2^2)} + 9\left(\frac{1.3}{2.4}\right)^2 \frac{(m^2-1^2)(m^2-3^2)}{m(m^2-2^2)(m^2-4^2)} + \ldots$$

$$+ (4k+1)\left\{\frac{1.3.5\ldots(2k-1)}{2.4.6\ldots(2k)}\right\}^2 \frac{(m^2-1^2)(m^2-3^2)\ldots\{m^2-(2k-1)^2\}}{m(m^2-2^2)(m^2-4^2)\ldots\{m^2-(2k)^2\}} \quad (70).$$

The resulting values of R, for $m = 3, 5, 7, \ldots 31$, are given in Table II*.

(ii) For practical purposes it is more convenient to express v_0 in terms of the u's, and also to replace fractional by decimal coefficients. I propose to deal elsewhere with this, and also with the comparison of the resulting formulae with some of the best known "graduation"-formulae.

19. *Interpolation.* We take U_θ, as in § 7, to be the value of U corresponding to $x = x_0 + \theta h$, where $0 < \theta < 1$.

(i) The ordinary (Lagrangian) formula for u_θ is obtained by taking a polynomial in x of degree j and equating it to $j+1$ consecutive u's. If, therefore, m u's are available, we have a choice of $m-j$ ordinary formulae. It will be found that the most accurate formula, i.e. the one which gives the least mean square of error for u_θ, is one of the j formulae which use both u_0 and u_1, but is different for different portions of the range (0 to 1) of values of θ. If we denote the roots of the equation in t

$$\sum_{g=1}^{g=j} (j, g-1)(j, g)(t-1, j)/(t-g) = 0$$

by $t_1, t_2, \ldots t_{j-1}$, and if we write $\theta_1 \equiv t_1 - 1$, $\theta_2 \equiv t_2 - 2$, $\ldots$ $\theta_{j-1} \equiv t_{j-1} - (j-1)$, then $\theta_1, \theta_2, \ldots \theta_{j-1}$ all lie between 0 and 1, and are in ascending order of magnitude; and they divide the interval (0 to 1) into j portions such that, if θ lies in the gth portion, the most accurate formula is the one which uses $u_{-g+1}, u_{-g+2}, \ldots u_{j-g+1}$.

(ii) The central-difference formula is therefore not the most accurate formula, except near the middle of the interval. It is, nevertheless, the best for ordinary purposes, since (1) its form is convenient for calculation, (2) it is the only formula (of this type) which throughout the whole interval makes the mean square of error of u_θ less than that of the given u's, and (3) the improvement effected by using any other formula is not much greater except at the extremities of the interval. Moreover, if

* By putting $m = 2k+1$, we obtain $R_k^2 = 1$, i.e.

$$1 = \frac{1}{2k+1} + 5\left(\frac{1}{1}\right)^2 \frac{k(k+1)}{(2k-1)(2k+1)(2k+3)} + 9\left(\frac{1.3}{1.2}\right)^2 \frac{(k-1)k(k+1)(k+2)}{(2k-3)(2k-1)\ldots(2k+5)} + \ldots$$

$$\ldots + (4k+1)\left\{\frac{1.3.5\ldots(2k-1)}{1.2.3\ldots k}\right\}^2 \frac{1.2.3\ldots(2k)}{1.3.5\ldots(4k+1)}.$$

This series contains $k+1$ terms. Inspection of Table II suggests that when $k = 2l+1$ the sum of the first $l+1$ terms is $\frac{1}{3}$, and the sum of the second $l+1$ terms $\frac{2}{3}$; but I have not been able to prove this.

we are subtabulating (i.e. constructing a table for which the interval is a submultiple of h), we obtain a smoother table by using the central-difference formula (or, if m is even, the mean of the two central-difference formulae) than by always using the most accurate formula. It should, however, be mentioned that, so long as we keep the original u's unaltered, smoothness and accuracy are antagonistic. The effect of using the most accurate formula for each interpolated value is to produce the maximum irregularity in passing from one interval to the next; and, conversely, the effect of using a special smoothing process, such as that of "osculatory interpolation," is to increase the inaccuracy.

(iii) Our new method uses m consecutive u's; and, if these include the $2j$ from u_{-j+1} to u_j, the resulting formula will be more accurate than any of those considered in (i). We therefore take as our basis the central-difference formula

$$(m=2n+1) \qquad U_\theta = U_0 + \theta\mu\delta U_0 + \theta^2/2!\,.\,\delta^2 U_0 + \ldots,$$

$$(m=2n) \qquad U_\theta = \mu U_{\frac{1}{2}} + (\theta - \tfrac{1}{2})\,\delta U_{\frac{1}{2}} + \theta(\theta-1)/2!\,.\,\mu\delta^2 U_{\frac{1}{2}} + \ldots,$$

and obtain the new formula by omitting terms after the difference of order j, and then replacing U by v throughout.

(iv) This formula is only suitable for improving isolated values. If we try to apply it to subtabulation, we shall introduce discontinuities, on account of pairs of values being found for quantities such as $U_0\,(m=2n+1)$ or $U_{\frac{1}{2}}\,(m=2n)$. For subtabulation, therefore, it is better to improve the data by the "graduation" method and then to apply the ordinary central-difference formula to each interval separately.

20. *Quadrature.* The ordinary formulae for quadrature are of two types, examples of which have been given in § 7. Both may be regarded as based on the formula for the "chordal" or the "tangential" area, which would be a correct formula if j were $=1$; for other values of j the formulae of the one type are obtained by altering the coefficients in a certain cyclical manner, while those of the other are obtained by introducing terminal corrections. Each type has certain advantages. It is sometimes stated that the "best" formula is the one obtained by treating u as a polynomial in x of the highest degree allowed by the data; but this does not rest on any clear definition of what is best. The most accurate formula, in our sense of the term, is obtained by the methods considered above.

There are four cases, according as the data are the bounding ordinates of the strips or the mid-ordinates, and according as their number is odd or even. In each case the formula is obtained, as before, by altering U's to v's.

(A.) Bounding ordinates given: m strips.

(i) $m=2n$.

$$\text{Area} = 2nh\,\{U_0 + (nh)^2\,U_0''/3! + (nh)^4\,U_0^{IV}/5! + \ldots\}$$

$$= 2nh\,\{U_0 + n^2/6\,.\,\delta^2 U_0 + (3n^4 - 5n^2)/360\,.\,\delta^4 U_0 + (3n^6 - 21n^4 + 28n^2)/15120\,.\,\delta^6 U_0 + \ldots\}.$$

(ii) $m = 2n + 1$.

$$\text{Area} = mh\{U_{\frac{1}{2}} + (\tfrac{1}{2}mh)^2 U_{\frac{1}{2}}''/3! + (\tfrac{1}{2}mh)^4 U_{\frac{1}{2}}^{IV}/5! + \ldots\}$$
$$= mh\{\mu U_{\frac{1}{2}} + (m^2 - 3)/24 . \mu\delta^2 U_{\frac{1}{2}} + (3m^4 - 50m^2 + 135)/5760 . \mu\delta^4 U_{\frac{1}{2}}$$
$$+ (3m^6 - 147m^4 + 1813m^2 - 4725)/967680 . \mu\delta^6 U_{\frac{1}{2}} + \ldots\}.$$

(B.) Mid-ordinates given: m strips.

(i) $m = 2n$.

$$\text{Area} = 2nh\{U_0 + (nh)^2 U_0''/3! + (nh)^4 U_0^{IV}/5! + \ldots\}$$
$$= mh\{\mu U_0 + (m^2 - 3)/24 . \mu\delta^2 U_0 + (3m^4 - 50m^2 + 135)/5760 . \mu\delta^4 U_0 + \ldots\}.$$

(ii) $m = 2n + 1$.

$$\text{Area} = mh\{U_{\frac{1}{2}} + (\tfrac{1}{2}mh)^2 U_{\frac{1}{2}}''/3! + (\tfrac{1}{2}mh)^4 U_{\frac{1}{2}}^{IV}/5! + \ldots\}$$
$$= mh\{U_{\frac{1}{2}} + m^2/24 . \delta^2 U_{\frac{1}{2}} + (3m^4 - 20m^2)/5760 . \delta^4 U_{\frac{1}{2}}$$
$$+ (3m^6 - 84m^4 + 448m^2)/967680 . \delta^6 U_{\frac{1}{2}} + \ldots\}.$$

The resulting formulae can all be adapted for numerical calculation in the manner explained in § 18 (ii).

21. *Other Applications.* The following are other applications; the method in all cases being the same, viz. to obtain the true formula in terms of central differences of the U's, and then to turn the U's into v's.

(i) *Differential Coefficients.* If U is the area measured up to the ordinate, U' is the ordinate. The ordinates enable us to trace the curve, and their differential coefficients show the position of the maximum ordinate, corresponding to the "modal" value of x. The formulae are well known.

(ii) *Moments.* The formulae for moments of areas can be improved in the same way. The U's may be either ordinates or areas, and the formulae will be different according as the moments are of odd or of even order. There are altogether 12 general formulae. (As to certain particular cases, see § 24.)

IV. Relation to Method of Least Squares and Method of Moments.

(A.) *Standard system (errors independent and equally variable).*

22. The results in §§ 18—21 may be regarded as obtained by treating U as a polynomial in x of degree j, and then, after performing any processes of integration, differentiation, etc., replacing the U's by the v's. But we should get the same result if we altered the order of these operations. It follows that, for any set of m consecutive u's, and for any given value of j, there is a polynomial

$$v \equiv A_0 + A_1 x + A_2 x^2 + \ldots + A_j x^j,$$

such that by taking $U = v$ throughout we obtain a less mean square of error for the resulting value of W than we should obtain by taking any other formula for U; and that this polynomial is the same whatever W may be, provided that W can be expressed as a linear compound of the m U's.

It may be shown that this polynomial is the same as is given by the method of least squares or the method of moments; it being understood that in the last-mentioned method the moments are not the moments of the area of which the u's are taken to be the ordinates, but the moments of the u's themselves. Hence these latter methods give the best results not only for the fit of the required polynomial to the given sequence as a whole, but also for the separate terms and for any quantity which can be expressed as a linear compound of them.

Although the new method and the old method (in either of its two forms) lead to the same result, they are based on quite different principles.

(i) The old method is a method of "fitting"; it considers the data as a whole, and finds the polynomial which fits them best. The new method deals separately with any particular u or with a quantity derived from some or all of the u's, and finds the best expression which can be substituted for it.

(ii) In the old method we consider that we are finding the real U, and that (e.g.) U_r is actually (taking $j=2$) of the form $A_0'+A_1'r+A_2'r^2$. In the new method we do not assume that v_r is identical with U_r; indeed, we say definitely that it is not, and we have in the coefficients of the "negligible" differences a means of measuring the extent of the mistake we make in substituting v_r for u_r.

(iii) In the new method the preliminary assumption (in the above case) is that v_r differs from u_r by a linear compound of differences of the 3rd order and upwards, and the fact that it is of the form $A_0+A_1r+A_2r^2$ is a deduction from this assumption. In the old methods this latter is the preliminary assumption, so that the process is reversed. To show that the latter assumption leads to the former, let

$$w_r \equiv A_0'+A_1'r+A_2'r^2 \quad \text{..........................(71)}$$

be the assumed value of U_r. Then, in the method of least squares, we make $\sum_r (w_r-u_r)^2$ a minimum. But

$$u_r = u_0 + r\,.\,\mu\delta u_0 + r^2/2!\,.\,\delta^2 u_0 + \phi(r),$$

where $\phi(r)$ is a definite linear compound of differences of u of order exceeding 2. Hence

$$w_r-u_r=(A_0'-u_0)+r(A_1'-\mu\delta u_0)+r^2(A_2'-\tfrac{1}{2}\delta^2u_0)-\phi(r).$$

Keeping w_r-u_r in this form, let us make $\sum_r (w_r-u_r)^2$ a minimum. Then we shall have a set of linear equations to determine $A_0'-u_0$, $A_1'-\mu\delta u_0$, and $A_2'-\frac{1}{2}\delta^2u_0$; and these equations will give these quantities as linear compounds of the ϕ's. In other words, we shall have

$$A_0'=u_0+\psi_0, \quad A_1'=\mu\delta u_0+\psi_1, \quad A_2'=\tfrac{1}{2}\delta^2u_0+\psi_2 \quad \text{........(72)},$$

where ψ_0, ψ_1, ψ_2 are linear compounds of the higher differences mentioned above. Hence, finally, w_r will be of the form

$$\left.\begin{aligned} w_r &= A_0'+A_1'r+A_2'r^2 \\ &= (u_0+r\mu\delta u_0+\tfrac{1}{2}r^2\delta^2u_0)+(\psi_0+r\psi_1+r^2\psi_2) \\ &= u_r+\chi(r) \end{aligned}\right\} \quad \text{...........(73)},$$

where $\chi(r)$ is a linear compound of the higher differences, with coefficients involving r.

23. Since the new method gives the same result as the old methods, our formulae for v_0, $\mu\delta v_0$, ... or for $\mu v_{\frac{1}{2}}$, $\delta v_{\frac{1}{2}}$, ... give an explicit solution of the equations which occur in the application of the old methods; and, conversely, these formulae may be expressed in terms of the moments, if on account of m being very great it is more convenient to use the latter.

To illustrate the use of the new formulae, let us take the example given in M. Merriman's *Method of Least Squares* (8th edition, page 131) of the velocities at successive tenths of the total depth of a river. If u is the velocity in feet per second, multiplied by 10,000, and if the unit of measurement of x is 0·1 of the depth, the differenced table is as follows:

x	u	δu	$\delta^2 u$	$\delta^3 u$	$\delta^4 u$	$\delta^5 u$	$\delta^6 u$	$\delta^7 u$	$\delta^8 u$	$\delta^9 u$
0	31950									
		+349								
1	32299		−116							
		+233		− 38						
2	32532		−154		+ 18					
		+ 79		− 20		+ 37				
3	32611		−174		+ 55		−229			
		− 95		+ 35		−192		+ 835		
4	32516		−139		−137		+606		−2372	
		−234		−102		+414		−1537		+5630
5	32282		−241		+277		−931		+3258	
		−475		+175		−517		+1721		
6	31807		− 66		−240		+790			
		−541		− 65		+273				
7	31266		−131		+ 33					
		−672		− 32						
8	30594		−163							
		−835								
9	29759									

The most laborious method is to take $x = 0$ as our zero; assuming U to be of the form $A_0' + A_1'x + A_2'x^2$, this leads to the equations

$$\left.\begin{aligned} 10A_0' + \quad 45A_1' + \quad 285A_2' &= M_0 = \ \ 317616 \\ 45A_0' + \ 285A_1' + \ 2025A_2' &= M_1 = 1408957 \\ 285A_0' + 2025A_1' + 15333A_2' &= M_2 = 8828813 \end{aligned}\right\},$$

the solution of which is

$$A_0' = (136M_0 - 57M_1 + 5M_2)/220 = 31951{\cdot}327,$$
$$A_1' = (-684M_0 + 437M_1 - 45M_2)/2640 = 442{\cdot}530,$$
$$A_2' = (12M_0 - 9M_1 + M_2)/528 = -76{\cdot}530.$$

A less troublesome method would be to take moments with regard to the middle of the range, i.e. to take the values of x to be $-4\frac{1}{2}, -3\frac{1}{2}, \ldots +4\frac{1}{2}$. To apply the new method, we take the values of x to be $-4, -3, \ldots +5$, and we find the values of $2\mu v_{\frac{1}{2}}$, $\delta v_{\frac{1}{2}}$, $2\mu\delta^2 v_{\frac{1}{2}}$ from the central differences of the interval (0, 1). The coefficients are given in Table I. B (ii), under $m = 10, j = 2$; and thus we obtain

$$\begin{aligned} 2\mu v_{\frac{1}{2}} &= (64798) - \tfrac{9}{5}(140) - \tfrac{16}{15}(-325) - \tfrac{9}{55}(886) \\ &= 64747{\cdot}68, \end{aligned}$$

$$\delta v_{\frac{1}{2}} = (-234) + \tfrac{12}{5}(-102) + \tfrac{9}{5}(414) + \tfrac{8}{15}(-1537) + \tfrac{3}{55}(5630)$$
$$= -246{\cdot}24,$$
$$2\mu\delta^2 v_{\frac{1}{2}} = (-380) + \tfrac{3}{2}(140) + \tfrac{2}{3}(-325) + \tfrac{1}{11}(886)$$
$$= -306{\cdot}12.$$

Since $2\mu v_{\frac{1}{2}} = v_1 + v_0$, $\delta v_{\frac{1}{2}} = v_1 - v_0$, this gives us the two middle v's; and thence we construct the table

x	v	δv	$\delta^2 v$
−4	31951·32		
		+366·00	
−3	32317·32		−153·06
		+212·94	
−2	32530·26		−153·06
		+ 59·88	
−1	32590·14		−153·06
		− 93·18	
0	32496·96		−153·06
		−246·24	
1	32250·72		−153·06
		−399·30	
2	31851·42		−153·06
		−552·36	
3	31299·06		−153·06
		−705·42	
4	30593·64		−153·06
		−858·48	
5	29735·16		−153·06
		−1011·54	
6	28723·62		

These values of v have to be divided by 10^4; and the number of significant figures can then be reduced.

In this example, I have followed Merriman in taking U to be of the 2nd degree in x; but it is doubtful whether this is justified by the data. The proper method of finding when differences of U become negligible is to take out, and difference, values of u at regular intervals, so as to increase the successive differences in increasing ratios. If, in our example, we take even values and odd values of x separately, we get the following results:

31950				32299			
	+ 582				+ 312		
32532		−598		32611		−641	
	− 16		− 95		− 329		− 46
32516		−693		32282		−687	
	− 709		+189		−1016		+196
31807		−504		31266		−491	
	−1213				−1507		
30594				29759			

The parallelism of the 3rd differences makes it very doubtful whether we are justified in assuming that 3rd differences of U are zero. If U can be represented by a polynomial in x, it seems more likely that this ought to be taken as being of the 4th degree in x. The calculation of $2\mu v_{\frac{1}{2}}$, $\delta v_{\frac{1}{2}}$, ... $2\mu\delta^4 v_{\frac{1}{2}}$ would then be as follows:

	64798 − 380 + 140 − 325 + 886	
$2\mu v_{\frac{1}{2}}$	1 + 0 − 0 + 8/33 + 9/143	= +64774·974
$2\mu\delta^2 v_{\frac{1}{2}}$	0 + 1 + 0 − 14/33 − 14/143	= − 328·862
$2\mu\delta^4 v_{\frac{1}{2}}$	0 + 0 + 1 + 8/11 + 18/143	= + 15·161

	− 234 − 102 + 414 − 1537 + 5630	
$\delta v_{\frac{1}{2}}$	1 + 0 − 1 − 16/33 − 9/143	= − 257·124
$\delta^3 v_{\frac{1}{2}}$	0 + 1 + 7/6 + 14/33 + 7/143	= + 4·534

This gives the following table:

x	v	δv	$\delta^2 v$	$\delta^3 v$	$\delta^4 v$
−4	31945·93				
		+361·06			
−3	32306·99		−134·81		
		+226·25		−18·21	
−2	32533·24		−153·02		+7·58
		+ 73·23		−10·63	
−1	32606·47		−163·65		+7·58
		− 90·42		− 3·05	
0	32516·05		−166·70		+7·58
		−257·12		+ 4·53	
1	32258·93		−162·17		+7·58
		−419·29		+12·11	
2	31839·64		−150·06		+7·58
		−569·35		+19·69	
3	31270·29		−130·37		+7·58
		−699·72		+27·27	
4	30570·57		−103·10		+7·58
		−802·82		+34·85	
5	29767·75		− 68·25		
		−871·07			
6	28896·68				

24. Since the results obtained by the new method are the same as would be given by equating moments, it follows that the method cannot be applied to the improvement of the moments themselves; in other words, if $M_g \equiv \sum_r r^g u_r$ is the gth moment, its mean square of error is already a minimum, and cannot be reduced by adding to M_g terms which involve differences of orders exceeding j. But it should be observed that (1) this only applies if $g \leqq j$, and (2) it only applies to the moment as defined above (cf. § 22). If the U's are ordinates of a figure, the moment of the area

of the figure is different from M_g, and our methods apply to the determination of its best value.

(B.) *Other systems.*

25. It has been assumed, throughout §§ 22—24, that the data belong to the standard system, under which the errors of the u's are independent and have all the same mean square; and it is only for this system that the method of least squares or the method of moments, as ordinarily used, will make the mean square of error of z a minimum. For any other system of errors, these methods will be correct, in the sense that they would give the true result if the data contained no error; but, in order that they may give the most accurate result, certain modifications are required.

(i) Denote the given u's by $u_f, u_{f+1}, \ldots u_{f+m-1}$; and let us find the best value for U_r by the new method. We take s or t to be one of the numbers $f, f+1, \ldots f+m-1$, and g or h to be one of the numbers 0, 1, 2, ... j.

If the best value for U_r is

$$v_r = \sum_s p_s u_s \qquad \text{(74)},$$

then (§ 9) the p's are found by making

$$R^2 \equiv \sum_s \sum_t \pi_{s,t} p_s p_t$$

a minimum, subject to certain conditions which may be written

$$\left.\begin{aligned} \sum_s p_s &= 1 \\ \sum_s s p_s &= r \\ \sum_s s^2 p_s &= r^2 \\ &\vdots \\ \sum_s s^j p_s &= r^j \end{aligned}\right\} \qquad \text{(75)}.$$

This (§ 10) gives m relations of the form

$$\sum_t \pi_{s,t} p_t = \sum_g \lambda_g s^g \qquad \text{(76)}.$$

Treating these as equations to give the p's, we find that

$$p_t = \sum_g \{\lambda_g F_g(t)\} \qquad \text{(77)},$$

where, if

$$P \equiv \begin{vmatrix} \pi_{f,f} & \pi_{f,f+1} & \pi_{f,f+2} & \cdots \\ \pi_{f+1,f} & \pi_{f+1,f+1} & \pi_{f+1,f+2} & \cdots \\ \pi_{f+2,f} & \pi_{f+2,f+1} & \pi_{f+2,f+2} & \cdots \\ \vdots & \vdots & \vdots & \end{vmatrix} \qquad \text{(78)},$$

$$P_{s,t} \equiv \text{cofactor of } \pi_{s,t} \text{ in } P \qquad \text{(79)},$$

then

$$F_g(t) = \sum_s s^g P_{s,t}/P \qquad \text{(80)}.$$

Substituting for p_t from (77) in the $(h+1)$th equation of (75) in the form

$$\sum_t t^h p_t = r^h,$$

we have

$$r^h = \sum_g \{g, h\} \lambda_g \qquad \text{(81)},$$

where
$$\{g, h\} \equiv \sum_t t^h F_g(t)$$
$$= \sum_s \sum_t s^g t^h P_{s,t}/P \quad \text{.........................(82).}$$

Substituting from (81) in (75), for $h = 0, 1, 2, \ldots j$, we have $j+1$ equations to determine the λ's; and, if we write

$$Q \equiv \begin{vmatrix} \{0, 0\} & \{1, 0\} & \{2, 0\} & \ldots \\ \{0, 1\} & \{1, 1\} & \{2, 1\} & \ldots \\ \{0, 2\} & \{1. 2\} & \{2, 2\} & \ldots \\ \vdots & \vdots & \vdots & \end{vmatrix} \quad \text{.........................(83),}$$

$$Q_{g,h} = \text{cofactor of } \{g, h\} \text{ in } Q \quad \text{.........................(84),}$$

these equations give
$$Q\lambda_g = Q_{g,0} + rQ_{g,1} + \ldots + r^j Q_{g,j} \quad \text{.....................(85).}$$

Hence, substituting in (77),
$$Qp_t = \sum_g \sum_h Q_{g,h} r^h F_g(t) \quad \text{..........................(86);}$$

and finally, by (74),
$$Qv_r = Q \sum_t p_t u_t = \sum_t \sum_g \sum_h Q_{g,h} r^h F_g(t) u_t \quad \text{.....................(87)}$$
$$= \sum_s \sum_t \sum_g \sum_h Q_{g,h} r^h s^g P_{s,t} u_t/P \quad \text{..........................(88).}$$

(ii) To adapt the method of least squares, we use the known property that the probability of occurrence of errors $e_f, e_{f+1}, \ldots e_{f+m-1}$ in the u's (each error being taken to a definite number of decimal places, so that we can omit the differentials de_f, $de_{f+1}, \ldots$) is proportional to

$$\exp\left(-\tfrac{1}{2} \sum_s \sum_t P_{s,t} e_s e_t/P\right) \quad \text{..........................(89),}$$

where P and $P_{s,t}$ are defined as in (i). This suggests that, if U_r is assumed to be of the form
$$U_r = w_r \equiv A_0 + A_1 r + A_2 r^2 + \ldots + A_j r^j \quad \text{.................(90),}$$

we should find the A's by making
$$\sum_s \sum_t P_{s,t} (w_s - u_s)(w_t - u_t) \quad \text{.......................(91)}$$

a minimum.

Differentiating (91) with regard to each of the A's, we have $j+1$ equations of the form
$$\sum_t F_g(t)(w_t - u_t) = 0 \quad \text{.............................(92),}$$

for $g = 0, 1, 2, \ldots j$. Hence, if we write
$$\sum_t F_g(t) u_t \equiv M_g \quad \text{.............................(93),}$$

we obtain
$$\sum_h \{g, h\} A_h = M_g \quad \text{.............................(94).}$$

Solving these equations in the A's, we find that
$$QA_h = \sum_g Q_{g,h} M_g \quad \text{.............................(95),}$$

and thence
$$Qw_r = \sum_h r^h \sum_g Q_{g,h} M_g$$
$$= \sum_t \sum_g \sum_h Q_{g,h} r^h F_g(t)\, u_t \quad \text{..........................(96).}$$

Comparing with (87), we see that this method gives the same result.

(iii) Also from (92) and (93) we see that the method of moments gives the same result, provided that in finding the gth moment we multiply u_t and w_t by $F_g(t)$, i.e. by $\sum_s P_{s,t} s^g$ (with any constant factor), instead of by t^g.

V. Treatment of Non-Standard Systems.

26. The formulae obtained in § 25 are so complicated that their practical application, as general formulae, would seem to be out of the question. We conclude that, except possibly in a few simple cases, the only practicable method of obtaining the best results is to convert our data to the standard system, in which the errors are independent and have all the same mean square.

A specially important case is that of data which represent frequency-distributions. Before considering this case, the following simpler cases may be noted.

(i) Suppose that the errors are independent but have not all the same mean square. Let the mean square of error of u_r be a_r^2.

(*a*) If a_r^2 is known, we can tabulate $a/a_r \,.\, u_r$, where a is a quantity which is the same for all the u's; and our formulae will then apply. But it will here, as in all other cases of this kind, be necessary that the fundamental condition as to existence of negligible differences (§ 1 (iii)) should continue to be satisfied.

(*b*) If a_r^2 is a definite function of U_r, let us aim at obtaining Y, a function of U. A small error e_r in the value of U_r will produce an error $(dY/dU)_r e_r$ in Y_r, and the mean square of this error will be $\{(dY/dU)_r\}^2 a_r^2$. We therefore choose Y to be such a function of U as to make $dY/dU = 1/a$, for all values of a; if y_r is the corresponding function of u_r, the mean square of error of y_r will be 1, and our method will apply to the sequence of values of y.

If, for instance, U is a physical measurement, the "probable error" of u might be proportional to U or to $\sqrt{U}$. In the former case we should have dY/dU proportional to $1/U$, so that we should tabulate $\log_{10} u$ in place of u, and then apply our formulae. In the latter case we should tabulate $\sqrt{u}$.

(ii) Suppose that the correlation of the errors is due to the fact that each error u_r consists of two portions ϵ_r and ζ_r, such that the ϵ's either are all equal or else form a sequence with regular differences, while the ζ's are independent. In such a case we cannot make any reduction in the ϵ's, and we must regard $U + \epsilon$ as the quantity whose value we are seeking, and ζ as its error. The case therefore comes under the category of independent errors.

Under this head would come a sequence of astronomical observations, all containing an error due to the "personal equation"; or a sequence of nautical observations based on dead-reckoning subject to an unknown current.

(iii) Suppose that 2nd differences of the U's are negligible, and that the mean product of errors of u_r and u_s is of the form

$$A + \tfrac{1}{2}B(r+s) - \tfrac{1}{2}C\,|\,r-s\,|,$$

the mean square of error of u_r being $A + Br$. We have then a choice of two methods.

(a) The mean square of error of $p_1u_1 + p_2u_2 + \dots + p_mu_m$ is a minimum when $p_2 = p_3 = \dots = p_{m-1} = 0$; the values of p_1 and p_m being then determined by the conditions of the case. If, for instance, $m = 2n+1$, and we are improving the u's, the improved value of u_{n+1} is $\tfrac{1}{2}(u_1 + u_{2n+1})$.

(b) The errors of the 1st differences are independent and have all the same mean square, so that we can (usually) treat these as the given quantities.

27. Next take the case of a frequency-distribution. Let N be the total number of measurements or observations, and let Nu_r be the number for which x is less* than x_r, so that u ranges in value from 0 to 1. The data will then ordinarily be arranged in the form of a table, the numbers in the compartments of the table being proportional to the 1st differences of the u's.

(i) The fundamental assumption is that the data are such as might be produced by random picking of N individuals out of an indefinitely great number for which the proportion classed as below x_r would be U_r.

(ii) The mean square of error of u_r is therefore $U_r(1-U_r)/N$, and the mean product of errors of u_r and u_s is $U_r(1-U_s)/N$ or $U_s(1-U_r)/N$ according as $r < s$ or $s < r$. Also, if y and y' are two of the 1st differences of the u's (i.e. if Ny and Ny' are the numbers in any two of the compartments of the table), and if Y and Y' are the corresponding 1st differences of the U's, the mean squares of error of y and of y' are $Y(1-Y)/N$ and $Y'(1-Y')/N$, and their mean product of errors is $-YY'/N$.

(iii) Hence, if the extreme values of x are x_0 and x_m, so that the observed u's are $u_0 \equiv 0, u_1, u_2, \dots u_{m-1}, u_m \equiv 1$, the probability of occurrence of errors $e_1, e_2, \dots e_{m-1}$ is proportional to

$$\exp\left(-\tfrac{1}{2}E^2/N\right) \quad\text{.................................(97)},$$

where

$$E^2 \equiv \frac{e_1^2}{U_1} + \frac{(e_2-e_1)^2}{U_2-U_1} + \frac{(e_3-e_2)^2}{U_3-U_2} + \dots + \frac{(e_{m-1}-e_{m-2})^2}{U_{m-1}-U_{m-2}} + \frac{e^2_{m-1}}{1-U_{m-1}} \quad\text{......(98)};$$

or, denoting the errors of the y's by $e'_{\frac{1}{2}}, e'_{\frac{3}{2}}, \dots e'_{m-\frac{1}{2}}$,

$$E^2 = \sum_{s=1}^{s=m} e'^2_{s-\frac{1}{2}}/Y_{s-\frac{1}{2}} \quad\text{..............................(99)}.$$

The form of this is the same as if the errors $e'_{\frac{1}{2}}, e'_{\frac{3}{2}}, \dots e'_{m-\frac{1}{2}}$ were independent; but they are not really independent, since their sum is zero.

(iv) It follows from (98) that (disregarding a constant multiplier)

$$F_g(s) = \{(s+1)^g - s^g\}/(U_{s+1}-U_s) - \{s^g - (s-1)^g\}/(U_s - U_{s-1}) \dots\text{(100)}.$$

Since, however, this involves the unknown U's, it is not of practical use, unless we proceed by successive approximations.

* In practice it is usually more convenient to make u range from 1 to 0, substituting "greater" here for "less." This involves some changes of sign.

(v) The correlated errors can be analysed into independent errors. If, for instance, we tabulate

$$\log(1-u_0) \equiv 0, \quad \log(1-u_1), \quad \log(1-u_2), \; \dots \; \log(1-u_{m-1}),$$

the errors of their 1st differences will be independent, provided that N is so large that, for each value of r, the error of u_r is small in comparison with $1-U_r$. This is useful for construction of control examples, but, for actual cases, the difficulty remains that the mean squares of error involve the U's.

(vi) If the number of compartments is large, so that YY' is always small in comparison with $Y(1-Y)$ and $Y'(1-Y')$, we see from (ii) that the correlation of the errors of the y's may be ignored, so that we can treat these errors as independent, the mean square of error of each y being $Y(1-Y)/N$. Hence (§ 26 (i) (*b*)) we can convert the system to the standard system by replacing the values of y by those of $2/\pi \, . \, \text{arc sin} \sqrt{y}$, or, sufficiently well for practical purposes, by those of $\sqrt{y}$.

(vii) This gives us a criterion for deciding what intervals should be used for the purpose of tabulation. It has been suggested* that it will generally be sufficient if the interval is so chosen that the whole number of classes lies between 15 and 25. But this will not usually be sufficient for our purposes, at any rate if N is large. We take the square roots of the 1st differences, and then apply our formulae; and the mistake so introduced into the quantity we are calculating ought to be small (say about one-fifth) in proportion to the probable error of the quantity. Roughly, in the ordinary case of a "double-tailed" distribution, with observations numbering (say) from 1000 to 4000, it will be sufficient for most purposes if the interval is such that no compartment contains more than about 7 per cent. of the total number.

VI. General Observations.

28. In deciding what order of differences we are to treat as negligible, we must remember that the terms which we introduce involve differences at the centre of the range, and that the conditions may be different from the conditions at the ends of the range. Suppose, for instance, that we require the gth moment of a double-tailed figure whose ordinates are the u's. We can take $\Sigma x^g u$ as an approximate formula. But the reason of this is that the true value differs from $\Sigma x^g U$ by an expression involving differences or differential coefficients of the extreme U's. These are very small, whatever their order may be. But at the centre the differential coefficients of low order will not necessarily be small.

29. The whole method depends on the fundamental assumption (§ 1 (iii)) that differences beyond those of a certain order are negligible. If this condition is not satisfied, the method fails, not because it is wrong, but because the interval of tabulation is not small enough, or because the independent variable is not suitably chosen. In such cases the only method is to substitute a new sequence, by a process analogous to that mentioned in § 26 (i) (*b*). This is a familiar process, of which the treatment of the tail of a distribution by taking logarithms is an example. We can then apply our method to the substituted data, though there may be some difficulty in finding the best formulae.

* G. U. Yule, *Introduction to the Theory of Statistics* (1911), p. 79.

TABLE I. *Formulae for v_0, etc. (standard system).*

This table relates to the standard system, in which the errors of the u's are independent and have all the same mean square. The highest order of differences of U that is not negligible is denoted by j.

The total number of u's involved in the formula being denoted by m, the cases of m odd and m even are dealt with in parts A and B of the table. Part A shows (for $j = 1$, 2, 3, or 4) (i) the coefficients of u_0, $\delta^2 u_0$, $\delta^4 u_0$, ... in the expressions for $v_0\,(m = 3, 5, 7, \ldots)$, $\delta^2 v_0\,(m = 5, 7, 9, \ldots)$, and $\delta^4 v_0\,(m = 7, 9, 11, \ldots)$, and (ii) the coefficients of $\mu\delta u_0$, $\mu\delta^3 u_0$, ... in the expressions for $\mu\delta v_0\,(m = 5, 7, 9, \ldots)$ and $\mu\delta^3 v_0\,(m = 7, 9, 11, \ldots)$; the u's being $u_{-n}, u_{-n+1}, \ldots u_n$. Part B shows (for $j = 1, 2, 3, 4$, or 5) (i) the coefficients of $\mu u_{\frac{1}{2}}$, $\mu\delta^2 u_{\frac{1}{2}}$, $\mu\delta^4 u_{\frac{1}{2}}$, ... in the expressions for $\mu v_{\frac{1}{2}}\,(m = 4, 6, 8, \ldots)$, $\mu\delta^2 v_{\frac{1}{2}}\,(m = 6, 8, 10, \ldots)$, and $\mu\delta^4 v_{\frac{1}{2}}\,(m = 8, 10, 12, \ldots)$, and (ii) the coefficients of $\delta u_{\frac{1}{2}}$, $\delta^3 u_{\frac{1}{2}}$, $\delta^5 u_{\frac{1}{2}}$, ... in the expressions for $\delta v_{\frac{1}{2}}\,(m = 4, 6, 8, \ldots)$, $\delta^3 v_{\frac{1}{2}}\,(m = 6, 8, 10, \ldots)$, and $\delta^5 v_{\frac{1}{2}}\,(m = 8, 10, 12, \ldots)$; the u's being $u_{-n+1}, u_{-n+2}, \ldots u_n$.

For an example of the use of the table, see § 23.

A. $m = 2n + 1$.

(i)			(ii)		
m	j	Co. u_0, $\delta^2 u_0$, $\delta^4 u_0$, ...	m	j	Co. $\mu\delta u_0$, $\mu\delta^3 u_0$, $\mu\delta^5 u_0$, ...
3	*0, 1*	$1+\frac{1}{3}$			
5	*0, 1*	$1+1+\frac{1}{5}$	**5**	*1, 2*	$1+\frac{2}{5}$
	2, 3	$1+0-\frac{3}{35}$			
		$0+1+\frac{2}{7}$			
7	*0, 1*	$1+2+1+\frac{1}{7}$	**7**	*1, 2*	$1+1+\frac{3}{14}$
	2, 3	$1+0-\frac{3}{7}-\frac{2}{21}$		*3, 4*	$1+0-\frac{5}{42}$
		$0+1+\frac{5}{7}+\frac{5}{42}$			$0+1+\frac{1}{3}$
	4, 5	$1+0-0+\frac{5}{231}$			
		$0+1+0-\frac{5}{66}$			
		$0+0+1+\frac{3}{11}$			
9	*0, 1*	$1+\frac{10}{3}+3+1+\frac{1}{9}$	**9**	*1, 2*	$1+\frac{9}{5}+\frac{9}{10}+\frac{2}{15}$
	2, 3	$1+0-\frac{9}{7}-\frac{2}{3}-\frac{1}{11}$		*3, 4*	$1+0-\frac{1}{2}-\frac{4}{33}$
		$0+1+\frac{9}{7}+\frac{1}{2}+\frac{2}{33}$			$0+1+\frac{7}{9}+\frac{14}{99}$
	4, 5	$1+0-0+\frac{5}{33}+\frac{5}{143}$			
		$0+1+0-\frac{7}{22}-\frac{28}{429}$			
		$0+0+1+\frac{7}{11}+\frac{14}{143}$			

(*Continued on page* 379) (*Continued on page* 380)

A (i) *continued.*

m	j	Co. u_0, $\delta^2 u_0$, $\delta^4 u_0$, ...
11	0, 1	$1+5+7+4+1+\frac{1}{11}$
	2, 3	$1+0-3-\frac{8}{3}-\frac{9}{11}-\frac{12}{143}$
		$0+1+2+\frac{4}{3}+\frac{4}{11}+\frac{5}{143}$
	4, 5	$1+0-0+\frac{20}{33}+\frac{45}{143}+\frac{6}{143}$
		$0+1+0-\frac{28}{33}-\frac{56}{143}-\frac{7}{143}$
		$0+0+1+\frac{12}{11}+\frac{54}{143}+\frac{6}{143}$
13	0, 1	$1+7+14+12+5+1+\frac{1}{13}$
	2, 3	$1+0-6-8-\frac{45}{11}-\frac{12}{13}-\frac{1}{13}$
		$0+1+\frac{20}{7}+\frac{20}{7}+\frac{100}{77}+\frac{25}{91}+\frac{2}{91}$
	4, 5	$1+0-0+\frac{20}{11}+\frac{225}{143}+\frac{6}{13}+\frac{10}{221}$
		$0+1+0-\frac{20}{11}-\frac{200}{143}-\frac{5}{13}-\frac{8}{221}$
		$0+0+1+\frac{18}{11}+\frac{135}{143}+\frac{3}{13}+\frac{9}{442}$
15	0, 1	$1+\frac{28}{3}+\frac{126}{5}+30+\frac{55}{3}+6+1+\frac{1}{15}$
	2, 3	$1+0-\frac{54}{5}-20-15-\frac{72}{13}-1-\frac{6}{85}$
		$0+1+\frac{27}{7}+\frac{75}{14}+\frac{25}{7}+\frac{225}{182}+\frac{3}{14}+\frac{1}{68}$
	4, 5	$1+0-0+\frac{50}{11}+\frac{75}{13}+\frac{36}{13}+\frac{10}{17}+\frac{15}{323}$
		$0+1+0-\frac{75}{22}-\frac{50}{13}-\frac{45}{26}-\frac{6}{17}-\frac{35}{1292}$
		$0+0+1+\frac{25}{11}+\frac{275}{143}+\frac{10}{13}+\frac{5}{34}+\frac{7}{646}$
17	0, 1	$1+12+42+66+55+26+7+1+\frac{1}{17}$
	2, 3	$1+0-18-44-45-24-7-\frac{18}{17}-\frac{21}{323}$
		$0+1+5+\frac{55}{6}+\frac{25}{3}+\frac{25}{6}+\frac{7}{6}+\frac{35}{204}+\frac{10}{969}$
	4, 5	$1+0-0+10+\frac{225}{13}+12+\frac{70}{17}+\frac{225}{323}+\frac{15}{323}$
		$0+1+0-\frac{35}{6}-\frac{350}{39}-\frac{35}{6}-\frac{98}{51}-\frac{1225}{3876}-\frac{20}{969}$
		$0+0+1+3+\frac{45}{13}+2+\frac{21}{34}+\frac{63}{646}+\frac{2}{323}$

(*Continued on page* 380)

A (i) *continued.*

m	j	Co. u_0, $\delta^2 u_0$, $\delta^4 u_0$, ...
19	*0, 1*	$1+15+66+132+143+91+35+8+1+\frac{1}{19}$
	2, 3	$1+0-\frac{198}{7}-88-117-84-35-\frac{144}{17}-\frac{21}{19}-\frac{8}{133}$
		$0+1+\frac{44}{7}+\frac{44}{3}+\frac{52}{3}+\frac{35}{3}+\frac{14}{3}+\frac{56}{51}+\frac{8}{57}+\frac{1}{133}$
	4, 5	$1+0-0+20+45+42+\frac{350}{17}+\frac{1800}{323}+\frac{15}{19}+\frac{20}{437}$
		$0+1+0-\frac{28}{3}-\frac{56}{3}-\frac{49}{3}-\frac{392}{51}-\frac{1960}{969}-\frac{16}{57}-\frac{7}{437}$
		$0+0+1+\frac{42}{11}+\frac{63}{11}+\frac{49}{11}+\frac{735}{374}+\frac{1764}{3553}+\frac{14}{209}+\frac{18}{4807}$

A (ii) *continued.*

m	j	Co. $\mu\delta u_0$, $\mu\delta^3 u_0$, $\mu\delta^5 u_0$, ...
11	*1, 2*	$1+\frac{14}{5}+\frac{12}{5}+\frac{4}{5}+\frac{1}{11}$
	3, 4	$1+0-\frac{4}{3}-\frac{8}{11}-\frac{15}{143}$
		$0+1+\frac{4}{3}+\frac{6}{11}+\frac{10}{143}$
13	*1, 2*	$1+4+\frac{36}{7}+\frac{20}{7}+\frac{5}{7}+\frac{6}{91}$
	3, 4	$1+0-\frac{20}{7}-\frac{200}{77}-\frac{75}{91}-\frac{8}{91}$
		$0+1+2+\frac{15}{11}+\frac{5}{13}+\frac{1}{26}$
15	*1, 2*	$1+\frac{27}{5}+\frac{135}{14}+\frac{55}{7}+\frac{45}{14}+\frac{9}{14}+\frac{1}{20}$
	3, 4	$1+0-\frac{75}{14}-\frac{50}{7}-\frac{675}{182}-\frac{6}{7}-\frac{5}{68}$
		$0+1+\frac{25}{9}+\frac{25}{9}+\frac{50}{39}+\frac{5}{18}+\frac{7}{306}$
17	*1, 2*	$1+7+\frac{33}{2}+\frac{55}{3}+\frac{65}{6}+\frac{7}{2}+\frac{7}{12}+\frac{2}{51}$
	3, 4	$1+0-\frac{55}{6}-\frac{50}{3}-\frac{25}{2}-\frac{14}{3}-\frac{175}{204}-\frac{20}{323}$
		$0+1+\frac{11}{3}+5+\frac{10}{3}+\frac{7}{6}+\frac{7}{34}+\frac{14}{969}$
19	*1, 2*	$1+\frac{44}{5}+\frac{132}{5}+\frac{572}{15}+\frac{91}{3}+14+\frac{56}{15}+\frac{8}{15}+\frac{3}{95}$
	3, 4	$1+0-\frac{44}{3}-\frac{104}{3}-35-\frac{56}{3}-\frac{280}{51}-\frac{16}{19}-\frac{1}{19}$
		$0+1+\frac{14}{3}+\frac{91}{11}+\frac{245}{33}+\frac{245}{66}+\frac{196}{187}+\frac{98}{627}+\frac{2}{209}$

B. $m = 2n$.

(i)

m	j	Co. $\mu u_{\frac{1}{2}}$, $\mu\delta^2 u_{\frac{1}{2}}$, $\mu\delta^4 u_{\frac{1}{2}}$, ...
4	*0, 1*	$1+\frac{1}{2}$
6	*0, 1*	$1+\frac{4}{3}+\frac{1}{3}$
	2, 3	$1+0-\frac{1}{7}$
		$0+1+\frac{5}{14}$
8	*0, 1*	$1+\frac{5}{2}+\frac{3}{2}+\frac{1}{4}$
	2, 3	$1+0-\frac{9}{14}-\frac{1}{6}$
		$0+1+\frac{6}{7}+\frac{1}{6}$
	4, 5	$1+0-0+\frac{5}{132}$
		$0+1+0-\frac{7}{66}$
		$0+0+1+\frac{7}{22}$
10	*0, 1*	$1+4+\frac{21}{5}+\frac{8}{5}+\frac{1}{5}$
	2, 3	$1+0-\frac{9}{5}-\frac{16}{15}-\frac{9}{55}$
		$0+1+\frac{3}{2}+\frac{2}{3}+\frac{1}{11}$
	4, 5	$1+0-0+\frac{8}{33}+\frac{9}{143}$
		$0+1+0-\frac{14}{33}-\frac{14}{143}$
		$0+0+1+\frac{8}{11}+\frac{18}{143}$
12	*0, 1*	$1+\frac{35}{6}+\frac{28}{3}+6+\frac{5}{3}+\frac{1}{6}$
	2, 3	$1+0-4-4-\frac{15}{11}-\frac{2}{13}$
		$0+1+\frac{16}{7}+\frac{12}{7}+\frac{40}{77}+\frac{5}{91}$
	4, 5	$1+0-0+\frac{10}{11}+\frac{75}{143}+\frac{1}{13}$
		$0+1+0-\frac{12}{11}-\frac{80}{143}-\frac{1}{13}$
		$0+0+1+\frac{27}{22}+\frac{135}{286}+\frac{3}{52}$

(*Continued on page* 382)

(ii)

m	j	Co. $\delta u_{\frac{1}{2}}$, $\delta^3 u_{\frac{1}{2}}$, $\delta^5 u_{\frac{1}{2}}$, ...
4	*1, 2*	$1+\frac{3}{10}$
6	*1, 2*	$1+\frac{4}{5}+\frac{1}{7}$
	3, 4	$1+0-\frac{5}{63}$
		$0+1+\frac{5}{18}$
8	*1, 2*	$1+\frac{3}{2}+\frac{9}{14}+\frac{1}{12}$
	3, 4	$1+0-\frac{5}{14}-\frac{5}{66}$
		$0+1+\frac{2}{3}+\frac{7}{66}$
	5, 6	$1+0-0+\frac{35}{1716}$
		$0+1+0-\frac{21}{286}$
		$0+0+1+\frac{7}{26}$
10	*1, 2*	$1+\frac{12}{5}+\frac{9}{5}+\frac{8}{15}+\frac{3}{55}$
	3, 4	$1+0-1-\frac{16}{33}-\frac{9}{143}$
		$0+1+\frac{7}{6}+\frac{14}{33}+\frac{7}{143}$
	5, 6	$1+0-0+\frac{56}{429}+\frac{21}{715}$
		$0+1+0-\frac{42}{143}-\frac{42}{715}$
		$0+0+1+\frac{8}{13}+\frac{6}{65}$
12	*1, 2*	$1+\frac{7}{2}+4+2+\frac{5}{11}+\frac{1}{26}$
	3, 4	$1+0-\frac{20}{9}-\frac{20}{11}-\frac{75}{143}-\frac{2}{39}$
		$0+1+\frac{16}{9}+\frac{12}{11}+\frac{40}{143}+\frac{1}{39}$
	5, 6	$1+0-0+\frac{70}{143}+\frac{35}{143}+\frac{7}{221}$
		$0+1+0-\frac{108}{143}-\frac{48}{143}-\frac{9}{221}$
		$0+0+1+\frac{27}{26}+\frac{9}{26}+\frac{33}{884}$

(*Continued on page* 383)

B (i) *continued.*

m	j	Co. $\mu u_{\frac{1}{2}}$, $\mu\delta^2 u_{\frac{1}{2}}$, $\mu\delta^4 u_{\frac{1}{2}}$, ...
14	*0, 1*	$1+8+18+\frac{120}{7}+\frac{55}{7}+\frac{12}{7}+\frac{1}{7}$
	2, 3	$1+0-\frac{54}{7}-\frac{80}{7}-\frac{45}{7}-\frac{144}{91}-\frac{1}{7}$
		$0+1+\frac{45}{14}+\frac{25}{7}+\frac{25}{14}+\frac{75}{182}+\frac{1}{28}$
	4, 5	$1+0-0+\frac{200}{77}+\frac{225}{91}+\frac{72}{91}+\frac{10}{119}$
		$0+1+0-\frac{25}{11}-\frac{25}{13}-\frac{15}{26}-\frac{1}{17}$
		$0+0+1+\frac{20}{11}+\frac{15}{13}+\frac{4}{13}+\frac{1}{34}$
16	*0, 1*	$1+\frac{21}{2}+\frac{63}{2}+\frac{165}{4}+\frac{55}{2}+\frac{39}{4}+\frac{7}{4}+\frac{1}{8}$
	2, 3	$1+0-\frac{27}{2}-\frac{55}{2}-\frac{45}{2}-9-\frac{7}{4}-\frac{9}{68}$
		$0+1+\frac{30}{7}+\frac{275}{42}+\frac{100}{21}+\frac{25}{14}+\frac{1}{3}+\frac{5}{204}$
	4, 5	$1+0-0+\frac{25}{4}+\frac{225}{26}+\frac{9}{2}+\frac{35}{34}+\frac{225}{2584}$
		$0+1+0-\frac{25}{6}-\frac{200}{39}-\frac{5}{2}-\frac{28}{51}-\frac{175}{3876}$
		$0+0+1+\frac{5}{2}+\frac{30}{13}+1+\frac{7}{34}+\frac{21}{1292}$
18	*0, 1*	$1+\frac{40}{3}+\frac{154}{3}+88+\frac{715}{9}+\frac{364}{9}+\frac{35}{3}+\frac{16}{9}+\frac{1}{9}$
	2, 3	$1+0-22-\frac{176}{3}-65-\frac{112}{3}-\frac{35}{3}-\frac{32}{17}-\frac{7}{57}$
		$0+1+\frac{11}{2}+11+\frac{65}{6}+\frac{35}{6}+\frac{7}{4}+\frac{14}{51}+\frac{1}{57}$
	4, 5	$1+0-0+\frac{40}{3}+25+\frac{56}{3}+\frac{350}{51}+\frac{400}{323}+\frac{5}{57}$
		$0+1+0-7-\frac{35}{3}-\frac{49}{6}-\frac{49}{17}-\frac{490}{969}-\frac{2}{57}$
		$0+0+1+\frac{36}{11}+\frac{45}{11}+\frac{28}{11}+\frac{315}{374}+\frac{504}{3553}+\frac{2}{209}$
20	*0, 1*	$1+\frac{33}{2}+\frac{396}{5}+\frac{858}{5}+\frac{1001}{5}+\frac{273}{2}+56+\frac{68}{5}+\frac{9}{5}+\frac{1}{10}$
	2, 3	$1+0-\frac{1188}{35}-\frac{572}{5}-\frac{819}{5}-126-56-\frac{72}{5}-\frac{189}{95}-\frac{4}{35}$
		$0+1+\frac{48}{7}+\frac{52}{3}+\frac{728}{33}+\frac{175}{11}+\frac{224}{33}+\frac{56}{33}+\frac{48}{209}+\frac{1}{77}$
	4, 5	$1+0-0+26+63+63+\frac{560}{17}+\frac{180}{19}+\frac{27}{19}+\frac{2}{23}$
		$0+1+0-\frac{364}{33}-\frac{784}{33}-\frac{245}{11}-\frac{6272}{561}-\frac{1960}{627}-\frac{96}{209}-\frac{7}{253}$
		$0+0+1+\frac{91}{22}+\frac{147}{22}+\frac{245}{44}+\frac{490}{187}+\frac{147}{209}+\frac{21}{209}+\frac{3}{506}$

B (ii) *continued.*

m	j	Co. $\delta u_{\frac{1}{2}}$, $\delta^3 u_{\frac{1}{2}}$, $\delta^5 u_{\frac{1}{2}}$, ...
14	*1, 2*	$1+\frac{24}{5}+\frac{54}{7}+\frac{40}{7}+\frac{15}{7}+\frac{36}{91}+\frac{1}{35}$
	3, 4	$1+0-\frac{30}{7}-\frac{400}{77}-\frac{225}{91}-\frac{48}{91}-\frac{5}{119}$
		$0+1+\frac{5}{2}+\frac{25}{11}+\frac{25}{26}+\frac{5}{26}+\frac{1}{68}$
	5, 6	$1+0-0+\frac{200}{143}+\frac{15}{13}+\frac{72}{221}+\frac{10}{323}$
		$0+1+0-\frac{225}{143}-\frac{15}{13}-\frac{135}{442}-\frac{9}{323}$
		$0+0+1+\frac{20}{13}+\frac{11}{13}+\frac{44}{221}+\frac{11}{646}$
16	*1, 2*	$1+\frac{63}{10}+\frac{27}{2}+\frac{55}{4}+\frac{15}{2}+\frac{9}{4}+\frac{7}{20}+\frac{3}{136}$
	3, 4	$1+0-\frac{15}{2}-\frac{25}{2}-\frac{225}{26}-3-\frac{35}{68}-\frac{45}{1292}$
		$0+1+\frac{10}{3}+\frac{25}{6}+\frac{100}{39}+\frac{5}{6}+\frac{7}{51}+\frac{35}{3876}$
	5, 6	$1+0-0+\frac{175}{52}+\frac{105}{26}+\frac{63}{34}+\frac{245}{646}+\frac{75}{2584}$
		$0+1+0-\frac{75}{26}-\frac{40}{13}-\frac{45}{34}-\frac{84}{323}-\frac{25}{1292}$
		$0+0+1+\frac{55}{26}+\frac{22}{13}+\frac{11}{17}+\frac{77}{646}+\frac{11}{1292}$
18	*1, 2*	$1+8+22+\frac{88}{3}+\frac{65}{3}+\frac{28}{3}+\frac{7}{3}+\frac{16}{51}+\frac{1}{57}$
	3, 4	$1+0-\frac{110}{9}-\frac{80}{3}-25-\frac{112}{9}-\frac{175}{51}-\frac{160}{323}-\frac{5}{171}$
		$0+1+\frac{77}{18}+7+\frac{35}{6}+\frac{49}{18}+\frac{49}{68}+\frac{98}{969}+\frac{1}{171}$
	5, 6	$1+0-0+\frac{280}{39}+\frac{35}{3}+\frac{392}{51}+\frac{2450}{969}+\frac{400}{969}+\frac{35}{1311}$
		$0+1+0-\frac{63}{13}-7-\frac{147}{34}-\frac{441}{323}-\frac{70}{323}-\frac{6}{437}$
		$0+0+1+\frac{36}{13}+3+\frac{28}{17}+\frac{315}{646}+\frac{24}{323}+\frac{2}{437}$
20	*1, 2*	$1+\frac{99}{10}+\frac{1188}{35}+\frac{286}{5}+\frac{273}{5}+\frac{63}{2}+\frac{56}{5}+\frac{12}{5}+\frac{27}{95}+\frac{1}{70}$
	3, 4	$1+0-\frac{132}{7}-52-63-42-\frac{280}{17}-\frac{72}{19}-\frac{9}{19}-\frac{4}{161}$
		$0+1+\frac{16}{3}+\frac{364}{33}+\frac{392}{33}+\frac{245}{33}+\frac{1568}{561}+\frac{392}{627}+\frac{16}{209}+\frac{1}{253}$
	5, 6	$1+0-0+14+\frac{147}{5}+\frac{441}{17}+\frac{3920}{323}+\frac{60}{19}+\frac{189}{437}+\frac{14}{575}$
		$0+1+0-\frac{84}{11}-\frac{784}{55}-\frac{2205}{187}-\frac{18816}{3553}-\frac{280}{209}-\frac{864}{4807}-\frac{63}{6325}$
		$0+0+1+\frac{7}{2}+\frac{49}{10}+\frac{245}{68}+\frac{490}{323}+\frac{7}{19}+\frac{21}{437}+\frac{3}{1150}$

TABLE II.

Values of reduction-ratio for central term (standard system).

[The table gives the value of R, where a^2 is the original mean square of error, and R^2a^2 is the reduced mean square of error, for different values of m and k; $m \equiv 2n+1$ being the number of terms used in the formula, and $2k$ or $2k+1$ being the highest order of differences that cannot be neglected.]

n	m = number of values of u in the formula $=2n+1$	Highest order of differences that cannot be neglected				
		1st ($k=0$)	2nd or 3rd ($k=1$)	4th or 5th ($k=2$)	6th or 7th ($k=3$)	8th or 9th ($k=4$)
1	3	·577	1·000	1·000	1·000	1·000
2	5	·447	·697	1·000	1·000	1·000
3	7	·378	·577	·753	1·000	1·000
4	9	·333	·505	·646	·787	1·000
5	11	·302	·455	·577	·690	·810
6	13	·277	·418	·528	·625	·721
7	15	·258	·389	·489	·577	·660
8	17	·243	·365	·459	·539	·614
9	19	·229	·345	·433	·508	·577
10	21	·218	·328	·411	·482	·547
11	23	·209	·313	·393	·460	·521
12	25	·200	·300	·376	·441	·498
13	27	·192	·289	·362	·424	·479
14	29	·186	·279	·349	·408	·461
15	31	·180	·270	·338	·395	·446

SUR UNE MÉTHODE D'INTERPOLATION EXPOSÉE PAR HENRI POINCARÉ, ET SUR UNE APPLICATION POSSIBLE AUX FONCTIONS DE SURVIE D'ORDRE n

PAR ALBERT QUIQUET.

Dans son *Calcul des Probabilités**, Henri Poincaré a exposé la méthode suivante d'interpolation. Je crois devoir la rappeler à la section actuarielle du Ve Congrès international des Mathématiciens, car elle m'a paru susceptible d'une application, que je ferai ensuite connaître, à la recherche des "fonctions de survie": ce nom peut être donné aux fonctions qui représentent analytiquement le nombre $l(x)$ des vivants à l'âge x, pour un nombre donné de naissances, ou pour un nombre donné d'individus ayant accompli en même temps un âge déterminé.

Voici succinctement l'exposé de Poincaré†.

Il s'agit d'appliquer la méthode des moindres carrés à la recherche d'une fonction inconnue $f(x)$.

On a mesuré m valeurs de cette fonction :

$$\begin{aligned} f(a_1) &= A_1, \\ f(a_2) &= A_2, \\ &\cdots\cdots \\ f(a_m) &= A_m. \end{aligned}$$

On aurait ainsi un certain nombre de points de la courbe

$$y = f(x).$$

On pourrait toujours par ces m points faire passer une courbe, mais cette solution ne serait pas la meilleure ; on fait passer une courbe *près* de ces points, aussi continue que possible.

On peut vouloir que cette courbe soit de degré h aussi petit que possible,

$$f(x) = C_0 + C_1x + \ldots + C_hx^h ;$$

* La deuxième édition (1912) de ce *Calcul des Probabilités* a été peut-être l'une des dernières œuvres du regretté maître. Il avait professé ce cours à la Sorbonne en 1893–1894, et, à cette époque, il avait bien voulu me laisser l'honneur de rédiger les leçons réunies dans la première édition. Près de vingt ans plus tard, avec une conscience dont je lui garde une respectueuse reconnaissance, il avait maintenu mon nom sur la couverture du volume, cependant revu et augmenté par lui.

† Chapitre XV. *Théorie de l'interpolation.*

h est plus petit que $m-1$, car si h était égal à $m-1$ on aurait une fonction satisfaisant exactement aux conditions.

Quelle valeur attribuer à h ?

Cette valeur est arbitraire. On la choisit d'abord assez petite, puis, si elle est insuffisante, on introduit un terme de plus dans le second membre, et ainsi de suite.

Supposons choisie une première valeur de h.

Nous déterminerons les coefficients du polynôme de telle façon que

$$\Sigma\,[f(a_i)-A_i]^2$$

soit minimum.

$f(x)$ est linéaire par rapport aux coefficients C.

Poincaré rattache cette question au développement en fraction continue du rapport

$$\frac{F'(x)}{F(x)},$$

où
$$F(x)=(x-a_1)(x-a_2)\dots(x-a_m).$$

Si, dans ce développement,

$$\frac{F'(x)}{F(x)}=\cfrac{1}{Q_1+\cfrac{1}{Q_2+\cfrac{1}{Q_3+\cdots+\cfrac{1}{Q_{m-1}+\cfrac{1}{Q_m}}}}},$$

on appelle D_i le dénominateur de la i^e réduite, D_i est un polynôme de degré i.

Ce sont ces polynômes D_i qui doivent retenir notre attention. Poincaré s'en sert pour écrire $f(x)$ sous une forme spéciale. Si $f(x)$ est un polynôme d'ordre h, il peut toujours, dit-il, être mis sous la forme

$$f(x)=C_0+C_1D_1(x)+C_2D_2(x)+\dots+C_hD_h(x).$$

Poincaré établit ensuite que l'on rendra minimum

$$\Sigma\,[f(a_i)-A_i]^2$$

si les coefficients C_i ont l'expression suivante :

$$C_i=\frac{A_1D_i(a_1)+A_2D_i(a_2)+\dots+A_mD_i(a_m)}{D_i^2(a_1)+D_i^2(a_2)+\dots+D_i^2(a_m)}.$$

Pour terminer l'exposé de la méthode, empruntons encore au *Calcul des Probabilités** l'explication de l'avantage des polynômes D.

"Je suppose que l'on ait essayé d'abord de représenter les observations par un polynôme de degré h : on a trouvé alors C_0, C_1, ..., C_h. On constate ensuite que la somme des carrés des erreurs commises est inadmissible : on se résigne alors à poursuivre avec un polynôme de degré $h+1$. Tout serait à recommencer, si l'on

* Page 290.

avait eu recours à un procédé quelconque; ici, au contraire, on n'a qu'à ajouter un terme $C_{h+1} D_{h+1}(x)$: les *précédents coefficients ne changent pas*, comme on le voit sur l'expression de C_i."

Les applications des mathématiques conduisent fréquemment à de semblables tâtonnements. La technique des assurances est peut-être une de celles qui ont le plus à y recourir: on conçoit sans peine l'intérêt que peuvent porter les actuaires à tout procédé capable d'abréger leurs essais, notamment quand ce procédé n'annule pas les efforts antérieurs. C'est précisément le cas des polynômes D, et c'est pourquoi il ne m'a pas semblé indifférent de montrer un exemple professionnel de leur utilité.

La base des opérations viagères, c'est-à-dire le nombre $l(x)$ des vivants, est rarement susceptible d'être figurée par un polynôme de degré h. Ce degré devrait en général être assez élevé, s'il s'agissait de représenter une table étendue de survie; et le nombre de termes qu'il entraînerait rendrait la formule peu maniable.

L'on recourt plutôt à certaines fonctions transcendantes, qui ont d'abord l'avantage de la concision, puis qui possèdent des propriétes appréciées, que l'analyse leur découvre, et dont les polynômes semblent moins bien pourvus. Je n'ai pas besoin de citer ici le prix, pour ainsi dire journalier, qui s'attache, dans les opérations courantes d'assurances, aux célèbres lois de deux actuaires anglais, Gompertz et Makeham.

Si les polynômes n'offrent qu'un intérêt secondaire pour l'interpolation de $l(x)$, ils ne sont pas cependant à écarter de tous les problèmes relatifs à la mortalité humaine. Le nombre des vivants n'est pas la seule fonction de l'âge x dont on fasse usage. Le taux de mortalité, qui dérive de $l(x)$, peut être cité comme une autre de ces fonctions. Pour nous borner à un seul cas d'application des polynômes, réalisé effectivement, de 0 à 25 ans le taux de mortalité des Assurés français (table AF) a été interpolé par une formule algébrique du sixième degré. Peut-être un essai des polynômes D eût-il indiqué si ce degré ne pouvait pas être abaissé.

Je me permettrai de citer encore une autre fonction de l'âge x, le logarithme de $l(x)$.

Dans la thèse que j'ai soutenue pour devenir membre agrégé de l'Institut des Actuaires français*, j'ai pu établir le parti qui est à tirer d'une certaine fonction z_x, la différence seconde de ce logarithme.

Lorsque z_x obéit à une relation de récurrence linéaire et d'ordre n,

$$B_0 z_x + B_1 z_{x+1} + \dots + B_n z_{x+n} = 0,$$

j'ai donné à $l(x)$ le nom de "fonction de survie d'ordre n." Ces fonctions forment une classe intéressante au point de vue de l'interpolation des tables de survie: elles comprennent notamment les lois de Gompertz, de Makeham, de Lazarus, de Janse, etc., et généralisent les propriétés de ces lois, en faisant ressortir les liens qui les unissent.

Or un cas, en quelque sorte limite, est à remarquer au point de vue de l'application des polynômes D. Ce cas est celui où l'échelle de récurrence des

* Albert Quiquet, *Représentation algébrique des Tables de Survie; généralisation des lois de Gompertz, de Makeham, etc.* Paris, 1893.

fonctions z_x n'admet qu'une racine multiple d'ordre n et égale à l'unité. z_x est alors un simple polynôme de degré $n-1$, et par suite le logarithme de $l(x)$ est un polynôme de degré $n+1$.

Cette forme particulière laisse cependant à $l(x)$ le caractère d'une "fonction de survie d'ordre n." On peut donc avoir intérêt à l'adapter à des observations brutes; pour savoir si cette adaptation est possible, on commencera naturellement par essayer les plus petites valeurs de n. Si elles ne conviennent pas, on passera à la suivante. C'est ainsi que s'introduiront les polynômes D: grâce à eux, cette recherche progressive s'opérera avec régularité et sans cesser d'utiliser, au fur et à mesure, les calculs successifs que l'on a laborieusement obtenus depuis l'origine.

Ces considérations sembleront assurément bien modestes dans un Congrès destiné surtout aux plus hautes spéculations des mathématiques pures. Mais notre section envisage les applications les plus immédiates des théories que nous transmettent nos maîtres. A l'heure où la science universelle déplore la perte récente de l'un d'eux, l'on voudra bien m'excuser, je l'espère, d'avoir apporté ici le souvenir des leçons qu'il nous a laissées, et d'avoir montré leur utilité possible, même dans des opérations aussi concrètes que les assurances sur la vie.

ON THE FITTING OF MAKEHAM'S CURVE TO MORTALITY OBSERVATIONS

By J. F. Steffensen.

Amongst all formulæ which have been proposed by actuaries for graduating mortality tables, none has a more established reputation than Makeham's Curve

$$y_x = \alpha + \beta c^x \qquad \text{..................................(1).}$$

Its authority rests less on the philosophical considerations on which it was originally founded than on the practical test to which it has been put, and on its well-known advantages for calculating complicated benefits.

In this paper I shall not deal with the question, what kind of mortality experience the formula may be presumed to fit; nor shall I discuss which are the best equations for determining the constants. I shall assume that the method of Least Squares has been chosen, and my main object is to find a practical and simple way of solving the resulting equations numerically.

Let σ_x be an observed value of y_x, and let Γ_x be proportional to the weight of σ_x. We have to determine α, β and c so, that the expression

$$\Sigma\,(y_x - \sigma_x)^2\,\Gamma_x \qquad \text{..................................(2)}$$

is a minimum, that is

$$\frac{\partial\Sigma}{\partial\alpha} = 0\,; \quad \frac{\partial\Sigma}{\partial\beta} = 0\,; \quad \frac{\partial\Sigma}{\partial c} = 0.$$

Replacing y_x by its value (1), these equations become

$$\left.\begin{aligned} \alpha\Sigma\Gamma_x + \beta\Sigma c^x\Gamma_x - \Sigma\sigma_x\Gamma_x &= 0 \\ \alpha\Sigma c^x\Gamma_x + \beta\Sigma c^{2x}\Gamma_x - \Sigma\sigma_x c^x\Gamma_x &= 0 \\ \alpha\Sigma x c^x\Gamma_x + \beta\Sigma x c^{2x}\Gamma_x - \Sigma\sigma_x x c^x\Gamma_x &= 0 \end{aligned}\right\} \qquad \text{..................(3).}$$

This system is linear with respect to α and β, but algebraic of a high degree with respect to c. Previously to the method proposed by me two other methods have chiefly been employed for dealing with these equations.

The first method is, in principle, the classical Gaussian one and requires that the equations be rendered linear by introducing approximate values of the constants. It is well known, that a strict application of this principle to the equations (3) involves considerable numerical work. This was, for the first time, carried through with consistency in a thorough paper by Joh. Karup*.

* "Die Ausgleichung der Sterblichkeitserfahrungen der Gothaer Bank nach der Gompertz-Makeham'schen Sterblichkeitsformel." *Rundschau der Versicherungen*, XXXIV. (1884). See also *Journal of the Institute of Actuaries*, XXIV. (1884), p. 70.

The second method seems to have been proposed simultaneously by Rosmanith* and Draminsky†. The guiding idea is to adopt a succession of trial values of c, and to determine the corresponding sets of α and β from the *two first* equations (3). For each set of constants thus obtained the value of the expression (2) is calculated, and it is evidently possible by continuing this process to get indefinitely close to the absolute minimum of (2). In order to obtain an actual solution of the three equations (3) it is, however, necessary (as pointed out by Rosmanith) to employ some sort of criterion for the approximation obtained to the absolute minimum, which complicates the calculation.

The method I have personally employed‡ seems more direct and less laborious than either of the two previous ones. I eliminate α and β from (3) and obtain a single equation in c

$$D \equiv \begin{vmatrix} \Sigma\Gamma_x & \Sigma c^x\Gamma_x & \Sigma\sigma_x\Gamma_x \\ \Sigma c^x\Gamma_x & \Sigma c^{2x}\Gamma_x & \Sigma\sigma_x c^x\Gamma_x \\ \Sigma x c^x\Gamma_x & \Sigma x c^{2x}\Gamma_x & \Sigma\sigma_x x c^x\Gamma_x \end{vmatrix} = 0 \quad \text{..............(4).}$$

To solve this equation numerically with any desired degree of accuracy presents no difficulty, remembering that $\log c = \cdot 04$ is generally a good first approximation. Three trial values of c will as a rule produce about five correct figures of this constant, by interpolation with Newton's divided differences; if it is further known that an increase of c means a decrease of D (as is the case in the experience dealt with below and may be presumed to hold good for similar experiences), we may often by means of only *two* trial values and linear interpolation between these be able to obtain c to about five figures. Having thus determined c, the values of α and β follow from any two of the equations (3).

For actual calculation we transform (4) as follows. We remember that

$$\Sigma x f(x) = \Sigma^2 f(x) \quad \text{................................(5),}$$

if the summation is commenced from below. It is here assumed, that the observations begin with $x=1$; otherwise Σ^2 must be interpreted as

$$\Sigma^2 + (t-1)\Sigma \quad \text{................................(6),}$$

where t stands for the first argument at which the observations begin. We may now instead of (4) write

$$D \equiv \begin{vmatrix} \Sigma\Gamma_x & \Sigma c^x\Gamma_x & \Sigma\sigma_x\Gamma_x \\ \Sigma c^x\Gamma_x & \Sigma c^{2x}\Gamma_x & \Sigma\sigma_x c^x\Gamma_x \\ \Sigma^2 c^x\Gamma_x & \Sigma^2 c^{2x}\Gamma_x & \Sigma^2\sigma_x c^x\Gamma_x \end{vmatrix} = 0 \quad \text{..............(7).}$$

Each new trial value of c requires the writing down of the following ten columns, which are readily formed, four-figured logarithms being as a rule sufficient for this part of the arithmetical work:

* *Mitteilungen des Oesterreichisch-Ungarischen Verbandes der Privatversicherungs-Anstalten*, vol. 2 (1906), p. 54; *Berichte des V internationalen Kongresses für Versicherungs-Wissenschaft*, vol. 2, p. 329. See also a recent paper by John S. Thompson in the *Transactions of the Actuarial Society of America*, XII. (1911), p. 225.

† *Dödelighed efter Forsikringsart og Forsikringstid*, Copenhagen, 1906, p. 99.

‡ *Dansk Forsikrings-Aarbog*, Copenhagen, 1909, p. 45.

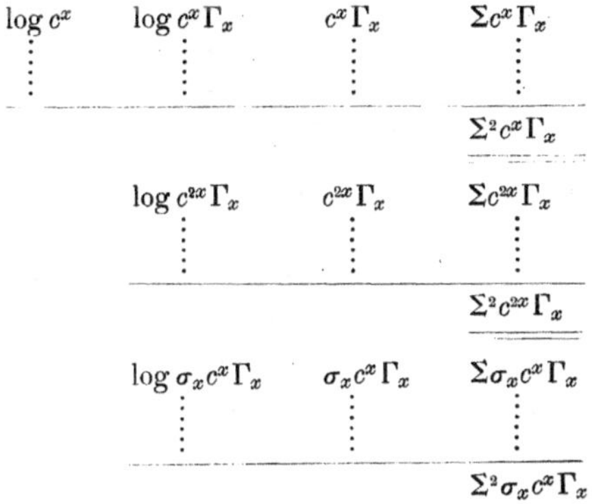

This calculation can, of course, be abbreviated by collecting the material into groups of, say, five years of age; I have, however, avoided this in the example below, as it was incumbent on me to show that a fully correct treatment of the material is possible without much increase of labour.

The further details of the method will best be understood from the example on which I have employed it. The table in question is the largest and most recent Danish experience, called $D^{M(5)}$ (healthy male insured lives, excluding the first five years of insurance), comprising 282,118 years under observation and 5065 deaths*. In the adjoined table E_x denotes as usual "Exposed to Risk" and θ_x "Died." The function to be represented by (1) is

$$y_x = \operatorname{col} p_x = \alpha + \beta c^x \quad \text{.............................(8)},$$

consequently we have

$$\sigma_x = -\log\left(1 - \frac{\theta_x}{E_x}\right) \quad \text{.............................(9)}.$$

As the observations would give $\sigma_{95} = \infty$, the observation for age 95 has been left out. For the weights w_x we have used the well-known approximate formula

$$w_x = \frac{p_x E_x}{q_x (\log e)^2} \quad \text{.............................(10)},$$

and Γ_x was taken as

$$\Gamma_x = \frac{p_x E_x}{q_x} \quad \text{.................................(11)}.$$

For reasons which we need not discuss here it is inexpedient to use the ungraduated values of p_x and q_x in the formulæ (10) and (11). A preliminary graduation should therefore be made; for this purpose it is usually sufficient to determine α, β and c from four ungraduated values of $\log l_x$. The preliminary formula used in this case was

$$\operatorname{col} p_x = \cdot000756 + 10^{\cdot039654x + \bar{5}\cdot68839} \quad \text{.................(12)}.$$

* *Dödlighetstabeller enligt nitton skandinaviska och finska Lifförsäkringsanstalters Erfarenheter*, Stockholm, 1906, p. 33.

The values of $\log \Gamma_x$ according to this formula are stated in column (5) of the adjoined table, while column (6) gives the mean error m_x or

$$m_x = \frac{\log e}{\sqrt{\Gamma_x}} \quad \text{..............................(13).}$$

As the first trial value for $\log c$ it was natural to take the result of the preliminary graduation or $\log c = \cdot 039654$; the value of D for this and two other trial values, namely $\log c = \cdot 039$ and $\log c = \cdot 040$, was calculated. Putting for convenience

$$\left. \begin{aligned} 10000 (\log c - \cdot 039) &= x \\ 10^{-22} D &= y \end{aligned} \right\} \quad \text{......................(14),}$$

the following table with divided differences was formed

x	y	δy	$\delta^2 y$
0·00	652		
		− 96·33	
6·54	22		−9·85
		−194·79	
10·00	−652		

The value of x for which y vanishes is found by Newton's formula*

$$y = y_0 + x \delta y_0 + x (x - x_1) \delta^2 y_0 = 0,$$

where $x_1 = 6{\cdot}54$. We find $x = 6{\cdot}68$ and finally

$$\log c = \cdot 039668.$$

For α and β we find thereafter by (3)

	α	β
From the 1st and 2nd equation	·00090332	·000046237
„ „ 1st „ 3rd „	90330	46237
„ „ 2nd „ 3rd „	90313	46238

The agreement between these values shows the accuracy of the process. As final values were adopted

$$\left. \begin{aligned} \alpha &= \cdot 0009033 \\ \log \beta &= \bar{5}{\cdot}66499 \\ \log c &= \cdot 039668 \end{aligned} \right\} \quad \text{............................(15),}$$

whence

$$\text{col}\, p_x = \cdot 0009033 + 10^{\cdot 039668 x + \bar{5}\cdot 66499} \quad \text{................(16).}$$

In order to test the success of this graduation I have first examined the column headed $(y_x - \sigma_x) : m_x$ with respect to the distribution of the signs + and −. If such a distribution is quite casual, the average number of isolated sequences of the rth

* As pointed out by Dr W. F. Sheppard, the solution of an equation for x may be avoided by putting $y = 0$ in the *inverse* formula:

$$x = x_0 + (y - y_0)\, \delta x_0 + (y - y_0)(y - y_1)\, \delta^2 x_0.$$

order is $n:2^{r+1}$, where n denotes the number of elements*. The average number of isolated sequences of all orders is $\frac{n}{2}$. In the present case we have $n=75$ and find

Order	Number of Sequences		
	calculated	observed	Difference
1	18·8	19	−0·2
2	9·4	9	+0·4
3	4·7	6	−1·3
4	2·3	1	+1·3
5	1·2	3	−1·8
6	0·6	0	+0·6
all	37·5	38	−0·5

The agreement is as close as can be wished.

I have secondly compared the magnitude of the deviations with the typical Law of Errors. In the small table below is indicated how many of the deviations $y_x - \sigma_x$ are found, according to observation and to theory, within k times the mean error. The table is easily verified by means of column (10) in the larger table, containing $(y_x - \sigma_x):m_x$. It is seen that the distribution is nearly typical, and would be still more so, if the observation for age 26 were left out. This observation is suspect *a priori*, as it has a deviation of more than six times the mean error. The occurrence

k	Number		Percentage	
	observed	calculated	observed	calculated
0·2	7	11·9	·093	·159
0·4	22	23·3	·293	·311
0·6	32	33·8	·427	·451
0·8	40	43·2	·533	·576
1·0	48	51·2	·640	·683
1·5	63	65·0	·840	·866
2·0	67	71·6	·893	·954
2·5	72	74·1	·960	·988
3·0	74	74·8	·987	·997

of this sort of abnormality in a mortality table cannot entirely be guarded against; it may be due to local connexions between observations, to clerical errors or to many other causes. On the whole I think the graduation may be considered successful, especially considering that the formula contains only three constants, to be determined by 75 observations.

Commutation columns at $3\frac{1}{2}$ °/₀ will be found in my Danish paper on the subject.

* Czuber: *Wahrscheinlichkeitsrechnung*, 2 Aufl. vol. I, p. 146.

Graduation of the $D^{M(5)}$ Table.

(1)	(2)	(3)	(4)	(5)	(6)	(7)	(8)	(9)	(10)
x	E_x	θ_x	σ_x	$\log \Gamma_x$	m_x	y_x	$y_x - \sigma_x$	$(y_x - \sigma_x)\,\Gamma_x$	$\frac{y_x - \sigma_x}{m_x}$
20	47	0	0·00000	4·2845	0·00313	0·001190	+ ·001190	+ 22·9	+0·38
21	75	1	582	4757	251	1218	− 4602	− 137·6	−1·83
22	111	0	000	6333	210	1248	+ 1248	+ 53·6	+0·60
23	179	1	244	8275	168	1281	− 1159	− 77·9	−0·69
24	260	3	504	9756	141	1317	− 3723	− 351·9	−2·64
25	337	0	0·00000	5·0735	0·00126	0·001357	+ ·001357	+ 160·7	+1·07
26	506	9	780	2341	105	1400	− 6400	−1097·2	−6·10
27	699	2	124	3578	091	1448	+ 0208	+ 47·4	+0·23
28	1027	5	212	5072	077	1500	− 0620	− 199·4	−0·81
29	1456	6	179	6404	066	1557	− 0233	− 101·8	−0·35
30	2245	10	0·00194	5·8090	0·00054	0·001619	− ·000321	− 206·7	−0·59
31	3112	12	168	9305	47	1688	+ 0008	+ 6·8	+0·02
32	4091	18	192	6·0281	42	1763	− 0157	− 167·5	−0·37
33	5071	22	189	0992	39	1845	− 0045	− 56·5	−0·12
34	6068	34	244	1542	36	1935	− 0505	− 720·2	−1·14
35	7034	31	0·00192	6·1944	0·00035	0·002034	+ ·000114	+ 178·3	+0·33
36	7873	36	199	2186	34	2142	+ 0152	+ 251·4	+0·45
37	8600	51	258	2313	33	2260	− 0320	− 545·0	−0·96
38	9157	58	275	2322	33	2390	− 0360	− 614·5	−1·08
39	9547	38	173	2230	34	2532	+ 0802	+1340·3	+2·39
40	9913	42	0·00185	6·2114	0·00034	0·002688	+ ·000838	+1363·3	+2·46
41	10108	55	237	1911	35	2859	+ 0489	+ 759·3	+1·40
42	10252	76	323	1678	36	3046	− 0184	− 270·8	−0·51
43	10236	69	294	1369	37	3251	+ 0311	+ 426·3	+0·84
44	10186	68	291	1041	39	3476	+ 0566	+ 719·3	+1·47
45	10040	83	0·00361	6·0664	0·00040	0·003722	+ ·000112	+ 130·5	+0·28
46	9816	89	396	0245	42	3991	+ 0031	+ 32·8	+0·73
47	9602	99	450	5·9823	44	4287	− 0213	− 204·5	−0·48
48	9241	84	397	9326	47	4610	+ 0640	+ 548·0	+1·36
49	8941	104	508	8846	50	4965	− 0115	− 88·2	−0·23
50	8442	106	0·00549	5·8256	0·00053	0·005353	− ·000137	− 91·7	−0·26
51	8077	112	607	7718	56	5779	− 0291	− 172·1	−0·52
52	7695	118	671	7156	60	6245	− 0465	− 241·6	−0·77
53	7332	123	735	6592	64	6756	− 0594	− 271·0	−0·92
54	6932	104	656	5992	69	7316	+ 0756	+ 300·4	+1·10
55	6557	117	0·00782	5·5388	0·00074	0·007929	+ ·000109	+ 37·7	+0·15
56	6198	148	1050	4779	79	8601	− 1899	− 570·9	−2·40
57	5802	110	0831	4125	85	9338	+ 1028	+ 265·8	+1·20
58	5460	153	1234	3489	92	0·010144	− 2196	− 490·4	−2·39
59	5113	120	1031	2830	99	11028	+ 0718	+ 137·7	+0·72
60	4707	123	0·01150	5·2094	0·00108	0·011997	+ ·000497	+ 80·5	+0·46
61	4403	141	1413	1426	117	13058	− 1072	− 148·9	−0·92
62	4099	128	1378	0733	126	14220	+ 0440	+ 52·1	+0·35
63	3863	152	1743	0092	136	15494	− 1936	− 197·7	−1·42
64	3584	119	1466	4·9380	148	16889	+ 2229	+ 193·2	+1·51
65	3300	136	0·01827	4·8633	0·00161	0·018418	+ ·000148	+ 10·8	+0·09
66	3057	133	1932	7911	175	20093	+ 0773	+ 47·8	+0·44
67	2777	165	2661	7101	192	21928	− 4682	− 240·2	−2·44
68	2521	138	2445	6286	211	23939	− 0511	− 21·7	−0·24
69	2291	121	2357	5473	231	26143	+ 2573	+ 90·7	+1·11
70	2075	126	0·02721	4·4645	0·00255	0·028557	+ ·001347	+ 39·3	+0·53
71	1849	126	3065	3743	282	31201	+ 0551	+ 13·1	+0·20
72	1640	132	3644	2818	314	34099	− 2341	− 44·8	−0·75
73	1447	128	4022	1870	350	37274	− 2946	− 45·3	−0·84
74	1252	116	4222	0833	395	40752	− 1468	− 17·8	−0·37
75	1090	109	0·04576	3·9821	0·00443	0·044564	− ·001196	− 11·5	−0·27
76	935	109	5383	8742	502	48739	− 5091	− 38·1	−1·01
77	774	86	5115	7506	579	53314	+ 2164	+ 12·2	+0·37
78	637	71	5132	6241	670	58327	+ 7007	+ 29·5	+1·05
79	535	57	4893	5063	767	63819	+ ·014889	+ 47·8	+1·94
80	443	59	0·06207	3·3818	0·00885	0·069837	+ ·007767	+ 18·7	+0·88
81	358	47	6112	2465	1034	76429	+ ·015309	+ 27·0	+1·48
82	295	56	9142	1191	1198	83653	− ·007767	− 10·2	−0·65
83	223	46	10033	2·9540	1448	91567	− 8763	− 7·9	−0·61
84	160	32	09691	7657	1799	0·100238	+ 3328	+ 1·9	+0·19
85	122	19	0·07352	2·6034	0·02168	109738	+ ·036218	+ 14·5	+1·67
86	95	33	18533	4496	2588	120147	− 65183	− 18·4	−2·52
87	52	9	08254	1422	3687	131552	+ 49012	+ 6·8	+1·33
88	36	11	15837	1·9361	4676	144047	− 14323	− 1·2	−0·31
89	23	8	18564	6944	6175	157738	− 27902	− 1·4	−0·45
90	15	5	0·17609	1·4609	0·08079	172737	− ·003353	− 0·1	−0·04
91	10	3	15490	2360	10466	189172	+ ·034272	+ 0·6	+0·33
92	6	1	07918	0·9644	14309	207178	+ ·127998	+ 1·2	+0·89
93	5	3	39794	8345	16619	226906	− ·171034	− 1·2	−1·03
94	1	0	00000	0835	39455	248521	+ ·248521	+ 0·3	+0·63

ON THE APPLICATION OF THE CALCULUS OF PROBABILITIES IN CALCULATING THE AMOUNT OF SECURITIES AND IN UTILISING STATISTICAL DATA FOR THE SAKE OF CLASSIFICATION OF LEGAL ACCIDENT-INSURANCE RISKS IN THE PRACTICE OF THE DUTCH STATE INSURANCE OFFICE*

By J. H. Peek.

The Dutch Act regulating the Insurance of labourers in case of accidents, allows, by provision of the articles 54 and following, private institutions and employers to undertake under certain restrictions the risk of insurance according to the provisions of that act. One of the conditions to be satisfied before the undertaking or transferring of such risk is permissible, is that the "Rijksverzekeringsbank" (State Insurance Office) shall hold from the future insurer a certain sum as a security for the obligations resulting from the law being observed. It has been thought desirable to fix these securities at a fraction of the average net value of the probable charge resulting from accidents to be expected per annum among the labourers at his risk, augmented by such a multiple of the mean deviation from that charge, that greater deviations have a smaller than an assigned probability.

Two problems arise from this:

(1) the determination of the mean deviation from the above said net value,

(2) that of the law relating to the distribution of deviations.

As these problems have proved to be of like importance in other cases daily occurring in the practice of legal accident-insurance, I think them of sufficient importance for communication.

The compensations provided by the Dutch Insurance Act consist chiefly of medical attendance, temporary weekly payments, which consist of a constant fraction of the weekly wages for the first six weeks after an accident, and afterwards of an amount depending on the degree of incapacity, and pensions in cases where permanent disablement results from the injury by accident (dependent upon the degree of incapacity) and, if death results, a pension to each of the dependents. If, in a group of insured persons of a constant number exposed to a constant danger, on the average $n_1 + n_2 + n_3 + \ldots + n_l$ accidents occur per unit of time (year) of which the cost is

* As far as no more recent results are concerned, this paper is an extract from an ample report, offered in 1902 to the Directors of the State Insurance Office by the author as Mathematical Adviser attached to the Institution.

respectively $m_1, m_2, m_3, \ldots m_l$, per accident, then the yearly premium due for the insurance of this group will be: $G = n_1 m_1 + n_2 m_2 + n_3 m_3 + \ldots + n_l m_l$.

Taking into consideration an indefinitely short period dt, the numbers $n_1 dt, n_2 dt, n_3 dt, \ldots n_l dt$, may be taken for the probabilities that during this period an accident may occur of which the value of the cost (pecuniary importance, in what follows briefly called "importance") will be $m_1, m_2, m_3, \ldots m_l$. The probabilities that two accidents may occur in so short a period being of second order, it seems permitted to neglect them. If during this period an accident occurs of importance m_s, the difference from the average value will be $m_s - Gdt$; if no accident occurs, the difference is: $-Gdt$. The probability of the first of these deviations is $n_s dt$, the probability of the second $1 - dt\Sigma n_s$.

Consequently the square of the mean deviation:

$$dt\Sigma (m_s - Gdt)^2 n_s + (Gdt)^2 (1 - dt\Sigma n_s)$$

or, neglecting powers of dt: $dt \,.\, \Sigma m_s^2 n_s$.

By integrating, with respect to t, from 0 to 1 we find the square of the mean deviation per annum to be

$$M = \Sigma\, m_s^2 n_s.$$

In order to facilitate as much as possible the calculation of this quantity we will write now

$$M = \Sigma\, m_s^2 n_s = a \Sigma\, p_s m_s^2;$$

where a denotes the number of persons belonging to the constant group (the number of Vollarbeiter; travailleurs-types; years of labour) and p_s the number of accidents per Vollarbeiter of importance m_s. Division by a gives the mean error per Vollarbeiter, denoted by

$$\Sigma\, m_s^2 p_s.$$

Let the cases be divided now into three groups:

Group I, cases of temporary incapacity, characterised by index 1

„ II, fatal accidents „ „ „ 2

„ III, accidents resulting in permanent incapacity characterised by index 3 at the left at the head of the symbol concerned; then

$$\Sigma p_s m_s^2 = \Sigma\, {}^1p_s m_s^2 + \Sigma\, {}^2p_s m_s^2 + \Sigma\, {}^3p_s m_s^2,$$

${}^np_s$ being understood as the product of two quantities, namely ${}^np$: the average number of accidents per annum of kind n per Vollarbeiter and ${}^nq_s$: the probability, that an accident of kind n will necessitate the expense m_s, then

$$\Sigma p_s m_s^2 = {}^1p \Sigma\, {}^1q_s \,.\, m_s^2 + {}^2p \Sigma\, {}^2q_s \,.\, m_s^2 + {}^3p \Sigma\, {}^3q_s m_s^2.$$

Here $\Sigma\, {}^1q_s \,.\, m_s^2$ is the average value of the square importance of a case of temporary incapacity, to be denoted by:

$${}^1M^2;$$

$\Sigma\, {}^2q_s m_s^2$ the average value of the square importance of a fatal accident to be denoted by ${}^2M^2$;

$\Sigma\, {}^3q_s m_s^2$ the average value of the square importance of a case of permanent incapacity to be denoted by ${}^3M^2$; so that

$$\Sigma p_s m_s^2 = {}^1p\, {}^1M^2 + {}^2p\, {}^2M^2 + {}^3p\, {}^3M^2.$$

The values of ${}^1M^2$, ${}^2M^2$ and ${}^3M^2$ depend on the amount and distribution of wages, the nature of accidents, the distribution of ages, the proportion of married and unmarried workmen and the formation of families. Consequently they might be quite different for different employers and that might be an impediment for application to practice. If we restrict application to larger groups of labourers we may expect less different values, especially if dividing by the square of the average annual wages L.

Let the quotient of ${}^rM^2$ and L^2 be denoted by ${}^r\mathfrak{M}^2$, then

$$\Sigma p_s m_s^2 = L^2({}^1p\,{}^1\mathfrak{M}^2 + {}^2p\,{}^2\mathfrak{M}^2 + {}^3p\,{}^3\mathfrak{M}^2).$$

Our first problem is now solved, if we know for every group of insured workmen the value of the average annual wages L, the numbers ${}^1p, {}^2p$ and ${}^3p$ (being the numbers of accidents per annum and per Vollarbeiter), and the values ${}^1\mathfrak{M}^2$, ${}^2\mathfrak{M}^2$ and ${}^3\mathfrak{M}^2$ (being the average square of the importance of an accident of the different groups per unit of average annual wages).

We have supposed (with respect to a previous calculation, executed before any experience had been acquired as to the effect of the Accident Insurance Act) that the values ${}^1\mathfrak{M}^2$ and ${}^2\mathfrak{M}^2$ are invariable. Moreover the term ${}^3p\,{}^3\mathfrak{M}^2$ was transformed into $\Sigma\,{}^3p_t\,.\,t^2\,{}^3\mathfrak{M}_1^2$ in which ${}^3p_t$ represents the number of accidents per Vollarbeiter, where a permanent incapacity of the fraction t of the total resulted, while ${}^3\mathfrak{M}_1^2$ represents the average square of the importance of a case of total incapacity per unit of average annual wages.

This transposition supposes that the average square of the expense of a case of permanent partial disablement of the fraction t may be represented by $t^2.{}^3\mathfrak{M}_1^2$. The quantity ${}^3\mathfrak{M}_1^2$ too was supposed to be invariable.

The calculation of the quantities $\mathfrak{M}^2$ can be effected most easily and most correctly from the expense of each of the individual accidents. As this information was not at hand when the previous calculation was made, we had to recur to the official Austrian statistics on the subject. These statistics relate to great numbers of cases and contain no data relating to the cost of each case individually. It was therefore impossible to obtain the quantities $\mathfrak{M}^2$ in a direct way, but to obtain them we could make use of the property, that the square of the mean deviation of an observed quantity from its average is equal to the average square minus the square of the average of the quantity. It may be well to recall the proof of this well-known property.

Let m_s denote the observed quantity, $\mathfrak{M}^2$ its average square, μ^2 the square of the mean deviation, and m the average of the individual values m_s; p_s the probability that the quantity m_s will be observed, then $\mu^2 = \Sigma(m_s - m)^2 p_s = \Sigma m_s^2 p_s - 2m\Sigma m_s p_s + m^2\Sigma p_s$, and $\Sigma m_s p_s = m$, $\Sigma p_s = 1$, therefore $\mu^2 = \mathfrak{M}^2 - m^2$ and $\mathfrak{M}^2 = \mu^2 + m^2$.

Now the square of the mean deviation may be approximated to by means of the property that this quantity taken over a cases is equal to a times the square of the deviation per case.

Consequently we had to calculate the value of the square of the mean deviation for the different territorial Austrian accidents insurance institutions from the abovementioned statistical publications, containing the number and the charge of accumulations of accidents belonging to each group, at that time (1901) disposable over five successive years. From the value obtained we could calculate the average square, the average value m being easily obtained. We will call this the *indirect* method.

If we dispose of individual data, then we do better in following the direct method by which we may obtain more accurate results by calculation of the desired average from the square of the importance of each accident.

So, as soon as a sufficient experience was at hand, according to this direct method, a calculation of $^2\mathfrak{M}^2$ and $^3\mathfrak{M}_1^2$ was made for several branches of industry, in order to verify whether these quantities for an approximate calculation could really be considered as invariable. For different branches of industry 5 values were calculated for $^2\mathfrak{M}$, the extremes of which were found to be 5·9 and 6·8; their arithmetical mean being 6·4. Though the statistical data were rather incomplete, the results were not so very different. For $^3\mathfrak{M}$, 12 values have been calculated for different branches of industry. The extreme values were 10·3 and 12·6, the arithmetical mean being 11·2. Again it appears that the above supposition was not altogether wrong. A similar calculation of $^1\mathfrak{M}$ has not been made. The importance varies here between 0 and thousands of guilders. The high results are comparatively scarce but have an important influence on the final result. Therefore the value of $^1\mathfrak{M}$ which will appear to be of little importance in the subsequent calculations has been obtained by the above explained indirect method from more ample statistical data than could have been taken into consideration, when calculating it in the direct way. It will be understood that the divergence of the different values acquired in this way, was far greater than that of the values $^2\mathfrak{M}$ and $^3\mathfrak{M}$, so much the more as we know the amount of the compensations of Group I to be in a high degree subject to personal estimation.

So far we have found the mean deviation to be: $\sqrt{L^2({}^1p.{}^1\mathfrak{M}^2 + {}^2p.{}^2\mathfrak{M}^2 + {}^3p.{}^3\mathfrak{M}^2)}$, in which formula $^2\mathfrak{M}$ and $^3\mathfrak{M}$ may practically be considered invariable, at least as far as cognate branches of industry are concerned. If we divide by L we shall find that in those branches the mean deviation only varies when $^1\mathfrak{M}$ (being of small importance) varies or when ^1p, ^2p and ^3p vary. Even when considering the same branch of industry, those values may differ a good deal, as will appear in due time.

Greater differences will disappear if we divide by the net-premium due for the insurance per 100 of annual wages.

This premium $n = 100\,({}^1p^1m + {}^2p^2m + {}^3p^3m)$; in which formula ^1m, ^2m and ^3m stand respectively for the average importance of cases of temporary incapacity, of fatal accidents, and of cases of permanent incapacity per unit of annual wages.

The result of the division is:

$$M^2 = \frac{{}^1p^1\mathfrak{M}^2 + {}^2p^2\mathfrak{M}^2 + {}^3p^3\mathfrak{M}^2}{100\,({}^1p^1m + {}^2p^2m + {}^3p^3m)}.$$

If we consider $^1\mathfrak{M}$, $^2\mathfrak{M}$, $^3\mathfrak{M}$ to be invariable for the same branch of industry as well as the quantities ^1m, ^2m and ^3m, then the value of M^2 appears to depend solely on the mutual proportions of ^1p, ^2p and ^3p. If these proportions were invariable too, in the same branch of industry, then M^2 might be considered invariable under the same circumstances. It has been proved already that even in much different branches of industry ${}^2\mathfrak{M}$ and ${}^3\mathfrak{M}$ vary very little. On investigation the same appeared to be the case with regard to the value of m. *A priori* it is not improbable that the mutual proportions of ${}^1p$, ${}^2p$ and ${}^3p$ should be nearly invariable. If the frequency of accidents should vary within the same branch of industry and in the same territory a difference in these proportions may arise from intensity of trade or economical depression, from different carefulness with regard to protection against accidents, etc. As long however as the technical character of a trade remains unaltered, these influences may have their effect on the frequency, but not on the issue of accidents. This, doubtless, holds true with respect to the proportion of the numbers of cases of permanent incapacity and of fatal accidents which are the most important.

Moreover if we pay special attention to the quantities in the formula for M^2, it is evident that on the whole ${}^3p\,{}^3\mathfrak{M}^2$ and ${}^3p.{}^3m$ are predominant because of the slightness of ${}^2p$ and of ${}^1\mathfrak{M}^2$ and ${}^1m$. So the differences in the mutual proportions of the values of p would have to be enormous to cause any important differences in the values of M^2.

The following table may prove that as a rule the mutual proportions of the heterogeneous cases never differ much, even when the numbers of accidents show great differences.

Branch of industry (name of the trade)	Number of accidents per one 1000 "Vollarbeiter" (years of labour)	Assuming the number of accidents causing permanent incapacity or death to be 100, we find for		
		the number of fatal accidents	the number of cases of permanent incapacity	the number of cases of temporary incapacity
Masons in the large cities	165	18	82	2497
" " other places	98	15	85	2782
Dock labourers in Rotterdam	538	14	86	2554
" " elsewhere	261	14	86	2584
Ship-builders in the north	142	13	87	4362
" " " " south	213	12	88	3542
Painters in the large cities	47·3	31	69	1540
" " other places	39·1	28	72	2881
Railway Company A	54·7	39	61	3105
" " B	61·0	48	52	3177
The building-trade in the large cities	148	15	85	2878
" " " " other places	142	12	88	3363
Inland navigation northern provinces	35·1	47	53	1100
" " southern provinces	53·5	51	49	877

When calculating the values of the quantity M for the railway companies A and B (where the proportions of the number of accidents show the greatest mutual differences), it appears how slight the influence of differences like these is on the value of M.

For company A, M amounts to 0·23, for B to 0·25. Consequently we may apply the values of M calculated from statistical data for the whole of a branch of industry to larger groups belonging to the same branch. We may even expect to find only slightly deviating values for different branches of industry.

Up to the present, in the development of $\Sigma\, m_s^2 n_s$, no allowance has been made for the so-called "multiple accidents" (those, by which more than one workman took injury).

The necessary information to make this allowance in calculation is wanting; an approximate calculation has proved that in the railway-trade (that is a trade where relatively perhaps more "multiple accidents" occur than in any other) these accidents add to the value of M much less than 3 °/₀ of the total value. So we are justified in calculating without these.

So far we have been operating exclusively according to what is called the combinatorial principle, though what may be called the physical principle has already been applied in calculating the quantities $\mathfrak{M}$ according to the indirect method. After in 1901 previous calculations had been made by application of the above explained combinatorial principle to Austrian statistics, it was felt to be necessary to compare the result to approximation of the same quantity. For this purpose we made use of the "Revidierte Unfall-Statistik" of 1890—1896, in which for several trades the total charge (B)* of all accidents, the total amount of wages (l), and the total number of "Vollarbeiter" (a) for each of the seven years (1890—1896) are communicated for each of the seven official territorial insurance-establishments. By totalizing the a and B for a certain establishment over all the seven years, we find the probable value of the charge per Vollarbeiter exposed to risk $\frac{\Sigma B}{\Sigma a}$. The deviation of B from the average charge is then $B - a\frac{\Sigma B}{\Sigma a}$. If a were invariable it would be easy to calculate from these the mean deviation for a workmen, from which the mean deviation per workman would be obtained by division by $\sqrt{a}$. Approximate equality of the number a could often be attained by adding the data for different establishments in opposite order of time. Perfect equality not being attainable, the square of the mean deviation per workman has been calculated by the formula

$$\frac{\Sigma\dfrac{\left(B - \dfrac{\Sigma B}{\Sigma a}\, a\right)^2}{a}}{n-1},$$

in which n denotes the resultant number of independent groupings.

By dividing the square of the result by $\left(\frac{\Sigma l}{\Sigma a}\right)^2$ and by $100\frac{\Sigma B}{\Sigma l}$ we finally obtain the value of M^2, comparable to the value obtained according to the combinatorial principle by means of the quantities ${}^1\mathfrak{M}^2$, ${}^2\mathfrak{M}^2$ and ${}^3\mathfrak{M}^2$. The average of the values obtained in this way for the same branch of industry, but for different institutions, was not considered to be the most probable, before the utmost care had been taken that these values should depend on statistical experience of the same extent, that deviations not depending on the number of observations (symptomatical deviations) were eliminated by inverse summations, and that the individual values should be as nearly as possible of equal weight.

* Charge is understood in this paper to be the value of all expenses resulting from accidents which occurred in one year of insurance, calculated at a rate of interest of 3½ per cent. per annum.

The following table of the values of M^2 proves that, if not always, still in the principal cases, where ample statistical information was at hand, the expected conformity as to order of greatness was indeed found to exist*.

Branch of Industry	Value of M^2 calculated after	
	the first method	the second method
Ironworks	0·074	0·091
Mechanical Brick-works	0·088	0·081
Glass-works	0·086	0·063
Mechanical forges	0·063	0·026
Engine-works	0·072	0·064
Gas-works	0·063	0·072
Wool-manufactories	0·082	0·032
Spinning-mills (cotton)	0·082	0·081
Paper-mills	0·088	0·067
Saw-mills	0·085	0·046
Sugar-factories	0·083	0·056
Builders	0·083	0·102
Mechanical printers	0·065	0·043
Average	0·078	0·063

Moreover we may conclude from the first column that, according to expectation, the value of M obtained by the first method will not differ very much for different kinds of trades.

If we know the value of M for a certain trade we may find the square of the average deviation for a certain group of persons in that trade by multiplying M^2 by the net-premium taken from that group per 100 guilders of wages, by the square of the average annual wages calculated for that group and by the number of labourers; that is according to the formula aL^2nM^2, if a denotes the number of labourers, $L =$ the average annual wages, and $n =$ the net annual premium for 100 guilders of wages.

As soon as sufficient statistical material concerning the legal insurance against accidents in the Netherlands was at hand, calculations have been repeated. This time calculations have been made exclusively according to the combinatorial principle, the quantities $\mathfrak{M}$ having been obtained according to the above called direct method. When applying the physical method the increase of the frequency of accidents which has been noted in our country as well as in Austria and Germany has a particularly disturbing influence. These symptomatic variations in the annual net-charge, though eliminated as is above said as much as possible, continue to be of so much importance that, at least in ample statistical material, the accidental deviations are quite overruled by them. This makes the results of

* Hitherto approximate conformity of combinatorial and physical results had only been obtained by Lexis as to the sex-ratio at birth (*Jahrbuch für National-Oekonomie*, 27 and 32), by the author as to the number of yearly deceased persons among the Dutch population (Das Problem vom Risiko in der Lebensversicherung, *Baumgartner's Zeitschrift*, 1899).

the physical method, where of course the statistical material should not be too extensive, unreliable in a great measure.

Throughout this series of calculations we used the same average value for ${}^{1}\mathfrak{M}^2$ and ${}^{2}\mathfrak{M}^2$, which, as far as the value of ${}^{2}\mathfrak{M}^2$ is concerned, seemed to be quite admissible. Besides the material was too small to derive a sufficient number of reliable values of ${}^{2}\mathfrak{M}^2$, while moreover fatal accidents have a comparatively small influence. That these values approximately might be considered invariable, does not hold good for ${}^{1}\mathfrak{M}^2$ as it does for ${}^{2}\mathfrak{M}^2$, but we noticed already that: (1) it is impossible to find reliable results for one single branch of industry, and that: (2) the term in ${}^{1}\mathfrak{M}^2$ does not influence the result very much.

As far as ${}^{3}\mathfrak{M}^2$ was concerned it has proved desirable to have at our disposal values that reflect the individual differences which, according to the statistical reports for different trades, appear to exist. It is true that the different values found for ${}^{3}\mathfrak{M}_1^2$ were only slightly different, but the introduction of this quantity rested upon a supposition which may only approximately be called correct, viz. the supposition that the average square of the importance of a case of permanent incapacity of the fraction t could be denoted by $t^2\,{}^{3}\mathfrak{M}_1^2$. A proportion as assumed here is certainly incorrect for the expenses of the first years after accident; for neither the expenses for medical treatment nor the payments during the period of sickness and recovery are at all connected with the percentages of resulting permanent incapacity. Moreover it has been proved that the older labourers are, the higher is the degree of incapacity, so that the average-age of the permanent invalids with a high percentage of incapacity is considerably higher than that of invalids with but slightly reduced capacity. Moreover, these circumstances being of great importance as to the result, it is desirable that not the average distribution of wages over the whole industry be contained in them, but the distribution of wages only over the branch of industry concerned, which certainly shows a closer dispersion. Therefore the value of ${}^{3}\mathfrak{M}^2$ is to be calculated directly from the material for every trade in particular.

The values found for M^2 in this way are given in the following table as far as they could be calculated from sufficiently extensive statistical material.

Industries	M
Mechanical tile-works and brick-works	0·23
Mechanical printers	0·22
Painters	0·27
Whitewashers and plasterers	0·25
Engineering works	0·27
Masons	0·25
Carpenters	0·23
Builders	0·24
Chemical industries	0·23
Saw-works	0·22
Forges	0·23
Engine-works	0·24
Non-mechanical ship-builders	0·26

Industries	M
Mechanical ship-builders	0·25
Paper-mills	0·27
Weaving-mills (cotton)	0·25
Distillers	0·23
Dock-labourers	0·23
Railways	0·26
Carriers and draymen	0·23

The extreme values are 0·22 and 0·27, which satisfactorily agrees with the supposition.

The law regarding the distribution of deviations of the real charges from the most probable or average value is

$$\frac{1}{2\pi}\int_{-\pi}^{+\pi}\cos(\theta\alpha)\cos\theta^3\frac{z}{3!}\cos\theta^5\frac{v}{5!}\dots e^{-\theta^2\frac{y}{2!}-\theta^4\frac{u}{4!}-\theta^6\frac{w}{6!}\dots}d\theta\,^*,$$

where α denotes the value of the deviation and the letters y, z, u, v, etc. consecutively the quantities $\Sigma m_s n_s$; $\Sigma m_s^2 n_s$; $\Sigma m_s^3 n_s$; $\Sigma m_s^4 n_s$, etc. In applying this general law we do not suppose, as we should do in applying *à priori* Gauss's law, that accidents of every importance are represented in large number. However, to apply it would give much trouble. It is evident that where very ample charges are concerned, combinatorial deviations will follow Gauss's law. Where only comparatively small charges are in question, we had at first to put up with what could be deduced about the probability of deviations from the Theorem of Tchebychef. However, from this theorem we obtain no information about the probability of deviations itself, but we obtain a deviation for which the probability of being surpassed remains below a given fraction.

It soon appeared that this theorem was by much too rough for practical purposes and that indeed the dispersion of deviations was by so much more restricted that research for another expedient was required. Mr K. Lindner, officer in chief of the Mathematical and Statistical department of the State Insurance Office, who has applied my theory developed above to the statistical results and further to the insurance-practice of that institution, has tried therefore to find an empirical frequency-curve, from the experience of the middle-sized charges commonly appearing in practice. For this purpose he calculated the values of M briefly communicated above and applied them to the data contained in the statistical reports of the financial results of the insurance of the years 1903—1907. In those statistical reports the total insured wages l, the total net charge B, and the number of "years of labour" a, are collected for a great number of trades and for each of the years 1903—1907. By means of these data could be calculated the average net charge per hundred $n=\frac{100\,\Sigma B}{\Sigma l}$, the average annual wages: $\frac{l}{a}$ and the mean deviation from the average charge of each year: $\sqrt{a\left(\frac{l}{a}\right)^2 nM^2}$.

* This formula is obtained according to the method explained in *Traité du calcul des probabilités* by H. Laurent, p. 247 sq., by considering accident-insurance as a succession of constant contracts, each of infinitesimal duration dt.

The deviation in the annual charge for each year could be calculated to be $B - \frac{1}{100}nl$ and this deviation could be expressed as a multiple or a fraction of the corresponding mean deviation. Out of 405 deviations, 190 were positive, 215 negative. The distribution as to the value was as follows:

Deviations per cent. of the mean deviation	Observed deviations	
	positive	negative
0— 20	31	30
21— 40	33	41
41— 60	23	33
61— 80	22	26
81—100	25	36
101—150	36	36
151—200	12	12
201—400	7	—
more than 400	1	1
Total amount	190	215

The asymmetry as to positive and negative deviations disappears for the greater part if we exclude observations dependent on smaller probable charges. The following table relates to the deviations from probable charges of 20,000 guilders at least.

Deviations per cent. of the mean deviation	Observed deviations		Distribution according to Gauss's law
	positive	negative	
0— 20	19	17	18
21— 40	20	26	18
41— 60	15	18	16
61— 80	17	15	14
81—100	13	19	13
101—150	24	15	21
151—200	5	6	10
201—400	3	—	5
more than 400	—	1	—
Total amount	113	117	115

For the purpose of a comparison of the dispersion with that according to Gauss's law, the latter is put in the third column next to the observed distribution. There appears to be sufficient concordance so that it seems to be admissible to assume this law as available in the case of not too small charges.

As rather satisfactory concordance with Gauss's law was found even for a charge of but 20,000 guilders, it was judged superfluous to draw up an empirical law. Moreover we may see in the obtained result another proof of the usefulness for practical purposes of the chosen approximate calculations of mean deviations. This is important for another question where the mathematical department of the State Insurance Office originally met with no small difficulties and for which an apparently sufficient solution has since been found by Mr Lindner. The State Office has to charge premiums from the branches of industry fairly; that is, counterbalancing as nearly as possible the charges of the insurance. The following rules prevail here. By Royal Resolution the different branches of industry are arranged in a great number of classes. Every class contains usually five (only in the lower classes three) degrees of risk, which are respectively about 10 and 20 °/₀ above or below the middle degree. The tariff contains for every degree of risk a proportional premium. With exception of a few special cases (to be mentioned hereafter) an establishment has to be arranged in the class in which the branch of industry practised is arranged. It depends on the individual character of the establishment as to which degree of the class premium shall be charged. Only when special reasons exist, an establishment can be arranged in another than the prescribed class as soon as special Royal Authorization has been obtained. At first it did not seem an improbable supposition that the risk in establishments belonging to the same branch of industry should vary between rather near limits, and from that idea the prescription for classification arose which was in a large measure borrowed from the Austrian Insurance Act. It soon appeared in giving effect to the Dutch Insurance Act that this plausible supposition was by no means supported by experience and that in the same trade the risks were often very much different. It is not exclusively the State Office that is an authorised insurer, as is the case with the Austrian official Accidents insurance institutions, and the existence of private insurance institutions prevents the State Office from expecting that evil risks will be compensated by good. Therefore it was felt necessary to profit as soon as possible by the statistical experience concerning the insurance of the different establishments individually.

Consequently we were soon seriously in want of a conclusive answer to the question when and in what circumstances the premium to be charged from an establishment may be deduced from statistical experience. In Austria the officials have been occupied in answering this question while attempting the general classification of industries. According to the official statistical report of the empire, entitled "Ergebnisse der zum Zwecke der Gefahrenclasseneinteilung überprüften Unfallstatistik der Jahre 1890—1896" (Seite VI), they go by the following criterion: "die aus den Unfalls- oder Belastungsziffern abgeleiteten Verhältniszahlen für die Unfallsgefährlichkeit der einzelnen Betriebsgattungen können nur dann als sichere, von Zufälligkeiten nicht mehr beeinflusste Ergebnisse angesehen werden, wenn die Beobachtung sich auf mehr als 10,000 Vollarbeiter erstreckt. Auch die aus Beobachtungen von mehr als 5000 Arbeitern abgeleiteten Verhältniszahlen können immerhin noch als massgebende Ergebnisse der Erfuhrung angesehen werden." Though we do not know on what considerations this criterion depends, it appears that for non-dangerous industries it demands insufficient security; for very dangerous industries an exorbitant one.

If the insurance of an average establishment causes great loss one year after the other, it is not right to go on charging the same premium from it until statistical experience finally runs over 5000 Vollarbeiter (years of labour), nor is it fair to classify an establishment belonging to a non-dangerous kind of industry satisfying the Austrian criterion, lower than normally, because of statistical reports though running over 2·5 millions of wages over perhaps no more than 5000 guilders of charge; in such a case of small charge the casual favourable issue of one accident may already have caused the deviation from the average value. Therefore we have tried to meet the difficulty in a different way.

Let b denote the statistical charge and l the corresponding total sum of wages, then the percentage of the premium: $\frac{100C}{l}$ may be derived from these data with an accuracy determined by the proportion of the mean deviation of the probable charge from the charge C.

The first thing that is to be done now is to make a reasonable choice of the degree of accuracy required here. In our case the choice was determined by certain legal prescriptions. Our purpose is to find the class in which the risk has to be arranged and, with reference to what has been communicated above, the question has to be answered, under what circumstances it is to be expected with sufficient probability, that the net premium necessary to meet the probable charge will be less than 20 °/ₒ higher or lower than the percentage of charges $\frac{100C}{l}$, deduced from statistical data. We assume a probability greater than $\frac{1}{2}$ to be sufficient, because then already there will be more chance for the establishment to be arranged in the right class than in a wrong. The mean deviation in the probable charge C may be approximated by the term $\sqrt{aL^2nM^2}$ if a denotes the number of labourers, L the average wages and n the charged net-premium $\frac{C}{aL}$.

If $aL = l$ be the amount of wages to which the charge C relates, then $aLn = ln$ is the hundred-fold of the probable charge C; so that we may write $\sqrt{L}.M\sqrt{100\,C}$ for $\sqrt{aL^2nM^2}$. Then the probable deviation may be expressed by $0{\cdot}675\sqrt{L}.M.\sqrt{100\,C}$, consequently greater deviations than these have a probability less than $\frac{1}{2}$, and such a deviation will coincide with the limits of the normal class, if it be equal to 20 °/ₒ of C. From this follows the condition that a statistical experience will begin to furnish sufficient indication of the class, if the expected charge satisfies the equation

$$0{\cdot}675\sqrt{L}.M\sqrt{100\,C} = \tfrac{1}{5}C,$$

from which $$C = 1139\,LM^2\,*.$$

If we know the average annual wages L, and the value of M, classification on exclusively statistical grounds will be allowed as soon as the expected net charge amounts to $1139\,LM^2$, in which condition, instead of the expected charge, the most probable value of it, the real net charge, must be substituted in case of application to practice.

Now the question still remains in what way allowances have to be made for statistical experience of smaller extent. If no other data were at hand the premium

* It must be observed that $1139LM^2$ is always large enough for an approximate application of Gauss's law, as follows from the preceding investigation, so that the factor 0·675, inferable from that law, was justly applied.

would have to be fixed at $\frac{100\,C}{l}$. In most other cases however we have other data at our disposal. If the classification of a separate establishment is concerned, an accessory datum may be found, for instance in the classification of the whole branch of industry that is in the average experience of all similar establishments together; if a branch of industry is concerned in most cases another branch may be found to which it may be compared and which may be classified on ampler statistical grounds. The experience of the State Insurance Office shows that this proceeding leads to results of less probability; as it has been proved that establishments appertaining to the same branch of industry may show the most different risks. Nevertheless, in case of scanty statistical experience we also have to operate after this method. Then a premium must be chosen between $\frac{100\,C}{l}$ and for instance the normal tariff-premium. It follows from the nature of this problem that whichever solution may be chosen here, it will remain comparatively arbitrary, as the probability of the applicability of normal classification in the case of a separate establishment is unknown; and even cannot be expressed numerically. For a practically serviceable solution the following consideration may be available.

Though *à priori* the statistical charge is the most probable value of the risk $\frac{100\,C}{l}$, we are safe in assuming that this value shows a deviation from the real risk of sign contrary to that of the difference between the real risk and the net tariff premium and of such an extent that greater deviations will have only a probability $\frac{1}{2}$, if by this supposition the risk should indeed be estimated right. This supposition takes for granted an influence of statistical data on classification increasing with experience; and only a few obvious restrictions are necessary always to obtain acceptable results.

It is evident, however, that any method will show equally serviceable results, if only this method admits of the charges $C < 1139\,LM^2$ having an influence on the net premium, dependent on the value of C. Our object was only to show that the theory developed in the preceding pages has in practice proved to be tenable and serviceable.

NOTES UPON THE CURVES OF CERTAIN FUNCTIONS INVOLVING COMPOUND INTEREST AND MORTALITY

By Robert Raynal Brodie.

It is well known that the rate of mortality is very high during the first few years of life and that it diminishes continuously till about age 10 or 12, thereafter increasing steadily right up to the end of life. This characteristic is common to all tables of mortality though the actual rates of mortality at different ages vary slightly according to different tables. If the rate of mortality were the same at all ages there would be no limit to the duration of life; and it would not be necessary for Life Assurance Offices to accumulate large reserves—the contributions of any year being simply applied to meet the claims of that year. Since, however, the rate of mortality does increase from year to year (omitting the first few years after birth) it is evident that a *level* periodic contribution for a sum payable at death must in the early years be greater and in the later years less than the amount required for the actual risk from year to year; in this way the effect of an increasing rate of mortality is to create a Fund out of the excess contributions of the early years so as to help to pay the excess claims due to high mortality in the later years of life. The amount of the Fund at any moment therefore must represent the difference between the contributions (with interest) received and the benefits paid and must be accumulated so as to help future contributions to pay future claims. Just as past contributions exceed past claims so future claims will sooner or later exceed future contributions. It is this feature of mortality that necessitates the accumulation of large Reserves by Life Assurance Societies; these Funds do not represent surplus; they must be on hand to meet the liabilities of the future.

In the case of benefits depending on mortality there are two forces operating on the reserve value that accrues from year to year, viz. Compound Interest tending to increase the Fund, and Mortality tending to reduce the Fund, and the progress of the Fund will depend upon the relative strength of these two forces. The object of the present note is to shew the resultant effect of these two forces upon the curve traced by such a Fund.

The following is the notation used in this paper:

$\bar{V}_t$ the reserve value at duration t arising out of past contributions in respect of each unit of benefit.

$\bar{\pi}$ the amount of yearly contribution payable by continuous instalments for each unit of benefit.

μ_t the intensity of mortality (per year) at moment t.

δ a year's interest (payable momently) per unit.

When a benefit becomes payable at duration t there will be $\overline{V}_t$ on hand to meet the claim in that particular case so that the general strain on the Funds of the Society will only be $1-\overline{V}_t$. The probability of death occurring within a year at moment t is μ_t. The expression $\mu_t(1-\overline{V}_t)$ is termed the "death strain" and represents the "loss" due to mortality. On the other hand the "gain" for the year due to premium and interest on reserve is $\overline{\pi}+\delta\overline{V}_t$. We therefore have the relation

$$\frac{d}{dt}\overline{V}_t=\overline{\pi}+\delta\overline{V}_t-\mu_t(1-\overline{V}_t) \quad \text{.........................(1).}$$

This relation is quite general and applies to all classes of benefit. It indicates separately the operation of the two forces of interest and mortality. Under any contract payable on death interest and mortality act in opposite directions. If, however, the benefit were payable only on survivance (such a benefit being known as a pure endowment) formula (1) would become

$$\frac{d}{dt}\overline{V}_t=\overline{\pi}+\delta\overline{V}_t-\mu_t(0-\overline{V}_t),$$

that is to say, the death strain in such a case becomes negative, and mortality instead of opposing the operation of interest acts in harmony with it. If under a pure endowment the accumulated contributions were repayable on death the element of mortality would be entirely eliminated and the death strain would be zero throughout.

Suppose that somewhere on the curve of $\overline{V}$ $\frac{d}{dt}\overline{V}_t$ passes through the value zero then at that point there is either a maximum or a minimum value, and there we have the relation

$$\overline{\pi}+\delta\overline{V}_t=\mu_t(1-\overline{V}_t) \text{..............................(2).}$$

At that point the forces of increment and decrement exactly counterbalance each other, and there the reserve curve has a turning value. If $\frac{d}{dt}\overline{V}_t$ pass from negative to positive the point will of course be a minimum, or if from positive to negative then it will be a maximum. If a benefit be payable in event of death only during a certain time the reserve value is zero at the expiration of the term, and since it must commence at zero, assuming a periodic contribution to be payable, it is evident that in such a case a maximum value will arise somewhere on the curve.

If contributions are paid in one sum we shall have in place of formula (1),—writing $\overline{A}_t$ instead of $\overline{V}_t$

$$\frac{d}{dt}\overline{A}_t=\delta\overline{A}_t-\mu_t(1-\overline{A}_t) \quad \text{...........................(3),}$$

and the condition for a maximum or minimum value will be

$$\delta\overline{A}_t=\mu_t(1-\overline{A}_t).$$

In the case of an assurance payable at death effected at a very early age the curve of the reserve value will pass through a minimum point, due to the rapid decrease in mortality. It is possible, however, by giving sufficient weight to the element of interest to eliminate this minimum value. Consider, for example, two single-payment assurances for £100 each, opened at age 3, one payable at death and the other payable at death or on attainment of age 21. The minimum value occurring about age 5 in the former case disappears under the endowment assurance, —the disappearance being caused in the latter case by the larger accumulative element. The following table shews the operation of the two forces of mortality and interest round about age 5, $\frac{d}{dx}\bar{A}$ in the former case passing from negative to positive and in the latter case being positive throughout. The figures are based on the Institute of Actuaries' Text Book 3 °/$_{\circ}$ Table.

TABLE I.

Age attained (x)	Assurance payable at death			Assurance payable at death or age 21			Age attained (x)
	$\delta\bar{A}_x$	$\mu_x(1-\bar{A}_x)$	$\frac{d}{dx}\bar{A}_x$	$\delta\bar{A}_{x:\overline{21-x}\rvert}$	$\mu_x(1-\bar{A}_{x:\overline{21-x}\rvert})$	$\frac{d}{dx}\bar{A}_{x:\overline{21-x}\rvert}$	
3	·782	1·314	−·532	1·808	·694	+1·114	3
4	·771	1·019	−·248	1·846	·518	+1·328	4
5	·766	·846	−·080	1·888	·413	+1·475	5
6	·767	·685	+·082	1·934	·320	+1·614	6
7	·771	·553	+·218	1·984	·246	+1·738	7
8	·779	·447	+·332	2·036	·189	+1·847	8
⋮							⋮

Differentiating both sides of the expression marked (1) we get

$$\frac{d^2\bar{V}_t}{dt^2} = \delta\frac{d}{dt}\bar{V}_t - \frac{d}{dt}\left[\mu_t(1-\bar{V}_t)\right] \quad \text{.....................(4).}$$

The condition for a point of inflexion on the curve of $\bar{V}$ is

$$\delta\frac{d}{dt}\bar{V}_t = \frac{d}{dt}\left[\mu_t(1-\bar{V}_t)\right].$$

If at that point $\frac{d^2\bar{V}_t}{dt^2}$ pass from positive to negative the curve thereabouts will be of the form ∫ or if it pass from negative to positive the curve will be of the form ⁄ . If the assurance is by single payment $\bar{A}_t$ will take the place of $\bar{V}_t$ in formula (4).

Most of the calculations made by Life Assurance Offices are based upon what are known as "Select Tables." These tables trace the mortality among a body of lives "selected" by medical examination. During the first few years succeeding the date of medical examination the mortality among such lives is much less than among lives of corresponding age in the general population, and the *proportional* increase in mortality due to the wearing-off of selection is correspondingly large. This rapid disappearance of the traces of selection tends to introduce a point of inflexion on the curve of $\overline{V}$. As an illustration shewing the difference between the curve of $\overline{V}$ according to a table where selection is present and a table where selection has worn off we may trace from ages 30 to 35 two single payment assurances for £1000 each payable at death on lives which entered at 20 and at 30. In the former case the effect of selection has worn off during the 10 years since date of entry. On the O[M] 3 °/₀ table the curve of the former will be of the form  and of the latter  as will be seen from the following figures:—

TABLE II.

Age attained (x)	Age 20 at entry			Age 30 at entry			Age attained (x)
	$\delta \frac{d}{dx} \overline{A}_x$	$\frac{d}{dx} [\mu_x (1 - \overline{A}_x)]$	$\frac{d^2 \overline{A}_x}{dx^2}$	$\delta \frac{d}{dx} \overline{A}_x$	$\frac{d}{dx} [\mu_x (1 - \overline{A}_x)]$	$\frac{d^2 \overline{A}_x}{dx^2}$	
⋮							
30	·210	·032	+·178	·315	2·604	−2·289	30
31	·215	·033	+·182	·267	·764	− ·497	31
32	·220	·037	+·183	·256	·345	− ·089	32
33	·226	·044	+·182	·255	·235	+ ·020	33
34	·231	·051	+·180	·256	·221	+ ·035	34
35	·236	·058	+·178	·257	·212	+ ·045	35
⋮							⋮

An increase in the rate of interest has always an accelerative effect on the ascent of the curve of $\overline{V}$, and it is evident therefore that a point of inflexion will be hastened or deferred by an increase in the rate of interest according as the curve thereabouts is of the form  or  .

We have seen that under a select table there are two opposing forces operating in the early years,—selection tending to give the curve the form  and interest giving the curve the form  and the resultant shape of the curve will depend upon the relative strength of these two forces. If the assurance is payable only at death it may not be possible to counteract the convex-upward shape of the curve in the early years resulting from selection, but under an endowment assurance payable at death or a certain age we can by making the term of the assurance sufficiently short give the accumulative element sufficient weight to smooth away the influence of selection upon the shape of the curve. This point may be made clearer by tracing for the first few years the reserves under three endowment assurances by single payment for £1000 each according to the $O^{[M]}$ 3 °/ₒ table, opened at age 30 and maturing at ages 60, 45 and 40 respectively.

TABLE III.

Duration	$\delta \frac{d}{dt} \bar{A}$	$\frac{d}{dt}[\mu(1-\bar{A})]$	$\frac{d^2\bar{A}}{dt^2}$
		maturing at 60	
0	·365	1·176	− ·811
1	·346	·872	− ·526
2	·338	·469	− ·131
3	·337	·243	+ ·094
		maturing at 45	
0	·540	·732	− ·192
1	·539	·427	+ ·112
2	·546	·204	+ ·342
3	·558	·043	+ ·515
		maturing at 40	
0	·630	·504	+ ·126
1	·638	·250	+ ·388
2	·652	·067	+ ·585
3	·669	− ·058	+ ·727

Here it will be seen that the effect of changing the maturity age from 60 to 45 is to hasten by about two years the point of inflexion resulting from selection; and by reducing the maturity age to 40 the point of inflexion is made to disappear altogether and all trace of selection upon the curve is eliminated.

It was already mentioned that under a temporary assurance with premiums payable throughout the term a maximum value arises. If such an assurance is effected by single payment it can easily be shewn that the reserve curves may assume any one of the following general shapes depending on the length of the term of the assurance and the nature of the mortality:

(1) no point of inflexion, no maximum value,

(2) a maximum value not preceded by any point of inflexion,

(3) a point of inflexion followed by a maximum value,

(4) two points of inflexion followed by a maximum value.

As an example of another type of curve we may take the case of an assurance for £1000 effected by single payment on a life aged 1 and expiring at age 25. No minimum value will arise under such a policy,—the expiry age is too low to admit of that,—and the reserve curve will descend throughout. In the early years $\frac{d^2\bar{A}_t}{dt^2}$

is positive; the curve then is at first convex-downwards and as the curves under such temporary assurances are invariably concave-downwards eventually a point of inflexion must arise. The position of the point of inflexion will be seen from the following figures and the curve will be of the type . The figures are based upon the Institute of Actuaries' Text Book 3 °/₀ table.

TABLE IV.

Age attained (x)	$\delta \frac{d}{dx} \bar{A}^{1}_{x:\overline{25-x}\rvert}$	$\frac{d}{dx}[\mu_x(1-\bar{A}^{1}_{x:\overline{25-x}\rvert})]$	$\frac{d^2}{dx^2} \bar{A}^{1}_{x:\overline{25-x}\rvert}$
⋮			
12	− ·0568	− ·2115	+ ·1547
13	− ·0539	− ·0794	+ ·0255
14	− ·0541	+ ·0538	− ·1079
15	− ·0606	+ ·1744	− ·2350
⋮			

In the case of assurances payable certain sooner or later (such as whole life or endowment assurances, as distinct from temporary assurances) it will be seen from the relation $\bar{V}_t = \frac{\bar{A}_t - \bar{A}_0}{1 - \bar{A}_0}$ that any peculiarities that may occur on the curve of the reserve-value where a single payment has been made will be reproduced at the identical points when a continuous premium is payable; that is to say, if a point of inflexion or a turning value has arisen in the former case it will occur also in the latter case at the same spot although the curves in the two cases will naturally ascend or descend at quite different rates. As regards temporary assurances it is only when the period of the assurance approaches the complement of life that the curves of benefits by single and continuous premium tend to have the same peculiarities.

On comparing the initial curve of the reserve arising under a pure endowment by single payment effected at an early age with the curve under a single payment assurance by a select table effected later in life it will be found that they are more or less of the same form, viz. convex-upwards. In the former case the shape of the curve is the result of rapidly *decreasing* mortality and a *negative* death strain; in the latter case it is the result of rapidly *increasing* mortality and a *positive* death strain.

CALCULATION OF MOMENTS OF AN ABRUPT FREQUENCY-DISTRIBUTION

By W. F. Sheppard.

1. The formulae adopted for calculation of the moments of a frequency-distribution—or, more generally, of a trapezette in respect of which our *data* are not ordinates but areas of strips—are not always correct. The formula for the case in which the figure is "double-tailed," i.e. has close contact at both extremities with the base, is fairly well known; the cases to be considered are those in which the figure, at one or both ends, is "abrupt," i.e. either ends with an ordinate which is not zero, or forms an angle between its upper boundary and the base.

2. The following notation will be used.

(i) y is the ordinate, corresponding to abscissa x, of the figure of frequency.

(ii) Observations are taken at intervals h in x.

(iii) A is the area of the strip of the figure which is bounded by ordinates corresponding to abscissae $x-\frac{1}{2}h$ and $x+\frac{1}{2}h$, and u is the average ordinate of this strip; i.e.

$$A \equiv hu \equiv \int_{x-\frac{1}{2}h}^{x+\frac{1}{2}h} y\,dx.$$

(iv) S is the total area from the ordinate y up to some definite ordinate; i.e.

$$S \equiv \int_x^C y\,dx.$$

(v) x_θ denotes $x_0+\theta h$; and y_θ, A_θ, u_θ, and S_θ are the corresponding values of y, A, u, and S.

(vi) The range of values of x is from $x_{-n} \equiv x_0 - nh$ to $x_m \equiv x_0 + mh$, where $m+n$ (but not necessarily m) is an integer; so that there are $m+n$ strips whose mid-ordinates correspond to the values of x_r for $r=-n+\frac{1}{2}, -n+\frac{3}{2}, \ldots m-\frac{1}{2}$. The *data* are the areas of these strips, i.e. are the values of A_r for these values of r.

(vii) Roman superscripts and D denote differentiation with regard to x, and δ and μ denote the central difference and the central mean for increment h in x; i.e.

$$y^{\mathrm{I}} \equiv Dy \equiv dy/dx,$$

$$\delta f(x) \equiv f(x+\tfrac{1}{2}h) - f(x-\tfrac{1}{2}h), \qquad \mu f(x) \equiv \tfrac{1}{2}\{f(x+\tfrac{1}{2}h) + f(x-\tfrac{1}{2}h)\},$$

so that, at any rate if we are dealing with polynomials in x,

$$\delta = 2\sinh\tfrac{1}{2}hD, \qquad \mu = \cosh\tfrac{1}{2}hD.$$

(viii) Σ denotes summation for the values of r mentioned in (vi) above.

(ix) $\triangleq$ denotes approximate equality.

3. The area and the successive moments of the figure are $\int y\,dx\,(=\Sigma A_r)$, $\int yx\,dx$, $\int yx^2dx, \ldots$, the limits of integration being $x = x_{-n}$ and $x = x_m$. But in actuarial work, and in statistical work generally, we have to deal with more general expressions of the form $\int y\phi(x)\,dx$, where $\phi(x)$ is a function of x which possibly is not given explicitly but is tabulated for a series of values of x. If, for instance, we are valuing an annuity fund, $\phi(x)$ may be such an expression as $(1+i)^{x-t}\,l_t/l_x$, $y\,dx$ being the total value at age t of the annuities conditionally payable to persons who are now between ages $x - \frac{1}{2}dx$ and $x + \frac{1}{2}dx$. The expression whose value we wish to find will therefore be denoted by

$$\int_{x_{-n}}^{x_m} y\phi(x)\,dx.$$

4. If the figure is double-tailed, the proper formula is

$$\int_{x_{-n}}^{x_m} y\phi(x)\,dx \triangleq \Sigma A_r \phi_1(x_r),$$

where

$$\phi_1(x) \equiv \phi(x) - \frac{1}{24}h^2\phi^{\text{II}}(x) + \frac{7}{5760}h^4\phi^{\text{IV}}(x) - \frac{31}{967680}h^6\phi^{\text{VI}}(x) + \ldots$$

$$= \mu\left\{\phi(x) - \frac{1}{6}h^2\phi^{\text{II}}(x) + \frac{7}{360}h^4\phi^{\text{IV}}(x) - \frac{31}{15120}h^6\phi^{\text{VI}}(x) + \ldots\right\}$$

$$= \left\{1 - \frac{1}{24}\delta^2 + \frac{3}{640}\delta^4 - \frac{5}{7168}\delta^6 + \ldots\right\}\phi(x)$$

$$= \mu\left\{1 - \frac{1}{6}\delta^2 + \frac{1}{30}\delta^4 - \frac{1}{140}\delta^6 + \ldots\right\}\phi(x);$$

the third form of $\phi_1(x)$ being used if $\phi(x)$ is tabulated for the mid-values $x_{-n+\frac{1}{2}}$, $x_{-n+\frac{3}{2}}, \ldots x_{m-\frac{1}{2}}$, and the fourth form if it is tabulated for the bounding values x_{-n}, $x_{-n+1}, \ldots x_m$.

For moments, the first or the second form of $\phi_1(x)$ is the more convenient according as we are taking moments with regard to a mid-ordinate or a bounding ordinate. Let ν_p be the true pth moment $\int_{x_{-n}}^{x_m} yx^p dx$; and let ρ_p be the ordinary "raw" pth moment $\Sigma A_r x_r^p$, obtained by massing the area of each strip along its mid-ordinate, and ρ_p' the alternative raw moment $\Sigma \frac{1}{2} A_r (x^p_{r-\frac{1}{2}} + x^p_{r+\frac{1}{2}})$, obtained by placing half the area of each strip along each of its bounding ordinates. Then the formulae are

$$\nu_p \triangleq \rho_p - \frac{1}{12}p_{(2)}h^2\rho_{p-2} + \frac{7}{240}p_{(4)}h^4\rho_{p-4} - \frac{31}{1344}p_{(6)}h^6\rho_{p-6} + \ldots,$$

$$\nu_p \triangleq \rho'_p - \frac{1}{3}p_{(2)}h^2\rho'_{p-2} + \frac{7}{15}p_{(4)}h^4\rho'_{p-4} - \frac{31}{21}p_{(6)}h^6\rho'_{p-6} + \ldots,$$

where $p_{(t)}$ denotes $p!/\{t!(p-t)!\}$. The terms after the first, in either expression, may be called the "corrections for massing." The two formulae give identical results.

5. The formula in § 4 is obtained by the following steps.

(A) If $A \equiv \delta F(x)$, then $y = F^{I}(x)$, and therefore

$$u = A/h = \delta/(hD) \,.\, y = y + \frac{1}{24} h^2 y^{II} + \frac{1}{1920} h^4 y^{IV} + \frac{1}{322560} h^6 y^{VI} + \dots,$$

$$y = hD/\delta \,.\, u = u - \frac{1}{24} h^2 u^{II} + \frac{7}{5760} h^4 u^{IV} - \frac{31}{967680} h^6 u^{VI} + \dots.$$

(B) $\int y\phi(x)\,dx = \int (hD/\delta \,.\, u)\,\phi(x)\,dx.$

(C) $\int_{x_{-n}}^{x_m} (hD/\delta \,.\, u)\,\phi(x)\,dx \triangleq \int_{x_{-n}}^{x_m} u\,\{hD/\delta \,.\, \phi(x)\}\,dx.$

(D) $\int u\,\{hD/\delta \,.\, \phi(x)\}\,dx = \int u\phi_1(x)\,dx,$

where $\phi_1(x)$ is defined as in § 4.

(E) $\int_{x_{-n}}^{x_m} u\phi_1(x)\,dx \triangleq h\Sigma u_r \phi_1(x_r).$

Of these steps, (A) and (B) are true absolutely, at any rate in the cases with which we are concerned; certain conditions as to continuity etc. being assumed with regard to y as a function of x. And (D) is true under similar conditions with regard to $\phi(x)$, which are of course satisfied if $\phi(x)$ is a power of x or a polynomial in x. But (C) and (E) only hold if the figure is double-tailed; (E) being the ordinary formula for quadrature by means of the "tangential area," and (C) being based on the fact that

$$\{hD/\delta \,.\, u\}\,\phi(x) - u\,\{hD/\delta \,.\, \phi(x)\}$$

consists of terms of the form (omitting powers of h and numerical coefficients)

$$u^{II}\phi(x) - u\phi^{II}(x), \qquad u^{IV}\phi(x) - u\phi^{IV}(x), \dots,$$

which are the derivatives of

$$u^{I}\phi(x) - u\phi^{I}(x), \qquad u^{III}\phi(x) - u^{II}\phi^{I}(x) + u^{I}\phi^{II}(x) - u\phi^{III}(x), \dots.$$

The formula cannot therefore be employed unless the figure is double-tailed.

6. But it does not follow that, in the cases in which the figure is not double-tailed—which are the cases to be considered in the remainder of this paper—, we can, as is sometimes suggested, content ourselves with treating $\int y\phi(x)\,dx$ as being equal to $\int u\phi(x)\,dx$, and then using the tangential formula for this latter; for this involves two separate mistakes, in that we first omit

$$\int (y-u)\,\phi(x)\,dx = \int \{hD/\delta - 1\}\,u \,.\, \phi(x)\,dx$$
$$= \int \left(-\frac{1}{24} h^2 u^{II} + \frac{7}{5760} h^4 u^{IV} - \dots\right) \phi(x)\,dx,$$

and then use an incorrect formula for calculating $\int u\phi(x)\,dx$. These two mistakes are of different kinds, and do not balance one another, except in some very special cases.

7. To express the complete formula, we use D_1 and D_2 to denote differentiation of u and of $\phi(x)$ respectively, so that, for differentiation of a product of u or any of its derivatives by $\phi(x)$ or any of its derivatives,

$$D = D_1 + D_2.$$

The formula is then*

$$\int_{x_{-n}}^{x_m} y\phi(x)\,dx \triangleq \Sigma A_r \{hD/\delta\,.\,\phi(x_r)\} + \Phi(x_m) - \Phi(x_{-n})$$

$$\triangleq \Sigma A_r \left\{\phi(x_r) - \frac{1}{24}h^2\phi^{\text{II}}(x_r) + \frac{7}{5760}h^4\phi^{\text{IV}}(x_r) - \frac{31}{967680}h^6\phi^{\text{VI}}(x_r) + \dots\right\} + \Phi(x_m) - \Phi(x_{-n}),$$

where†

$$\Phi(x) \equiv \left\{\frac{1}{hD}\frac{\frac{1}{2}hD_1}{\sinh\frac{1}{2}hD_1} - \frac{1}{2\sinh\frac{1}{2}hD}\frac{\frac{1}{2}hD_2}{\sinh\frac{1}{2}hD_2}\right\}A\phi(x)$$

$$= \frac{\frac{1}{2}hD_2}{2\sinh\frac{1}{2}hD_1}\left\{\left(\coth\tfrac{1}{2}hD - \frac{1}{\frac{1}{2}hD}\right) - \left(\coth\tfrac{1}{2}hD_2 - \frac{1}{\frac{1}{2}hD_2}\right)\right\}A\phi(x).$$

Performing the differentiations, we find that

$$\Phi(x) = \left\{\frac{1}{12}h\phi^{\text{I}}(x) - \frac{1}{240}h^3\phi^{\text{III}}(x) + \frac{1}{6048}h^5\phi^{\text{V}}(x) - \dots\right\}A$$

$$+\left\{-\frac{1}{240}h^2\phi^{\text{II}}(x) + \frac{1}{3024}h^4\phi^{\text{IV}}(x) - \dots\right\}hA^{\text{I}}$$

$$+\left\{-\frac{7}{1440}h\phi^{\text{I}}(x) + \frac{61}{120960}h^3\phi^{\text{III}}(x) - \frac{13}{362880}h^5\phi^{\text{V}}(x) + \dots\right\}h^2A^{\text{II}}$$

$$+\left\{\frac{41}{120960}h^2\phi^{\text{II}}(x) - \frac{31}{725760}h^4\phi^{\text{IV}}(x) + \dots\right\}h^3A^{\text{III}}$$

$$+\left\{\frac{31}{161280}h\phi^{\text{I}}(x) - \frac{1051}{29030400}h^3\phi^{\text{III}}(x) + \frac{1547}{383201280}h^5\phi^{\text{V}}(x) - \dots\right\}h^4A^{\text{IV}}$$

$$+\left\{-\frac{103}{5806080}h^2\phi^{\text{II}}(x) + \frac{161}{47900160}h^4\phi^{\text{IV}}(x) - \dots\right\}h^5A^{\text{V}}$$

$+$ etc.

With regard to this formula, the following points should be noted.

(i) If $\phi(x)$ is tabulated at intervals h, and its derivatives would have to be calculated from the differences of the tabulated values, it is best to express the coefficients of A, hA^{I}, ... in terms of these differences. The coefficients of the (central) differences in the successive portions of $\Phi(x)$ are given in Table I.

(ii) The values of $A, hA^{\text{I}}, h^2A^{\text{II}}, \dots$ for x_{-n} and for x_m have to be calculated from the given values of A for $x_{-n+\frac{1}{2}}, x_{-n+\frac{3}{2}}, \dots$ and for $x_{m-\frac{1}{2}}, x_{m-\frac{3}{2}}, \dots$. Usually we can get fairly good values by using an auxiliary function, as in the example given in § 9 below; when this is not possible, we have to use the actual differences of the A's, and these will not generally give such good results. In these latter cases, the alternative method considered in § 11 may be regarded as simpler.

8. For moments, the formulae of § 7 give

$$\nu_1 \triangleq \rho_1 + \Phi_1(x_m) - \Phi_1(x_{-n})$$

$$\triangleq \rho_1' + \Phi_1(x_m) - \Phi_1(x_{-n}),$$

* Where the symbol $\triangleq$ is used, in the remainder of this paper, the expressions which it connects are absolutely equal if y and $\phi(x)$ are polynomials in x.

† This differs from the expression originally given by me in *Proc. Lond. Math. Soc.* XXIX. 358—9, in being expressed in terms of the A's instead of in terms of the y's.

TABLE I. Coefficients of products of A, hA^{I}, h^2A^{II}, ... by $\delta\phi(x)$, $\mu\delta^2\phi(x)$, $\delta^3\phi(x)$, ..., or by $\mu\delta\phi(x)$, $\delta^2\phi(x)$, $\mu\delta^3\phi(x)$, ..., in $\Phi(x)$ (§ 7).

	A	hA^{I}	h^2A^{II}	h^3A^{III}	h^4A^{IV}	h^5A^{V}
$\delta\phi(x)$	$+\frac{1}{12}$	0	$-\frac{7}{1440}$	0	$+\frac{31}{161280}$	0
$\mu\delta^2\phi(x)$	0	$-\frac{1}{240}$	0	$+\frac{41}{120960}$	0	$-\frac{103}{5806080}$
$\delta^3\phi(x)$	$-\frac{11}{1440}$	0	$+\frac{19}{26880}$	0	$-\frac{2567}{58060800}$	0
$\mu\delta^4\phi(x)$	0	$+\frac{29}{24192}$	0	$-\frac{47}{414720}$	0	$+\frac{10817}{1532805120}$
$\delta^5\phi(x)$	$+\frac{521}{483840}$	0	$-\frac{1009}{8294400}$	0	$+\frac{290113}{30656102400}$	0
$\mu\delta\phi(x)$	$+\frac{1}{12}$	0	$-\frac{7}{1440}$	0	$+\frac{31}{161280}$	0
$\delta^2\phi(x)$	0	$-\frac{1}{240}$	0	$+\frac{41}{120960}$	0	$-\frac{103}{5806080}$
$\mu\delta^3\phi(x)$	$-\frac{13}{720}$	0	$+\frac{53}{40320}$	0	$-\frac{1981}{29030400}$	0
$\delta^4\phi(x)$	0	$+\frac{41}{60480}$	0	$-\frac{103}{1451520}$	0	$+\frac{3709}{766402560}$
$\mu\delta^5\phi(x)$	$+\frac{241}{60480}$	0	$-\frac{2351}{7257600}$	0	$+\frac{14941}{766402560}$	0

$$\nu_2 \triangleq \rho_2 - \frac{1}{12}h^2\rho_0 + \Phi_2(x_m) - \Phi_2(x_{-n})$$

$$\triangleq \rho_2' - \frac{1}{3}h^2\rho_0' + \Phi_2(x_m) - \Phi_2(x_{-n}),$$

$$\nu_3 \triangleq \rho_3 - \frac{1}{4}h^2\rho_1 + \Phi_3(x_m) - \Phi_3(x_{-n})$$

$$\triangleq \rho_3' - h^2\rho_1' + \Phi_3(x_m) - \Phi_3(x_{-n}),$$

$$\nu_4 \triangleq \rho_4 - \frac{1}{2}h^2\rho_2 + \frac{7}{240}h^4\rho_0 + \Phi_4(x_m) - \Phi_4(x_{-n})$$

$$\triangleq \rho_4' - 2h^2\rho_2' + \frac{7}{15}h^4\rho_0' + \Phi_4(x_m) - \Phi_4(x_{-n}),$$

$$\nu_5 \doteq \rho_5 - \frac{5}{6} h^2 \rho_3 + \frac{7}{48} h^4 \rho_1 + \Phi_5(x_m) - \Phi_5(x_{-n})$$

$$\doteq \rho_5' - \frac{10}{3} h^2 \rho_3' + \frac{7}{3} h^4 \rho_1' + \Phi_5(x_m) - \Phi_5(x_{-n}),$$

etc.,

where $\Phi_1(x)$, $\Phi_2(x)$, ... are functions whose values are given by Table II.

The terms due to the Φ's may be called the "corrections for abruptness"; the terms which precede them (after the first) being still called the "corrections for massing," though really the two sets of corrections together form the complete correction for massing.

If the figure is single-tailed, it will usually be simplest to take the abrupt end as the zero for x. The given A's will then be $A_{\frac{1}{2}}$, $A_{\frac{3}{2}}$, ...; and from these we determine A_0, hA_0^{I}, $h^2A_0^{II}$, ... The coefficients of A_0, hA_0^{I}, $h^2A_0^{II}$, ... in the quantities to be added to ρ_1/h, $\rho_2/h^2 - 1/12 . \rho_0$, ... in order to obtain ν_1/h, ν_2/h^2, ... are given in Table III.

TABLE II. Coefficients of A, hA^{I}, h^2A^{II}, ... in $\Phi_1(x)/h$, $\Phi_2(x)/h^2$, ... (§ 8).
($\xi \equiv x/h$.)

	co. A	co. hA^{I}	co. h^2A^{II}	co. h^3A^{III}	co. h^4A^{IV}	co. h^5A^{V}
$\Phi_1(x)/h$	$+\frac{1}{12}$	0	$-\frac{7}{1440}$	0	$+\frac{31}{161280}$	0
$\Phi_2(x)/h^2$	$+\frac{1}{6}\xi$	$-\frac{1}{120}$	$-\frac{7}{720}\xi$	$+\frac{41}{60480}$	$+\frac{31}{80640}\xi$	$-\frac{103}{2903040}$
$\Phi_3(x)/h^3$	$+\frac{1}{4}\xi^2$ $-\frac{1}{40}$	$-\frac{1}{40}\xi$	$-\frac{7}{480}\xi^2$ $+\frac{61}{20160}$	$+\frac{41}{20160}\xi$	$+\frac{31}{53760}\xi^2$ $-\frac{1051}{4838400}$	$-\frac{103}{967680}\xi$
$\Phi_4(x)/h^4$	$+\frac{1}{3}\xi^3$ $-\frac{1}{10}\xi$	$-\frac{1}{20}\xi^2$ $+\frac{1}{126}$	$-\frac{7}{360}\xi^3$ $+\frac{61}{5040}\xi$	$+\frac{41}{10080}\xi^2$ $-\frac{31}{30240}$	$+\frac{31}{40320}\xi^3$ $-\frac{1051}{1209600}\xi$	$-\frac{103}{483840}\xi^2$ $+\frac{161}{1995840}$
$\Phi_5(x)/h^5$	$+\frac{5}{12}\xi^4$ $-\frac{1}{4}\xi^2$ $+\frac{5}{252}$	$-\frac{1}{12}\xi^3$ $+\frac{5}{126}\xi$	$-\frac{7}{288}\xi^4$ $+\frac{61}{2016}\xi^2$ $-\frac{13}{3024}$	$+\frac{41}{6048}\xi^3$ $-\frac{31}{6048}\xi$	$+\frac{31}{32256}\xi^4$ $-\frac{1051}{483840}\xi^2$ $+\frac{1547}{3193344}$	$-\frac{103}{290304}\xi^3$ $+\frac{161}{399168}\xi$

TABLE III. Coefficients of A_0, hA_0^{I}, $h^2A_0^{\mathrm{II}}$, ... in $-\Phi_1(x_0)/h$, $-\Phi_2(x_0)/h^2$, ... for $x_0=0$ (§ 8).

	co. A_0	co. hA_0^{I}	co. $h^2A_0^{\mathrm{II}}$	co. $h^3A_0^{\mathrm{III}}$	co. $h^4A_0^{\mathrm{IV}}$	co. $h^5A_0^{\mathrm{V}}$
$-\Phi_1(0)/h$	$-\cdot08333\ 3333$	0	$+\cdot00486\ 1111$	0	$-\cdot00019\ 2212$	0
$-\Phi_2(0)/h^2$	0	$+\cdot00833\ 3333$	0	$-\cdot00067\ 7910$	0	$+\cdot00003\ 5480$
$-\Phi_3(0)/h^3$	$+\cdot02500\ 0000$	0	$-\cdot00302\ 5794$	0	$+\cdot00021\ 7221$	0
$-\Phi_4(0)/h^4$	0	$-\cdot00793\ 6508$	0	$+\cdot00102\ 5132$	0	$-\cdot00008\ 0668$
$-\Phi_5(0)/h^5$	$-\cdot01984\ 1270$	0	$+\cdot00429\ 8942$	0	$-\cdot00048\ 4445$	0

9. As a simple example, take the half of the ordinary figure of error; h being half the standard deviation. Taking h as the unit, the equation to the curve, if the total area is 10,000, is

$$y=10^4\sqrt{\{1/(2\pi)\}}\,.\,\exp.\,(-x^2/8).$$

This gives the values of A shown below on the left, and the values of $2\mu A$ that may be used for calculating the raw moments (ρ_1', ρ_2', ...) with regard to $x=0$.

x	A
$\frac{1}{2}$	3829
$1\frac{1}{2}$	2998
$2\frac{1}{2}$	1837
$3\frac{1}{2}$	881
$4\frac{1}{2}$	331
$5\frac{1}{2}$	97
$6\frac{1}{2}$	22
$7\frac{1}{2}$	4
$8\frac{1}{2}$	1
	10000

x	$2\mu A$
0	3829
1	6827
2	4835
3	2718
4	1212
5	428
6	119
7	26
8	5
9	1
	20000

To find the values of A, hA^{I}, h^2A^{II}, ... for $x=0$, we take logarithms:

x	$\log_{10}A$	1st diff.	2nd diff.
$\frac{1}{2}$	3·5831		
		-1063	
$1\frac{1}{2}$	3·4768		-1064
		-2127	
$2\frac{1}{2}$	3·2641		-1064
		-3191	
$3\frac{1}{2}$	2·9450		-1061
⋮	⋮	⋮	⋮

We have therefore, approximately,

$$\log_{10}A=3\cdot5964-\cdot0532x^2,$$

whence, by differentiation,

$$\begin{aligned} A^{I} &= -\cdot 2450 A x, \\ A^{II} &= -\cdot 2450 A^{I} x - \cdot 2450 A, \\ A^{III} &= -\cdot 2450 A^{II} x - \cdot 4900 A^{I}, \\ A^{IV} &= -\cdot 2450 A^{III} x - \cdot 7350 A^{II}, \\ A^{V} &= -\cdot 2450 A^{IV} x - \cdot 9800 A^{III}. \end{aligned}$$

Hence, for $x = 0$, we have

$$A = 3948, \ A^{I} = 0, \ A^{II} = -967, \ A^{III} = 0, \ A^{IV} = 711, \ A^{V} = 0, \ \ldots.$$

The following is a comparison of the raw moments as ordinarily used (ρ_1, ρ_2, ...) with the corrected values after allowing for massing and for abruptness, and with the true values; these latter being found by calculating the true moment of each strip of the original figure, and then altering this moment in the ratio of the integral number taken as the area of the strip to the true area of the strip.

	Moments				
	1st	2nd	3rd	4th	5th
Raw (ρ)	16292·00	40840·00	131717·00	501019·00	2158528·25
Corrected for massing	16292·00	40006·67	127644·00	480890·67	2051140·00
Corrected for massing and for abruptness	15958·16	40006·67	127745·78	480890·67	2051057·17
True...	15958·13	40006·43	127741·31	480830·04	2050388·68

Hence we deduce the following values of the mean and of the moments with regard to the mean.

	Mean	Moments with regard to mean			
		2nd	3rd	4th	5th
From raw moments (ρ's) ...	1·629200	14297	18595	81695	266434
For raw moments corrected for massing	1·629200	13464	18595	74839	260938
From raw moments corrected for massing and for abruptness ...	1·595816	14540	17495	82189	255326
From true moments	1·595813	14540	17492	82154	255038

10. With regard to the above example, the following points should be noted:

(i) The "corrections for massing," taken alone, do not affect the third moment with regard to the mean. This is always the case; for this moment as deduced from the raw values (ρ's) is

$$\rho_3 - 3(\rho_1/\rho_0)\rho_2 + 2(\rho_1/\rho_0)^3\rho_0,$$

while, as deduced from the raw values corrected for massing, it is

$$\left(\rho_3 - \frac{1}{4}h^2\rho_1\right) - 3\left(\frac{\rho_1}{\rho_0}\right)\left(\rho_2 - \frac{1}{12}h^2\rho_0\right) + 2\left(\frac{\rho_1}{\rho_0}\right)^3\rho_0,$$

and these are equal.

(ii) The corrections for abruptness are specially important for the purpose of determining the mean; and, for the moments generally, the most important part of these corrections is that which is due to A_0. We can, in fact, usually get fairly good results by using A_0 alone, which can be determined with tolerable accuracy, and ignoring hA_0^{I}, $h^2A_0^{\text{II}}$, ...

(iii) For the second and fourth moments with regard to the mean, we certainly get better results by omitting both sets of corrections than by the "corrections for massing" alone. This may seem to justify the ordinary practice; but it is due to the nature of the example. The peculiarity would be still more marked if the curve descended steadily from the initial ordinate. The more common case is that in which the bounding ordinate of the figure is appreciably smaller than the maximum ordinate; and then the correction for massing becomes important. Taking only the leading corrections, the fully corrected second moment with regard to the mean is

$$\rho_2 - \frac{1}{12}h^2\rho_0 - \left(\rho_1 - \frac{1}{12}hA_0\right)^2\Big/\rho_0,$$

so that the correction is approximately

$$-\frac{1}{12}h^2\rho_0 + \frac{1}{6}hA_0\frac{\rho_1}{\rho_0}.$$

The two portions of this are of opposite signs, so that they tend to counteract one another; but, if A_0 is so small that $\rho_1 A_0$ is less than $\frac{1}{4}h\rho_0^2$, it would even be better to make the "correction for massing" alone than to make no correction at all. The corrections for massing are, of course, more important if the raw moments are ρ_1', ρ_2', ... than if they are ρ_1, ρ_2, ...

11. If the values of A, hA^{I}, h^2A^{II}, ... would have to be determined directly from the differences of the given A's, no suitable auxiliary function being available, it may be simpler, for calculation of a few moments, to use an alternative method. Let S be the area defined in § 2 (iv), so that $S^1 = -y$. Then we know the values of S for $x = x_{-n}, x_{-n+1}, \ldots x_m$. The general formula is

$$\int_{x_{-n}}^{x_m} y\phi(x)\,dx = -\int_{x_{-n}}^{x_m} S^{\mathrm{I}}\phi(x)\,dx$$

$$= \int_{x_{-n}}^{x_m} S\phi^{\mathrm{I}}(x)\,dx - \{S_m\phi(x_m) - S_{-n}\phi(x_{-n})\}$$

$$\simeq h\Sigma\tfrac{1}{2}\{S_{r-\frac{1}{2}}\phi^{\mathrm{I}}(x_{r-\frac{1}{2}}) + S_{r+\frac{1}{2}}\phi^{\mathrm{I}}(x_{r+\frac{1}{2}})\} - \Theta(x_m) + \Theta(x_{-n}),$$

where $$\Theta(x) \equiv S\phi(x) + \left\{\frac{1}{12}hD - \frac{1}{720}h^3D^3 + \frac{1}{30240}h^5D^5 - \dots\right\} Sh\phi^{\mathrm{I}}(x).$$

For moments, let C_k be the "chordal" (trapezoidal) area of the figure whose ordinate is Sx^k, i.e.

$$C_k \equiv h\Sigma\tfrac{1}{2}\{S_{r-\frac{1}{2}}x^k{}_{r-\frac{1}{2}} + S_{r+\frac{1}{2}}x^k{}_{r+\frac{1}{2}}\}.$$

Then, for $k > 0$, the kth moment with regard to the ordinate at the lower extremity is

$$\nu_k \simeq kC_{k-1} - \Theta_k(x_m) + \Theta_k(x_{-n}),$$

where $$\Theta_k(x) \equiv Sx^k + k\left\{\frac{1}{12}hD - \frac{1}{720}h^3D^3 + \frac{1}{30240}h^5D^5 - \dots\right\} Shx^{k-1}$$

$$= Sx^k + k\left\{\frac{1}{12}\mu\delta - \frac{11}{720}\mu\delta^3 + \frac{191}{60480}\mu\delta^5 - \dots\right\} Shx^{k-1}.$$

For a single-tailed figure, we should usually take the abrupt end as the zero for x, and the other end as the zero for S; and ν_k would then be formed by adding to kC_{k-1} the value of

$$k\left\{\frac{1}{12}\mu\delta - \frac{11}{720}\mu\delta^3 + \frac{191}{60480}\mu\delta^5 - \dots\right\} Shx^{k-1}$$

at the abrupt end.

In the example considered in § 9, for instance, we have the following data for the calculation of the second moment:—

x	S	Sx	1st diff.	2nd diff.	3rd diff.	4th diff.
0	10000	0				
			+6171			
1	6171	6171		−5996		
			+ 175		+3483	
2	3173	6346		−2513		− 820
			−2338		+2663	
3	1336	4008		+ 150		−1825
			−2188		+ 838	
4	455	1820		+ 988		−1084
			−1200		− 246	
5	124	620		+ 742		− 165
			− 458		− 411	
6	27	162		+ 331		+ 180
			− 127	⋮	− 231	⋮
7	5	35	⋮		⋮	
8	1	8				
9	0	0				

Taking -820 to be the constant 4th difference at the commencement of the table, we obtain

x	Sx	1st diff.	2nd diff.	3rd diff.	4th diff.
		+16470		+5123	
0	0		−10299		−820
		+ 6171		+4303	
1	6171		− 5996		−820
		+ 175		+3483	
2	6346		− 2513		−820

and thence
$$\nu_2 = 2\,.\,19170 + 1/12\,.\,22641 - 11/720\,.\,9426$$
$$= 40083.$$

This is not so good as the value obtained in § 9, but it is better than the ordinary raw moment.

The objection to the method, when several moments have to be calculated, is that the values of Shx^{k-1} have to be differenced separately for each value of k.

12. By performing the differentiations in $\Theta(x)$ we obtain a formula analogous to the formula of § 7, but involving the derivatives of S instead of those of A. It will be found that, in the general formula of § 11,

$$\begin{aligned}\Theta(x) = &\left\{\phi(x) + \frac{1}{12}h^2\phi^{II}(x) - \frac{1}{720}h^4\phi^{IV}(x) + \dots\right\}S \\ &+ \left\{\frac{1}{12}h\phi^{I}(x) - \frac{1}{240}h^3\phi^{III}(x) + \frac{1}{6048}h^5\phi^{V}(x) - \dots\right\}hS^{I} \\ &+ \left\{-\frac{1}{240}h^2\phi^{II}(x) + \frac{1}{3024}h^4\phi^{IV}(x) - \dots\right\}h^2S^{II} \\ &+ \left\{-\frac{1}{720}h\phi^{I}(x) + \frac{1}{3024}h^3\phi^{III}(x) - \frac{1}{34560}h^5\phi^{V}(x) + \dots\right\}h^3S^{III} \\ &+ \left\{\frac{1}{6048}h^2\phi^{II}(x) - \frac{1}{34560}h^4\phi^{IV}(x) + \dots\right\}h^4S^{IV} \\ &+ \left\{\frac{1}{30240}h\phi^{I}(x) - \frac{1}{57600}h^3\phi^{III}(x) + \frac{1}{380160}h^5\phi^{V}(x) - \dots\right\}h^5S^{V} \\ &+ \left\{-\frac{1}{172800}h^2\phi^{II}(x) + \frac{1}{570240}h^4\phi^{IV}(x) - \dots\right\}h^6S^{VI} \\ &+ \text{etc.}\end{aligned}$$

For the moments, this gives

$$\begin{aligned}\nu_1 &\doteq C_0 \;\; - \Theta_1(x_m) + \Theta_1(x_{-n}),\\ \nu_2 &\doteq 2C_1 - \Theta_2(x_m) + \Theta_2(x_{-n}),\\ \nu_3 &\doteq 3C_2 - \Theta_3(x_m) + \Theta_3(x_{-n}),\\ \nu_4 &\doteq 4C_3 - \Theta_4(x_m) + \Theta_4(x_{-n}),\\ \nu_5 &\doteq 5C_4 - \Theta_5(x_m) + \Theta_5(x_{-n}),\\ &\text{etc},\end{aligned}$$

where $\Theta_1(x)$, $\Theta_2(x)$, ... are functions whose values are given by Table IV.

TABLE IV. Coefficients of S, hS^{I}, h^2S^{II}, ... in $\Theta_1(x)/h$, $\Theta_2(x)/h^2$, ... (§ 12).
($\xi \equiv x/h$.)

	co. S	co. hS^{I}	co. h^2S^{II}	co. h^3S^{III}	co. h^4S^{IV}	co. h^5S^{V}	co. h^6S^{VI}
$\Theta_1(x)/h$	$+\xi$	$+\frac{1}{12}$	0	$-\frac{1}{720}$	0	$+\frac{1}{30240}$	0
$\Theta_2(x)/h^2$	$+\xi^2+\frac{1}{6}$	$+\frac{1}{6}\xi$	$-\frac{1}{120}$	$-\frac{1}{360}\xi$	$+\frac{1}{3024}$	$+\frac{1}{15120}\xi$	$-\frac{1}{86400}$
$\Theta_3(x)/h^3$	$+\xi^3+\frac{1}{2}\xi$	$+\frac{1}{4}\xi^2$ $-\frac{1}{40}$	$-\frac{1}{40}\xi$	$-\frac{1}{240}\xi^2$ $+\frac{1}{504}$	$+\frac{1}{1008}\xi$	$+\frac{1}{10080}\xi^2$ $-\frac{1}{9600}$	$-\frac{1}{28800}\xi$
$\Theta_4(x)/h^4$	$+\xi^4+\xi^2$ $-\frac{1}{30}$	$+\frac{1}{3}\xi^3$ $-\frac{1}{10}\xi$	$-\frac{1}{20}\xi^2$ $+\frac{1}{126}$	$-\frac{1}{180}\xi^3$ $+\frac{1}{126}\xi$	$+\frac{1}{504}\xi^2$ $-\frac{1}{1440}$	$+\frac{1}{7560}\xi^3$ $-\frac{1}{2400}\xi$	$-\frac{1}{14400}\xi^2$ $+\frac{1}{23760}$
$\Theta_5(x)/h^5$	$+\xi^5+\frac{5}{3}\xi^3$ $-\frac{1}{6}\xi$	$+\frac{5}{12}\xi^4$ $-\frac{1}{4}\xi^2$ $+\frac{5}{252}$	$-\frac{1}{12}\xi^3$ $+\frac{5}{126}\xi$	$-\frac{1}{144}\xi^4$ $+\frac{5}{252}\xi^2$ $-\frac{1}{288}$	$+\frac{5}{1512}\xi^3$ $-\frac{1}{288}\xi$	$+\frac{1}{6048}\xi^4$ $-\frac{1}{960}\xi^2$ $+\frac{1}{3168}$	$-\frac{1}{8640}\xi^3$ $+\frac{1}{4752}\xi$

The combination of kC_{k-1} with the terms involving S gives the same result as the raw moment with the "corrections for massing"; and the remaining terms are the "corrections for abruptness." In fact, if we omit the first line of $\Theta(x)$, the remainder, with sign changed, is identical with $\Phi(x)$ of § 7, A being equal to $-\delta S$.

13. For the purpose of the formulae in this paper, it has been assumed that the extreme values of $A_{-n+\frac{1}{2}}$ and $A_{m-\frac{1}{2}}$ actually represent the areas of the corresponding strips. Sometimes this is not the case; the value of A may be measured from or to an ordinate lying inside the strip, or the curve of frequency may cut the base of the strip, so that the true (algebraical) area of the strip is less than the area represented by the given value of A. In such cases special methods have to be employed.

14*. A special application of the method of § 8 is to calculation of the first moment, all deviations being taken to be positive. Let K be the true first moment with regard to y, and L_t the raw first moment with regard to y_t, both moments being calculated in this way; i.e. (replacing x, y by ξ, η), let

* This section has been added as a result of discussion at the Congress.

$$K \equiv \int_{x_{-n}}^{x_m} \eta\,|\,\xi - x\,|\,d\xi = \int_{x_{-n}}^{x} \eta\,(x-\xi)\,d\xi + \int_{x}^{x_m} \eta\,(\xi - x)\,d\xi,$$

$$L_t \equiv \Sigma A_r\,|\,x_r - x_t\,| = \sum_{r=-n+\frac{1}{2}}^{r=t-\frac{1}{2}} A_r\,(x_t - x_r) + \sum_{r=t+\frac{1}{2}}^{r=m-\frac{1}{2}} A_r\,(x_r - x_t).$$

Then it will be found that

$$K_t \triangleq L_t + \Phi_1(x_m) - 2\Phi_1(x_t) + \Phi_1(x_{-n}),$$

where $\Phi_1(x)$ has the value stated in § 8. In the case of a double-tailed figure, if the differences near A_t are fairly regular, K_t will be approximately equal to

$$L_t - \frac{1}{6}\mu h A_t + \frac{11}{360}\mu\delta^2 h A_t - \frac{191}{30240}\mu\delta^4 h A_t + \dots.$$

For calculating K for intermediate values of x, we have

$$K_t^{\mathrm{I}} = S_t' - S_t,$$

$$K_t^{\mathrm{II}} = 2y_t, \quad K_t^{\mathrm{III}} = 2y_t^{\mathrm{I}}, \dots,$$

where S_t' is the total area from y_{-n} to y_t, and S_t is the total area from y_t to y_m. If we use an auxiliary function, we have

$$hK_t^{\mathrm{II}} = 2A_t - \frac{1}{12}h^2 A_t^{\mathrm{II}} + \frac{7}{2880}h^4 A_t^{\mathrm{IV}} - \dots,$$

$$h^2 K_t^{\mathrm{III}} = 2hA_t^{\mathrm{I}} - \frac{1}{12}h^3 A_t^{\mathrm{III}} + \frac{7}{2880}h^5 A_t^{\mathrm{V}} - \dots,$$

etc.

Or, if we use central differences,

$$hK_t^{\mathrm{II}} = 2\mu A_t - \frac{1}{3}\mu\delta^2 A_t + \frac{1}{15}\mu\delta^4 A_t - \dots,$$

$$h^2 K_t^{\mathrm{III}} = 2\delta A_t - \frac{1}{6}\delta^3 A_t + \frac{1}{45}\delta^5 A_t - \dots,$$

$$h^3 K_t^{\mathrm{IV}} = 2\mu\delta^2 A_t - \frac{1}{2}\mu\delta^4 A_t + \dots,$$

$$h^4 K_t^{\mathrm{V}} = 2\delta^3 A_t - \frac{1}{3}\delta^5 A_t + \dots,$$

etc.

For the median ordinate, K is a minimum. Let the corresponding value of x be $x_{t+\theta}$. Then the minimum value of K is

$$K_t + h(S_t' - S_t)\,\theta + 2\,\{h^2 y_t \theta^2/2! + h^3 y_t^{\mathrm{I}} \theta^3/3! + \dots\},$$

θ being given by

$$0 = \frac{1}{2}(S_t' - S_t) + hy_t\theta + h^2 y_t^{\mathrm{I}}\theta^2/2! + h^3 y_t^{\mathrm{II}}\theta^3/3! + \dots.$$

The successive approximations to the value of K, supposing hy_t^{I} to be small in comparison with y_t, are

$$K_t,$$

$$K_t - \frac{1}{4}(S_t' - S_t)^2/y_t,$$

$$K_t - \frac{1}{4}(S_t' - S_t)^2/y_t - \frac{1}{24}(S_t' - S_t)^3\, y_t^{\mathrm{I}}/y_t^3.$$

A METHOD OF REPRESENTING STATISTICS BY ANALYTICAL GEOMETRY

By F. Y. Edgeworth.

I. *Introduction.* The statistics which it is here attempted to represent are of a kind that commonly occurs in biology, sociology and other statistical sciences. This general type is characterised by simple curves of frequency, such that if the abscissa denote the size of an object or attribute, and the ordinate the frequency of each size, the ordinate has only one maximum, and becomes zero at each extremity. (Compare Pearson, *Mathematical Contributions to the Theory of Evolution,* No. XIV, pp. 4, 5; where an additional property is attributed to the general type.)

The proposed method is a development of that which has been presented by the present writer under the title of "Method of Translation" (see *Journal of the Royal Statistical Society* for Dec. 1898, March, June and September, 1899, and March, 1900). The method consists in the transformation of a normal (Gaussian) curve of error by substituting for each of the magnitudes grouped according to such a law some assigned *function* of that magnitude. If, as will usually be the case, the origin from which the magnitudes to be operated on are measured is well outside the sensible area of the normal curve which is to be "translated"—say at a distance from the centre of the normal curve of more than twice its *modulus,* or thrice its *standard deviation*—then any points on the abscissa of the transformed curve, P, Q, R ... divide its area into the same proportions (form the same percentiles) as the corresponding points p, q, r ... on the abscissa of the generating normal curve. Accordingly, the median of the transformed curve may be taken as coincident with the centre of the generating normal curve, say 0; and to every abscissa of the normal curve, say ξ, measured from 0, there will correspond, measured from the same point, an abscissa of the transformed curve, say X, being a certain function of ξ. If $X = a\xi$ where a is a constant, the transformed curve is still normal, the scale only or size of the modulus being altered. Since the forms with which we have to deal do not for the most part deviate widely from the normal type, there is a propriety in assuming for the assigned function a Taylorian expansion of the form

$$a\xi + k\xi^2 + l\xi^3 + \ldots,$$

where k and l are small with respect to a. Or, if for the small quantities k/a, l/a we put respectively κ and λ, we may write the operator

$$a(\xi + \kappa\xi^2 + \lambda\xi^3).$$

The modulus of the generating normal curve may be taken to be unity without loss of generality. Thus each element or small rectangular strip of the generating curve at the point ξ with base $\Delta\xi$, and of height $\frac{1}{\sqrt{\pi}}\exp. -\xi^2$ is translated to the distance $a(\xi + \kappa\xi^2 + \lambda\xi^3 + \ldots)$; its base being changed from $\Delta\xi$ to

$$\Delta X = \frac{dX}{d\xi}\Delta\xi = a(1 + 2\kappa\xi + 3\lambda\xi^2 + \ldots)\Delta\xi,$$

and its height altered in the inverse ratio. It will not in general be necessary to proceed beyond the third power of ξ in the expansion of X. The transformation may be also conceived as the *reverse* of the following transformation. Consider the curve which represents the *integral* of the represented curve; and transform this integral by simply writing for its abscissa X (measured from the central point, 0) ξ multiplied by the factor $(a + k\xi + l\xi^2)$. Whence (by reversion of series) the abscissa ξ (relating to the integral of the normal curve) may be equated to an expression of the form

$$A + BX + CX^2 + \ldots.$$

The rationale of this method consists not merely, and not principally, in its exactly representing those cases in which each member of the frequency group under consideration is a definite function of some member of a group distributed according to the normal law of error. For instance if the velocities of winds, or of any other objects, are distributed normally, then the corresponding *energies* will be distributed according to a frequency curve which is an exact translation of a normal curve. So if the diameters of oranges or other spherical bodies vary normally, the solid contents, proportional to the cubes of the diameters, will be distributed according to a law which is given by putting X in the above written formula $3b^2c(\xi + \kappa\xi^2 + \frac{1}{3}\kappa^2\xi^3)$; where b is the mean diameter (the average of the normal group of diameters), c is the modulus of that normal group; and $\kappa = c/b$, a small fraction in the case of many natural objects (as pointed out by the present writer, *Philosophical Magazine*, 1892, Vol. XXXIV, p. 435). The cases in which translation is *the* formula are doubtless not uncommon (compare *Journal of the Statistical Society*, Vol. LXI, 1898, p. 678). But the reason of the method lies deeper. It consists in the affinity of the formula to that universal law which is ever and everywhere approximately fulfilled throughout the whole vast realm of Statistics—Statistics proper as distinguished from mere arithmetic by sporadic or fortuitous dispersion. This "generalised law of error" may be written (in the simple case of a single variable, x)

$$y = y_0 - \frac{1}{3!}k_1\frac{d^3}{dx^3}y_0 + \frac{1}{4!}k_2\frac{d^4}{dx^4}y_0 + \ldots$$

where y_0 is the normal error function $\frac{1}{\sqrt{\pi}c}\exp. -x^2/c^2$; k_1 and k_2 are closely related to the important coefficients designated by Professor Pearson as β_1 ("Contributions to the Mathematical Theory of Evolution," No. II, *Transactions of the Royal Society*, 1895, *et passim*), and η (*Biometrika*, Vol. IV, p. 177). Provided that the essential attribute of Statistics, plurality of independent agencies, is present in some degree, it may be expected that this law will be approximately fulfilled for a portion of any frequency group, a central body as distinguished from the extremities which obey no general law. (See the present writer's exposition of the "Generalised Law of Error" in the *Journal of the Royal Statistical Society for* 1906, referring to his paper on the "Law of Error" in the *Cambridge Philosophical Transactions for* 1905.) For example, suppose a group of magnitudes to be formed by batches of twenty-five digits taken at random from mathematical tables, or the developments of a constant such as π. The sum of the twenty-five (generally n) digits may conveniently be divided by 5 (generally $\sqrt{n}$). Here, the group being symmetrical,

$$\beta_1 (= \mu_3^2/\mu_2^3) = 0, \quad \eta (= \mu_4^2/\mu_2^2 - 3) = - \cdot 0489\dot{9}\dot{6}.$$

Accordingly the generalised law of error applied to the data gives the equation

$$y = \frac{1}{\sqrt{\pi}\sqrt{16\cdot 5}} e^{-\frac{x^2}{16\cdot 5}} \left(1 - \frac{\cdot 048\dot{9}\dot{6}}{4}\left[\frac{1}{2} - 2\frac{x^2}{16\cdot 5} + \frac{2}{3}\frac{x^4}{(16\cdot 5)^2}\right]\right);$$

neglecting higher terms of expansion affected with the coefficients k_4, k_6, etc.; where x is the number of digits in excess of the mean number (in a batch of 25/5), viz. 22·5. (If instead of 25 any other number n of digits is taken and the sum divided by $\sqrt{n}$, the equation remains unaltered, except as to the factor outside the square brackets, which is inversely proportionate to n.) According to the method of translation, a normal curve with unit modulus being placed with its centre at the point (on the abscissa) 22·5, it is transformed by the operator $4\cdot 08 (\xi - \cdot 004 \xi^3)$; the constants being determined by one of the methods presently to be described (below, p. 430). There are, then, to be distinguished three loci: (1) the particular real form, in the present case a polynomial of unmanageable complexity, in general unknowable; (2) the generalised law of error, which is approximately true up to a certain extent of the abscissa, and ultimately becomes unmeaning, if not false; (3) the particular representative curve, which is approximately true up to a certain extent of the abscissa, and ultimately becomes false, if not unmeaning. Thus the above written expressions both for the generalised law of error and for the particular representative break down long before the extremity of the real locus is reached, the point 45 (measured from zero), corresponding to the probability of all the twenty-five digits proving to be 9's!

The contrast between the generalised law and the particular representative is more significant when η is positive and large. The use of such a representative particular, true only as to the main body of the frequency group, but with fictitious extremities—as it were handles of inferior material—seems to be a requirement of mathematical statistics. As a good way of meeting that requirement the Method of Translation is recommended.

II. *Determination of the Constants by Moments.* For the determination of the constants in the formula for translation there is available the use of *moments*; which Professor Pearson has shown both by theoretical considerations and practical success to be not only serviceable for the fitting of curves in general, but also particularly appropriate to frequency groups of the kind here considered (see his "Observations on the Fitting of Curves" in *Biometrika*, Vol. I, Part III, Vol. II, Part I, and his *Contributions to the Mathematical Theory of Evolution*, Nos. I, II, XIV, *et passim*). There seems indeed a peculiar propriety in the use of moments in the case of a method which claims affinity with the (generalised) law of error. For the simplest and perhaps most fundamental proof of the validity of that law is afforded by the coincidence between the moments of the real and those of the approximate locus. (See the present writer's "Law of Error," *Cambridge Philosophical Transactions*, 1905; and compare Poincaré, *Calcul des Probabilités* Leçon XV, the priority of which—in respect of the *normal* law of error—not known to the writer in 1905, has been acknowledged by him in the *Bulletin of the International Institute of Statistics*, 1909). This proof virtually assumes that a locus which has the same moments as another *is* the same locus.

The observed moments are connected with the sought constants a, κ, λ, by the following simple propositions. If $Y=f(X)$ is the required frequency-curve, then by construction $Y=f(\xi)\dfrac{d\xi}{dX}$; where $f(\xi)$ is the normal error-function with unit constant, and $X=a(\xi+\kappa\xi^2+\lambda\xi^3)$. Therefore $\int_{-\infty}^{\infty} X^n f(X)\,dX$ (the symbol ∞ being used to denote an extreme limit not necessarily at an infinite distance from the origin), the nth moment of the curve $Y=f(\xi)$ about its median

$$=\int_{-\infty}^{\infty} a^n(\xi+\kappa\xi^2+\lambda\xi^3)^n f(\xi)\,d\xi.$$

The moments about the median, say M_1, M_2, M_3 etc., can thus be deduced in terms of the required constants from the well-known values of the moments of the normal error-curve about its centre; and the moments of the translated curve about its centre of gravity, say μ_1, μ_2, μ_3, can be deduced from the moments about the median by a process which Professor Pearson has made familiar. We have thus

$$M_1 = a\,\tfrac{1}{2}\kappa,$$
$$M_2 = a^2(\tfrac{1}{2}+\tfrac{3}{4}\kappa^2+\tfrac{3}{2}\lambda+\tfrac{15}{8}\lambda^2),$$
$$M_3 = a^3(\tfrac{9}{4}\kappa+\tfrac{15}{8}\kappa^3+\tfrac{45}{4}\kappa\lambda+\tfrac{315}{16}\kappa\lambda^2),$$
$$\ldots\ldots\ldots\ldots\ldots\ldots\ldots\ldots\ldots\ldots$$

Whence
$$\mu_1 = M_1 = \tfrac{1}{2}\kappa,$$
$$\mu_2 = M_2 - M_1^2 = a^2\tfrac{1}{2}(1+\kappa^2+3\lambda+\tfrac{15}{4}\lambda^2),$$
$$\mu_3 = M_3 - M_1^3 - 3M_1\mu_2 = a^3\kappa(\tfrac{3}{2}+\kappa^2+9\lambda+\tfrac{135}{8}\lambda^2),$$
$$\ldots\ldots\ldots\ldots\ldots\ldots\ldots\ldots\ldots\ldots\ldots\ldots\ldots\ldots$$

We have accordingly for the constant β_1, hereinafter written simply β, the equation

$$\beta\,(=\mu_3^2/\mu_2^3)=\frac{8\kappa^2(\tfrac{3}{2}+\kappa^2+9\lambda+\tfrac{135}{8}\lambda^2)^2}{(1+\kappa^2+3\lambda+\tfrac{15}{4}\lambda^2)^3};$$

an expression which may be simplified by putting $\chi=\kappa^2$. By parity we may deduce the expression for $\eta\,(=\mu_4/\mu_2^2-3)$, and thus obtain the fundamental equations connecting the observed moments with (two of) the required constants:—

$$\text{(I)}\quad 8\chi(\tfrac{3}{2}+\chi+9\lambda+\tfrac{135}{8}\lambda^2)^2-\beta(1+\chi+3\lambda+\tfrac{15}{4}\lambda^2)^3=0,$$

$$\text{(II)}\quad 4(6\chi+3\lambda+3\chi^2+54\chi\lambda+27\lambda^2+135\chi\lambda^2+\tfrac{405}{4}\lambda^3+\tfrac{1215}{8}\lambda^4)$$
$$-\eta(1+\chi+3\lambda+\tfrac{15}{4}\lambda^2)^2=0.$$

These equations are not so formidable as they appear at first sight; owing mainly to the incident that the required values of χ and λ are usually small fractions. Accordingly the higher powers of the variable may often be neglected for a first approximation. For example, some statistics of barometric heights which seem to be fairly typical of a moderately skew frequency-curve (the statistics discussed in the papers already referred to in the *Journal of the Royal Statistical Society*, 1898–9) present for χ and λ respectively ·13 and ·34. These values being substituted in the fundamental equations we obtain from the abridged equations

$$\text{(I)}\quad 18\chi - \cdot 13\,(1 + 3\chi + 9\lambda) = 0,$$

$$\text{(II)}\quad 24\chi + 12\lambda - \cdot 34\,(1 + 2\chi + 6\lambda) = 0,$$

neglecting all but the *first* powers of the variables, the approximate solution $\chi = \cdot 0084$, $\lambda = \cdot 0146$. And these values remain correct to the second place of decimals, when the omitted terms are replaced in the equations. Commonly it is sufficient, for a first approximation at least, to take account only of the *second* (in addition to the first) powers of the variables. Abridged methods of solution will no doubt suggest themselves to the practical statistician. Difficulty from the occurrence of imaginary roots is not to be apprehended, as there is reason to believe that real values of χ and λ may be found, if not for every conceivable pair of values for β and η that is consistent with the general description of the frequency-curves with which we have to do, at least for all probable and usual data. For instance approximate roots to the equations are readily found in the case of all the examples of the Pearsonian *Types* given by Mr Palin Elderton in his *Frequency-curves*, and in the case of the examples given by Professor Pearson in his second *Contribution* (except Nos. X–XII, which are to be treated according to one of the methods prescribed below for exceptional cases, and Nos. IX and XV, for which the moments up to the fourth are not explicitly given). The whereabouts of the required roots having been ascertained, it is not very troublesome to proceed to nearer approximation by means of the expressions for $\frac{d\beta}{d\kappa}$, $\frac{d\beta}{d\lambda}$ and $\frac{d\eta}{d\chi}$, $\frac{d\eta}{d\lambda}$. In fine, if the method of translation were generally adopted, mathematicians with more leisure and industry than the present writer would soon construct a Table showing for every assigned pair of values of β and η (within the limits of ordinary practice) the corresponding values of $\chi\,(=\kappa^2)$ and η. It is hardly necessary to add that when the values of κ and η have been obtained, the value of a is given by the equation

$$a^2 = 2\mu_2/(1 + \kappa^2 + 3\lambda + \tfrac{15}{4}\lambda^2).$$

III. *Determination of the Constants by Percentiles.* A start on the quest for the values of κ and λ may be obtained from a summary use of *percentiles.* Thus, suppose the position of the *median* to have been observed; and that the point on the axis of X below the median, between which and the median are intercepted ·3414 of the total area (the point corresponding to the standard deviation on the generating normal curve), is at the distance S from the median. Likewise let the *quartile* above the median be at the distance Q from the median, and at the distance R let there occur the *decile* above which there is just a tenth of the area (corresponding to 1·28155 times the standard deviation—or ·906 the modulus—of the generating curve). We thus obtain the three linear equations for the three quaesita:

$$a\,(\cdot 707 - \cdot 707^2\kappa + \cdot 707^3\lambda) = S,$$

$$a\,(\cdot 477 + \cdot 477^2\kappa + \cdot 477^3\lambda) = Q,$$

$$a\,(\cdot 906 + \cdot 906^2 \;\; + \cdot 906^3\lambda) = R.$$

The solution thus easily obtained is inadequate, both because it does not utilise all the data, and because it does not take into account the difference in weight and the *correlation* of the errors incident to the several equations. For a more perfect solution the required constants must be connected with the correlated errors incident to the whole system of percentiles (as to which correlation see W. F. Sheppard, "On the Application of the Theory of Error," *Transactions of the Royal Society*, 1898, Vol. 192 A, p. 114 *et seq.*). Beginning with the outermost percentile, say at the upper extremity of the group, cutting off to the right the fraction v of the total area, we know that its error is due to the incident that the area to the right of the true, the ideal percentile, is not, as *a priori* most probably, v, but v *plus* (or *minus*) a deviation therefrom, say e_v (on the principle underlying Laplace's "Method of Situation," as pointed out by the present writer, *Philosophical Magazine*, 1886, Vol. XXII, p. 374. Compare Sheppard, *loc. cit.*). Likewise the error of the percentile next in order depends on the deviation of the area to the right of the corresponding true percentile from the true area. Thus the errors of portions of area corresponding to the percentiles form a system analogous to the errors of the numbers or proportions which are presented when from an indefinitely large *mélange* of differently coloured balls there is extracted at random a large sample numbering N. To fix the ideas, suppose that there are mixed up balls of *seven* colours, say those of the prism, violet, indigo, blue, green, yellow, orange, red; in the proportions v, i, b, g, y, o, r; Nv being the number of violet balls in the supposed indefinitely large bag or jar, Ni the number of indigo balls, and so on. The probability that the number of violet balls in the sample numbering N will exceed (the most probable number) Nv by Ne_v is approximately proportionate to exp. $-Ne_v^2/2v(1-v)$. The probability that when $N(v+e_v)$ violet balls are drawn, there will be a certain excess $N\epsilon_i$ of indigo balls above that number of indigo balls which is now most probable, viz.

$$N(1-v-e_v)\frac{i}{1-v},$$

is proportional to $\quad \text{exp.} - N(1-v-e_v)\,\epsilon_i^2/2i(1-v-i).$

This probability, multiplied by the above written probability of e_v, gives the probability of the double event e_v and ϵ_i. Now put $\epsilon_i = e_i + e_v i/(1-v)$, so that the deviation of the number of indigo balls from the originally most probable number thereof, viz. Ni, becomes Ne_i; and we find for the probability of the two errors e_v and e_i:

$$\text{Const.} \times \text{exp.} - \tfrac{1}{2}N\left(\frac{e_v^2}{v(1-v)} + \frac{(1-v)(e_i + e_v i/(1-v))^2}{i(1-v-i)}\right);$$

neglecting in the numerator of the latter part of the expression terms which are of the *third* order of magnitude. (Compare as to the orders of magnitude "Law of Error," *Cambridge Philosophical Transactions*, 1905, p. 58.) Multiplying out, we have for the probability of the double event e_v and e_i

$$\text{Const.} \times \text{exp.}\ \tfrac{1}{2}N\left(\frac{e_v^2(1-i)}{v} + 2e_ve_i + \frac{e_i^2(1-v)}{i}\right)\Big/(1-v-i).$$

Proceeding similarly with the remaining colours up to the sixth inclusive—the number of the seventh, red, is given when the errors of the other six numbers are assigned—we find for the system of six errors $e_v, e_i \ldots, e_0$ the probability

$$\text{Const.} \times \text{exp.} - \tfrac{1}{2}N\left(\frac{e_v^2(r+v)}{v} + \frac{e_i^2(r+i)}{i} + \ldots + 2e_ve_i + \ldots\right)\Big/r.$$

This expression will be found to agree with that obtained by Professor Pearson in connexion with his Criterion of Random Sampling (*Phil. Mag.* July, 1900).

If green is so to speak the *median* colour, viz.

$$v+i+b<\tfrac{1}{2},\quad v+i+b+g>\tfrac{1}{2},$$

we might write for g, u_0 and for the proportions y, o above the median colour, or class u_1, u_2, etc. and for the proportions below u_{-1}, u_{-2}, etc. *mutatis mutandis*; and for the corresponding errors e_0; e_1, e_2; e_{-1}, e_{-2}.

We have now to connect these errors of area with errors of the percentiles involving the sought constants. The error pertaining to any compartment, of which the content is ideally u_1, is made up of two errors, one at its upper and one at its lower limit, each of which may be regarded as a little rectangle of which the base is the deviation of the corresponding percentile from its true position, and the height is the ordinate of the translated curve at the point. Thus the rectangle error at that boundary of the tth compartment which is nearer the median has for base the difference between the observed distance from the median, say Ω_t, and the ideal distance $a\xi_t+k\xi_t^2+l\xi_t^3$, if ξ_t is the abscissa of the normal error-curve corresponding to the position of the lower limit of u_t (the point above which the area of the normal curve $=u_t+u_{t+1}+u_{t+2}+\ldots$). Also the ordinate of the transformed curve at the point under consideration is $f(\xi_t)\left(\frac{d\xi}{dX}\right)_{\xi=\xi_t}$, where $f(\xi_t)$ is the ordinate of the normal curve with unit modulus for abscissa ξ_t, say ζ_t. Therefore the transformed ordinate $=\zeta_t/(a+2k\xi_t+3l\xi_t^2)$, and the error of the tth compartment, viz. e_t,

$$=-\frac{(a\xi_t+k\xi_t^2+l\xi_t^3-\Omega_t)\,\zeta_t}{a+2k\xi_t-t+3l\xi_t^2},$$

plus a similar product which is formed by substituting $t+1$ for t in this expression (and changing the sign of the expression). The expression thus formed is to be substituted for e_t in the above written expression for the probability of the system of errors; and similarly formed expressions are to be substituted for all the other e's. Let the logarithm of the resulting function of a, k, l be called $-P$. Then the simultaneous equations,

$$\frac{dP}{da}=0,\quad \frac{dP}{dk}=0,\quad \frac{dP}{dl}=0,$$

give the values of a, k and l which make the probability of the observed complex event a maximum, provided that the second term in the expansion of P fulfils the condition of a minimum. The equations become simplified when, as very commonly, k and l are small with respect to a, and accordingly it is allowable as above to put, for X, $a(\xi+\kappa\xi^2+\lambda\xi^3)$, where κ and λ are small fractions. The exposition may be completed by considering one or two simplified examples. First suppose that κ and λ are each zero. The problem then reduces to the simple and familiar one (treated in papers above referred to): to determine the modulus (or standard deviation) from observed percentiles. In this simple case the ordinate of the transformed curve reduces to ζ_t/a, where a is the sought modulus, and the equivalent of e_t becomes

$$-\frac{\zeta_t(a\xi_t-\Omega_t)}{a}+\frac{\zeta_{t+1}(a\xi_{t+1}-\Omega_{t+1})}{a}.$$

In the expression for P made up of squares and products of e's of this type, it is proper to equate to zero the differential coefficient of P with respect to a; the *complete* differential, the a in the denominator not being treated as constant (in accordance with the principles of Inverse Probability as stated by the present writer—after Laplace and Pearson—in the *Journal of the Royal Statistical Society*, 1908, Vol. LXXI, p. 388, and context, p. 499, etc.). The resulting value for a, and the expression for the error to which it is liable, will be found to be in substantial agreement with Mr Sheppard's determination of the weights which make the mean square of error of the standard deviation a *minimum*.

An introduction to the general method, and estimate of the trouble it is likely to require, is afforded by the case in which only one of the constants, viz. λ, is supposed to be zero. For further simplicity it may be supposed that there are only three percentiles, the median and the two quartiles. It will be found convenient in this case, and generally, to treat the observed median provisionally as the true one, to measure the ξ's in absolute quantity on either side of the median, and to form the expressions for E_1, E_{-1}, in general E_t, E_{-t}, where E_t is the error incident to the whole area cut off to the right by the tth percentile, and E_{-t} is the error of the area cut off to the left by the percentile designated as $-t$. The E's can readily be obtained from the e's above defined. It will be seen that the error introduced by treating the observed median as the true one tends to disappear by compensation.

When the values which have been determined for k and l (or *mutatis mutandis* κ and λ) have been substituted in P (above defined) the probability of any system of errors, say Δa, Δk, Δl, in those values is given by the normal (hyper-) surface

$$z = \text{Const.} \times \exp. - \left(\frac{d^2P}{da^2}\Delta a^2 + 2\frac{d^2P}{da\,dk}\Delta a \Delta k + \ldots\right).$$

This surface may be compared in respect of tenuity with that which pertains to the determination of the moments (as to which see Pearson, "Contributions," No. IV, *Transactions of the Royal Society*, Vol. 191 A, p. 266 *et seq.*), or to the determination of κ and λ as functions of the moments. It is not to be hastily assumed that in this comparison the moments will have the advantage, because this is so in the case of the constant pertaining to the normal curve, as shown by Mr Sheppard. For in the case of the normal curve it is given by Inverse Probability that the most probable value of the modulus (or standard deviation) is a function of (the first and second) moments (cp. *Journal of the Statistical Society*, 1908, *loc. cit.* p. 388). As the (normal) curve of error for that determination rises to a higher apex, so it spreads less than the curve pertaining to other determinations. But the most probable values of the constants entering into the formula for translation are not simple functions of low moments. The method of moments which has been exhibited is a good method of determining the constant, but not necessarily the very best in respect of accuracy.

In the practical choice of the method for determining the constants, such a difference of probable error as may exist in the abstract between the method of moments and that of percentiles will not appear decisive to those who hold with a recent writer that "there are very few arguments made from probable error which would be seriously affected if the probable error were altered by 25 per cent." (Rhind, *Biometrika*, Vol. VII, pp. 127 and 386). Much more serious differences arise from the imperfections occurring in concrete statistics. Suppose the data are not finely graduated, say only four or five degrees or percentiles given; or suppose that at the extremities there occur heterogeneous measurements, such as dwarfs and giants may be in the case of anthropometry. In such cases the method of percentiles can utilise all the sound data, and omit those that are suspicious, without manipulation; but the method of moments has to be eked out by hypotheses.

Percentiles, too, admit of abridged methods which may often be convenient and sufficient even when the data admit of greater precision. In the light of the theory practical statisticians could no doubt devise a procedure intermediate in respect of facility and accuracy between the very summary method mentioned at the beginning of this section and the very elaborate method afterwards set forth. For instance, to constitute each of the three equations employed in the summary method *two* percentiles instead of one might be employed. Thus in the barometric statistics already referred to, in the neighbourhood of the point on the upper arm of the curve which has above it a tenth of the total area, there occur two percentiles with the corresponding proportions of the modulus unity obtained from the ordinary tables,

$$\xi = {\cdot}8958, \quad \xi = 1{\cdot}0796.$$

Whence we obtain the following two equations:

$$a\ {\cdot}8958 + k\ {\cdot}8958^2 + l\ {\cdot}8958^3 = 4{\cdot}5,$$
$$a1{\cdot}0796 + k1{\cdot}0796^2 + l1{\cdot}0796^3 = 5{\cdot}5.$$

By adding these two equations we obtain a compound equation presumably better than either of the components; which is to be combined with two other equations each similarly compounded. But caution must be exercised lest by taking in additional observations *unweighted* we should make the result *worse* (as pointed out by the present writer, *Philosophical Magazine*, 1893; where, as noticed by Mr Sheppard, *loc. cit.*, p. 135, there was committed the error of treating the percentiles as uncorrelated. The matter had been rightly stated by the writer in an earlier paper, *Philosophical Magazine*, 1886, Vol. XXII, p. 375).

IV. *Exceptional Cases.* Percentiles are not so available as moments for the treatment of exceptional cases in which the generating normal curve is not only extended (or contracted), but also in part folded over in the process of translation. For instance, suppose the coefficients a and l to become both zero, the operator taking the form $X = k\xi^2$. Then the lower (left) half of the normal curve will be swung round the axis of Y at the centre of the (original) curve so as to be coincident with the upper half; while both are distorted in such wise that the initial ordinate now becomes infinite. The infinity of a single ordinate, it may be remarked, is not fatal to the representative character of a curve. To determine β we have now the following equations (X being as before measured from the centre of the generating normal curve, as origin, and $f(\xi)$ denoting the normal error-function with unit modulus):

$$M_1 = 2\int_0^\infty XF(X)\,dX = 2\int_0^\infty k\xi^2 f(\xi)\,d\xi = \tfrac{1}{2}k;$$
$$M_2 = 2\int_0^\infty X^2F(X)\,dX = 2\int_0^\infty k^2\xi^4 f(\xi)\,d\xi = \tfrac{3}{4}k^2;$$
$$M_3 = 2\int_0^\infty X^3F(X)\,dX = 2\int_0^\infty k^3\xi^6 f(\xi)\,d\xi = \tfrac{15}{8}k^3.$$

Whence

$$\mu_2 = M_2 - M_1^2 = \tfrac{1}{2}k^2;$$
$$\mu_3 = M_3 - (M_1^3 + 3M_1\mu_2) = k^3;$$
$$\beta = \mu_3^2/\mu_2^3 = 1/\tfrac{1}{8} = 8.$$

The same result may be deduced from the fundamental equations (above, p. 430). For now, λ vanishing, k becomes infinite. Accordingly, the fraction which is equated to β becomes in the limit equal to 8. The corresponding value of η obtained in either way is 12.

Now let l receive a positive value, a being still zero. And first let the value of l be small relatively to k. Then the elements of the normal on the left of the central axis will continue to be swung round to the right up to the point at which $\frac{dX}{d\xi}$ becomes zero, that is when $2k\xi = 3\lambda\xi^2$. Thus practically the whole of the area on the left is swung round to the right when $\frac{2}{3}k/l > 2$, say, or 2·12 (corresponding to 3 times the standard deviation), or $l/k < \frac{1}{3}$. The transformed curve will, however, have lost the typical character of uni-modality (cp. above, p. 427) at an earlier point, the point at which the expression for $\frac{dY}{dX}$ changes from negative to positive (becoming infinite at the point $\xi = \frac{2}{3}k/l$). By varying the values of k and l within the limits indicated the formula may be adapted to a considerable variety of data of the sort instanced by Professor Pearson in the tenth, eleventh and twelfth examples of his second Contribution (*Transactions of the Royal Society*, 1895, Vol. 196 A). Further latitude is obtained by supposing a to have a small value (relatively to k or l). Thus, suppose that a/k is a small fraction, say, while l is zero, then the whole portion of the area on the left of the point ξ defined by the condition $a - 2k\xi = 0$ is swung to the right round the axis of Y at the point distant from the origin on the negative side. The resulting curve will have a mode other than that formed by the infinite ordinate at the extreme left, which, if a/k be sufficiently small, may well fall within the limits of the first of the given compartments—often comprising a considerable percentage of the whole area—and so escape notice.

By assigning different relations to the constants a great variety of cases, perhaps "magis mathematicae quam naturales," can be manufactured. It must suffice here to specify one more anomaly. Suppose l is negative, while k is zero: the operator is then $a(\xi - \lambda\xi^3)$. And the expression for $\frac{dY}{dX}$ becomes

$$\frac{6l\xi^3 + \xi(6\lambda - 2)}{(1 - 3\lambda\xi^2)^3},$$

ξ being measured positively, or in absolute magnitude, to the left (as well as to right) of the origin. If $\lambda > \frac{1}{3}$, this expression is *positive* in the immediate neighbourhood of the origin, and continues *positive* up to the point at which $\xi^2 = 1/3\lambda$. For instance, if $\lambda = \frac{1}{2}$, $\frac{dY}{dX}$ is positive on either side of the origin up to the point $\xi = \cdot 816$, where Y becomes infinite. We have thus a symmetrical **U**-shaped curve. If, now, k becomes sensible, we have an unsymmetrical inverted curve, provided that the values of k and λ are such and so related that the cubic equation (corresponding to $\frac{dY}{dX} = 0$), $6\lambda\xi^3 - 4k\xi^2 + \xi(6\lambda - 2) - 2k = 0$, has only one real root, corresponding to a single *inverse mode.*

To him who grounds the use of a particular representative curve on its affinity to the generalised law of error these freaks of analysis are of secondary interest. They serve, however, to dispel the suspicion that the method is deficient in the power of adaptation.

V. *Frequency-surfaces.* The analogue of the preceding method in cases where the given statistics relate to two (or more) characters, say X and Y, utilises, in addition to the mean *powers* of those variables, mean *products*, such as the sum-integrals of X^2Y and XY^2. Constants corresponding to these additional data are presented by adding to the expression for X hitherto employed terms involving η (as well as ξ) the coordinate of a normal surface with unit moduli and coefficient of correlation r. Thus for powers and products of the third dimension we may write

$$X = a(\xi + \kappa_1\xi^2 + \kappa_1'\xi\eta),$$
$$Y = b(\eta + \kappa_2\eta^2 + \kappa_2'\eta\xi).$$

For the fourth dimension, as there are *three* additional data, it would seem proper to introduce three new "λ's," λ_1' the coefficient of $\xi^3\eta$, λ_2' the coefficient of $\eta^3\xi$, and λ'' the coefficient of $\xi^2\eta^2$ in the expressions for *both* X and H. Substituting these expressions in the values of the mean powers and products given by observation, we obtain as many equations as there are variables; of which the solution may be practicable, if the coefficients of the type κ and λ are small, in which case the coefficient of correlation, r, differs by a small quantity, say ρ, from what it would have been if κ, λ, ... were all zero, say r' $(= \mu_{11}/\sqrt{\mu_{2v}\mu_{v2}})$. Thus we may substitute $r' + \rho$ for r wherever r occurs in the expressions equated to our data—for instance in the integral of $\xi^2\eta^2$. We thus obtain equations comparable in respect of the smallness of the required roots with equations (I) and (II) of Section II, which usually prove manageable.

The solution may be simplified in cases where we know that mean products of the type X^2Y, XY^2, X^3Y may be ignored (perhaps even we only do not know the contrary), for then the distribution will depend only on the mean *powers*. Let there be obtained from the third and fourth (together with the first and second) mean powers (moments) the coefficients of translation for the two curves which represent the frequency of X irrespectively of Y (with equation $Z = \int_{-\infty}^{\infty} F(X, Y)\,\Delta Y$) and the corresponding frequency of Y irrespectively of X. Say the coefficients are respectively κ_1, λ_1 and κ_2, λ_2, then the given surface may be regarded as generated by *translating* a normal surface, say $z = f(\xi, \eta)$, with unit moduli and coefficient of correlation, r, as follows. For every element (small rectangular column) of the generating surface standing on base $\Delta\xi\Delta\eta$ there is substituted an element of equal content at the point $X = a(\xi + \kappa_1\xi^2 + \lambda_1\xi^3)$, $Y = b(\eta + \kappa_2\eta^2 + \lambda_2\eta^3)$; and with area $\Delta X\Delta Y$ correspondingly modified. The coefficient r is thus connected with the observed product-sum $\Sigma\Sigma XY$ (and the other data). By construction $\Sigma\Sigma XY$, say

$$\mu_{12} = \int_{-\infty}^{\infty}\int_{-\infty}^{\infty} a(\xi + \kappa_1\xi^2 + \lambda_1\xi^3)\, b(\eta + \kappa_2\eta^2 + \lambda_2\eta^3)\, d\xi d\eta$$
$$= \int_{-\infty}^{\infty}\int_{-\infty}^{\infty} ab(\xi\eta + \kappa_1\kappa_2\xi^2\eta^2 + \lambda_1\xi^3\eta + \lambda_2\xi\eta^3 + \lambda_1\lambda_2\xi^3\eta^3)\, d\xi d\eta,$$

omitting integrals which reduce to zero. Assigning to those which are retained their respective values (for a normal surface with unit moduli) we have:

$$\mu_{12} = ab \{r + \kappa\kappa_2 \tfrac{1}{4}(1 + 2r^2) + (\lambda_1 + \lambda_2)\tfrac{3}{4}r + \lambda_1\lambda_2(\tfrac{9}{8}r + \tfrac{3}{4}r^3)\}.$$

There is thus obtained for r a cubic equation, with of course a real root. But if this root is greater than unity, the solution is not admissible.

VI. *Comparison with the received method.* The method of translation labours under a heavy disadvantage in that it has not been accepted by the highest living authority on mathematical statistics. Professor Pearson's objections, it must be admitted, had force when directed against the method as first presented. But they are not equally applicable to the method as now developed. It cannot be now objected—if it ever was against the present writer, as distinct from one with whom he was bracketed—that the curve proposed is not a "graduation formula," or that it is not "determined by constants of which the probable errors are easily deducible." (See K. Pearson's "Rejoinder," *Biometrika*, Vol. IV, pp. 201, 202, 211.) Nor can there now be urged the objection: "What is x of which the observed character X is a function?...What is the character which obeys the normal law?... The function x has no real existence as a biological entity" (*loc. cit.* p. 199). This objection might be applicable if the proposed form was advocated as a particular real curve (above, p. 429), related to some real attribute distributed normally, say, as energy is to velocity. But the objection is not equally applicable to the position now taken. The best defence of this position is that it is the same as Professor Karl Pearson's. For his Types, as here interpreted, are but particular representative curves (above, p. 429), formed by a judicious divergence from the normal law of error (a divergence well indicated by himself, *loc. cit.*, pp. 179 and 210 (VII)). In the Pearsonian system the normal curve is transformed by substituting, for the abscissa, say in our terminology ξ, of the *differential* coefficient of the logarithm of a normal curve, an abscissa X *divided* by a quadratic expression, say

$$b_0 + b_1X + b_2X^2;$$

in the system here proposed, the normal curve is transformed by substituting for the abscissa ξ of the *integral* of the normal curve an abscissa X *multiplied* by a quadratic expression, say $a_0 + a_1X + a_2X^2$ (above, p. 428). Is there anything to choose in respect of *rationale* between the two transformations? Are they not both equally true or equally false (as maintained in a former paper, *Journal of the Statistical Society*, 1906, Vol. LXIX, p. 512 and context)? The question for the practical statistician is whether they are equally useful.

A use special to the method of translation is to confirm some of the results obtained by the received method; not by a servile repetition, but as an independent witness giving a similar but not an identical version. This would be a strange mode of verifying ordinary mathematical formulae; but in dealing with Probabilities, it may be appropriate. Thus Professor Pearson has pointed out, with special reference to his Type III which does not utilise the *fourth* moment, that commonly the distance of the mean from the median is about a third of what it is from the mode (*Contributions*, No. II, p. 375. Compare Yule, *Introduction to the Theory of Statistics*, p. 121: "there is an approximate relation between mean, median and mode that appears to hold good with surprising closeness for moderately asymmetrical distributions...expressed by the equation Mode = Mean − 3 (Mean − Median)"). Now the case in which the fourth moment is not required on our system is when $\lambda = 0$. In that case, the mode is given by the equation

$$\xi^2 + \frac{1}{2\kappa}\xi + \frac{1}{2} = 0.$$

This equation gives for the distance of the mode, in the negative direction, from the median $\xi = -\kappa$ nearly,

$$X = a(\xi + \kappa\xi^2) = -a\kappa = -k \text{ (nearly)},$$

a small fraction. But the distance of the mean from the median in a positive direction is found to be $\frac{1}{2}k$. Therefore the distance of the median from the mean is about a third of the distance of the mode from the mean. The Pearsonian Type III is situate so to speak on the dividing line between two types which utilise the fourth moment with *range* on the one side finite, on the other side unlimited. The transition is marked by the *sign* of the "criterion," $6 + 3\beta_1 - 2\beta_2$ (where β_1 is our β and $\beta_2 = \mu_4^2/\mu_2^2 =$ our $\eta + 3$). The transition corresponds in our system to the passage of λ through zero. But the criterion of this transition is not so simple as the above-written one. It is the relation between β and η which is given by putting $\lambda = 0$ in the fundamental equations I and II (above, p. 430) and eliminating χ. In the common case, however, when χ $(= \kappa^2)$ is small, this eliminant is much simplified. We have then, neglecting powers of χ above the first:

$$\text{(I)}\ \chi(18 - 3\beta) = \beta, \qquad \text{(II)}\ \chi(24 - 2\eta) = \eta.$$

Whence $$24\beta - 18\eta + \eta\beta = 0.$$

Neglecting the product $\eta\beta$ as of the second order of magnitude and writing

$$\eta = \beta_2 - 3,$$

and dividing by a constant, we have the Pearsonian criterion $6 + 3\beta_1 - 2\beta_2$. As for the *range*, there is a distinction on our system, too, between limited and unlimited; and the dividing line is exactly at the vanishing point of λ. When λ is zero, the representative curve may be considered to stop or break down at the point where $\frac{dY}{dX}$ changes sign a *second* time as we move in a negative direction from the origin, the median; the point defined by giving to the radical in the root of the above-written equation for the distance of the mode from the origin, the *negative* sign, that is $\frac{1}{2k}$ nearly $= a/2\kappa$—a point at which the curve may be regarded as no longer sensible when $1/2k > 2{\cdot}1$ (above, p. 436), say $\kappa < \frac{1}{4}$. But when λ is not zero, the position of the mode is given as a root of the cubic equation

$$6\lambda\xi^3 + 4k\xi^2 + \xi(2 + 6\lambda) + 2\kappa = 0$$

(the above-stated quadratic *plus* terms involving λ). The criterion distinguishing limited from unlimited range is the condition that this cubic equation should have two impossible roots; and it will be found that there are or are not impossible roots according as λ is negative or positive. The distinction, then, between λ positive or negative, marks a real difference in the data. But of course by those who consider that our general knowledge of approximate forms terminates with the generalised law of error—refers to the body and not the tail of a frequency-curve—exact propositions about range must be taken *cum grano*. Other features of the Pearsonian types may be traced in the variant representation now proposed. When they cannot be so traced, it is suggested that they are of secondary importance.

With regard to the general purposes of the statistician, the advantage seems by no means to be all on the side of the received method. It is true that in that method the connexion between the observed moments and the required constants is particularly simple. By a stroke of singular power and felicity, Professor Pearson has expressed that relation by a series of simple equations which contrast favourably with our many-dimensioned system (above, equations I and II). On the other hand, the relation between the constants and the *form*, or (integral) equation, of the representative frequency-curve is not so simple in the Pearsonian system. A slight variation of quantity in the constants is attended with convulsive changes in the form of the functions. Thus in dealing with statistics which seem to be in *pari materia*, frequency groups nearly coincident with the normal error-curve (such as the anthropometric measurements dealt with by Miss Fawcett in the first volume of *Biometrika*), a single step may plunge the practitioner from one unfamiliar form to another, from Type I to the stupendous Type IV. But on the plan here proposed the corresponding changes are quite gentle: presumably from a small positive value of the constant λ to a small negative value thereof, with corresponding small changes in the constants a and κ. Data which exercise the mathematician on one system present no significant peculiarity on the other. Thus the example of Types V and VI given by Professor Pearson in his contribution No. X (*Transactions of the Royal Society*, 1901, Vol. CXCVII) differs in our system from ordinary cases principally in having peculiarly large values for κ and λ. No doubt the distinctions marked by the Pearsonian criterion κ_2 do correspond to real differences in the statistical material as well as the distinctions marked by his criterion κ. And for some purposes it may be useful to take account of those differences; but for other purposes it may be useful to ignore them. Not only is the analytic form employed in the method of translation uniform, simple and familiar, namely the error-function (together with a rational algebraic expression of the third degree), but it is particularly useful as connecting the percentiles of the given group with the constants (when ascertained by moments) of the representative curve, by simple relations, with the aid of ordinary tables.

The method of translation has also an advantage in being capable of application (as pointed out already, p. 435), when the data are imperfect. Even when the data are not imperfect it may often be convenient to roughly characterise a given frequency-curve without employing the laborious apparatus of moments and criteria.

One deficiency indeed, in the method of translation, is felt and acknowledged; one thing it yet lacks—*authority*.

CONTRIBUTION TO LAPLACE'S THEORY OF THE GENERATING FUNCTION

By D. Arany.

I. The expression

$$Q(x) \equiv l!/\{x!(l-x)!\} . p^x q^{l-x} \quad (p+q=1) \qquad \text{.....................(1)}$$

is only defined for integral values of x, and is arbitrary for other values.

In the Calculus of Probabilities it is convenient to replace $Q(x)$ by an expression that shall be equal to it if x is an integer and shall have a definite meaning whatever the value of x may be.

The ordinary method is to use Stirling's theorem

$$n! = n^n e^{-n} \sqrt{(2\pi n)} \exp\{B_1/(1.2).n^{-1} - B_2/(3.4).n^{-3} + \ldots\}, \qquad \text{.........(2)}$$

where B_1, B_2, ... are Bernoulli's numbers.

Substituting in (1), we obtain

$$Q(x) = l^l/\{x^x(l-x)^{l-x}\} . \sqrt{\{l(2\pi)^{-1} x^{-1}(l-x)^{-1}\}} . p^x q^{l-x} .$$
$$\exp\{B_1/(1.2).(l^{-1} - x^{-1} - (l-x)^{-1}) - \ldots\}. \ \text{...(3)}$$

Writing $x = lp + t$, $l - x = lq - t$, this becomes

$$Q(x) = \frac{\exp\left\{-\frac{1}{1.2}\frac{t^2}{l}\left(\frac{1}{p}+\frac{1}{q}\right)+\ldots\right\}}{\sqrt{2\pi lpq\left(1+\frac{t}{lp}\right)\left(1-\frac{t}{lq}\right)}} \exp\left\{\frac{B_1}{1.2}\left(\frac{1}{l}-\frac{1}{lp+t}-\frac{1}{lq-t}\right)-\ldots\right\} \text{...(6)}$$

$$= \frac{\exp\left(-\frac{t^2}{2lpq}\right)}{\sqrt{2\pi lpq}} \text{ approximately. } \qquad \text{..(7)}$$

II. It can be shown that both (6) and (7) can be obtained by extending a method due to Laplace, called the method of Generating Functions. As the original method is applicable to every function of x which is defined for integral values of x, we will give the extension of the method in the general case, and then check the result by applying it to $Q(x)$.

Let $f(x)$ be a function of x whose values are given for $x = 0, 1, 2, \ldots l$. Then we define $f(x)$ for other values by the equation

$$2\pi i f(x) = \int_{u=a}^{u=a+2\pi i} e^{-xu} \left\{\sum_{t=0}^{t=l} f(t) e^{tu}\right\} du, \qquad \text{....................(8)}$$

which can be proved to hold for any integral value of x and for any value of a. We then write

$$\phi(u) \equiv \sum_{t=0}^{t=l} f(t)\, e^{tu},$$

and call this the *generating function* of $f(t)$. For a, which is quite arbitrary, we choose a value which makes

$$\phi(a) = \phi(a + 2\pi i) = 0.$$

We then, for any particular value of x, define a value s of u which makes the first derivative of $e^{-xu}\phi(u)$ zero. This gives

$$\phi'(s) - x\phi(s) = 0, \quad x = \phi'(s)/\phi(s). \quad \text{......................(11)}$$

We also define a new variable y related to u by the equation (for this value of x)

$$e^{-xu}\,\phi(u) = e^{-xs}\,\phi(s)\, e^{-y^2}. \text{..............................(12)}$$

We then have

$$2\pi i\, f(x) = e^{-xs}\,\phi(s) \int_{y=-\infty}^{y=\infty} e^{-y^2} \left(\frac{du}{dy}\right) dy. \text{......................(13)}$$

It then remains to develop $\left(\dfrac{du}{dy}\right)$ in a Maclaurin-series in powers of y.

This is done by introducing certain functions $\psi(u)$ and $\theta(u)$, explained in the paper*, such that

$$\left.\begin{aligned} (du/dy)_{y=0} &= i\,\{\theta(s)\}^{-\frac{1}{2}} \\ (d^2u/dy^2)_{y=0} &= i^2\left[d\,\{\theta(u)\}^{-\frac{2}{2}}/du\right]_{u=s} \\ (d^3u/dy^3)_{y=0} &= i^3\left[d^2\,\{\theta(u)\}^{-\frac{3}{2}}/du^2\right]_{u=s} \\ &\text{etc.} \end{aligned}\right\} \text{........................(20)}$$

Substituting in (13), we obtain

$$f(x) = \frac{e^{-xs}\,\phi(s)}{2\sqrt{\pi}} \left\{\theta^{-\frac{1}{2}}(s) - \frac{1}{2\,!}\frac{1}{2}\left[\frac{d^2\{\theta(u)\}^{-\frac{3}{2}}}{du^2}\right]_{u=s} + \frac{1}{4\,!}\frac{1\,.\,3}{2\,.\,2}\left[\frac{d^4\{\theta(u)\}^{-\frac{5}{2}}}{du^4}\right]_{u=s} - \dots\right\} \text{....(22)}$$

We can prove that

$$\left.\begin{aligned} \theta(s) &= \frac{0\,!\,dx}{2\,!\,ds} \\ \theta'(s) &= \frac{1\,!\,d^2x}{3\,!\,ds^2} \\ \theta''(s) &= \frac{2\,!\,d^3x}{4\,!\,ds^3} \\ &\text{etc.} \end{aligned}\right\}, \text{..................................(26)}$$

* $\psi(u) = \log \dfrac{\phi(u)}{\phi(s)}$

$$\theta(u) = \frac{\psi_2(s)}{2\,!} + \frac{\psi_3(s)}{3\,!}(u-s) + \frac{\psi_4(s)}{4\,!}(u-s)^2 + \text{etc.}$$

where

$$\psi_k(s) = \left\{\frac{d^k\,\psi(u)}{du^k}\right\}_{u=s}.$$

and therefore (22) becomes approximately

$$f(x) = \frac{e^{-xs}\,\phi(s)}{\sqrt{2\pi\, dx/ds}} \qquad \text{.................................(27)}$$

We can transform this into

$$f(x) = B\,\frac{e^{-\int s\,dx}}{\sqrt{2\pi\, dx/ds}}. \qquad \text{..............................(29)}$$

For

$$f(x) = Q(x),$$

the value of s is

$$s = \frac{t}{l}\left(\frac{1}{p} + \frac{1}{q}\right) - \frac{1}{2}\left(\frac{t}{l}\right)^2\left(\frac{1}{p^2} - \frac{1}{q^2}\right) + \dots.$$

Also $B = 1$, and therefore

$$Q(x) = \frac{e^{-\frac{t^2}{2lpq}}}{\sqrt{2\pi lpq}},$$

as in (7).

RÉPERTOIRE BIBLIOGRAPHIQUE DES SCIENCES MATHÉMATIQUES

Par A. Gérardin.

A la mémoire de notre regretté Président Henri Poincaré.

Le Répertoire Bibliographique des Sciences Mathématiques est destiné à épargner aux travailleurs de longues et pénibles recherches. Il contient les titres des Mémoires relatifs aux Mathématiques pures et appliquées de 1800 à 1900 inclusivement, ainsi que des travaux relatifs à l'histoire des Mathématiques depuis 1600 jusqu'à 1889 inclus.

Le Répertoire parait par séries de 100 fiches 8,5 × 13,5 sur papier fort, chaque fiche renfermant en moyenne neuf titres complets. Actuellement, vingt séries sont parues, comprenant 2000 fiches et l'indication de 18523 articles. La série 21, à la composition, est réservée à la Théorie des Nombres, à la bibliographie complète du dernier théorème de Fermat, et à la Géométrie.

Après avoir rendu un juste hommage au labeur de mes prédécesseurs, MM. Humbert, C. A. Laisant, Raffy, je vais donner ici par matière, mais sans entrer dans le détail, les chiffres des articles classés.

Classe A. (1355 articles.)

Algèbre élémentaire, théorie des équations algébriques et transcendantes; groupes de Galois; fractions rationnelles; interpolation.

Classe B. (752 articles.)

Déterminants; substitutions linéaires; élimination; théorie algébrique des formes; invariants et covariants; quaternions; équipollences et quantités complexes.

Classe C. (779 articles.)

Principes du calcul différentiel et intégral; applications analytiques; quadratures; intégrales multiples; déterminants fonctionnels; formes différentielles; opérateurs différentiels.

Classe D. (1279 articles.)

Théorie générale des fonctions et son application aux fonctions algébriques et circulaires; séries et développements infinis, comprenant en particulier les produits infinis et les fractions continues considérées au point de vue algébrique; nombres de Bernoulli; fonctions sphériques et analogues.

CLASSE E. (220 articles.)

Intégrales définies, et en particulier intégrales eulériennes.

CLASSE F. (207 articles.)

Fonctions elliptiques avec leurs applications.

CLASSE G. (191 articles.)

Fonctions hyperelliptiques, abéliennes, fuchsiennes.

CLASSE H. (1297 articles.)

Equations différentielles et aux différences partielles; équations fonctionnelles; équations aux différences finies; suites récurrentes.

CLASSE I. (1163 articles.)

Arithmétique et Théorie des Nombres; analyse indéterminée; théorie arithmétique des formes et des fractions continues; division du cercle; nombres complexes, idéaux, transcendants.

CLASSE J. (650 articles.)

Analyse combinatoire; calcul des probabilités; calcul des variations; théorie générale des groupes de transformations [en laissant de côté les groupes de Galois (A), les groupes de substitutions linéaires (B) et les groupes de transformations géométriques (P)]; théorie des ensembles de M. Cantor.

CLASSE K. (1040 articles.)

Géométrie et Trigonométrie élémentaires (étude des figures formées de droites, plans, cercles et sphères); Géométrie du point, de la droite, du plan, du cercle et de la sphère; Géométrie descriptive; Perspective.

CLASSE L. (526 articles.)

Coniques et surfaces du second degré.

CLASSE M. (617 articles.)

Courbes et surfaces algébriques; courbes et surfaces transcendantes spéciales.

CLASSE N. (117 articles.)

Complexes et congruences; connexes; systèmes de courbes et de surfaces; géométrie énumérative.

CLASSE O. (912 articles.)

Géométrie infinitésimale et géométrie cinématique; applications géométriques du calcul différentiel et du calcul intégral à la théorie des courbes et des surfaces; quadrature et rectification; courbure; lignes asymptotiques, géodésiques, lignes de courbure; aires; volumes; surfaces minima; systèmes orthogonaux.

Classe P. (213 articles.)

Transformations géométriques; homographie, homologie et affinité; corrélation et polaires réciproques; inversion; transformations birationnelles et autres.

Classe Q. (246 articles.)

Géométrie, divers; géométrie à n dimensions; géométrie non euclidienne; *analysis situs*; géométrie de situation.

Classe R. (1319 articles.)

Mécanique générale; Cinématique; Statique comprenant les centres de gravité et les moments d'inertie; dynamique; mécanique des solides; frottement; attraction des ellipsoïdes.

Classe S. (953 articles.)

Mécanique des fluides; hydrostatique; hydrodynamique; thermodynamique.

Classe T. (2444 articles.)

Physique Mathématique; élasticité; résistance des matériaux; capillarité; lumière; chaleur; électricité.

Classe U. (1065 articles.)

Astronomie; Mécanique céleste et géodésie.

Classe V. (851 articles.)

Philosophie et histoire des sciences mathématiques; Biographie.

Classe X. (327 articles.)

Procédés de calcul; tables; Nomographie; calcul graphique; planimètres; instruments divers.

En résumé; Analyse Mathématique 7893 articles; Géométrie, 3671, et 6959 pour les Mathématiques appliquées.

La classification détaillée et aussi complète que possible de toutes les questions du domaine mathématique a été établie indépendamment des méthodes.

COMMUNICATIONS

SECTIONS IV (*a*) AND IV (*b*). JOINT MEETING

(PHILOSOPHY, HISTORY, DIDACTICS)

THE PRINCIPLES OF MATHEMATICS IN RELATION TO ELEMENTARY TEACHING

By A. N. Whitehead.

We are concerned not with the advanced teaching of a few specialist mathematical students, but with the mathematical education of the majority of boys in our secondary schools. Again these boys must be divided into two sections; one section consists of those who desire to restrict their mathematical education, the other section is made up of those who will require some mathematical training for their subsequent professional careers, either in the form of definite mathematical results or in the form of mathematically trained minds.

I shall call the latter of these two sections the mathematical section, and the former the non-mathematical section. But I must repeat that by the mathematical section is meant that large number of boys who desire to learn more than the minimum amount of mathematics. Furthermore most of my remarks about these sections of boys apply also to elementary classes of our university students.

The subject of this paper is the investigation of the place which should be occupied by modern investigations respecting Mathematical Principles in the education of both of these sections of schoolboys.

To find a foothold from which to start the enquiry, let us ask, why the non-mathematical section should be taught any mathematics at all beyond the barest elements of arithmetic. What are the qualities of mind which a mathematical training is designed to produce when it is employed as an element in a liberal education?

My answer, which applies equally to both sections of students, is that there are two allied forms of mental discipline which should be acquired by a well designed mathematical course. These two forms though closely allied are perfectly distinct.

The first form of discipline is not in its essence logical at all. It is the power of clearly grasping abstract ideas, and of relating them to particular circumstances. In other words, the first use of mathematics is to strengthen the power of abstract thought. I repeat that in its essence this has nothing to do with logic, though as a matter of fact a logical discipline is the best method of producing the desired effect. It is not the philosophical theory of abstract ideas which is to be acquired, but the habit and the power of using them. There is one and only one way of acquiring the habit and the power of using anything, that is by the simple commonplace method of habitually using it. There is no other short cut. If in education we desire to produce a certain conformation of mind, we must day by day, and year by year, accustom the students' minds to grow into the desired structural shape. Thus to teach the power of grasping abstract ideas and the habit of using them, we must select a group of such ideas, which are important and are also easy to think about because they are clear and definite.

The fundamental mathematical truths concerning geometry, ratio, quantity, and number, satisfy these conditions as do no others. Hence the fundamental universal position held by mathematics as an element of a liberal education.

But what are the Fundamental Mathematical truths concerning Geometry, Quantity and Number?

At this point we come to the great question of the relation between the modern science of the Principles of Mathematics and a Mathematical Education.

My answer to the question as to these Fundamental Mathematical truths is that, in any absolute sense, there are none. There is no unique small body of independent primitive unproved propositions which are the necessary starting-points of all mathematical reasoning on these subjects. In mathematical reasoning the only absolute necessary presuppositions are those which make logical deduction possible. Between these absolute logical truths and so-called Fundamental truths concerning Geometry, Quantity and Number, there is a whole new world of mathematical subjects concerning the logic of propositions, of classes, and of relations.

But this subject is too abstract to form an elementary training ground in the difficult art of abstract thought.

It is for this reason that we have to make a compromise and start from such obvious general ideas as naturally occur to all men when they perceive objects with their senses.

In Geometry, the ideas elaborated by the Greeks and presented by Euclid are, roughly speaking, those adapted for our purpose, namely, ideas of volumes, surfaces, lines, of straightness and of curvature, of intersection and of congruence, of greater and less, of similarity, shape, and scale. In fact, we use in education those general ideas of spatial properties which must be habitually present in the mind of anyone who is to observe the world of phenomena with understanding.

Thus we come back to Plato's opinion that for a liberal education, Geometry, as he knew it, is the queen of sciences.

In addition to Geometry, there remain the ideas of quantity, ratio, and of number. This in practice means, Elementary Algebra. Here the prominent ideas are those of 'any number,' in other words, the use of the familiar x, y, z, and of the dependence of variables on each other, or otherwise, the idea of functionality. All this is to be gradually acquired by the continual use of the very simplest functions which we can devise, of linear functions, graphically represented by straight lines, of quadratic functions graphically represented by parabolas, and of those simple implicit functions graphically represented by conic sections. Thence, with good fortune and a willing class, we can advance to the ideas of rates of increase, still confining ourselves to the simplest possible cases.

I wish here emphatically to remind you that both in Geometry and in Algebra a clear grasp of these general ideas is not what the pupil starts from, it is the goal at which he is to arrive. The method of progression is continual practice in the consideration of the simplest particular cases, and the goal is not philosophical analysis but the power of use.

But how is he to practise himself in their use? He cannot simply sit down and think of the relation $y = x + 1$, he must employ it in some simple obvious way.

This brings us to the second power of mind which is to be produced by a mathematical training, namely, the power of logical reasoning. Here again, it is not the knowledge of the philosophy of logic which it is essential to teach, but the habit of thinking logically. By logic, I mean deductive logic.

Deductive logic is the science of certain relations, such as implication, etc., between general ideas. When logic begins, definite particular individual things have been banished. I cannot relate logically this thing to that thing, for example this pen to that pen, except by the indirect way of relating some general idea which applies to this pen to some general idea which applies to that pen. And the individualities of the two pens are quite irrelevant to the logical process. This process is entirely concerned with the two general ideas. Thus the practice of logic is a certain way of employing the mind in the consideration of such ideas; and an elementary mathematical training is in fact nothing else but the logical use of the general ideas respecting Geometry and Algebra which we have enumerated above. It has therefore, as I started this paper by stating, a double advantage. It makes the mind capable of abstract thought, and it achieves this object by training the mind in the most important kind of abstract thought, namely, deductive logic.

I may remind you that other choices of a type of abstract thought might be made. We might train the children to contemplate directly the beauty of abstract moral ideas, in the hope of making them religious mystics. The general practice of education has decided in favour of logic, as exemplified in elementary mathematics.

We have now to answer the further question, what is the *rôle* of Logical Precision in the Teaching of Mathematics? Our general answer to the implied question is obvious, Logical Precision is one of the two objects of the teaching of mathematics, and it is the only weapon by which the teaching of mathematics can achieve its other object. To teach mathematics is to teach logical precision. A mathematical teacher who has not taught that, has taught nothing.

But having stated this thesis in this unqualified way, its meaning must be carefully explained; for otherwise its real bearing on the problem of education will be entirely misunderstood.

Logical precision is the faculty to be acquired. It is the quality of mind which it is the object of the teaching to impart. Thus the habit of reading great literature is the goal at which a literary education aims. But we do not expect a child to start its first lesson by reading for itself Shakespeare. We recognize that reading is impossible till the pupil has learnt its alphabet and can spell, and then we start it with books of one syllable.

In the same way, a mathematical education should grow in logical precision. It is folly to expect the same careful logical analysis at the commencement of the training as would be appropriate at the end. It is an entire misconception of my thesis to construe it as meaning that a mathematical training should assume in the pupil a power of concentrated logical thought. My thesis is in fact the exact opposite, namely that this power cannot be assumed, and has got to be acquired, and that a mathematical training is nothing else than the process of acquiring it. My whole groundwork of assumption is that this power does not initially exist in a fully developed state. Of course like every other power which is acquired, it must be developed gradually.

The various stages of development must be guided by the judgment and the genius of the teacher. But what is essential is, that the teacher should keep clearly in his mind that it is just this power of logical precise reasoning which is the whole object of his efforts. If his pupils have in any measure gained this, they have gained all.

We have not yet however fully considered this part of our subject. Logical precision is the full realization of the steps of the argument. But what are the steps of the argument? The full statement of all the steps is far too elaborate and difficult an operation to be introduced into the mathematical reasoning of an educational curriculum. Such a statement involves the introduction of abstract logical ideas which are very difficult to grasp, because there is so rarely any need to make them explicit in ordinary thought. They are therefore not a fit subject-ground for an elementary education.

I do not think that it is possible to draw any theoretical line between those logical steps which form a theoretically full logical investigation, and those which are full enough for most practical purposes, including that of education. The question is one of psychology, to be solved by a process of experiment. The object to be attained is to gain that amount of logical alertness which will enable its possessors to detect fallacy and to know the types of sound logical deduction. The objects of going further are partly philosophical, and also partly to lay bare abstract ideas whose investigation is in itself important. But both these objects are foreign to education.

My opinion is that, on the whole, the type of logical precision handed down to us by the Greek mathematicians is roughly speaking what we want. In Geometry, this means the sort of precision which we find in Euclid. I do not mean that we should use his famous *Elements* as a textbook, nor that here and there a certain compression in his mode of exposition is not advisable. All this is mere detail. What I do mean is that the sort of logical transition which he made explicit, we should make explicit, and that the sort of transition which he omits, we should omit.

I doubt however whether it is desirable to plunge the student into the full rigour of Euclidean Geometry without some mitigation. It is for this reason that the modern habit, at least in England, of laying great stress, in the initial stages, on training the pupil in simple constructions from numerical data is to be praised. It means that after a slight amount of reasoning on the Euclidean basis of accuracy, the mind of the learner is relieved by doing the things in various special cases, and noting by rough measurements that the desired results are actually attained.

It is important however that the measurements be not mistaken for the proofs. Their object is to make the beginner apprehend what the abstract ideas really mean.

Again in algebra, the notation and the practical use of the symbols should be acquired in the simplest cases, and the more theoretical treatment of the symbolism reserved to a suitable later stage. In fact my rule would be initially to learn the meaning of the ideas by a crude practice in simple ways, and to refine the logical procedure in preparation for an advance to greater generality. In fact the thesis of my paper can be put in another way thus, the object of a mathematical education is to acquire the powers of analysis, of generalization, and of reasoning. The two processes of analysis and generalization were in my previous statement put together as the power of grasping abstract ideas.

But in order to analyse and to generalize, we must commence with the crude material of ideas which are to be analysed and generalized. Accordingly it is a positive error in education to start with the ultimate products of this process, namely the ideas in their refined analysed and generalized forms. We are thereby skipping an important part of the training, which is to take the ideas as they actually exist in the child's mind, and to exercise the child in the difficult art of civilizing them and clothing them.

The schoolmaster is in fact a missionary, the savages are the ideas in the child's mind; and the missionary shirks his main task if he refuse to risk his body among the cannibals.

At this point I should like to turn your attention to those pupils forming the mathematical section. There is an idea, widely prevalent, that it is possible to teach mathematics of a relatively advanced type—such as differential calculus, for instance—in a way useful to physicists and engineers without any attention to its logic or its theory.

This seems to me to be a profound mistake. It implies that a merely mechanical knowledge without understanding of ways of arriving at mathematical results is useful in applied science. It is of no use whatever. The results themselves can all be found stated in the appropriate pocket-books and in other elementary works of reference. No one when applying a result need bother himself as to why it is true. He accepts it and applies it. What is of supreme importance in physics and in engineering is a mathematically trained mind, and such a mind can only be acquired by a proper mathematical discipline.

I fully admit that the proper way to start such a subject as the Differential Calculus is to plunge quickly into the use of the notation in a few absurdly simple cases, with a crude explanation of the idea of rates of increase. The notation as thus known can then be used by the lecturers in the Physical and Engineering Laboratories. But the mathematical training of the applied scientists consists in making these ideas precise and the proofs accurate.

I hope that the thesis of this paper respecting the position of logical precision in the teaching of mathematics has been rendered plain. The habit of logical precision with its necessary concentration of thought upon abstract ideas is not wholly possible in the initial stages of learning. It is the ideal at which the teacher should aim. Also logical precision, in the sense of logical explicitness, is not an absolute thing: it is an affair of more or less. Accordingly the quantity of explicitness to be introduced at each stage of progress must depend upon the practical judgment of the teacher. Lastly, in a sense, the instructed mind is less explicit; for it travels more quickly over a well-remembered path, and may save the trouble of putting into words trains of thought which are very obvious to it. But on the other hand it atones for this rapidity by a concentration on every subtle point where a fallacy can lurk. The habit of logical precision is the instinct for the subtle difficulty.

LE RAISONNEMENT LOGIQUE DANS L'ENSEIGNEMENT MATHÉMATIQUE SECONDAIRE ET UNIVERSITAIRE

PAR R. SUPPANTSCHITSCH.

Il y a une dizaine d'années, l'on aurait jugé inutile de dire un mot sur le raisonnement logique dans l'enseignement mathématique tant secondaire qu'universitaire. Tout enseignement mathématique, sauf le primaire dans tous les cas, était logique, rien que logique, ayant pour but principal, presque unique, le développement de la faculté de raisonner. Les maîtres suivaient alors un chemin bien tracé et sans crainte possible d'abérrations. On pourrait envier à ces anciennes écoles le calme parfait de leur œuvre pédagogique que ne troublaient pas encore les discussions publiques. Mais regardons le revers de la médaille! Les anciens maîtres connaissaient fort bien leur tâche, ils l'accomplissaient tous les ans; mais cela leur avait enlevé la faculté de regarder au dehors. On aurait dit qu'ils portaient avec plaisir les œillères qu'on trouve inutiles aujourd'hui même pour les êtres inférieurs. Puis, au commencement de ce siècle vint la grande réforme, d'origine française. Alors, on s'est aperçu que le fondement soi-disant logique des éléments de mathématiques n'était rien moins qu'irréprochable, qu'il y avait là un problème trop difficile pour l'enseignement secondaire, que maintes démonstrations, d'une sécheresse horrible pour les enfants, étaient dépourvues de toute valeur. En même temps, on commençait dans quelques cas à faire largement appel à l'intuition des élèves, et les résultats obtenus étaient des plus surprenants. Certes, un jeune homme apprend plus vite un nombre considérable des théorèmes qui sont vérifiés dans des cas simples, qu'une seule démonstration rigoureuse; il peut même saisir parfaitement leur portée et acquérir le don de s'en servir à propos. Ainsi, il accumulera très vite des connaissances utiles pour la vie, et son instruction mathématique, étendue quoique superficielle, lui permettra d'aborder des problèmes d'une certaine difficulté. Mais par là, je ne veux pas dire que cette acquisition rapide des notions mathématiques puisse avoir, pour le développement des facultés les plus nobles de l'esprit, autant de valeur que l'enseignement approfondi d'un seul théorème même très restreint.

L'enseignement logique une fois abandonné, on croyait ne point pouvoir délivrer assez les programmes des anciennes méthodes qui, cependant, avaient contribué pendant des siècles à la formation des esprits élevés. Si rien n'est plus dangereux pour l'enseignement qu'une immobilité prolongée, cependant une évolution brusquée produit souvent des effets aussi fâcheux. On croirait, dans certains pays, l'enseignement mathématique élémentaire transformé en un véritable jeu. Ce fait ne serait que des plus heureux, s'il n'avait deux inconvénients. D'abord, la vie devient toujours plus difficile, elle nous impose des efforts grandissants, elle rejette avec une cruauté impitoyable tous ceux qui n'ont pas acquis dès leur jeunesse les forces nécessaires pour cette épouvantable lutte: tandis que nous n'avons plus le souci de former nos enfants à l'école puisque nous ne cherchons qu'à leur éviter tout travail sérieux. Ensuite, ne croyons point que nos enfants puissent trouver amusantes ces soi-disant récréations mathématiques. L'esprit railleur de la jeunesse saisit très vite que l'expérience faite dans une leçon de physique est liée étroitement avec les recherches sérieuses. Au contraire, l'expérience mathématique, appelée à éclairer une certaine vérité, n'est souvent qu'un moyen infaillible d'embrouiller ceux qui n'ont encore que des notions imparfaites et d'ennuyer les autres. Prenons un exemple: voici un cercle et un point A situé en dehors. Je mène plusieurs droites passant par le point A et coupant le cercle. J'obtiens une corde sur chacune d'elles et je vois que la corde située sur le diamètre passant par A est la plus longue et que les cordes décroissent, si j'éloigne la droite du centre...etc. C'est absolument simple. Cependant, il me semble qu'on l'a jugé assez compliqué, car autrement je ne comprendrais pas comment on a pu imaginer l'appareil suivant pour apprendre aux enfants ces propriétés des cordes de cercle. On marque le point A sur un carton gris et assez fort et on y trace le cercle. Puis, on fabrique avec des ciseaux une espèce d'aiguille en carton rouge p.e., évidée dans toute sa longueur. On y marque le point A et on attache l'aiguille en carton gris avec un œil de manière que les deux points A se superposent et que l'aiguille puisse tourner autour de l'œil. Mettons l'appareil en mouvement! Alors, on voit dans l'ouverture de l'aiguille les variations de la corde limitée par le cercle...etc. Les élèves sont tenus, quelquefois, de se faire eux-mêmes cet appareil. Maintenant, il faut avoir le courage de le dire: si un élève ne saisit pas immédiatement et à l'aide d'un dessin grossier les propriétés de la corde mobile, il sera entièrement inapte aux études mathématiques. Rien n'est plus inutile que cet essai d'une expérience à laquelle il est toujours nécessaire de consacrer de 30 minutes à une heure. Or, dans l'enseignement, le superflu est nuisible. Et l'effet moral est pis encore: ou bien, on risque de suggérer aux élèves l'idée qu'on se livre en mathématiques à des expériences fastidieuses, ou bien qu'on est assez peu averti de la vigueur et de la vivacité de l'esprit chez les jeunes gens.

Je vous prie de ne pas me considérer comme un adversaire acharné des expériences mathématiques; mais, selon moi, il est indispensable d'attendre que les choses deviennent assez compliquées pour qu'on ait peine à les imaginer, nettement et longtemps, sans se servir des modèles. Je ne cite qu'un seul exemple: les beaux modèles cinématiques qui mettent en évidence l'engendrement des cyclides. Mais alors, les expériences sont réservées aux classes les plus élevées de l'enseignement secondaire et surtout à l'enseignement supérieur.

Je n'ai pas encore parlé de la logique dans l'enseignement secondaire. J'y reviens. Mais ne craignez pas que j'aille plaider pour le rétablissement des erreurs abandonnées. Bien communiquer les meilleures méthodes est aussi, à mon avis, un enseignement logique, qui cesse de l'être seulement lorsqu'on présente toujours des cas singuliers, exprès imaginés, on le dirait, pour cacher la vraie nature de problèmes qui autrement seraient facilement saisis dans toute leur généralité. Est-il nécessaire de traiter en géométrie descriptive p.e., l'intersection de deux plans, en partant des remparts, qui sont bien plus compliqués que ces notions géométriques élémentaires?

Est-il logique d'imposer aux élèves la nécessité toujours indispensable de voir enfin par eux-mêmes les lois cachées dans ces nombreux cas logiquement identiques, il est vrai, mais très différents en apparence ? L'enseignement devient logique, selon moi, dès que l'on réduit ces points de départ pris à la vie pratique, éternellement renouvelés, à l'importance qui leur convient et qu'on fait comprendre aux élèves que la connaissance des bonnes méthodes et la faculté de s'en assimiler de nouvelles sont les qualités les plus précieuses pour la plupart des hommes, sauf peut-être pour les grands génies. La logique infiniment subtile, nécessaire aujourd'hui pour inventer ou même pour bien comprendre une démonstration vraiment correcte, n'est plus à la portée de nos élèves de collèges, mais ils auront besoin toujours de la logique de méthodes qui ne dépasse pas leur intelligence. Je veux dire qu'ils doivent apprendre à pouvoir se servir bien vite d'une méthode enseignée, à voir si son application est possible dans des cas relativement compliqués, à critiquer les différentes méthodes et à choisir la plus pratique dans un cas donné. Cette faculté précieuse ne s'acquiert nullement, je le crois du moins, quand on n'habitue pas les élèves de bonne heure à des procédés généraux. Il ne faut pas s'illusionner encore sur la complexité de la vie pratique : jamais, dans les collèges, on ne pourra enseigner la géométrie au point qu'un arpentage exécuté dans les conditions de la réalité puisse donner des résultats sérieux. C'est un exemple seulement, qu'il serait aisé de multiplier. Rappelons-nous seulement combien même les élèves sortant de Polytechnique ou de Centrale ont encore à apprendre dans la vie pratique. Les exemples techniques de l'enseignement secondaire ne servent donc qu'à lui donner plus de vie, et cela est très juste. Mais n'exagérons pas ! Nous ne voulons point que nos élèves ne voient dans les mathématiques qu'un moyen et non une science. Il faut, au contraire, leur apprendre à gouter la beauté de cette science. A ce sujet, il est indispensable de leur faire de temps à autre une démonstration rigoureuse, bien choisie, qui leur donnera le goût de la rigueur et de la critique. Je veux insister encore sur un point important. Les études mathématiques, comme celle de la grammaire, forcent les jeunes gens à s'exprimer avec précision et simplicité. Rien ne favorise plus l'habitude des expressions claires que la nécessité de rendre dans le langage ordinaire un théorème compliqué, mais que l'on a compris parfaitement. Vous savez tous l'énoncé exact des conditions nécessaires et suffisantes pour qu'une suite de nombres ait une limite ; vous savez aussi que les commençants sont souvent tentés de formuler ce théorème d'une manière très incomplète. J'entre d'autant moins dans la discussion des erreurs possibles que quelques-unes d'entre elles sont très subtiles. Cependant, je citerai un exemple simple qui, très à la portée des élèves, les forcerait à réfléchir sur l'énoncé de ce théorème. Prenons la suite de nombres :

$$+\tfrac{2}{3},\ -\tfrac{3}{4},\ +\tfrac{4}{5},\ -\tfrac{5}{6},\ +, \ldots$$

Le nombre 1 est évidemment tel qu'il y a des nombres dans cette suite qui s'en rapprochent arbitrairement, mais il n'est pas *la limite* de ces nombres.

L'esprit de la critique, la faculté de s'exprimer avec clarté—voilà encore des qualités bien nécessaires à un bon citoyen. On ne peut guère imaginer, en effet, un véritable homme d'état dénué d'esprit critique. C'est pour cette raison, sans doute, qu'on a toujours jugé les études des mathématiques pures très utiles pour la culture nationale.

Sur l'enseignement universitaire, j'ai peu de chose à dire. Il est divisé par sa nature en un enseignement purement scientifique, donné aux futurs savants, et un enseignement donné aux candidats du professorat secondaire. Le premier, confié toujours aux plus grands savants du pays, suit par ce fait même les derniers progrès de la science. Dans le second, on est obligé déjà de choisir les méthodes et la matière en vue d'un but pratique. Les grandes découvertes modernes de la science ont amené de fréquentes réformes des programmes de ces études dans les différents pays. D'autre part, la révision des programmes de l'enseignement secondaire exige de nouvelles qualités du maître, surtout une connaissance plus complète des applications techniques. La nécessité d'études pratiques pourrait nuire à la haute précision à laquelle on est habitué depuis longtemps dans l'enseignement universitaire. Un second danger menace l'enseignement donné aux futurs professeurs, c'est celui d'idées pédagogiques, mal conçues et défendues par des partisans manquant d'une instruction mathématique très solide. Ici, il faut toutefois excepter la France, où à l'Ecole Normale Supérieure un enseignement pédagogique parfait est donné par des professeurs des mathématiques. Mais je ne crains pas non plus un abaissement de l'enseignement universitaire dans les autres pays, car dans l'intérêt grandissant qu'on prête à l'étude critique des principes, je vois un remède aux deux dangers. Je peux résumer ainsi mon opinion sur le meilleur enseignement universitaire à donner aux candidats du professorat: on doit ne pas toucher sauf quelques allégements au fond de cet enseignement, donner une place bien plus large aux applications techniques, étudier celles-ci, autant que possible, dans une école technique, faire toujours donner par des professeurs de mathématiques aux candidats à la fin de leurs études des connaissances pédagogiques indispensables, et enfin, souligner, le cas échéant, la valeur de la rigueur. Ainsi, on obtiendra des maîtres qui réuniront au gout absolument indispensable de la rigueur celui des applications pratiques. Que c'est là une chose possible, les grands mathématiciens sortis de l'école polytechnique en France nous le prouvent. En passant de vue l'enseignement supérieur, je n'ai parlé que de l'enseignement universitaire. En effet, l'enseignement mathématique dans les écoles techniques proprement dites, j'exclue formellement l'école polytechnique de France, doit être donné, selon moi, d'une manière tout à fait différente. Dans ces établissements, il s'agit d'enseigner surtout et avant tout des théorèmes et des méthodes qui sont applicables aux problèmes posés aux ingénieurs par la vie pratique. Ces problèmes étant bien souvent trop difficiles, vu l'état actuel de nos connaissances, il est encore nécessaire d'habituer les élèves à simplifier les données réelles de telle façon que le résultat final ne soit pas trop modifié. Cet enseignement, malgré son but pratique, n'est pas moins scientifique; en outre, il doit contribuer à la formation définitive d'esprit des futurs ingénieurs. Nous voyons que l'enseignement mathématique aux écoles techniques exige de la part du professeur non seulement des connaissances mathématiques approfondies, mais aussi une grande expérience et un don particulier d'enseigner. Je ne peux que signaler ici ces grandes difficultés et je renvoie ceux qui désireraient de plus amples renseignements au très beau rapport de M. Czuber sur les écoles polytechniques en Autriche.

COMMUNICATIONS

SECTION IV (*a*)

(PHILOSOPHY AND HISTORY)

A SET OF POSTULATES FOR ARITHMETIC AND ALGEBRA

BY HENRY BLUMBERG.

In the following paper, I wish to present a set of postulates for arithmetic and algebra that, I believe, combines very well the advantages of the so-called "genetic" and "axiomatic" methods for foundations in mathematics. This set has grown out of conversations that I had with Prof. Zermelo and it is our intention to publish our results in greater detail in the near future. To-day I shall confine myself to indicating the principal points.

We start with a set of postulates defining the most general field ("Körper") for which the commutative law of multiplication need not hold. Such a field will be henceforth simply termed a "non-commutative" field. By the addition of one postulate, we obtain a second set of postulates defining the most general non-commutative field that contains as a sub-field the field of rational numbers*. By the further addition of one postulate, we obtain a third set defining the most general non-commutative field that contains as a sub-field the field of real numbers. Finally, by the addition of one postulate to this third set, we obtain a fourth set of postulates defining the most general non-commutative field that contains as a sub-field the field of ordinary complex numbers.

Certain concepts and principles of logic and the theory of aggregates will be presupposed without being expressly formulated. The notion of "finite number" is however *not* presupposed. In the proofs, therefore, it will not be permissible to use the expression "a finite number of times" or the expression "and so on" when reference is implicitly made to the series of ordinal numbers. The development of arithmetic and algebra on the basis of our postulates includes, therefore, a rigorous development of the theory of finite numbers. In this connection, I may refer to a paper of Prof. Zermelo, read at the Fourth International Congress of Mathematicians and entitled, "Ueber die Grundlagen der Arithmetik."

I also wish to call attention to the fact that there are no postulates of order.

For other sets of postulates for arithmetic and algebra to be found in mathematical literature, I refer to Hilbert's *Grundlagen der Geometrie* (third edition) and, in particular, to the articles of Prof. Huntington in the *Transactions of the American Mathematical Society* and in the *Annals of Mathematics.* In these articles of Huntington you will also find extensive bibliographies.

* More accurately, "that contains as a sub-field a field isomorphic with the field of rational numbers with respect to addition and multiplication." A similar remark is to be made later with regard to the real and the complex numbers.

I may add, by way of explanation, that the term "equality" has the same significance below as the term "identity." The logical properties of the latter concept are presupposed.

Let $\mathfrak{F}$ be an aggregate of objects; "addition," denoted by the symbol $+$, and "multiplication," denoted by the symbol $\times$, two operations. The scheme of our postulates is then as follows:

Postulate A_1. If a and b are elements of $\mathfrak{F}$, then $a+b$ is a uniquely determined element of $\mathfrak{F}$.

Postulate A_2. The associative law of addition holds for $\mathfrak{F}$; i.e. if a, b, c are elements of $\mathfrak{F}$, then

$$(a+b)+c=a+(b+c).$$

Postulate A_3. There exists an element z (zero) in $\mathfrak{F}$, such that for every element a of $\mathfrak{F}$ (1) $a+z=a$, and (2) an element $\bar{a}$ exists such that $a+\bar{a}=z$.

Postulate M_1. If a and b are elements of $\mathfrak{F}$, then $a\times b$ is a uniquely determined element of $\mathfrak{F}$.

Postulate M_2. The associative law of multiplication holds for $\mathfrak{F}$; i.e. if a, b, c are elements of $\mathfrak{F}$, then

$$(a\times b)\times c=a\times(b\times c).$$

Postulate M_3. There exists an element u (unity) in $\mathfrak{F}$, such that for every element a of $\mathfrak{F}$ that satisfies the inequality $a+a\neq a$ (1) $a\times u=a$, and (2) an element a' exists such that $a\times a'=u$.

Postulate D. The distributive laws of addition and multiplication hold for $\mathfrak{F}$; i.e. if a, b, c are elements of $\mathfrak{F}$, then

$$a\times(b+c)=(a\times b)+(a\times c)$$

and

$$(b+c)\times a=(b\times a)+(c\times a).$$

These seven postulates constitute the first set of postulates mentioned above; i.e. they define the most general non-commutative field. Among the important deductions from these postulates are:

(1) The uniqueness of the zero element z and the unit element u.

(2) The uniqueness of the inverse elements $\bar{a}$ and a' of an element a, the former with respect to addition and the latter with respect to multiplication.

(3) The commutative law of addition.

(4) The law that a product equals z when and only when at least one of its factors equals z.

I now wish to indicate how the integral and rational numbers are defined and ordered.

Definition: An E-aggregate is a system of elements of $\mathfrak{F}$ that contains

(α) the element u,

(β) with every element x the element $x+u$.

That such aggregates exist is shown by the fact that $\mathfrak{F}$ itself is an E-aggregate.

The greatest common divisor of all the E-aggregates is again an E-aggregate. This special E-aggregate we denote by N. Its elements play the rôle of natural or positive integral numbers. N may also be defined as follows: N is an E-aggregate such that every sub-aggregate of it that is also an E-aggregate is identical with N.

For N the principle of mathematical induction holds. This principle may be formulated as follows:

Principle of induction for N: If A_x is a statement regarding an undetermined object x such that

(α) *A_u is true,*

and (β) *the truth of A_x implies the truth of A_{x+u},*

then A_x is true for every element x of N.

By means of this principle, we can prove the

Theorem: If a and b belong to N, then $a+b$ and $a \times b$ belong to N.

The following theorem holds:

Theorem: N is infinite or not according as z is not or is contained in N. (The positive definition of infinity by means of correspondence is employed.)

Definition: $\mathfrak{R}$ is the greatest common divisor of all non-commutative fields that are sub-aggregates of $\mathfrak{F}$. ($\mathfrak{F}$ itself is also to be considered a sub-aggregate of $\mathfrak{F}$.)

The following theorem holds:

Theorem: If N is not infinite, it is identical with $\mathfrak{R}$.

We are now in a position to appreciate the meaning of the next postulate.

Postulate R. Every non-commutative sub-field of $\mathfrak{F}$ is infinite.

By virtue of this postulate, we obtain on the basis of the preceding theorems the

Theorem: z is not contained in N.

From this follows the

Theorem: If a is contained in N, then $\bar{a}$ is not contained in N.

We now define the aggregate G as follows: An element x of $\mathfrak{F}$ belongs then and only then to G, when either $x=z$ or x belongs to N or $\bar{x}$ belongs to N.

The following theorem holds:

Theorem: The sum, difference and product of any two elements of G also belong to G.

We are now ready to introduce the notion of order for the elements of G. When the two elements a and b of G are such that $a+\bar{b}$ belongs to N, we say, "a is greater than b" and write, "$a > b$." It may then be shown that for any two elements a and b of G, one and only one of the three relations, $a > b$, $a = b$, $b > a$, holds. All the well-known theorems of order for integral numbers may then be deduced by the aid of the foregoing theorems. In particular, for example, the fundamental theorem may be proved, that every sub-aggregate of N possesses a smallest element.

We then show:

Theorem: Every element of $\mathfrak{R}$ *has the form* $m \times n'$, *where* m *and* n *belong to* N.

By means of this theorem we order the elements of $\mathfrak{R}$ on the basis of the order of the elements of N.

By means of the last theorem we are also able to prove the

Theorem: If a *is any element of* $\mathfrak{F}$ *and* r *any element of* $\mathfrak{R}$, *then* $a \times r = r \times a$.

In particular, then, *the commutative law of multiplication holds for* $\mathfrak{R}$.

We thus see that these eight postulates define the most general non-commutative field containing as a sub-field the field of rational numbers.

If it is our desire to obtain a set of postulates defining the rational numbers alone, we need only add the following postulate to those already given.

Postulate P. $\mathfrak{F}$ is primitive with respect to the foregoing postulates; i.e. every sub-aggregate of $\mathfrak{F}$ for which the foregoing postulates hold is identical with $\mathfrak{F}$. (This postulate may be called the "postulate of primitivity.")

It may be remarked, in passing, that on the supposition of the consistency of the postulates A_1—A_3, M_1—M_3, D and R, the consistency of the entire set of nine postulates may be proved. For, by means of the first eight, we were able to show the existence of $\mathfrak{R}$, which in fact satisfies all of our nine postulates.

In the ordered aggregate $\mathfrak{R}$, we may define a Dedekind's section in the usual way. The sum and the product of two sections are then also defined, as certain definite sections, in the usual way. Our next postulate is as follows:

Postulate C. To every section S of $\mathfrak{R}$ there corresponds uniquely an element s of $\mathfrak{F}$ such that, when S_1 and S_2 are two sections and s_1 and s_2 the corresponding elements, $s_1 + s_2$, $s_1 \times s_2$ correspond respectively to $S_1 + S_2$, $S_1 \times S_2$; the corresponding elements are furthermore not all identical.

The set of postulates A_1—A_3, M_1—M_3, D, R and C defines the most general non-commutative field containing as a sub-field the field of real numbers.

If we denote by $\mathfrak{C}$ the aggregate that consists of $\mathfrak{R}$ and all the elements of $\mathfrak{F}$ that correspond to sections of $\mathfrak{R}$, it may be shown, after the manner of Dedekind, that $\mathfrak{C}$ possesses all the properties of the ordinary real numbers. In particular, I mention the fact that the commutative law of multiplication can be proved to hold for $\mathfrak{C}$.

To obtain a set of postulates defining the real numbers alone, we need only add to the 9 postulates A_1—A_3, M_1—M_3, D, R and C the postulate of primitivity P unaltered.

To obtain a set of postulates for the most general non-commutative field containing as a sub-field the field of ordinary complex numbers, we add to the postulates already given (the postulate of primitivity being of course excepted) the following:

Postulate I. There exists in $\mathfrak{F}$ an element i such that $i \times i = \bar{u}$ and for every element c of $\mathfrak{C}$ $i \times c = c \times i$.

It may be shown that the requirement that $i \times c = c \times i$ for every element c of $\mathfrak{C}$ is not redundant.

In a similar manner as above, we obtain a set of postulates for complex numbers alone by adding to the foregoing postulates the postulate of primitivity P.

The field of quaternions may also be dealt with from our point of view; but I do not wish to enter upon its discussion here.

I also do not wish to discuss the question of independence. Our postulates are however essentially independent.

To sum up, I append a list of the sets of postulates considered.

For the most general non-commutative field:

$$A_1, A_2, A_3, M_1, M_2, M_3, D.$$

For the most general non-commutative field containing as a sub-field the field of rational numbers:

$$A_1, A_2, A_3, M_1, M_2, M_3, D, R.$$

For the most general non-commutative field containing as a sub-field the field of real numbers:

$$A_1, A_2, A_3, M_1, M_2, M_3, D, R, C.$$

For the most general non-commutative field containing as a sub-field the field of complex numbers:

$$A_1, A_2, A_3, M_1, M_2, M_3, D, R, C, I.$$

For the field of rational numbers:

$$A_1, A_2, A_3, M_1, M_2, M_3, D, R, P.$$

For the field of real numbers:

$$A_1, A_2, A_3, M_1, M_2, M_3, D, R, C, P.$$

For the field of complex numbers:

$$A_1, A_2, A_3, M_1, M_2, M_3, D\ R, C, I, P.$$

A SET OF POSTULATES FOR ABSTRACT GEOMETRY, EXPRESSED IN TERMS OF THE SIMPLE RELATION OF INCLUSION

By Edward V. Huntington.

The following abstract gives the main results of a paper which will be published in full in the *Mathematische Annalen**.

The fundamental concepts in our discussion are two: (1) a *symbol* K, which can stand for any *class of elements* A, B, C, ..., and (2) a *symbol* R, which can stand for any *relation* between two elements of the class K. For example:

(1) K may denote the class of all men, and "$A\ R\ B$" may mean "A is the son of B"; or

(2) K may denote a class consisting of the following six numbers: 2, 3, 5, 6, 15, 30, while "$A\ R\ B$" means "A is a factor of B"; or

(3) K may denote the class of all spheres with diameter not less than one inch, while "$A\ R\ B$" means "A is inside of B," or "B contains A"; etc., etc.

These symbols K and R are thus variables, which may take on any desired values; any particular determination of K and R, as in the examples just given, may be called a "system (K, R)."

Instead of considering all possible systems (K, R), however, we agree to confine our attention to such systems (K, R) as satisfy certain *conditions*, or *postulates*, mentioned below. These conditions, or postulates, are all independent of one another, so that no one of them is superfluous (see below); furthermore, the restrictions which they impose are so extensive that we have the following theorem:

There is one and only one system (K, R) *which satisfies all the postulates*; or, more precisely, all the systems (K, R) which satisfy the postulates are logically identical, in the sense that they are all isomorphic with one another with respect to the variables K and R.

The simplest example of a system that satisfies all the postulates is the system (compare example (3) above) in which K is the class of all "concrete" spheres in perceptual space, while R is the relation of inclusion (so that "$A\ R\ B$" means "A

* For statements in regard to the relation of this paper to earlier papers, the reader is referred to the completed article.

within B")*. Since all other systems (K, R) that satisfy the postulates are logically identical with this system, as far as theorems expressible in terms of K and R are concerned, it will be convenient to employ even for the general case a terminology which will suggest this familiar geometrical interpretation. We shall therefore call any system (K, R) which possesses all the prescribed properties an *abstract geometry* (understanding under geometry, Euclidean geometry of three dimensions); the elements A, B, C, ..., of such a system we shall call *abstract spheres*, and the relation R the relation of *abstract inclusion*. It must not be forgotten, however, that these abstract spheres and this abstract relation R are concepts which are defined only through the postulates; that is, K and R are symbols which have no properties except those which are explicitly mentioned in the postulates. All theorems which are deducible from our postulates will be valid for *any* system (K, R) which satisfies the postulates, and do not depend in any way on the special interpretation which may be given to the variables K and R.

The postulates fall into two groups: the "general laws" 1—18, and the "existence postulates" E 1—E 7.

POSTULATE 1. If $A\ R\ B$ and $B\ R\ C$, then $A\ R\ C$.

POSTULATE 2. If $A\ R\ B$, then A and B are distinct.

These postulates require the relation R to be "transitive," "non-reflexive," and hence "non-symmetrical," and exclude at once such systems (K, R) as example (1) above.

In the statements of the further postulates, certain combinations of the fundamental symbols K and R are of such frequent occurrence that we shall replace them by single terms, introduced by definition as occasion requires. Also, for the sake of brevity, we shall usually omit the word "abstract," and speak of our "abstract spheres" simply as "spheres."

DEF. If A is a sphere which contains no other sphere within it, then A is called an *abstract point*, or simply a *point*.

There is nothing in this definition which requires our "points" to be "small"; thus in example (3) above, the "abstract points" of the system are simply the "inch spheres."

POSTULATE 3. (*a*) If the class of spheres which contain the point A is the same as the class of spheres which contain the point B, then $A = B$. (*b*) If the class of points within a sphere S is the same as the class of points within a sphere T, then $S = T$.

The following definition is new, and is of fundamental importance for the whole discussion.

DEF. Let A and B be any given points. If X is a point such that every sphere which contains A and B also contains X, then X is said to belong to the *segment* [AB]. The segment [AB], or [BA], is thus a class of point uniquely determined by A and B.

* The question whether the "concrete" or "perceptual" spheres of our spacial experience do actually possess all the prescribed properties, is a question of psychology or natural science, into which we do not enter here.

DEF. The "prolongation" of a segment $[AB]$ beyond A is the class of points X such that the segment $[XB]$ contains A; similarly, the prolongation beyond B is the class of points X such that $[AX]$ contains B. The "line AB" is then defined as the class of points consisting of the segment $[AB]$ together with its two prolongations.

POSTULATE 4. If a point X belongs to a segment $[AB]$, then $[AB]$ is the "simple sum" of the two segments $[AX]$ and $[BX]$; that is, every point of $[AB]$ belongs either to $[AX]$ or to $[BX]$, and these two segments have no common point except X.

POSTULATE 5. If two lines have two distinct points in common, they coincide.

From these postulates 1—5 *all the formal laws of order for points on a line can be deduced.*

DEF. The definition of a *triangle* $[ABC]$, and its six extensions, and the definition of the *plane* ABC, are analogous to the definitions of segment and line. (For a full statement of this analogy, see the completed article.)

POSTULATE 6. If a point X belongs to a triangle $[ABC]$, then $[ABC]$ is the "simple sum" of the three triangles $[ABX]$, $[ACX]$, and $[BCX]$. (Compare Postulate 4.)

POSTULATE 7. If the segment $[XY]$ intersects the segment $[AB]$, then the triangles $[ABX]$ and $[ABY]$ have no point in common except the points of $[AB]$.

POSTULATE 8. If two planes have three non-collinear points in common, they coincide. (Compare Postulate 5.)

DEF. If a plane having all the properties stated in Postulates 6—8 contains two lines AB and CD which have no point in common, then these lines are said to be *parallel*, and we write $AB \parallel CD$. If two lines AB and CD are either parallel or coincident, we shall write $AB \sim CD$.

POSTULATE 9. If two distinct lines are parallel to a third line, they are parallel to each other.

POSTULATE 10. If AB and CD are parallel lines, then no one of the four points A, B, C, D lies within the triangle formed by the other three.

POSTULATE 11. Let A, B, C, D be any set of four points, no three of which are collinear, and A', B', C', D' any other set of four points, no three of which are collinear; and consider the two sets of six lines each, which these points determine. If now $AB \sim A'B'$, $AC \sim A'C'$, $AD \sim A'D'$, $BC \sim B'C'$, and $BD \sim B'D'$, then also $CD \sim C'D'$.

From these postulates 1—11 *all the non-metrical formal laws for the plane can be deduced.*

DEF. In particular, we can show that if we consider all the parallelograms that have a given segment $[AB]$ as one diagonal, then the other diagonal of each of these parallelograms will intersect $[AB]$ in a point M which depends only on A and B. This point is called the *mid-point* of the segment $[AB]$, denoted by $M = \text{mid } AB$.

The following definitions lead to the metrical properties.

Def. If A and B are points within a sphere S, and if all points of the prolongations of $[AB]$ are outside of S, then the segment $[AB]$ is called a *chord* of the sphere S, and the points A and B are said to lie *on the surface* of the sphere.

Def. If O is a point within a sphere S, and if every pair of chords which intersect at O are the diagonals of a parallelogram, then O is called the *centre* of the sphere S.

The theorem that a sphere cannot have more than one centre follows from Postulates 1—11.

Def. Two segments $[AB]$ and $[CD]$ are called *congruent* (in symbols: $AB \equiv CD$) when and only when one of the following conditions is satisfied: (1) if the two segments are on the same or parallel lines, we must have mid $AD =$ mid BC or mid $AC =$ mid BD (translation); (2) if they have a common end point, but do not lie in the same line, then they must be radii of the same sphere (rotation); (3) if they do not lie on the same or parallel lines, nor have a common end point, then there must be two segments $[OX]$ and $[OY]$ which are congruent to the given segments according to (1) and congruent to each other according to (2).

Postulate 12. If $AB \equiv CD$ and $CD \equiv EF$, then $AB \equiv EF$.

That is, the relation of congruence must be "transitive."

Postulate 13. If the surfaces of two concentric spheres are cut by one radius in A and X and by another radius in B and Y, then $AX \equiv BY$.

That is, the portions of two radii intercepted between the surfaces of two concentric spheres are congruent.

Postulate 14. Let A, B, C, X be a set of four points of which the first three are collinear, and let A', B', C', X' be another set of four points of which the first three are collinear; and consider the two sets of six segments each, which these points determine. If $AB \equiv A'B'$, $AC \equiv A'C'$, $BC \equiv B'C'$; $AX \equiv A'X'$, and $BX \equiv B'X'$; then also $CX \equiv C'X'$.

From these postulates 1—14, together with a postulate demanding that all the points of our system shall lie in one plane, all the *formal laws of plane geometry* can be deduced. For geometry of three dimensions the following further postulates are required.

Def. The definition of the *tetrahedron* $[ABCD]$, with its 14 extensions, and the definition of the *space* $ABCD$, are analogous to the definitions of triangle and plane.

Postulate 15. If a point X belongs to a tetrahedron $[ABCD]$, then $[ABCD]$ is the "simple sum" of the four tetrahedra $[ABCX]$, $[ABDX]$, $[ACDX]$, and $[BCDX]$. (Compare Postulate 6.)

Postulate 16. If the segment $[XY]$ intersects the triangle $[ABC]$, then the tetrahedra $[ABCX]$ and $[ABCY]$ have no point in common except the points of $[ABC]$. (Compare Postulate 7.)

Postulate 17. If $ABCD$ is a space, then every point belongs to this space.

This postulate limits the system to three dimensions. The following postulate is analogous to Postulate 10:

Postulate 18. If a line XY is parallel to a plane ABC, then no one of the five points A, B, C, X, and Y belongs to the tetrahedron formed by the other four.

These postulates 1—18 *are independent of one another, and constitute the fundamental "formal laws" of geometry.* It remains to state the postulates which provide the "existence theorems."

Postulate E 1. There are, in the class K, at least two distinct points.

This, and the following postulates, are not to be understood as stating the absolute existence of any concrete object; they mean simply that among all possible systems (K, R) we agree to consider only such as satisfy these conditions. For example, in view of Postulate E 1, a system (K, R) which did not contain at least two distinct points would not be classed as a geometric system.

Postulate E 2. If AB is a line, there is a point X not on that line.

Postulate E 3. If AB is a line, and C a point not on that line, then there is a point X such that CX is parallel to AB.

A system in which this postulate is satisfied may be called a system in which parallels can be freely drawn.

Postulate E 4. If $[AB]$ is any segment in a system in which parallels can be freely drawn, then on any half-line OP there is a point X such that $[OX] \equiv [AB]$.

Postulate E 5. If S_1, S_2, S_3, ... is an infinite sequence of spheres, each of which lies within the preceding one, then there is a point X which lies within them all.

Postulate E 6. If any sphere has a centre, then every sphere has a centre, provided, of course, that it is not itself a point.

Postulate E 7. If ABC is a plane, there is at least one point not in that plane.

These existence postulates E 1—E 7 *are independent of each other and of the general laws* 1—18; that is, for each of these existence postulates E 1—E 7, there is a system (K, R) which satisfies all the general laws 1—18, and all the existence postulates except the one in question.

The two groups of postulates 1—18 and E 1—E 7 taken together form *a complete set of postulates for Euclidean geometry of three dimensions.* To obtain a corresponding *set of postulates for two dimensions*, we have merely to omit Postulates 15—18, and replace Postulate E 7 by its opposite; in this case the most natural interpretation of "abstract sphere" would be *circle* instead of *sphere.*

The deductions from the postulates are carried sufficiently far, in the completed article, to establish the theorem of isomorphism mentioned above. The most important element in the demonstration is of course the introduction of coordinates.

The previous authors to which the writer is chiefly indebted are Peano, Russell, Hilbert, Veblen and Schur.

LA VALEUR ET LES RÔLES DU PRINCIPE D'INDUCTION MATHÉMATIQUE

PAR ALESSANDRO PADOA.

§ 1. *La valeur.*

On sait que M. H. Poincaré—le savant dont la perte vient de frapper douloureusement le monde scientifique—s'est occupé beaucoup du principe d'induction mathématique, qu'il considérait comme "le raisonnement mathématique par excellence", enveloppant "une infinité de syllogismes" et "irréductible au principe de contradiction"*. Or, si je vais réfuter encore une fois ces assertions, et d'une manière que j'ose croire définitive, j'espère que personne ne voudra m'accuser d'un manque de respect dont je ne me sens pas coupable. Je pense que les moindres erreurs des grands sont si funestes qu'il faut se hâter à les révéler en toute lumière; d'autant plus lorsqu'on les rencontre dans une œuvre puissante, même étonnante, telle que celle de M. Poincaré.

Mais j'ai réfléchi qu'un esprit si aigu et si profond n'aurait insisté sur des assertions qui me semblent fausses, si d'abord elles n'avaient été l'expression ambiguë d'une pensée exacte; c'est pourquoi, sous l'erreur qui paraît dans ses mots, j'ai tâché de découvrir la vérité partielle qui s'était présentée à son esprit.

D'ailleurs, pour vaincre définitivement un malentendu et empêcher qu'il puisse persister ou se réproduire, il faut tâcher d'en découvrir l'origine.

1. D'abord—si par "raisonnement", ou mieux par "forme de raisonnement", on entend une "proposition de logique déductive pure"—le principe d'induction mathématique *n'est pas un raisonnement,* mais *une propriété des nombres entiers absolus* (qu'ici j'appellerai "nombre", tout court). La voici :

"*Si* une classe renferme le zéro et *si* cette classe renferme le successif de chaque individu qui lui appartient, *alors* cette classe renferme tous les nombres".

Dans plusieurs traités on énonce ce principe de cette autre manière, qui est plus proche de ses applications :

"Pour démontrer qu'une proposition est vraie lorsqu'une des lettres qui y sont employées désigne un nombre quelconque, il suffit de s'assurer que cette proposition est vraie lorsque ce nombre est zéro, et que, si elle est vraie pour un certain nombre, elle doit être vraie aussi pour son successif".

* H. Poincaré : "Sur la nature du raisonnement mathématique", ap. *Revue de Métaphysique*, t. II, p. 371; *La science et l'hypothèse*, p. 19.—L. Couturat, *Les principes des mathématiques*, p. 62.

Mais l'équivalence de ces deux énoncés est évidente; car il suffit de relire le second, en portant l'attention sur *la classe des valeurs de la lettre considérée pour lesquelles la proposition donnée résulte vraie,* pour retrouver immédiatement le premier énoncé.

Or, si un mathématicien a pu classer le principe dont il est question parmi les "formes de raisonnement", il n'a fait que suivre la tradition de la plupart des traités de Logique, dont les auteurs, hommes de lettres, se plaisant à montrer qu'ils n'ignorent pas ce principe, le placent parmi les méthodes générales de preuve et de recherche.

Toutefois, lorsqu'il disait que c'est "le raisonnement mathématique par excellence", l'attention de M. Poincaré devait se porter sur le mot "mathématique" justement opposé à "logique", et sur la phrase "par excellence", qui en affirmait "l'importance" et "le caractère particulier"; mais après, le seul mot qui avait été mal choisi, "raisonnement", l'emporta sur les autres et l'égara.

2. Si, lorsqu'on dit d'une assertion qu'elle enveloppe "une infinité de syllogismes", on veut nier le droit d'en affirmer la conclusion avant d'avoir examiné un nombre infini d'implications—le principe d'induction mathématique *n'enveloppe point une infinité de syllogismes*; car (ainsi que nous venons de le voir) c'est *une proposition dont la prémisse est l'affirmation simultanée de deux seules conditions* [1].

D'ailleurs, une proposition qui impliquerait une infinité de syllogismes ne saurait avoir *aucune application*; tandis que notre principe est très fréquemment employé par tout mathématicien.

Doit-on faire un crime à ce principe de son mérite, c'est-à-dire de servir à *démontrer* une proposition quelconque se rapportant à *tout* nombre, sans devoir la *vérifier* pour *chaque* nombre? ce serait comme si quelqu'un, après avoir franchi un obstacle, ne voulait pas le croire et se doutait même de la possibilité de le franchir.

Doit-on faire un tort à ce principe du fait qu'il se rapporte à un *ensemble infini,* tel que celui des *nombres finis?* mais alors on devrait faire ce reproche à la plupart des propositions scientifiques; car, par exemple, lorsqu'en Logique on parle d'*une classe quelconque,* ou qu'en Géométrie on parle d'*un triangle quelconque,* c'est bien de *toute classe* et de *tout triangle* qu'on entend parler, bien que l'ensemble des classes et que l'ensemble des triangles soient *infinis.*

Or—si un mathématicien, tout en connaissant par conviction et par expérience la bonté et la puissance de ce principe, a pu dire qu'il enveloppe "une infinité de syllogismes"—c'est qu'au lieu de se rapporter au principe même, il se rapportait à *l'argumentation vulgaire* qui le remplace dans la plupart des traités d'Arithmétique pour les écoles.

En effet, très souvent—au lieu d'admettre *explicitement* notre principe, c'est-à-dire de lui donner une place à part parmi les *propositions* de l'Arithmétique—la première fois qu'il se présenterait l'occasion de l'appliquer, après avoir établi les deux prémisses [1], on dit:

"donc, notre proposition étant vraie pour *zéro,* doit être vraie pour *un*; mais alors, étant vraie pour *un,* elle doit être vraie pour *deux*; mais alors, étant vraie pour *deux,* elle doit être vraie pour *trois*; et ainsi de suite sans fin".

Cette phrase "et ainsi de suite sans fin", rigoureuse au point de vue *substantiel*, ne l'est pas au point de vue *formel*; car elle n'énonce qu'une *induction simple*. C'est pourquoi l'esprit critique de M. Poincaré s'arrête et remarque que, pour la rendre légitime, il faudrait autant de constatations qu'il y a de nombres, à savoir une infinité.

Mais lorsque Maurolico, ou quelque autre savant, a énoncé explicitement le principe d'induction mathématique, c'est précisément qu'il avait compris que, cette chaîne d'implications étant infinie, si on devait chaque fois se croire autorisé à en affirmer la conclusion après s'être arrêté à un quelconque de ses anneaux, autant valait-il d'affirmer une fois pour toutes le droit de tirer la conclusion des deux prémisses [1], sans même commencer à parcourir cette chaîne infinie d'implications.

Par conséquent, je crois que la vérité qui s'était présentée à l'esprit de M. Poincaré est celle-ci: "l'argumentation vulgaire, moyennant laquelle on croit se passer du principe d'induction mathématique envisagé comme proposition à part, en confirme la nécessité"; mais il a énoncé cette vérité de manière à faire croire à tout le monde, jusqu'à lui-même, qu'il voulait mettre en doute la légitimité de ce principe.

3. Pour ce qui se rapporte à la dernière des trois opinions citées, c'est-à-dire l'irréductibilité de notre principe à celui de contradiction, à un certain point de vue on pourrait la considérer comme une vérité de La Palisse, si jamais ce brave capitaine, auquel on fait un reproche de la naïveté de ses admirateurs, avait étudié Logique.

En effet, je ne connais pas deux principes dont l'un soit *réductible* à l'autre, sans l'intervention de quelque *autre* principe. Mieux encore: récemment, dans ma *Logique déductive*, j'ai même analysé trois *formes* différentes du *principe de contradiction*, irréductibles entre elles sans l'intervention d'autres principes*.

Cela n'empêche pas que dans plusieurs traités de Logique on rencontre des *pseudo-réductions* de *tout principe logique* au principe de contradiction; pseudo-réductions dues aux pièges tendus par le langage ordinaire aux auteurs, qui ne se sont pas aperçus d'avoir recours aux principes *mêmes* qu'ils voulaient démontrer ou à d'*autres* encore, qu'ils n'avaient pas même analysés.

Je pense donc que M. Poincaré—ayant accepté sans examen la prétendue réductibilité de toute "forme de raisonnement" au principe de contradiction—a voulu dire tout simplement que *le principe d'induction mathématique n'est pas démontrable par les seuls principes de logique pure,* ce qui est vrai.

Or, comme cela arrive pour toute proposition qui appartient à une Science particulière, par ex. à la Géométrie, ce fait n'a rien d'étonnant pour nous; mais il devait paraître étonnant à M. Poincaré, qui avait classé le principe d'induction mathématique parmi les formes de raisonnement [1].

Et alors, il a énoncé la vérité dont je viens de parler de manière à faire croire à tout le monde, jusqu'à lui-même, que cette "irréductibilité au principe de contradiction" était une *tare* du principe d'induction mathématique.

* *La logique déductive dans sa dernière phase de développement* par A. Padoa avec une *Préface* de G. Peano—Gauthier-Villars, Paris, 1912. Voir les formules (II) à la p. 90, (II') à la p. 91, et P. 113 à la p. 94.

En résumant, tous les égarements à ce sujet prennent naissance du faux classement de notre principe parmi les *raisonnements*: faux classement qui existait avant M. Poincaré, que son langage n'a pas voulu renier, mais dont son esprit ne pouvait accepter les conséquences. De là sa manière caractéristique de traiter la question, par des phrases ambiguës et des argumentations incertaines entre le paradoxe et le sophisme.

§ 2. *Les rôles.*

4. Maintenant, je vais m'occuper d'une autre question. D'abord, je précise que pour moi une *science particulière* est envisagée comme un *ensemble non ordonné* (mais fini) de propositions, se rapportant à un vocabulaire particulier et dont chacune énonce un *fait*, c'est-à-dire une vérité complète; tandis qu'une *théorie déductive* est envisagée comme une *succession* (ensemble ordonné) de propositions, dont chacune y a un de ces rôles: *postulat* ou *définition* ou *théorème.*

On voit d'ici la possibilité de tirer d'une *même science* particulière *plusieurs théories* déductives, dans lesquelles une *même proposition* peut avoir des *rôles différents.* Cela dépend, jusqu'à un certain point, de la volonté de l'auteur; de manière que, pourvu qu'il soit rigoureux, on ne pourrait lui adresser aucune critique au sujet du *rôle* qu'il a voulu confier à chaque proposition dont il s'agit, sans entrer dans le domaine de la Psychologie ou de la Pédagogie, ce qui en ce moment n'est pas dans mon intention.

Par conséquent, si après avoir reconnu que le principe d'induction mathématique est une proposition arithmétique [1], on demandait *quel est son rôle dans l'Arithmétique,* envisagée comme *science particulière,* on ferait une demande dépourvue de signification. On peut seulement examiner quel est le *rôle* de cette proposition dans *chaque théorie déductive* dans laquelle on l'énonce *explicitement* comme proposition à part, prendre note des différentes *possibilités actuelles* et en chercher *d'autres.*

5. Dès 1889, dans ses *Arithmetices principia, nova methodo exposita*—ouvrage classique, qui donne le premier exemple d'une théorie déductive complète dans laquelle l'idéographie logique remplace tout à fait le langage ordinaire—M. G. Peano confia au principe d'induction mathématique [1] le rôle de *postulat,* avec quatre autres dont voici la traduction en langage courant:

(1) *zéro est un nombre,*

(2) *le successif de chaque nombre est un nombre déterminé,*

(3) *zéro n'est le successif d'aucun nombre,*

(4) *les successifs de nombres différents entre eux sont aussi différents entre eux.*

6. Comme toute définition doit avoir la forme d'une *égalité,* le principe d'induction mathématique [1] *ne peut prendre le rôle de définition*; mais il peut entrer *explicitement* dans une définition, de manière à le pouvoir considérer après comme *conséquence immédiate d'une définition.*

C'est-ce qui arrive dans *The principles of Mathematics* par M. B. Russell, parus en 1903, où l'on rencontre une *définition* qu'on peut résumer ainsi:

on appelle nombre (*inductif*) *la classe qui est contenue en toute classe vérifiant le principe d'induction mathématique.*

7. Enfin, nous allons voir que notre principe peut prendre aussi le *rôle* de *théorème*; mais cette dernière possibilité [4] se partage en *plusieurs possibilités*, selon *le différent choix* des principes moyennant lesquels on *démontre* ce théorème.

Naturellement, en vous parlant de ses *démonstrations* je laisserai de côté les très nombreux exemples de *paralogismes* dans lesquels le principe même entre subrepticement.

Souvenons-nous que l'auteur d'une théorie déductive [4] doit confier à quelques-unes des idées de la science dont il s'agit le *rôle* d'*idées primitives*. Celles-ci ne seront pas définies dans la théorie en question; mais, reliées entre elles moyennant les postulats, elles serviront après à définir toutes les autres idées propres à cette science. Par exemple, dans la théorie de M. Peano il y a *trois idées primitives*, à savoir: "zéro", "nombre" et "le successif de". Or, en 1900, c'est-à-dire trois ans avant que M. Russell donnât la dite définition de "nombre" moyennant "zéro" et "le successif de" [6], j'avais remarqué que le système des idées primitives de M. Peano [5] est *réductible* *.

Et j'avais donné cette *définition* de "zéro":

"zéro est le nombre qui n'est le successif d'aucun nombre".

En même temps que je réduisais ainsi de *trois* à *deux* les *idées primitives* de M. Peano, j'en réduisais de *cinq* à *quatre* les *postulats*; et précisément, tout en conservant ces deux [5]:

"le successif de chaque nombre est un nombre déterminé" et

"les successifs de nombres différents entre eux sont aussi différents entre eux",

je remplaçais les trois autres par les deux suivants:

"*il y a au moins un nombre qui n'est le successif d'aucun nombre*", et

"*si* une classe renferme un nombre qui n'est le successif d'aucun nombre et *si* cette classe renferme le successif de chaque individu qui lui appartient, *alors* cette classe renferme tous les nombres".

Cette proposition est à peu près le principe d'induction mathématique [1]; toutefois, ainsi qu'on peut voir dans ma *Théorie des nombres entiers absolus* †, avant de le pouvoir énoncer dans la forme exacte que nous avons vue au commencement, il faut *déduire* des postulats énoncés que:

"*il n'y a pas plus d'un nombre qui ne soit le successif d'aucun nombre*";

après quoi, ayant *admis l'existence* et *démontré l'unicité* du "nombre qui n'est le successif d'aucun nombre", on peut légitimement énoncer la dite *définition* de "zéro", qui permet de tirer le principe d'induction mathématique du quatrième de mes postulats.

8. Après quelques années, en 1908, M. Mario Pieri publia une note très remarquable *Sopra gli assiomi aritmetici* ‡, dans laquelle, en adoptant *mon système d'idées primitives* [7], il forme *un autre système de postulats*, qui se compose encore de *quatre* propositions.

* Voir mes leçons sur *L'Algebra e la Geometria quali teorie deduttive* (Roma).

† *Revue de Mathématique*, Turin, 1902 (t. VIII, p. 45—54).

‡ Boll° dell' Accademia Gioenia di Scienze naturali di Catania (Italie).

Une d'entre elles est commune aux systèmes à M. Peano, à moi et à M. Pieri; c'est:

"le successif de tout nombre est un nombre déterminé";

une autre est celle qui tout à l'heure avait pour moi le rôle de théorème:

"il n'y a plus qu'un nombre qui ne soit le successif d'aucun nombre";

les deux autres sont:

"*il y a au moins un nombre*", et

"*toute classe, qui renferme au moins un nombre, en renferme un qui n'est le successif d'aucun nombre renfermé par cette classe*".

Le mérite le plus grand de M. Pieri est celui d'avoir augmenté, pour ainsi dire, la *distance déductive* entre les postulats et le principe d'induction mathématique, envisagé comme *théorème*: distance qui était encore très petite dans mon système.

Sa manière de déduire le principe d'induction mathématique de son système de postulats me sembla en même temps si ingénieuse et si simple que je me suis forcé de l'adapter à un nouveau système, que j'ai construit il y a presque deux années et dont j'ai donné notice à la Société *Mathesis* à Gênes, mais que je n'ai pas encore publié*.

9. En remontant jusqu'à la frontière qui, à mon avis, sépare l'Arithmétique de la Logique, j'ai fixé comme *domaine caractéristique* de l'Arithmétique celui qui prend naissance d'*une seule idée primitive*, qui correspond au mot "couple"†.

En effet—ainsi qu'il résulte des études plus récentes dues à l'école de M. Peano d'accord en cela avec la tradition depuis Aristote—la Logique se réduit à la considération des relations d'*égalité*, d'*appartenance* et d'*inclusion* entre des *individus* et des *classes* quelconques, ou qui découlent de classes quelconques moyennant les opérations de *réunion*, d'*intersection* et de *négation*. Dans toutes ces relations et opérations les *individus* et les *classes* sont considérés tacitement comme *coexistants* (dans *l'espace* et dans *le temps*); et par suite tout concept d'*ordre* ou de *succession* est tacitement exclu du domaine de la Logique.

Or, tandis qu'il n'y a qu'*une classe* à laquelle appartiennent *deux individus* donnés, avec *eux* on peut former *deux couples*, selon qu'on place *avant* (dans *l'espace* ou dans *le temps*) l'un ou l'autre de deux individus donnés. Le "couple" est donc le premier exemple de *succession*, l'expression primordiale du concept d'*ordre*.

D'ailleurs—comme mon analyse prouve que le concept de "couple" est *suffisant* à développer, par de simples définitions, toute la terminologie actuellement employée dans l'Arithmétique—c'est lui *seul* qui forme tout mon *système d'idées primitives* [5].

10. Tandis que les *Arithmetices principia* par M. Peano, s'inspirant à l'ouvrage classique de M. Dedekind *Was sind und was sollen die Zahlen?*, envisagent les nombres au point de vue *ordinal*, ainsi que ma première théorie et celle de M. Pieri—*The principles of Mathematics*, d'abord par M. Russell et récemment par MM. Russell et

* *I principii dell' Aritmetica*, Séances du 15 et du 22 février 1911.

† A. Padoa, *D'où convient-il de commencer l'Arithmétique?* Communication au IVe Congrès de Philosophie, à Bologne, 1911; publiée par la *Revue de Métaphysique et de Morale*, Paris, juillet 1911.

Whitehead, s'inspirant au *Grundlagen der Arithmetik* par M. Frege et à la célèbre *Théorie des ensembles* de M. Georg Cantor, envisagent les nombres (cardinaux) au point de vue *cardinal*, ainsi que ma nouvelle théorie.

Mais, chose d'abord étonnante, MM. Russell et Whitehead affirment que *tous* les *termes arithmétiques* sont capables d'être *définis* au moyen de *termes logiques*; et, de même, toutes les *propositions arithmétiques* sont capables d'être *démontrées* au moyen de *propositions logiques* et des dites définitions. De manière que, selon ces auteurs, il n'y a plus *aucune idée primitive* et *aucun postulat* pour l'Arithmétique, et celle-ci n'est qu'un prolongement de la Logique.

Ce fait s'explique en remarquant que pour ces auteurs le domaine de la Logique est plus étendu qu'il n'est pour moi [9], de manière que les propriétés du "couple" se trouvent déjà dans leurs prolégomènes à l'Arithmétique.

Mais moi, voulant séparer la Logique de l'Arithmétique, je fonde celle-ci sur le "couple", envisagé comme *idée primitive*, et sur *un postulat* qui en établit la propriété essentielle.

11. En représentant par l'écriture $x\,;y$ le couple [9] dont x est *l'antécédent* et y est *le conséquent*, j'énonce comme *postulat* que :

$$\text{si } x\,;y = r\,;s, \text{ alors } x = r \text{ et } y = s\,;$$

d'où, en remplaçant r par y et s par x, et en simplifiant :

$$\text{si } x\,;y = y\,;x, \text{ alors } x = y\,;$$

c'est-à-dire que :

$$\text{si } x \text{ n'est pas égal à } y, \text{ alors } x\,;y \text{ n'est pas égal à } y\,;x,$$

ce qui distingue le couple $x\,;y$ de la classe renfermant x et y (qui est toujours égale à celle qui renferme y et x).

Après quoi je donne les *définitions* suivantes :

dire que $x\,;y$ est *disjoint* de $r\,;s$ signifie que x n'est pas égal à r et que y n'est pas égal à s;

j'appelle : "*correspondance biunivoque*" tout ensemble de couples tel que, si $x\,;y$ et $r\,;s$ lui appartiennent, alors $x;y$ est disjoint de $a;b$;

"correspondance biunivoque entre les a et les b" toute correspondance biunivoque telle que a et b sont respectivement l'ensemble des antécédents et des conséquents des couples qui lui appartiennent*;

"*autocorrespondance* biunivoque entre les a" toute correspondance biunivoque entre les a et les a;

"autocorrespondance biunivoque *irréductible* entre les a" toute autocorrespondance biunivoque entre les a qui ne renferme aucune autocorrespondance biunivoque entre une *partie* des a (cette partie étant différente de la totalité des a et du *rien*)†;

* Ce qui n'exclut pas que *chacune* des classes a et b ne renferme *aucun* objet.

† Il s'ensuit que cette correspondance n'est pas l'*identité*, sauf si la classe considérée est un *élément*, c'est-à-dire une classe qui renferme *un seul* objet.

"autocorrespondance biunivoque *ouverte*" toute autocorrespondance biunivoque dont on peut tirer un ensemble *partiel* de couples, tel que, s'il renferme x ; y, il doit aussi renfermer y ; z*;

"classe *numérotable*" toute classe a telle qu'il existe (au moins) une *autocorrespondance biunivoque irréductible et non ouverte* entre les a.

D'après M. Frege et ma théorie de *l'abstraction mathématique*†, je *définis*:

"*le nombre* des objets d'une classe numérotable" comme "l'abstraction de cette classe par rapport à la possibilité d'une correspondance biunivoque";
d'où l'on *déduit* que:

si a et b sont des classes numérotables, "le nombre des a est le même que le nombre des b" seulement si "l'on peut établir une correspondance biunivoque entre a et b."

Après quoi, j'énonce les *définitions*:

"nombre" (comme abréviation de "nombre entier absolu") est toute valeur prise par "le *nombre* des x" lorsque x est une classe *numérotable*;

si x et y sont des nombres, "x est *plus petit* que y" signifie que "si a et b sont des classes numérotables dont les nombres sont respectivement x et y, alors on peut établir une correspondance biunivoque entre x et une partie de y (différente de y)";
d'où il *résulte* (après avoir démontré que aucune classe numérotable n'est infinie dans l'acception de M. G. Cantor) que

jamais chacun de deux nombres est plus petit que l'autre ....(1).

Après quoi je *définis* "le *minimum* d'un groupe de nombres" comme "le nombre (s'il existe) qui *appartient* à ce groupe et qui est *plus petit* que tout autre nombre du même groupe" ....(2);
et j'énonce comme second *postulat* que:

tout groupe de nombres (qui renferme quelque nombre) *a son minimum*...(3).

Ensuite, je *définis* le "zéro" comme "le nombre des objets d'un groupe qui ne renferme *aucun* objet," et je *démontre* que:

"si a est un nombre quelconque, le groupe, qui renferme a et tout nombre plus petit que a, est une classe numérotable";
en *nommant* ce nombre "le *successif* de a", il *résulte* que

"le successif de chaque nombre est un nombre déterminé".

* Même une autocorrespondance biunivoque *irréductible* peut être *ouverte*. En effet, si a est l'ensemble des nombres entiers relatifs (signés) et si u est l'ensemble des couples r ; $r+1$, où r est un a, c'est-à-dire l'ensemble des couples

... -3 ; -2 -2 ; -1 -1 ; 0 0 ; 1 1 ; 2 2 ; 3 ...,

u est autocorrespondance biunivoque *irréductible*. Or, par exemple, l'ensemble v des couples s ; $s+1$, où s est un a qui n'est pas plus petit que 5, est une *partie* de u (car v ne renferme pas, par ex., 2 ; 3); et s'il renferme x ; y, il renferme aussi y ; z. Donc u est *ouverte*.

D'ailleurs, l'ensemble des couples

0 ; 1   1 ; 2   2 ; 3   3 ; 4   4 ; 5   5 ; 6   6 ; 7   7 ; 8   8 ; 9   9 ; 0

nous donne un exemple de autocorrespondance biunivoque *irréductible* et *non ouverte*; ce qui justifie la définition suivante.

† Voir ma communication à ce sujet au Congrès national de Philosophie, à Parme (Italie), 1907.

Après avoir *défini* "le *précedent* d'un nombre quelconque" comme "le nombre (s'il existe) dont le *successif* est derechef le nombre *donné*" ....(4),
je *démontre* que :

le précédent de chaque nombre, différent de zéro, est un nombre déterminé, qui est plus petit que le nombre donné ....(5).

D'ailleurs, moyennant mes postulats et les dites définitions je *démontre* toute propriété connue des concepts arithmétiques que je viens de définir.

12. Enfin (pour ce qui a rapport à ce discours), on peut énoncer *le principe d'induction mathématique* [1] et en donner cette *démonstration par absurde* [8]:

Soit A un groupe de nombres, tel que :

(1) A renferme "zéro",

(2) A renferme le *successif* de chaque nombre qu'il renferme.

Il s'agit de démontrer que A renferme *tous les nombres*; en d'autres termes, si l'on désigne par *B l'ensemble des nombres n'appartenant pas* à A (s'il y en avait) il s'agit de démontrer que B ne renferme *aucun nombre.*

En effet, autrement, B aurait son *minimum* (3), qui appartiendrait à B (4) et serait *différent de zéro*, puisque (par la *première prémisse*) "zéro" appartient à A. Par conséquent, le *précédent* de ce minimum serait un *nombre déterminé* (5), qui, étant *plus petit* que ce minimum (5), ne pourrait pas appartenir à B (1) (2). Donc, "le *précédent* du *minimum* de B" devrait appartenir à A.

Mais alors (par la *seconde prémisse*), le *successif* de ce nombre, c'est-à-dire le *minimum* de B (4), appartiendrait aussi bien à A que à B (2), ce qui est *absurde.*

Comme il est *absurde* que B ait son *minimum*, il résulte que B ne renferme *aucun nombre* (3), c'est-à-dire que A renferme *tous les nombres.* *c.q.f.d.*

13. Pour finir, permettez-moi une petite remarque psychologique.

L'*argumentation vulgaire* par laquelle on remplace le principe d'induction mathématique [2] est très accessible; mais le principe lui-même ne l'est pas autant. Il vaut donc mieux le *différer* jusqu'à ce qu'on doive l'employer *pour démontrer d'autres propositions.*

Or cela—qui n'était pas possible en lui confiant le *rôle* de *postulat,* ainsi que fait M. Peano [5], ou *en le renfermant dans la définition de nombre,* ainsi que font MM. Russell et Whitehead [6]—est possible seulement en lui confiant le *rôle* de *théorème,* ainsi que l'on vient de voir.

SUR LES LOIS GÉNÉRALES DE L'ALGORITHME DES SYMBOLES DE FONCTION ET D'OPÉRATION

PAR C. BURALI-FORTI.

Les symboles de fonction et d'opération dont on fait usage dans *l'Algèbre générale*, sont sujets à des lois fixes et logiquement coordonnées entre elles. A ce fait on doit la puissance et la diffusion de l'algorithme algébrique.

La notation *complète** des *Quaternions*, telle qu'elle est donnée par Hamilton, est aussi admirable pour la précision des conceptions et de la forme, de sorte que les Quaternions de Hamilton forment, dans leur petit domaine†, un système parfait.

Les notations vectorielles des autres auteurs sont sujettes à des lois différentes de celles de l'Algèbre ordinaire et de celles qui ont été établies par Hamilton, et elles sont en outre contradictoires entre elles. Voilà la cause de la diffusion limitée du calcul vectoriel et de son inapplicabilité à plusieurs questions de physique, de mécanique et de géométrie.

Le sujet de cette communication a, donc, une réelle importance pratique et théorique. M. Marcolongo et moi‡, suivant les lois que je vais exposer dans cette communication, nous avons pu obtenir un calcul vectoriel *intrinsèque*§ qui est supérieur pour la *puissance*, la *simplicité*, et la *précision de conception et de forme*, à tous les calculs vectoriels ordinaires. Nous pouvons appliquer notre calcul n'importe à quelle question physico-mécanico-géométrique‖.

Mon exposition peut être aisément traduite avec les symboles de logique mathématique de M. Peano ou de M. Russell. Pour l'étude de la question j'ai fait usage de symboles logiques, mais je vais l'exposer en langage commun, pour en rendre aisée la lecture à qui que ce soit. Les exemples sont relatifs au calcul vectoriel; mais on pourrait facilement en choisir dans d'autres domaines de la mathématique.

* C'est-à-dire avec les opérateurs $\mathbf{I}$, $\mathbf{I}^{-1}$.

† C. Burali-Forti et R. Marcolongo, *Eléments de calcul vectoriel* (A. Hermann, Paris, 1910, Edit. ital. Zanichelli, Bologna, 1909); *Omografie vettoriali* (G. B. Petrini, Torino, 1909); *Analyse vectorielle générale* I. *Transformations linéaires* (Mattei, Pavia, 1912).

‡ Voir note (†).

§ C.-à-d. indépendant des coordonnées de quelque espèce que ce soit.

‖ Voir note (†), *Analyse vectorielle générale*.

Ma communication a un double but: d'abord de faire connaître que, s'il existait une question relative aux notations vectorielles, cette question, aujourd'hui, est déjà résolue; en second lieu de démontrer d'une manière évidente que la logique mathématique symbolique n'est point, comme erronément quelques-uns le pensent, un tachygraphe des idées, mais bien un puissant instrument d'analyse des idées mêmes.

1. *Opérateurs.* Les symboles de fonction monodrome

$$\sin, \cos, \ldots, \log, \ldots$$

dont on fait usage dans l'Analyse, les symboles

$$\mathbf{I}, \mathbf{V}, \mathbf{D}, \mathbf{K}, \mathbf{C}, \mathbf{R}, \frac{d}{dP}, \text{div}, \text{rot}, \text{grad}, \text{Rot}, \Delta, \Delta',$$

qui ont une très grande importance dans le *Calcul Vectoriel,* peuvent être appelés *opérateurs* (à *gauche,* sous-entendu) et sont tous soumis à des *propriétés formelles* qui sont suffisantes pour donner la *définition formelle* de la classe *opérateur,* sous la forme logique suivante.

Si u, v sont des classes quelconques nous écrivons

$$(* u)\, v$$

au lieu de "opérateur (à gauche) entre les u et les v,"

et nous disons que "f est un élément de la classe $(* u)\, v$"

lorsque:

1° f est un symbole, simple ou composé, qui placé à gauche d'un élément arbitraire x de la classe u, est tel que la notation

$$fx$$

indique un élément, et un seul, de la classe v, qu'on peut déduire de x suivant une loi déterminée;

2° x, x' étant des éléments de la classe u, la condition $x = x'$ entraîne la condition $fx = fx'$.

2. De la définition formelle d'opérateur que nous venons de donner, il résulte qu'avec les symboles ordinaires d'opération (n. 9) on peut former des opérateurs. Voici des exemples:

Si a est un nombre (en général complexe d'ordre n),

$$a+,\ a-,\ a\times \textit{ sont des } (* \text{nombre}) \text{ nombre},$$

$$a/ \textit{ est un } (* \text{nombre non nul}) \text{ nombre},$$

Si $\mathbf{u}$ est un vecteur,

$$\mathbf{u}+,\ \mathbf{u}-,\ \mathbf{u}\wedge,\ \textit{sont des } (* \text{vecteur}) \text{ vecteur},$$

$$\mathbf{u}\times \textit{ est un } (* \text{vecteur}) \text{ nombre réel},$$

$$\mathbf{u}+,\ \mathbf{u}- \textit{ sont des } (* \text{point}) \text{ point}.$$

3. Pour les notations générales nous faisons les remarques suivantes:

Comme Lagrange, Abel, Hamilton, ...nous écrivons simplement fx et non $f(x)$; les parenthèses étant inutiles, sauf dans le cas où x est exprimé par deux ou plusieurs symboles. Si f est formé par deux ou plusieurs symboles, on peut écrire $(f)\,x$, les parenthèses ayant aussi en ce cas le rôle ordinaire de grouper.

Les éléments de la classe u peuvent être des éléments simples $(x, x', \ldots)$, ou des couples $((x, y), (x', y'), \ldots)$, ou des ternes $((x, y, z), (x', y', z'), \ldots)$, etc. Nous écrivons toujours

$$fx,\ f(x, y),\ f(x, y, z),\ \ldots$$

et f peut être appelé respectivement, opérateur *simple, binaire, ternaire, ...*et correspond à l'ordinaire *symbole des fonctions d'une, deux, trois, ...variables.*

Lorsqu'il n'est point nécessaire d'indiquer la classe v nous disons que "f est un opérateur pour les u" et la classe "opérateur pour les u" peut être indiquée par la notation $(* u)$.

Si f est un $(*u)\,v$, alors (n. 1), x étant un élément arbitraire de la classe u, il existe un et un seul élément y de la classe v tel que $fx = y$. Si, y étant un élément arbitraire de la classe v, l'élément x de la classe v tel que $fx = y$ existe, et s'il est unique, alors l'opérateur f est appelé *opérateur réversible.* Si f est un $(*u)\,v$ réversible et si $fx = y$, alors l'$(*v)\,u$ qui appliqué (à gauche) à y donne x, sera indiqué (Hamilton) par la notation f^{-1}. Il s'ensuit que les deux conditions

$$fx = y,\quad x = f^{-1}y$$

expriment, sous des formes différentes, une même propriété, et la transformation de l'une dans l'autre est soumise aux règles algébriques ordinaires des puissances.

4. *Domaines d'un Opérateur.* Un symbole f peut être opérateur pour plusieurs classes $u, u', u'', \ldots$. Ces classes sont appelées *domaines de l'opérateur f.*

Si la classe u est un des domaines de l'opérateur f, alors toute classe u', contenue dans la classe u, est encore un des domaines de l'opérateur f (n. 1). C'est-à-dire que, l'opérateur f étant défini dans le domaine u, nous pouvons rétrécir son domaine à une classe u' quelconque contenue dans la classe u.

5. Mais nous pouvons encore étendre le domaine de l'opérateur f de la manière logique suivante :

Soit u' une classe qui contient la classe u et telle qu'il existe des u' qui ne sont pas des u ; et soit v' une classe qui contient la classe v, l'identité entre les classes v, v' n'étant pas exclue. Cela posé, et y étant un élément de la classe u' qui n'appartient pas à la classe u, nous définissons (n. 1) fy comme un élément bien déterminé de la classe v'.

Il s'ensuit alors que la classe u' (plus étendue que la classe u) est un des domaines de l'opérateur f. L'élément fx, où x est un u', est un des anciens éléments de la classe v, si x est un u ; c'est un des nouveaux éléments de la classe v', si x est un u' qui n'est pas un u.

Le procédé d'extension que nous venons d'indiquer a été souvent appliqué en Mathématique. Voici des exemples :

Une fois définis les opérateurs sin, cos dans les domaines des *nombres réels,* on étend le domaine aux *nombres complexes,* et sin, cos sont encore des opérateurs sujets aux lois que nous venons d'établir au n. 1. Pour l'opérateur log on suit en Analyse un procédé bien différent; log est, dans le domaine des nombres complexes, un opérateur *polydrome*, tandis que dans le domaine des nombres réels il est un opérateur *monodrome*.

L'opérateur différentiel vectoriel grad a pour domaine *ordinaire* la classe *nombre réel*. M. Marcolongo et moi, nous avons étendu le domaine de l'opérateur grad à la classe *homographie vectorielle*, qui contient la classe *nombre réel*.

De même pour l'opérateur Δ dont le domaine ordinaire est la classe *nombre réel*.

6. Parmi les domaines de l'opérateur f existe-t-il *un domaine maximum*?

Le procédé logique que nous venons de donner au n. 5 pour étendre à la classe u' le domaine de l'opérateur f, ne limite pas l'extension de la classe u', et la classe u' pourrait bien être la *classe totale*. Mais alors un seul symbole serait suffisant pour indiquer plusieurs opérateurs avec des domaines différents entre eux.

Donc l'opérateur f doit nécessairement avoir un *domaine pratique maximum* qui évidemment dépend de l'état de nos connaissances, état effectif ou supposé. Nous allons fixer bientôt (n. 8) un criterium général pour établir le domaine pratique maximum d'un opérateur. Avant tout nous devons examiner comment les propriétés d'un opérateur dépendent du domaine dans lequel l'opérateur même est appliqué, c.-à-d. dépendent du *domaine maximum relatif* que nous imposons à l'opérateur.

7. Soit f un $(*u)v$ et u' une classe contenue dans la classe u. Nous indiquons par la notation

$$f \mid u' \quad \ldots\ldots\ldots\ldots\ldots\ldots\ldots\ldots (1)$$

l'opérateur ϕ tel que:

1° ϕ est un $(*u')v$, tel que si x est un élément arbitraire de la classe u', on ait toujours $\phi x = fx$;

2° la classe u' est le domaine maximum relatif que nous imposons à l'opérateur ϕ.

Les propriétés de l'opérateur (1) dépendent de l'opérateur f et de la classe u'. Voici des exemples:

Exemple 1er. Des opérateurs

$$\sin \mid -\frac{\pi}{2} \mapsto \frac{\pi}{2} \qquad \sin \mid 0 \mapsto \pi,$$

le premier est réversible, le second ne l'est pas.

Exemple 2ème. L'opérateur vectoriel différentiel du second ordre

$$\text{rot grad}$$

est toujours nul si son domaine maximum est formé par les *nombres réels* fonctions d'un point. Au contraire il n'est pas nécessairement nul,

$$\text{rot grad} = \text{grad Rot},$$

si son domaine maximum est formé par les homographies vectorielles fonctions d'un point. C.-à-d.

$$\text{grad rot} \mid \text{nombre réel} = 0$$

$$\text{grad rot} \mid \text{homographie vectorielle} \neq 0, \text{ en général.}$$

Exemple 3ème. Si s est un nombre réel et $\mathbf{u}$ est un vecteur, posons

$$\alpha = s + \mathbf{u} \wedge .$$

L'opérateur α ¦ vecteur normale au vecteur $\mathbf{u}$

est un *quaternion* de Hamilton, et précisément le quaternion β tel que

$$\mathbf{S}\beta = s, \quad \mathbf{V}\beta = \mathbf{u}.$$

Au contraire l'opérateur α ¦ vecteur

n'est pas un quaternion mais une homographie dont la *dilatation* est le nombre s et le *vecteur* est $\mathbf{u}$; c'est-à-dire l'homographie γ telle que

$$\mathbf{D}\gamma = s, \quad \mathbf{V}\gamma = \mathbf{u}.$$

On en tire aisément que l'homographie γ n'est pas un quaternion en observant que

$$\gamma^2 = \{(s^2 - \mathbf{u}^2) + (2s\mathbf{u}) \wedge\} + \mathbf{H}(\mathbf{u}, \mathbf{u})$$

n'est pas un quaternion dans le domaine maximum "vecteur," tandis que le carré d'un quaternion est toujours un quaternion.

Exemple 4ème. Pour les opérateurs Δ, Δ' que M. Marcolongo et moi nous avons introduits sous cette forme, on a la correspondance suivante avec la notation complète de Hamilton :

$$\Delta \text{ ¦ nombre réel} = -\nabla^2 \text{ ¦ nombre réel},$$

$$\Delta' \text{ ¦ vecteur} = -(\mathbf{I}\nabla^2\mathbf{I}^{-1}) \text{ ¦ quaternion droit}$$

(ce qui prouve que Δ et Δ' sont des opérateurs bien distincts !). L'opérateur

$$\Delta \text{ ¦ homographie}$$

n'a pas de correspondant dans les notations quaternionnelles de Hamilton.

Les exemples que nous venons d'exposer prouvent que, pour établir quelles sont les propriétés d'un opérateur en général, il est nécessaire de fixer le domaine dans lequel nous voulons appliquer l'opérateur même.

Il n'est point nécessaire de faire un usage systématique de la notation (1); il suffit d'en connaître l'existence et de l'employer dans les cas douteux.

8. *Un même symbole ne doit jamais être employé pour indiquer deux ou plusieurs opérateurs qui n'ont pas en commun l'algorithme formel.* Voilà un criterium général pour la formation du domaine maximum d'un opérateur. Voici des exemples.

Exemple 1er.—Soit π un plan protectif et f l'opérateur réversible de la classe

(∗ point de π) droite de π

qu'on appelle habituellement *polarité.*

Dans la géométrie projective on indique par le même symbole f la *polarité inverse,* qui est un élément de la classe

(∗ droite de π) point de π.

Le domaine (maximum) de l'opérateur f est donc donné par la classe

"point de π" *vel* "droite de π" ....(1), et dans ce domaine l'opérateur f est réellement donné par deux opérateurs bien distincts dont l'un est l'inverse de l'autre, car

$$f \mid \text{droite de } \pi = (f \mid \text{point de } \pi)^{-1}.$$

Toutefois, en vertu du *principe de dualité*, l'algorithme formel de l'opérateur

$$f \mid \text{point de } \pi$$

est identique à l'algorithme formel de l'opérateur inverse, et il n'y a pas d'inconvénients à donner à l'opérateur f le domaine maximum (1).

Ce que nous venons de dire est applicable à la *polarité* dans une étoile (de droites et de plans) et à l'opérateur *index*, de Grassmann.

Exemple 2ème.—Soient π, π' des plans (classes de points), O un point qui n'appartient pas à la classe

$$\pi \text{ vel } \pi' \quad \ldots\ldots\ldots\ldots\ldots\ldots\ldots\ldots\ldots(1),$$

et f *l'*opérateur $(\ast\,\pi)\,\pi'$ qui, appliqué à un point x de π, donne le point projection de x fait du point O sur le plan π' (prospectivité de centre O).

Dans la Géométrie projective on indique par le même symbole f l'opérateur f^{-1}; c.-à-d. que f réprésente un opérateur dans le domaine (1) tel que

$$f \mid \pi' = (f \mid \pi)^{-1}.$$

Dans ce cas, étendre le domaine de l'opérateur f à la classe π' représente une acception primitive et incomplète du concept de fonction, car par ex. les opérateurs $f \mid \pi, f \mid \pi'$ ont, chacun, *une droite limite*, tandis que $f \mid (\pi \text{ vel } \pi')$ a *deux droites limites* qui ont des propriétés géométriques bien différentes.

Exemple 3ème.—Si dans l'exemple précédent nous considérons le cas limite $\pi = \pi'$, $f \mid (\pi \text{ vel } \pi')$, identique à $f \mid \pi$, est une homologie et on a

$$f \mid \pi = (f \mid \pi)^{-1}:$$

ce qui est absurde, car f n'est pas, en général, une *homologie harmonique*.

Dans ce cas l'acception primitive et incomplète du concept de fonction conduit à un absurde, ce qui n'empêche pas que cet absurde est habituel dans la Géométrie projective.

Exemple 4ème.—On indique, d'ordinaire, avec un même symbole (Δ_2, Δ^2, ...) l'opérateur qui a pour tachygraphe cartésien l'*opérateur de Laplace*.

M. Marcolongo et moi nous avons décomposé cet opérateur en deux opérateurs distincts Δ et Δ' tels que

Δ *est un* ($\ast$ homographie) homographie

Δ' „ ($\ast$ vecteur) vecteur,

car les propriétés formelles de Δ et Δ' sont bien différentes, comme il résulte des formules suivantes

$$(\Delta\alpha)\,\mathbf{x} = \text{grad}\left\{\frac{d(\alpha\mathbf{x})}{dP} - 2\alpha\frac{d\mathbf{x}}{dP}\right\} + \alpha \text{ grad}\frac{d\mathbf{x}}{dP},$$

$$\Delta'\,\mathbf{u} = \text{grad}\frac{d\mathbf{u}}{dP} = \text{grad div } \mathbf{u} - \text{rot}^2\,\mathbf{u},$$

$$\text{rot}\,\Delta' = \Delta'\,\text{rot} = -\,\text{rot}^3$$

$$\text{Rot}\,\Delta = \Delta\,\text{Rot} = -\,\text{Rot}^3$$

$$\text{grad}\,\Delta = \Delta'\,\text{grad} = \text{grad}\,\frac{d}{dP}\,\text{grad}$$

$$\Delta'^2 = \text{grad}\,\Delta\,\text{div} - \text{rot}\,\Delta'\,\text{rot}$$

Δ^2 n'a pas de formule correspondante.

L'opérateur unique (Δ_2, Δ^2, ...) correspond aussi dans ce cas à une acception primitive et incomplète du concept de fonction.

Nous croyons que les exemples que nous venons de donner sont suffisants et qu'il n'est pas indispensable d'indiquer les aberrations formelles qui ont conduit des auteurs à indiquer par les notations

$$\sin(\mathbf{uv}), \quad \cos(\mathbf{uv}), \quad \frac{\mathbf{v}}{\mathbf{u}}$$

le produit *vectoriel* et *scalaire* du vecteur **u** par le vecteur **v**, la dyade **H** (**u**, **v**), ou à indiquer par

$$(\mathbf{u}\,\text{grad})\,\mathbf{v}$$

le vecteur

$$\frac{d\mathbf{v}}{dP}\,\mathbf{u}$$

qui n'a pas de relation avec le gradient.

9. *Symboles d'operation.* Si u, v, w sont des classes quelconques par la notation

$$(u * v)\,w \ldots\ldots\ldots\ldots\ldots\ldots\ldots\ldots(1)$$

nous indiquons la classe "symbole d'opération entre les u et les v qui produit un w."

La *définition formelle* de la classe (1) est la suivante; La classe $(u * v)\,w$ est formée par tout symbole $\circ$ tel que:

1°. Quel que soit l'élément x de la classe u et l'élément y de la classe v, le symbole $\circ$ placé entre x et y donne un élément bien déterminé de la classe w, qui reste indiqué par la notation

$$x \circ y$$

et qui dépend de x et y selon une loi déterminée.

2°. Si x, x' sont des éléments de la classe u et y, y' des éléments de la classe v les conditions

$$x = x', \quad y = y'$$

entraînent la condition

$$x \circ y = x' \circ y'.$$

Par exemple:

$+$, $-$, $\times$ *sont des* (nombre $*$ nombre) nombre,

$/$ *est un* (nombre $*$ nombre non nul) nombre,

$+$, $-$, $\wedge$ *sont des* (vecteur $*$ vecteur) vecteur,

$\times$ *est un* (vecteur $*$ vecteur) nombre réel.

Lorsqu'il n'est point nécessaire d'indiquer explicitement la classe w nous disons que $\circ$ est un "symbole d'opération entre les u et les v": la classe "symbole d'opération entre les u et les v" peut être indiquée par la notation $(u * v)$.

10. Pour abréger l'écriture, nous écrivons, avec M. Peano,

$$u \,\vdots\, v$$

au lieu de "classe des couples (x, y), x et y variant, respectivement, dans les classes u, v."

Si f est un opérateur (binaire) de la classe

$$\{*(u \,\vdots\, v)\}\, w \ldots\ldots\ldots\ldots\ldots\ldots\ldots\ldots\ldots\ldots(1)$$

alors il existe un élément $\circ$ de la classe

$$(u * v)\, w \ldots\ldots\ldots\ldots\ldots\ldots\ldots\ldots\ldots\ldots(2)$$

et un seul, tel que $x \circ y = f(x, y)$.

Réciproquement, si $\circ$ est un élément de la classe (2) il existe un, et un seul, élément f de la classe (1) tel que

$$f(x, y) = x \circ y.$$

On peut donc établir une correspondance réversible entre les éléments des deux classes

$$\{*(u \,\vdots\, v)\}\, w, \quad (u * v)\, w:$$

c.-à-d. *que dans un algorithme quelconque on peut, logiquement, substituer à un symbole d'opération, un symbole de fonction binaire, et réciproquement.*

11. De la remarque que nous venons de faire découle la *question pratique* suivante:

Doit-on, dans un algorithme, donner la préférence aux **symboles d'opération,** *ou aux* **opérateurs binaires** *?*

Il n'est pas possible de donner une réponse générale. Il faut examiner les cas particuliers qui se présentent pratiquement.

(*a*) Au lieu des symboles d'opération $+$, $\times$ dont on fait usage en Algèbre, mettons les opérateurs binaires σ, π, c.-à-d. écrivons, x, y étant des nombres,

$$\sigma(x, y) = x + y, \quad \pi(x, y) = x \times y.$$

Les formules ordinaires et très simples de l'Algèbre, par ex.

$$(x + y) + z = x + (y + z), \quad (x \times y) \times z = x(y \times z),$$
$$(x + y) \times z = x \times z + y \times z$$

prennent la forme suivante:

$$\sigma\{\sigma(x, y), z\} = \sigma\{x, \sigma(y, z)\}, \quad \pi\{\pi(x, y), z\} = \pi\{x, \pi(y, z)\},$$
$$\pi\{\sigma(x, y), z\} = \sigma\{\pi(x, z), \pi(y, z)\},$$

qui est presque illisible.

On peut conclure: *Si des symboles d'opération ont un algorithme identique ou semblable à celui des symboles* $+$, $-$, $\times$, $/$, *de l'Algèbre, la substitution des opérateurs binaires aux symboles d'opération donne lieu à des formules presque illisibles; cette substitution n'est donc pas à faire.*

Pour cette raison aussi, M. Marcolongo et moi nous avons refusé les notations **S (uv)**, **V (uv)** pour le produit intérieur et vectoriel, lorsqu'on ne fait pas usage des quaternions. Il faut observer que, dans la théorie des quaternions, **S** et **V** sont des opérateurs mais non des opérateurs binaires ; d'où l'absurdité de la notation usuelle.

De même pour les notations à *système cellulaire*

$$(\mathbf{uv}), \quad [\mathbf{uv}],$$

avec la circonstance agravante que le symbole de fonction binaire est formé par deux parenthèses, contrairement à toutes les règles très répandues de l'algorithme algébrique, qui donne aux parenthèses le seul rôle de grouper.

(*b*) Soit $\circ$ un des symboles d'opération $(u * v)\,w$ et soit a un élément arbitraire de la classe u.

Par la définition formelle (n. 1) d'opérateur il résulte que le symbole composé

$$a\,\circ$$

est un élément de la classe $(* v)\,w$. Le symbole $\circ$ donne donc naissance a une infinité d'opérateurs entre les v et les w ; un élément arbitraire a de la classe u détermine un de ces opérateurs.

Au contraire, si f est un opérateur de la classe $\{*(u \,\vdots\, v)\}\,w$ et

$$f(a, y) = a \circ y,$$

alors l'opérateur binaire f ne donne pas naissance à un opérateur explicite comme $a\,\circ$.

Par ex. M. Marcolongo et moi nous avons obtenu par les symboles d'opération $\wedge$, $\times$ les opérateurs

$$\mathbf{u}\wedge, \quad \mathbf{u}\times,$$

le premier desquels donne la forme générale de l'*homographie axiale* qui est d'un usage continu dans les applications physico-mécaniques. Voila une autre des puissantes raisons pour lesquelles nous avons refusé les notations **S**(**uv**), **V**(**uv**), (**uv**), [**uv**].

On peut conclure : *Si le symbole d'opération* $\circ$ *donne naissance à des opérateurs* $a\circ$ *qui ont une grande importance dans un algorithme, on doit donner la préférence au symbole d'opération plutôt qu'au symbole de fonction binaire.*

(*c*) Au contraire : *si des symboles d'operation ont un algorithme très différent de celui des symboles* $+$, $-$, $\times$, $/$ *de l'Algèbre, et s'ils ne donnent pas naissance à des opérateurs importants, on aura, en général, avantage à remplacer de tels symboles par des symboles de fonction binaire.*

Par ex. M. Marcolongo et moi nous avons indiqué par la notation

$$\mathbf{H}(\mathbf{u}, \mathbf{v}),$$

avec le symbole de fonction binaire **H**, la *dyade* que Gibbs indique par la notation

$$\mathbf{vu}$$

produit-dyadique de u par v, le symbole d'opération étant sous-entendu. Il est facile de reconnaître que notre notation est plus avantageuse dans le calcul vectoriel que celle de Gibbs.

12. En Analyse on sous-entend, presque toujours, le symbole $\times$ entre deux nombres x, y et on écrit simplement xy au lieu de $x \times y$.

En général: *est-il possible logiquement, est-il avantageux pratiquement, de sous-entendre le symbole générique* $\circ$ *d'opération et d'écrire simplement* xy *au lieu de* $x \circ y$?

Soit $\circ$ un symbole de la classe

$$(u * v)\, w$$

et convenons d'écrire toujours xy au lieu de $x \circ y$. De la définition formelle d'opérateur (n. 1) il résulte alors que x, élément arbitraire de la classe u, est un $(* v)\, w$, c.-à-d. que la classe u est contenue dans la classe $(* v)\, w$.

Donc: *La définition formelle d'opérateur* (n. 1) *une fois admise, et* u *n'étant pas contenue dans la classe* $(* v)\, w$, *il n'est pas logiquement possible d'indiquer, tout court, avec la notation* xy *l'élément* $x \circ y$ *de la classe* w.

La condition "u n'est pas contenue dans la classe $(* v)\, w$" est nécessaire. Soit par exemple f un $(* v)\, w$ et pour le "produit fonctionnel, fx, de f par x," introduisons le symbole $\circ$ tel que

$$f \circ x = fx.$$

Il s'ensuit que

$$f \circ = f;$$

et comme $f \circ$ et f sont des éléments de la classe $(* v)\, w$, l'égalite $f \circ = f$ n'est pas absurde; bien plus, c'est une identité. En d'autres termes le symbole d'opération pour le produit fonctionnel de f par x est inutile.

Soit à présent u une classe qui n'est pas contenue dans la classe $(* v)\, w$. Le symbole $\circ$ une fois donné nous pouvons **convenir,** pour abréger l'écriture, de sous-entendre le symbole $\circ$; bien entendu **sous-entendre** et **non** *supprimer* d'une manière absolue. Le symbole $\circ$ doit être toujours à notre disposition pour l'écrire dans tous les cas douteux ou toutes les fois qu'il se présentera comme formellement nécessaire ainsi que dans les exemples du n. 2.

Il est évident qu'une fois que l'on aura fait la convention de sous-entendre un symbole déterminé de la classe

$$(u * v)\, w$$

il ne sera plus possible de sous-entendre aucun autre symbole de la même classe; ni même de sous-entendre un des symboles quelconque de la classe $(u * v)\, w'$.

La possibilité logique, au moins formelle, étant admise, l'utilité pratique est évidente pour un symbole $\circ$ qui paraît continuellement dans les calculs.

En Algèbre les symboles $+$, $\times$ sont presque aussi souvent employés l'un que l'autre. A quoi attribuer la suppression de $\times$ et non de $+$? Probablement au fait que dans des expressions contenant $+$ et $\times$ le symbole $+$ est un séparateur et $\times$ un agrégatif; par exemple

$$x + y \times z = x + (y \times z) \neq (x + y) \times z$$

$$x \times y + z \times u = (x \times y) + (z \times u) \neq x \times (y + z) \times u.$$

Dans le calcul géométrique de Grassmann on peut sous-entendre le symbole du *produit alterné* qui est d'un usage continu. L'usage des parenthèses dans la notation allemande est absolument inutile, tandis que quelquefois il est nécessaire d'avoir à sa disposition le symbole du produit alterné (symbole d'opération).

De même suivant Hamilton on peut sous-entendre le symbole du produit fonctionnel entre deux opérateurs (n. 13).

Les cas que nous venons de citer sont les seuls dans lesquels (à part la possibilité logique) on puisse sous-entendre le symbole d'opération sans se heurter à de gros obstacles formels.

13. *Produits des opérateurs.* Soient u, v, w des classes, α un opérateur de la classe $(*\, u)\, v$ et β un opérateur de la classe $(*\, v)\, w$.

On appelle produit (fonctionnel) de α par β, l'opérateur γ de la classe $(*\, u)\, w$ tel que, quel que soit l'élément x de la classe u, on ait

$$\gamma x = \beta\,(\alpha x) \qquad \text{.................................}(1).$$

L'élément γ de la classe $(*\, u)\, w$ reste donc défini par la relation (1) comme une fonction binaire des opérateurs β, α. On peut donc définir un *opérateur* f tel que

$$\gamma = f\,(\beta,\, \alpha),$$

ou bien définir un *symbole d'opération* $\circ$ tel que

$$\gamma = \beta \circ \alpha.$$

Dans les deux cas la relation (1) donne

$$\{f\,(\beta,\, \alpha)\}\; x = \beta\,(\alpha x) \qquad \text{..............................}(2),$$

$$(\beta \circ \alpha)\, x = \beta\,(\alpha x) \qquad \text{..............................}(3).$$

On voit aisément que les premiers membres des relations (2), (3) sont plus compliqués que le deuxième; le premier membre de la relation (3) est plus simple que le premier membre de la relation (2).

Il vaut donc mieux donner la préférence au symbole d'opération. Mais le produit fonctionnel est d'un usage si continu et si étendu qu'il semble préférable de sous-entendre le symbole $\circ$ et d'écrire simplement $\beta\alpha$, c.-à-d. de mettre

$$(\beta\alpha)\, x = \beta\,(\alpha x)$$

et, par conséquent, d'écrire $\qquad \beta\alpha x$

sans parenthèses, car les deux groupements $(\beta\alpha)\, x$, $\beta\,(\alpha x)$ ont la même signification.

Nous *convenons* donc de *sous-entendre toujours le symbole d'opération pour le produit* (*fonctionnel*) *des opérateurs.*

Cette convention faite, il résulte par ex. que β est un élément des deux classes (n. 12)

$$(*\, u)\, v, \quad [*\, \{(*\, u)\, v\}]\, \{(*\, u)\, v\}\,;$$

ce qui est absurde, en général.

14. La convention que nous venons de faire donne naissance à de nouveaux opérateurs qui ont une grande importance formelle.

Soient u, v, w des classes quelconques; $\circ$ un symbole d'opération de la classe $(u * v)\, w$; α un opérateur de la classe $(*\, u)\, v$. Si a est un élément de la classe u, alors $a \circ$ (11 (b)) est un symbole d'opération de la classe $(*\, v)\, w$. Étant

α un élément de $(*\, u)\, v$

$a \circ$ „ $(*\, v)\, w$

on peut considérer (n. 13) le produit (fonctionnel) de l'opérateur α par l'opérateur $a\,\mathrm{o}$. Un tel produit, en vertu de la convention que nous venons d'établir au n. 13, reste indiqué par la notation

$$(a\,\mathrm{o})\,\alpha$$

ou, plus simplement encore, par la notation

$$a\,\mathrm{o}\,\alpha,$$

car le groupement $a\,(\mathrm{o}\,\alpha)$ n'a point de signification.

Si $\mathbf{u}$ est un *vecteur* et α est une *homographie vectorielle*, alors

$$\left.\begin{array}{l}\mathbf{u} \wedge \alpha \textit{ est une } \text{homographie vectorielle} \\ \mathbf{u} \times \alpha \textit{ est un } (* \text{ vecteur}) \text{ nombre réel}\end{array}\right\} \ldots\ldots\ldots\ldots\ldots\ldots (1).$$

L'homographie $\mathbf{u} \wedge \alpha$ est d'un usage continu dans les applications physico-mécaniques.

Gibbs obtient les éléments (1), non comme produit (fonctionnel) de deux opérateurs, mais par le moyen d'une définition directe du produit d'un vecteur par une dyadique. Le calcul qui en résulte est bien compliqué, car il n'est pas sujet aux lois formelles du produit des opérateurs établies d'une manière géniale, complète et simple par Hamilton.

ON ISOID RELATIONS AND THEORIES OF IRRATIONAL NUMBER

By Philip E. B. Jourdain.

I call a relation which is reflexive, symmetrical, and transitive an *isoid relation.* Such relations come prominently forward in the work of Peano* and his school. The closely connected subject of "definition by abstraction" has been especially treated by Burali-Forti and more recently (1907) by Padoa; the writer last named develops some of the conclusions previously reached by Russell.

Russell† pointed out a means whereby, given that a stands in an isoid relation to b, an entity can always be defined which has a one-many relation to all such b's as stand in the isoid relation spoken of to a. Russell called the principle which is based on this construction the "principle of abstraction." I wish to point out that Russell uses, in a manner apparently not quite free from arbitrariness, different methods for the construction of such entities according as he deals with integer numbers (cardinal or ordinal) or real numbers, and to initiate a discussion as to why these particular forms should be used in preference to others. In connexion with this, I will discuss certain points in the theories of irrational number.

I. Consider the case of cardinal numbers. Where u, v, ... are classes which have the isoid relation of what Cantor called "equivalence," and what Dedekind and Russell called "similarity," to each other, Russell constructed a many-one relation between the similar classes and an entity, z, which is, in conformity with the views of Frege and Weber, "the cardinal number" of all these classes. But Russell also remarked that other particular S's are possible: "...la démonstration donne une façon dont ceci peut se faire, sans prouver qu'il n'y a pas d'autres façons de la faire." In fact, we shall, in the case of real numbers, meet with a different construction of S and z.

But Russell's proof of his proposition 6. 2 does not make the nature of the existence-theorem clear. The nature of existence-theorems is, in essentials, the same throughout mathematics; in the present case what we do is‡: We show that,

* Cf., for example, my paper on "The Development of the Theories of Mathematical Logic and the Principles of Mathematics," *Quart. Journ. of Math.* 1912, pp. 272, 280, 282–3, 284, 293–4, 302–4.

† "Sur la logique des relations," *Rev. de Math.* VII, 1902, propositions 6. 1, 6. 2; *The Principles of Mathematics*, Cambridge, 1903, pp. 166, 219–20, 305, 314, 499, 519; cf. L. Couturat, *Les Principes des Mathématiques*, Paris, 1905, pp. 49—51, 77.

‡ I use Russell's notation of the paper referred to.

if xRy, by choosing z to be $\breve{\rho}x$ (which $= \rho y$), the relation ϵ is an example of an S. In Russell's proof this does not seem as clear as it might be.

II. Russell's second construction of an entity which justifies the principle of abstraction is in his theory of real numbers. Here he takes a definite member out of the class of classes between which an isoid relation subsists, so that the relation between the classes in isoid relation and the new entity obtained by the principle of abstraction is the same isoid relation as that which correlates the classes spoken of, and is not ϵ, as it was before. In the case of real numbers, the isoid relation is the relation which is practically the same as that which Cantor has called "Zusammengehörigkeit" and which Russell* has translated "coherency." The entity chosen by Russell is what Pasch and Peano called a "segment" of rational numbers.

III. It may be remarked, by the way, that the chief advantages of the principle of abstraction seem to be the avoidance of the necessity (1) of introducing new indefinables and (2) of redefining equality; and the advantage, mentioned in the book of 1903†, of enabling the existence-theorems to be proved is comparatively unimportant.

If we apply this second method to the theory of cardinal numbers, we need, among all the classes similar to u, one specialized one. Thus we should have the results that u is similar to its cardinal number, instead of, as in the theory of Russell, u is a member of its cardinal number. We will return to the consideration of this theory.

IV. If, now, we attempt to define the real number belonging to a class of rational numbers by a method analogous to that which Russell has employed to define the cardinal number of a class, we would arrive at the definition as the class of those classes of rational numbers which are coherent to u. It seems to me that there is no particular advantage in defining a real number as a class of such classes over the definition as a segment, or *vice versa*.

V. Let us now consider if there is, among all classes which are similar to a given class u, a specialized one which can therefore be used as a definition of "the cardinal number of u." The method which immediately suggests itself is to start with the null-class and define the cardinal number 1 as the class whose only member is the null-class. Then 2 may be defined as the class whose members are the null-class and 1; and so on. It is visible that, in this way, we would get a one-one correlation of cardinal numbers with ordinal numbers; but, on the other hand, there is a rise of type at each step, so that it would appear that we would be unable to define any of the transfinite cardinal or ordinal numbers. However, it must be remarked that the rise of type is of a different nature from the rise of type which is treated at length by Russell. There would appear to be no objection to the definition of types whose order is a limit of lower orders.

The distinction between the hierarchy of types which I have indicated and the hierarchy which Russell has treated is, in a way, analogous to the distinction that there is between limits in the case of ordinal numbers and those in the case of

* *Principles*, p. 283.

† P. ix. Cf. Whitehead and Russell, *Principia Mathematica*, Cambridge, 1912, p. 4.

cardinal numbers; or again between ω^ω in Cantor's notation and the cardinal number $\aleph_0$ exponentiated by $\aleph_0$, or the ordinal type α exponentiated by the ordinal type β (Hausdorff and myself). It seems that there would be no difficulty, then, in extending the series of Alephs beyond $\aleph_\omega$ and the series of ordinal numbers beyond ω_ω. It is known that Russell's theory of types puts a limit to the series of Alephs and ordinal numbers at these points.

VI. It has appeared from somewhat careful historical investigations of mine that irrational "numbers" were based by Cauchy on geometrical foundations, but that later there grew up, in some text-books, a would-be arithmetical theory in which the irrational numbers are defined as the limits of series of rational numbers which have no rational limits and which are convergent. Since the "convergence" of a series was usually defined by means of a presupposed limit, the circle was obvious, and it was the avoidance of this circle which seems to have given rise to the theories of Weierstrass and Cantor*, and perhaps also of Méray. When, however, the "convergence" of a sequence (s_ν) is defined without assuming the existence of a limit, by means of the well-known criterion

$$|s_{\nu+\mu} - s_\nu| < \epsilon,$$

there is no circle, but the assumption of existence appears even more clearly. In the analogous case of the Dedekind-Peano theory, this assumption was rightly emphasized by Russell†. On the other hand, Russell‡ seems to neglect the advance made in the avoiding of the circle. This advance lay, perhaps, rather in the fact that mathematicians began to see the necessity of attending to fundamental questions than in the fact that an obvious logical error was not committed. Still, almost certainly the error did not seem so obvious about 1870 as it does now.

In Russell's theory§, which was anticipated by Pasch||, the selection of the entity defined as the real number appears somewhat arbitrary. There is, quite ready to hand, another entity, which is formed in a manner analogous to that in which the entity which Frege and Russell called "the cardinal number" of classes which are in the isoid relation of similarity is formed. It would, further, appear that this was the entity which Frege¶ indicated as the real number belonging to certain classes of rationals. In any case, there is a difficulty in admitting that Russell's entity or this entity is *the* real number corresponding to a given segment of rationals. Many entities, no one of which seems to have any special advantage over the others, can be defined by "the principle of abstraction."

VII. It may be remarked that theories of irrational numbers do not depend upon limits** in the sense that irrationals are defined as limits. If limits are defined

* See Cantor, *Math. Ann.* XXI, 1883, pp. 566, 568.

† *Principles*, 1903, pp. 270, 280—285.

‡ See *Ibid.* pp. 278, 280—285.

§ *Ibid.* pp. 271—275, 285, 286.

|| *Einleitung in die Differential- und Integralrechnung*, Leipzig, 1882; see especially p. 11. It is a curious fact that Pasch seemed so unconscious of the merits of this definition that later (*Grundlagen der Analysis*, Leipzig and Berlin, 1909, p. 98) he abandoned his definition by defining (or postulating) an irrational number as a new "thing," somewhat as Peano did.

¶ *Die Grundlagen der Arithmetik*, Breslau, 1884, pp. 114—115.

** Cf. Russell, *op. cit.* pp. 277—278.

before irrationals are introduced (as they usually are, for pedagogical reasons, because some limits are rational), to define an irrational as a limit would be to commit the circle referred to. In Peano's introduction of the conception of a limit, on the other hand, a "limit" and a "real number" are merely two names for the same thing; thus limits are not introduced at first for rationals alone, but are "defined by abstraction," as Peano and Burali-Forti would say, as new entities*, and this "definition" leads to Russell's definition or to the one given in the fourth section.

VIII. In his remarks on Dedekind's axiom, Russell† quoted his criticisms‡ on Couturat's account of Dedekind's axiom, given in Couturat's *De l'infini mathématique*§. Now these criticisms are very much to the point *à propos* of Couturat's mistaken version of the axiom, in which the point dividing the Dedekindian sections is said to be *both* greater than all the numbers of the lower section, *and* less than all the numbers of the upper section. But it does not seem always to be relevant to Dedekind's own statement.

IX. The following points seem worthy of notice in the theories of irrationals:

(1) Dedekind's theory had not for its object to prove the existence of irrationals‖: it showed the necessity, as Dedekind thought, for the mathematician to create them. In the idea of the creation of numbers, Dedekind was followed by Stolz; but Weber and Pasch showed how the supposition of this creation could be avoided: Weber defined real numbers as sections (*Schnitte*) in the series of rationals; Pasch (like Russell) as the segments which generate these sections;

(2) In Weierstrass's theory, irrationals were defined as classes of rationals. Hence Russell's¶ objections do not hold against it, nor does Russell seem to credit Weierstrass and Cantor with the avoidance of quite the contradiction that they did avoid;

(3) The real objection to Weierstrass's theory, and one of the objections** to Cantor's theory, is that equality has to be re-defined;

(4) In the various arithmetical theories of irrational numbers there are three tendencies: (*a*) the number is defined as a logical entity—a class or an operation—, as with Weierstrass††, Weber, Pasch, Russell and Pieri; (*b*) it is "created," or, more frankly, postulated, as with Dedekind, Stolz, Peano, and Méray; (*c*) it is defined as a sign (for what, is left indeterminate), as with Heine, Cantor‡‡, Thomae,

* Cf. *Archiv der Math. und Phys.* (3), XVI, pp. 36—37.

† *Op. cit.* pp. 279—280.

‡ *Mind*, N. S., VI, 1897, p. 117.

§ Paris, 1896, p. 416.

‖ Cf. Russell, *op. cit.* p. 280.

¶ *Ibid.* p. 282.

** Not made by Russell, *ibid.* p. 284; cf. p. 285.

†† I have referred elsewhere (*Archiv der Math. und Phys.* (3), XIV, pp. 306, 311) to the merits of Weierstrass's theory, which were unappreciated by Frege (*Grundgesetze der Arithmetik*, II, Jena, 1903, pp. 149—154).

‡‡ It is a remarkable fact that Cantor was a nominalist as regards real numbers (*Math. Ann.* XXI, 1883, pp. 589—590), and rejected with scorn the nominalistic thesis in the case of integers (*Zur Lehre vom Transfiniten*, Halle, 1890, pp. 16—18).

Pringsheim. I will not consider here the geometrical theories, in which, as with Paul du Bois-Reymond, a real number is a sign for a length;

(5) In Russell's theory it appears to be equally legitimate to define a real number in various ways. I do not propose the alternative form I have suggested as in any way superior to Russell's own, but merely to show the need of some further discussion.

Finally, I would remark that Russell's later work seems to show that the advantage claimed in 1903 for the definition of certain numbers as classes, viz. that an "existence-theorem" can be asserted of them, cannot be maintained. Russell's "existence," in fact, which does not apply to relations such as the rationals unless a relation is a class of couples, does not relate to the question of what mathematicians usually call the "existence" of real numbers.

DELLE PROPOSIZIONI ESISTENZIALI

DI G. PEANO.

La logica matematica fece negli ultimi anni grandi progressi; molto si deve ai professori Burali-Forti e Padoa. Queste teorie sparse, assursero ad un corpo scientifico completo nell'opera: Whitehead and Russell, *Principia mathematica*, t. 1, 1910, t. 2, 1912.

Ivi tutte le teorie sono esposte in modo completo, e con critica profonda; le questioni più difficili ed astruse vi sono esaminate e sviluppate.

Gli autori introducono numerosi simboli, per rappresentare le varie idee di logica che essi studiano; e questi simboli sono utili e necessarii per una teoria completa delle relazioni, per l'analisi della mutua dipendenza delle proposizioni logiche, per la distinzione delle medesime in proposizioni primitive e in derivate, e per lo studio di tutte le questioni relative alla intricata teoria degli insiemi, e dei paradossi o antinomie, che si incontrano in questa teoria.

Da alcuni anni, io mi occupo della parte più elementare della logica matematica, e precisamente delle sole idee di logica che si incontrano nella matematica classica, aritmetica, algebra, calcolo infinitesimale. La logica matematica si presenta nel mio Formulario mathematico, non come scienza a sè, ma come strumento per esprimere ed analizzare le proposizioni di matematica. Varie questioni, non sono affrontate, ma quando fu possibile, furono girate ed eliminate. Per questa via il Formulario mathematico contiene solo una ventina di simboli logici, e quelli di uso comune si riducono a 10.

Per dare un esempio di questo doppio modo si trattare le questioni, o col vincerle, o coll' eliminarle, considererò le proposizioni esistenziali.

La parola *esiste* del linguaggio comune ha più valori, a seconda dei casi. Noi ci accorgiamo del diverso valore d'una stessa parola, dalle diverse sue proprietà, e modo di comportarsi nel calcolo logico. Nelle due proposizioni:

"*Esiste* un numero quadrato somma di due quadrati,"
"*Esiste* l'integrale d'una funzione continua"

i due *esiste* hanno valori diversi; il primo si riferisce ad una classe, il secondo ad un individuo. Anche nel linguaggio comune, essi sono formalmente distinti, poichè il primo è accompagnato dall' articolo *un*, ed il secondo da *il*.

Perciò, volendosi la corrispondenza univoca fra idee e simboli, non si possono esprimere ambe le proposizioni esistenziali con uno stesso simbolo.

Gli autori dei *Principia mathematica* introdussero perciò due simboli $\exists$ e E per rappresentare le due idee.

Il Formulario Mathematico invece, usa il simbolo $\exists$ nel primo significato. Le proposizioni di matematica in cui comparisce il secondo *esiste*, sono trasformate in modo da risultare espresse per simboli più elementari. Darò esempi dei due casi.

Il simbolo Cls si legge "classe." E siccome una classe è determinata da una proprietà, o condizione, il simbolo Cls si potrà anche leggere "proposizione condizionale," o proposizione contenente una o più variabili reali.

Λ è il simbolo della classe nulla, o vuota, o della proprietà posseduta da nessun ente, o d'una condizione assurda. Quindi $a = \Lambda$ significa "la classe a è nulla," "non esistono a," "la condizione a è assurda."

$a \cap b$ indica la classe comune agli a e b. Quindi

$$a \cap b = \Lambda$$

significa "non esistono enti comuni alle classi a e b," ovvero "nessun a è b."

$\sim$ (leggasi *non*) indica la negazione, e si scrive davanti ad una classe o ad una proposizione, o al segno di relazione. Quindi

$$a \sim = \Lambda$$

significa "la classe a non è nulla," ossia "esistono degli a." E

$$a \cap b \sim = \Lambda$$

significa "esistono degli individui che sono ad un tempo a e b," cioè più comunemente "qualche a è b," proposizione particolare affermativa.

Ora in logica matematica, si suol dapprima parlare dei simboli:

ϵ, che indica la proposizione individuale: $x \epsilon a$ significa "x è un a."

$\supset$, che indica la proposizione universale affermativa: "$a \supset b$" significa "ogni a è b," ovvero "ogni individuo che ha la proprietà a ha la proprietà b," ovvero "gli individui che verificano la condizione a, verificano pure la b," ovvero "da a segue b," o "se a, allora b."

Poi i simboli $\cap$ (et), $\cup$ (aut), $\sim$ (non) già visti, come pure Λ (nulla).

Allora tanto la proposizione "esistono degli a," quanto l'altra "qualche a è b," si possono esprimere coi simboli precedenti, e non c' è necessità di un simbolo nuovo.

Però, un po' per brevità, e specialmente per uniformarsi al linguaggio comune, il Formulario mathematico usa spesso il simbolo $\exists$, che si legge "esiste," e che si può definire:

$$\exists a \,.=.\, a \sim = \Lambda \quad \text{Def.}$$

"Se a è una classe, allora noi scriveremo $\exists a$ (che si legge "esistono degli a") invece di "la classe a non è nulla."

Daremo alcuni esempi dall' aritmetica. Perciò conviene dire che nel Formulario, il simbolo N significa "numero intero, della serie 1, 2, 3,…." E per convenzioni quasi evidenti, N^2 significa "numero quadrato" e $N^2 + N^2$ vale "somma di due quadrati." Allora si ha la proposizione aritmetica, tutta scritta in simboli

$$\exists N^2 \cap (N^2 + N^2)$$

"esistono numeri quadrati, somma di due quadrati." Invece

$$\sim \exists N^3 \cap (N^3 + N^3),$$

che si può anche scrivere:

$$N^3 \cap (N^3 + N^3) = \Lambda$$

"dire che un numero è cubo, ed è la somma di due cubi, è assurdo."

Altro esempio. La proposizione di Euclide "dato un numero m, si può determinare un numero primo maggiore di m," si potrà scrivere:

$$m \epsilon N \,.\, \supset \,.\, \exists \mathrm{Np} \cap (m + N).$$

Il simbolo aritmetico "Np" vale "numero primo."

Ora daremo esempi della seconda specie di "esiste."

Le relazioni, indicate dai simboli $=$, $\supset$, ϵ, sono differenti. In vero, da $x = y$ segue $y = x$; ciò che non è vero per le relazioni $x \supset y$ e $x \epsilon y$. La relazione ϵ è distributiva al segno $\cup$ (aut):

$$x \epsilon a \cup b \,.=.\, x \epsilon a \,.\cup.\, x \epsilon b.$$

"Se x è a o b, allora x è a, o x è b, e viceversa." Ciò che non è vero pel segno $\supset$. Ma le relazioni $= \supset \epsilon$ sono legate da numerose relazioni. Noi possiamo decomporre il segno $=$ nel segno ϵ e in un altro segno ι (leggasi eguale o ἴσος). Basta che noi poniamo la definizione:

$$\iota x = y \ni (y = x).$$

Allora $y \epsilon \iota x$ vale $y = x$; e $\iota x \supset a$ vale $x \epsilon a$. E ιx è la classe composta dal solo individuo x.

Se a è una classe che contiene un solo individuo, per $\imath a$ noi indichiamo questo individuo. Definizione della differenza:

$$a \epsilon N \,.\, b \epsilon a + N \,.\, \supset \,.\, b - a = \imath N \cap x \ni (a + x = b).$$

"Se a è un numero, e b è un numero superiore ad a, allora $b - a$ indica quel numero x che sodisfa la condizione $a + x = b$."

La prima proposizione che si presenta nella teoria della differenza, è che nell' ipotesi $a \epsilon N \,.\, b \epsilon a + N$, *esiste* la differenza $b - a$; ciò è che $b - a$ indica un numero determinato; e noi scriviamo:

$$a \epsilon N \,.\, b \epsilon a + N \,.\, \supset \,.\, b - a \epsilon N.$$

Definizione del massimo:

$$u \epsilon \mathrm{Cls}'N \,.\, \supset \,.\, \max u = \imath u \cap x \ni (y \epsilon u \,.\, \supset_y \,.\, y \leqq x).$$

"Se u è una classe di numeri, allora $\max u$ (leggasi: massimo degli u), indica quell' individuo della classe u, e x tale che, per ogni y nella classe u, sempre è $y \leqq x$."

Se $u \epsilon \mathrm{Cls}'N$, non sempre esiste $\max u$. Per l'esistenza, noi dobbiamo aggiungere un altra condizione:

$$u \epsilon \mathrm{Cls}'N \,.\, \exists u \,.\, m \epsilon N \,.\, \sim \exists u \cap (m + N) \,.\, \supset \,.\, \max u \epsilon u.$$

"Se u è una classe di numeri, e se esiste u (ciò è, se la classe u non è vacua, od esistono individui nella classe u), e se m è un numero, e non esiste alcun u superiore ad m, allora $\max u$ indica un elemento della classe u (ciò è, esiste $\max u$ nella classe u)."

La definizione del *limite* d'una funzione (reale di variabile reale) può assumere più forme, ove sempre si presenta il segno ɿ. Data una funzione ad arbitrio, non sempre il limite esiste. Noi consideriamo un teorema molto semplice, che afferma l'esistenza del limite:

$$x \in (qFN) \text{ cres} \,.\, \supset .\, \lim x \in q \cup \iota (+ \infty).$$

"Se x è una successione crescente di quantità, allora $\lim x$ è una quantità finita o vale $+\infty$; ciò è, $\lim x$ esiste nel campo delle quantità finite o infinite."

La somma d'una serie, la derivata di una funzione, sono definite come limiti. Dunque le proposizioni "la somma della serie esiste, o la serie è convergente," "la derivata della funzione esiste," sono sempre indicate per "la somma della serie $\in q$," "la derivata $\in q$."

Così per l'integrale.

In conseguenza, la parola *esiste,* col secondo valore, sempre è con un'espressione x che o indica un individuo, o è senza senso. Noi esprimiamo questo *esiste* in simboli, coll' affermazione $x \in u$, ove u è una classe conveniente. Quindi il Formulario mathematico adotta nessun segno speciale per il secondo *esiste.* Anche il primo *esiste* può essere eliminato, mediante il segno Λ, come noi già vedemmo. L'uso o non di simboli per indicare le due specie di *esiste,* è questione di commodità in pratica, non di necessità in logica.

ÜBER EINE ANWENDUNG DER MENGENLEHRE AUF DIE THEORIE DES SCHACHSPIELS

VON E. ZERMELO.

Die folgenden Betrachtungen sind unabhängig von den besonderen Regeln des Schachspiels und gelten prinzipiell ebensogut für alle ähnlichen Verstandesspiele, in denen zwei Gegener unter Ausschluss des Zufalls gegeneinander spielen; es soll aber der Bestimmtheit wegen hier jeweilig auf das Schach als das bekannteste aller derartigen Spiele exemplifiziert werden. Auch handelt es sich nicht um irgend eine Methode des praktischen Spiels, sondern lediglich um die Beantwortung der Frage: kann der Wert einer beliebigen während des Spiels möglichen Position für eine der spielenden Parteien sowie der bestmögliche Zug mathematisch-objektiv bestimmt oder wenigstens definiert werden, ohne dass auf solche mehr subjektiv-psychologischen wie die des "vollkommenen Spielers" und dergleichen Bezug genommen zu werden brauchte? Dass dies wenigstens in einzelnen besonderen Fällen möglich ist, beweisen die sogenannten "Schachprobleme," d. h. Beispiele von Positionen, in denen der Anziehende *nachweislich* in einer vorgeschriebenen Anzahl von Zügen das Matt erzwingen kann. Ob aber eine solche Beurteilung der Position auch in anderen Fällen, wo die genaue Durchführung der Analyse in der unübersehbaren Komplikation der möglichen Fortsetzungen ein praktisch unüberwindliches Hindernis findet, wenigstens theoretisch denkbar ist und überhaupt einen Sinn hat, scheint mir doch der Untersuchung wert zu sein, und erst diese Feststellung dürfte für die praktische Theorie der "Endspiele" und der "Eröffnungen," wie wir sie in den Lehrbüchern des Schachspiels finden, die sichere Grundlage bilden. Die im folgenden zur Lösung des Problems verwendeten Methode ist der "Mengenlehre" und dem "logischen Kalkül" entnommen und erweist die Fruchtbarkeit dieser mathematischen Disziplinen in einem Falle, wo es sich fast ausschliesslich um *endliche* Gesamtheiten handelt.

Da die Anzahl der Felder, sowie die der ziehenden Steine endlich ist, so ist es auch die Menge P der möglichen Positionen $p_0, p_1, p_2 \ldots p_t$, wobei immer Positionen als verschieden aufzufassen sind, je nachdem Weiss oder Schwarz am Zuge ist, eine der Parteien schon rochiert hat, ein gegebener Bauer bereits verwandelt ist u. s. w. Es sei nun q eine dieser Positionen, dann sind von q aus "Endspiele" möglich $\mathfrak{q} = (q, q_1, q_2 \ldots)$, nämlich Folgen von Positionen, die mit q beginnen und im Einklang mit den Spielregeln auf einander folgen, sodass jede Position q_λ aus der vorhergehenden $q_{\lambda-1}$ abwechselnd durch einen zulässigen Zug von Weiss oder Schwarz hervorgeht. Solch ein mögliches Endspiel $\mathfrak{q}$ kann entweder in einer "Matt" oder

"Patt" Stellung sein natürliches Ende finden, oder aber auch—theoretisch wenigstens—unbegrenzt verlaufen, in welchem Falle die Partie zweifellos als unentschieden oder "remis" zu gelten hätte. Die Gesamtheit Q aller dieser zu q gehörenden "Endspiele" $\mathfrak{q}$ ist stets eine wohldefinierte, endliche oder unendliche Untermenge der Menge $P^{\mathfrak{a}}$, welche alle möglichen abzählbaren Folgen gebildet aus Elementen p von P umfasst.

Unter diesen Endspielen $\mathfrak{q}$ können einige in r oder weniger "Zügen" (d. h. einfachen Positionswechseln $p_{\lambda-1} \longrightarrow p_\lambda$, nicht etwa Doppelzügen) zum Gewinn von Weiss führen, doch wird dies in der Regel auch noch vom Spiel des Gegners abhängen. Wie muss aber eine Position q beschaffen sein, damit Weiss, wie Schwarz auch spielt, in höchstens r Zügen den Gewinn *erzwingen* kann? Ich behaupte, die notwendige und hinreichende Bedingung hierfür ist die Existenz einer nicht verschwindenden Untermenge $U_r(q)$ der Menge Q von folgender Beschaffenheit:

1. Alle Elemente $\mathfrak{q}$ von $U_r(q)$ enden in höchstens r Zügen mit dem Gewinn von Weiss, sodass keine dieser Folgen mehr als $r+1$ Glieder enthält und daher $U_r(q)$ jedenfalls endlich ist.

2. Ist $\mathfrak{q} = (q, q_1, q_2 \ldots)$ ein beliebiges Element von $U_r(q)$, q_λ ein beliebiges Glied dieser Reihe, welches einem ausgeführten Zuge von Schwarz entspricht, also entweder immer ein solches gerader oder eines ungerader Ordnung, je nachdem bei q Weiss oder Schwarz am Zuge ist, sowie endlich q'_λ eine mögliche Variante, sodass Schwarz von $q_{\lambda-1}$ aus ebensogut nach q'_λ wie nach q_λ hätte ziehen können, so enthält $U_r(q)$ noch mindestens ein Element der Form $\mathfrak{q}'_\lambda = (q, q_1, \ldots q_{\lambda-1}, q'_\lambda, \ldots)$, welches mit $\mathfrak{q}$ die ersten λ Glieder gemein hat. In der Tat kann in diesem und nur in diesem Falle Weiss mit einem beliebigen Elemente $\mathfrak{q}$ von $U_r(q)$ beginnen und jedesmal, wo Schwarz q'_λ statt q_λ spielt, mit einem entsprechenden $\mathfrak{q}'_\lambda$ weiterspielen, also unter allen Umständen in höchstens r Zügen gewinnen.

Solcher Untermengen $U_r(q)$ kann es freilich mehrere geben, aber die Summe je zweier ist stets von derselben Beschaffenheit, und ebenso auch die Vereinigung $\overline{U}_r(q)$ aller solchen $U_r(q)$, welche durch q und r eindeutig bestimmt ist und jedenfalls von 0 verschieden sein, d. h. mindestens ein Element enthalten muss, sofern überhaupt solche $U_r(q)$ existieren. Somit ist $\overline{U}_r(q) \neq 0$ die notwendige und hinreichende Bedingung dafür, dass Weiss den Gewinn in höchstens r Zügen erzwingen kann. Ist $r < r'$ so ist stets $\overline{U}_r(q)$ Untermenge von $\overline{U}_{r'}(q)$, weil dann jede Menge $U_r(q)$ sicher auch die an $U_{r'}(q)$ gestellten Anforderungen erfüllt, also in $\overline{U}_{r'}(q)$ enthalten sein muss, und dem kleinsten $r = \rho$, für welches noch $\overline{U}_r(q) \neq 0$ ist, entspricht der gemeinsame Bestandteil $U^*(q) = \overline{U}_\rho(q)$ aller solchen $\overline{U}_r(q)$; dieser umfasst alle solche Fortsetzungen, mit denen Weiss in der kürzesten Zeit gewinnen muss. Nun besitzen aber diese Minimalwerte $\rho = \rho_q$ ihrerseits ein von q unabhängiges Maximum $\tau \leqq t$, wo $t+1$ die Anzahl der möglichen Positionen ist, sodass $U(q) = \overline{U}_\tau(q) \neq 0$ die notwendige und hinreichende Bedingung dafür darstellt, dass in der Position q irgend ein $\overline{U}_r(q)$ nicht verschwindet und Weiss überhaupt "auf Gewinn steht." Ist nämlich in einer Position q der Gewinn überhaupt zu erzwingen, so ist er es auch, wie wir zeigen wollen, in höchsten t

Zügen. In der Tat müsste jedes Endspiel $\mathfrak{q} = (q, q_1, q_2 \ldots q_n)$ mit $n > t$ mindestens eine Position $q_\alpha = q_\beta$ doppelt enthalten, und Weiss hätte beim ersten Erscheinen derselben ebenso weiter spielen können wie beim zweiten Male und jedenfalls schon früher als beim n ten Zuge gewinnen, also $\rho \leqq t$.

Ist andererseits $U(q) = 0$, so kann Weiss, wenn der Gegner richtig spielt, höchstens remis machen, er kann aber auch "auf Verlust stehen" und wird dann versuchen, dass "Matt" möglichst hinauszuschieben. Soll er sich noch bis zum s^{ten} Zuge halten können, so muss eine Untermenge $V_s(q)$ von Q existieren von folgender Beschaffenheit:

1. In keinen der in $V_s(q)$ enthalten Endspiele verliert Weiss vor dem s^{ten} Zuge.

2. Ist $\mathfrak{q}$ ein beliebiges Element von $V_s(q)$ und in $\mathfrak{q}$ durch einen erlaubten Zug von Schwarz q_λ ersetzbar durch q'_λ, so enthält $V_s(q)$ noch mindestens ein Element der Form

$$\mathfrak{q}'_\lambda = (q, q_1, q_2, \ldots q_{\lambda-1}, q'_\lambda \ldots),$$

welches mit $\mathfrak{q}$ bis zum λ^{ten} Gliede übereinstimmt und dann mit q'_λ weitergeht. Auch diese Mengen $V_s(q)$ sind sämtlich Untermengen ihrer Vereinigung $\overline{V}_s(q)$, welche durch q und s eindeutig bestimmt ist und die gleiche Eigenschaft besitzt wie V_s selbst, und für $s > s'$ wird jetzt $\overline{V}_s(q)$ Untermenge von $\overline{V}_{s'}(q)$. Die Zahlen s, für welche $\overline{V}_s(q)$ von O verschieden ausfällt, sind entweder unbegrenzt oder $\leqq \sigma \leqq \tau \leqq t$, da der Gegner, wenn überhaupt, den Gewinn in höchstens τ Zügen müsste erzwingen können. Somit kann Weiss dann und nur dann mindestens remis machen, wenn $V(q) = \overline{V}_{\tau+1}(q) \neq 0$ ist, und im anderen Falle kann er vermöge $V^*(q) = \overline{V}_\sigma(q)$ den Verlust noch mindestens $\sigma \leqq \tau$ Züge hinausschieben. Da jedes $U_r(q)$ gewiss auch den an $V_s(q)$ gestellten Anforderungen genügt, so ist jedes $\overline{U}_r(q)$ Untermenge jeder Menge $\overline{V}_s(q)$, und $U(q)$ Untermenge von $V(q)$. Das Ergebnis unserer Betrachtung ist also das folgende:

Jeder während des Spiels möglichen Position q entsprechen zwei wohldefinierte Untermengen $U(q)$ und $V(q)$ aus der Gesamtheit Q der mit q beginnenden Endspiele, deren zweite die erste umschliesst. Ist $U(q)$ von O verschieden, so kann Weiss, wie Schwarz auch spielt, den Gewinn erzwingen und zwar in höchstens ρ Zügen vermöge einer gewissen Untermenge $U^*(q)$ von $U(q)$, aber nicht mit Sicherheit in weniger Zügen. Ist $U(q) = 0$ aber $V(q) \neq 0$, so kann Weiss wenigstens remis machen vermöge der in $V(q)$ enthaltenen Endspiele. Verschwindet aber auch $V(q)$, so kann Weiss, wenn der Gegner richtig spielt, den Verlust höchstens bis zum σ^{ten} Zuge hinausschieben vermöge einer wohldefinierten Menge $V^*(q)$ von Fortsetzungen. Auf alle Fälle sind nur die in U^*, bezw. V^* enthaltenen Partieen im Interesse von Weiss als "korrekt" zu betrachten, mit jeder anderen Fortsetzung würde er, wenn in Gewinnstellung, bei richtigem Gegenspiel den gesicherten Gewinn verscherzen oder verzögern, sonst aber den Verlust der Partie ermöglichen oder beschleunigen. Ganz analoge Betrachtungen gelten natürlich auch für Schwarz, und als "korrekt" zu Ende geführte Partieen hätten diejenigen zu gelten, welche *gleichzeitig* den beiderseitigen Bedingungen entsprechen, sie bilden also in jedem Falle wieder eine wohldefinierte Untermenge $W(q)$ von Q.

Die Zahlen t und τ sind von der Position unabhängig und lediglich durch die Spielregeln bestimmt. Jeder möglichen Position entspricht eine τ nicht überschreitende Zahl $\rho = \rho_q$ oder $\sigma = \sigma_q$, je nachdem Weiss oder Schwarz in ρ bezw. σ Zügen, aber nicht in weniger, den Gewinn erzwingen kann. Die spezielle Theorie des Spiels hätte diese Zahlen, soweit dies möglich ist, zu bestimmen oder wenigstens in Grenzen einzuschliessen, was bisher allerdings nur in besonderen Fällen, wie bei den "Problemen" oder den eigentlichen "Endspielen" gelungen ist. Die Frage, ob die Anfangsposition p_0 bereits für eine der spielenden Parteien ein "Gewinnstellung" ist, steht noch offen. Mit ihrer exacten Beantwortung würde freilich das Schach den Charakter eines Spieles überhaupt verlieren.

ON SUPERPOSITION AS A BASIS FOR GEOMETRY—ITS LOGIC, AND ITS RELATION TO THE DOCTRINE OF CONTINUOUS QUANTITY

By R. F. Muirhead.

Abstract.

I.

This memoir aims at the re-establishment of Geometry on a purely geometrical basis.

The introduction points out that the notion of *number* is no more fundamental or exact than the notion of *measurement* of space or extended magnitudes, and suggests that the attempts to base the doctrine of measurement on that of number, which have occupied so much of the attention of mathematicians in recent years, are to be regarded as mistaken, provided it can be shown that a purely geometrical doctrine based on Superposition can be logically established.

It is pointed out that the most fundamental part of geometry consists in the elementary principles of Topologics* (German, *Topologie*), which deal with those conceptions of bodies which are anterior to the ideas of *shape* and *measurement*, and that these form the natural and proper basis of geometrical theory.

The chief characteristics of the theory of Geometry developed in this memoir are these:

The theory rejects the notion of space as distinct from bodies in space, and deals with the relations of one body to another, 'space' being simply a body singled out from other bodies as a body of reference, but having no essential difference from the other bodies considered. Such bodies are conceived as capable of coinciding or "occupying the same space" in the sense that any number of them may be wholly or partially coincident, whether they be unidimensional or polydimensional.

It takes the view that the boundary of a body or space (which is a body or space of one dimension lower than that which it bounds) is purely a boundary, and not an element of the space which it bounds. Thus, e.g. a line does not consist of points, but any point in a simple bounded line is the common boundary of two simple bounded lines which are parts of the whole. If this point be removed, no element is removed from the whole, but only the separation of the whole into parts is annulled. And

* Sir G. Greenhill suggests *Topics* as the name for this part of geometrical theory.

the end-point of a line cannot be removed without removing part of the line itself. It is suggested that to conceive a line as made up of points leads to unnecessary intricacy and obscurity.

Another characteristic of the present theory is that it begins with a purely topological treatment of space, such notions as straightness and metrical relations being introduced later after appropriate postulates have been added to the topological first principles.

In developing the theory much use is made of the idea of a displacement-sequence of points. (See "On a Method of Studying Displacement," *Edin. Math. Soc. Proc.* Vol. XV. pp. 119—128, and "On the Transformation Sequence," *ibid.* Vol. XVI. pp. 70—76.)

II. Geometry of the Simple Line.

The notion of a simple bounded line AB with boundary-points or end-points A and B is an indefinable.

Postulates. 1. AB may be divided into two parts AP and PB by a point P such that AP and PB are simple bounded lines. Every such point is a point in AB, and the boundary points A and B, as well as points *in* AB are points *of* AB.

2. Every point of AB is either a point of AP or a point of PB, and of these P is the only point common to AP and PB.

3. Every point of AP and of PB is a point of AB.

Postulates of Order.

Sense of a simple bounded line (line).

The symbol $|AB|$ denotes the line AB without reference to *sense*.

Postulate of Production of a Line. Two cases: (1) the line, however far produced, never closes (infinite line); (2) the line when sufficiently produced becomes closed, i.e. the end-points coincide forming a single point which ceases to be a boundary point, and so a *simple closed line* is formed, having no boundary.

Definition of Continuity of a simple bounded line AB. Every point P in it is a common end-point or boundary of two simple bounded lines AP and PB.

Def. Superposability of two lines l_1 and l_2. Two simple bounded lines are mutually superposable if they can be placed with a segment or set of segments l_1' of l_1 coincident with a segment or set of segments l_2' of l_2 so that every point of l_1' coincides with one and only one point of l_2' and *vice versa*, and that segments of l_1' of like sense coincide with segments of l_2' having like sense.

Definition of Relative Continuity of two superposable lines l_1 and l_2. This implies that l_1' and l_2' are both continuous and have their common end-points identical with end-points of l_1 or l_2.

Postulate of Relative Mobility. Two lines l_1, l_2 are movable along one another in such a manner as to remain superposed and relatively continuous, so that any point of l_1 can be made to coincide with any point of l_2.

Postulate of Relative Equability. Two lines l_1, l_2 are relatively equable if when P_1 a point of l_1 coincides with P_2 a point of l_2, the correspondence by coincidence of all the remaining pairs of points is fixed, however l_1 and l_2 may move relatively to any other line on which they are superposed.

Cor. If l_1 and l_2 are relatively equable, and also l_2 and l_3, then l_1 and l_3 are relatively equable.

Coincidence of *Tracts, Congruent* Tracts.

Congruent point-figures.

Definition of Equality of Tracts. Two tracts are equal if they are congruent to the same tract.

Magnitude of a tract defined to be a tract starting from a point of reference $(0)_0$ in a line of reference l_0 which is equal to the tract in question. Thus all equal tracts have *the same* magnitude.

Proof of the rules of Equality:

(i) $a = a$,

(ii) if $a = b$, $b = a$,

(iii) if $a = b$ and $b = c$ then $a = c$,

as applied to tracts.

Notation $|AB|$ to denote a segment or line AB without reference to *sense*. The *absolute magnitude* or *sense-less magnitude* of AB is a segment of the *line of reference* l_0, equal to AB or BA, starting from $(0)_0$ and ending on the same side of $(0)_0$ as $(1)_0$.

Displacement-Sequence of Points, and *Sequence of Displacements.* While l_2 moves along l_1 until A_2 falls on B_1, B_2 which originally coincided with B_1 will fall on a point C_1, and C_2 which originally coincided with C_1 will fall on a point D_1, and so on.

The series of points $A_1B_1C_1D_1 \ldots$ forms a Displacement-Sequence of Points. So does the series $A_2B_2C_2D_2 \ldots$.

Prop. A further displacement of A_2 to C_1 will bring B_2 to D_1, C_2 to E_1, etc.

Cor. If $A_1A_2A_3 \ldots A_n \ldots$ is a displacement-sequence of points then

$$A_rA_{r+1} \ldots A_{r+p}$$

is a figure congruent with the figure $A_sA_{s+1}A_{s+2} \ldots A_{s+p}$.

Definition of Relative Quantitativeness of relatively Equable lines. If any number of lines on each of which a tract is marked, are superposed simultaneously on the line of reference l_0 in such a manner that the first point of the $r+1$th tract coincides with the last point of the rth, etc., and the first point of the first tract coincides with a fixed point of l_0, then the last point of the last tract will coincide with a point of l_0 which is independent of the order in which the various tracts are placed, if the lines are *quantitative relatively to* l_0.

Cor. If any number of lines are quantitative relatively to l_0, they are quantitative relatively to one another.

An illustration is given, showing that *quantitativeness* is a property not implied in the postulate of *Equability,* and that *Equability* is not implied in the postulates which preceded it.

Definition of Addition in Tracts. The sum of any number of tracts $a, b, c, \ldots k$, is that tract of l_0 which has the first point of a for its first point, and the last point of k for its last point, when these tracts are superposed on l_0 so that the first point of b coincides with the last of a, and the first of c with the last of b, and so on.

Proof of formal laws:

$$\text{(i)} \quad (a+b)+c=a+(b+c),$$

$$\text{(ii)} \quad a+b=b+a, \text{ etc.}$$

Definition of symbol $-a$, where a is a tract, and of symbol 0.

Definition of $a-b$ as $a+(-b)$.

Definition of "greater" and "less" applied to tracts.

Deduction of rules for symbols $>, <, =, +, -$.

Definition of "greater" and "less" applied to sense-less line segments, and of addition and subtraction of such.

Definition of Equality of Ratio, by the aid of *Uniformly strainable* lines.

Alternative definition of Equality of Ratio after the manner of Euclid.

Magnitude of a Ratio.

Multiplication and *Division* of one magnitude by another.

Formal laws relating to the symbols $=, +, -, \times, \div$.

Definition of Limiting Point.

Regular Sequence of Points.

Scale of Magnitude. If we take the displacement-sequence of points $\ldots(-3)_0$, $(-2)_0$, $(-1)_0$, $(0)_0$, $(1)_0$, $(2)_0 \ldots$ on the line of reference l_0 then the tracts $(0)_0(1)_0$, $(0)_0(2)_0$, $(0)_0(3)_0$, etc., are the positive integers and are denoted by 1, 2, 3, etc., while the tracts $(0)_0(-1)_0$, $(0)_0(-2)_0$, etc., are the negative integers, and are denoted by -1, -2, -3, etc.

All real numbers are *magnitudes*, i.e. tracts of l_0 with $(0)_0$ as first point.

The Simple Closed Line.

Sense of a simple closed line.

A simple closed line having sense is a circuit.

Displacement-Sequence of Points in a closed line. Two cases: (1) no two points of the sequence coincide, however far it may be continued; (2) A_n coincides with A_0 no intermediate points coinciding; (2α) none of the tracts A_1A_2, $A_2A_3 \ldots A_{n-1}A_n$ contains A_0; (2β) $r-1$ of these tracts contain A_0.

III. The Simple Surface.

Generation of Simple Bounded Surface S whose boundary is a simple closed line c.

Topological Postulates.

Postulate of Production of Simple Surface.

Simple *Infinite* Surface.

Simple *Closed* Surface. The generating closed line grows from a point and contracts to a point.

Postulate of Relative Superposability of Simple Surfaces.

Postulate of Relative Mobility of Simple Surfaces.

Latent Points.

Postulate of Relative Equability of Simple Surfaces. A' being fixed on A, and B' (not a latent point for A' fixed) being fixed on B, then P' is fixed on P.

Congruent Figures. Notation $F \equiv F'$.

Definition of Equality of Distances. Distance from A to B denoted by (A, B).

Postulate $(P, Q) = (Q, P)$.

Laws of *Equality of Distances.*

Surface of Reference.

Displacement-Sequence of Points and *Sequence of Displacements* in a Surface.

Associated Displacement-Sequences of Points.

Latent Points for Displacements.

Finite Displacement-Sequences of Points.

Definition of Straight Displacement-Sequence of Points. If $A'B'C'D'\ldots$ be a sequence in S' coincident with $ABCD\ldots$ in S and if S' be moved in S so that B', C' go to C, B, and if then D' falls on A, the sequences are *straight.*

Postulate of Rotation. *Circles.*

Coincident circles are movable in one another.

Two concentric circles cannot have a common point.

Generation of Simple Surface by circular line.

(1) Infinite Surface.

(2) Closed Surface.

In (2) every circle has two centres—"opposite points."

Of two opposite points, each is the anti-point of the other.

Every pair of opposite points is congruent with every other pair.

If a displacement has a latent point K, every sequence of points associated with it lies on a circle, centre A.

In a simple closed surface, if a displacement has one latent point, it has two, which are opposites.

In an infinite surface every rotation has but one latent-point, and every circle but one centre, and no displacement can have more than one latent-point.

Finite Displacement-Sequence.

Three distinct successive points of a displacement-sequence determine a displacement completely.

Definition of "greater" and "less" as applied to *distances.*

Laws of the symbols $=$, $>$, $<$ as applied to distances.

Definition of Relative Quantitativeness of two Equable Surfaces. If every l' belonging to S' which can *move* in any l belonging to S is *quantitative* relatively to l, then S and S' are relatively quantitative.

Ratio between arcs of a circle.

If a straight sequence has one latent-point, it has two, which are opposites.

No straight sequence in an infinite surface can have a latent-point, or lie on a circle.

Construction of a Straight Displacement-Sequence of Points.

Postulate of the Straight Line.

Two straight lines which have two common points which are not opposites, coincide completely.

Cor. 1. All straight lines are mutually *congruent.*

Cor. 2. In an infinite surface two straight lines that do not coincide cannot have more than one common point.

Angle subtended by two points at a third.

Definition of magnitude of angle when the two points are equidistant from the third.

General definition of magnitude of angle subtended by two points at a third.

Sense of Angles. Angles greater than 1.

Two equal circles intersect in two points at most.

Cor. The diagonals of a rhombic four-point bisect one another.

If on two equal circles, two arcs AB and A_1B_1 are equal, then $(A, B)=(A_1, B_1)$.

Conversely, if, for points on two equal circles C and C_1 $(A, B)=(A_1, B_1)$ then A and B divide C into two arcs which are respectively equal to the two arcs into which C_1 is divided by A_1 and B_1.

If $(A, B)=(A_1B_1)=(A, C)=(A_1, C_1)$ and $(B, C)=(B_1, C_1)$ the three-point ABC is congruent either with $A_1B_1C_1$ or with $A_1C_1B_1$.

If $(A, B)=(A_1, B_1)$, $(B, C)=(B_1, C_1)$ and $(C, A)=(C_1, A_1)$ then ABC is either congruent to $A_1B_1C_1$ or *anticongruent* to $A_1B_1C_1$.

If $(A, B)=(A_1, B_1)$ and $(A, C)=(A_1, C_1)$ and if the angle subtended by B, C at A is equal to that subtended by B_1, C_1 at A_1, then ABC is congruent to $A_1B_1C_1$.

Postulate of Euclidean Plane Geometry.

All Displacement-Sequences of points associated with a straight sequence are congruent to it.

IV. Application of Similar Methods to the Geometry of Solid Space.

Generation of simple solid space by a simple closed surface S which grows from a point.

Case 1. S never reduces to a point.

Case 2. S reduces to a point. (Finite solid space without boundary.)

Topological Postulates.

Postulates of Superposability, of Equability, and of Mobility.

Postulates of Pivotal Motion about a fixed point, and of *Rotation* about two points which are not opposites.

Postulate of Relative Quantitativeness.

Mode of picturing the finite Solid Space that has no boundary, in ordinary Euclidean Space.

V.

Less simple spaces, in which some of the preceding Postulates do not hold good.

RÉFLEXIONS SUR LA LOI DE L'ATTRACTION

PAR VICOMTE ROGER DU BOBERIL.

Le mouvement des planètes dans leurs orbites est un mouvement composé de deux forces : la première est une force de projection dont l'effet s'exercerait dans la tangente de l'orbite, si l'effet continu de la seconde cessait un instant ; cette seconde force tend vers le soleil, et, par son effet, précipiterait les planètes vers le soleil, si la première force venait à son tour à cesser un seul instant.

La première de ces forces peut être regardée comme une impulsion dont l'effet est uniforme et constant, et qui a été communiqué aux planètes dès la formation du système planétaire. La seconde peut être considérée comme une attraction vers le soleil, et se doit mesurer, comme toutes les qualités qui partent d'un centre, par la raison inverse du carré de la distance, comme en effet on mesure les quantités de lumière, d'odeur, etc., et toutes les autres quantités ou qualités qui se propagent en ligne droite et se rapportent à un centre. Or, il est certain que l'attraction se propage en ligne droite, puisqu'il n'y a rien de plus droit qu'un fil à plomb et que, tombant perpendiculairement à la surface de la terre, il tend directement au centre de la force et ne s'éloigne que très peu de la direction du rayon au centre. Donc on peut dire que la loi de l'attraction doit être la raison inverse du carré de la distance uniquement parce qu'elle part d'un centre ou qu'elle y tend, ce qui revient au même.

Mais comme ce raisonnement préliminaire, quelque bien fondé que je le croie, pourrait être contredit par les gens qui font peu de cas de la force des analogies, et qui ne sont accoutumés à se rendre qu'à des démonstrations mathématiques ; Newton a cru qu'il valait beaucoup mieux établir la loi de l'attraction par les phénomènes mêmes que par toute autre voie ; et il a en effet démontré géométriquement que si plusieurs corps se meuvent dans des cercles concentriques et que les carrés des temps de leurs révolutions soient comme les cubes de leurs distances à leur centre commun, les forces centripètes de ces corps sont réciproquement comme les carrés des distances, et que si les corps se meuvent dans les orbites peu différentes d'un cercle, ces forces sont aussi réciproquement comme les carrés des distances pourvu que les apsides de ces orbites soient immobiles. Ainsi les forces par lesquelles les planètes tendent aux centres et aux foyers de leurs orbites suivent en effet la loi du carré de la distance ; et la gravitation étant générale et universelle la loi de cette gravitation est constamment celle de la raison inverse du carré de la distance ; et je ne crois pas que personne doute de la loi de Kepler, et que l'on puisse nier que cela ne soit ainsi pour Mercure, pour Vénus, pour la Terre, pour Mars, pour Jupiter et pour Saturne, surtout en les considérant à part, et comme ne pouvant se troubler les uns les autres et en ne faisant attention qu'à leur movement autour du soleil.

Toutes les fois donc qu'on ne considérera qu'une planète ou qu'un satellite, se mouvant dans son orbite autour du soleil ou d'une autre planète, ou qu'on aura que deux corps tous deux en mouvement, ou dont l'un est en repos et l'autre en mouvement, on pourra assurer que la loi de l'attraction suit exactement la raison inverse du carré de la distance, puisque, par toutes les observations, la loi de Kepler se trouve vraie tant pour les planètes principales que pour les satellites de Jupiter et de Saturne. Cependant on pourrait dès ici faire une objection tirée des mouvements de la lune, qui sont irréguliers, au point que M. Halley l'appelle sidus contumax, et principalement du mouvement de ses apsides qui ne sont pas immobiles, comme le demande la supposition géométrique sur laquelle est fondé le résultat qu'on a trouvé de la raison inverse du carré de la distance pour la mesure de la force d'attraction dans les planètes.

A cela il y a plusieurs manières de répondre. D'abord on pourrait dire que la loi s'observant généralement dans toutes les autres planètes avec exactitude, un seul phénomène où cette même exactitude ne se trouve pas, ne doit pas détruire cette loi ; on peut le regarder comme une exception dont on doit chercher la raison particulière. En second lieu, on pourrait répondre, comme l'a fait M. Cotes, que quand même on accorderait quelle loi d'attraction n'est pas exactement dans ce cas en raison inverse du carré de la distance et que cette raison est un peu plus grande, cette différence peut s'estimer par le calcul, et qu'on trouvera qu'elle est presque insensible, puisque la raison de la force centripète de la lune, qui de toutes est celle qui doit être la plus troublée, approche soixante fois plus près de la raison du carré que de la raison du cube de la distance.

"Responderi potest, etiamsi concedamus hunc motum tardissimum exinde profectum quod vis centripetæ proportio aberret aliquantulùm à duplicatâ, aberrationem illam per computum mathematicum inveniri posse et plané insensibilem esse: ista enim ratio vis centripetæ lunaris, quæ omnium maximè turbari debet, paululùm quidem duplicatam superabit; ad hanc vero sexaginta ferè vicibus propiùs accedet quam ad triplicatam. sed verior erit responsio, etc." (Editoris præf. in édit. 2ème Newton auctore Roger Cotes.)

Et, en troisième lieu, on doit répondre plus positivement que ce mouvement des apsides ne vient point de ce que la loi d'attraction est un peu plus grande que dans la raison inverse du carré de la distance, mais de ce qu'en effet le soleil agit sur la lune par une force d'attraction qui doit troubler son mouvement et produire celui des apsides, et que par conséquent cela seul pourrait bien être la cause qui empêche la lune de suivre exactement la règle de Kepler. Newton a calculé, dans cette vue, les effets de cette force perturbatrice, et il a tiré de sa théorie les équations et les autres mouvements de la lune avec une telle précision, qu'ils répondent très exactement, et à quelques secondes près, aux observations faites par les meilleurs astronomes ; mais pour ne parler que du mouvement des apsides, il fait sentir, dès la XLVe proposition du premier livre, que la progression de l'apogée de la lune vient de l'action du soleil ; en sorte que jusqu'ici tout s'accorde, et sa théorie se trouve aussi vraie et aussi exacte dans tous les cas les plus compliqués, comme dans ceux qui le sont le moins.

Cependant un de nos grands géomètres* a prétendu que la quantité absolue du mouvement de l'apogée ne pouvait pas se tirer de la théorie de la gravitation, telle qu'elle est établie par Newton, parce qu'en employant les lois de cette théorie, on trouve que ce mouvement ne devrait s'achever qu'en dix-huit ans, au lieu qu'il s'achève en neuf ans. Malgré l'autorité de cet habile mathématicien, et les raisons qu'il a données pour soutenir son opinion, j'ai toujours été convaincu, comme je le suis encore aujourd'hui, que la théorie de Newton s'accorde avec les observations; je n'entreprendrai pas ici de faire l'examen qui serait nécessaire pour prouver qu'il n'est pas tombé dans l'erreur qu'on lui reproche, je trouve qu'il est plus court d'assurer la loi de l'attraction telle qu'elle est, et de faire voir que la loi que M. Clairaut a voulu substituer à celle de Newton n'est qu'un supposition qui implique contradiction.

Car admettons pour un instant ce que M. Clairaut prétend avoir démontré, que, par la théorie de l'attraction mutuelle, le mouvement des apsides devrait se faire en dix-huit ans, au lieu de se faire en neuf ans, et souvenons-nous en même temps, qu'à l'exception de ce phénomène, tous les autres, quelques compliqués qu'ils soient, s'accordent dans cette même théorie, très exactement avec les observations: à en juger d'abord par les probabilités, cette théorie doit subsister puisqu'il y a un nombre très considérable de choses où elle s'accorde parfaitement avec la nature; qu'il n'y a qu'un seul cas où elle en diffère et qu'il est aisé de se tromper dans l'énumérations des causes d'un seul phénomène particulier. Il me paraît donc que la première idée qui doit se présenter, est qu'il faut chercher la raison particulière de ce phénomène singulier; et il me semble qu'on pourrait en imaginer quelqu'une. Par exemple, si la force magnétique de la Terre pouvait, comme le dit Newton, entrer dans le calcul, on trouverait peut-être qu'elle influe sur le mouvement de la Lune, et qu'elle pourrait produire cette accélération dans le mouvement de l'apogée; et c'est dans ce cas où en effet il faudrait employer deux termes pour exprimer la mesure des forces qui produisent le mouvement de la lune. Le premier terme de l'expression serait toujours celui de la loi de l'attraction universelle, c'est à dire la raison inverse et exacte du carré de la distance, et le second terme représenterait la mesure de la force magnétique.

Cette supposition est sans doute mieux fondée que celle de M. Clairaut, qui me paraît beaucoup plus hypothétique et sujette d'ailleurs à des difficultés invincibles. Exprimer la loi de l'attraction par deux ou plusieurs termes, ajouter à la raison inverse du carré de la distance une fraction du carré-carré, au lieu de $\frac{1}{xx}$ mettre $\frac{1}{xx}+\frac{1}{mx^4}$, me paraît n'être autre chose que d'ajuster une expression de telle façon qu'elle corresponde à tous les cas. Ce n'est plus une loi physique que cette expression représente; car en se permettant une fois de mettre un second, un troisième, un quatrième terme, etc., on pourrait trouver une expression qui, dans toutes les lois d'attraction, représenterait les cas dont il s'agit, en l'ajustant en même temps aux

* M. Clairaut.

mouvements de l'apogée de la lune et aux autres phénomènes, et par conséquent cette supposition, si elle était admise, non seulement anéantirait la loi de l'attraction en raison inverse du carré de la distance, mais même donnerait entrée à toutes les lois possibles et imaginables. Une loi en physique n'est loi que parce que sa mesure est simple, et que l'échelle qui la représente est non seulement la même mais encore qu'elle est unique, et qu'elle ne peut être représentée par une autre échelle; or, toutes les fois que l'échelle d'une loi ne sera pas représentée par un seul terme, cette simplicité et cette unité d'échelle, qui fait l'essence de la loi, ne subsiste plus, et par conséquent il n'y a plus aucune loi physique.

Comme ce dernier raisonnement pourrait paraître n'être que de la métaphysique, et s'il y a peu de gens qui le sachent apprécier, je vais tâcher de la rendre sensible en m'expliquant davantage. Je dis donc, que toutes les fois qu'on voudra établir une loi sur l'augmentation ou la diminution d'une qualité ou d'une quantité physique, on est strictement assujetti à n'employer qu'un terme pour exprimer cette loi; ce terme est la représentation de la mesure qui doit varier, comme en effet la quantité à mesurer varie; en sorte que si la quantité n'étant d'abord qu'un pouce devient ensuite un pied, une aune, une toise, une lieue, etc., le terme qui l'exprime devient successivement toutes ces choses ou plutôt les représente dans le même ordre de grandeur, et il en est de même de toutes les autres raisons dans lesquelles une quantité peut varier.

De quelque façon que nous puissions donc supposer qu'une qualité physique puisse varier, comme cette qualité est une, sa variation sera simple et toujours exprimable par un seul terme, qui en sera la mesure; et dès qu'on voudra employer deux termes on détruira l'unité de la qualité physique, parce que ces deux termes représenteront deux variations différentes dans la même qualité, c'est à dire deux qualités au lieu d'une. Deux termes sont en effet deux mesures, toutes deux variables et inégalement variables; et dès lors elles ne peuvent être appliquées à un sujet simple, à une seule qualité; et si on admet deux termes pour représenter l'effet de la force centrale d'un astre, il est nécessaire d'avouer qu'au lieu d'une force il y en a deux, dont l'une sera relative au premier terme, et l'autre relative au second terme; d'où on voit évidemment qu'il faut, dans le cas présent, que M. Clairaut admette nécessairement une force différente de l'attraction, s'il emploie deux termes pour représenter l'effet total de la force centrale d'une planète.

Je ne sais pas comment on peut imaginer qu'une loi physique, telle qu'est celle de l'attraction, puisse être exprimée par deux termes par rapport aux distances; car s'il y avait, par exemple, une masse M dont la vertu attractive fut exprimée par $\frac{aa}{xx} + \frac{b}{x^4}$, n'en résulterait-il pas le même effet que si cette masse était composée de deux matières différentes comme, par exemple, de $\frac{1}{2}M$ dont la loi d'attraction fut exprimée par $\frac{2aa}{xx}$, et de $\frac{1}{2}M$ dont l'attraction fut $\frac{2b}{x^4}$? Cela me paraît absurde.

Mais, indépendamment de ces impossibilités qu'implique la supposition de M. Clairaut, qui détruit aussi l'unité de loi sur laquelle est fondée la vérité et la belle simplicité du système du monde, cette supposition souffre bien d'autres difficultés que M. Clairaut devait, ce me semble, se proposer avant que de l'admettre, et commencer au moins par examiner d'abord toutes les causes particulières qui pourraient produire le même effet. Je sens que si j'eusse résolu, comme M. Clairaut, le problème des trois corps, et que j'eusse trouvé que la théorie de la gravitation ne donne en effet que la moitié du mouvement de l'apogée, je n'en aurais pas tiré la conclusion qu'il en tire contre la loi de l'attraction; aussi est-ce cette conclusion que je contredis, et à laquelle je ne crois pas qu'on soit obligé de souscrire, quand même M. Clairaut aurait pu démontrer l'insuffisance de toutes les autres causes particulières.

Newton dit (page 547, tome III): "In his computationibus attractionem magneticam terræ non consideravi, cujus itaque quantitas perparva est et ignoratur; si quando vero hæc attractio investigari poterit, et mensura graduum in meridiano, ac longitudines pendulorum isochronorum in diversis parallelis, legesque motuum maris et parallaxis lunæ cum diametris apparentibus solis et lunæ ex phænomenis determinatæ fuerint licebit calculum hunc omnem accuratiùs* repetere."—Ce passage ne prouve-t-il pas bien clairement que Newton n'a pas prétendu avoir fait l'énumération de toutes les causes particulières, et n'indique-t-il pas en effet que si on trouve quelques différences avec sa théorie et les observations, cela peut venir de la force magnétique de la terre, ou de quelque autre cause secondaire? et par conséquent, si le mouvement des apsides ne s'accorde pas aussi exactement avec sa théorie que le reste, faudra-t-il pour cela ruiner sa théorie par le fondement, en changeant la loi générale de la gravitation? ou plutôt ne faudra-t-il pas attribuer à d'autres causes cette différence, qui ne se trouve que dans ce seul phénomène? M. Clairaut a proposé une difficulté contre le système de Newton; mais ce n'est tout au plus qu'une difficulté qui ne doit ni ne peut devenir un principe; il faut chercher à la résoudre, et non pas à en faire une théorie dont toutes les conséquences ne sont appuyées que sur un calcul; car, comme je l'ai dit, on peut tout représenter avec un calcul et on ne réalise rien; et si on se permet de mettre un ou plusieurs termes à la suite de l'expression d'une loi physique, comme l'est celle de l'attraction, on ne nous donne plus que de l'arbitraire, au lieu de nous représenter la réalité.

Au reste il me suffit d'avoir établi les raisons qui me font rejeter la supposition de M. Clairaut, celles que j'ai de croire que, bien loin qu'il ait pu donner atteinte à la loi de l'attraction, et renverser l'astronomie physique, elle me paraît, au contraire, demeurer dans toute sa vigueur, et avoir des forces pour aller encore bien loin; et celà, sans que je prétende avoir dit, à beaucoup près, tout ce qu'on peut dire sur cette matière, à laquelle je désirerais, qu'on donnât, sans prévention, toute l'attention qu'il faut pour la bien juger.

Je me suis borné à démontrer que la loi de l'attraction par rapport à la distance ne peut être exprimée que par un terme, et non par deux ou plusieurs termes; que par conséquent l'expression que M. Clairaut a voulu substituer à la loi du carré des distances, n'est qu'une supposition qui renferme une contradiction; c'est là le seul point auquel je me suis attaché: mais comme il paraît, par sa réponse, qu'il ne m'a pas assez entendu, je vais tâcher de rendre mes raisons plus intelligibles en les traduisant en calcul; ce sera la seule réplique que je ferai à sa réponse.

La loi de l'attraction par rapport à la distance ne peut pas être exprimée par deux termes.

* Buffon, tome 1, p. 279, Paris, Pourrat frères, 1837.

Première Démonstration.

Supposons que $\frac{1}{x^2} \pm \frac{1}{x^4}$ représente l'effet de cette force par rapport à la distance x; ou, ce qui revient au même, supposons que $\frac{1}{x^2} \pm \frac{1}{x^4}$, qui représente la force accélératrice, soit égale à une quantité donnée A pour une certaine distance; en résolvant cette équation, la racine x sera ou imaginaire, ou bien elle aura deux valeurs différentes: donc, à différentes distances, l'attraction serait la même, ce qui serait absurde; donc la loi de l'attraction, par rapport à la distance, ne peut pas être exprimée par deux termes. (Ce qu'il fallait démontrer.)

Seconde Démonstration.

La même expression $\frac{1}{x^2} \pm \frac{1}{x^4}$, si x devient très grand, pourra se reduire à $\frac{1}{x^2}$, et si x devient très petit, elle se réduira à $\pm \frac{1}{x^4}$, de sorte que si $\frac{1}{x^2} \pm \frac{1}{x^4} = \frac{1}{x^2}$, l'exposant n doit être un nombre compris entre 2 et 4; cependant ce même exposant n doit nécessairement renfermer x, puisque la quantité d'attraction doit, de façon ou d'autre, être mesurée par la distance; cette expression prendra donc alors une forme comme $\frac{1}{x^2} \pm \frac{1}{x^4} = \frac{1}{x^x}$ ou $= \frac{1}{x+r}$; donc une quantité, qui doit être nécessairement un nombre compris entre 2 et 4, pourrait cependant devenir infinie, ce qui est absurde, donc l'attraction ne peut pas être exprimée par deux termes. (Ce qu'il fallait démontrer.)

INTORNO AI METODI USATI DAGLI ANTICHI GRECI PER ESTRARRE LE RADICI QUADRATE

Di Gino Loria.

I procedimenti adoperati dagli antichi Greci per effettuare le operazioni aritmetiche sono tuttora avvolti, se non in impenetrabile mistero, almeno in una discreta penombra, non trovandosi essi esposti in alcuna opera speciale; la ragione di questo fatto, indubbiamente deplorevole, risiede nell' opinione allora diffusa che la Logistica (Aritmetica pratica) fosse un soggetto troppo umile per dare materia ad un trattato *ad hoc*, bastando a ciò la viva voce del maestro. Tuttavia, esaminando con cura tutta la letteratura superstite, si riesce, senza gravi stenti, a ricostruire le vie lastricate e battute dai nostri progenitori scientifici per esseguire *le prime quattro operazioni aritmetiche* ed a concludere che esse non differiscono da quelle che noi oggi seguiano, se non nei minuti particolari, provenienti dalle differenze che passano fra i sistemi di numerazione in uso allora ed oggi. Ben maggiori ostacoli incontra chi tenti di divinare i metodi in uso sotto il cielo dell' Ellade per estrarre le *radici quadrate dei numeri* (sottinteso *interi*)! Nel nobile scopo di superarli furono istituite vaste e profonde ricerche, fondate sopra ipotesi, molte delle quali sono, non soltanto geniali, ma anche plausibili: i risultati di tali studî sono tanto noti ai cultori della storia della scienza antica, che è superfluo io mi attardi ad enumerarli ed esaminarli quì*.

Vi sono però alcuni documenti atti a meglio lumeggiare la questione, che benchè pubblicati da quasi tre lustri, non vennero ancora ritenuti meritevoli di studio da parte degli storici della matematica: alludo a quelli fatti conoscere dal celebre ellenistadanese J. L. Heiberg in occasione del primo giubileo di M. Cantor†. Tali documenti appartengono all' epoca bizantina, cioè ad un periodo di tempo nel quale l' astromatematico dei Greci aveva cessato di brillare di luce sua propria; onde è estremamente probabile (si potrebbe dire certo) che quanto è ivi insegnato altro non sia che un semplice relitto del patrimonio aritmetico accumulato durante il periodo aureo della matematica greca. La parte per noi veramente essenziale risulta di due Tabelle di valori approssimati delle radici quadrate di una lunga serie di numeri, espresse una volta in *interi e frazioni ordinarie*, un' altra volta col *sistema astronomico* (cioè in interi e frazioni aventi per denominatori 60, 60^2 ecc.). Quest' ultima presenta, salvo errore, mediocre interesse perchè non è che il compendio

* Mi limito a ricordare l' ottima monografia di S. Günther, *Die quadratischen Irrationalitäten und deren Entwickelungsmethoden* (Abh. zur Gesch. der Mathematik, IV. Heft, 1882).

† *Byzantinische Analekten* (Abh. zur Gesch. der Mathematik, IX. Heft, 1899).

di risultati ottenuti applicando l' algoritmo insegnato da TEONE ALESSANDRINO in un notissimo passo del suo *Commento all' Almagesto* di CLAUDIO TOLOMEO, algoritmo che, in fondo, coincide con quello tuttora in uso. Ben più importante è la prima di quelle TABELLE, come agevolmente si comprende riflettendo che, mentre la semplice conoscenza del risultato di un calcolo *esatto* di regola nulla insegna intorno al procedimento usato per conseguirlo, l' indagine accurata della genesi di un valore *approssimato* può guidare alla scoperta della strada percorsa da chi lo ha ottenuto; onde, con la pubblicazione del documento in questione, l' HEIBERG ha notevolmente accresciute le molte benemerenze che egli da tempo si è acquistate presso i cultori della storia delle matematiche. Dallo studio di esse che ci accingiamo a fare trarremo un metodo uniforme*, basato sopra considerazioni elementarissime, mediante cui è possibile ottenere, non soltanto i valori approssimati ivi registrati, ma anche altri che s' incontrano in ARCHIMEDE ed ERONE.

1. Trascriviamo anzitutto, in simboli moderni, la prima delle TABELLE pubblicate dall' HEIBERG:

$$\sqrt{2}=1+\frac{5}{12},\qquad \sqrt{19}=4+\frac{29}{81},\qquad \sqrt{98}=9+\frac{18}{20},$$

$$\sqrt{3}=1+\frac{11}{15},\qquad \sqrt{20}=4+\frac{17}{36},\qquad \sqrt{99}=9+\frac{19}{20},$$

$$\sqrt{5}=2+\frac{15}{64},\qquad \sqrt{21}=4+\frac{21}{36},\qquad \sqrt{101}=10+\frac{1}{20},$$

$$\sqrt{6}=2+\frac{9}{20},\qquad \sqrt{22}=4+\frac{29}{42},\qquad \sqrt{102}=10+\frac{2}{20},$$

$$\sqrt{7}=2+\frac{38}{59},\qquad \sqrt{23}=4+\frac{43}{54},\qquad \sqrt{109}=10+\frac{37}{84},$$

$$\sqrt{8}=2+\frac{53}{64},\qquad \sqrt{24}=4+\frac{9}{10},\qquad \sqrt{110}=10+\frac{41}{84},$$

$$\sqrt{10}=3+\frac{5}{31},\qquad \sqrt{26}=5+\frac{1}{10},\qquad \sqrt{111}=10+\frac{45}{84},$$

$$\sqrt{11}=3+\frac{19}{60},\qquad \sqrt{27}=5+\frac{12}{61},\qquad \sqrt{112}=10+\frac{49}{84},$$

$$\sqrt{12}=3+\frac{13}{28},\qquad \sqrt{28}=5+\frac{16}{55},\qquad \sqrt{115}=10+\frac{76}{105},$$

$$\sqrt{13}=3+\frac{37}{61},\qquad \sqrt{29}=5+\frac{22}{57},\qquad \sqrt{119}=10+\frac{20}{22},$$

$$\sqrt{14}=3+\frac{43}{58},\qquad \sqrt{30}=5+\frac{21}{44},\qquad \sqrt{120}=10+\frac{21}{22},$$

* Giova tenere presente che *l' uniformità* in una serie di calcoli approssimati congeneri consiste nella costanza del *concetto* che li informa, non nell' *identità di particolari* in tutte le applicazioni, che non si può pretendere quando si ignora se furono tutti eseguiti della stessa persona e nel medesimo intento.

$\sqrt{15} = 3 + \frac{48}{55}$, $\sqrt{31} = 5 + \frac{25}{44}$, $\sqrt{122} = 11 + \frac{1}{22}$,

$\sqrt{17} = 4 + \frac{1}{8}$, $\sqrt{35} = 5 + \frac{11}{12}$, $\sqrt{123} = 11 + \frac{2}{22}$,

$\sqrt{18} = 4 + \frac{7}{29}$, $\sqrt{37} = 6 + \frac{1}{12}$, $\sqrt{147} = 12 + \frac{1}{8}$.

2. Esponiamo poi qualche considerazione generale atta a servici di guida nel nostro saggio di divinazione del procedimento a cui si attenne l' anonimo costruttore di questa TABELLA.

Il teorema che dà il quadrato di un binomio (EUCLIDE, *Elementi* II, 4) mostra subito che

$$\sqrt{a^2 + b} < a + \frac{b}{2a} \qquad \text{.............................(1)};$$

giova notare che tale relazione verrà sempre applicata nell' ipotesi che a^2 sia il massimo quadrato contenuto nel numero $a^2 + b$.

Per trovare un altra relazione analoga, ma di senso opposto, si paragoni $\sqrt{a^2 + b}$ alla frazione $a + \frac{b}{2a + x}$, ossia $a^2 + b$ a $\left(a + \frac{b}{2a + x}\right)^2$; si vedrà che, assumendo $x = \frac{b}{2a}$, risulta

$$a + \frac{b}{2a + \frac{b}{2a}} < a + \frac{b}{2a},$$

ossia

$$a + \frac{2ab}{4a^2 + b} < \sqrt{a^2 + b} \qquad \text{.............................(2)};$$

rileviamo per incidenza che per $a = b = 1$ si ottiene $1 + \frac{2}{5} = \frac{7}{5}$, valore approssimato di $\sqrt{2}$ adoperato da ARISTARCO DA SAMO.

Siccome i due termini della frazione $\frac{2ab}{4a^2 + b}$ risultano spesso così grandi da riuscire praticamente inservibili, così spesso giova ricorrere, invece che alla (2), alla formola

$$a + \frac{b}{2a + 1} < \sqrt{a^2 + b}, \qquad \text{.............................(3)},$$

valida per $b < 2a + 1$, come si vede applicando il succitato teorema di EUCLIDE.

La (1) si è ottenuta partendo dal valore a di $\sqrt{a^2 + b}$, approssimato *per difetto*; prendendo invece le mosse dal numero $a + 1$, approssimato *per eccesso*, si trova similmente

$$\sqrt{(a + 1)^2 - c} < (a + 1) - \frac{c}{2(a + 1)} \qquad \text{.......................(4)}.$$

La conoscenza da parte dei Greci di limitazioni di tal fatta venne generalmente ammessa da coloro che si occuparono della questione che stiamo trattando, nè può essere revocata in dubbio da chiunque conosca il livello matematico raggiunto da un popolo che era in grado di risolvere tutte le questioni riducibili ad equazioni lineari e quadratiche.

Altrettanto può ripetersi riguardo ad altre semplicissime proposizioni, che servono a restringere i limiti entro cui, in forza delle formole precedenti, cade una qualunque radice quadratica: i ragionamenti che servono ad ottenerle, tradotti nella simbolica moderna, si possono presentare sotto la seguente forma:

Se
$$\frac{a}{b} < \frac{c}{d},$$
sarà pure, qualunque siano i numeri (positivi) m e n,
$$\frac{ma}{mb} < \frac{nc}{nd}$$
e
$$\frac{ma}{nc} < \frac{mb}{nd}, \qquad \frac{nc}{ma} > \frac{nd}{mb};$$
queste relazioni non si alterano aggiungendo l' unità a ciascun membro, onde
$$\frac{ma+nc}{nc} < \frac{mb+nd}{nd}, \qquad \frac{ma+nc}{ma} > \frac{mb+nd}{mb}$$
ossia
$$\frac{ma+nc}{mb+nd} < \frac{c}{d}, \qquad \frac{ma+nc}{mb+nd} > \frac{a}{b},$$
vale a dire
$$\frac{a}{b} < \frac{ma+nc}{mb+nd} < \frac{c}{d} \;\ldots\ldots\ldots\ldots\ldots\ldots\ldots\ldots(5):$$
questa doppia diseguaglianza ci riuscirà utilissima.

Utile è pure l' osservazione seguente: si supponga di essere giunti ad una limitazione del seguente tipo:
$$\frac{a}{b} < N < \frac{c}{d};$$
qualunque siano i numeri μ, ν sarà pure
$$\frac{a\mu}{b\mu} < N < \frac{\left(\frac{c\mu}{bd}\right)}{b\mu} \;\ldots\ldots\ldots(6), \qquad \frac{\left(\frac{a\nu}{bd}\right)}{d\nu} < N < \frac{c\nu}{d\nu} \;\ldots\ldots\ldots(7);$$
ora è chiaro che si otterrà per N un valore più approssimato di quanto siano i numeri $\frac{a}{b}$, $\frac{c}{d}$ scegliendo una frazione che abbia per denominatore $b\mu$ e per numeratore un intero compreso fra $a\mu$ e $\frac{c\mu}{bd}$, oppure per denominatore $d\nu$ e per numeratore un intero compreso fra $\frac{a\nu}{bd}$ e $c\nu$.

Mostreremo ora come queste semplici osservazioni siano sufficienti ad ottenere *tutti gli elementi* della precedente TABELLA, dopo di avere notato in generale che i valori ivi registrati sono tanto più approssimati quanto più piccoli sono i numeri radicandi.

3. Applicando la formola (1) e scrivendo $\doteq$ invece di $<$, per indicare che si tratta sempre di eguaglianze approssimate, si trova:
$$\sqrt{17} = \sqrt{4^2+1} \doteq 4 + \frac{1}{8},$$
$$\sqrt{26} = \sqrt{5^2+1} \doteq 5 + \frac{1}{10}.$$

$$\sqrt{37} = \sqrt{6^2+1} \doteq 6 + \frac{1}{12},$$

$$\sqrt{101} = \sqrt{10^2+1} \doteq 10 + \frac{1}{20},$$

$$\sqrt{102} = \sqrt{10^2+2} \doteq 10 + \frac{2}{20},$$

$$\sqrt{122} = \sqrt{11^2+1} \doteq 11 + \frac{1}{22},$$

$$\sqrt{123} = \sqrt{11^2+2} \doteq 11 + \frac{2}{22},$$

$$\sqrt{147} = \sqrt{12^2+3} \doteq 12 + \frac{3}{24} = 12 + \frac{1}{8},$$

tutti valori registrati nella riferita TABELLA*.

Applicando invece la formola (4), con la stessa avvertenza di prima riguardo al segno $\doteq$, si ottiene

$$\sqrt{24} = \sqrt{5^2-1} \doteq 5 - \frac{1}{10} = 4 + \frac{9}{10},$$

$$\sqrt{35} = \sqrt{6^2-1} \doteq 6 - \frac{1}{12} = 5 + \frac{11}{12},$$

$$\sqrt{98} = \sqrt{10^2-2} \doteq 10 - \frac{2}{20} = 9 + \frac{18}{20},$$

$$\sqrt{99} = \sqrt{10^2-1} \doteq 10 - \frac{1}{20} = 9 + \frac{19}{20},$$

$$\sqrt{119} = \sqrt{11^2-2} \doteq 11 - \frac{2}{22} = 10 + \frac{20}{22},$$

$$\sqrt{120} = \sqrt{11^2-1} \doteq 11 - \frac{1}{22} = 10 + \frac{21}{22},$$

che pure si trovano in quella TABELLA.

4. Servendosi ora delle formole (1), (2) si ottiene

$$1 + \frac{2}{5} < \sqrt{1+1} < 1 + \frac{1}{2},$$

cioè

$$\frac{7}{5} < \sqrt{2} < \frac{3}{2},$$

onde la (5) porge il seguente valore approssimato:

$$\sqrt{2} \doteq \frac{2.7+3}{2.5+2} = \frac{17}{12} = 1 + \frac{5}{12},$$

conformemente a quanto registra la TABELLA ed usa ERONE.

* Si può aggiungere che lo stesso metodo dà $\sqrt{50} = \sqrt{7^2+1} \doteq 7 + \frac{1}{14}$, che s' incontra in ERONE.

Similmente si ha:

$$1+\frac{2}{3}<\sqrt{1+2}<1+1, \qquad \frac{5}{3}<\sqrt{3}<2, \qquad \sqrt{3}\doteq\frac{4.5+3.2}{4.3+3.1}=\frac{26}{15}=1+\frac{11}{15},$$

valore che s' incontra, non soltanto nell' anonima TABELLA, ma anche in ARCHIMEDE.

Analogamente:

$$2+\frac{8}{18}<\sqrt{2^2+2}<2+\frac{2}{4}, \qquad \frac{22}{9}<\sqrt{6}<\frac{5}{2}, \qquad \sqrt{6}\doteq\frac{2.22+5}{2.9+2}=\frac{49}{20}=2+\frac{9}{20},$$

$$2+\frac{12}{19}<\sqrt{2^2+3}<2+\frac{3}{4}, \qquad \frac{50}{19}<\sqrt{7}<\frac{11}{4}, \qquad \sqrt{7}\doteq\frac{25.50+14.11}{25.19+14.4}=\frac{9.156}{9.59}=2+\frac{38}{59},$$

$$2+\frac{16}{20}<\sqrt{2^2+4}<2+\frac{4}{4}, \qquad \frac{14}{5}<\sqrt{8}<3, \qquad \sqrt{8}\doteq\frac{11.14+9.3}{11.5+9.1}=\frac{181}{64}=2+\frac{53}{64},$$

$$3+\frac{12}{38}<\sqrt{3^2+2}<3+\frac{2}{6}, \qquad \frac{63}{19}<\sqrt{11}<\frac{10}{3}, \qquad \sqrt{11}\doteq\frac{3.63+10}{3.19+3}=\frac{199}{60}=3+\frac{19}{60},$$

$$3+\frac{18}{39}<\sqrt{3^2+3}<3+\frac{3}{6}, \qquad \frac{45}{13}<\sqrt{12}<\frac{7}{2}, \qquad \sqrt{12}\doteq\frac{2.45+7}{2.13+2}=\frac{97}{28}=3+\frac{13}{28},$$

$$5+\frac{50}{135}<\sqrt{5^2+5}<5+\frac{5}{10}, \qquad \frac{115}{21}<\sqrt{30}<\frac{11}{2}, \qquad \sqrt{30}\doteq\frac{2.115+11}{2.21+2}=\frac{241}{44}=5+\frac{21}{44},$$

tutti valori che si leggono anche in quella TABELLA.

5. Ad altri si perviene col mezzo di un' opportuna combinazione delle formole (1), (3), (5). Citiamo anzitutto un secondo valore di $\sqrt{3}$ registrato in margine della TABELLA. Servendosi delle citate formole (1), (3) si trova anzitutto

$$\frac{5}{3}<\sqrt{3}<\frac{7}{4},$$

donde

$$\sqrt{3}\doteq\frac{56.5+153.7}{56.3+153.4}=\frac{1351}{780}=1+\frac{571}{780}.$$

Similmente si trova:

$$3+\frac{4}{7}<\sqrt{3^2+4}<3+\frac{4}{6}, \qquad \frac{25}{7}<\sqrt{13}<\frac{11}{3}, \qquad \sqrt{13}\doteq\frac{11.25+15.11}{11.7+15.3}=\frac{220}{61}=3+\frac{37}{61},$$

$$4+\frac{6}{9}<\sqrt{4^2+6}<4+\frac{6}{8}, \qquad \frac{14}{3}<\sqrt{22}<\frac{19}{4}, \qquad \sqrt{22}\doteq\frac{10.14+3.19}{10.3+3.4}=\frac{197}{42}=4+\frac{29}{42},$$

$$5+\frac{2}{11}<\sqrt{5^2+2}<5+\frac{2}{10}, \qquad \frac{57}{11}<\sqrt{27}<\frac{26}{5}, \qquad \sqrt{27}\doteq\frac{57+10.26}{11+10.5}=\frac{317}{61}=5+\frac{12}{61},$$

tutto ciò in conformità con la riportata TABELLA.

Un' altra coppia di numeri registrati nella stessa si ottengono assumendo a come valore approssimato per difetto di $\sqrt{a^2+b}$, nell' ipotesi che a^2 sia il massimo quadrato contenuto in a^2+b, e $a+\frac{b}{2a}$ come valore approssimato per eccesso e quindi applicando le (5); quali siano i numeri in questione risulta da ciò che segue:

$$3<\sqrt{10}<\frac{19}{6}, \qquad \sqrt{10}\doteq\frac{3+5.19}{1+5.6}=\frac{98}{31}=3+\frac{5}{31},$$

$$4<\sqrt{18}<\frac{17}{4}, \qquad \sqrt{18}\doteq\frac{4+7.17}{1+7.4}=\frac{123}{29}=4+\frac{7}{29}.$$

Attenendosi allo stesso schema di calcolo, ma applicando le formole (2) e (4), si giunge ad un' altro elemento della TABELLA; infatti:

$$3+\frac{30}{41}<\sqrt{3^2+5}<4-\frac{2}{8}, \quad \frac{153}{41}\sqrt{14}<\frac{15}{4}, \quad \sqrt{14}\doteq\frac{2.153+23.15}{2.41+23.4}=\frac{217}{58}=3+\frac{43}{58}.$$

Servendosi invece delle (3), (4) se ne ottengono altri quattro, come emerge dai seguenti calcoli:

$$3+\frac{6}{7}<\sqrt{3^2+6}=\sqrt{4^2-1}<4-\frac{1}{8}, \quad \frac{27}{7}<\sqrt{15}<\frac{31}{8}, \quad \sqrt{15}\doteq\frac{27+6.31}{7+6.8}=\frac{213}{55}=3+\frac{48}{55},$$

$$4+\frac{3}{9}<\sqrt{4^2+3}=\sqrt{5^2-6}<5-\frac{6}{10}, \quad \frac{13}{3}<\sqrt{19}<\frac{22}{5}, \quad \sqrt{19}\doteq\frac{17.13+6.22}{17.3+6.5}=\frac{353}{81}=4+\frac{29}{81},$$

$$4+\frac{7}{9}<\sqrt{4^2+7}=\sqrt{5^2-2}<5-\frac{2}{10}, \quad \frac{43}{9}<\sqrt{23}<\frac{24}{5}, \quad \sqrt{23}\doteq\frac{43+9.24}{9+9.5}=\frac{259}{54}=4+\frac{43}{54},$$

$$5+\frac{4}{11}<\sqrt{5^2+4}=\sqrt{6^2-7}<6-\frac{7}{12}, \quad \frac{59}{11}<\sqrt{29}<\frac{65}{12}, \quad \sqrt{29}\doteq\frac{3.59+2.65}{3.11+2.10}=\frac{307}{57}=5+\frac{22}{57}.$$

6. Tutti i rimanenti valori registrati nella TABELLA, che stiamo studiando si possono ottenere applicando il concetto algebricamente espresso dalle formole (6), (7), come passiamo a dimostrare.

Col mezzo delle (1), (2) si trova anzitutto

$$\frac{34}{17}<\sqrt{5}<\frac{9}{4},$$

ovvero

$$\frac{143+\frac{1}{7}}{64}<\sqrt{5}<\frac{144}{64};$$

onde approssimativamente si può assumere:

$$\sqrt{5}\doteq\frac{143}{64}=2+\frac{15}{64}.$$

Similmente:

$$\frac{76}{17}<\sqrt{20}<\frac{9}{2}, \qquad \frac{161-\frac{1}{7}}{36}<\sqrt{20}<\frac{162}{36}, \qquad \sqrt{20}\doteq\frac{161}{36}=4+\frac{17}{36}.$$

Applicando invece le (1), (3) si trova

$$\frac{41}{9}<\sqrt{21}<\frac{37}{8},$$

ossia

$$\frac{164}{36}<\sqrt{21}<\frac{165+\frac{2}{3}}{36},$$

onde si presenta spontanea la posizione

$$\sqrt{21}\doteq\frac{165}{36}=4+\frac{21}{36}.$$

In modo analogo si trova:

$$5+\frac{3}{11}<\sqrt{28}<5+\frac{3}{10}, \qquad 5+\frac{15}{55}<\sqrt{28}<5+\frac{16+\frac{1}{2}}{55}, \quad \sqrt{28}\doteq5+\frac{16}{55},$$

$$\frac{61}{11}<\sqrt{31}<\frac{28}{5}, \qquad \frac{244}{44}<\sqrt{31}<\frac{246+\frac{2}{5}}{44}, \quad \sqrt{31}\doteq\frac{245}{44}=5+\frac{25}{44},$$

$$\frac{219}{21} < \sqrt{109} < \frac{209}{20}, \qquad \frac{876}{84} < \sqrt{109} < \frac{877 + \frac{4}{5}}{84}, \qquad \sqrt{109} \doteq \frac{877}{84} = 10 + \frac{37}{84},$$

$$\frac{220}{21} < \sqrt{110} < \frac{21}{2}, \qquad \frac{880}{84} < \sqrt{110} < \frac{882}{84}, \qquad \sqrt{110} \doteq \frac{881}{84} = 10 + \frac{41}{84},$$

$$\frac{221}{22} < \sqrt{111} < \frac{211}{20}, \qquad \frac{884}{84} < \sqrt{111} < \frac{886 + \frac{1}{5}}{84}, \qquad \sqrt{111} \doteq \frac{885}{84} = 10 + \frac{45}{84},$$

$$\frac{222}{21} < \sqrt{112} < \frac{53}{5}, \qquad \frac{888}{84} < \sqrt{112} < \frac{890 + \frac{2}{5}}{84}, \qquad \sqrt{112} \doteq \frac{889}{84} = 10 + \frac{49}{84},$$

$$\frac{225}{21} < \sqrt{115} < \frac{43}{4}, \qquad \frac{1125}{105} < \sqrt{115} < \frac{1128 + \frac{3}{5}}{105}, \qquad \sqrt{115} \doteq \frac{1126}{105} = 10 + \frac{76}{105}.$$

Quanto al valore
$$\sqrt{11} \doteq 3 + \frac{181}{390},$$
che una mano ignota scrisse in margine della TABELLA, non mette conto occuparcene chè il valore $\sqrt{11} = 3 + \frac{1}{3} = 3 + \frac{130}{390}$ è più approssimato ed espresso in numeri più piccoli, onde, per una doppia ragione, è preferibile.

Notiamo che l' ispezione della riportata TABELLA mostra che chi la compose ignorò essere $\sqrt{m^2 n} = m\sqrt{n}$; tutt' almeno non seppe utilizzare tale identità.

Sarei lieto che le osservazioni esposte richiamassero l' attenzione degli studiosi sopra una questione di grande interesse per la conoscenza della matematica greca e sopra documenti atti, a mio credere, di projettare grande luce su di essa.

NOTE ON FOURIER'S INFLUENCE ON THE CONCEPTIONS OF MATHEMATICS

By Philip E. B. Jourdain.

It is well known that Fourier had precise notions on the convergence of series, and demonstrated the convergence of the series and integrals named after him in a manner which, at bottom, is exact, and is, in fact, the way which was later put into the form of a model of mathematical deduction by Dirichlet. I wish here to emphasize that Fourier gave, in other directions as well, the most important stimulus, perhaps, given in the nineteenth century to the development of the conceptions of pure mathematics. As we may see in many cases, lack of interest in what mathematics had not immediate physical applications prevented his cultivation of pure mathematics—which, in this case, is the logical discussion of the methods used in mathematical physics.

I.

In the first place, since it was mainly due to Fourier's work that the older Eulerian distinction between functions which are "continuous" and "discontinuous" *in their analytical expression* had to be given up, this work must be regarded as one of the chief causes of Cauchy's new (1821) conception of the continuity of functions*, and also of the connected reform with respect to the limits of validity of formulae. Then also, the extension of the conception of function, which was brought about principally by Fourier's work, necessitated a return to the conception of an integral as a sum—or rather as the limit of certain sums. Fourier always used the term "discontinuity" in Euler's sense, so that with him also a "discontinuous" function had not a differential quotient at certain points. Nevertheless, the integrals of such functions appeared as coefficients in the Fourier's series. Thus, such integrals could not be defined as inverses of differential quotients, though it seemed plain that they had a geometrical interpretation as areas†. This reinstatement of the original conception of Leibniz seems to me very important, since, if we have not taken into account this character of Fourier's work, there appears to be an unfilled gap between the older conception of the integral in Euler's works and the somewhat shortly expressed conception of Cauchy‡.

* Cf. A. Brill and M. Noether, "Die Entwicklung der Theorie der algebraischen Functionen in älterer und neuerer Zeit," *Jahresber. der Deutschen Math.-Ver.* III, 1894, pp. 160—163.

† Cf. Fourier's prize memoir of 1811, printed in the *Mém. de l'Inst.* IV, 1819-20, pp. 301—302; *Théorie analytique de la Chaleur*, Paris, 1822, art. 220; *Œuvres*, I, p. 210; A. Freeman's translation (*The Analytical Theory of Heat*, Cambridge, 1878), p. 186.

‡ Cf. my papers in *Bibl. Math.* (3), VI, 1905, p. 193 (note), and *Archiv der Math. und Phys.* (3), X, 1906, p. 261 (note).

II.

In the second place, it appears that, if Fourier did not actually discover the conception of non-uniform convergence, he came very close to it, as the following extract will show.

Fourier considers the equation

$$\frac{\pi}{4} = \cos x - \frac{1}{3}\cos 3x + \frac{1}{5}\cos 5x - \dots,$$

and remarks* that: "It is easy to see now that the result necessarily holds if the arc x is less than $\frac{1}{2}\pi$. In fact, attributing to this arc a definite value X as near to $\frac{1}{2}\pi$ as we please, we can always give to m a value so great, that the term $\frac{k}{2m}(\sec x - \sec 0)$, which completes the series, becomes less than any quantity whatever." Now, Fourier† explicitly says that the series referred to is *always* convergent, "that is to say that writing instead of y any number whatever, and following the calculation of the coefficients, we approach more and more to a fixed value, so that the difference of this value from the sum of the calculated terms becomes less than any assignable magnitude." However, he leaves the proof of this assertion to the reader, and remarks, in the next article, that, when x is $\frac{1}{2}\pi$, y is the straight line perpendicular to the axis of x which joins parallel straight lines above and below this axis. It is quite possible that Fourier, like so many later writers, thought that y became many-valued at points such as $\frac{1}{2}\pi$, though perhaps his attention was not directed to this point, since it would naturally be of greater importance for the physicist to consider the curves formed by the taking of a finite number of terms of the series in order to get a close approximation to states of distribution of heat which may be observed in nature.

* *Mém. de l'Inst.* IV, p. 279; *Théorie*, art. 188; Freeman, p. 153. In the *Théorie*, but not in the memoir of 1811, is added to the above quotation: "But the exactness of this conclusion is based on the fact that the term $\sec x$ acquires no value which exceeds all possible limits, whence it follows that the same reason cannot apply to the case in which the arc x is not less than $\frac{1}{2}\pi$."

† *Mém. de l'Inst.* IV, p. 270; *Théorie*, art. 177; Freeman, p. 143.

WALTHER VON DYCK

spricht über den Mechaniker und Ingenieur GEORG VON REICHENBACH.

Das Deutsche Museum in München beabsichtigt, eine Reihe von Lebensbeschreibungen und Urkunden zur Geschichte der Naturwissenschaften und der Technik herauszugeben, vorzugsweise gestützt auf die in den reichen Sammlungen des Museums aufbewahrten historischen Objekte und Dokumente. Die Lebensbeschreibung Georg von Reichenbach's bildet den ersten Band dieser Veröffentlichungen, welchen der Verfasser dem Congress überreicht.

Die Bedeutung Reichenbachs liegt in seinen bekannten Schöpfungen auf dem Gebiete des Baues astronomischer und geodätischer Instrumente und in dem Bau seiner Wassersäulenmaschinen für den Salinenbetrieb in Berchtesgaden und Reichenhall. Die im Deutschen Museum aufbewahrten Dokumente haben darüber hinaus sehr interessante Arbeiten über die Construction von Dampfmaschinen mit hochgespanntem Dampf zu Tage gefördert, die hier zum ersten Mal veröffentlicht sind.

Die Vorlage des Buches auf dem gegenwärtigen Congress in Cambridge ist gerechtfertigt durch die mannigfachen Beziehungen der Arbeiten Reichenbachs zu den technischen Leistungen in England. Einmal sind es die Studien, die Reichenbach als junger Ingenieur im Auftrag des Kurfürsten Karl Theodor von der Pfalz in Soho, in der Fabrik von Boulton und Watt gemacht hat, wie sein späterer Aufenthalt auf den grossen Eisenwerken des Lord Balcarres. Weiter hat Reichenbach den Instrumentenbau Englands (Ramsden, Cary, Troughton) vorzugsweise auf den dortigen Sternwarten kennen gelernt. Der grosse Erfolg seiner späteren Arbeiten auf diesem Gebiet liegt in der Verbesserung der Kreiseinteilungsmethoden, durch welche es ihm gelang, die damals ungeheuren Dimensionen der Instrumente zu reduciren und ihre Genauigkeit wesentlich zu erhöhen.

Auch den Brückenbau, den Schiffsbau und ganz besonders auf artilleristischem Gebiet die Geschütz- und Geschosskonstruktion hat Reichenbach mit neuen Ideen befruchtet.

MITTEILUNGEN ÜBER DIE EULERAUSGABE

Von Ferdinand Rudio.

Hochgeehrte Versammlung,

Es sind jetzt 15 Jahre verflossen, seit auf dem ersten Internationalen Mathematiker-Kongress in Zürich die Forderung gestellt wurde, es sollten sich die Mathematiker der ganzen Welt zur Herausgabe der Werke Leonhard Eulers vereinigen. Damit wurde eine Forderung wiederholt, die auf eine jahrzehntelange Geschichte zurückblicken konnte, eine Geschichte, in der die Namen Fuss und Jacobi mit besonderem Glanze hervorleuchten. Auch auf dem dritten Internationalen Mathematikerkongress, der 1904 in Heidelberg stattfand und der dem Andenken Jacobis gewidmet war, bot sich Gelegenheit, von neuem an die Ehrenschuld zu erinnern, die schon längst hätte getilgt sein sollen.

Aber erst das Jahr 1907, das Jahr, in dem der zweihundertste Geburtstag Eulers von der ganzen wissenschaftlichen Welt festlich begangen wurde, führte von den Worten zur Tat. Unter dem Eindruck der Bewegung, die sich bei der Feier dieses Geburtstages der mathematischen Welt bemächtigte, unter dem Eindruck namentlich der erhebenden Geburtstagsfeier in Basel beschloss die Schweizerische Naturforschende Gesellschaft in ihrer Jahresversammlung zu Freiburg eine Eulerkommission niederzusetzen mit dem Auftrage, *die Mittel und Wege zu studieren, die zu einer Gesamtausgabe der Werke Eulers erforderlich sind.*

Dieser Beschluss rief überall freudigen Wiederhall hervor. So fasste auf Veranlassung der Deutschen Mathematiker-Vereinigung der vierte Internationale Mathematiker-Kongress in Rom, April 1908, einstimmig folgende Resolution:

> "Der vierte Internationale Mathematiker-Kongress in Rom betrachtet eine Gesamtausgabe der Werke Eulers als ein Unternehmen, das für die reine und angewandte Mathematik von der grössten Bedeutung ist. Der Kongress begrüsst mit Dank die Initiative, die die Schweizerische Naturforschende Gesellschaft in dieser Angelegenheit ergriffen hat, und spricht den Wunsch aus, dass das grosse Unternehmen von dieser Gesellschaft in Gemeinschaft mit den Mathematikern der andern Nationen ausgeführt werde. Der Kongress bittet die Internationale Assoziation der Akademien und insbesondere die Akademien zu Berlin und Petersburg, deren glorreiches Mitglied Euler gewesen ist, das genannte Unternehmen zu unterstützen."

Mit dieser Resolution war das Eulerunternehmen unter den Schutz der Internationalen Mathematiker-Kongresse gestellt worden und die Mitglieder des fünften Kongresses dürfen daher mit Recht erwarten, dass ihnen über die Ereignisse der letzten vier Jahre Bericht erstattet werde.

Bei dem, was Ihnen allen bekannt ist, will ich freilich nicht lange verweilen. Sie wissen, dass es der Schweizerischen Naturforschenden Gesellschaft innerhalb eines einzigen Jahres gelang, eine den ganzen Erdball umfassende Begeisterung zu entflammen, wie sie in der Geschichte der Wissenschaft ohne Beispiel dasteht.

Bis zum Herbst 1909 war durch freiwillige Beiträge und durch Subskriptionen, insbesondere durch die Beteiligung der Akademien von Berlin, Paris und Petersburg, die Summe von rund einer halben Million Franken gesichert, und so durfte denn am 6. September 1909 die Schweizerische Naturforschende Gesellschaft auf ihrer Jahresversammlung zu Lausanne—der Stadt, wo einst die *Introductio* und die *Methodus inveniendi lineas curvas* erschienen waren,—jenen denkwürdigen Beschluss fassen, dem wir in absehbarer Zeit die Gesamtausgabe der Werke Leonhard Eulers werden zu verdanken haben.

Mit der Durchführung des Unternehmens wurde ein Redaktionskomitee betraut, das aus Adolf Krazer, Paul Stäckel und dem Sprechenden besteht.

Es bedarf keiner weiteren Ausführung, dass für ein so gewaltiges Unternehmen zunächst umfangreiche Vorarbeiten zu erledigen waren. Insbesondere handelte es sich um die Ausarbeitung eines ausführlichen Redaktionsplanes, um eine genaue Revision des Stäckelschen Entwurfes einer Einteilung der Werke Eulers und sodann um die Gewinnung der Herausgeber der einzelnen Bände. Dank der freundlichen Unterstützung, die dem Redaktionskomitee von zahlreichen Kollegen zu teil wurde, konnten diese Vorarbeiten bis Mitte 1910 abgeschlossen werden. Für jeden Band waren ein oder mehrere Herausgeber gewonnen worden, im ganzen 37, die sich auf Deutschland, England, Frankreich, Italien, Oesterreich, Russland, Schweden und die Schweiz verteilen.

Inzwischen hatte die rührige Firma B. G. Teubner, in deren Hände die technische Durchführung des Unternehmens gelegt worden war, durch zahlreiche Druckvorlagen dafür gesorgt, dass definitive Festsetzungen über die Art und die Grösse der Schrift und über die Anordnung des Satzes getroffen werden konnten. Dabei stellte sich heraus, dass die ursprünglich in Aussicht genommene Korpusschrift, die auch allen Berechnungen zu Grunde gelegen hatte, dem monumentalen Charakter, den eine Eulerausgabe beanspruchen darf, nicht völlig genügen würde und dass eine grössere und würdigere Schrift gewählt werden müsse. Wer die bereits erschienenen Bände gesehen hat, wird der vom Redaktionskomitee getroffenen Wahl die Zustimmung wohl kaum versagen.

Noch im Jahre 1910 konnte mit dem Satz von 3 Werken, der *Algebra*, der *Dioptrik* und der *Mechanik*, begonnen werden. Dabei sei gleich hier bemerkt, dass jede Fahne und jeder Bogen in allen Korrekturen nicht nur vom Herausgeber, sondern auch vom Generalredaktor und noch einem andern Mitgliede der Redaktion gelesen werden, sodass für die denkbar grösste Korrektheit der Eulerausgabe garantiert werden kann.

Am 1. August 1911, dem Tage der Schweizerischen Bundesfeier, konnte der Schweizerischen Naturforschenden Gesellschaft bei ihrer Jahresversammlung in Solothurn der erste Band der Eulerausgabe vorgelegt werden. Der stattliche, über 93 Bogen umfassende Band enthält ein *Vorwort zur Gesamtausgabe der*

Werke von Leonhard Euler, die Fuss'sche *Lobrede auf Herrn Leonhard Euler* und sodann als Hauptinhalt die *Algebra* Eulers mit den *Additions* von Lagrange. Der Band ist geschmückt mit einer Reproduktion des Stiches, das der Basler Kupferstecher Christian von Mechel nach dem von dem Basler Maler Emanuel Handmann im Jahre 1756 gemalten Oelbilde Eulers gestochen hatte. Dass der erste Band der Eulerausgabe von Heinrich Weber herausgegeben worden ist, möge als ein besonders freundliches Omen angesehen werden.

Wenige Wochen nach dem ersten Bande war die Redaktion in der Lage, der Versammlung deutscher Naturforscher und Aerzte in Karlsruhe als zweiten Band die erste Hälfte der *Dioptrik* zu überreichen, und im Frühjahr dieses Jahres lag mit dem dritten Bande das grosse Werk der *Dioptrik*, herausgegeben von Emil Cherbuliez, fertig vor. Im *Jahresbericht der Deutschen Mathematiker-Vereinigung* ist in zwei Nummern ein ausführliches Referat über die drei ersten Bände der Eulerausgabe erschienen und es ist darin mit besonderem Nachdruck auf die geradezu klassische Schönheit der oft verkannten Dioptrik hingewiesen und die Erwartung ausgesprochen worden, dass das Werk in dem neuen Gewande eine wahre Auferstehung feiern werde.

Und heute habe ich die Ehre, Ihre Aufmerksamkeit auf zwei weitere Bände lenken zu dürfen, deren Abschluss noch gerade rechtzeitig für den diesjährigen Kongress hat erzielt werden können. Es handelt sich um die von Paul Stäckel herausgegebene *Mechanik*. Wie die Originalausgabe von 1736 erscheint sie in *zwei* Bänden, obwohl in dem ursprünglichen Entwurf die Ausgabe in *einem* Bande vorgesehen war. Dieser eine Band hätte aber—im wesentlichen zufolge der gewählten grösseren Schrift—nicht weniger als 111 Bogen umfasst, die zum Preise von 25 Fr. zu liefern ein Ding der Unmöglichkeit gewesen wäre—ganz abgesehen von der Monstrosität einer solchen Publikation.—Um dies noch genauer zu begründen, darf wohl auf die Herstellungskosten der drei ersten Bände hingewiesen werden, die in dem Berichte der Eulerkommission für das Jahr 1911/12 mitgeteilt sind. Der erste Band, die Algebra, hat allein rund 22,000 Fr. gekostet, denen aus dem Abonnement nur 9450 Fr. Einnahmen gegenüberstehen. Dieser eine Band hat also ein Defizit von über 12,000 Fr. verursacht. Günstiger stellt sich die Rechnung bei den zwei dünneren Dioptrikbänden, die mit rund 31,000 Fr. Ausgaben und 19,000 Fr. Einnahmen den Eulerfond zusammen mit 12,000 Fr. belasten. Aus diesen Zahlen ergibt sich die ernste Mahnung, der sich kein Einsichtiger wird verschliessen können, dass die Bände im Durchschnitt nicht über 60 Bogen umfassen dürfen, wenn nicht das ganze Unternehmen schwer gefährdet werden soll.

Doch ich kehre zur Mechanik zurück. Der Herausgeber hat dem ersten Bande ein ausführliches, an historischen Notizen reiches Vorwort vorausgeschickt. Ausserdem ziert diesen Band ein nach dem Stiche von Friedrich Weber hergestelltes Porträt Eulers. Die Petersburger Akademie hatte zwar die in ihrem Besitze befindliche Originalplatte in liberalster Weise zur Verfügung gestellt, da aber das Format zu gross war, so mussten wir uns mit einer Reproduktion begnügen.

Den jetzt vorliegenden fünf Bänden werden sich in wenigen Monaten drei weitere angeschlossen haben. Als sechster Band wird noch im Laufe dieses Jahres die erste Hälfte der Abhandlungen über die elliptischen Integrale, herausgegeben von Adolf Krazer, erscheinen. Der Band ist bereits fertig gesetzt und korrigiert. Auch von dem folgenden Bande, der die zweite Hälfte der genannten Abhandlungen bringen wird, ist bereits ein grosser Teil gesetzt. Beide Teile werden über 90 Bogen umfassen und mussten daher in zwei Bänden untergebracht werden. Von einem weiteren Bande, den von Gerhard Kowalewski herausgegebenen *Institutiones calculi differentialis*, ist ebenfalls ein Teil gesetzt. Die Eulerausgabe schreitet also rüstig vorwärts.

Ich darf nicht schliessen, ohne des gewaltigen handschriftlichen Materiales zu gedenken, das die Petersburger Akademie dem Redaktionskomitee zur Verwertung überlassen und nach Zürich gesandt hat. Das gedruckte Verzeichnis dieser kostbaren Manuskripte umfasst auf 13 Quartseiten 209 Nummern und hat seitdem noch wesentliche Ergänzungen erfahren. Einen besonders wertvollen Bestandteil dieses Materials bilden natürlich die Briefe von und an Euler, von denen bis jetzt nur ein Teil veröffentlicht ist. Freilich ist mit den Petersburger Manuskripten der Briefwechsel noch lange nicht erschöpft, und es wird daher eine der nächsten Aufgaben des Redaktionskomitees sein, das Material, soweit nur irgend möglich, zu vervollständigen. Ich darf wohl die Gelegenheit benutzen, die hohe Versammlung zu bitten, das Redaktionskomitee bei dieser Arbeit nach Kräften zu unterstützen. Zu grossem Danke sind wir bereits der Royal Society verpflichtet, die auf Veranlassung von Sir J. Larmor die in ihrem Besitze befindlichen Briefe Eulers hat kopieren lassen und uns die Kopien zugestellt hat.

Mit der Sichtung des gesamten handschriftlichen Materiales ist Gustaf Eneström beschäftigt, dem wir das treffliche *Verzeichnis der Schriften Leonhard Eulers* verdanken. Herr Eneström hat zur Durchführung dieser Aufgabe schon zweimal Aufenthalt in Zürich genommen und wird die Arbeit voraussichtlich im nächsten Jahre soweit gefördert haben, dass er den in Aussicht gestellten Bericht erstatten kann. Es darf aber schon jetzt gesagt werden, dass das Manuskriptenmaterial, auch abgesehen von den Briefen, noch viel Interessantes und Wertvolles zu liefern verspricht.

Viribus unitis! Mit diesem Rufe habe ich vor 15 Jahren auf dem ersten Internationalen Mathematiker-Kongress auf die Ehrenpflicht einer Herausgabe der Werke Leonhard Eulers hingewiesen. Dank der Mitwirkung der ganzen mathematischen Welt hat die Schweizerische Naturforschende Gesellschaft es unternehmen können, an die Lösung dieser schweren und verantwortungsvollen Aufgabe heranzutreten. Aus meinen Mitteilungen werden Sie aber entnommen haben, dass es noch immer grosser und allgemeiner Anstrengungen bedarf, um das gewaltige Werk zu einem glücklichen Ende zu führen. *Viribus unitis* sei daher auch heute wieder unsere Losung! Mit diesem Rufe stelle ich von neuem das Unternehmen der Eulerausgabe unter den Schutz der Internationalen Mathematiker-Kongresse.

THE GEOMETRY OF THALES

By Percy J. Harding.

In this paper there will, of course, be found no new facts; but merely a review of well-known historical statements and an attempt to connect them.

It has often been stated that Thales introduced an abstract Geometry of lines and angles; whereas the Egyptians dealt with a practical Geometry of areas and volumes, and possessed only slight beginnings of any more advanced views.

This seems to be true when we compare the papyrus of Ahmes with the theorems attributed to Thales. But the latter, like men of all ages, must have made his advances gradually and with difficulty; and to understand his position we ought to avoid ascribing to him too great or too sudden a bound forward into abstract science. The gradual nature of his advances is clearly pointed out by the practical measurements that are related of him, forming, as these do, a natural link between the geometry of the Egyptians and his own more abstract work.

One of the most remarkable of the theorems attributed to Thales is that the angle in a semicircle is a right angle; and it is in connection with this theorem that I wish to enquire with what amount of geometrical knowledge he should be credited.

Allman and Cantor consider that Thales knew not only the theorem just mentioned, but also that the sum of the angles of a triangle equals two right angles; and further that he succeeded in deducing one of these theorems from the other with the help of his knowledge that the angles at the base of an isosceles triangle are equal. They differ from one another as to which of these two theorems he deduced from the other; but by each, in his conjectured proof, Thales is made to treat angles very freely as magnitudes. In each case he adds angles together, regards one angle of a triangle as equal to the sum of the other two, or to half the sum of all three. In fact a veritable algebraical equation is worked with angles as the magnitudes concerned; and Allman even goes so far as to say that in this he laid the foundations of algebra.

The question for this paper is whether Thales had such a complete view of angles as magnitudes that he could deal with them in the way just mentioned.

Before answering this question, I propose to consider the different stages that can be traced in the development of the complete notion of a plane angle—stages that every individual has to go through, and that every day present difficulties to beginners.

(i) An angle originally is simply a corner. In this stage there is no clear idea of an angle as a magnitude. Corners may be of different shapes; but one is hardly regarded as greater than another. The corner of a square or rectangle is one of a special shape rather than of special size.

(ii) In the next stage an angle is a slope or inclination. Here there is some notion of magnitude attached to an angle, but not a very complete one. Two lines may have the same or different slopes; at most the inclination of one of them to a base line may be regarded as greater than that of another. But in this stage angles are not thought of as pure magnitudes that may be the subjects of addition or subtraction, or of any form of equation.

(iii) In the third stage we have the fully developed view of an angle as a magnitude, generally explained with the help of the notion of revolution.

It is generally allowed that, in an early stage of thought, an angle may be regarded only from the first of the preceding points of view. But surely there may be a period in the development of thought in which an angle is looked at from the point of view of (ii); whilst (iii) is not understood at all. I think the Egyptians were in this stage, and my reason for thinking so is to be found in their calculations of *seqt*.

In the *seqt* we find a *cosine* in embryo. The *seqt* was not a true ratio, but was sufficient to determine the slope of a slanting edge of a pyramid on a square base. The *seqt* seems to have been used to enable the builders of a pyramid to give the required slope to its edges; from which it follows that the designers felt that a pyramid might be built with an edge either too steep or not steep enough. That is, they felt that the idea of greater and less could be applied to the inclination of an edge to the base, though they may not verbally have reached quite so far.

That stages corresponding to the three just mentioned occur generally in the development of the idea of magnitude, and that it is possible for this development to be arrested at the end of stage (ii), may be verified by watching the thoughts of children. But a very striking case of this arrested development is to be seen in Euclid's Fifth Book.

It is often discussed* whether, in the Fifth Book, Euclid regarded a ratio as a magnitude; and the different answers given suggest that he hesitated in the matter. The correct answer seems to be that he left this thought where he found it, i.e. in stage (ii). In the fifth and sixth definitions he speaks of the *sameness* of two ratios: in the seventh definition one ratio may be greater than another. But ratios are never said to be equal, nor are two ratios added together.

The avoidance of the use of the word "equal" with regard to two quantities of the same kind, although some notion of magnitude is clearly present, seems to be characteristic of stage (ii). If the word "equal" is found in any edition of a work

* See Prof. Hill on the Theory of Proportion in the *Mathematical Gazette* for July 1912, Arts. 12 and 13.

that is really in this stage, it is probably merely an alteration made almost unconsciously by an editor who is himself in a more advanced stage of thought.

An illustration of this is that some editors of Euclid's Fifth Book, though apparently intending to keep pretty close to his words, use the word "equal," or let it slip in, without any note or qualifying remark*.

Having now seen the level of the Egyptian view of an angle, the view that we may suppose came before Thales through his Egyptian teachers, we may ask whether, in considering an angle as a magnitude, he himself ever reached stage (iii).

I submit that historical evidence answers this question in the negative; and that whoever, in reconstructing proofs for Thales, represents him as regarding an angle from the point of view of stage (iii), is reading his own more advanced thoughts into Thales' work.

In the first place Proclus says that in stating the theorem that the angles at the base of every isosceles triangle are equal he did not use the Greek word for equal; but "in archaic fashion he phrased it, *like* (ὅμοιαι)†."

Next, if, disregarding the angle in a semicircle and the sum of the three angles of a triangle, we look at the other cases, either of measurement or theory, in which Thales deals with angles, we can see that, although he looked at angles in a more abstract way than the Egyptians and had a better notion of their use and importance, yet there is no suggestion of the addition of angles, but only a better view of slope and *seqt*.

But if Thales knew the theorem of the sum of the angles of a triangle his thought must have been in stage (iii). And if he was in stage (iii), he might easily have seen a connection between this theorem and that of the angle in a semicircle. But, as other things suggest that he was only in stage (ii), direct historical evidence is certainly required to satisfy us that he knew the theorem of the sum of the angles of a triangle. There is no such direct evidence; but only an inference from a statement of Geminus. All the rest is pure conjecture.

Now, the statement of Geminus is, that the ancient geometers proved the theorem about the sum of the angles of a triangle for three species of triangles, viz. equilateral, isosceles and scalene, separately; and as further Proclus states that this theorem was first proved in a general manner by the Pythagoreans, it is inferred that the older geometers mentioned were none other than Thales and his successors in the Ionic school.

To this I say that it is not certain, and not even probable, that Thales was one of the older geometers referred to.

First Proclus ascribes the general proof to the Pythagoreans rather than to Pythagoras. Hence its date may be after the death of the latter; and if so the older geometers mentioned might have been decidedly younger than Thales.

Next, all the attempts that have been made to reconstruct the proof of these older geometers have required a perpendicular to be drawn from the vertex of a

* See Hill's quotation from Simson in the same article.

† See Allman's *Greek Geometry from Thales to Euclid*, Chap. I.

triangle to its base: a construction that ought not to be supplied so lightly. For this construction is conspicuous by its absence from the Rhind papyrus; and that this is not a mere accidental omission may be inferred from the fact that the priests of the temple of Edfu, and even later land measurers, were no further advanced in this particular. It is this absence of the perpendicular that spoils the Egyptian measurements.

What could cause the area of an isosceles triangle to be found by taking the product of the lengths of one side and half the base, except unfamiliarity with the perpendicular mentioned?

The statement that Oinopides was the first to draw a perpendicular from an external point on to a straight line is generally received with some hesitation or qualification. It is supposed that the passage means that he was the first to give the pure ruler and compass method. This may be the meaning. But the very fact that the perpendicular was in any way coupled with the name of Oinopides seems to suggest that it was at least not common before his time.

But the Egyptians were familiar with the carpenter's square and could draw right angles and rectangles. True, but to draw a right angle as a corner between the bounding lines of a figure under construction, or as part of an ornamental drawing, is a very different thing from introducing an internal perpendicular into a triangle for the sake of discovering or proving some further property. In fact, after the bounding lines of a figure had been drawn, the introduction of further lines preliminary to a proof—a proceeding common enough in Euclid—would not suggest itself very readily to people who had little or no experience of a proof in any form.

Further, as the ancient historians have given us a fairly detailed list of constructions and theorems due to Thales, is it possible that, if he had discovered the theorem of the sum of the angles of a triangle, no one of the ancients should have recollected to attribute to so celebrated a man a discovery that would certainly have been his most remarkable accomplishment?

I conclude that it is not even probable that Thales was one of the older geometers who knew, through some archaic proof, that the sum of the angles of any triangle is equal to two right angles; and hence, to render it probable that he did not advance beyond stage (ii) in the notion of an angle as a magnitude, it only remains to deal with the theorem of an angle in a semicircle.

The historical foundation for the opinion that Thales discovered this theorem is a passage from Diogenes Laertius in which, on the authority of Pamphila, it is stated that Thales was the first to describe a right-angled triangle in a circle.

The objections to the acceptance of this passage as evidence of Thales' knowledge of the theorem in question are, first,—either Diogenes Laertius, or Pamphila (perhaps both) did not quite understand their own tale and confused a theorem with a problem; secondly,—it is strange that, if, in ancient times, this theorem was generally attributed to Thales, Proclus should have omitted all mention of it in his account of Thales' work.

As far as this paper is concerned, it may be said that if this theorem is not due to Thales the question as to his view of an angle is at an end. There is nothing left in his reported work that lifts him out of stage (ii).

But if it is considered that Pamphila's statement is evidence that Thales knew that the angle in a semicircle is a right angle, we have to consider in what way Thales could have arrived at this theorem, and whether it was necessary, or even likely, that he should take a stage (iii) view of an angle.

The following method seems to me probable. The Egyptians were experts in ornamental Geometrical drawing. They were familiar with squares and with surfaces mapped out into squares by horizontal and vertical lines. They must also have been familiar with rectangles, for a row of these squares would produce a rectangle. And also, though the author of the papyrus of Ahmes does not express himself very clearly, he plainly, in measuring areas, had squares and rectangles in his mind.

Further, squares with their inscribed circles are found in Egyptian decorations. The most natural way for them to obtain these would be to draw the diagonals of a square* and so obtain the centre from which they could draw the inscribed circle.

Though I do not know that figures of squares and rectangles with their circumscribed circles have been discovered in Egyptian drawings, it seems difficult to suppose that the Egyptians stopped short of these; but even if they did, that is if no such wall decoration met the gaze of Thales, it would have been no great step in advance for him to have drawn such a figure himself.

And, bearing in mind that Proclus says that Thales attempted some things in a general or more abstract manner, we may regard him as lifting the whole subject away from wall decoration. He might thus have before him the abstract figure of a rectangle, its two diagonals, and the circumscribing circle with the intersection of the diagonals as its centre.

Proclus further says that Thales first demonstrated that the circle was bisected by its diameter. We may, of course, erase the word "demonstrated," as belonging to a later time and substitute "observed" or "stated." Then applying this to the figure of a rectangle and its circumscribed circle, and omitting from consideration one of the diagonals of the rectangle, Thales might easily have observed that the other diagonal was a diameter of the circle, and that the figure on one side of it was a semicircle in which was an angle that was first drawn as an angle of a rectangle.

If next, the figure were to be drawn in a different order,—the circle first, then a diameter and then an angle in the semicircle on one side of the diameter, I think that the statement of Pamphila is quite justified and fairly applicable to the thought of a time before Pythagoras. The criticism of Proclus comes from a mind of a later date.

My final conclusion is that, though Thales dealt with angles in a far more advanced way than the Egyptians, there is not sufficient reason for supposing him to have reached stage (iii) or to have known the theorem about the sum of the angles

* The introduction of the diagonals of a square for the purpose of further construction is very different from the introduction of a perpendicular into a triangle for the purpose of proof.

of a triangle. The different statements made about him form a much more consistent whole without these suppositions.

Moreover, the statement of Apollodorus that it was Pythagoras who first drew a right angle in a circle is not, if the preceding view is adopted, decidedly contradictory of Pamphila. For the Pythagoreans, with their deductive method, and after the general proof of the theorem of the sum of the angles of a triangle, could easily advance to a deductive proof of the theorem of the angle in a semicircle, and might hence be reported to be its discoverers.

Finally, I may add that since writing this paper I have seen in Zeuthen's *History of Mathematics* a suggestion that Thales might have reached the theorem of the angle in a semicircle in the way just given; but as what is here written was sketched out in my own mind long before the publication of Zeuthen's book, I have thought it better to leave this paper as it is.

NOTE HISTORIQUE SUR LA THÉORIE DES NOMBRES

PAR A. GÉRARDIN.

J'ai déjà commencé à publier les œuvres inédites d'Ed. Lucas, que M. C. A. Laisant a bien voulu m'envoyer. Malheureusement, ce dossier ne contient aucune note manuscrite sur les problèmes journaliers étudiés par l'auteur, ni sur les 22 questions d'Algèbre Supérieure retrouvées dans les papiers inédits de Fermat. Je possède actuellement la liste complète de tous les travaux connus d'Ed. Lucas sur la théorie des nombres, et la fin de ce travail sera publiée dans quelques mois.

Les œuvres de Proth sont malheureusement perdues d'une façon définitive, et je ne puis publier que divers renseignements inédits sur sa vie, et le recueil des travaux de ce jeune arithmologue mort à la fleur de l'âge.

J'adresse tous nos remerciements aux personnes qui veulent bien nous aider dans cette tâche ingrate de recherches, en particulier à M. le Dr Pein, au sujet de Proth.

Pour le P. Pépin et G. de Rocquigny, je ne puis rien publier d'inédit, mais j'indique les travaux intéressants peu ou pas connus; pour le dernier auteur cité, j'ai donné plus d'une centaine de solutions choisies parmi les 3 ou 400 problèmes non résolus posés par lui.

Je désirerais recevoir tous renseignements inédits ou non, lettres, papiers, notes, articles...de tous arithmologues, modernes ou anciens, vivants ou morts, illustres ou inconnus. Dès que j'aurai réuni la plus grande somme possible de documents sérieux, même en prêt, je compte éditer une série d'intéressantes monographies historiques ayant spécialement trait à la Théorie des Nombres.

Bibliographie de mes articles parus dans "Sphinx-Œdipe" à ce jour.

Ed. Lucas, t. IV, 1909, p. 93 ; t. V, 1910, p. 33, 60, 89, 121 ; t. VI, 1911, p. 38—39, 164—173, 177—185 ; t. VII, 1912, p. 103—107.
Proth, t. VI, 1911, p. 36—37, 79, 151—156, 185 ; t. VII, 1912, p. 50—51.
Pépin, t. III, 1908, p. 188 ; t. IV, 1909, p. 27, 30 ; t. V, 1910, p. 1—16, 35, 56.
G. de Rocquigny, t. VI, 1911, p. 40—43, 65—70, 81—86, 97—101, 129—132.

Divers articles de Fermat, Euler, Genocchi, Lamé, Seelhoff, Schlömilch, Réalis, E. de Jonquières, Catalan, Prestet, Cesaro, etc.

THE IDEAS OF THE "FONCTIONS ANALYTIQUES" IN LAGRANGE'S EARLY WORK

By Philip E. B. Jourdain.

It does not seem to have been previously remarked that the ideas which were given by Lagrange in a paper published in 1774*, and later in the well-known *Théorie des Fonctions Analytiques* of 1797, were present in a memoir by Daviet de Foncenex† on the fundamental principles of mechanics, which was published in 1761. It is known that Lagrange told Delambre‡ that he wrote for Foncenex, "among other things, a new theory of the lever in the third part of a memoir by Foncenex in the *Miscellanea Taurinensia* for 1759," and it is apparently to the part described below that Lagrange meant to refer. In the treatment of the lever we have the development of a function of x in a Taylor's series, and the notation of the differential quotients of this function by dashes, which is mentioned as something habitually done. Thus, Reiff§ is certainly wrong when he says that the notation of ψ' for $\frac{d\psi}{dx}$ first appears in Lagrange's|| paper of 1770 on the resolution of literal equations by series. This question of notation seems to be important because it indicates a wish to avoid the use of infinitesimals, at least in principle.

In a letter written on the 24th of November, 1759, to Euler, Lagrange¶ said: "I have also composed the elements of mechanics and the differential and integral calculus for the use of my pupils, and I believe that I have developed the true metaphysics of their principles as far as it is possible." It would appear then, that Vivanti** is mistaken in saying that Lagrange, at the beginning of his career, attempted to vindicate the rigour of the infinitesimal calculus, and only *later* arrived at the thought of replacing this calculus by a more rigorous one. Indeed, Lagrange

* "Sur une nouvelle espèce de calcul relatif à la différentiation et à l'intégration des quantités variables," *Nouveaux Mémoires de l'Acad. de Berlin*, 1772 (published 1774); *Œuvres*, III, pp. 441—476.

† "Sur les principes fondamentaux de la Méchanique," *Miscellanea Taurinensia*, II, 1760—1761, pp. 299—322. The part considered here is the fourth part (pp. 319—322) and entitled "Du Levier."

‡ Cf. *Œuvres de Lagrange*, I, p. xi. Cf. also E. Netto in M. Cantor's *Vorlesungen über Geschichte der Mathematik*, IV, Leipzig, 1908, pp. 308—309.

§ *Geschichte der unendlichen Reihen*, Tübingen, 1889, p. 147.

|| "Nouvelle méthode pour résoudre les équations littérales par le moyen des séries," *Histoire de l'Acad. de Berlin*, 1768 (published 1770); *Œuvres*, III, pp. 5—73.

¶ *Œuvres*, XIV, p. 173.

** In Cantor, *op. cit.* IV. p. 644.

seems to have convinced himself, at an early date, that the use of infinitesimals was fundamentally rigorous, and to have used both the infinitesimal method and the method of derived functions side by side during his whole life*.

In Foncenex's memoir the action of force applied to move a lever is $p\,.\,\xi x$, where p is the magnitude of the force, and x is the distance of the point of application from the fulcrum, and ξ is the sign of a function. Two forces, each equal to $\frac{p}{2}$ in magnitude, and applied at points E and D of the lever, have the same effect as p applied at the point A, where $AD = AE = z$. Now the action of the first on the fulcrum is $\frac{p}{2}\xi(x+z)$, and the action of the second is $\frac{p}{2}\xi(x-z)$. Thus we have the equation

$$p\,.\,\xi x = \frac{p}{2}\xi(x+z) + \frac{p}{2}\xi(x-z),$$

or

$$2\,.\,\xi x = \xi(x+z) + \xi(x-z),$$

which must be true whatever z may be. If, then, we make z infinitely small, and develop the function ξ by known methods, we find, where $\xi' x$, $\xi'' x$, $\xi''' x$, ... denote, as is the usual custom, the successive differences of ξx divided by dx†,

$$\xi(x+z) = \xi x + \frac{z\,.\,\xi' x}{2} + \frac{z^2\,.\,\xi'' x}{2\,.\,3} + \frac{z^3\,.\,\xi''' x}{2\,.\,3\,.\,4} + \dots,$$

$$\xi(x-z) = \xi x - \frac{z\,.\,\xi' x}{2} + \frac{z^2\,.\,\xi'' x}{2\,.\,3} - \frac{z^3\,.\,\xi''' x}{2\,.\,3\,.\,4} + \dots.$$

Consequently

$$\frac{z^2\,.\,\xi'' x}{3} + \frac{z^4\,.\,\xi^{\text{iv}} x}{3\,.\,4\,.\,5} + \frac{z^6\,.\,\xi^{\text{vi}} x}{3\,.\,4\,.\,5\,.\,6\,.\,7} + \dots = 0;$$

and, dividing this equation by $\frac{z^2}{3}$, we find separately $\xi'' x = 0$, $\frac{z^2\,.\,\xi^{\text{iv}} x}{4\,.\,5} = 0$, and so on. By integrating the first twice we get

$$\xi x = Hx + K,$$

and we must have $\xi x = 0$ when $x = 0$. Hence the law of the lever.

* Contrary to the statement of Rouse Ball, *A Short Account of the History of Mathematics*, London, 1908, p. 410.

† "$\xi' x$, $\xi'' x$, $\xi''' x$, etc. dénotant selon l'usage ordinaire les différences successives de ξx divisées par dx"; *op. cit.* p. 321. Evidently this infinitesimal explanation had, at that time, to be added for the benefit of other mathematicians.

COMMUNICATIONS

SECTION IV (*b*)

(DIDACTICS)

ON THE TEACHING OF THE THEORY OF PROPORTION

By M. J. M. Hill.

ABSTRACT.

Art. 1. Objects of the paper.

(1) To explain in a simple and direct manner the Theory of Proportion when the magnitudes concerned are incommensurable.

(2) To discuss the position of the subject in the Mathematical Curriculum.

Art. 2. Specification of the characteristics of magnitudes *of the same kind.*

Art. 3. Propositions relating to Magnitudes and their Multiples.

(In what follows, magnitudes are denoted by capital letters and positive integers by small letters.)

Prop. I. $n(A+B+C+\ldots)=nA+nB+nC+\ldots$.

Prop. II. $(a+b+c+\ldots)N=aN+bN+cN+\ldots$.

Prop. III. $(r(s))A=r(sA)=s(rA)=(s(r))A$.

Prop. IV. If $A>B$, then $r(A-B)=rA-rB$.

Prop. V. If $a>b$, then $(a-b)R=aR-bR$.

Prop. VI. $rA \gtreqless rB$ according as $A \gtreqless B$; and $A \gtreqless B$ according as $rA \gtreqless rB$.

Prop. VII. $aR \gtreqless bR$ according as $a \gtreqless b$; and $a \gtreqless b$ according as $aR \gtreqless bR$.

Prop. VIII. If A, B, C are magnitudes *of the same kind*, and $A>B$, then integers n, t exist such that $nA>tC>nB$.

Art. 4. Definition of the ratio of Commensurable Magnitudes.

It is shown that if A, B, C are commensurable magnitudes, and

(i) If $A>B$, then $A:C>B:C$,

(ii) If $A=B$, then $A:C=B:C$,

(iii) If $A<B$, then $A:C<B:C$.

Conversely, (iv) If $A:C>B:C$, then $A>B$,

(v) If $A:C=B:C$, then $A=B$,

(vi) If $A:C<B:C$, then $A<B$.

ART. 5. Definition of the irrational number by means of the system of rational numbers.

ART. 6. Fundamental hypotheses on which the definition of the ratio of incommensurable magnitudes is based.

It is *assumed* that, if A, B, C be any magnitudes *of the same kind*, whether they have a common measure or not, and

$$\text{if } A > B, \text{ then } A : C > B : C,$$
$$\text{if } A = B, \text{ then } A : C = B : C,$$
$$\text{if } A < B, \text{ then } A : C < B : C.$$

Prop. IX. If A, B, C be any magnitudes *of the same kind*, whether they have a common measure or not, and

$$\text{if } A : C > B : C, \text{ then } A > B,$$
$$\text{if } A : C = B : C, \text{ then } A = B,$$
$$\text{if } A : C < B : C, \text{ then } A < B.$$

Prop. X. To prove that

$$A : B \gtreqless \frac{s}{r} \text{ according as } rA \gtreqless sB,$$

and conversely that $$rA \gtreqless sB \text{ according as } A : B \gtreqless \frac{s}{r}.$$

It is shown that, if A and B are incommensurable, then $A : B$ determines an irrational number.

ART. 7. Two ratios are defined to be equal when no rational fraction lies between them; but unequal when it can be shown that a single rational fraction lies between them.

ART. 8. The Test for Equal Ratios.

$$A : B = C : D,$$

if (1) all values of r, s which make $rA > sB$ also make $rC > sD$,

if (2) all values of r, s which make $rA = sB$ also make $rC = sD$,

and if (3) all values of r, s which make $rA < sB$ also make $rC < sD$.

ART. 9. Prop. XI. Proof of theorem due to Stolz that the 2nd alternative in the preceding article is involved in the 1st and 3rd, and is therefore superfluous.

ART. 10. The test for distinguishing the larger from the smaller of two unequal ratios.

ART. 11. Properties of Equal Ratios (First Group).

These propositions are independent of one another, and may therefore be taken in any order. They do not depend on Prop. VIII. They depend on the Test for Equal Ratios.

Prop. XII. If $A : B = C : D$, then $rA : sB = rC : sD$.

Prop. XIII. If $A : B = C : D$, then $B : A = D : C$.

Prop. XIV. If $A : B = C : D = E : F$, and if all the magnitudes are *of the same kind*, then $A : B = A + C + E : B + D + F$.

Prop. XV. $A : B = nA : nB$.

Prop. XVI. If $A : B = X : Y$, then $A + B : B = X + Y : Y$.

Prop. XVII. If $A + B : B = X + Y : Y$, then $A : B = X : Y$.

ART. 12. Properties of Equal Ratios (Second Group).

These propositions are independent of one another. They all involve Prop. VIII. They depend on the Test for Equal Ratios.

Prop. XVIII. If A, B, C, D be four magnitudes *of the same kind*, and if $A : B = C : D$, then $A : C = B : D$.

Corollary. $B \gtreqless D$ according as $A \gtreqless C$.

Prop. XIX. If A, B, C be three magnitudes *of the same kind*; if T, U, V be three magnitudes *of the same kind*; if $A : B = T : U$, and if $B : C = U : V$, then $A : C = T : V$.

Corollary. $T \gtreqless V$ according as $A \gtreqless C$.

Prop. XX. If A, B, C be three magnitudes *of the same kind*; if T, U, V be three magnitudes *of the same kind*; if $A : B = U : V$, and if $B : C = T : U$, then $A : C = T : V$.

Corollary. $T \gtreqless V$ according as $A \gtreqless C$.

ART. 13. Properties of Equal Ratios (Third Group).

These propositions cannot be conveniently deduced *directly* from the Test for Equal Ratios. They are however readily deduced by the aid of propositions in the two preceding groups, and therefore depend ultimately on the Test for Equal Ratios.

Prop. XXI. If $A + C : B + D = C : D$, then $A : B = C : D$.

Prop. XXII. If $A : C = X : Z$, and if $B : C = Y : Z$, then

$$A + B : C = X + Y : Z.$$

Prop. XXIII. If A, B, C, D are four magnitudes *of the same kind*, if $A : B = C : D$, and if A be the greatest of the four magnitudes, then $A + D > B + C$.

ART. 14. *The existence of the Fourth Proportional.*

Prop. XXIV. If A and B be two magnitudes *of the same kind*, and if C be a third magnitude, then a fourth magnitude Z *of the same kind* as C exists, such that $A : B = C : Z$.

Except in the case where A, B, C are segments of straight lines (or cases readily reducible to this case) the proof of this depends on an Axiom corresponding to the Cantor-Dedekind Axiom.

ART. 15. *The position of the subject in the Mathematical Curriculum.*

Propositions I—XXIII can be explained without difficulty to persons of average mathematical capacity having a knowledge of Elementary Algebra. On the other hand Prop. XXIV (except in the special case mentioned above) can only be explained to persons with more than the average capacity. A knowledge of Props. I—XXIII should be obtained before the study of the Theory of Similar Figures is commenced.

Euclid's Treatment in his Fifth Book is intricate and difficult because he uses his Seventh Definition (the Test for distinguishing between Unequal Ratios) when the Fifth Definition (the Test for Equal Ratios) is sufficient. All the proofs given in this paper depend on the Fifth Definition only.

Objects of the Paper.

1. The primary object of this paper is to set forth the Theory of Proportion in a simple and direct manner, which is applicable when the magnitudes concerned have no common measure.

I set out the propositions to be proved and give demonstrations of some of them as specimens, references being given in regard to the rest (Arts. 1—14).

The secondary object of the paper is to discuss the position of the subject in the mathematical curriculum (Art. 15).

I do not therefore make any attempt to go deeply into first principles, and I avoid altogether the difficulties of non-Archimedean Geometries.

Magnitudes of the same kind.

2. The first step in the argument is to specify the characteristics of magnitudes *of the same kind.*

These will be readily admitted if we consider the magnitudes to be segments of lines, or areas, or volumes, or weights.

I follow in essentials Stolz's account of the properties of absolute magnitudes in his *Allgemeine Arithmetik*, Erster Theil, page 69.

A system of magnitudes is said to be *of the same kind* when the magnitudes possess the following characteristics:

(1) Any two magnitudes *of the same kind* may be regarded as equal or unequal.

In the latter case one of them is said to be the smaller, and the other the larger of the two.

(2) Two magnitudes *of the same kind* can be added together. The resulting magnitude is a magnitude *of the same kind* as the original magnitudes.

The Commutative and Associative Laws apply to the Addition. So that

$$A + B = B + A, \quad (A + B) + C = A + (B + C).$$

This property makes it possible to form multiples of a magnitude.

(3) If A and B be two magnitudes *of the same kind*, and A be greater than B, then another magnitude X *of the same kind* as A and B exists such that

$$B + X = A.$$

This may also be written

$$X = A - B.$$

(4) If A be any magnitude, and n any positive integer whatever, then a magnitude X *of the same kind* as A exists such that

$$nX = A.$$

(5) If A be greater than B, a multiple of B exists which is greater than A.

The fifth characteristic is known as the Axiom of Archimedes. It is not a consequence of the preceding four characteristics.

The following deduction from the above is specially useful in the Theory of Proportion:

If A and B are two magnitudes *of the same kind*, and any multiple whatever of A, say rA, is chosen, and any multiple whatever of B, say sB, is chosen, then one and only one of the alternatives

$$rA > sB, \quad rA = sB, \quad rA < sB$$

always exists, and it is assumed to be possible to determine which one of these alternatives exists.

Propositions concerning Magnitudes and Multiples of Magnitudes.

3. In what follows, magnitudes will be denoted by capital letters and positive whole numbers by small letters.

Prop. I. (Euc. V. 1.)

$$n(A + B + C + \dots) = nA + nB + nC + \dots.$$

The simplest case of this is

$$n(A + B) = nA + nB.$$

For a rigid deduction of this from the Associative and Commutative Laws I refer to my edition of *Euclid V and VI*, 2nd edition, pp. 125—6. It is tedious, and I do not think the beginner should be stopped at this stage with it. I think it sufficient to say that the effect of the Associative and Commutative Laws is this, that when any number of magnitudes are to be added together, they may be arranged in any order and grouped in any way, the magnitudes in each group may be first added together, and then finally the sum of the groups can be found, and that the result so obtained will always be the same.

Thus $n(A + B)$ is the sum of n groups, each of which is $A + B$.

The magnitude A occurs n times; and therefore taking these together, their sum is nA.

The magnitude B occurs n times; and taking these together, their sum is nB.

The sum of the two groups is $nA + nB$.

$$\therefore\ n(A+B) = nA + nB.$$

From this the complete proposition may be deduced.

Prop. II. (Euc. v. 2.)

$$(a+b+c+\ldots)N = aN + bN + cN + \ldots\ .$$

The simplest case of this is

$$(a+b)N = aN + bN.$$

Now $(a+b)N$ means that N is taken $a+b$ times.

Group the first a N's together. Their sum is aN.

Group the remaining b N's together. Their sum is bN.

$$\therefore\ (a+b)N = aN + bN.$$

From this the complete proposition can be deduced.

For a rigid deduction of the proposition from the Associative and Commutative Laws see my *Euclid V and VI*, 2nd edition, p. 127.

Prop. III. $(r(s))A = r(sA) = s(rA) = (s(r))A.$

In Prop. I, suppose that each of the magnitudes $B, C, \ldots$ is equal to A, and that there are, including A, altogether s magnitudes.

Then $n(A+B+C+\ldots)$ is $n(sA)$,

and $nA + nB + nC + \ldots$ is $s(nA)$.

$$\therefore\ n(sA) = s(nA).$$

Or replacing n by r, $r(sA) = s(rA)$ ....(I).

Next in Prop. II suppose that each of the integers $a, b, c, \ldots$ is equal to s; and that there are r such integers.

Then $(a+b+c+\ldots)N$ becomes $(r(s))N$,

and $aN + bN + cN + \ldots$ becomes $r(sN)$.

$$\therefore\ (r(s))N = r(sN),$$

or replacing N by A, $(r(s))A = r(sA)$ ....(II).

Interchanging s and r, $(s(r))A = s(rA)$..............................(III).

Then from (I), (II), (III) it follows that

$$(r(s))A = r(sA) = s(rA) = (s(r))A.$$

Corollary: $s[(n(r))A] = r[(n(s))A].$

Prop. IV. If $A > B$, then $r(A-B) = rA - rB$. (Euc. v. 5.)

Since $A > B$, then by Art. 2 (3) a magnitude C exists such that

$$A = B + C,$$

$$\therefore\ rA = rB + rC,$$

$$\therefore\ rC = rA - rB,$$

but $$C = A - B,$$

$$\therefore\ r(A - B) = rA - rB.$$

Prop. V. If $a > b$, then $(a - b) R = aR - bR$. (Euc. v. 6.)

This is proved in the same way as Prop. IV by putting $a = b + c$.

Prop. VI. If $A > B$, then $rA > rB$,

If $A = B$, then $rA = rB$,

If $A < B$, then $rA < rB$.

Conversely, If $rA > rB$, then $A > B$,

If $rA = rB$, then $A = B$,

If $rA < rB$, then $A < B$.

If $A > B$, then, as in Prop. IV,

$$rA = rB + rC,$$

$$\therefore\ rA > rB, \text{ and so on.}$$

The converse proposition follows from the original proposition as a logical consequence.

Prop. VII. According as $a \gtreqless b$, so is $aR \gtreqless bR$.

Conversely, according as $aR \gtreqless bR$, so is $a \gtreqless b$.

The proofs are similar to those in Prop. VI.

Prop. VIII. If A, B, C are three magnitudes *of the same kind*, and if $A > B$, then integers n, t exist such that

$$nA > tC > nB.$$

Since $A > B$, $A - B$ is a magnitude *of the same kind* as A and B and therefore *of the same kind* as C.

Hence by Archimedes' Axiom an integer n exists such that

$$n(A - B) > C,$$

$$\therefore\ nA > nB + C.$$

Since $nA > C$, it is possible that $nA > 2C, 3C, \ldots$.

Suppose that tC is the greatest multiple of C which is less than nA.

Then $(t + 1) C$ must be either greater than nA or equal to nA.

$$\therefore\ (t + 1) C \geqq nA,$$

but $$nA > nB + C,$$

$$\therefore\ (t + 1) C > nB + C,$$

$$\therefore\ tC > nB.$$

But $$nA > tC,$$

$$\therefore\ nA > tC > nB.$$

This is a very important proposition in the Theory. Its proof constitutes almost the whole of the demonstration of Euc. v. 8.

The ratios of Commensurable Magnitudes.

4. The next step is the introduction of the idea of ratio. I begin by considering the case of magnitudes which have a common measure.

Let there be three magnitudes A, B and C of the same kind, all of which are measured by G.

Let $A = aG$, $B = bG$, $C = cG$ where a, b, c are positive integers.

Then the ratio of A to C, i.e. the ratio of aG to cG, is defined to be the rational fraction $\frac{a}{c}$.

Similarly the ratio of B to C is $\frac{b}{c}$.

Now according as $$A \gtreqless B,$$
so is $$aG \gtreqless bG,$$
so is $$a \gtreqless b,$$
so is $$\frac{a}{c} \gtreqless \frac{b}{c},$$
so is $$A : C \gtreqless B : C.$$

It has therefore been proved that if A, B, C have a common measure, then according as
$$A \gtreqless B,$$
so is $$A : C \gtreqless B : C.$$

Conversely (see Prop. IX below) it will follow from the above that if A, B, C have a common measure, then according as
$$A : C \gtreqless B : C,$$
so is $$A \gtreqless B.$$

The Irrational Number.

5. The next step is to explain the manner in which an irrational number is defined.

Let some rule be given by means of which the system of all the rational numbers can be separated into two classes such that every number in one class (called the lower class) is less than every number in the other class (called the upper class); then the following cases have to be distinguished.

(i) The lower class has no greatest number, and the upper class has no least number. Then between the two classes there is a gap. This gap is filled by the creation of a number. It cannot be a *rational* number, because every *rational* number falls by hypothesis into one of the two classes. Consequently it is called an *irrational* number; and it separates the whole system of *rational* numbers into two classes, such that every number of the lower class is less than every number of the upper class.

It is regarded as known because all its properties can be inferred from a knowledge of all the rational numbers which are less than it, and a knowledge of all the rational numbers which are greater than it.

(ii) The case in which the lower class contains a greatest number. This greatest number is a rational number, because it belongs to the lower class. Calling it $\frac{r}{s}$, the upper class contains all the rational numbers greater than $\frac{r}{s}$, and therefore can contain no least number.

In this case the number $\frac{r}{s}$ separates the whole system of rational numbers into two classes such that each number in the lower class is less than each number in the upper class.

The lower class contains $\frac{r}{s}$ and all rational numbers less than $\frac{r}{s}$; the upper class contains all the numbers greater than $\frac{r}{s}$.

(iii) The case in which the upper class contains a least number.

This least number is a rational number because it belongs to the upper class. Calling it $\frac{r}{s}$, the lower class contains all the numbers less than $\frac{r}{s}$ and therefore can contain no greatest number.

In this case the number $\frac{r}{s}$ separates the whole system of rational numbers into two classes such that each number in the lower class is less than each number in the upper class.

The upper class contains $\frac{r}{s}$ and all numbers greater than $\frac{r}{s}$: the lower class contains all the numbers less than $\frac{r}{s}$.

Cases (ii) and (iii) are regarded as not essentially distinct.

The ratios of Incommensurable Magnitudes.

6. If A and B have no common measure, then the definition of ratio given in Art. 4 is inapplicable, and it is necessary to say either that A cannot have a ratio to B or else to give a new definition of the ratio of A to B.

This second alternative will be chosen.

This new definition must be of such a nature that when it is applied to two magnitudes A and B which have a common measure, it will give the same value for the ratio of A to B as that which would be given by the definition of Art. 4. (Otherwise there would be a conflict between the two definitions.)

In order to construct a definition of ratio which shall be applicable whether the magnitudes concerned have a common measure or not, we look for a definition which satisfies the following conditions:

Suppose that A, B, C are any magnitudes *of the same kind*, they may or may not have a common measure, then the conditions are as follows:

(i) If A be greater than B, then the ratio of A to C is greater than that of B to C;

(ii) If A be equal to B, then the ratio of A to C is equal to that of B to C.

To these may be added for symmetry a condition which is included in (i).

(iii) If A be less than B, then the ratio of A to C is less than that of B to C.

These statements may be abbreviated into the following:

(i) If $A > B$, then $A : C > B : C$;

(ii) If $A = B$, then $A : C = B : C$;

(iii) If $A < B$, then $A : C < B : C$.

It is obvious that between these conditions, which will in future be referred to as fundamental assumptions, and the results obtained in Art. 4 there can be no conflict.

From these assumptions follows as a logical consequence

Prop. IX. (i) If $A : C > B : C$, then $A > B$;

(ii) If $A : C = B : C$, then $A = B$;

(iii) If $A : C < B : C$, then $A < B$.

Suppose, if possible, $A : C > B : C$, but $A \not> B$.

Then either $A = B$ or $A < B$.

But if $A = B$, then by one of the fundamental assumptions

$$A : C = B : C,$$

which is contrary to the hypothesis.

And if $A < B$, then by one of the fundamental assumptions

$$A : C < B : C,$$

which is contrary to the hypothesis.

Hence $A > B$.

The other cases may be proved in like manner.

Prop. X. To prove that according as $rA \gtreqless sB$, so is $A : B \gtreqless \frac{s}{r}$; and conversely that according as $A : B \gtreqless \frac{s}{r}$, so is $rA \gtreqless sB$.

Suppose that $rA > sB$.

Then by Art. 2 (4) a magnitude Q exists such that

$$B = rQ,$$

$$\therefore\ rA > s(rQ),$$

$$\therefore\ rA > r(sQ),$$

$$\therefore\ A > sQ,$$

$$\therefore\ A : B > sQ : B,$$

by one of the fundamental assumptions,

$$\therefore\ A : B > sQ : rQ,$$

$$\therefore\ A : B > \frac{s}{r}.$$

Similarly if $$rA < sB,$$

$$A : B < \frac{s}{r}.$$

If $rA = sB$, taking $B = rQ$, it follows that $A = sQ$,

$$\therefore\ A : B = \frac{s}{r}.$$

It is important to notice this last case. When any relation of the form $rA = sB$ exists, then A and B have a common measure.

Conversely, suppose $$A : B > \frac{s}{r}.$$

Take $$B = rQ.$$

Then $$\frac{s}{r} = sQ : rQ = sQ : B,$$

$$\therefore\ A : B > sQ : B.$$

Hence $$A > sQ \qquad \text{(Prop. IX)},$$

$$\therefore\ rA > r(sQ),$$

$$\therefore\ rA > s(rQ),$$

$$\therefore\ rA > sB.$$

Similarly if $$A : B = \frac{s}{r}, \quad \text{then} \quad rA = sB,$$

and if $$A : B < \frac{s}{r}, \quad \text{then} \quad rA < sB.$$

It is now possible to decide whether $A : B$ is greater than, equal to, or less than any given rational fraction $\frac{s}{r}$.

Now when A and B are magnitudes *of the same kind*, so also are rA and sB, and it is assumed in Art. 2 that it is possible to determine whether they are equal or unequal, and if unequal which of them is the greater. Hence it is always possible to decide whether

$$rA > sB, \quad \text{or} \quad rA = sB, \quad \text{or} \quad rA < sB.$$

Now $$\text{if } rA > sB, \quad \text{then} \quad A : B > \frac{s}{r},$$

$$\text{if } rA = sB, \quad \text{then} \quad A : B = \frac{s}{r},$$

$$\text{if } rA < sB, \quad \text{then} \quad A : B < \frac{s}{r}.$$

Let us now separate the whole system of rational numbers into two classes, such that every number in the lower class is less than every number in the upper class in accordance with the following rules:

(i) If $A : B > \frac{s}{r}$, let $\frac{s}{r}$ be put into the lower class;

(ii) If $A : B < \frac{s}{r}$, let $\frac{s}{r}$ be put into the upper class.

From this step onwards we must take separately the cases

(I) A and B have no common measure,

and (II) A and B have a common measure.

In case (I) A and B have no common measure, and therefore by a result obtained in Prop. X no relation of the form $rA = sB$ can hold, and therefore $A : B$ can never be equal to any rational fraction, and therefore it will be possible by means of rules (i) and (ii) to determine the class to which every rational fraction is to be assigned.

In this case then $A : B$ lies between the rational fractions in the upper class and those in the lower class. It is therefore the irrational number which separates the whole system of rational numbers into these two classes.

In case (II) A and B have a common measure, say G. Let $A = aG$ and $B = bG$.

Now $$A : B = \frac{a}{b}.$$

If $\frac{s}{r}$ is less than $\frac{a}{b}$, it is less than $A : B$, and must therefore be assigned to the lower class.

If $\frac{s}{r}$ is greater than $\frac{a}{b}$, it is greater than $A : B$, and must therefore be assigned to the upper class.

With regard to $\frac{a}{b}$ itself, it is immaterial whether it is assigned to the lower class or to the upper class. In either case $\frac{a}{b}$ determines the separation of all the rational numbers into two classes such that every number in the lower class is less than every number in the upper class.

Hence again $A : B$ is equal to the number, in this case rational, which separates the whole system of rational numbers into two classes such that each number in the lower class is less than each number in the upper class.

To determine the value of $A : B$ approximately the process of finding the Greatest Common Measure is the most convenient.

Another method which is more readily explained to beginners is this:

Take Q so that $B = rQ$.

If Q measure A, so that $A = sQ$, then

$$A : B = sQ : rQ = \frac{s}{r}.$$

But if Q do not measure A, suppose A falls between tQ and $(t+1)Q$,

$$\therefore\ tQ < A < (t+1)Q,$$

$$\therefore\ tQ : B < A : B < (t+1)Q : B,$$

$$\therefore\ tQ : rQ < A : B < (t+1)Q : rQ,$$

$$\therefore\ \frac{t}{r} < A : B < \frac{t+1}{r}.$$

Therefore $A : B$ differs from $\frac{t}{r}$ by less than $\frac{1}{r}$, and therefore taking r to be a large number, $\frac{t}{r}$ will be an approximate measure of $A : B$.

Definition of Equal and Unequal Ratios.

7. The preceding theory can now be applied to find tests for determining

(*a*) when two ratios are equal;

(*b*) when two ratios are unequal.

In Art. 6 a particular case of unequal ratios, viz.:—the case in which both ratios had the same second term, was considered, viz.:

If A, B, C are magnitudes *of the same kind*, and if $A > B$, then it was assumed as fundamental that $A : C > B : C$.

It will now be proved that a rational fraction exists which lies between the unequal ratios $A : C$ and $B : C$.

Now A, B, C being magnitudes *of the same kind* and $A > B$, it is known by Prop. VIII that integers n, t exist such that

$$nA > tC > nB,$$

and
$$\therefore\ A : C > \frac{t}{n} > B : C.$$

Hence the rational fraction $\frac{t}{n}$ lies between the unequal ratios $A : C$ and $B : C$.

Suppose now we take any two ratios $A : B$ and $C : D$, and a rational fraction $\frac{s}{r}$ lies between them, say $A : B > \frac{s}{r} > C : D$; then if we separate the whole system of rational numbers into two classes as previously described, $\frac{s}{r}$ will be in the lower class determined by $A : B$, but in the upper class determined by $C : D$.

Hence if a rational fraction lie between two ratios $A : B$ and $C : D$, then $A : B$ and $C : D$ separate the whole system of rational numbers into two classes in different ways, and we say that $A : B$ and $C : D$ are unequal.

If, however, no rational fraction lie between $A : B$ and $C : D$, then $A : B$ and $C : D$ separate the whole system of rational numbers into two classes in the same way, and we say that $A : B = C : D$.

We say therefore that

(1) two ratios are equal if no rational fraction lies between them;

(2) two ratios are unequal if any single rational fraction lies between them.

The Test for Equal Ratios

(Euclid's Fifth Definition).

8. I proceed to obtain Euclid's conditions from the definition given that two ratios are equal if no rational fraction lie between them.

Suppose that $A : B = C : D$.

Take any rational fraction $\frac{r}{s}$.

There are three possible alternatives

$$\text{(i) } A : B > \frac{r}{s}; \quad \text{(ii) } A : B = \frac{r}{s}; \quad \text{(iii) } A : B < \frac{r}{s}.$$

Now (i) If $A : B > \frac{r}{s}$, then in order that $\frac{r}{s}$ may not lie between $A : B$ and $C : D$, it is necessary that $C : D > \frac{r}{s}$.

But if $A : B > \frac{r}{s}$, then $sA > rB$ and conversely.

And if $C : D > \frac{r}{s}$, then $sC > rD$ and conversely.

Hence if any integers r, s exist such that $sA > rB$, then must also $sC > rD$.

(ii) If $A : B = \frac{r}{s}$, then in order that no rational fraction may lie between $A : B$ and $C : D$, it is necessary that $C : D = \frac{r}{s}$.

Hence if any integers r, s exist such that $sA = rB$, then must also $sC = rD$.

(iii) Similarly if $A : B < \frac{r}{s}$, then must $C : D < \frac{r}{s}$.

Hence if any integers r, s exist such that $sA < rB$, then must also $sC < rD$.

Hence if $A : B = C : D$, then when the integers r, s are such that

$sA > rB$, then must also $sC > rD$ ....(i),

but if the integers r, s are such that

$sA = rB$, then must also $sC = rD$ ....(ii),

and if the integers r, s are such that

$sA < rB$, then must also $sC < rD$ ....(iii).

These are Euclid's conditions for the equality of the ratios $A : B$ and $C : D$.

Simplification of the Test for Equal Ratios.

9. It is important to prove (as was, I believe, first proved by Stolz, *l.c.* p. 87)

Prop. XI. If all values of r, s which make $sA > rB$ also make $sC > rD$ ...(I),

and if all values of r, s which make $sA < rB$ also make $sC < rD$ ...(II),

and if any values of r, s exist, say $r = r_1$, $s = s_1$ which make $s_1A = r_1B$, then must also $s_1C = r_1D$.

By hypothesis $s_1A = r_1B$.

Suppose if possible $s_1C \neq r_1D$.

Hence either

(i) $s_1C > r_1D$.

$\therefore s_1C - r_1D$ is a magnitude *of the same kind* as D.

$\therefore$ An integer n exists such that

$$n(s_1C - r_1D) > D,$$

$$\therefore ns_1C > (nr_1 + 1)D,$$

but $s_1A = r_1B$,

$$\therefore ns_1A = nr_1B < (nr_1 + 1)B.$$

Hence $ns_1A < (nr_1 + 1)B$,

but $ns_1C > (nr_1 + 1)D$,

and $\therefore$ putting $s = ns_1$, $r = nr_1 + 1$, it is seen that for these values of r, s the hypothesis (II) is not satisfied.

$$\therefore s_1C \not> r_1D.$$

Or

(ii) $s_1C < r_1D$.

$\therefore r_1D - s_1C$ is a magnitude *of the same kind* as D.

$\therefore$ An integer n exists such that

$$n(r_1D - s_1C) > D,$$

$$\therefore ns_1C < (nr_1 - 1)D,$$

but $s_1A = r_1B$,

$$\therefore ns_1A = nr_1B > (nr_1 - 1)B.$$

Hence $ns_1A > (nr_1 - 1)B$,

but $ns_1C < (nr_1 - 1)D$,

and $\therefore$ putting $s = ns_1$, $r = nr_1 - 1$, it is seen that for these values of r, s the hypothesis (I) is not satisfied.

$$\therefore s_1C \not< r_1D.$$

It has now been shown that s_1C is neither greater nor less than r_1D.

$$\therefore s_1C = r_1D.$$

Hence the second set of conditions in Euclid's Definition is involved in the first and third sets of conditions. It is therefore superfluous.

Hence $A : B = C : D$,

if all values of r, s which make $sA > rB$ also make $sC > rD$,

and if all values of r, s which make $sA < rB$ also make $sC < rD$.

And $\therefore$ if $A : B > \frac{r}{s}$, it is necessary that $C : D > \frac{r}{s}$,

but if $A : B < \frac{r}{s}$, it is necessary that $C : D < \frac{r}{s}$,

whatever integers r, s may be, and it is not necessary to show also that

$$\text{if } A : B = \frac{r}{s}, \quad \text{then } C : D = \frac{r}{s}.$$

The Test for distinguishing between Unequal Ratios

(Euclid's Seventh Definition).

10*. Two ratios are defined to be unequal if any single rational fraction lies between them.

Let $\frac{r}{s}$ lie between $A : B$ and $C : D$.

Suppose $$A : B > \frac{r}{s} > C : D.$$

Now if $A : B > \frac{r}{s}$, then $sA > rB$ and conversely.

And if $C : D < \frac{r}{s}$, then $sC < rD$ and conversely.

Hence if any single pair of integers r, s exist such that

$$sA > rB, \quad \text{but} \quad sC < rD,$$

then $$A : B > C : D.$$

It may be noted that $A : B > C : D$ in the two following cases:

If integers r, s exist such that

$$\text{(i)} \quad sA = rB, \quad \text{but } sC < rD,$$

or $$\text{(ii)} \quad sA > rB, \quad \text{but } sC = rD.$$

Euclid includes the second in his Seventh Definition; and Saccheri and Commandinus note that he might also have included the first.

In either of these cases it can be proved that integers r', s' exist such that

$$s'A > r'B, \quad \text{but } s'C < r'D,$$

which is the form first given.

Let us take (ii).

Since $$sA > rB,$$

$$\therefore\ sA - rB \text{ is a magnitude } \textit{of the same kind} \text{ as } B,$$

$$\therefore \text{ an integer } n \text{ exists such that } n(sA - rB) > B,$$

$$\therefore\ nsA > (nr + 1)B,$$

but $$sC = rD, \quad \therefore\ nsC = nrD < (nr + 1)D.$$

Hence putting $$r' = nr + 1, \quad s' = ns,$$

integers r', s' exist such that $s'A > r'B$, but $s'C < r'D$.

$$\therefore\ A : B > \frac{r'}{s'} > C : D.$$

Properties of Equal Ratios. First Group of Propositions.

11. It is now possible to prove the properties of Equal Ratios without employing the Test for Unequal Ratios, or any propositions depending thereon.

* This article is not required in the subsequent exposition. It is given here only on account of its close connection in thought with the preceding article.

For the actual proofs, excepting the few here given as specimens, I refer to my *Euclid V and VI*, second edition.

Prop. XII. If $A : B = C : D$, then $rA : sB = rC : sD$. (Euc. v. 4.)

Prop. XIII. If $A : B = C : D$, then $B : A = D : C$. (Euc. v. Corollary to 4.)

Prop. XIV. If $A : B = C : D = E : F$, and if all the magnitudes are *of the same kind*, then $A : B = A + C + E : B + D + F$. (Euc. v. 12.)

Prop. XV. $A : B = nA : nB$. (Euc. v. 15.)

Prop. XVI. If $A : B = X : Y$, then $A + B : B = X + Y : Y$. (Euc. v. 18.)

Prop. XVII. If $A + B : B = X + Y : Y$, then $A : B = X : Y$. (Euc. v. 17.)

The proofs of these propositions may be conducted on the same lines. They are independent of one another and may be taken in any order.

I give as examples the proofs of Props. XII and XVI.

Prop. XII. If $A : B = C : D$, to show that $rA : sB = rC : sD$.

Let $\frac{p}{q}$ denote any rational fraction, then it follows from Art. 9 that it is only necessary to consider the following alternatives:

Either $rA : sB > \frac{p}{q}$,

$\therefore q(rA) > p(sB)$,

$\therefore (q(r))A > (p(s))B$,

$\therefore A : B > \frac{p(s)}{q(r)}$.

But $A : B = C : D$,

$\therefore C : D > \frac{p(s)}{q(r)}$,

$\therefore (q(r))C > (p(s))D$,

$\therefore q(rC) > p(sD)$,

$\therefore rC : sD > \frac{p}{q}$.

Hence if $rA : sB > \frac{p}{q}$,

then $rC : sD > \frac{p}{q}$.

Or $rA : sB < \frac{p}{q}$,

$\therefore q(rA) < p(sB)$,

$\therefore (q(r))A < (p(s))B$,

$\therefore A : B < \frac{p(s)}{q(r)}$.

But $A : B = C : D$,

$\therefore C : D < \frac{p(s)}{q(r)}$,

$\therefore (q(r))C < (p(s))D$,

$\therefore q(rC) < p(sD)$,

$\therefore rC : sD < \frac{p}{q}$.

Hence if $rA : sB < \frac{p}{q}$,

then $rC : sD < \frac{p}{q}$.

Hence no rational fraction can lie between $rA : sB$ and $rC : sD$.

$$\therefore rA : sB = rC : sD.$$

Prop. XVI. If $A : B = X : Y$, then $A + B : B = X + Y : Y$.

Compare $A + B : B$ with any rational fraction $\frac{r}{s}$.

Then by Art. 9 it is necessary only to consider the alternatives:

Either $A+B:B>\frac{r}{s}$,

$\therefore\ s(A+B)>rB.$

It is necessary to take separately the cases $s<r$, $s=r$, $s>r$.

(i) $s<r$, $sA>(r-s)B$,

$$A:B>\frac{r-s}{s},$$

but $A:B=X:Y$,

$$\therefore\ X:Y>\frac{r-s}{s},$$

$$\therefore\ sX>(r-s)Y,$$

$$\therefore\ s(X+Y)>rY,$$

$$\therefore\ X+Y:Y>\frac{r}{s}.$$

(ii) $s=r$, $sY=rY$,

$$\therefore\ s(X+Y)>rY,$$

$$\therefore\ X+Y:Y>\frac{r}{s}.$$

(iii) $s>r$, $sY>rY$,

$$\therefore\ s(X+Y)>rY,$$

$$\therefore\ X+Y:Y>\frac{r}{s}.$$

Hence if $A+B:B>\frac{r}{s}$,

then $X+Y:Y>\frac{r}{s}$.

Or $A+B:B<\frac{r}{s}$,

$\therefore\ s(A+B)<rB.$

In this case s must be $<r$,

$$\therefore\ sA<(r-s)B,$$

$$A:B<\frac{r-s}{s},$$

but $A:B=X:Y$,

$$\therefore\ X:Y<\frac{r-s}{s},$$

$$\therefore\ sX<(r-s)Y,$$

$$\therefore\ s(X+Y)<rY,$$

$$\therefore\ X+Y:Y<\frac{r}{s}.$$

Hence if $A+B:B<\frac{r}{s}$,

then $X+Y:Y<\frac{r}{s}$.

Hence no rational fraction lies between $A+B:B$ and $X+Y:Y$.

$$\therefore\ A+B:B=X+Y:Y.$$

Properties of Equal Ratios. Second Group of Propositions.

12. I take next a group of propositions, which depend on Prop. VIII above.

Prop. XVIII. If A, B, C, D be four magnitudes *of the same kind*, and if $A:B=C:D$, then $A:C=B:D$. (Euc. v. 16.) From this it follows as a Corollary that according as $A \gtreqless C$, so is $B \gtreqless D$. (Euc. v. 14.)

Prop. XIX. If $A:B=T:U$, and if $B:C=U:V$, then $A:C=T:V$. (Euc. v. 22.) From this it follows as a Corollary that according as $A \gtreqless C$, so is $T \gtreqless V$. (Euc. v. 20.)

Prop. XX. If $A : B = U : V$, and if $B : C = T : U$, then $A : C = T : V$. (Euc. v. 23.) From this it follows as a Corollary that according as $A \gtreqless C$, so is $T \gtreqless V$. (Euc. v. 21.)

Euclid proves the propositions here given as Corollaries first, and then deduces the main propositions from them.

I give as an example the proof of

Prop. XX. Compare $A : C$ with any rational fraction $\frac{r}{s}$.

It is necessary to consider the two following alternatives only:

Either $\quad A : C > \frac{r}{s}$,

$$\therefore\ sA > rC.$$

B is a magnitude of the same kind as sA, rC.

Hence by Prop. VIII integers n, t exist such that

$$n(sA) > tB > n(rC).$$

Now $\quad (n(s))A = n(sA) > tB$,

$$\therefore\ A : B > \frac{t}{n(s)},$$

$$U : V = A : B > \frac{t}{n(s)},$$

$$\therefore\ (n(s))U > tV \ldots\ldots(\text{I}).$$

Also $\quad tB > n(rC)$,

$$\therefore\ tB > (n(r))C,$$

$$\therefore\ B : C > \frac{n(r)}{t}.$$

Now $\quad T : U = B : C > \frac{n(r)}{t}$,

$$\therefore\ tT > (n(r))U \ \ldots(\text{II}).$$

We have to eliminate U between (I) and (II),

$$s(tT) > s[(n(r))U] \text{ from (II)},$$

$$s[(n(r))U] = r[(n(s))U]$$

Cor. to Prop. III,

$$r[(n(s))U] > r(tV) \text{ from (I)},$$

$$\therefore\ s(tT) > r(tV),$$

$$\therefore\ t(sT) > t(rV),$$

$$\therefore\ sT > rV,$$

$$\therefore\ T : V > \frac{r}{s}.$$

Or $\quad A : C < \frac{r}{s}$,

$$\therefore\ sA < rC.$$

B is a magnitude of the same kind as sA, rC.

Hence by Prop. VIII integers n, t exist such that

$$n(sA) < tB < n(rC).$$

Now $\quad (n(s))A = n(sA) < tB$,

$$\therefore\ A : B < \frac{t}{n(s)},$$

$$U : V = A : B < \frac{t}{n(s)},$$

$$\therefore\ (n(s))U < tV \ \ldots(\text{III}).$$

Also $\quad tB < n(rC)$,

$$\therefore\ tB < (n(r))C,$$

$$\therefore\ B : C < \frac{n(r)}{t}.$$

Now $\quad T : U = B : C < \frac{n(r)}{t}$,

$$\therefore\ tT < (n(r))U \ \ldots(\text{IV}).$$

We have to eliminate U between (III) and (IV),

$$s(tT) < s[(n(r))U] \text{ from (IV)},$$

$$s[(n(r))U] = r[(n(s))U]$$

Cor. to Prop. III,

$$r[(n(s))U] < r(tV) \text{ from (III)},$$

$$\therefore\ s(tT) < r(tV),$$

$$\therefore\ t(sT) < t(rV),$$

$$\therefore\ sT < rV,$$

$$\therefore\ T : V < \frac{r}{s}.$$

If	$\therefore\ A : C > \frac{r}{s}$,	If	$\therefore\ A : C < \frac{r}{s}$,
then	$T : V > \frac{r}{s}$.	then	$T : V < \frac{r}{s}$.

Hence no rational fraction lies between $A : C$ and $T : V$.

$$\therefore\ A : C = T : V.$$

Properties of Equal Ratios. Third Group of Propositions.

13. I take next the propositions

Prop. XXI. If $A + C : B + D = C : D$, then $A : B = C : D$. (Euc. V. 19.)

Prop. XXII. If $A : C = X : Z$, and if $B : C = Y : Z$, then $A + B : C = X + Y : Z$. (Euc. V. 24.)

Prop. XXIII. If A, B, C, D are four magnitudes *of the same kind*, if $A : B = C : D$, and if A be the greatest of the four magnitudes, then $A + D > B + C$. (Euc. V. 25.)

The proofs of these propositions can be obtained without using the preceding propositions, as I have shown in the *Cambridge Philosophical Transactions*, Vol. XIX. Part II, pp. 166—170. But these proofs are involved and difficult and altogether unsuitable to an elementary course of instruction. These propositions are best deduced from earlier* propositions in Euclid's manner.

Prop. XXI depends on Props. XVII and XVIII.

Prop. XXII depends on Props. XIII, XVI and XIX.

Prop. XXIII depends on Props. XVII and XVIII.

The existence of the Fourth Proportional.

14. I come now to a very important proposition, which is far too difficult to be included in an elementary course, except in the special case in which the magnitudes concerned are segments of straight lines or cases readily reducible thereto.

Euclid assumed it, and as Simson points out without necessity, in the Eighteenth Proposition of the Fifth Book. The proposition, as will be seen, rests for its validity on an axiom corresponding to the Cantor-Dedekind Axiom. It is fundamental in the Theory of Ratio. It is as follows:

Prop. XXIV. If A and B be magnitudes *of the same kind*, and if C be any third magnitude, then there exists a fourth magnitude Z *of the same kind* as C such that

$$A : B = C : Z.$$

Now if $A : B = C : Z$,

then $Z : C = B : A$.

* As these earlier propositions have now been established by the aid of the Fifth Definition without using the Seventh Definition, the demonstrations are no longer dependent on the Seventh Definition.

(1) Suppose that A and B have a common measure, and

$$\therefore \text{ integers } r, s \text{ exist such that } B : A = \frac{r}{s}.$$

Then $$Z = \frac{rC}{s}.$$

For $$\frac{rC}{s} : C = rC : sC = \frac{r}{s} = B : A.$$

(2) Suppose B and A have no common measure.

Let $B : A$ be equal to the irrational number ρ.

Let $p_1, p_1', p_1'', \ldots$ represent rational numbers in the lower class determined by ρ in *ascending* order of magnitude.

Let $p_2, p_2', p_2'', \ldots$ represent rational numbers in the upper class determined by ρ in *descending* order of magnitude.

Construct the magnitudes $p_1C, p_1'C, p_1''C, \ldots$, and call them magnitudes of the form p_1C.

Construct the magnitudes $p_2C, p_2'C, p_2''C, \ldots$, and call them magnitudes of the form p_2C.

$$\therefore\ p_1C < p_1'C < p_1''C < \ldots < p_2''C < p_2'C < p_2C.$$

Let us separate all the magnitudes *of the same kind* as C into two classes by the following rules:

(i) Put all those into the upper class which are greater than every one of the magnitudes of the form p_1C.

Call any one of these a magnitude Y.

(ii) Into the lower class put all those magnitudes which are not greater than every one of the magnitudes of the form p_1C.

Call any one of these magnitudes a magnitude X.

It will be proved in the first place that

Every magnitude Y is greater than every magnitude X.

From the definition of the magnitudes X, it appears that any X, say X_1, does not exceed every magnitude of the form p_1C.

Suppose $$X_1 \leqq p_1'C.$$

But every Y exceeds every magnitude of the form p_1C.

$$\therefore\ Y > p_1'C,$$

$$\therefore\ Y > X_1,$$

$$\therefore \text{ every } Y \text{ exceeds every } X.$$

It will be proved in the second place that

The set of magnitudes X includes no greatest magnitude.

The characteristic of the magnitudes X is that they are not greater than every magnitude of the form p_1C.

Suppose X' is one of the magnitudes X.

Let $$X' \gtrless p_1'C.$$

Now there is no greatest p_1.

Suppose $$p_1' < p_1'',$$
$$\therefore\ p_1'C < p_1''C.$$

It is then permissible to take $$X'' = p_1''C,$$
$$\therefore\ X' < X'',$$

and so on; other magnitudes X can be found continually increasing in order of magnitude.

Therefore the set of magnitudes X includes no greatest magnitude.

Now the magnitudes X and Y together include all the magnitudes *of the same kind* as C. In regard to these magnitudes we assume an axiom corresponding to the Cantor-Dedekind Axiom for the straight line, as follows:

If all the magnitudes *of the same kind* as C be separated into two classes such that every magnitude of the one class is less than every magnitude of the other class, then there is one and only one magnitude *of the same kind* as C which produces this separation, and it is either the greatest magnitude of the one class, or the least magnitude of the other class.

Now it has been proved that the magnitudes X include no greatest magnitude.

Therefore the magnitudes Y include a least magnitude.

Call this least magnitude Z.

Since Z is a magnitude Y, it is greater than every magnitude of the form p_1C.

Write this thus: $$Z > \text{every } p_1C,$$
$$\therefore\ Z : C > \text{every } p_1.$$

It will be proved next that $$Z : C < \text{every } p_2,$$
$$\text{i.e. } Z < \text{every } p_2C.$$

Suppose if possible $$Z \geqq \text{some } p_2C,$$
say $$Z \geqq p_2''C.$$

Then since the rational numbers p_2 include no least rational number, take
$$p_2''' < p_2''.$$

Then $$p_2'''C < p_2''C,$$
$$\therefore\ p_2'''C < Z.$$

Now $$\text{every } p_2 > \text{every } p_1,$$
$$\therefore\ p_2''' > \text{every } p_1,$$
$$\therefore\ p_2'''C > \text{every } p_1C,$$
$$\therefore\ p_2'''C \text{ is a magnitude } Y.$$

But $$p_2'''C < Z.$$

Hence there is a magnitude Y which is less than Z.

But this is contrary to the definition of Z, viz. that it was the least of the magnitudes Y.

Hence $$Z < \text{every } p_2 C,$$

$$\therefore\ Z : C < \text{every } p_2.$$

And it was proved that $$Z : C > \text{every } p_1.$$

Hence $Z : C$ determines the same separation of the whole system of rational numbers as the irrational number ρ, i.e. the same separation as $B : A$,

$$\therefore\ Z : C = B : A,$$

$$\therefore\ A : B = C : Z.$$

The position of the study of the Theory of Proportion in the Mathematical Curriculum.

15. It is clear that the properties of similar figures cannot be discussed without an adequate theory of proportion, something in fact to replace the treatment in Euclid's Fifth Book.

And now the question arises whether the theory of proportion of incommensurable magnitudes can be explained in a manner which is simple enough to be intelligible to those who are commencing the study of the Theory of Similar Figures.

I am prepared to admit at once that Euclid's Theory as given in his Fifth Book does not satisfy this condition. That is the result of all experience in teaching.

It seems to me that the definition of the ratio of Commensurable Magnitudes should be given at a very early stage in the study of Geometry.

The following propositions can then be proved:

(1) If two triangles (parallelograms) have the same altitude and commensurable bases, then their areas are proportional to their bases.

(2) If two straight lines a, b be cut by four parallel straight lines c, d, e, f, and if the segment on a formed by c and d be commensurable with the segment on a formed by e, f, then these segments are proportional to the corresponding segments on b.

(3) If in equal circles there are two angles at the centre which have a common measure, then the angles are proportional to the arcs which they subtend.

I have found from several years' experience that any person with average mathematical capacity, prepared in this way, when he begins the study of Similar Figures, is able to understand the Theory of Proportion explained in Arts. 1—13.

I will now state the difference between this treatment and that given in Euclid's Fifth Book.

Euclid has two fundamental definitions, viz.: the Fifth, which is the test for determining when two ratios are equal, and the Seventh, which is the test for distinguishing the greater from the smaller of two unequal ratios.

He proves properties of Equal Ratios in two different ways. Some of them he deduces directly from the Fifth Definition. To prove others he employs the Seventh Definition, or propositions depending thereon, as well as the Fifth.

Now arguing *a priori* it might be expected that if the Fifth Definition is a full and complete one, it should be sufficient to prove all the properties of Equal Ratios without the aid of the Seventh. This is certainly the case for all the propositions dealing with Equal Ratios in the Fifth Book.

It might therefore be expected that the use of the Seventh Definition for purposes for which the Fifth is adequate would introduce complexity into the argument.

In fact it is not at all clear to any one trying to follow the course of Euclid's argument whether it leads naturally to the employment of the Fifth Definition or to that of the Seventh. Euclid's proofs are difficult and intricate. They do not appear to run on the same lines.

It must not be supposed that I am suggesting any departure from the spirit of Euclid's method. All that I do is to assimilate the proofs of those propositions dealing with properties of Equal Ratios in which Euclid employs the Seventh Definition to the proofs of those other propositions in which he relies only on the Fifth.

When this is done the difficulty and the complexity of the argument disappear, and it is much easier to grasp as a whole.

The general case of Prop. XXIV, Art. 14, I do not attempt to explain in an elementary course. It is, however, a very good example of the use of Dedekind's Theory of the separation of the system of rational numbers into two classes, such that every number in the one class is less than every number in the other class.

Any one, who has studied the far easier examples of the same kind of separation in the other propositions in Arts. 11—13, will be in a much better position to study this theory when he commences the study of irrational numbers in the Calculus; and will be far better equipped for the study of the Calculus than he would have been if he had gone through no previous training in the subject. Whether I am right in thinking that the theory I have set out in Arts. 1—13 should be explained to those who are commencing the study of the Theory of Similar Figures or not, I have no doubt in my mind that the subject should form a part of the mathematical curriculum; and that, at whatever stage the subject be studied, the treatment I suggest is much easier to grasp than Euclid's, and that the reasoning is no less rigid than his.

SYSTEMATISCHE REKREATIONSMATHEMATIK IN DEN MITTLEREN SCHULEN

Von N. Hatzidakis.

Die antikonservative Strömung, die man zur Zeit in allen Gebieten des menschlichen Denkens und Handelns leicht bemerken kann—sei es Politik oder Wissenschaft oder Litteratur oder Kunst—hat auch *unsere* Wissenschaft mitgerissen. Alte mathematische Gebäude, die man früher, so zu sagen, in der öffentlichen mathematischen Meinung, für nicht weniger felsenfest als diese uralten, felsenfesten Cambridger Colleges-Gebäude hielt, wurden fast plötzlich niedergestürzt und ein allgemeines Zweifelsfieber hat unsere Gehirnsadern erobert. Und zwar hat sich dieses Niederstürzen des Alten und Überlieferten nach zwei wesentlich verschiedenen Richtungen verbreitet, nach der Richtung der wissenschaftlichen Forschung und nach der Richtung der Verbreitung des Wissens, spezieller des Unterrichts.

Ist schon im Gebiete *der Forschung* dieses Hasten, lass uns besser sagen, diese *Mode* des Wegschaffens der alten prinzipiellen Begriffe und der Grundlegung einer immer grösseren, ungefähr in geometrischer Progression wachsenden Unzahl neuer Zweige—z. B. Geometrien!—trotz des grossen Vorteiles des tieferen Eingreifens in das Wesen unserer Wissenschaft—nicht ganz frei von Gefahr, so steht doch die Frage, wie weit *im Unterricht* Altes wegzujagen und Neues hinzuzunehmen ist, bedeutend mehr kompliziert vor uns, ist sogar vielleicht—wenigstens was den mittleren Unterricht anbetrifft—eine theoretisch unlösbare Frage, zu deren Lösung man praktisch durch immer feinere Approximationen immer näher—doch immer *asymptotisch*!—gelangen kann, oder wenigstens zu gelangen versucht. Die *Internationale Mathematische Unterrichtskommission* befindet sich vorläufig in der ersten Stufe ihres Zieles, nämlich in der Periode der Untersuchung, *wie* es eigentlich mit dem mittleren mathematischen Unterricht in den verschiedenen Kulturländern steht; das Material, was schon bis heute eingesammelt worden ist, ist selbstverständlich bedeutend gross und wird immer reichhaltiger, so dass endgiltige Resultate—ja, *überhaupt* Resultate—nicht in voller Klarheit vorliegen können. Man kann aber—ich glaube—ganz abgesehen vom Material—diese verschiedenen Fragen und Zweifel über Zweck und Wesen des mittleren mathematischen Unterrichts—ja, sogar des gesammten mittleren Unterrichts—im Allgemeinen in *zwei* Kategorien teilen: erstens, *wieviel* und *was* in den verschiedenen Schulen und Klassen gelesen werden soll, und zweitens, *wie* das gelesen wird, mit anderen Worten: die Kategorie *der Quantität* und *Qualität* des zu unterrichtenden Materials und die Kategorie *der*

Methode des Unterrichts. Ich glaube nun, dass man in den bisherigen verschiedenen Versuchen eine deutliche Vorliebe für die *erste* Kategorie spüren kann. Speziell ist, von deutscher Seite her, ein grosser Umschwung vielfach diskutiert, teilweise auch durchgeführt worden, nämlich die Einführung der ersten prinzipiellen Begriffe der höheren Mathematik—natürlich in möglichst vereinfachter Form!—in die mittleren Schulen. Vielleicht hat man auch schon versucht—oder wird versuchen!—auch die Begriffe der nicht-euklidischen Geometrie in den niederen Unterricht einzuführen—ja, es ist sogar anlockend, auch die mehrdimensionalen, geradlinigen oder gekrümmten, Räume in die Oberrealschulen hineinzudrängen, u.s.w. Es ist hier nicht mein Zweck, in irgend eine Diskussion über die Vorteile oder Nachteile dieser Neuerungen einzugehen—das würde übrigens, wie es so oft geschehen ist, zu endlosen Wegen führen!—Ich wende mich dagegen kurz dem eigentlichen Kerne meiner heutigen Betrachtungen zu, und dieser Kern ist *der Schüler* selbst. Wir sind ja alle einmal Schüler gewesen, und, wenn auch nicht gerade *von uns* gesagt werden kann, dass wir —*in der Mathematik!*—durchschnittliche Schüler waren—sonst hätten wir ja sie sehr wahrscheinlich nicht weiter als Spezialfach getrieben!—so begreifen wir doch leicht, dass—abgesehen von den mathematisch Begabten, die ja immer eine kleine Minorität ausmachen!—dem *durchschnittlichen* Schüler Mathematik mehr oder weniger ziemlich schwer, oder jedenfalls nicht besonders interessant vorkommt. Der Grund dieser Erscheinung liegt teils in der grossen Abstraktion, die die Mathematik überhaupt verlangt, teils in den langen logischen Ketten von Syllogismen, die den noch nicht stark entwickelten kindlichen Geist ermüden, etwa so wie eine lange, steile Treppe des Kindes Körper, ja, es giebt vielleicht auch andere Gründe mehr komplizierter Natur—uns interessiert jetzt jedenfalls nur das häufige Dasein eines solchen schlaffen Interesses. Natürlich kann der Lehrer sagen: "der Schüler *muss* lernen." Das tut er auch wirklich im Allgemeinen: er lernt die Mathematik, die man ihn zu lernen zwingt, lernt aber sehr oft *nicht*, die Mathematik zu lieben! Liebe überhaupt setzt eine Freiheit voraus und die Freiheit des Denkens lassen wir dem Schüler *nicht* zu. Indem wir nämlich die Lehrbücher des Unterrichts nach unserem eigenen Gutdünken verfassen, tun wir eigentlich nichts andres als *dies*: wir verlangen, dass der Geist des Kindes unserem schon stark und methodisch denkenden Geist—dem Geist des Erwachsenen—folgt und sich danach formt. Und doch liegt der Natur des kindlichen Geistes gerade eine Systematisierung, eine schablonenmässige Anordnung und Anhäufung seiner Ideen fern; das Kind denkt vielmehr auf seine eigene Weise, und die ist der *sprungweise* Gedankengang. Der kindliche Geist ist ein Schmetterling, der von Blume zu Blume niedrig und sprungweise fliegt, der des Erwachsenen dagegen ein Adler, der hoch und lange in der Luft schwebt. Ich will nun damit nicht gerade sagen, dass dieser Freiheit wegen wir dem Kinde auch *die* Freiheit geben sollen, mit uns über das zu unterrichtende Material zu diskutieren, nicht nur weil das Kind über die Weise seines Denkens nichts ahnt, sondern auch aus *dem* Grunde, dass wir Lehrer dann vielleicht dieselbe unangenehme Behandlung seitens der Kinder bekommen würden, wie die jetzigen englischen Minister seitens der Stimmrecht verlangenden englischen Damen! Ich meine aber, dass, anstatt *das Kind* zu unserem wissenschaftlichen Gedankengang emporzuheben zu versuchen, *wir* dagegen der kindlichen Seele näher kommen sollen, so dass wir in ihr den Keim des Interesses sanft und unbemerkt pflanzen. Es schadet auch meines Erachtens *nicht* so sehr, wenn wir zu diesem Zwecke ab und zu von dem von vornherein mit *strenger* Logik durchdachtem Wege abweichen und dem Schüler in einen kleinen Seitenpfad, wonach er Lust bekommt, folgen. Wenn wir ausserdem in Betracht ziehen, dass ein anderes karakteristisches Kennzeichen der Kinderseele die bestimmte Neigung zum Konkreten ist, so kommen wir, was die Mathematik anbetrifft, zu dem Resultat, dass dem Kinde solche Mathematik, oder wenigstens *auch* solche Mathematik, gegeben werden soll, die so viel wie möglich *sprungweise* verfährt und so viel wie möglich *konkrete* Begriffe behandelt. So gelangen wir zu den verschiedensten mathematischen Spielen und Unterhaltungen, die wir kurz mit dem Stichwort *Rekreations-* oder *Unterhaltungsmathematik* bezeichnen können. Dann kommt aber gleich die Frage nach einem geeigneten Buche, denn dass der Lehrer von sich selbst solche einzelne Unterhaltungsaufgaben giebt, ist ja sehr schwierig und auch nicht zweckmässig. Vielmehr handelt es sich um eine *systematische* Anordnung des Stoffes, so dass, *ohne dass der Schüler es bemerkt*, trotz des Wandelns in den Seitenpfäden, er doch schliesslich ein bestimmtes Ziel erreicht und seinen Weg nicht ganz in der Wildniss verliert. Bücher über Unterhaltungsmathematik giebt es nun viele, ich brauche nur an Schubert's "Mathematische Mussestunden" oder an das englische von Ball oder das französische von Lucas zu erinnern, allein keines von diesen Büchern ist für den Schüler geeignet, denn sie sind ja nicht für ihn, sondern für den gebildeten Laien geschrieben. Das Buch für den Schüler sollte ganz anders verfasst werden. Vor allem kommen aber die Fragen: Ist ein solches Buch überhaupt möglich? In welchen mathematischen Gebieten und in welcher Ausdehnung kann die Unterhaltungsmathematik dem Lehrer als ein Hilfswerkzeug zur mathematischen Ausbildung des Schülers dienen? Das alles sind Fragen, die der Antwort vielleicht noch lange harren werden, die Experimente in verschiedenen Gebieten und verschiedenen Völkern verlangen, Fragen also, die nicht im voraus theoretisch gelöst werden können, geschweige denn in diesen kurzen Betrachtungen, die Sie die Güte gehabt haben so aufmerksam zu hören. Bevor ich aber meinen kleinen Vortrag schliesse, bitte ich Sie mir zu erlauben, dessen Hauptpunkte kurz zusammenzufassen und zwar unter dem Stichworte:

Pro discipulo.

(1) Es wäre wünschenswert, in den Betrachtungen, den Erörterungen und überhaupt in der ganzen Reformbewegung für die niederen und die mittleren Schulen, den Faktor *Kind* nicht als ein mathematisches *Unendlichkleines* zu betrachten.

(2) Es wäre wünschenswert, die Psychologie der Kinderseele nach der mathematischen Seite hin durch geeignete Betrachtungen seitens der Lehrer mit mehr experimentalem Material zu bereichern.

(3) Es wäre wünschenswert, die Frage, *ob* und *wie* und in *welcher* Ausdehnung die Unterhaltungsmathematik in die Schulen eingeführt werden kann und soll, gründlich zu studieren.

(4) Und, was ja Alles vorige zusammenfasst, es wäre wünschenswert, dem Kinde das Recht einzuräumen, *ZUERST* EIN KIND, UND *DANN* EIN SCHÜLER ZU SEIN.

SUR QUELQUES NOUVELLES MACHINES ALGÉBRIQUES

Par A. Gérardin.

Les nouvelles machines dont je dois vous parler ici très brièvement sont destinées à résoudre en nombres entiers les équations indéterminées. Cette opération peut se faire automatiquement et l'on arrive déjà à des résultats tout à fait intéressants. Outre mes machines personnelles je dois en citer d'autres qui s'appuient sur le même principe, mais qui diffèrent beaucoup dans l'application ; ce sont celles de M. Kraïtchik, jeune ingénieur russe à Liège, et celle de MM. Carissan, deux bons collaborateurs, à qui je souhaite pleine réussite.

Mais ces machines destinées en principe à la résolution de problèmes ardus d'arithmétique supérieure, peuvent être présentées sous une forme neuve et servir ainsi à de jeunes élèves de 13 à 15 ans pour la décomposition des nombres. Un concours de ce genre à été ouvert en juillet au collège de Lesneven (France), après autorisation de M. le Recteur de Rennes.

Ce concours, dirigé par le professeur M. P. Carissan, et basé sur l'étude de mon procédé, a donné de bons résultats, et a permis à ces jeunes élèves de décomposer en quelques minutes des nombres de 6 ou 7 chiffres, au lieu de demander plusieurs journées de travail par les méthodes classiques.

Tous les problèmes qui nous intéressent sont résolus par des équations de la forme

$$ax^2 + bx + c = y^2,$$

en nombres entiers, et il suffit de connaître le tableau des résidus quadratiques pour pouvoir appliquer immédiatement mon procédé.

Je représente un résidu par une case blanche sur une bande quadrillée et un non-résidu par une case noire. Ces bandes sont périodiques, et pour avoir la solution, il suffit de chercher, sans fatigue, une ligne *entièrement blanche.* On peut aussi remplacer la case noire par le chiffre 1, et la blanche par un zéro, ou bien encore utiliser l'Initiateur Mathématique de M. J. Camescasse, jeu de cubes colorés destiné à apprendre les premiers éléments du calcul, et qui est lui-même une application de l'Initiation Mathématique de M. C. A. Laisant.

Je vais exposer un exemple très simple. Soit à décomposer 263 069 ; j'écris $N = x^2 - y^2 = 263\ 069$. Je vois facilement que y est forcément de la forme $4l + 2$, d'où pour le module 4, la bande périodique 1 1 0 1, ... de même pour les autres modules, et l'on obtient la figure suivante établie en 3 minutes.

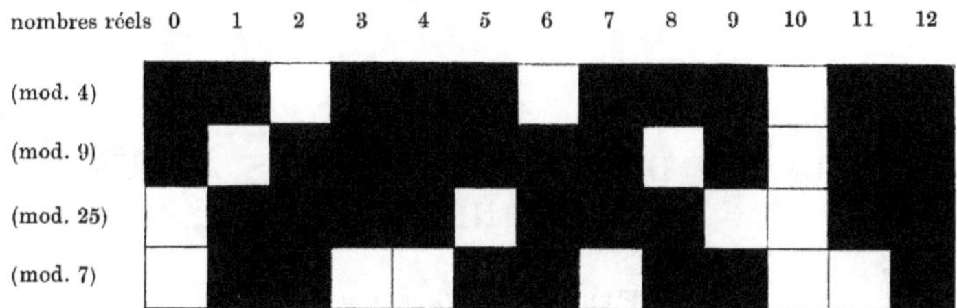

La solution cherchée est $y = 10$, d'où $N = 503 \times 523$; le groupe de 4 bandes données ci-dessus est valable pour tous les nombres de la forme $6300m + 2939$ sans aucune retouche.

Nous espérons voir mettre à l'étude cette question par nos collègues de divers pays, qui voudront bien me tenir au courant de leurs résultats.

L'ENSEIGNEMENT DES MATHÉMATIQUES À L'USAGE DES INGÉNIEURS

PAR HENRY LE CHATELIER.

Les mathématiques sont indispensables pour toutes les études de science pure ou appliquée, surtout quand il s'agit de sciences arrivées à un haut degré de perfection, comme celles dont font usage les ingénieurs: la mécanique, la physique et même la chimie. Le seul objet de la science est de préciser les relations mutuelles des différents phénomènes naturels; or toutes ces relations, comme les lois de Mariotte et de Ohm, le principe de Carnot, s'expriment au moyen de formules numériques qui relient entre elles certaines grandeurs mesurables, c'est-à-dire au moyen de véritables expressions algébriques. Toutes les déductions à tirer de ces formules en vue des applications doivent nécessairement être obtenues par des calculs, par des transformations algébriques.

Les calculs nécessaires pour l'établissement des ponts et des fers métalliques s'appuient sur les formules de la mécanique rationnelle et sur la résistance des matériaux. Ces calculs deviennent plus compliqués dans le cas des pièces de machines, en raison des phénomènes d'inertie mis en jeu par le mouvement même de ces pièces et des phénomènes de résonnance, si fréquents dans les déplacements périodiques à grande vitesse. Enfin la complexité des calculs atteint son maximum avec les machines utilisant le mouvement des fluides: turbines, ventilateurs, aéroplanes. Les théories de l'hydrodynamique et de l'aérodynamique doivent leur complexité non seulement à la nature de leurs formules algébriques, mais encore aux incertitudes des hypothèses sur lesquelles s'appuient ces formules.

En électricité les exemples sont tout à fait analogues; la simple application des lois de Ohm aux circuits multiples des réseaux urbains soulève déjà des problèmes intéressants qui se compliquent rapidement dans le cas des courants alternatifs en raison de l'obligation de tenir compte de la capacité des lignes, de leur induction sur les circuits voisins et enfin dans certains cas des phénomènes de résonnance entre les canalisations et les machines. La construction des dynamos pose des questions dont l'étude mathématique est toujours très compliquée, parfois même impossible. Il faut alors faire intervenir, en plus des lois de l'électricité, celles du magnétisme et tenir compte des formes toujours compliquées de machines, nécessitées par les conditions de réfroidissement extérieur.

Pendant longtemps en chimie l'expérience brutale, le pur empirisme a suffi aux industries les plus variées; mais aujourd'hui le rôle de plus en plus important joué par la mécanique chimique oblige les chimistes à s'occuper parfois de différentielles et d'intégrales, de logarithmes; sans parler de la nécessité, absolue pour un chimiste qui prétend au rang d'ingénieur, d'être en même temps mécanicien et électricien.

Il semble pourtant à première vue que les connaissances mathématiques, indispensables pour un ingénieur moyen, ne doivent pas être très étendues; je parle ici d'un ingénieur de capacité ordinaire, qui ne se propose pas de renouveler les découvertes de Lord Kelvin, mais se content d'appliquer des méthodes de calcul connues, en s'inspirant d'exemples donnés par ses prédecesseurs. Pour cet usage les géométries euclidienne et descriptive, le calcul algébrique, la résolution des équations des deux premiers degrés; quelques éléments du calcul différentiel et intégral et enfin un peu de trigonométrie suffisent largement. Mais alors, dira-t-on, le travail nécessaire pour acquérir ces connaissances se réduit à fort peu de choses. Il n'en est rien; il faut pour acquérir les connaissances élémentaires indispensables à un ingénieur, une somme de travail beaucoup plus considérable qu'on ne le suppose communément.

Le but poursuivi par un ingénieur n'est pas d'arriver à se tirer plus ou moins brillamment d'un examen, après avoir repassé à satiété pendant quelques jours les questions du programme; il doit être capable d'appliquer sans hésitation et sans erreur les méthodes qu'il a apprises à des cas tout différents de ceux qui lui ont été enseignés dans ses cours, et il lui faut garder cette aptitude pendant toute la durée de son existence. Ses connaissances mathématiques doivent donc s'être incrustées dans son esprit de façon à en être devenu parties intégrantes et ne pas être restées à l'état de vernis superficiel que les années auraient vite fait d'effacer.

En fait pour graver dans l'esprit les notions élémentaires de mathématiques nécessaire à l'ingénieur, il faut lui demander un travail semblant à première vue tout à fait disproportionné avec le but poursuivi. J'ai souvent entendu dire à propos de l'École Polytechnique de Paris: A quoi bon tant de mathématiques pour former des ingénieurs et des artilleurs? quel usage feront-ils jamais des fonctions elliptiques ou abeliennes? Les défenseurs de ces études répondent parfois que les hautes mathématiques sont nécessaires pour la résolution de certains problèmes de physique transcendante. Cet argument ne vaut rien, car le rôle des ingénieurs n'est pas de se livrer à des spéculations plus ou moins creuses sur la constitution de la matière. Si réellement ces mathématiques supérieures n'avaient pas d'autre utilité, il faudrait les sacrifier sans hésitation. Les choses rares et curieuses regardent les artistes et les virtuoses, mais pas les ingénieurs dont le rôle est d'agir et non de rêver. En réalité l'utilité pour les ingénieurs de l'étude des mathématiques supérieures est de donner à leur esprit un entraînement qui leur permette de retrouver sans hésitation et d'employer en toute certitude les méthodes plus simples, dont ils auront journellement à se servir. Un enfant n'arrive à faire la règle de trois sans hésitation qu'après avoir appris à résoudre les équations algébriques; il ne sait réellement extraire une racine carrée qu'après avoir étudié le calcul algébrique, la multiplication des polynomes. De même pour savoir employer les différentielles et les intégrales les plus simples, il est indispensable de s'être frotté à des fonctions plus complexes. Les mathématiques transcendantes sont utiles à tout ingénieur par les clartés qu'elles projettent indirectement sur les mathématiques élémentaires.

Un second point de vue à prendre en considération est la différence nécessaire des méthodes d'éducation convenant à des ingénieurs ou à des futurs savants. Dans la science pure, plus les méthodes et les formules deviennent générales, plus elles groupent de faits différents et par suite plus elles sont abstraites, plus grande est leur perfection. La théorie des vecteurs a permis de grouper ainsi toute une série de chapitres distincts de la science : problèmes de géométrie, composition des forces et des mouvements, composition des couples de rotation, etc. Son intérêt est indiscutable pour le savant, peut-être n'en est-il pas de même pour le jeune ingénieur. Celui-ci a un intérêt de premier ordre à rattacher le plus directement possible tout résultat aux principes intuitifs ou aux faits expérimentaux suivant de point de départ ; il faut supprimer tout intermédiaire inutile, toute formule abstraite, dont la trace dans l'esprit est d'autant moins profonde qu'elle s'éloigne davantage des réalités concrètes ; par exemple, pour les résolutions des équations du premier et du second degré, il ne faut pas insister sur les formules générales qui donnent les valeurs des racines en fonctions des paramètres de ces équations. Tout au plus, faut-il citer ces formules générales à titre de curiosité ; dans tout problème on doit exiger les calculs complets conduisant à la résolution directe des équations ; dans le cas de deux équations du premier degré à deux inconnues on fera indifféremment employer la méthode de résolution par substitution ou par coefficient indéterminé ; ces méthodes une fois bien possédées ne s'oublient plus pour toute la durée de l'existence, tandis qu'il suffit de quelques mois pour perdre complètement de vue les formules générales de résolution et s'exposer à les appliquer à faux si on veut les rechercher dans un livre.

Indépendamment de leur utilité pratique pour la résolution de tel ou tel problème l'étude des mathématiques est encore indispensable pour la formation générale de l'esprit, en raison de leur valeur éducative. A ce point de vue général on devrait particulièrement retenir deux branches des mathématiques : la géométrie descriptive, c'est-à-dire la représentation de solides placés dans l'espace au moyen de leur projection orthogonale sur deux plans perpendiculaires, et la géométrie analytique, c'est-à-dire la résolution de problèmes de géométrie en combinant l'algèbre avec les coordonnées cartésiennes. Ces deux sciences ont depuis peu perdu en France la place considérable qu'elles occupaient autrefois dans l'enseignement mathématique. Cela est très regrettable, mais le discrédit dans lequel elles sont tombées s'explique par le fait qu'elles se sont laissées envahir par des théories générales et abstraites, qui leur ont enlevée une grande partie de leur valeur éducative.

La géométrie descriptive est extraordinairement utile pour le développement de l'imagination géométrique, de la faculté de voir simultanément dans l'espace des ensembles de points, de lignes et de surfaces ; cela permet de raisonner sur les corps solides, en particulier sur les machines ; c'est donc une formation indispensable pour l'ingénieur. Mais pour atteindre ce but la descriptive ne doit être qu'une application directe de la géométrie ordinaire, ne comportant aucune méthode particulière en dehors de l'emploi des deux plans de projection ; chaque problème doit être traité pour lui-même sans recourir à aucun artifice dispensant de reprendre les raisonnements d'une façon complète et d'arriver ainsi plus vite au résultat, comme cela est indispensable pour les épures données aux examens.

De même la géométrie analytique devrait être une simple application de l'algèbre à la géométrie euclidienne, sans théories particulières, sans fonctions spéciales à telle ou telle classe de surface. Son but doit être d'habituer au maniement du calcul algébrique; en prenant comme sujet d'exercice des relations concrètes de figures géométriques on facilite la succession des raisonnements et on habitue l'esprit à passer d'un fait concret à sa représentation algébrique ou à suivre la marche inverse, deux opérations que comportent toutes les applications des mathématiques aux problèmes de l'art de l'ingénieur.

Il existe enfin un dernier point de vue à faire entrer en ligne de compte, dont l'importance prime toutes les considérations précédentes et dont l'usage n'est pas restreint aux mathématiques mais s'applique à toutes les sciences, n'est pas réservé à la formation des ingénieurs mais intéresse aussi le savant désintéressé; c'est la nécessité d'obliger l'étudiant à faire œuvre personnelle au cours de ses études, à lui demander un effort pour l'acquisition de connaissances qui lui sont nécessaires et de ne pas lui permettre de se contenter, comme le dit si judicieusement F. W. Taylor, de jouer le rôle d'une éponge absorbant sans peine et sans refléxion les matières qui lui sont offertes dans les cours oraux. Les devoirs écrits, les calculs, la lecture et le résumé de certains mémoires originaux devraient représenter la majeure partie du travail des étudiants. En se contentant de prendre des notes à ses cours, de les apprendre par cœur, puis de les répèter à ses examinateurs, ils acquièrent seulement un vernis superficiel bien vite disparu.

Les avantages du travail personnel sont trop évidents pour être sérieusement contestés; mais pour obtenir ce travail, les méthodes à mettre en œuvre sont délicates et les professeurs reculent devant une tentative très hasardeuse. On se heurte à la répugnance de la majorité des jeunes gens pour tout effort personnel et surtout pour tout effort intellectuel. Les plus paresseux copient les devoirs de leurs camarades plus actifs. Cette supercherie n'a pas seulement l'inconvénient d'annuler tout travail pour une catégorie d'étudiants, elle fausse en même temps les classements et décourage les meilleurs élèves; leur activité baisse quand leur travail ne reçoit plus une sanction efficace. Cette difficulté d'ordre pédagogique n'est certainement pas insoluble; elle est cependant très difficile à résoudre, surtout avec l'organisation politique et sociale des états démocratiques. Les élèves des écoles scientifiques ou techniques se considèrent comme les possesseurs d'un privilège inviolable; leurs parents et leurs protecteurs en jugent de même. Il n'est pas possible d'éliminer d'une école des jeunes gens par le seul motif qu'ils se refusent à tout effort personnel. Les influences politiques, la concurrence des établissements similaires, paralysent trop souvent les efforts des corps enseignants et les détourne de s'écarter des vieilles routes; il n'en subsiste pas moins que le point de beaucoup le plus important pour une bonne formation mathématique des ingénieurs serait d'obtenir d'eux une plus grande somme d'effort personnel au cours de leurs études.

Pour me résumer je conclurai ainsi: L'éducation mathématique nécessaire au futur ingénieur doit embrasser un champ d'études deux fois plus étendu environ que celui qui lui sera réellement utile pour la pratique de son métier. Cet enseignement différera de celui de la science pure, étudiée en dehors de toute préoccupation utilitaire, par le développement considérable donné à la géométrie descriptive et à la géométrie analytique et en même temps par la suppression de toutes les théories générales, de toutes les formules abstraites qui servent à réduire le développement d'une démonstration particulière, tout en augmentant considérablement la longueur de la chaîne reliant les conclusions à leurs prémisses; enfin le travail personnel des étudiants devra être développé dans la mesure où les circonstances le permettront.

THE PLACE OF DEDUCTION IN ELEMENTARY MECHANICS

By G. St L. Carson.

A science consists of a definite class or classes of entities, a set or sets of postulates relating them, and a series of deductions which are logical consequences of these postulates. In the earlier stages of its evolution it may be—it usually has been—that the sets of postulates contain redundancies. Only when the logical consistence of these sets has been investigated, and the number of independent postulates established, can the science be termed complete, for then only is it certain that the number of assumptions has been reduced to a minimum, and that no one of them conflicts with any other. Although the concept of a perfect science was attained by Euclid in connection with Geometry, the first approximately successful presentation of a science in this form came as late as Newton, and then in connection with Mechanics; Geometry, on the other hand, has only been completed within the last generation, and this after struggles extending over two thousand years.

This historic distinction between Geometry and Mechanics implies a corresponding didactic distinction. It is the peculiarity of Geometry, as opposed to other physical sciences, that its postulates, and many of the deductions which can be made from them, are and have long been the common property of civilised mankind. Who doubts, whether he has learnt Geometry or not, that all right angles are equal, that only one parallel can be drawn through a point to a given line, or that any diameter divides a circle into two identical parts? Most, if not all, of the postulates of Mechanics are, on the other hand, known only to those who have consciously adopted them. Indeed, many persons, otherwise educated, will dispute their truth. The problems before the teacher are, therefore, entirely different in the two cases. In Geometry he steps into an inheritance of pre-acquired space concepts, crude perhaps, but formed beyond doubt; he can develope deductions with little trouble concerning postulates. But in Mechanics he must set up the entities and develope the postulates from the commencement; he steps into no such inheritance as does the teacher of Geometry.

With methods of exhibiting mechanical entities, and the choice of postulates for use in a first treatment of the subject, this paper is not concerned. It is assumed, however, in accordance with modern practice in most schools, that the number of postulates is more than the minimum. My first concern is to point out that, treat it how we may, the direct evidence which can be brought before a boy in support of any of these postulates is singularly narrow and unconvincing. And it is an essential part of the argument that this weakness should be exposed at the outset. Take for instance the triangle of forces; the pupil may fairly ask

Within what degree of accuracy has it been demonstrated?

Is it true whatever be the body on which the forces act?

Is it true at all places, and at all times?

Is it true under any circumstances of motion?

Is it true for forces of all kinds, electrical, magnetic, or any other?

Or again, take the proportionality between force and acceleration, if the subject be developed so that this is a postulate. The pupil may be convinced that, in his own locality and for some substances, the statement is approximately true. But does the functional relation between force and acceleration involve no other variables, e.g. temperature? And is its form the same for all kinds of matter? May not the force be proportional to the square of the acceleration for some substances other than those used in the experiments? Pupils should be trained, and trained from the outset, to question in this manner the degree of accuracy of every measurement, and the generality of the circumstances under which each experiment is performed. It is scientific method, and education of the most practical and valuable character.

But then, the pupils may say, what is the use of going further? Must not some better evidence be obtained? There are of course many reasons which can and should be given for going on in faith, but one of the most illuminating and interesting illustrations is contained in Faraday's verification of Coulomb's law of electrostatic attraction. Few boys fail to find interest in the picture of Faraday, basing highly complex calculations on Coulomb's crude experiments, testing them inside the highly charged iron box, and emerging from it with a demonstration that the law corresponds really closely with observed facts. The pupil thus realises the true importance of deduction as an aid to his very imperfect powers of observation. In place of building complacently on a foundation whose imperfections have been glossed over, too often with pulleys mounted on ball bearings and other viciously misleading trivialities, he has a sane idea of what he is doing. He sees that, on these meagre foundations, he is to erect a structure which will come into contact with practical experience at many points, and must be judged by its degree of accordance with such experience at all these points.

The structure having been erected on a number of these very dubious supports, it becomes necessary to examine their possible interconnection, to ascertain whether the truth or falsity of any one of these assumptions involves of logical necessity the truth or falsity of any others. To illustrate the process suggested, I have dealt with the postulates of statics, but the method is equally applicable to dynamics, or to any combination of mechanical postulates.

The customary assumptions in a first treatment of statics—suggested, and rightly so, by crude experiment—are three in number, viz., the triangle of forces, the principle of the lever, and the principle of moments for two forces acting along intersecting lines. All three have been adopted, but any or all may be true or false. Thus the possibilities are eight in number, as shewn in the following table, and among them must the truth be sought:

	Triangle of Forces	Principle of Lever	Principle of Moments
1	true	true	true
2	false	true	true
3	true	false	true
4	true	true	false
5	true	false	false
6	false	true	false
7	false	false	true
8	false	false	false

Now it is possible to show, by logical deduction, that any two of these assumptions are necessary consequences of the third*. When this is done (and the proofs are well within the comprehension of a boy of 17), the alternatives 2 to 7 disappear, and the three assumptions have become one. The only possibility now being the simultaneous truth or falsity of all, the rough experimental results acquire greater import. If all three were false it is trebly unlikely that they should every one accord fairly well with experiment, and the only alternative to the falsity of all is the truth of all. The probability of this truth has thus been strongly reinforced by processes purely logical in nature.

It is worthy of notice that the conventional deductive method does not give an equal amount of strength to the hypothesis. Postulating the triangle of forces, the principle of the lever and the principle of moments are obtained by logical deduction, the possible alternatives being left thus:

Triangle of Forces	Principle of Lever	Principle of Moments
true	true	true
false	true	true
false	true	false
false	false	true
false	false	false

five in number. The proofs that *any* two of these postulates can be deduced from the third are not given, and the full power of deduction to reinforce assumption is not exhibited.

Treating other groups of postulates in the same manner, the structure is seen to be based, not on weak isolated supports, but on interlinked groups of such supports, the strength of the foundation being greatly increased by these interconnections. And finality is reached when the supports have been interlinked into groups, between

* If this is done, selecting any one assumption only, the consistence of the three assumptions follows at once, and the demonstration of consistence should not be overlooked.

which it can be demonstrated that no such logical interconnections are possible. This, I conceive, is the true aspect of Newton's achievement in the statement of his three laws of motion; he stated a number of consistent and independent hypotheses, and developed the whole subject from these by purely logical processes. I do not imagine that Newton had any such general concept of a science as is set out at the beginning of this paper but he perceived, intuitively or sub-consciously, that Mechanics could be based on three sets of postulates, each set referring to a different class of entities. In their statement his logic was defective, but this is trifling compared with the greatness of his achievement; he formed an ideal for Mechanics similar to that of Euclid for Geometry, and he attained practical success in its elaboration, in contrast with Euclid's decided failure.

We have now before us three distinct didactic treatments of Mechanics; the old method, the method now current, and a development such as has here been suggested. The old method presents the science in its complete form, with no indication of its evolution. Three postulates are laid down, to be accepted in blind faith, and from them the subject is developed by logical processes; it is a course in applied deduction. The current method presents a number of mechanical assumptions, based on foundations whose strength is hardly discussed, and uses them in application to various problems. There is little or no attempt to discuss their logical interconnection, and certainly no suggestion of its scientific meaning; it is a course in applied computation. But if this current method were followed by a logical discussion, exhibiting Mechanics as based on independent supports, each consisting of interlinked assumptions as has been described, I venture to suggest that it might fairly be called a course in applied mathematics.

The tendency of modern education, as it seems to me, is to lay undue stress on direct sensation as the one and only basis for faith. Undoubtedly education must find its origins, and these as widespread as possible, in direct sensation; but a false and very dangerous ideal is left, unless finally these origins are linked together by logical process, so as to give them their maximum strength and expose their ultimate weakness. Final contentment with a set of postulates, which may or may not be inconsistent or redundant, and for which there has appeared little real justification, is vicious; vicious also is the attempt in a first course to develope any science from a minimum of hypothesis. The one method is an undue suppression of perception, the other an undue glorification. The first step of the teacher should be to develope a wide spirit of enquiry; the second should be to breed a "divine discontent" with the imperfections of perceptual evidence. Recognising its essential nature, our inevitable bondage to it, we may yet liberate ourselves, so far as may be, in each branch of knowledge. It is the peculiar function of mathematics to point the way to this freedom in each science, and it is here that modern developments of mathematical thought may yet find application in other sciences.

THE CALCULUS AS A SUBJECT OF SCHOOL INSTRUCTION

By T. P. Nunn.

I do not seek in this paper to state the case for teaching the calculus in schools. I assume that the strength of this case is so widely recognised that there will be an audience for a practical discussion of ways and means. What follows may, perhaps, serve as a starting point for such a discussion. At the same time, since many who exclude the calculus from the school curriculum do so not on principle but simply because they doubt the possibility of teaching it effectively, the recommendations which I propose to develop will incidentally be arguments in favour of the general position here to be assumed.

I. *Preliminary: The Difficulties of the Calculus.*

The difficulties which the ordinary or "non-specialist" pupil finds in understanding the calculus fall under two main heads. Some are logical difficulties—that is, difficulties in grasping the broad ideas of the subject and in following the reasonings. Others are technical difficulties—that is, difficulties in conducting arguments by means of the conventional notation. The practical teacher may insist that there is a third and more fundamental category—the boy's difficulty is in understanding the "good" of the calculus and "what it is all about." But the difficulties under this head are, I suggest, really compounded of difficulties of the logical and technical kinds.

(a) *Logical difficulties.*

To begin with the logical difficulties. In one way or another these all cluster round the notion of a *limit* which is, of course, the foundation upon which the whole of the calculus is built. I wish here to urge that the natural difficulty of this notion is gratuitously increased by ancient misconceptions to which, in England at any rate, text-books and teachers cling with unfortunate persistence. Most of them spring from the twin doctrines that a differential coefficient is the ratio of two "infinitely small" quantities, and that an integral is the sum of an "infinite" number of such "infinitely small" quantities. The parentage of these notions is well known: they derive from the *minima indivisibilia* of Leibniz. For my own part I hold that conclusive proof of the non-existence of these "infinitesimals" has been given by authors of whom I may name Mr Bertrand Russell as typical. I venture, therefore, to suggest that the teacher who wishes his pupils to have clear ideas about limits should avoid the word "infinite" in this connexion altogether. So long as it is retained in his exposition the really simple and definite features of the notion are bound to be obscured by a halo of mysticism.

A logical error related to the foregoing one is often covered by the use of the "blessed word" *ultimately*. Thus we are told that when the sides of a regular polygon are incessantly increased in number the figure "ultimately" becomes a circle. The teacher who has already determined to eschew the word "infinite" in talking about limits will do well to add "ultimately" to his *index expurgatorius*. It is the only way to reconcile his teaching to the common-sense of his pupil who sees perfectly clearly that however many sides a polygon may assume it must eternally remain a polygon and can never, without the aid of magic, be translated into a circle.

Examined more closely these modes of statement are seen to be faulty for one and the same reason: they treat the limit of a series of terms as if it were itself a term of the series. Consider a very simple instance. The fraction

$$\frac{0^2+1^2+2^2+\ldots+l^2}{l^2+l^2+l^2+\ldots+l^2}=\frac{1}{3}+\frac{1}{6l}$$

for all positive integral values of l. The larger l is the nearer the fraction approaches in value to $\frac{1}{3}$; but it cannot properly be said that it will ever become exactly $\frac{1}{3}$. Yet $\frac{1}{3}$ is quite properly said to be the limit of the values of the fraction. Now if $\frac{1}{3}$ were only one of the numbers obtained by assigning particular values to l it would not be the limit of those values. It is the limit because although *not* one of the series of values of the fraction it bears the following unique relation to that series: there is no number so little greater than $\frac{1}{3}$ that a term of the series cannot be found whose value is still nearer to $\frac{1}{3}$. To say that the value of the fraction is $\frac{1}{3}$ when l is infinite or to say that the fraction ultimately becomes $\frac{1}{3}$ is simply to darken counsel. At best such statements are unilluminating and ambiguous paraphrases of the truth about the fraction. They should most certainly be avoided in teaching beginners.

(*b*) *Difficulties of notation.*

To turn to the difficulties of the ordinary notation. These difficulties all flow from our representing as an algebraic fraction a quantity which—as we diligently teach —is not a fraction at all. The confusion produced by this procedure becomes worse confounded when, in spite of our protest, we *do* treat dy/dx as a fraction as in the following argument:

$$\begin{aligned} dy/dx &= 3x^2, \\ dy &= 3x^2dx, \\ y &= \int 3x^2dx \\ &= x^3 + C. \end{aligned}$$

I do not suggest that this notation has no convenience, nor that it may not be employed profitably by the mathematician for whom the ideas which it originally expressed are as dead as the dodo. But we are dealing with the beginner. For him notation is not a means of concealing thought but a means of fixing volatile conceptions and of giving coherence to elusive trains of reasoning. He cannot make use of a piece of notation like the foregoing without being affected by its obvious presuppositions, no matter how firmly he is ordered to ignore them. These presuppositions are, of course, those referred to in the previous section. They regard dy and dx as "infinitely small" increments of the variables whose ratio has reached a definite final value—here the value $3x^2$. This being so, the value of the infinitesimal dy is of course to be obtained by multiplying the value of dx by $3x^2$. The integration is simply the process of summing an infinite number of infinitely small products of this form. Now the notation quoted undoubtedly expresses these ideas and operations perfectly clearly. But for this very reason it is bound to confuse the boy who has learnt that if dx and dy are magnitudes at all they are both zero, that $3x^2dx$ means $3x^2 \times h$ when h is zero and is therefore itself zero, and that no one can ever make anything by the addition of a series of zeros.

It is probable that few students of the calculus escape this confusion of ideas in the early stages of their progress. It can be avoided only in one way—namely, by avoiding the notation which creates it. Fortunately this way is not only simple; in addition there is nothing in it which need offend the most delicate sense of propriety. The dy/dx notation is not the only one with a distinguished pedigree. If it can claim the great Leibniz as its first ancestor, the dot notation can point in a similar way to the great Newton. There is also the D notation which finds a place in the writings of most modern mathematicians. Finally we may remember that good early work in the calculus was done practically without any special notation at all. We have therefore several alternative modes of procedure at our command, each legitimised by historical prestige or distinguished modern usage. It may be found that by the use of one of these or of a combination the difficulties involved in the ordinary notation may be avoided.

II. *Methodological suggestions.*

The way is perhaps now clear for some practical proposals. Of the incidental adumbration of the main ideas of the calculus in early graphical exercises I will say nothing. This part of the pedagogy of the subject is of great importance but it has been frequently treated before. You will permit me to begin at the point where calculations definitely based upon algebraic form first make their appearance.

(*a*) *The general method of procedure.*

The first question that arises has reference to the general spirit in which the work should be undertaken. Is the doctrine of the calculus to be developed systematically as a separate "subject" or is it to emerge from the study of topics that arise from time to time in a general course of algebra? I venture to urge as forcibly as possible the superiority of the second of these methods. I do not, I hope, underestimate the disciplinary value of "purism" in mathematics. It is undoubtedly an excellent thing that a boy should realise the conception of a perfect piece of mathematical architecture compact of homogeneous materials, and fulfilling a single harmonious design. But in actual history a subject rarely grows up from the first in this form. The perfect structure is the result of a process of selection and criticism working upon a wide range of material. A boy will not as a rule gain real discipline from the "pure" logical development of a subject unless the historical conditions of that development are in the main reproduced in his instruction. Mere instruction *ab initio* in accordance with perfect logical form will have little more effect upon his mind than the marbles of the Vatican had upon the aesthetic sensibility of the dogs in William James's famous argument. To be really effective, instruction in accordance with the ideal of purism must be largely a purified reconstruction of knowledge which, in some form or another, the boy already possesses. Whether he can undertake this process with advantage depends upon whether he can, before the close of his school life, accumulate a sufficient knowledge of the subject upon the lower logical plane.

I suggest, therefore, that the boy's initiation into the doctrines of the calculus should, in principle, reproduce the circumstances of their emergence in history. To the pioneers Wallis and Mercator and to a certain extent Newton, the new methods were simply developments of algebra called into existence to meet problems for whose solution the existing methods were inadequate. It was only with the development of a special technique that they became a new branch of mathematics with a separate individuality and a distinctive architectonic. It is, I submit, in accordance with the soundest maxims of pedagogy that instruction in the calculus should at first take this unspecialised form in the case of all pupils. The non-specialist will probably never pursue it beyond this stage. In that case he may still acquire a perfectly sound knowledge of the essentials of the method and become acquainted with its most important applications. The specialist will reach the point at which the subject cannot be further developed without the aid of a specialised technique. His progress will be all the more sound and rapid because in the earlier stage his attention was not diverted from the study of first principles by the necessity of mastering simultaneously a novel notation.

Lastly, this method of approach represents the calculus in its true light—namely, as the doctrine of a certain relation between functions. The pupil has already learnt that if the length of the side of a (variable) square can be represented by the function $ax+b$ its area is represented by the function $a^2x^2+2abx+b^2$. He has now to learn that if the velocity of a point after time x can be represented by the function ax^2 the distance it has travelled is represented by the function $\frac{1}{3}ax^3$. The only novelty is the novelty of the means by which the new kind of connexion is discovered or demonstrated. A technical notation may be convenient in the one case, as in the other, for the purpose of specifying concisely the kind of connexion which exists between the related functions; but it is in neither case necessary in principle.

(*b*) *The first stage.*

The relation just spoken of may be looked at in two ways. From one point of view it is the relation between a function and its differential coefficient; from the other it is the relation between a function and its integral. From which end shall our pupils look at it first? To this question I reply that although eventually it is best to regard the result of differentiation as the "derived" function yet there is much to be said for adopting the opposite procedure at first. This might be expected from the fact that integration preceded differentiation in history. The questions which arise most naturally and are the simplest to state are, in fact, questions of integration. It is, for example, more natural to ask how far a point will travel with a specified form of movement than to ask what law of speed has produced a given movement. Even when the second question is the one actually proposed—as Galileo proposed it to himself in his famous investigation—the attempt to solve it most naturally takes the form of testing suggested laws—that is, it takes the form of integration. In general, we may say that the characteristic method of the mathematical sciences is the deduction of integral consequences from laws expressed in differential equations. The fact that such equations cannot, in general, be solved without a preliminary collection of material by differentiation is, so to speak, merely a mathematical accident which must not be allowed to affect our exposition unduly.

I suggest, then, that our inquiries should begin with integration and that in the main they should follow the direction taken by John Wallis in his *Arithmetica Infinitorum* (1655). Wallis's methods have recently been described in some detail in the *Mathematical Gazette* and illustrations of their use in school teaching are given in the exhibition attached to this Congress. It will be sufficient to say here that his problems largely took the form of determining the areas of curves or volumes of solids whose ordinates or cross-sections follow given laws. His methods were both simple and ingenious and led him to the momentous conclusion that if the ordinate (or the cross-section) follows the law y (or A) $= kx^{n-1}$ the area (or the volume) follows the law A (or V) $= \frac{1}{n} kx^n$. In the course of his attempt to make this rule as wide as possible Wallis invented fractional and negative indices. It is perhaps not too late to commemorate his brilliant and fertile work by christening his generalisation "Wallis's Law."

The fraction which has already been considered in connexion with the doctrine of limits may be used to indicate the essentials of Wallis's procedure. Let the rectangles in the figure (fig. 1) have equal breadth and let their areas in succession be

Fig. 1.

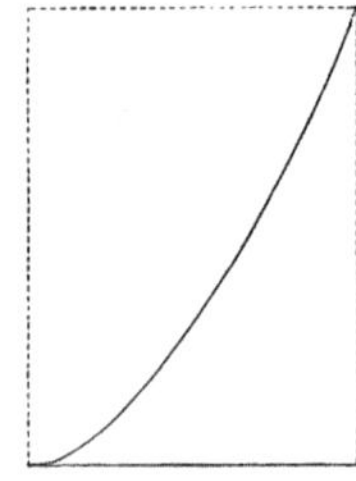

Fig. 2.

$1^2, 2^2, 3^2, \ldots l^2$, the area of the first being taken arbitrarily as unity. Let the complete dotted rectangle be built up of $l+1$ rectangles each equal to the largest of the former series. Then the ratio of the increasing series to the complete dotted rectangle is

$$\frac{0^2 + 1^2 + 2^2 + \ldots + l^2}{(l+1)\, l^2} = \frac{1}{3} + \frac{1}{6l}.$$

As l increases the fraction approximates to $\frac{1}{3}$ and the series of rectangles approximates to the area under the parabolic segment. We conclude, then, that this area is just one-third of the area of the rectangle contained by the abscissa and ordinate of its final point.

To the logic of this argument we must shortly return. In the meantime let us note that the method, though evolved in connexion with problems of mensuration, is readily applicable to many other kinds. In order that this application may be made it is sufficient that the problem be capable of representation in one of the graphical forms to which the method was originally applied. For example, the series of rectangles may represent the spaces covered in equal times by a point whose speed increases by sudden spurts so that during successive intervals 0^2, 1^2, 2^2, 3^2 ... (arbitrary) units of distance are covered. *Then* the area of the corresponding parabolic figure will give the space covered in the same total time if the speed increased uniformly in such a way that at any moment it was proportional to the time since the movement began. Wallis's Law assures us that in this case the point will have gone one-third of the distance which it would have covered if it had moved for the same time with the maximum actual speed.

The point to be reached by the pupil at the end of this first stage may now be indicated. Given a curve whose ordinates follow such a law as $y = 3 - 2x + 4{\cdot}2x^2$ he should be able to write down at once the corresponding law for the area under the curve: $A = 3x - x^2 + 1{\cdot}4x^3$. Given that the distance covered by a moving point in time t is given by the law $s = 2t + 1{\cdot}3t^2 - 0{\cdot}8t^3$ he should be able to write down the law of its speed: $v = 2 + 2{\cdot}6t - 2{\cdot}4t^2$. Moreover he should, on cross-examination, be able to justify the rules which he has applied to these problems.

(c) *The second stage; limits.*

The logic of the methods just described is readily seen to be defective at two points. In the first place it is evident that the tops of the rectangles, however numerous they may be, can never really become a parabola, and, similarly, that the value of the fraction can never really be exactly $\frac{1}{3}$. Practically this objection to our conclusion is of no importance. Our knowledge of the ways of Nature forbids us to suppose that in a case of this kind she acts *per saltum*. Even if our lives depended upon the exact truth of our result we might reaffirm it with a perfectly serene mind. But Ideal Reason demands something more than the strongest practical convictions; she will not let us rest until those convictions become unassailable deductions from an impeccable theory. It is necessary therefore to demonstrate explicitly that if the ordinate follows the law $y = kx^2$ the area is given *exactly* by the law $A = \frac{1}{3}kx^3$ without any qualification whatsoever.

In the second place we assumed (in the second example) that a boy knows what he means by saying that the speed of a point increases incessantly and that it is given at the moment t by the law kt^2. No doubt we may safely credit him with as much knowledge of these matters as is possessed by the policeman whose kinematical statements on oath are as a rule accepted by the magistrate. Here, again, however, Ideal Reason demands precision and explicitness greater than are required to meet the ordinary needs of practice.

The attempt to add logical perfection to the practical certainty of our previous work marks the second stage of our progress.

Consideration may first be given to the exact meaning of the term *rate* when only a single moment of time is in question.

Let the area of the single rectangle of fig. 3 represent the distance covered in

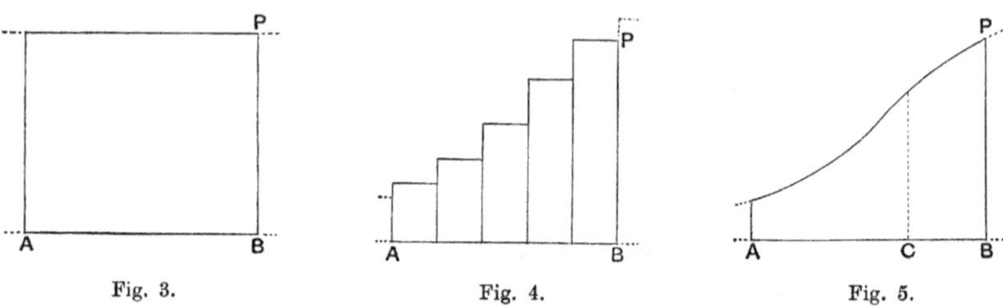

Fig. 3. Fig. 4. Fig. 5.

unit time (AB) by a point which moves uniformly—that is, so that in $1/n$th of the time it covers $1/n$th of the distance. There is no difficulty in the statement that its velocity or rate of movement is throughout v centimetres per second. It is obvious also that the height of PB as well as the area AP will be v. In fig. 4 let the point move in n equal spurts, the velocity during each spurt being uniform. Then if during the last nth of the second the distance covered is $1/n$th of v centimetres, the velocity during that last nth of a second may obviously be described as v cms./sec. as before. Moreover, although the area of the last rectangle is reduced to v/n, the height PB still remains v, and this is true however large n may become. Finally, let the distance accomplished at any moment be such that it is represented by the area under the corresponding section of the continuous curve of fig. 5. Also let the form of this curve be such that for the moment represented by B the ordinate PB is still v. In this case it is clear that there is no part of the interval AB during which the velocity has any constant value. But figs. 3, 4 and 5 have this in common, namely that the ordinate PB is in each case v. While, therefore, we cannot say that the velocity *during* any time in fig. 5 is v, we can, by agreement, adopt a new form of words and say that the velocity *at* the moment B is v cms./sec. We are now in a position to define, at any rate in graphical terms, exactly what we mean by the rate of movement of a point *at* a given moment. Draw a curve such that the area under it up to a given ordinate is equal to the space traversed up to the moment represented by the abscissa. Then the height of the ordinate is the velocity of the point at that moment.

In view of previous observations it is important to understand the true connexion between the velocity thus defined and the average velocity of the point during a given interval. During the interval CB (fig. 5) the average velocity is the area CP divided by CB. As C is taken nearer to B this quotient will become more nearly equal to v, but it can never become exactly equal to it. PB is a line, *not* an "infinitely thin rectangle," and the velocity at B is, in consequence, *not* the average velocity during any "infinitely short time" before B. It occupies, in fact, with reference to those average velocities the unique position which we have characterised by the term *limit.*

The term itself may be introduced at this stage, but there is much to be said for deferring its use and for contenting oneself for the present with the graphical symbol of the idea which it represents.

Meanwhile the method just employed may be permitted to suggest a mode of improving the demonstration of Wallis's Law. In fig. 6 let the curve of which AB is a segment have the property: that the area under it from the origin to any ordinate such as PQ whose abscissa is x is kx^3. The problem is to determine the exact height of PQ. Take two other ordinates CD and EF one on each side of PQ and at any distance h from it. Draw the rectangles on DQ and QF whose areas shall be equal respectively to the slices under the curve between CD and PQ and PQ and EF. Let the curve cross the upper ends of these rectangles at p and p'.

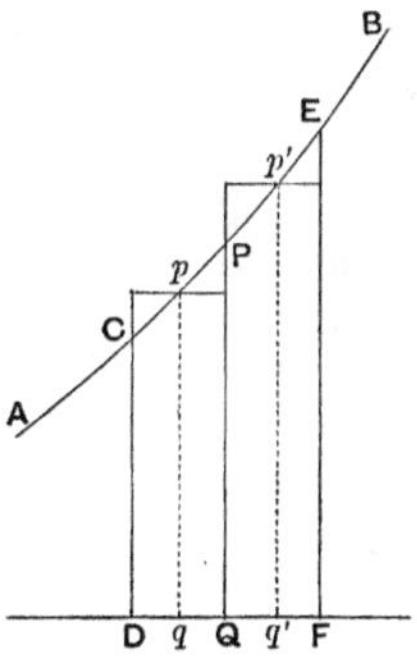

Fig. 6.

It is easy to calculate the heights of the ordinates pq and $p'q'$. For the area of the rectangle above DQ is

$$k\{x^3 - (x-h)^3\}$$

and that of the rectangle above QF $k\{(x+h)^3 - x^3\}$. Dividing each of these quantities by h we find that

$$pq = k\{3x^2 - h(3x - h)\} \text{ and } p'q' = k\{3x^2 + h(3x+h)\}.$$

Now if h is smaller than $3x$, pq must be less than $3kx^2$ and $p'q'$ greater. The smaller h becomes—that is, the nearer pq and $p'q'$ approach to PQ—the nearer their heights are to $3kx^2$. If we name any height however little different from $3kx^2$, either pq or $p'q'$ can be made to occupy a position such that its height is still nearer to $3kx^2$. It is clear, then, that there is only one value left which can possibly be attributed to PQ—namely the value $3kx^2$ itself. It will be noted that the result of this argument is no longer an approximation. In essence it is an application of the logical method of arriving at the truth by exclusions. The ordinate PQ must have some value, and it is shown that no other value than $3kx^2$ can be attributed to it without inconsistency. The argument though indirect is, in principle, simple and conclusive.

(*d*) *The generalisation of results: the D notation.*

The foregoing problems have been considered in detail partly in the hope of showing that a treatment of the calculus may be very elementary and yet avoid gross errors, and partly to illustrate the expository value of a certain graphical device. It is usual to teach differentiation by reference to the slope of a curve. There are undoubtedly good reasons for the practice but I feel confident that the plan here adopted is more effective. The very fact that a tangent to a curve cannot be drawn with conviction is a serious drawback to its use in early stages. Moreover the tangent cannot conveniently be used to illustrate the notion of integration. On the other hand the relation between the ordinate of a curve and the area under it illustrates with equal directness the processes both of integration and of differentiation, while the ease with which ordinates can be drawn to a given curve or a curve added to given ordinates gives the method considerable practical superiority over its rival. So great, in fact, are its tactical advantages, that in my own teaching I have sometimes based upon it a useful though transitory nomenclature. The term *ordinate-function* is given to the function which defines the height of the ordinate of a curve and the term *area-function* to the function which gives the area under the curve from the origin up to the ordinate at x. In these terms, integration is the determination of the area-function corresponding to a given ordinate-function, differentiation the determination of the ordinate-function corresponding to a given area-function. It will be seen that the nomenclature has at least the advantage of bringing out clearly that in the calculus we are constantly engaged with a definite kind of relation between functions.

You will note that upon the definition given all integrals are definite integrals. For example, if the ordinate-function is $\sin x$ the area-function is $1 - \cos x$. But this consequence instead of being a defect is in elementary work an advantage; for the integrals are given at once in the form required for the solution of problems. But while the relation between the ordinate and the curve may well remain the standard graphic expression to which calculus problems can be reduced in case of need, yet the pupil's problems will show him the fundamental relation of functions in a variety of settings. In addition to the relation, already mentioned, of the cross-sectional area to the volume of a solid there is the relation between the ordinate and the slope at a given point of a curve. The important uses of this relation for expository and illustrative purposes are too well known to need mention. From these and other non-graphical contexts the boy will soon arrive at a sound general notion of a purely formal relation between functions. It is now time to introduce a final nomenclature and a definite notation. In accordance with convention a function B which stands to a function A as an ordinate-function stands to its area-function is called the *first derived function* of A. The relation is denoted by the symbolism $B = D(A)$. If C is another function standing to B as the "slope-function" of a curve stands to its ordinate-function and standing to A as the slope-function stands to the area-function, then we write $C = D(B)$, $C = D^2(A)$. When the converses of these relations are in view either we replace the positive indices by negative ones or (borrowing a suggestion from the formal logicians) we invert the D. Thus in the instances just given we write either $B = D^{-1}(C)$, $A = D^{-2}(C)$ or $B = \reflectbox{$D$}(C)$, $A = \reflectbox{$D$}^2(C)$.

(*e*) *The programme.*

The length to which this paper has already extended forbids any discussion of the range of topics from the calculus which should be included in the school programme. I must be contented with a brief confession of faith. The programme should be largely determined by the character of the courses in physics, etc., with which the mathematical work is correlated. I am inclined to treat at an early stage a wider range of functions than is admitted by some important modern writers. The sine and cosine would come early on my list because the problems in which they are involved are so important and so interesting. The logarithmic function, also, cannot be omitted, in spite of the difficulties which Mr C. S. Jackson has signalised in his recent excellent report. The calculus should be applied to the determination of certain simple expansions, algebraic and trigonometrical. Finally the programme should include a demonstration that these expansions are all special cases of the generalisations known as Taylor's and Maclaurin's Theorems.

LA COMMISSION INTERNATIONALE DE L'ENSEIGNEMENT MATHÉMATIQUE DE 1908 À 1912

RAPPORT PRÉSENTÉ À LA SÉANCE DU VENDREDI 23 AOÛT 1912

PAR H. FEHR
Secrétaire-général de la Commission.

COMPTE RENDU SOMMAIRE

A. **Introduction.**

La Commission internationale de l'Enseignement mathématique a été instituée par le 4e Congrès international des mathématiciens tenu à Rome du 6 au 11 avril 1908. Dans sa séance du 11 avril le Congrès adopta la résolution* suivante :

" *Le Congrès ayant reconnu l'importance d'un examen comparé des méthodes et des plans d'études de l'enseignement mathématique dans les écoles secondaires des différentes nations, confie à MM.* KLEIN, GREENHILL *et* FEHR, *le mandat de constituer une Commission internationale qui étudiera ces questions et présentera un rapport d'ensemble au prochain Congrès.*"

Le comité de trois membres désigné par le Congrès a pris le nom de *Comité central*; il s'est constitué de la manière suivante :

Président : M. le Prof. F. Klein, G.R.R., Göttingue.

Vice-président : Sir George Greenhill, F.R.S., Londres.

Secrétaire-général : M. le Prof. H. Fehr, Genève.

Dans une réunion tenue à Cologne, en septembre 1908, le Comité central établit le *Rapport préliminaire* destiné à renseigner les délégués sur l'organisation de la Commission et à leur fournir des indications générales concernant le plan des travaux.

Il convient de rappeler ici les principaux points qui ont servi de base à l'organisation de la Commission et à l'élaboration des nombreux travaux rédigés dans les principaux pays.

* Cette résolution fut proposée par la section *Philosophie, Histoire et Enseignement*, à la suite d'une série de rapports sur l'enseignement mathématique dans les principaux pays. Sur l'initiative de M. le prof. Dav.-Eug. Smith, auteur du rapport concernant les Etats-Unis, elle décida de soumettre au Congrès une résolution tendant à créer une Commission internationale chargée de faire une étude d'ensemble des progrès de l'enseignement mathématique dans les différentes nations. Cette proposition avait déjà été formulée par le savant professeur de New-York, en 1905, dans sa réponse à une enquête sur les "réformes à accomplir" entreprise par M. H. Fehr dans la Revue internationale *L'Enseignement mathématique* Vol. VII, 1905 (p. 469).

B. Organisation de la Commission.

I. Les Délégations.

(*a*) La Commission est formée par des délégués représentant les pays qui ont pris part au moins à deux des Congrès internationaux des mathématiciens avec une moyenne d'au moins deux membres. Chacun de ces pays a droit à un délégué. Les pays qui ont eu une moyenne d'au moins dix représentants peuvent avoir deux ou trois délégués. Dans les votations et les discussions de la Commission, chaque pays n'a cependant qu'une voix.

Les pays, dits *pays participants*, appelés à prendre part aux travaux de la Commission, sont les suivants:

Allemagne (3 délégués)
Autriche (3)
Belgique (1)
Danemark (1)
Espagne (1)
Etats-Unis d'Amérique (3)
France (3)
Grèce (1)
Hollande (1)
Hongrie (3)
Iles britanniques (3)
Italie (3)
Japon (1)
Norvège (1)
Portugal (1)
Roumanie (1)
Russie (3)
Suède (1)
Suisse (3)

Les pays qui ne répondent pas aux conditions ci-dessus, mais qui par leurs institutions peuvent contribuer aux progrès de la science, ont été invités à se faire représenter par un délégué qui suivrait les travaux de la Commission, sans toutefois prendre part aux votations.

Ces pays sont dits *pays associés*; le Comité central s'est adressé aux pays suivants:

Argentine (Rép.)
Australie
Brésil
Bulgarie
Canada
Chili
Chine
Colonie du Cap
Egypte
Indes anglaises
Mexique
Pérou
Serbie
Turquie

et il a pu obtenir des représentants pour l'Australie, le Canada, la Colonie du Cap, le Mexique et la Serbie. Des pourparlers se poursuivent pour quelques Etats, et nous espérons qu'à l'occasion du Congrès du Cambridge il sera possible de les faire aboutir définitivement*.

Voici la liste des membres de la Commission qui ont fonctionné pendant la période de quatre ans qui s'est écoulée entre les deux Congrès.

* Pendant le Congrès, le Comité central a enregistré l'adhésion du Brésil, représenté par M. E. R. Gabaglia (Rio de Janeiro), et de la Bulgarie, représentée par M. A.-v. Sourek (Sofia).

Délégués des pays participants :

Allemagne : MM. F. KLEIN (Gœttingue), P. STÆCKEL (Carlsruhe), P. TREUTLEIN (Carlsruhe).

Autriche : MM. E. CZUBER, W. WIRTINGER, R. SUPPANTSCHITSCH.

Belgique : M. J. NEUBERG (Liège).

Danemark : M. P. HEEGAARD (Copenhague).

Espagne : M. Z.-G. de GALDEANO (Saragosse).

Etats-Unis : MM. Dav.-Eug. SMITH (New-York), W. OSGOOD (Cambridge, Mass.), J.-W.-A. YOUNG (Chicago).

France : MM. A. de SAINT-GERMAIN, C.-A. LAISANT et C. BOURLET.

Grèce : M. C. STÉPHANOS (Athènes).

Hollande : M. J. CARDINAAL (Delft).

Hongrie : MM. M. BEKE, C. RADOZ, RATZ (Budapest).

Iles Britanniques : Sir George GREENHILL, Prof. E.-W. HOBSON, Mr C. GODFREY.

Italie : MM. G. CASTELNUOVO (Rome), Fr. ENRIQUES (Bologne), G. SCORZA (Palerme).

Japon : M. R. FUJISAWA (Tokio).

Norvège : M. ALFSEN (Christiania).

Portugal : M. Gomes TEIXEIRA (Porto).

Roumanie : M. G. TZITZEICA (Bucarest).

Russie : MM. N. v. SONIN, KOJALOVIC, K. W. VOGT (St-Pétersbourg).

Suède : M. H. v. KOCH (Stockholm).

Suisse : MM. H. FEHR (Genève), C.-F. GEISER (Zurich), J.-H. GRAF (Berne).

Délégués des pays associés :

Australie : M. CARSLAW (Sidney) ; suppléant en Europe : Prof. BRAGG (Leeds).

Canada : M. BOVEY (Londres).

Colonie du Cap : M. HOUGH (Capetown).

Mexique : M. Valentin GAMA (Tacuyaba).

Serbie : M. Michel PETROVITCH (Belgrade).

Décès.—La Commission a eu le regret d'enregistrer les décès de trois de ses membres. Ce fut d'abord M. G. VAILATI, l'un des délégués italiens, qui a été remplacé par M. SCORZA (Palerme). Puis au cours de la présente année elle a été privée du concours de M. BOVEY, recteur du Collège Impérial technique de South-kensington à Londres qui s'était spécialement chargé de nous renseigner sur l'enseignement mathématique au Canada, et de M. P. TREUTLEIN, G.H.R., membre de la délégation allemande ; celle-ci perd en lui un collaborateur très actif et fort apprécié pour ses ouvrages didactiques. M. Treutlein est décédé subitement le 26 juillet dernier à l'âge de 67 ans. Sur la proposition de la Société mathématique allemande, le Comité central l'a remplacé dans la Commission par M. le Prof. Dr A. THÆR (Hambourg).

Démission.—M. le Prof. Z.-G. DE GALDEANO (Saragosse) a désiré se retirer de la Commission à la fin de cette première période. Le Comité central lui a exprimé ses vifs remerciements pour tout l'intérêt qu'il n'a cessé de témoigner à la Commission. M. C. J. RUEDA (Madrid) a été désigné comme délégué espagnol.

II. SOUS-COMMISSIONS NATIONALES.

Les différentes délégations ont été invitées à s'adjoindre des Sous-commissions nationales, comprenant des représentants des divers degrés de l'enseignement mathématique dans les établissements d'instruction générale ou dans les écoles techniques ou professionnelles. Ces Sous-commissions ont apporté un concours très précieux aux délégués pour la préparation des rapports. C'est à leurs membres que l'on doit en grande partie les nombreuses publications qui ont été entreprises sur l'initiative de la Commission.

III. DISPOSITIONS FINANCIÈRES.

Le 4me Congrès international n'ayant fourni aucun subside les gouvernements des *pays participants* ont été invités à mettre à la disposition de leur délégation une somme permettant de couvrir entièrement les frais de la délégation et de la sous-commission nationale et de contribuer aux frais généraux de la Commission.

Pour subvenir aux frais généraux de la Commission (comprenant notamment les frais du secrétariat-général et du Comité central), il a été constitué un fonds formé par des contributions annuelles de cent francs par *pays participant.*

IV. ORGANE OFFICIEL DE LA COMMISSION.

PUBLICATION DES RAPPORTS DES SOUS-COMMISSIONS.

La Revue internationale *L'Enseignement mathématique,* dirigée par MM. LAISANT et FEHR, sert d'organe à la Commission. Elle publie les Rapports du Comité central et rend régulièrement compte des travaux de la Commission et des Sous-commissions.

Les Sous-commissions publieront leurs rapports suivant leur propre convenance. Le Comité central a toutefois exprimé le désir que ces rapports soient imprimés suivant le format de *L'Enseignement mathématique* et que les délégations des divers pays en adressent 75 exemplaires au secrétariat-général qui les fait distribuer aux membres de la Commission.

V. LANGUES OFFICIELLES.

La correspondance et les rapports doivent être rédigés dans l'une des quatre langues admises aux Congrès internationaux des mathématiciens, au gré des auteurs. Ces langues sont : l'allemand, l'anglais, le français et l'italien.

C. **Objet des travaux de la Commission.**

Dans le texte même de la résolution du Congrès de Rome, il n'est question que de l'enseignement mathématique dans les écoles secondaires. Mais étant donné que le but de ces écoles et la durée de leurs études sont très variables d'un Etat à un autre, le Comité central a jugé utile de faire porter son travail sur l'ensemble du champ d'instruction mathématique depuis la première initiation jusqu'à l'enseignement supérieur. En outre il ne s'est pas borné aux établissements d'instruction générale conduisant à l'Université, mais il a également fait étudier l'enseignement mathématique dans les écoles techniques ou professionnelles. Ces établissements ont en effet une importance croissante; il y avait donc lieu d'accorder une attention toute spéciale à l'enseignement mathématique qui s'y donne.

Le Comité central a donc entrepris une étude d'ensemble de l'enseignement mathématique dans les différents types d'écoles et à ses divers degrés. Il s'agit d'une *étude objective destinée à présenter l'état actuel et les tendances modernes de cet enseignement.* Comme on l'a dit dans les Réunions de Bruxelles et de Milan, la Commission ne cherche nullement à uniformiser l'enseignement mathématique, mais avant tout à mettre en lumière les tendances modernes. La Commission ne peut et ne veut rien imposer, mais ses travaux permettront aux professeurs de savoir ce qui se fait dans les nations voisines et ils les renseigneront aussi sur l'organisation de son propre pays. La comparaison des documents et l'étude des expériences faites ailleurs contribueront à réaliser de nouveaux progrès dans tous les domaines de l'enseignement mathématique.

Le *plan général des travaux* élaboré par le Comité central était destiné à servir de guide aux délégués et aux membres des Sous-commissions nationales, afin de leur donner des indications sur les principaux points à prendre en considération. Toutefois, en raison de la diversité même de l'organisation dans les différents pays, il n'était pas possible d'imposer un plan unique, s'adaptant à la fois aux conditions des divers pays. La plus grande liberté a donc été laissée aux rapporteurs.

Voici les titres des principaux objets signalés dans le Rapport préliminaire:

PREMIÈRE PARTIE: *Etat actuel de l'organisation et des méthodes de l'instruction mathématique.*—I. Les divers types d'écoles.—II. But de l'instruction mathématique et branches d'enseignement.—III. Les examens.—IV. Les méthodes d'enseignement.—V. Préparation des candidats à l'enseignement.

DEUXIÈME PARTIE: *Les tendances modernes de l'enseignement mathématique.*—I. Les idées modernes concernant l'organisation scolaire.—II. Les tendances modernes concernant le but de l'enseignement et les branches d'études.—III. Les examens.—IV. Les méthodes d'enseignement.—V. La préparation des candidats à l'enseignement.—Remarque générale.

Les rapports sommaires que présenteront les délégués en déposant leurs publications indiqueront les points caractéristiques des travaux des Sous-commissions nationales.

D. **Séances du Comité central et de la Commission.**

Le *Comité central* s'est réuni pour la première fois à Cologne, en septembre 1908; puis à Carlsruhe, au commencement d'avril 1909; à Bâle, fin décembre 1909; à Bruxelles en août 1910; à Milan en septembre 1911 et enfin dans les premiers jours de juillet à Hahnenklee, dans le Harz, auprès de son président, que des raisons de santé empêchent malheureusement d'assister au Congrès de Cambridge. Son absence sera vivement regrettée non seulement par les membres de la Commission, mais par tous les Congressistes.

En outre de nombreux pourparlers ont eu lieu entre le Secrétaire-général et les membres de la Commission.

Réunion de Bruxelles (9–10 août 1910).—Le Comité central a saisi l'occasion de l'Exposition universelle de Bruxelles pour organiser une réunion partielle de la Commission, à laquelle elle avait tout particulièrement invité les délégués des pays voisins. Son appel a rencontré le meilleur accueil auprès de la plupart des délégations. Onze pays se trouvaient représentés par plus de trente membres des sous-commissions nationales. Nous nous bornons à rappeler ici la brillante conférence de M. C. BOURLET (Paris) sur la pénétration réciproque des mathématiques pures et des mathématiques appliquées dans l'enseignement secondaire. Elle a été reproduite *in extenso* dans le Compte rendu détaillé* publié par le Secrétaire-général dans la *Circulaire nº* 3, qui comprend en outre un résumé des conférences organisées à l'exposition du 11–16 août 1910.

Réunion de Milan (18–21 septembre 1911).—La Commission a tenu sa première réunion plénière† à Milan, en septembre 1911. En dehors des séances du Comité central et des Commissions spéciales, la réunion qui, en réalité, avait pris l'ampleur d'un véritable congrès international de l'enseignement mathématique, comprenait quatre séances, dont la première était consacrée à la présentation des rapports des Sous-commissions nationales. Pour les deux séances suivantes, le Comité central a estimé qu'il était utile de concentrer le débat sur deux questions importantes concernant l'une l'enseignement moyen, l'autre l'enseignement supérieur. Les questions mises à l'ordre du jour à Milan étaient les suivantes:

A. I. Les mathématiques dans l'enseignement moyen: *Dans quelle mesure peut-on tenir compte, dans les écoles moyennes* (lycées, collèges, gymnases, écoles réales, etc.), *de l'exposé systématique des mathématiques?* II. *La question de la fusion des différentes branches mathématiques dans l'enseignement moyen.*—Rapport de la Sous-commission A; rapporteurs MM. CASTELNUOVO et BIOCHE.—Discussion.

B. *L'enseignement mathématique théorique et pratique destiné aux étudiants en sciences physiques et naturelles.*—Rapport de la Sous-commission B; rapporteur: M. TIMERDING.—Discussion.

Mentionnons également la séance générale publique avec les belles conférences de M. le sénateur COLOMBO et de M. le Prof. F. ENRIQUES.

A la suite de l'extension considérable qu'ont pris ses travaux la Commission ne voit pas la possibilité de donner à Cambridge une étude comparée des différents rapports nationaux. Pour plusieurs pays les rapports ne sont du reste pas encore terminés. Dans sa réunion de Milan la Commission a donc estimé nécessaire de soumettre au Vme Congrès une proposition tendant à renouveler son mandat jusqu'au congrès suivant.

Réunion de Cambridge (*août* 1912) *et réunions ultérieures.*—Les séances de la Commission ont été organisées sur le même plan que celles de Milan. Comme suite

* Voir l'*Ens. mathém.* du 15 septembre 1910.

† Le compte rendu détaillé fait l'objet de la *Circulaire nº* 5 (75 p.), *Ens. math.* du 15 nov. 1911.

à la *question A*, l'ordre du jour comprend une discussion sur "*l'intuition et l'expérience dans l'enseignement mathématique des écoles moyennes*" (rapporteur: M. D.-E. SMITH). Tandis que pour la *question B* il convenait, après la discussion générale de Milan, de se limiter plus particulièrement à "*la préparation mathématique des physiciens*" (rapporteur: M. C. RUNGE).

Au cas où le mandat de la Commission serait prolongé, le Comité central aborderait ensuite l'étude d'autres questions d'une importance fondamentale; elles seraient discutées dans des réunions à placer entre le 5me et 6me Congrès international. Il apportera notamment une attention toute spéciale à la préparation théorique et pratique des professeurs de mathématiques.

E. **Publications concernant la Commission.**

Grace au concours dévoué des membres des Sous-commissions nationales, la Commission se trouve en possession d'un ensemble de documents fort précieux. Si nous nous bornons aux rapports proprement dits sur l'enseignement mathématique dans les différents pays, leur nombre dépasse 280, répartis sur plus de 150 fascicules ou volumes et représentant actuellement un ensemble de plus de 9000 pages in-8°.

Nous donnons ci-après [p. 642] la liste complète des publications parues jusqu'à ce jour ou actuellement en préparation.

Les rapports sont terminés dans les pays suivants: Suède, Hollande, France, Suisse, Autriche, Japon, Etats-Unis, Iles britanniques, Danemark (9 pays).

Sont en cours de publication les rapports concernant l'Allemagne, la Belgique, l'Espagne, la Hongrie, l'Italie, la Norvège, la Roumanie et la Russie (8 pays).

Dans d'autres Etats, il se prépare également des rapports. Nous pouvons déjà mentionner l'Australie, où notre délégué M. H. C. CARSLAW a entrepris une étude sur les mathématiques dans les écoles moyennes et dans l'enseignement supérieur.

Cette vaste enquête sur l'état actuel et les tendances modernes de l'enseignement mathématique une fois terminée, il s'agira d'en tirer parti en la faisant connaître au corps enseignant et aux autorités intéressées. Dans une étude comparée de différents rapports qui les concernent, les conférences ou sociétés de professeurs examineront les vœux et conclusions à transmettre aux autorités respectives, dans le but de faire progresser l'enseignement des mathématiques.

THE MATHEMATICAL TRAINING OF THE PHYSICIST IN THE UNIVERSITY.

REPORT PRESENTED BY C. RUNGE.

The international commission on the Teaching of Mathematics requested me to make an enquiry into the state of the mathematical studies of the students of physics at the universities of different countries. A circular letter was sent out during the winter of 1911–1912 to collect the necessary information, and a number of answers from Italy, Austria, Germany, Switzerland, Holland, England and the United States were received. It is however extremely difficult to form an adequate idea of the state of things in another country with which you are not familiar, because the same term describing the mathematical education may mean very different things when used by different men, so that a description does not convey a very definite idea to the reader. Some countries have developed their educational conditions in so widely different a manner, that it is almost impossible to understand them except by visiting the country and seeing for yourself. In the announcement of his lectures on mathematical physics Helmholtz used to invite the students by saying no other previous knowledge was wanted besides the elements of the calculus. But his so-called elements were by no means considered elementary by his students. In the same way the terms "analytic geometry," "integral and differential calculus," no doubt have a different meaning with my correspondents from different countries. Though vague the report may however help us to get acquainted with the state of affairs. But I beg to consider it, incomplete and imperfect as it is, only as an introduction leading up to the discussion, which is to follow. I shall follow here the order of the questions in my circular letter. The original text is given in the notes.

I*. The subjects of mathematical lectures attended by students of physics in their regular course of studies seem to be very much the same almost everywhere. They pass through a fairly complete course of instruction in mathematics, which however has no particular connection with the special work that a student of physics as such ought to do. Analytic geometry, differential and integral calculus, elements

* *Question* I: Welche mathematischen Fächer gehören zum regelrechten Studium eines Physikers? Wird in den Anforderungen an die mathematische Ausbildung der Physiker ein Unterschied gemacht zwischen Physikern einer mehr experimentellen und einer mehr theoretischen Richtung? Wird von den Mathematik-Professoren besondere Rücksicht auf die Bedürfnisse der Physiker genommen? Sind besondere mathematische Kurse für Physiker eingerichtet? Wie weit und in welchem Sinne beteiligen sich die Mathematik-Professoren an den Vorlesungen (*a*) über Mechanik (*b*) über sonstige insbesondere moderne Gebiete der mathematischen Physik?

of ordinary differential equations, dynamics are taught, the mathematical professors in general paying little attention to the particular needs of physical students. No distinction is made between students of experimental physics and of theoretical physics and there are on these mathematical subjects no special mathematical courses for physicists.

The lectures on dynamics are almost always given by mathematical professors and sometimes also certain other lectures on subjects of mathematical physics. Some of my correspondents bitterly complain of the mathematical training of students of physics in consequence of the professors of pure mathematics ignoring some mathematical theorems and methods, that are of greatest importance to the physicist. Green's and Stokes' theorem for instance ought to be taught in the integral calculus and Fourier's series ought to be treated in a more practical manner and not only as furnishing interesting instances of conditions of convergence. Vector analysis ought to be taught quite regularly by the mathematical professors, so that students should be perfectly familiar with it in their studies in dynamics and mathematical physics. Instead of that some of the mathematical professors spend too much time on the logical foundations of the calculus. Their subtilty is lost on the student who is only beginning to grasp the power of the calculus and ought to be trained to use it. This applies equally to the student of pure mathematics. On the whole there seems to be no need for special mathematical courses for students of physics nor does it seem necessary to compel them to attend more mathematical lectures. But a need seems to be strongly felt for mathematicians and physicists to draw closer together. The spirit of the mathematical teaching should be altered, so as to make it more practical and easier to apply to physical problems. At present the gap is very wide and is not tending to close up.

II*. Modern graphical methods seem to be very slightly introduced except in France and a few other universities. On the other hand descriptive geometry is now provided for, nearly everywhere. In some countries it is even cultivated to such an extent that opposition seems to be awakening. Indeed the work in descriptive geometry has a tendency to overgrow other mathematical branches, if it is not carefully pruned down. Though exceedingly useful in its way as training the power of geometrical intuition the student may become too fond of it and neglect the calculus. Besides there is scarcely any development going on in descriptive geometry, so that little research work is done in connection with it. Whatever the value of descriptive geometry may be, there can be little doubt that graphical methods showing how to represent functions of one and of more than one variable and how to handle them, have at least the same importance for any students of natural philosophy. The physicist in a great many cases has to deal with empirical functions that are conveniently and with sufficient accuracy represented graphically. To approximate them by means of analytical expressions becomes in many cases so tedious that it is out of

* *Question* II: Wie weit sind auf den Universitäten die modernen graphischen Methoden über graphische Integration und Nomographie verbreitet? Lernen die Physik-Studierenden darstellende Geometrie, numerisches Rechnen, numerische Auflösung von Gleichungen, numerische Auflösung von Differentialgleichungen, Methode der kleinsten Quadrate? Lernen sie den Gebrauch mathematischer Apparate, des Rechenschiebers, der Rechenmaschine, des Planimeters? Geschicht dies in besonderen mathematischen Vorlesungen oder Uebungen oder nur beiläufig im physikalischen Praktikum?

the question. When these functions enter into mathematical considerations, all operations are best carried on graphically. Engineers have long been in the habit of doing so. But it is only since Massau (Gand) and d'Ocagne (Paris) worked out the methods systematically, that they have become a branch of mathematics. It appears that except in France teachers of mathematics have not yet paid sufficient attention to this remarkable growth of our mathematical power. The methods of graphical integration of any given function of a real or complex variable, of ordinary differential equations and of some partial differential equations, should be considered essential parts of the integral calculus. Due attention should be paid to them in every course on the calculus. And there is this second advantage, that it makes the calculus easier. The difficulty of performing the analytical integration of a given function has led to an undue preponderance of differentiation in the teaching of the calculus. The graphical methods restore the natural order. With an empirical function the process of integration is performed easily and accurately, while the process of differentiation is much more uncertain. The methods of nomography, if not taught in a special course of lectures, might be included in the lectures on coordinate geometry; but it will not do to leave them out altogether.

The *numerical* methods of handling an empirical function are older than the graphical methods and in consequence better known. Nevertheless they are also too much neglected in the mathematical education of the student of physics. The method of least squares is generally taught, but greater stress ought to be laid on the calculus of finite differences, on the numerical solution of equations, on numerical calculation of integrals and the numerical solution of differential equations. The difficulty is that many professors of mathematics have never been in the habit of calculating numerically and seem to have an aversion to teaching their students. Many are not familiar with the handling of mathematical instruments, with the slide rule, the integraph, the planimeter, the calculating machine, and little mention is made of them in the mathematical lectures. The student thus forms a wrong idea of the possibility of carrying out a mathematical operation. As long as he is only interested in mathematical theorems, this does not much matter. But a physicist or an engineer cannot be satisfied with existence-theorems, he wants the actual numerical result for the given data he has before him. Of course he may learn to calculate in his practical work in the laboratory, and his teachers in physics will to a certain extent take care of this part of his mathematical education. Nevertheless a decided advance could be made, if the mathematical teachers better understood the wants of their students in physics and did not rest satisfied with the proof of a system of theorems. The solution of a mathematical problem is not complete unless it supplies an answer to any intelligent question raised in connection with the problem. It is not sufficient that the answer may be found "by a finite number of operations"; but one must be able to find it without an unreasonable expenditure of work.

The use of the slide rule is commonly learned incidentally in the physical laboratory or in engineering courses and the calculating machine or planimeter may be learned in the same way. But if they were handled in the mathematical courses it would react favourably on the teaching, and the students of pure mathematics would profit by it as well.

III*. This however is only possible with individual teaching, which in most of the universities and in most of the mathematical courses is not insisted upon. The organisation of the mathematical exercises ought to be on a similar plan to the exercises in a chemical or physical laboratory. As to descriptive geometry this is recognized everywhere. Nearly everywhere special rooms are provided for the exercises on the drawing board and the professor enjoys the help of assistants to look after each student individually to correct his drawing and to discuss with him his difficulties. The same plan ought to be carried out for exercises in numerical and graphical calculation and for general mathematical exercises. The difficulty that a student finds in his mathematical studies is not so much that of understanding the proof of a mathematical theorem but more that of grasping the contents of it, of seeing its application in a variety of cases, of knowing how to make use of it. For the student of physics or of engineering the power to use his mathematics is of primary importance. Without it he might almost as well dispense with the knowledge of mathematical theorems altogether. They will be an unnecessary burden to his intellect. This power can only be acquired by exercise. Mathematics cannot be learned by lectures alone, any more than piano playing can be learnt by listening to a player. It does not seem advisable to leave the exercises to the physical student himself, although he will himself feel the necessity of them, and although his physical studies will bring him in contact with problems where his mathematics are wanted. The better pedagogic no doubt is to combine the theorem with its application, to help the student to find his way back from the generalisation to the single case. It is the single case that he wants to understand, the abstraction has its value only as a means to understand the single case.

In mathematical exercises organised after the usual plan of laboratory work, the teacher is able to help his students in a far more efficient manner than by lectures alone. Each student has his particular way of looking at things and should not be forced to another way unless for good reasons. Written solutions of problems need not be excluded, but the personal intercourse, the discussion of individual difficulties, ought to be insisted on. Besides graphical work and the use of mathematical instruments must be shown to each student separately and must be practised for some time under the eye of the teacher.

IV†. Most of my correspondents agree in considering this the only way for the organisation of mathematical exercises. Some express their opinion that the difficulty is not that the students of physics do not learn enough mathematics but rather that they do not learn to apply their mathematics to concrete problems. "There often exists a curious disproportion of which they are well aware between their mathematical ability and the depth and wideness of their mathematical studies. This evil can hardly be remedied by lectures but only by prolonged practice."

* *Question* III: Wie sind die mathematischen Uebungen der Physik-Studierenden beschaffen? Werden sie in der Weise von Laboratoriumsübungen abgehalten durch Ausführen der Aufgaben an Ort und Stelle und steht der Docent oder seine Assistenten in persönlicher Beziehung zu den einzelnen Studierenden?

† *Question* IV: Was sind Ihre eignen Ansichten über die Zweckmässigkeit des gegenwärtigen Unterrichts? Schlagen Sie Aenderungen vor in Bezug auf Ausdehnung oder Einschränkung des mathematischen Unterrichts oder in Bezug auf Unterscheidung der Physik-Studierenden nach Gruppen oder in Bezug auf Unterrichtseinrichtungen?

Among the exercises the more elementary ones are naturally of greater importance than the more advanced. While only a small minority of students of physics will draw advantage from the highest mathematical studies, all of them will profit by the full possession of the graphical and numerical methods and the ability to use mathematical instruments. These exercises ought to accompany the mathematical lectures of the first two years, and ought to constitute an integral part of the mathematical education of the physicist.

To recapitulate, the general opinion seems to be that the mathematical instruction for physical students is very much in need of reform. Large portions of the theoretical matter intended for the pure mathematician might be left out and the attention concentrated on those subjects which are of continual application in mathematical physics. Improvement is to be looked for in the progressive adaptation of the teaching of applied mathematics to the more modern views of Physics on Electricity and Matter. As to the breaking up of students into groups it seems that it is both impracticable and undesirable. There is a limit to the indefinite multiplication of classes owing to the limited number of teachers and the limited amount of funds available.

The main danger is the gap between physicists and pure mathematicians that seems to be widening. Mathematics is suffering from over-specialisation and is cutting itself off from Natural Philosophy and Experimental Science. But very little can be done by regulations. What is really wanted is that Mathematical teachers should understand the problems and the needs of the Physicist.

Questionnaire. Nous croyons utile de rappeler, dans son texte français, le questionnaire distribué l'hiver dernier par M. Runge au nom de la *Sous-commission B*, composée de MM. Klein (Gœttingue), président, Bourlet (Paris), Fehr (Genève), Sir G. Greenhill (Londres), Coialovich (St Pétersbourg), Levi-Civita (Padoue), Runge (Gœttingue), Timerding (Braunschweig), Webster (Worcester, E. U.), Wirtinger (Vienne).

(*Extrait de la circulaire* de M. le prof. C. Runge.) Objet: *Les mathématiques dans les études universitaires des physiciens.*

1. Quelles sont les branches mathématiques qui appartiennent à un enseignement régulier destiné au physicien? Dans la préparation mathématique des physiciens fait-on une différence entre les étudiants qui suivent une direction plutôt expérimentale et ceux qui suivent une voie plus théorique?

Les professeurs de mathématiques tiennent-ils particulièrement compte des besoins des physiciens?

Y a-t-il des cours de mathématiques spécialement destinés aux physiciens?

Dans quelle mesure et à quel point de vue les mathématiciens participent-ils aux cours (*a*) de mécanique; (*b*) à d'autres cours et particulièrement à ceux qui se rattachent au domaine moderne de la physique mathématique?

2. Jusqu'à quel point les méthodes graphiques modernes d'intégration et de nomographie sont-elles répandues dans les universités?

Les étudiants en physique sont-ils appelés à apprendre la géométrie descriptive, le calcul numérique, la résolution numérique des équations différentielles et la méthode des moindres carrés?

Apprennent-ils le maniement d'instruments mathématiques tels que la règle à calcul, la machine à calculer et les planimètres?

Y a-t-il des cours ou des exercices spéciaux à cet effet ou cet enseignement se fait-il dans les travaux pratiques de physique?

3. Quelle est l'organisation des exercices mathématiques destinés aux physiciens? Ces exercices ont-ils lieu suivant le mode habituel des travaux de laboratoire? Le professeur ou ses assistants entrent-ils en relation personnelle avec les différents étudiants?

4. Quelle est votre opinion personnelle sur l'opportunité de l'organisation actuelle de cet enseignement?

Avez-vous des propositions à faire au sujet d'une extension ou d'une réduction de l'enseignement mathématique ou au sujet d'une distinction des étudiants en physique en divers groupes ou encore pour ce qui concerne l'organisation de l'enseignement?

DISCUSSION

Sir J.-J. Thomson remercie M. Runge de sa belle conférence. Elle a été suivie d'une discussion d'un grand intérêt à laquelle ont pris part MM. P. Stæckel, C. Bourlet, F. Enriques, Sir G. Greenhill, A.-G. Webster, E. Borel, Sir J. Larmor, C. Bioche, A.-E.-H.-Love, E.-W. Hobson, G.-A. Gibson, Sir J.-J. Thomson et C. Runge. En outre, afin de ne pas allonger le débat, M. Lanchester nous a adressé par écrit les remarques qu'il comptait présenter à la séance.

Quelques orateurs ont insisté sur les dangers qu'il y aurait de négliger le côté logique dans l'enseignement supérieur, même si celui-ci s'adresse uniquement à des physiciens, ce qui n'est généralement pas le cas. Tout en reconnaissant la nécessité d'accorder une juste place au côté pratique, à l'aide de problèmes bien choisis, ils estiment que le côté utilitaire ne doit pas prédominer. D'autres font remarquer que les étudiants en mathématiques eux-mêmes retireraient un grand bénéfice d'un enseignement par lequel on les initierait davantage aux applications d'ordre pratique. C'est de ce côté qu'on devrait diriger l'effort sans porter préjudice à l'enseignement théorique bien approprié. Le temps nécessaire se trouvera aisément; on pourrait le prendre sur les heures de laboratoire de la physique, dont le nombre, au dire même des physiciens, est parfois exagéré. Il suffirait de renoncer à des manipulations souvent sans intérêt et sans aucune portée théorique ou pratique.

On lira ci-après le résumé de la seconde partie de cette séance qui a été l'une des plus intéressantes du Congrès.

Prof. Staeckel (Karlsruhe) ist überzeugt, dass die Vorschläge Runges durchgeführt zu werden verdienen, und möchte nur eine gewisse Schwierigkeit beleuchten, die darin liegt, dass die Ausbildung der Physiker und Mathematiker zum Teil gemeinsam erfolgt und dafür gesorgt werden muss, dass die Pflege der logischen Seite der Mathematik nicht vernachlässigt wird.

M. Bourlet (Paris) présente les excuses de la Sous-commission française de n'avoir pas répondu au questionnaire de M. le Prof. Runge. La raison est que, devant une question aussi vaste, la Sous-commission a décidé de faire plus que de donner des renseignements sommaires, et qu'elle publiera tout un volume.

M. Bourlet indique à grands traits ce que sera cet ouvrage.

Comme il l'a déjà expliqué dans des réunions antérieures, l'enseignement est donné en France dans un esprit très général dans les classes de *Mathématiques spéciales* et dans les *Cours de Mathématiques générales* des Universités. En France, on essaie de donner aux jeunes gens qui se destinent aux sciences mathématiques, physiques et naturelles, une éducation générale commune. C'est ainsi que les élèves de l'Ecole normale Supérieure suivaient jadis en première année *tous* les *mêmes* cours.

En dehors de l'étude de l'état actuel de l'enseignement oral, la Commission étudiera les ouvrages en usage pour donner un tableau complet de ce qui se fait en France.

Mais M. Bourlet estime que le travail ne devrait pas s'arrêter là et que le rôle de la Commission internationale est de chercher des directions communes au moyen d'une enquête parmi les mathématiciens, les physiciens et les ingénieurs. Cette enquête pourra porter sur deux points: les *matières* qu'il faut enseigner, les *méthodes* qu'il y a lieu d'employer. M. Bourlet estime que c'est surtout le choix des matières qu'il y a lieu de faire avec soin ; quant à la méthode, il est d'avis qu'il ne peut pas y en avoir deux, une pour les mathématiciens et l'autre pour les physiciens, et que l'enseignement des mathématiques aux physiciens doit conserver son caractère de rigueur et de logique. On n'enseignera pas les fonctions monogènes de M. Borel, mais ce que l'on enseignera on devra le faire avec toute la précision nécessaire. La Commission internationale devant se réunir vraisemblablement à Pâques 1914 à Paris, M. Bourlet espère que le volume sera prêt quelques mois d'avance.

M. F. Enriques (Bologne) remarque que M. Runge s'est placé à un point de vue utilitaire en se demandant quels instruments il faudra donner au physicien. Il y a lieu de poser une autre question qui n'a pas moins un intérêt pratique, c'est-à-dire comment on peut attirer les jeunes gens à l'étude de la physique.

M. Enriques remarque que, dans plusieurs pays au moins, il y a une tendance des étudiants les *mieux doués* à préférer l'étude des questions de mathématiques pures à celles de physique. Cela dépendrait-il de ce que l'on néglige le plus souvent de présenter aux élèves les côtés de la physique théorique qui sont les plus beaux et les plus intéressants, en expliquant le but même de la science qui consiste à mettre d'accord les faits avec les représentations de notre esprit?

Il n'est pas douteux que les progrès de la physique exigent de se tenir près des faits d'expérience et que l'expérience demande un long apprentissage. Cependant on ne peut méconnaître que certains esprits élevés ne sauraient se soumettre à l'effort de cet apprentissage s'ils ne voient d'abord l'intêret des questions qu'on est appelé à résoudre par cette voie. D'autres aussi qui n'auront jamais la faculté de fournir eux-mêmes un travail expérimental utile, pourraient apporter une contribution aux questions théoriques grâce à une intuition heureuse.

M. Enriques pense que l'enseignement actuel de la physique mathématique n'est pas fait assez dans le sens de développer cette intuition et parfois même qu'il tend à repousser les esprits qui ont une plus grande intuition par un excès développement des algorithmes de l'analyse où le moyen semble être pris comme une fin.

Sir G. Greenhill (London). The Report we have had in our hand for preliminary study, and the exposition of it by Prof. Runge, have covered the whole ground of the state of mathematical training of the student of Physics.

But a participator in the discussion must confine himself to some detail, where his experience encourages him to offer some remarks.

In my own case I endorse heartily what is said of the overgrowth of Descriptive Geometry (see Question II).

The subject is too often easy and soft, and the student grows too fond of it, and cannot be brought back to harder study of the Calculus.

We found this at Coopers Hill College, where I was one of the original staff. Reports from India complained of the young engineer we sent out; that he knew no mathematics, but he could *draw*.

Give him a task on the drawing board with his instruments and he is pleased to work for ever. In fact his instructors at Coopers Hill used to groan at his output of drawings.

So although we envy Prof. Gutzmer his Mathematical Institute at Halle, there is one point of his equipment we do not admire so much—the elaborate laboratory for drawing and descriptive Geometry.

For my part I would rather see a Mathematical Workshop in its place, with a carpenter's bench and tools, an anvil and some large hammers, and other simple apparatus for illustrating elementary dynamical principles, and making them familiar to the muscular sense.

Prof. Webster (Clark University, United States). I find myself in general agreement with most of the gentlemen who have spoken, especially with the ideas of Professor Runge, as expressed in his admirable report. I regret very much the distance that often separates the mathematician from the physicist, especially the lack of perception that often exists of what the physical student needs. I should like to point out a very important practical consideration in the words "Ars longa, vita brevis." In other words so much time must be taken for the education of the physicist that it is totally wrong to waste precious time on what is not useful. I fully agree that the teaching should be thorough—the best is none too good, but it is foolish to spend time in pointing out to the student imaginary difficulties that he will never encounter. As an example, the physicist will need to integrate many functions, but he will not require a rigid discussion of the conditions of integrability of a function, since all the functions that he will meet will probably be integrable. He will have to make frequent use of Fourier's series, but it would be wrong to devote several months of his time to the study of all sorts of convergence and non-convergence of series in general. I believe that every mathematician should know the main principles of physics, just as much as the physicist should be acquainted with the leading methods of mathematics. It is also quite possible to illustrate mathematical ideas, by physical examples which are interesting, illuminating and perfectly logical. I need only suggest the notions of velocity and acceleration, and the employment of such physical ideas as are the possession of everyone.

Sir Joseph Larmor (Cambridge) said that what he had heard induced him to ask them to consider the position of the mathematicians. In mathematical teaching they ought not to be expected to give up everything that was interesting, and confine themselves to routine work and to providing mere tools to be used in more attractive occupations. What we were suffering from was over-specialisation. In his view mathematics included the whole of theoretical physics; there was no essential difference between reasoning on the doctrine of thermodynamics and reasoning on the doctrine of limits. Mathematics was not all technical algebraical analysis, any more than physics was all experimental dexterity.

M. E. Borel (Paris) se déclare entièrement d'accord avec M. Runge. Nous devons tendre à rendre plus simple l'enseignement des mathématiques. Puis il donne des renseignements sur ce qui se fait à l'Ecole normale supérieure. En première année l'enseignement est commun aux physiciens et aux mathématiciens, sauf de rares exceptions. La méthode et le choix des matières dépendent souvent des questions d'actualité, de mode et de la personnalité du maître.

M. Ch. Bioche (Paris) estime que les physiciens ou naturalistes doivent avoir quelques connaissances générales des mathématiques, en dehors des connaissances pratiques qu'ils utilisent journellement, de façon à ce qu'ils puissent savoir qu'à l'occasion de telle question les intéressant ils trouveront auprès d'un mathématicien professionnel les indications qui leur seraient utiles.

Lorsqu'on veut traiter un problème d'ordre pratique il ne suffit pas d'en avoir une solution théorique; il faut en avoir une solution commode. Il importe que, dans l'enseignement mathématique, on se préoccupe de discuter les solutions des problèmes au point de vue de la simplicité dans les applications.

Prof. Love (Oxford) wished to associate himself with Sir J. Larmor's view. The ideal thing would be, every mathematician a physicist and every physicist a mathematician. To gain time for the necessary training it would be desirable to discard obsolete things in mathematics and to devise some scheme by which a general knowledge of physics could be brought within the reach of all students of mathematics. In regard to the proposed choice of mathematical matter, he thought there might be a tendency to make the subject too narrow, with too little regard to the possible future application of branches of analysis which are now studied for their own sake.

Prof. Hobson (Cambridge) said that perhaps too large a share of the blame for the undoubted evils of the present divorce of Mathematical and Physical teaching had been put upon the Mathematicians. If the mathematical teacher spent too much time on matters which would be better omitted, and in the teaching of merely analytical expertness, it might be said that the Physicist spent too much time on the merely manipulative work of the Laboratory. Mathematical teachers were expected to teach only what was directly useful to the Physicist, and also what was directly useful to the Engineer, or to the Chemist. The main object of the Mathematical teaching ought to be that of developing habits of clear and logical thinking in the pupils, and that could not be hampered by a too close adherence to what was required only for application in other departments. Dexterity in such matters as the use of the slide rule was not part of what the Mathematician should be expected to teach. He would lose his self-respect if he became a mere provider of tools for use in subjects outside his own.

Sir J. J. Thomson (Cambridge), l'illustre physicien, reconnaît que de nos jours on exagère parfois le temps accordé aux travaux pratiques dans l'enseignement supérieur où ils prennent une trop grande place. Ce serait nuisible de donner aux physiciens un enseignement mathématique spécialisé, par contre il faut qu'ils soient familiarisés avec les applications pratiques.

Prof. Runge. I believe that the introduction of numerical and graphical exercises does not interfere in any way with the logical deduction of pure mathematical theorems. My idea is that practical mathematics are indispensable not only to the physicist but also to the pure mathematician in so far as they enable him to work out examples for the application of pure mathematical theorems, and in this way enlighten him and aid his logical deductions.

Remarks of F. W. Lanchester (Birmingham). The question of the mathematical training of the physicist in the university has perhaps two aspects—that is from the point of view of the mathematician, and from the standpoint of the work which the student in after life finds it incumbent upon him to undertake. Whether trained primarily as engineer or physicist a large number of those who are trained in applied mathematics either enter by intention or by accident into the engineering profession. The following remarks may be taken to relate more particularly to this class of student.

The weakness commonly exhibited by the engineering student, who having graduated in a university takes a post as an engineer or in an engineering work, is found more often than not to lie in his inability to deduce from the complex maze of facts with which he has to deal, a definite problem in what may be termed "examination-question" form *. In my experience there is usually ample knowledge of mathematical principles and mathematical method, but there is a very obvious deficiency in the development of the mathematical student's reasoning powers, which are the essential counterpart to any purely mathematical training, before it can serve him usefully in the responsible carrying out of engineering work. It is naturally very questionable whether it is possible by any modification or supplementing of the university education to make good this deficiency entirely; in fact it stands to reason that for the elastic application of his mathematical training the latter will require to be supplemented by considerable actual experience before it can be fully utilised. Notwithstanding this, I consider that there are certain subjects (which in the discussion of the paper were belittled by some of the speakers present) which a little reflection will show to be of vital importance. As an example (and in my opinion a very important example) it is impossible to overrate the utility from an educational point of view of descriptive geometry. There is no subject that teaches a student to *think in three dimensions* so well as descriptive geometry, and the inability to think in three dimensions is one of the most common failings of the university mathematically trained engineer. This aspect of descriptive geometry is quite apart from its obvious utility in training a man to read drawings of a complicated character such as are constantly before him in the course of his duties. I would call particular attention incidentally to the difference between a study such as descriptive geometry which is a real educating factor and enlarges the capacity of the mind, and those mathematical studies which cannot be regarded in the same

* It is as if some one accustomed to dining off a well cooked and nicely served steak were shown the carcase of a freshly slaughtered ox and told to get his dinner out of that!

light, but of which one must think as analogous to tools in the hands of a workman. I do not wish to suggest that all mathematical work is merely a matter of providing a mental tool equipment, but only that a great deal of the ordinary mathematical curriculum can be looked upon from this point of view, rather than be considered as a real educative factor.

I am convinced that the mathematical investigation of three-dimensional problems is in many cases detrimental to the expansion of the mind, inasmuch as the mathematician rather than think in three dimensions employs a machine process to evade mental gymnastics; for instance in fluid dynamics he thinks once and for all in three dimensions in an imaginary infinitesimal unit cube of the fluid. From this he makes equations and from these equations he obtains his results. From the time these equations are formulated he may forget all about three-dimensional space until he comes to the interpretation of his results. I think that in the mathematical training of a young engineer such dangerous methods should not be introduced until he has had a very thorough grounding in descriptive geometry. There is truly the risk pointed out by Professor Greenhill that after he has devoted a great deal of time to descriptive geometry he will be liable to shirk, or ignore, or neglect the more rigid mathematical method; but even if this undesirable consequence results (which in nowise appears necessary) I think that it may be taken as the lesser of two evils.

Briefly, from a more general point of view, I think that in the discussion of the circular there has been too much tendency to dwell on detail questions as to what should be taught and what should not be taught, and to ignore the principle that true education ought to be a process of developing and expanding the mind in those directions in which such development and expansion is most likely to serve the ends in view, and that the provision of a mental tool equipment is or should be considered as merely an incidental.

BERICHT ÜBER DIE HERAUSGABE EINER BIBLIOGRAPHIE DES MATHEMATISCHEN UNTERRICHTS 1900—1912

VON KARL GOLDZIHER.

Die Reformkommission der ungarischen Mittelschullehrer-Vereinigung legte besonderes Gewicht darauf die Vorarbeiten und die ausländische Entwickelung unserer Bewegung zu beachten. So entstand im Jahre 1907 ein ausführlicher Bericht, der auch in der deutschen Bearbeitung des ungarischen Reformwerks Platz gefunden hat und in seiner ungarischen Ausgabe mit einer vorläufigen Bibliographie ergänzt worden ist. Wir waren der Ansicht, dass man die vielseitige und internationale Entwickelung der Reformgedanken nur dann übersehen und studieren kann, wenn eine möglichst vollständige Bibliographie zur Verfügung steht. Eine solche Arbeit erbringt auch vom allgemein pädagogischen Gesichtspunkt den "statistischen" Beweis für die natürliche Entstehung und Berechtigung unserer wichtigsten Forderungen, indem sie zeigt, dass die Reformideen schon seit 40 Jahren literarisch angebahnt sind, und in fast allen Kulturstaaten allmählich Anklang finden. Dies ist ja der Grund, dass die I. M. U. K. in so erfolgreicher Weise ihr Unternehmen ausführen konnte; ein Unternehmen, das wohl die erste grössere pädagogische Bestrebung ist, die durch einheitliche internationale Arbeit vollführt wird. Es ist somit für die Weiterführung und für den Abschluss der Tätigkeit der I. M. U. K. von besonderem Werte, dass eine historisch angelegte, alle Einzelheiten verfolgende bibliographische Zusammenstellung der Reformarbeiten zustande komme.

Das Studium der Entwickelung in den einzelnen Ländern, die Ausgestaltung spezieller Probleme, die Auswahl entsprechender Lehrbücher erfordert vorwiegend eine solche Sammlung; was besonders dürch den Umstand ersichtlich wird, welch' überaus reichliche Fülle von Aufsätzen, offiziellen Berichten, Sammelwerken und Lehrbüchern in den letzten 12 Jahren auf dem Gebiete des mathematischen Unterrichtswesens erschienen ist. Diese Production kann noch in nächster Zeit, in der die Wirkung der Bewegung erst die rechten Früchte tragen wird, noch beträchtlicher werden.

Diese Umstände veranlassten Herrn Prof. D. E. Smith in New York und ganz unabhängig von ihm den Verfasser dieses Berichtes zur Sammlung aller literarischen Ergebnisse der Reformbewegung. Ein glücklicher Zufall wollte es, dass wir auf diesem Arbeitsfelde zusammentrafen und das Bureau of Education der amerikanischen Regierung zur Herausgabe dieses für den mathematischen Schulunterricht wertvollen Materiales gewinnen konnten. Unser gemeinschaftliches Material erstreckte sich ursprünglich auf die allgemeinen Probleme, auf die Vorgeschichte (deren Beobachtung in den einzelnen Staaten besonders interessant ist), auf die speziellen Reformgedanken nach Gegenständen geordnet und auf die Entwickelung in den einzelnen Staaten. Wir versuchten weiterhin die Liste der wichtigsten neuen Lehrbücher zusammenzustellen. Das ursprüngliche Material umfasste ca 5000 Titel, eine Fülle also, die wohl kaum von anderen Lehrfächern übertroffen werden könnte.

Die Sonderung des Materials ergab jedoch mancherlei Schwierigkeiten, so dass wir uns entschliessen mussten, vorerst nur eine Auswahl zu treffen, um diesem Kongress an der Hand der wichtigsten Erscheinungen die Berechtigung eines solchen Unternehmens vorzuführen. Das Werk, das demnächst (wir hoffen im Herbst 1912) erscheinen soll, und von der amerikanischen Regierung jedem, der sich dafür interessiert, zugeschickt werden wird, umfasst ca 1800 Titel aus den Jahren 1900—1912. Die Zusammenstellung ist folgende:

I. Allgemeine Fragen des mathematischen Unterrichts.

II. Der math. Unterricht an den Volksschulen.

III. Der math. Unterricht an den höheren Schulen.

IV. Der math. Unterricht an den technischen Lehranstalten.

V. Der math. Unterricht an den Colleges und Universitäten.

VI. Der arithmetische Unterricht.

VII. Der algebraische Unterricht.

VIII. Der geometrische Unterricht.

IX. Der Unterricht in der Trigonometrie.

X. Der Unterricht in der analytischen Geometrie.

XI. Der Unterricht in der Infinitesimalrechnung.

XII. Der Unterricht in der zeichnenden und in der darstellenden Geometrie.

XIII. Graphische Methoden.

XIV. Logarithmus und Rechenschieber.

XV. Funktionsbegriff.

XVI. Wechselbeziehungen und Anwendungen der Mathematik.

XVII. Geschichte der Mathematik.

XVIII. Ausbildung der Lehrkräfte.

Die Redaktion hat Prof. Smith übernommen, der darin von seinen Assistenten am Teachers College in New York unterstützt wurde.

Vor allem mussten des Raummangels wegen die Aufsätze vor 1900, also die ältern Versuche wegbleiben; dann mussten wir uns in Bezug auf die Entwickelung in den einzelnen Staaten nur auf die wichtigsten Erscheinungen beschränken. Eine ganz lückenlose Zusammenstellung in der letztgenannten Richtung wäre nur mit

Hülfe der Fachgenossen aus den einzelnen Ländern möglich gewesen. Wir mussten endlich von unserer Liste der modernen Lehrbücher ganz absehen, weil wir die Zeit für eine solche Zusammenstellung heute noch nicht für entsprechend hielten.

Unser kleines Werk ist somit nur eine Vorarbeit auf diesem Gebiete, eine Probe für ein grösseres Unternehmen, von dessen Wichtigkeit und praktischer Brauchbarkeit wir vollständig überzeugt sind. Wir sind der amerikanischen Regierung dafür zu grossem Danke verpflichtet, dass sie die Herausgabe dieser ersten Bibliographie übernommen hat.

Prof. Smith und ich kamen nun nach gründlicher Erwägung der Frage zu dem Ergebnis, bei Gelegenheit dieses Kongresses dem Centralcomité der I. M. U. K. die folgenden Vorschläge zur Diskussion und gefl. Annahme zu empfehlen:

(1) Das Centralcomité möge in das Programm seiner weiteren Tätigkeit die Zusammenstellung einer möglichst vollständigen Bibliographie aufnehmen; als Zeitpunkt des Abschlusses erscheint der 1. Jan. 1915 am zweckmässigsten.

(2) Bei dieser Arbeit möge die Mitarbeit von entsprechenden wissenschaftlichen Kräften aus allen Kulturstaaten, die sich an der Bewegung beteiligen, herangezogen werden, damit eine Vollständigkeit nach allen Richtungen gesichert werde.

(3) Das Centralcomité möge einen Sonderausschuss zur Durchführung dieser Arbeit erwählen.

(4) Das Centralcomité möge an das Bureau of Education der amerikanischen Regierung in offizieller Form das Ansuchen richten, auch die Herausgabe des grösseren Werkes übernehmen zu wollen.

INTUITION AND EXPERIMENT IN MATHEMATICAL TEACHING IN THE SECONDARY SCHOOLS*

Report presented by David Eugene Smith.

1. *Method of investigation.*

In the year 1911 the Central Committee appointed a subcommittee known as "Subcommittee A" charged with the duty of investigating the rôle of intuition in the teaching of secondary mathematics. It assigned to Dr W. Lietzmann of Barmen, Germany, the work of preparing a questionnaire that should facilitate the work of securing the data upon which to base a report, which labour was diligently performed. The questionnaire was sent out during the winter of 1911–1912 and replies were received by April 1, 1912, from representative teachers of Austria, England, France, Germany, Switzerland, and the United States. These replies were held until July 1, and no others having been received by that date, to the writer of this paper was assigned the duty of preparing the present report.

In the several countries the method of investigation differed materially. In some cases the "rapporteur" sent out local questionnaires; in others he referred to the printed reports of the International Commission; while in others a committee considered the replies. The results seem to show that it was a matter of no particular significance which plan was adopted.

* The German topic as assigned was "Anschauung und Experiment im mathematischen Unterricht der höheren Schulen." This was translated into French as "L'intuition et l'expérience dans l'enseignement mathématique des écoles moyennes." The translation is not exact, Anschauung being neither exactly Intuition nor exactly Perception, and neither Experiment nor Experience being a satisfactory equivalent of the German word Experiment.

This report is based upon data collected by Dr Walther Lietzmann, Oberlehrer an der Oberrealschule in Barmen, Germany. The data were secured through replies to a questionnaire prepared by Dr Lietzmann, these replies having been sent by the following gentlemen:

Austria, Prof. Dr Erwin Dintzl, Vienna;

England, Charles Godfrey, M.A., headmaster of the Royal Naval College, Osborne;

France, M. Ch. Bioche, Professeur au lycée Louis-le-Grand, Paris;

Germany, The late Prof. Dr P. Treutlein, G. H. R., Direktor des Real- und Reform-Gymnasiums, Karlsruhe; and Dr W. Lietzmann, Barmen.

Switzerland, M. H. Fehr, Professeur à l'Université de Genève, Geneva;

United States of America, Professors David Eugene Smith, Teachers' College, Columbia University, New York City, and J. W. A. Young, The University of Chicago, Chicago, Illinois.

2. *Types of school considered.*

The questionnaire having been prepared in the German language, and naturally under the influence of the German school system, the terms used do not exactly meet the situation as it appears in all of the other countries. It is therefore necessary to define with some care the terms employed, and to give the English interpretation to the several questions.

The Gymnasien and "Realanstalten" of Austria and Germany correspond rather closely in years to the lycées of France, and to the Gymnasien and "Kantonschulen" of Switzerland. They correspond somewhat less closely to the so-called "public schools" of England, while they are radically different from the public "high schools" of the United States. It therefore becomes necessary to define "Unterricht der höheren Schulen" as the teaching in those school years that correspond in the age of pupils with the school years of the Gymnasien, of the lycées, of the "public schools" (in England), and of the last part of the "elementary school," all of the "high school," and the first two years of "college" in the United States. This means that the study includes the work of pupils from about the age of 10 to about the age of 19 years. The French translation of *écoles moyennes* seems best to describe the schools in question.

It should also be noted that the replies have reference to the work done in schools of a general type, including both the classical and the non-classical, and usually not to that done in such special schools as those devoted chiefly to agriculture, mechanical arts, navigation, mining, and the like. Austria, for example, expressly limits the report to the Gymnasium, Realgymnasium, Reformrealgymnasium, and Realschule. England considers what are described loosely as "public schools," those which supply the greater number of students to the ancient universities of Oxford and Cambridge, and other general secondary schools. Germany and Switzerland report only upon the work of the Gymnasien, Realgymnasien, and Oberrealschulen, and France chiefly upon that of the lycées and collèges. In the United States, owing to the fact that the school system differs materially from that of Europe, the report refers to the upper classes in the "elementary school," to all of the work of the general "high school," and to the first two years of the "college*." In all the reports, however, emphasis has been laid upon the general type of school rather than the special.

3. *The question of elementary and higher schools.*

It is evident that the scope of the inquiry might well have been enlarged so as to include the important question of intuition and experience in the Kindergarten, the Écoles maternelles, the primary school, and even the home. As Dr Montessori is now bringing into prominence in Italy and America, and as thousands of successful teachers everywhere have long been proving, intuition and experience play a very important part in the first stages of a child's education in general, and with respect to

* The "elementary school" generally has 8 years or Grades, the child entering Grade I at about the age of 6–7, and finishing Grade VIII at the age of 13–14. The "high school" has 4 years, the pupil entering at the age of 14–15, and leaving at the age of 17–18. The "college" follows, the entering age being 17–18, and the 4 years being completed at the age of 21–22, the bachelor's degree being then conferred. The strictly-speaking university courses are entered at the age of 22–23, and the degree of doctor of philosophy may be obtained at the age of 25–26.

mathematics in particular. M. Laisant has brought this important question to the attention of his countrymen, and Lietz and other well-known educators have done the same in other countries. It was felt by the committee, therefore, that it was better to call the attention of teachers to the importance of the question in the *écoles moyennes* at this time, and to reserve for further consideration the question of the primary schools. It is hoped that, in case the Central Committee is continued for another period of four years, this question will be considered with the care that it deserves.

As to higher institutions of learning it should be observed that the topic assigned to this committee is closely related to that assigned to Subcommittee B, *The Mathematical Training of the Physicist in the University*, and this relationship appears the more evident as we consider the report that has already been presented by Professor Runge at the session of August 26, 1912.

4. *Method followed in this Report.*

It seems to the committee that it is advisable, in this report, to give a summary of the reports prepared by the representatives of the various countries, rather than to submit these reports in full. This is especially the case because the representatives of some of the countries have merely referred to pages of published monographs instead of giving the information directly. It also seems advisable to give this résumé under each of the general rubrics suggested by the committee rather than to attempt any other arrangement.

5. *The general situation.*

Before proceeding to consider the questions in detail it is desirable to say a few words concerning the general situation in the various countries. The first thing that must strike any observer is that we are passing through a period of great change in all that pertains to secondary education, including the field assigned to this committee. A subject like descriptive geometry, for example, can hardly be said to occupy any definite position, so rapidly changing are the views concerning its importance, its significance, and the schools in which it should be taught. Austria makes much of it in the Realschule, less of it in the Realgymnasium, and almost nothing of it in the Gymnasium. This is what we might expect, but what descriptive geometry will do in the Reformrealgymnasium it is rather early to say, since this type of school is only beginning to appear in that country. In England there is seen the same spirit of unrest, and the training of boys in special lines is having its influence in the general fields as well. The teaching of boys who are going into the army, engineering, or surveying, tends to assume a more intuitional and experimental character than that of other boys, and the army requirements during the last 10 years have no doubt had a powerful reaction on the character of the teaching throughout the English public schools. On the other hand, as regards the teaching of boys other than specialists, probably the prevalent view in England is that the proper field for methods of intuition and experiment is in the middle and lower classes rather than the upper. These methods are viewed with suspicion by many masters who are concerned with the more able mathematical boys, and in many cases practical methods learnt by the boys in the lower classes are allowed to rest in the upper. Many teachers consent to the postponement of abstract methods during the earlier teaching only on the understanding that the abstract character of the higher teaching shall be preserved.

In the United States the spirit of unrest in all educational matters is also manifest, and in particular with respect to this whole question of intuition and experiment in mathematics. The question has been agitated for the past ten years, and many kinds of experiments have been tried, varying from an extreme laboratory method with a minimum of mathematics to the most abstract kind of work in which intuition and experiment played almost no part. It seems quite impossible to report as to any fixed policy or general consensus of opinion upon the subject.

We shall now take up *seriatim* the various topics set forth in the questionnaire that was sent out by Dr Lietzmann last winter. Since it will not be profitable to present orally the report in full, the details being better assimilated from the printed page, a brief summary of the results will now be given.

In the work of measuring and estimating, a more practical form of mensuration seems to be developing, especially in Austria, Germany and Switzerland. England, France, and the United States seem to have given the matter less attention, or at least to have secured less definite results. An elementary trigonometry is more commonly found at an early period in the first three countries, thus allowing for outdoor work with simple instruments at an earlier stage.

In the matter of geometric drawing and graphic representation of solids, the various countries seem to be in a transition stage between the period in which this was considered part of the duties of the art teacher and that in which it is to be taken over by the department of mathematics. The tendency is general to consider this work as part of mathematics. The nature of the work is not, however, at all settled; even the term "descriptive geometry" has no well-defined meaning. It may be said in a general way that the subject is taught in schools of the Oberrealschule type, but not in those of the Gymnasium type, and that the tendency is to increase the work both in amount and quality in the former.

Graphic methods of representing functions have become universal in the last generation. From the idea of a line representing an equation the tendency is at present to that of a graphic representation of a function. Just how much the pupil is acquiring of the function concept seems often to be questioned, and the whole subject is in the experimental stage at present. Of the value of squared millimetre paper there is no question anywhere, but it seems equally true that its use has been abused by the over-extensive treatment of equations and by its application to proving the obvious.

The contracted methods of computation that were prominently advocated fifty years ago do not seem to have advanced materially, owing to the feeling that they are not really practical. On the other hand the use of logarithms seems on the increase, and the slide rule has come into great favour in the technical schools where approximate calculation is prominent. Graphic methods of computation and of the approximation of roots of numerical higher equations (as worked out by Professor

Runge, for example)*, have found but little place in the schools of the type under investigation. It is probably too early to say what their success will be in this kind of school, or when, if ever, any large body of teachers will be found capable of treating the subject.

In general it may be said that more progress towards the recognition of the rôle of intuition and experiment in secondary mathematics seems to have been made of late in Austria, Germany, and Switzerland than in England, France, and the United States†. A comparison of the work done in these several countries, particularly in the way of applying mathematics seriously to the problems of life, and of visualizing the work in mensuration, will be one of the valuable results of the present international movement.

The greatest questions of all, however, relate to the nature of geometry and to the treatment of the function concept. Other questions are important, but here is the dominant issue that must be met in the next few years.

The first of these questions is this: How much of the geometry of the secondary schools shall be inductive, and how much deductive? Few are ready to assert to-day that it is best to begin geometry with a study of Euclid or Legendre. There must be a preparatory stage, and this must be characterized by intuition and experiment. But how much time shall be assigned to this stage? and exactly what ground shall be covered? and, what is more important, to what extent shall intuition replace deduction of the Euclid type?‡ Must we have two or even three years of *Anschauungslehre*, as some have advocated, or is a year or a half-year enough?§ How much must the rigorist in geometry be compelled to concede to the demands of the non-mathematical mind? Was Newton right when he expressed the opinion that all this intuitive work is pretty, but that it is not geometry? Must the real geometry of the past go by the board, as went the mediaeval logic, or will its position be strengthened by this propaedeutic work? Is the value of the Euclidean type of geometry such as to save it from destruction, or is it to be so diluted by this consideration of pictures, of models, and of simple mensuration as to be unrecognizable? Is the over-powering force of to-day, the force of modern industry, the cause of the growth of intuition and experiment in geometry? And if so, what will the over-powering force of to-morrow, the force of social considerations, demand? These are questions that we hear about us, but they are questions which we are not yet able scientifically to answer, and fortunately they are not within the province of this committee to attempt to answer.

In general it may be said that it is the plan of the Teutonic countries to mix the intuitional and the deductive work from the outset, while in France, and now in England, the plan is to let an inductive cycle precede a deductive one. The United States is only beginning to talk about the question, whatever tendency there is being towards the Anglo-French plan. Now is either of these plans better than the other? and can this be proved? Or is the question one of racial habit? Would the German

* In his Columbia University (New York) lectures.

† At least this is the deduction from the reports submitted in reply to the questionnaire.

‡ Readers are referred to the recent works by Treutlein and Timerding, published by Teubner.

§ See Lietzmann, *Stoff und Methode des Raumlehreunterrichts in Deutschland*, 1912.

plan succeed in the United States, or is the French plan better adapted to such a conglomeration of races? Would the English scheme of arranging the work in three stages, with intuition and experiment the initial one, be better for Germany than the totally different scheme that characterizes most of the recent text-books of that country? Many people look upon the German plan as unscientific, attributing its apparent success to the excellently trained teachers who carry it out, while others look upon the cycle arrangement as ultra-scientific. Which is right? or is neither right? It is questions of this kind that the reports of the Commission will help us to answer in the next few years; and for the present it suffices to state the problem and to call attention to its importance.

The second important question relates to the treatment of the function concept. Here the rôle of intuition, in the first steps, is more clearly defined, since we have no well-tried body of knowledge to be set aside. The chief argument for the elaboration of the function concept seems to be that the calculus has already found place in the schools under our consideration, and if it is to hold this place and continue to grow in strength, we must cease to impose it merely from above—we must prepare for it from below. The notions of limit, invariability, rate, function, and graph must be so gradually introduced and must become so clearly understood that when the calculus is reached they will be met as we meet familiar friends. How to do this economically is one of the problems relating to intuitional mathematics. It is one of the interesting facts of present education that teachers are demanding the elimination of the incommensurable quantity from elementary demonstrative geometry, only to find themselves face to face with a demand for the study of limits, functions, and rate of change. The movement in favour of the elaboration of the function concept, however, is too recent to judge of its permanence in secondary education. Starting in France within the last twenty years, and vigorously advocated in Germany within the last decade, it has much to commend it if reasonably treated*.

6. *Measuring and estimating*†.

Austria reports that much is now being made of this work in the lower and middle classes, and that this will have its influence in the higher classes. Some idea of the nature of this work may be obtained from the recently published *Praktisch-geometrische Schülerübungen für die unteren Klassen*, by Fr. Schiffner. As an illustration of the nature of this work, Dr Matter, of the Stiftsgymnasium in Seitenstetten, relates that the children in the second class (numbering from the lowest in the school) measure the height of a tower by the simple use of a measuring tape and a large protractor. They find two angles and the included side of a vertical triangle, and then lay this off on the horizontal plane, and finally draw it to scale and measure the resulting figure. Similar work is also done in the classes in geography. In order to ascertain how much work of this general nature is done in the upper classes of the Gymnasien, Realgymnasien, and the Realschulen, Dr Erwin Dintzl of Vienna, who reports for Austria, secured information from 38 institutions. He finds

* See Schimmack, *Die Entwicklung der mathematischen Unterrichtsreform in Deutschland*, 1911.

† Messen und Schätzen (Mesure et estimation des grandeurs).

that in Vienna there is difficulty in carrying out a scheme of field work in connection with trigonometry owing to a lack of proper instruments, the crowding of the school curriculum, and the proper demands of out-door play. Occasionally, however, facilities for this work are found, and he mentions a satisfactory Taschen-Universal-Instrument made by Neuhöfer for the use of students and at the price of 170 Kroner (say a little over £7, 178 fr., or $34·50). In other cities a satisfactory amount of geodetic work is being done in the upper classes, the proximity to the country making this feasible.

In England, 58 °/₀ of the "public schools" investigated by Mr Godfrey attempt no work in practical geodetic measurements, but in the remaining 42 °/₀ such work is done, but as a rule only by special classes of students, such as those preparing to be surveyors, engineers, or army officers. Out of the schools reporting, 14 °/₀ have a theodolite, and others have plane tables or other instruments. In the other secondary schools 56 °/₀ do work of this kind. Apparently more work of this nature is done here than in the older type of "public school," theodolites being found in 23 °/₀ of the schools, and simpler instruments in numerous others.

In France it is rare that a lycée or collège does any work in geodetic or astronomical measurement. In the *Écoles des Arts et Métiers,* where the pupils are about 17 years of age, elaborate courses in surveying are given, and pupils become expert in the use of such instruments as the theodolite. This work is done in the department of mathematics. The teacher of geometric design is usually the teacher of mathematics, and he relates the work in surveying to that in drawing, as in the case of profiles, plots and the like.

In Germany the work varies in the different states. Dr Lietzmann states that in the Prussian schools the theodolite is usually found, and that along with it are seen simple instruments for angle measure, angle mirrors and prisms, measuring rods, and the like. Simple instruments are often made by the pupils, particularly instruments for the measure of angles*.

Much is made of out-door work in the classes in geometry and trigonometry, in the measuring of heights and distances the pupils making use of the instruments†. In Mecklenburg and Oldenburg about two-thirds of all the secondary schools give systematic field work, and even in the rest this work is not neglected. At least in certain parts of Germany apparatus and models for the teaching of geometry play some part, and a helpful list of materials is given in a recent monograph by Thaer, Geuther, and Böttger‡, and in one by Cramer§. The testimony of teachers there as elsewhere,

* For the rôle of intuition in the class-room in geometry, and for the conduct of a class period, see Dr W. Lietzmann, *Die Organisation des mathematischen Unterrichts an den höheren Knabenschulen in Preussen*, p. 65.

† Ibid., p. 161. See also Dr H. Wieleitner, *Der mathematische Unterricht...im Königreich Bayern*, pp. 42, 62; Dr E. Geck, *Der mathematische Unterricht...im Königreich Württemberg*, p. 25; Prof. H. Cramer, *Der mathematische Unterricht im Grossherzogtum Baden*, p. 35; Prof. J. Wirz, *Der mathematische Unterricht...im Elsass-Lothringen*, p. 9.

‡ Thaer, Geuther, und Böttger, *Der mathematische Unterricht in den Gymnasien...Mecklenburgs und Oldenburgs*, pp. 22, 24, 27, 78.

§ *Der mathematische Unterricht im Grossherzogtum Baden*, p. 30.

however, is that there is a danger in the extensive use of models in geometry, although a reasonable use of apparatus in field work has a value that cannot be doubted.

With respect to mathematico-astronomical work in the German schools, it is coming to be recognized that such work is feasible and desirable. Numbers of Höherenschulen in Prussia have telescopes for astronomical observation, although those that are suited to measuring are not usually found*. The subject has recently been treated in a monograph by Professor Hoffmann, and to this article reference must be made for detailed information†. The extent to which he has carried this work in his school is indicative of the German *Lehrfreiheit*, a freedom that is wanting in free America and in most other countries.

The helpful table that Dr Brandenberger has prepared to show at a glance the work done in the Gymnasien and Realschulen of Switzerland‡ gives the impression that relatively little field work in geometry, and even less in astronomy, is done in the Gymnasien. In Canton Bern surveying is given in the summer; in Unterwalden some practical work is done with the theodolite and in the measuring of lengths and areas; a few Gymnasien give an hour a week to surveying during one year, usually with pupils of the age of 16–17 years; a few others possess observatories with sufficient apparatus for mathematical work. Of 25 Realschulen reporting, however, 12 give surveying as a definite topic in their curricula§ and the rest make more or less of the subject in connection with trigonometry. A number of the schools are supplied with such instruments as the theodolite, cross staff, angle mirror, angle prism, and plane table. In certain schools, as in the technical classes at Lugano and St Gall, very satisfactory courses are given, including triangulation, the measurement of base lines, the making of profile maps, and the finding of altitudes.

In the United States there are, generally speaking, no required courses in trigonometry or geodesy in the "high schools,"—that is in those 4-year schools that are intermediate between the 8-year elementary schools and the 4-year college. In the third or fourth year when the pupil is 16–17 years old, elective work is offered in trigonometry by the better class of high schools. Occasionally the theodolite or some similar instrument is used, but in general the "high schools" do not possess such apparatus. Such work is regularly taken up in the first year in college, and then as an elective study. It may thenceforth be carried on as an elective by students of general mathematics, but it is required of engineers.

In certain special schools, however, some interesting work may be found. For example, Principal Stark studied with the writer not long ago the subject of primitive mathematical instruments and their relation to present-day teaching. The results were put in practice in the Ethical Culture School in New York, the children making astrolabes from paper protractors and deriving much interest from their out-door measurements with these and other instruments. In the Horace Mann School, connected with Teachers' College, New York, there is at present a class in vocational

* Lietzmann, *Die Organisation*, p. 43.

† Prof. B. Hoffmann, *Mathematische Himmelskunde und niedere Geodäsie an den höheren Schulen.*

‡ Dr K. Brandenberger, *Der mathematische Unterricht an den Schweizerischen Gymnasien und Realschulen*, pp. 13–25; p. 57.

§ Ibid., pp. 60, 62, 119.

mathematics, well supplied with simple instruments which they use out of doors in their eighth school year, preparatory to the more scientific work that follows*.

Astronomical work requiring the use of instruments may be begun in the first year of college, and may be continued thereafter. It is elective throughout. The student who pursues the courses will be prepared to do scientific photographic work in the senior year (the fourth year in college, about 21–22 years of age). Colleges of the better class are supplied with material for work of this character.

The educational problem may be stated generally as follows: What simple, inexpensive instruments may advantageously be used to increase the interest in the early stages of mathematics? In particular, what can be done to make the inductive cycle or phase of geometry more real and interesting, without weakening the deductive side? Then, what further apparatus can be used to advantage in the later geometry and trigonometry?

7. *Geometric drawing and graphic representation*†.

The questionnaire asked first concerning the work done in descriptive geometry (oblique parallel projection), preparation of plans and elevations, central projection, and the theory of shadows; concerning the fusion of this teaching and the teaching of stereometry; and concerning the teaching of the subject in the department of mathematics or the department of drawing.

In Austria, in the Realschulen, the subject is nominally in the hands of a special teacher examined and appointed for the purpose. Professor Dintzl raises the question as to the wisdom of this arrangement, saying that to the qualifications of a teacher of descriptive geometry must be joined those of a teacher of mathematics if we are to have successful results. At present, in about a third of the schools, the teacher of mathematics is also the teacher of descriptive geometry in the three upper classes (V–VII). There is a well-defined line of demarcation between stereometry and descriptive geometry as the work is carried on, although the latter is employed in representing the solids met in the former. It appears, therefore, that certain definite work in descriptive geometry is done in the upper classes of the Realschulen. In the Gymnasien descriptive geometry is not a separate object of study. In the work in stereometry, however, plans and elevations are drawn, and orthogonal projections of parallelepipeds, octahedrons, pyramids, and the like are prepared‡.

In England descriptive geometry was formerly taught only to boys who needed it for special examinations, and then generally by the art master, the result being mere work by rule with great attention to artistic ink-work. Two or three causes are operating to change this position: (1) the Army examiners now treat the subject as part of mathematics, and do not require inking-in, the result being that mathematical masters are not discouraged from teaching the subject by consciousness

* The class is conducted by Mr J. C. Brown who has contributed to the psychology of elementary drill work in mathematics.

† Zeichnen und Darstellen (Dessin et représentation graphique).

‡ See also Adler, *Der Unterricht in der darstellenden Geometrie an den Realschulen, Gymnasien, Realgymnasien und Reformgymnasien*, and Müller, *Der Unterricht in der darstellenden Geometrie in den technischen Hochschulen*, in Heft 9 of the Austrian reports.

of their technical weakness on the artistic side. (2) Descriptive geometry is now required for an honours degree in mathematics at Cambridge, and this circumstance will result in an increased output of mathematical masters interested in the subject and will have its direct reaction on school work. (3) Boys intending to study engineering at the University are encouraged to master descriptive geometry and perspective at school. Oblique parallel perspective is little studied in English schools. Descriptive geometry according to Monge (with work in plans, elevations, rabattements, etc.) is taught in 77 °/₀ of the "public schools" reporting, and perhaps in half of these it now forms part of the school course for the upper classes, in other cases being confined to specialists. Of the other secondary schools, 60 °/₀ have it in their courses. It seems not to have established itself, however, because it does not enter into the ordinary school examinations. Where it is taught two-thirds of schools correlate it with solid geometry, and in the majority of the "public schools" cases it is taught by a mathematician. It is seldom that the "public schools" do what is often done in other secondary schools, namely correlate descriptive geometry with manual work in the carpenter's shop. Perspective, being more technical than descriptive geometry, has not been so extensively taken over by the department of mathematics. In 63 °/₀ of the "public schools" reporting, it is taught in connection with drawing and usually by the drawing teacher. Accurate drawing in connection with projective geometry, including the construction of conics by anharmonic (cross) ratio properties, is not done at all in 69 °/₀ of the schools, and only a few of the most advanced boys in the other schools are encouraged to reach this comparatively advanced work.

France reports that descriptive geometry is taught in the lycées and collèges only in 1re C and D (the pupils being 16 years of age), in "Mathématique A et B," and in the classes preparing for the higher scientific schools*.

The teaching of descriptive geometry is generally distinct from that of stereometry. Usually geometric drawing is not taught by the teacher of mathematics, but at present there is an active effort being made to put the direction of the work in the department of mathematics†.

Germany, like Austria, seems to be taking a leading position in this line of work. Models for the study of descriptive geometry seem to be not uncommon. In the Realgymnasien and Oberrealschulen especially there is given definite work in this line. Oblique parallel projection is taught in Prussia in the Obersekunda (the seventh year of this type of school, or the pupil's tenth school year), and descriptive geometry as a special subject is taught in Oberprima (the last, ninth year‡). The tendency in Northern Germany has been to make descriptive geometry in its various phases an elective subject in the domain of art instruction. In Saxony and the Southern states it has more generally been required§. The tendency throughout Germany seems to be in favour of putting geometric drawing, under whatever name,

* See the report of M. Rousseau and that of M. Blutel (vol. II).

† The report as sent to this committee was very brief, and for further information the reader is, therefore, referred to the printed reports of the Commission.

‡ See particularly, Lietzmann, *Die Organisation*, etc., p. 139; and Zühlke, *Der Unterricht im Linearzeichnen und in der darstellenden Geometrie an den deutschen Realanstalten.*

§ Thaer, Geuther, und Böttger, *loc. cit.*, p. 81.

in the hands of the teachers of mathematics*. In South Germany the work has for a long time been in a satisfactory condition, but it was quite neglected in North Germany until 1898. In general it may be said that geometric drawing is commonly given in the Gymnasien and Oberrealschulen; that, as might be expected, it is found chiefly in Sekunda and Prima; that in the Oberrealschulen it has much more attention than in the Gymnasien; that it was formerly related to the art work, but now it is being taken over by the mathematical department; that in the non-technical schools the descriptive geometry is related more to the representation of the usual geometric solids by projection than directly to the making of the working drawings of the artisan; and that more attention is being paid to the subject than is the case in the non-technical schools of any other country except Austria and Switzerland.

In Switzerland† the Gymnasien devote little attention to descriptive geometry, most of them not mentioning it‡. In the Realanstalten, however, it is found almost without exception in the last three years§. In the published curricula Switzerland makes very clear the nature of the work that is done under the general terms of "Geometrisches Zeichnen" and "Darstellende Geometrie." For example, in the curriculum of Bern the course includes work in geometric ornament, such as parquetry flooring; orthogonal projection of geometric solids; drawing of conics and other plane curves; drawing of machine models; shadow constructions; axonometry; polar perspective; solids of rotation; and plans and elevations. Such work, in a non-technical school, is found in the reports of only the three countries above named, among those that have replied to the questionnaire. Moreover it appears that in the large majority of the schools the practical bearing of all this work upon industry is fully recognized. It also appears that the work in descriptive geometry is looked upon as belonging to the teacher of mathematics rather than to the teacher of art||.

In the United States few general "high schools" offer work in geometric drawing of any kind. Formerly some such work was occasionally given by the teacher of drawing, but the awakening of an interest in art a few years ago led to the substitution of more free-hand drawing and colour work for the mechanical drawing then somewhat in vogue. The change was justified, for the mechanical drawing was badly done as a rule and the country needed all of the improvement in art appreciation that it could get. In the technical "high schools" that are beginning to appear in the larger cities, work of this kind is being begun, but not as yet with the thoroughness that characterizes the Austrian, German, and Swiss schools. In the technical courses of the colleges it is begun in the freshman year (about 18 years of age) and is taken only by students of architecture or engineering.

* Thaer, Geuther, und Böttger, pp. 31, 32, 36, 56, 70, 73; Wieleitner, *loc. cit.*, pp. 29, 35, 43, 56; Witting, *loc. cit.*, pp. 27, 29, 33, 36, 57, 64; Cramer, *loc. cit.*, pp. 15, 20; Schimmack, *loc. cit.*, pp. 13, 15, 20, 43; Wirz, *loc. cit.*, pp. 9, 17, 25, 31, 50; Zühlke, *loc. cit.*

† See Brandenberger, *loc. cit.*, pp. 14–25, 129–137.

‡ Ibid., pp. 14–20, 129.

§ Ibid., pp. 20–25, 130. One of the clearest statements of the nature of the work to be found in any of the reports appears on pp. 130–135.

|| Ibid., pp. 60, 62, 118, 137.

These courses are rarely given by the mathematician, being left to the engineer or architect.

The educational problem may, therefore, be stated as follows: Are we making enough of drawing in connection with geometry? In particular, are we recognizing the practical value, not merely of the working drawings of the artisan, but also of geometric design, of cartography, of topographical drawing, and of geometric construction?

8. *Graphic methods**.

Graphic methods of one form or another are now found in the courses in mathematics, at least in the Realanstalten, in all countries, having gradually made their way from engineering, through thermodynamics and general physics, to pure mathematics. The extent to which these methods are used varies, however.

In Austria millimetre paper is used for all of the simple functions, such as the elementary ones of algebra, and the exponential, logarithmic, and circular functions. The value of the sine is found to 2 decimal places by measuring on the squared paper, and computations are often verified graphically. The planimeter is rarely used. Vector geometry is used for the purpose of representing the complex number, and for elucidating certain parts of mechanics. For the work in complex numbers it appears in the sixth class (reviewed in the eighth), and for mechanics it is introduced in the sixth (Realschule) or seventh (Gymnasium) class. The real consideration of the scalar field is hardly undertaken, but it is touched upon in the study of magnetic and electric fields, the concept of the potential, and the like, in the upper classes.

In England about 90 °/₀ of the schools state that the graphical study of statistics is given. It seems to begin mainly in the lower and middle classes of the schools. Probably it involves little beyond the plotting of tables of statistics, and for a new subject it has made rapid progress, having both the encouragement of the mathematicians and an abundant opportunity for application. The graphical representation of functions is taught in all of the "public schools." In 27 °/₀ of the schools it is begun in the lower classes, in 58 °/₀ in the middle classes, and in 15 °/₀ in the upper classes. Similar results are found in the other secondary schools. The work is connected with the plotting of equations and with the approximation of the roots. Whether the idea of functionality is really grasped in this work seems open to question, but the graph forms a basis from which further advance may be made. The use of vectors is found in a large majority of the schools, in connection with mechanics (velocities, accelerations, forces), this latter subject being part of the mathematical course in England. In some schools the vector is used in connection with complex numbers. Graphical statics is taught generally. Areas are estimated by squared paper in most schools, but the planimeter is rarely used.

* Graphische Methoden. (Darstellung von Funktionen auf Millimeterpapier. Vektorendarstellung. Skalare Felder. Graphisches Rechnen, insbesondere graphische Statik. Flächenauswertung mit Millimeterpapier oder Planimeter.)

Les méthodes graphiques. (Représentation de fonctions sur du papier millimétrique. Représentation des vecteurs. Champ scalaire. Calcul graphique et spécialement statique graphique. Evaluation de surfaces à l'aide du papier millimétrique ou du planimètre.)

In France the notion of coordinates is introduced when the pupil is about 14 years of age. The graph is used in the study of equations, as apparently in all other countries. The technical schools make use of graphic mechanics. In the third year of the École des Arts et Métiers (age of pupil, 20 years) and in the higher scientific schools graphical statics is taught. The planimeter is not used.

In Germany the custom is growing of having, at least in the class-rooms of Tertia or Sekunda, a portion of the blackboard ruled in squares, usually about 5 cm. on a side, for the graphic representation of functions*. In some of the Gymnasien† there is carefully organized work in such representation of functions from Obertertia on. In Obertertia such functions as

$$y = \frac{b}{a}\sqrt{a^2 - x^2}, \quad y = \frac{1}{x}, \quad y = \sqrt{a^2 + x^2}, \quad \text{and} \quad y = ax^3 \quad \text{or} \quad ax^4$$

are treated graphically. In Untersekunda this is extended to the exponential, logarithmic, and trigonometric functions‡, and in the Obersekunda and Prima to functions of degree exceeding two§. In all cases the function concept is mentioned as prominent, and the use of the graph merely for purposes of solving an equation is not in evidence‖. Graphic aids in computation are also employed to some extent in the German schools**.

In Switzerland the graphic representation of equations and functions is general, as in other countries††, and is extended to the treatment of limits. The question of the function concept is not settled there, and it does not appear to be settled anywhere. Exactly what can be done, and how it is to be begun, are matters still awaiting the results of experience. Dr Brandenberger calls attention, for example, to the fact that in the course of study of one Realschule (where it would be expected) the word does not appear at all, while in a certain Gymnasium (where mathematics is less in evidence) it plays a prominent part. It would seem that the practice does not differ so much as the printed statements, and that the notion of function is introduced generally when there is a demand for it. Some schools emphasize it early, others late, and experience seems not as yet to have given any definite verdict.

In the United States the graphic representation of simple algebraic functions, with millimetre paper, is commonly begun in the first year of the "high schools" (about 14 years of age), and continued, chiefly with curves of the second degree, in the third year (age 16 years). The graphic treatment of vectors and scalars is reserved for the college, where it commonly begins in the algebra work of the freshman year (age 18 years). It is not taken up with any thoroughness unless the student elects a course in vector analysis, which is frequently offered in the junior or senior year (age 20, 21 years), and is frequently relegated to the graduate years (about 22–24 years).

* Lietzmann, *Die Organisation*, p. 37.

† Ibid., p. 161.

‡ Ibid., pp. 161, 170.

§ See also Thaer, Geuther, und Böttger, *loc. cit.*, pp. 6, 9, 79; Wieleitner, *loc. cit.*, pp. 20, 42; Witting, *loc. cit.*, p. 36; Geck, *Der mathematische Unterricht...im Königreich Württemberg*, pp. 21, 25, 59; Schnell, *Der mathematische Unterricht...im Grossherzogtum Hessen*, pp. 20, 29, 39, 41.

‖ See particularly Schimmack, *loc. cit.*, pp. 19, 22, and Wirz, *loc. cit.*, p. 45.

** Timerding, *Die Kaufmännischen Aufgaben*, p. 35.

†† Brandenberger, *loc. cit.*, pp. 95, 101, 103, 104–109.

9. *Numerical computation**.

The inquiry of the committee related to abridged computation, to the use of the slide rule and tables, and to graphical and numerical methods of approximation of the roots of equations.

In Austria abridged methods of computation are introduced in the third class (age about 13) in connection with the mensuration of surfaces†. The slide rule has not yet found general acceptance in the secondary schools. The reason for this seems to be the expense of the instrument, the cheaper ones not being accurate enough to be of value; but the question of time to acquire the necessary facility is also a serious one. In the upper classes the numerical computation is performed almost exclusively by the aid of logarithms. The new programme requires the elements of probabilities in the upper class, and the simple problems of life insurance will require the use of mortality tables.

In England the mechanical rules for contracted operations with decimals have long been in use, but it seems that at present these formal rules are losing ground. The omission of useless figures and the presentation of results to a given number of significant figures must become increasingly general as practical and laboratory work in the lower classes gains ground. On the other hand many masters say that logarithms prove more useful than contracted methods, and that contracted rules are learnt but are seldom used afterwards. In 30 °/₀ of the "public schools" and in 66 °/₀ of the other secondary schools the slide rule is not used; it is generally used by pupils preparing for special examinations. The common use of the 4-figure logarithmic table is doubtless the cause for the little use of the slide rule. Tables of squares and square roots are not used in 40 °/₀ of the "public schools," and tables of cube roots are not used in 55 °/₀ of the "public schools." The other secondary schools make less use of them. Tables of logarithms are used in all schools, 48 °/₀ introducing them in the upper classes, and 37 °/₀ in the middle classes. Tables of trigonometric functions are used in all schools, in about half of them being introduced in the upper classes, and in about 20 or 25 in the middle classes. Mortality tables are not generally used. In about 65 °/₀ of the schools 4-figure tables are used. Graphical methods of treating transcendental equations are not introduced systematically, and are found in 47 °/₀ of the "public schools," and in 16 °/₀ of the other secondary schools.

In France the abridged methods that became prominent in the middle of the 19th century are no longer taught. They seem not to be demanded in practical work. The slide rule is not used in the lycées and collèges except in classes preparing for technical schools. In the scientific classes of the lycées and technical schools the pupils learn the use of logarithms, usually with tables to five decimals. Trigonometric tables, with the logarithms of the functions, are common, as are tables‡ of square and cube roots and of $\log(1+r)$. The approximate solution of numerical higher equations, either graphically or by numerical computation, is given only in the *classes de mathématiques spéciales*.

* Rechnen und Berechnen. Calculs et évaluations numériques.

† See the Austrian reports, Heft 3, S. 40, and Heft 6, S. 9.

‡ For example, Combet's tables, published by Belin, Paris.

In Germany contracted operations (abgekürztes Rechnen) are not required in the Prussian programme, although they are given in particular institutions*. The question of the real value of the work seems unsettled. In Bavaria it has some place in the lower classes of the Gymnasium and Realschule†. In Saxony the particular need for the subject is found in the Obertertia or Untersekunda of the Gymnasium‡. In Baden and Elsass-Lothringen it is assigned place in Quarta of the Oberrealschule§, and in general it may be said to have at least a nominal position in the various states of the Empire. The slide rule seems to be making its way very slowly in Germany as elsewhere‖. The use of tables is general there as in other countries.

Five-place tables are used in the majority of schools, but about a third use four-place tables, interpolations being made mentally instead of by the use of proportional parts. In the lower classes the method of computing the tables is mentioned incidentally, the exercises in actual computation by means of series being relegated to the higher classes of the Oberrealschulen. The graphic treatment and the numerical approximations of the roots of numerical higher equations has place in the modern reform movement**. In the Realanstalten, however, the use of approximate methods (such as Newton's rule and the *Regula falsi*) is found in Oberprima.

In Switzerland the contracted operations are emphasized in the technical schools. The slide rule is found in the programmes of 6 of the 25 Realschulen, and in 2 Gymnasien††. Logarithms are universal, the majority of schools using 5-place tables, although about 30 °/₀ use 7-place tables. Some advance is being made in the direction of decimalizing the angle, tables having been prepared for such divisions. Tables of roots and mortality are in use. In 25 °/₀ of the Gymnasien the subject of infinite series is introduced, and applications are made to the theory of computing tables of logarithms. Methods of approximation of the roots of numerical higher equations are given. Graphic methods, the *Regula falsi*, and the Newtonian approximation are given in all Realschulen, and certain schools add Horner's, Lagrange's, or Gräffe's method.

In the United States contracted methods with decimals are rarely taught, either in the elementary schools or the general "high schools." In the "technical high schools" (a relatively new type, parallel to the older type of general "high school"), and in the colleges where engineering courses are given, these methods are taught incidentally in the courses where they would naturally be used. The slide rule is coming into extensive use in the "technical high schools," and it is used in all engineering courses in the colleges. In the better types of general high schools it is

* Lietzmann, *Die Organisation*, pp. 104, 159; *Stoff und Methode...Unterricht*, p. 85.

† Wieleitner, *loc. cit.*, pp. 24, 41.

‡ Witting, *loc. cit.*, p. 16.

§ Cramer, *loc. cit.*, p. 18; Wirz, *loc. cit.*, pp. 29, 33.

‖ Lietzmann, *Stoff und Methode...Unterricht*, p. 70 seq.; Timerding, *Die Kaufmännischen Aufgaben*, p. 35.

** Lietzmann, *Die Organisation*, pp. 161, 173, 178, 182, 197; Thaer, *loc. cit.*, p. 73; Wieleitner, *loc. cit.*, p. 31; Witting, *loc. cit.*, p. 24.

†† Brandenberger, *loc. cit.*, pp. 60, 90.

occasionally explained to the classes in trigonometry. Mathematical tables are not used in the "high schools" until trigonometry, which is an elective study, is reached. In general, 5-place tables are used. In some arithmetics and in some official courses the table of compound interest is used in the eighth grade (age 13 years). Mortality tables are rarely seen in the secondary school. The method of computing tables is not taken up in the elementary classes, but in the calculus and in higher algebra (college studies, age 18–19) the student is shown how certain functions ($\sin x$, e^x, $\log x$, etc.) are expanded in series and how these series may be used in the making of tables. Graphic and arithmetic solutions of numerical higher equations are given in the college course in algebra (age about 18 years), the arithmetic solutions being effected usually by Horner's method. Graphic methods of solving such equations are beginning to be more highly appreciated.

The bearing of this type of computation upon the industrial problems of the present is important. In our time the typewriter has replaced the pen to a great extent, and the computing cash register has replaced the old cash drawer. Japan has kept her soroban* because she believes that the era of mechanical calculation is to return. In the Western World the cheapening of the calculating machine is testifying to Japan's forethought. What should the schools do in view of this change? Already the attempt to train the "lightning calculator" is past, but are we entering upon a period of practical graphical computation? of the replacing of logarithmic tables by the slide rule? and of inexpensive machines that shall perform the ordinary calculations, as they already perform the extraordinary ones? If so, are we to make the mistake that we made with logarithms, of using 7-place tables when 4-place ones would be better, or can we select with scientific forethought the material that is really practical?

This committee was charged with the duty of reporting upon present conditions and of propounding questions, but not of making recommendations as to the future. Since the stating of a problem is quite as important as the solution, it is hoped that some of the questions that have been raised will serve a good purpose. That certain of these problems deserve more than a passing notice need hardly be asserted†.

Questionnaire. Rappelons ici le questionnaire qui a servi de base à l'étude de M. Smith. Il avait été rédigé et distribué par M. Lietzmann, au nom de la Sous-commission A.

Extrait de la circulaire de M. le Dr W. Lietzmann.—Objet: *L'intuition et l'expérience dans l'enseignement mathématique des écoles moyennes.*

Délimitation du sujet. L'intuition et l'expérience jouent un rôle prépondérant dans l'enseignement géométrique des écoles élémentaires et des cours complémentaires (Fortbildungschule), ainsi que dans l'enseignement propédeutique des écoles moyennes. Dans la suite, il ne sera pas question de tout cela. Nous allons de même laisser de côté les cas multiples où l'intuition et l'expérience sont destinées à compléter ou à remplacer les développements logiques et déductifs, dans l'enseignement systématique de la géométrie dans le domaine des éléments d'Euclide. Nous aurons peut-être l'occasion, dans une séance ultérieure de la Commission, d'examiner ces questions extrêmement

* See the report of Professor Fujisawa.

† After this report was ready for the press the note by Messrs Cardinaal and Barrow, relating to the schools of Holland, appeared in *L'Enseignement mathématique*, July, 1912, p. 327. The plea for the freedom of the teacher in the matters referred to in this report is worthy of careful consideration.

importantes, en tenant compte du point de vue psychologique. Afin de bien délimiter le sujet, il conviendra donc à Cambridge de s'en tenir au rôle de l'intuition et de l'expérience dans les classes supérieures des écoles moyennes.

Pour organiser les travaux préparatoires, il est désirable d'avoir un tableau de l'état actuel de ce qui se fait dans les différents pays. Nous nous permettons à cet effet de vous soumettre les questions suivantes:

1. *Mesure et estimation des grandeurs.* Dans quels établissements, gymnase, école réale supérieure, etc., dans quelle étendue et dans quelles classes (âge des élèves):

(*a*) Procède-t-on à des mesures *géodésiques* pratiques pour les utiliser ensuite numériquement? Usage du théodolite, de la chaîne d'arpenteur, etc.

(*b*) Fait-on des observations et des mesures *astronomiques* avec des problèmes qui s'y rattachent? Usages d'appareils photographiques, instruments universels.

2. *Dessin et représentation graphique.* Dans quels établissements, dans quelle étendue et dans quelles classes présente-t-on:

(*a*) La géométrie descriptive (Projection oblique?—Plan et élévation?—Projection centrale?—Théorie des ombres?).

Y a-t-il fusion entre cet enseignement et l'enseignement de la stéréométrie? L'enseignement est-il donné par le maître de mathématiques ou par le maître de dessin?

(*b*) Les méthodes graphiques (Représentation de fonctions sur du papier millimétrique?—Représentation des vecteurs?—Champ scalaire?—Calcul graphique et spécialement statique graphique?—Evaluation de surfaces à l'aide du paper millimétrique ou du planimètre?).

3. *Calculs et évaluations numériques.*

(*a*) Calcul abrégé à l'aide de fractions décimales?

(*b*) Emploi de la règle à calcul?

(*c*) Tables numériques (nombre de décimales pour le calcul logarithmique et les fonctions trigonométriques?—Emploie-t-on aussi des tables de racines carrées ou de racines cubiques et des tables de mortalité?—Est-ce que l'on montre à l'aide d'exemples comment on peut calculer les valeurs des logarithmes et des fonctions trigonométriques?)

(*d*) Résolution numérique et graphique des équations par approximation (Règle de Newton?—Regula Falsi?—Méthodes nomographiques?).

Les deux publications ci-après permettent d'orienter le lecteur sur quelques-unes de ces questions et sur les réponses concernant l'Allemagne:

P. Zühlke: *Der Unterricht im Linearzeichnen und in der darstellenden Geometrie* (III. Bd., Heft 3 der Abhandl.).

B. Hoffmann: *Astronomie, Vermessungswesen, mathematische Geographie an den höheren Schulen* (III. Bd., Heft 4 der Abhandl.), actuellement sous presse.

DISCUSSION.

M. Fujisawa, qui présidait la première partie de la séance, remercia M. Smith, au nom de l'assemblée, pour son exposé si documenté, puis il ouvrit la discussion sur cet intéressant rapport.

M. Laisant estime qu'il eût été préférable de faire porter l'enquête sur l'ensemble de l'enseignement, depuis la première initiation, tandis que la Sous-commission a cru devoir limiter l'étude à l'enseignement moyen. Puis MM. Thaer et Lietzmann développent et complètent certains passages du rapport de M. Smith pour ce qui concerne tout particulièrement l'Allemagne, M. E. Dintzl pour l'Autriche, M. Bioche pour la France, MM. Siddons et Carson pour l'Angleterre, et M. Goldziher pour la Hongrie. Envisageant la question dans son ensemble, M. v. Dyck (Munich) tient à faire constater que les discussions soulevées par la Commission montrent le rôle utile qu'elle a déjà exercé jusqu'ici en appelant les mathématiciens du monde entier à réfléchir sur les questions si importantes des méthodes et des plans d'étude de l'enseignement mathématique.

M. C.-A. Laisant (Paris).—Je ne voudrais à aucun prix entraver, ni même retarder, la discussion de l'intéressant et consciencieux rapport de notre collègue M. D.-E. Smith. Mais il est de mon devoir de vous dire les motifs pour lesquels il m'est impossible de prendre part à cette discussion, tout en reconnaissant qu'elle pourra produire quelques résultats utiles.

La question me semble mal posée, ou pour mieux dire posée d'une façon incomplète, ce qui, par cela même, la rend insoluble. Le rôle de l'intuition et de l'expérience dans l'éducation en général, dans l'enseignement mathématique notamment, est l'un des problèmes capitaux de la pédagogie. Mais restreindre l'examen de ce problème à la considération des écoles moyennes (ou secondaires) c'est le rendre insoluble, et méconnaître les conditions psychologiques et physiologiques du développement cérébral.

Suivant, en effet, qu'un enfant qui débute dans l'enseignement secondaire aura subi une préparation antérieure de telle ou telle nature, les procédés pédagogiques que vous pourrez lui appliquer utilement différeront du tout au tout. Si chez cet enfant, depuis le début, on a fait appel à l'intuition et à l'expérience, la continuation sera toute naturelle, et il n'y aura pour ainsi dire qu'à mettre l'élève en garde contre le danger des surprises, si le raisonnement logique n'exerce pas un contrôle assez sévère.

Si votre élève, au contraire, n'a acquis ses notions mathématiques antérieures que dogmatiquement et à force de mémoire — comme cela arrive trop souvent hélas — l'intuition et l'expérience exigeront de sa part de nouveaux efforts pour être mises en œuvre, en outre ; il les considérera avec une sorte de dédain, comme des procédés manuels bons pour un apprentissage, mais indignes de lui. Et je n'hésite même pas à dire, moi partisan irréductible de l'intuition et de l'expérience introduites dans l'enseignement, que, dans certains cas particuliers, cette introduction peut devenir dangereuse.

La nature ne connaît pas nos divisions artificielles, nos classifications souvent arbitraires. Il est impossible de négliger le passé en s'occupant du présent, sous peine de grosses déceptions. Un architecte qui s'occuperait des matériaux à employer pour la construction des étages d'un édifice, sans se soucier de l'examen des fondations, serait d'une imprudence sans pareille.

Je crois donc que la Commission s'est trompée, en ne donnant pas à la question l'ampleur qu'elle comporte. Mais je suis sûr que le débat sera repris ultérieurement dans les conditions qu'entraîne fatalement la nécessité même des choses. Soit au prochain congrès, soit dans des réunions préparatoires qui auront lieu auparavant, la question se posera dans toute son unité, comme elle doit être posée. La discussion actuelle, je le répète, sera un élément utile, mais elle ne saurait avoir un caractère définitif.

Je regrette l'erreur commise par la Commission. Je serais cependant désolé qu'on pût attribuer à ma brève intervention une pensée d'obstruction qui est bien loin de moi.

Allemagne: *Remarques* de M. A. Thaer (Hambourg).—Mehrfach ist dem Schmerze über den Tod des Herrn P. Treutlein wohltuender Ausdruck verliehen worden, ganz besonders wird er aber heute bei der Besprechung von "Anschauung und Experiment im mathematischen Unterricht" vermisst werden, denn er ist seit mehr als einem Menschenalter der Führer dieser Bewegung in Baden gewesen und durch die Unterstützung des Herrn F. Klein für ganz Deutschland geworden. Es ist als ein Glück zu bezeichnen, dass es ihm noch vergönnt war sein Buch "Der geometrische Anschauungsunterricht als Unterstufe eines zweistufigen geometrischen Unterrichts an unseren höheren Schulen" zu vollenden. Das Werk ist zu sehr der Ausdruck der hinreissenden Persönlichkeit des Herrn Treutlein, als dass es einem andern möglich wäre eine einigermassen entsprechende Schilderung zu geben. Er empfiehlt darin die in Österreich-Ungarn seit einem halben Jahrhundert mit Erfolg erprobte Methode der Zweistufigkeit des geometrischen Unterrichts. Aber auch wer in diesem Punkte von ihm abweicht, wird dem Buche ausserordentlich viel entnehmen, vor allem die Richtigkeit der feinen Charakterisierung, die Herr D. E. SMITH von der deutschen Art des Reform-Unterrichts gibt, "To mix the intuitional and the deductive work." Herr F. Klein hat in dem Einführungswort zu dem Buche des Herrn P. Treutlein jedes Missverständnis über den deutschen Anschauungsunterricht durch folgende Worte auszuschliessen versucht: "Der Kunst des Lehrers bleibt es überlassen, von den anschauungsmässigen Elementen schon während des vorbereitenden Unterrichts allmählich zur logischen Erfassung der Zusammenhänge überzuleiten, sodass nicht nur eine vorläufige Kenntnisnahme der grundlegenden räumlichen Vorstellungen, sondern unmittelbar eine tragfähige Grundlage für den höheren geometrischen Unterricht gewonnen ist, der, wie es Max Simon einmal ausdrückt, eine chemische Verbindung von Anschauung und Logik darstellen soll."

Aber nicht bloss in diesem Punkt sondern auch in den übrigen Beziehungen gibt der Bericht des Herrn D. E. Smith bei aller Kürze ein treues und klares Bild über den gegenwärtigen Stand der Reformbewegung in Deutschland, soweit er sich aus den umfassenden und eingehenden I. M. U. K. Berichten ersehen lässt und die Abschnitte "Anschauung und Experiment" betrifft.

Es könnte also genügend erscheinen, wenn ich hier nur noch einmal zur näheren Aufklärung auf die I. M. U. K. Abhandlungen hinweise, die neben den Einzelberichten die hier behandelten Fragen in mehr zusammenfassender Darstellung bringen, das sind die Arbeiten der Herren Hoffmann, Lietzmann, Schimmack, Zühlke.

So nahe es liegt, den Stand der Reform aus den Lehrbüchern zu beurteilen, sie können, wie auch schon die Berichte der Herren Lietzmann, Schimmack und Zühlke hervorgehoben haben, nicht als unbedingt treues Bild gelten, selbst die ausdrücklich als Reformlehrbücher bezeichneten, so treffliche Wegweiser sie sind. Denn bei der Kürze der Zeit basieren sie zu sehr auf der subjektiven Ansicht der Verfasser und der im günstigsten Fall kurzfristigen Erfahrung eines ersten Experiments. Der lebendige Austausch von Erfahrungen beim Besuch von Schulen und auf Versammlungen ist jedenfalls als Ergänzung wünschenswert und tatsächlich von den oben genannten Herren benutzt worden, ebenso wie die eingehenden Antworten in den versandten Fragebogen. Auch ich kann natürlich nur subjektive Eindrücke mitteilen, aber ich möchte betonen, dass sie nur zum kleinsten Teil dem eigenen Unterricht ihren Ursprung verdanken.

Der Ruf nach Anschauung hat zunächst zur Herstellung zahlreicher künstlicher Anschauungsmittel geführt. Mir scheint die herrschende Tendenz jetzt dahin gehen zu wollen, die natürlichen Anschauungsmittel mehr zu benutzen, *den Schüler anzuleiten mathematische Begriffe aus seiner Umgebung zu abstrahieren und wieder auf sie anzuwenden.* Wichtig ist, dass man sich nicht auf den Schulraum beschränkt, deshalb ist der *Feldmessunterricht* so fruchtbar, weil er ins Freie führt. Möglich ist er auch in der Grossstadt. Neben der allerdings nicht ganz zu vernachlässigenden Benutzung genauerer Instrumente—Messstange und Sextant verdienen vielleicht den Vorzug vor Bandmass und Theodolit—ist die einfache Zählung der Schritte, das aus dem Zeichenunterricht wohlbekannte Verfahren der Bestimmung scheinbarer Grössen mit Hilfe des ausgestreckten Armes und eines Massstabes zu pflegen. Benutzt man als Massstab den logarithmischen Rechenstab (sliding rule), so kann man die Winkel besonders bis 5° überraschend genau bestimmen.

Was die *Astronomie* betrifft, so möchte ich feststellen, dass ich unter den etwa 20 neu gebauten Schulen, die ich in Deutschland gesehen, auch höhere Mädchenschulen, keine ohne astronomisches Observatorium gefunden habe, zum Teil auch mit Präzisionsinstrumenten ausgestattet, was mir aber nicht das Wesentliche erscheint. Die Hauptsache bleibt ja, dass der durch das Copernikanische System hochmütig auf alle unmittelbare Anschauung herabblickende Schüler erst einmal wieder einigermassen im Ptolemäischen heimisch wird, die Drehung des Himmelsgewölbes, die Bewegung von Mond und Sonne, einiger Planeten, der Jupitermonde roh messend verfolgt. Ideal ist gewiss ein Unterricht, wie ihn Herr Hoffmann schildert, ich fürchte nur, nicht 5 % der Schulen werden ihn adoptieren, und mancher wird sich ganz abhalten lassen, weil er den Vorwurf der Unwissenschaftlichkeit fürchtet, wenn er es nicht so vollendet wie Herr Hoffmann macht. Das beste astronomische Schulobservatorium ist ein flaches Dach, zugleich eine vorzügliche Basis für die Pothenotsche Aufgabe und, wenn ein sonstiger Turm zur Verfügung steht, für die Hansensche.

Für *Linearzeichnen und Darstellende Geometrie* möchte ich auf Herrn Zühlkes Arbeit verweisen und nur erwähnen, dass man neuerdings versucht, die Zentralperspektive nicht an den Schluss, sondern an den Anfang der Körperdarstellung zu stellen, älteren Anregungen des Württembergers Hertter und des Berliners Martus folgend. Für die Versuche spricht, dass dem Schüler aus dem Zeichenunterricht die Praxis geläufig ist und dass die Bilder anschaulicher und wohlgefälliger sind als die axonometrischen. Dagegen spricht, dass die Theorie schwieriger ist. Für die Darstellung der Kugel, besonders zur korrekten zeichnerischen Lösung astronomischer Aufgaben, gewinnt die stereographische Projektion etwas mehr Freunde. Da vielfach Mathematiker sich jetzt in Deutschland auch die Lehrbefähigung in Geographie erwerben, wird Kartenprojektion und Kartenlesen auch im mathematischen Unterricht gepflegt.

Das *Zeichnen von Kurven* und graphischen Lösungen wird nach der Ansicht vieler bereits übertrieben. Vor allem sind statistische Kurven vielfach eher geeignet, den Begriff der Funktion zu verschleiern, als ihn zu enthüllen. In der Geographie sind sie gewiss nützlich, aber in der Mathematik eigentlich erst dann, wenn sie zur Feststellung einer funktionalen Abhängigkeit führen. Bei Kursschwankungen eines Industriepapiers dürfte das schwer sein, aber z. B. aus der Gold- und Silberproduktion auf den Preis des Silbers, soweit er nicht durch Börsenspekulationen alteriert ist, zu schliessen, ist für einen Schüler der Mittelklassen nicht zu schwer.

Es ist richtig, was Herr D. E. Smith sagt, dass der *logarithmische Rechenstab* in Deutschland bis vor Kurzem noch wenig verbreitet war, Schuld war erst der hohe Preis. Noch vor 10 Jahren kostete ein guter Stab $12\frac{1}{2}$ M. Dann kamen billige und minderwertige Fabrikate, die erst recht den Rechenstab diskreditierten. Seit der Preis für gute Stäbe mit trigonometrischer Einrichtung auf 5 M. herabgegangen ist, nimmt ihr Gebrauch ausserordentlich zu. In Bayern ist er durch Herrn v. Dyck für alle Realanstalten obligatorisch gemacht. In einzelnen Schulen hat er aus den Oberklassen alle logarithmischen, trigonometrischen, Quadrat- und sonstigen Tafeln verdrängt. Für physikalische und Versicherungs-Aufgaben sind allerdings noch Tafeln nötig. Eine grosse Rolle spielt der Rechenstab auch gerade bei der graphischen Lösung numerischer Gleichungen, algebraischer sowohl wie transzendenter.

Wenn natürlich auch der Schüler die Kurven selbst zeichnen soll, möchte ich doch erwähnen, dass ein von Herrn J. Schröder in Hamburg konstruierter Apparat gute Dienste tun kann. Die Kurve wird aus Draht in einer ebenfalls aus Drähten gebildeten Ebene hergestellt. Macht man die Ebene um die Gerade $x-y=0$ drehbar, so verwandelt sich jede Kurve durch Drehung um 180° im Raum in die inverse, also e^x in log nat x, sin x in arc sin x, tg x in arc tgx, u. s. w.

Quadratisches Papier wird sehr stark benutzt. Die Flächenbestimmung durch Abzählung findet z. B. Verwendung bei der Integralrechnung in der Auswertung bestimmter Integrale. Hier ist es nicht nur zur Prüfung eines erhaltenen Resultats nützlich, sondern es zeigt dem Schüler auch einen Ausweg, wo die ihm bekannten Formeln versagen.

Autriche: *Remarques de* M. E. Dintzl (Vienne).—Wenn ich zunächst dem zur Diskussion gestellten Programm die Frage herausgreife, wie der aus Unterricht in der *Darstellenden Geometrie* an den höheren Lehranstalten betrieben wird, so geschieht es deshalb, weil diese Disziplin in *Oesterreich* insbesondere an den Realschulen schon seit langem mit grosser Sorgfalt gepflegt wird und auch in der gegenwärtigen Form ganz charakteristische Merkmale aufweist. Ich könnte mich zwar damit begnügen, auf die ausführlichen Berichte von Prof. Müller und Adler in den Veröffentlichungen der österreichischen Subkommission hinzuweisen. Es gibt aber noch ein plastischeres Mittel, die Unterrichtsmethode in diesem Gegenstande vorzuführen, d. i. an der Hand von Zeichnungen, welche von den Schülern in den Unterrichtsstunden wirklich ausgeführt worden sind. Ich habe darum solche Schülerzeichnungen von vier Wiener Realschulen (2. Staatsrealschule im 7., 13. und 18. Bezirk) und einem Wiener Realgymnasium (2. Bezirk) mitgebracht und lade Sie ein, dieselben zu besichtigen. Was Ihnen bei der Durchsicht dieser Zeichnungen vor allem auffallen wird, das sind die vielen Darstellungen technischer Objekte, z. B. von Gerüstteilen, Säulen, Gesimsen, ja von ganzen Häusern und Kapellen. Dies ist ein besonderes Merkmal des modernen Unterrichtes und hängt aufs engste mit den Reformideen zusammen, welche Profes. Dr Emil Müller von der technischen Hochschule in Wien inbezug auf den Unterricht in der Darstellenden Geometrie an den technischen Hochschulen vertritt. Solche Darstellungen einfacher technischer Objekte erwecken, wie von Seite der Lehrer stets versichert wird, das lebhafteste Interesse der Schüler und zwar aus einem doppelten Grunde. Einmal ist es klar, dass Wirklichkeitsaufgaben, hier also einfache Beispiele aus der Architektur, den Schüler mehr fesseln, als theoretische Aufgaben. Dazu kommt aber noch ein psychologisches Moment, d. i. die ständige Kontrolle, welche der Schüler aus sich selbst, aus seiner unmittelbaren Anschauung heraus auszuüben vermag. Das Bild eines technischen Objektes ist, wie einmal gesagt wurde, einem Porträt vergleichbar, an dem man bei jedem Striche, den man zeichnet, sieht, ob derselbe richtig ist oder nicht. Sie dürfen aber daraus nicht den Schluss ziehen, als ob der Unterricht in einer systemlosen Aneinanderreihung solcher Aufgaben bestünde. Dies widerlegen am besten die mitgebrachten Zeichnungen.

Ich wende mich nun den übrigen Fragen zu. Hiebei kann ich auf Details um so leichter verzichten, als ja Prof. D. E. Smith in seinem ausgezeichneten Berichte dieselben für unser Land in klarer Weise beantwortet. Hier nur einige Bemerkungen allgemeiner Natur. Es ist wahr, dass an den österreichischen Mittelschulen Anschauung und Experiment eine grosse Rolle spielen und zwar nicht bloss auf der dreijährigen Unterstufe, welche wir für eine ungemein wertvolle Institution halten, sondern auch in den mittleren und oberen Klassen, wo sie in der Arithmetik in einer vorwiegend graphischen Behandlung des Zahl- und Funktionsbegriffes, in der Geometrie in der Pflege des zeichnerischen Momentes zum Ausdrucke kommen. Die Bewegung schreitet in dieser Richtung gegenwärtig eher vor als zurück. Es ist augenblicklich schwer möglich, irgend eine Kritik auszuüben und auf die beiden Hauptfragen, welche Prof. D. E. Smith in seinem Referate aufgeworfen hat, präzise Antworten zu geben, da ja die Reformbewegung in Oesterreich noch nicht so langen Datums ist. Aber eines kann heute schon gesagt werden. Wir stehen vor einer Schwierigkeit, um nicht zu sagen vor einer Gefahr, nämlich davor, dass durch das beständige Hereinziehen immer neuer Gegenstände der angewandten Mathematik—wie sie durch das praktische Leben gefordert werden—der systematische Unterricht mehr, als gut ist, eingeengt und bedrückt wird. Wie soll man nun dieser Gefahr wirksam entgegnen? In dieser Hinsicht hat vor kurzem Professor Sobotka von der böhmischen Universität in Prag Ideen entwickelt, welche sehr beachtenswert sind und für die künftige Entwicklung in Oesterreich von Bedeutung sein können. Sobotka spricht zunächst von der Bedeutung des Experimentes für den eigentlichen mathematischen Unterricht in Arithmetik und Geometrie und sagt, dass "das Experiment (Demonstration von Modellen, graphische Darstellungen, etc.) von Lehrer und Schüler zu pflegen sei, aber nur soweit es das selbständige Erfassen und die Ausbildung des Raumanschauungsvermögens unterstützt, zum Nachdenken zwingt oder dem Gegenstande der Betrachtung neue Seiten abzugewinnen geeignet ist, somit als *Behelf* und nicht als *ausschliessliches Mittel zum Zweck*." "Was die *Anwendung auf technische Probleme* oder solche, die dem praktischen Leben überhaupt entnommen sind, anlangt, so sei vorderhand—ähnlich wie es in der Physik teilweise der Fall ist, ein besonderes Praktikum der angewandten Mathematik in jeder Mittelschule einzuführen. Hier käme das Experiment in ausgedehntem Masse und in systematischer Anwendung zur Geltung. In weiterer Folge sei jedoch anzustreben, dass angewandte Mathematik mit praktischen Uebungen als Lehrgegenstand, wenn nicht allgemein, so doch wenigstens an den Realschulen, *obligatorisch* eingeführt werde."

Das sind in aller Kürze die Ideen von Sobotka, soweit sie mit den zur Diskussion gestellten Fragen in Zusammenhang stehen. Man mag gegen diese Vorschläge vielleicht vom Standpunkt der Fusion Bedenken erheben. Dieselben scheinen aber nicht schwerwiegend zu sein—um so weniger als wir ja in dem Unterrichte in der Darstellenden Geometrie und in seinem Verhältnis zum systematischen Unterricht in der Stereometrie ein gutes Vorbild besitzen. Sicher haben diese Ideen das eine für sich, dass auf diese Weise der Intuition und dem Experimente eine genau fixierte Stellung zugewiesen ist, eine Stellung, welche auch der so wichtigen Erziehung der jungen Leute im logischen, deduktiven Denken genügend Raum zur Entwicklung lässt.

France. M. Ch. Bioche (Paris) insiste sur l'importance qu'on donne en *France* à la Géométrie descriptive et au Dessin géométrique.

L'exécution d'un dessin géométrique, ou d'une épure, oblige l'élève

1° à constater la nécessité d'un raisonnement logique pour établir les faits géométriques, par exemple l'existence d'un cercle passant par 3 points;

2° à se préoccuper de déduire de la solution théorique d'un problème une construction qui soit à la fois simple, précise et susceptible d'être effectuée dans les limites de la feuille de papier.

Les professeurs de mathématiques ont fréquemment demandé que l'enseignement du dessin géométrique soit confié aux professeurs de mathématiques pour que cet enseignement facilite et complète l'enseignement théorique.

Iles britanniques. Mr G. St L. Carson (Tonbridge) said that, in *England* at any rate, there appeared to be confusion between the terms experiment and intuition. Results such as the angle properties of parallel lines, which are at once accepted by a child of 12 if expressed in non-technical terms, are made the subject of numerical measurement with a protractor and yet referred to as intuitive, instead of experimental. He was strongly of opinion that only intuitions in the proper sense of the term should be taken as the basis of a first course of formal Geometry, and that all possible intuitions should be taken as the basis. In this way only could the appearance of proving the obvious, so dangerous and destructive to the logical sense of a young child, be avoided.

He demurred most strongly to the suggestion that logic might be less rigorous at an early age. More postulates than the minimum necessary might and should be accepted, but the processes of deduction based on these postulates should be entirely rigorous from the outset. Any want of rigour in deduction is detected sooner or later, and with bad results; whereas analysis of the interconnection of the set of postulates which have been assumed is, at a later stage, a natural and interesting course for the pupil.

METHODS OF INTUITION AND EXPERIMENT IN SECONDARY SCHOOLS

By C. Godfrey.

One of the questions proposed for discussion at the International Congress of Mathematicians (Cambridge, 1912) is indicated at the head of the present report. In order to ascertain the prevalence of such methods in England, a questionnaire was sent out to 551 schools; namely to 112 schools represented on the Headmasters' Conference, and to 439 schools represented on the Incorporated Association of Headmasters. Replies were received from 74 and 295 schools in the two groups respectively, which give percentages of 66 and 67. The result of the enquiry should therefore be very fairly representative of the work done in these schools in England and Wales, but throws no light on the conditions in Scotland and Ireland as few replies were received from these parts.

The constitution of the Headmasters' Conference provides that "as a general rule there should be 100 boys at least in any school represented at the Conference, and about 10 resident undergraduates at the Universities who have gone direct from the school." It is therefore fairly accurate to describe Conference Schools as large schools that prepare boys for the Universities—mainly, of course, the ancient Universities of Oxford and Cambridge.

The school life is usually 13 to 18 or 19. These schools are described loosely as "public schools."

The schools whose Headmasters have joined the Incorporated Association of Headmasters are secondary schools, varying greatly in size and character, and including both grammar schools of ancient foundation and new municipal secondary schools. The leaving age is as a rule lower than in the public schools; perhaps about 16 is practically the upper limit in the majority of cases. The age of entry varies greatly; the lower limit is eight or nine, but many boys enter from elementary schools up to the age of 14 or so. Generally speaking these schools do not prepare boys for Oxford and Cambridge; on the other hand they supply a large proportion of students to the numerous modern Universities.

The following questionnaire was sent out:—

A. Name of School.

B. Does the Headmaster object to the name of your school appearing in the report?

C. *To what extent and in what classes (upper, middle or lower school, or Army class) do you introduce the following?*

1. Practical fieldwork.

   Are measurements made and numerical results deduced? Do you use theodolite (or equivalent instrument)? plane-table? other instruments?

2. Practical astronomy.
3. Descriptive geometry (plan and elevation).
4. Perspective drawing.
5. Is the teaching of descriptive geometry and perspective associated with that of solid geometry?
6. Is descriptive geometry and perspective taught by a mathematical or an art master?
7. Accurate drawing work in connection with projective geometry: construction of conics by anharmonic ratio properties, etc.
8. Graphical study of statistics, etc.
9. Graphical representation of functions.
10. Graphical treatment of vectors.
11. Graphical statics.
12. Formal methods of contracted multiplication and division.
13. Use of slide rule (*a*) front scales, (*b*) trigonometrical scales.
14. Tables of squares.
15. Tables of square roots.
16. Tables of cube roots.
17. Tables of logarithms.
18. Tables of trigonometrical functions.
19. To how many places are the tables you use calculated?
20. Graphical work in solution of transcendental equations

$$\left(e.g.\ \log x = 1 + \frac{1}{x}\right),$$

   in conjunction with Newton's rule, processes of trial and error, etc.

21. Estimation of area by means of (*a*) squared paper, (*b*) planimeter.

....

Signature of informant.

In public schools the ages of lower classes may be taken as 12 to 15; of middle classes, 14 to 17; of upper classes, 16 to 18 or 19. But of course it may happen that a clever boy of 14 is in the highest mathematical class, and corresponding anomalies of age may be found in the other classes.

Public schools are usually reclassified for Mathematics: thus, "upper classes" means "upper mathematical classes" and so forth. The boys in the highest mathematical classes are usually working for mathematical scholarships at the Universities. In the larger schools there are also, as a rule, a number of boys preparing for the competitive examination that admits to the Army colleges, and these boys are often grouped as a separate class. If numbers permit, there may be a separate class for future engineers or surveyors. These classes—Army, engineers, surveyors, are grouped together in the present report as "specialists."

It is difficult to estimate the ages corresponding to "lower," "middle," "upper" classes in other secondary schools, on account of the great variety in the character of these schools. Perhaps in the majority of the large secondary day schools the ages are—lower, 8 to 13; middle, 12 to 15; upper, 14 to 16. Army classes do not usually exist in these schools. Reclassification for Mathematics is by no means universal. The number of boys working for scholarships is in most cases too small to affect the general style of the work.

In all schools the teaching of specialists (as defined above) tends to be of a more intuitional and experimental character than that of other boys, a result ultimately connected with the requirements of their chosen professions, and in the case of Army boys caused more directly by the style of the papers set in the entrance examination. The Army requirements during the last ten years have no doubt had a powerful reaction on the character of the teaching throughout English public schools.

The following is an analysis of the answers received. Percentages for each class of school have been calculated on the number of replies received from schools of that class.

1. *Practical Fieldwork, in which Measurements are made and Numerical Results deduced:*

Public Schools.—Not done, 58 per cent. In the remaining 42 per cent. such work is done, but as a rule only by special classes of students (officers' training corps, surveyors, engineers, etc.). In only two schools is this stated to be part of Army class work; in two others it is classed with geography. On the whole it is more frequently done in upper classes than in middle and lower. As to instruments, theodolite is specified in 14 per cent., simplified angle-instruments in 5 per cent., sextant in 7 per cent., prismatic compass in 1 per cent., plane table in 9 per cent., level in 4 per cent.

Other Secondary Schools.—Not done, 44 per cent. In the remaining 56 per cent., upper, middle and lower classes are mentioned with equal frequency. Apparently more generally done than in public schools, and less confined to specialists. Classed with geography in 14 per cent. Theodolite 23 per cent., simplified angle-instruments 9 per cent., sextant 10 per cent., compass 3 per cent., plane table 25 per cent., level 3 per cent.

2. *Practical Astronomy.*—Very rarely attempted, viz., in 15 per cent. of public schools, 8 per cent. of other secondary schools. Perhaps the climate is not very suitable. Four schools in all are provided with observatories. Where the work is done, it is concerned with determination of latitude, direction, and time.

3. *Descriptive Geometry, after Monge (Plan and Elevation, etc.)*.—Schools are in a transitional state in the treatment of this subject. Formerly it was only taught to boys who needed it for special examinations, and then generally by the art master; this commonly meant working by rule with great attention to artistic ink-work. Two or three causes are at work to change this position: (1) The Army examiners now treat the subject as part of Mathematics, and do not require inking-in; mathematical masters are, therefore, not discouraged from teaching the subject by consciousness of their technical weakness on the artistic side; (2) Descriptive geometry is now required for an Honours degree in Mathematics at Cambridge; this circumstance will result in an increased output of mathematical masters interested in descriptive geometry; and, further, will have a direct reaction on schoolwork. (3) Boys intending to study engineering at the University are encouraged to master descriptive geometry and perspective at school.

The figures are:—

Public Schools.—Not done, 23 per cent.; done in upper classes, 38 per cent.; middle, 8 per cent.; lower, 3 per cent.; Army class, 38 per cent.; specialists other than Army class, 23 per cent. In some of the above schools the subject is being taught in more than one place, *e.g.*, both in upper classes and in the Army class. Where taught, it is associated with solid geometry in 66 per cent., taught by a mathematician in 42 per cent., by an art master in 58 per cent.

Other Secondary Schools.—Not done, 40 per cent.; done in upper classes, 14 per cent.; middle, 21 per cent.; lower, 18 per cent.; specialists, 3 per cent., the rest unclassified. A feature in this group of schools is the teaching of plan and elevation in connection with manual work, generally carpentry (18 per cent.); this may explain the fact that it is taught lower in the school than is the case in the public schools. Where taught, it is associated with solid geometry in 54 per cent., taught by a mathematician in 35 per cent., by an art master in 44 per cent., and by a manual instructor in 20 per cent.

4. *Perspective.*—This subject has not been annexed by the mathematician to the same extent as descriptive geometry; it is still mainly an art subject. The correlation with Mathematics probably extended no further, in most cases, than an explanation to the more advanced pupils of the mathematical principles of perspective.

Public Schools.—Not done, 38 per cent.; done in upper classes, 22 per cent.; middle, 14 per cent.; lower, 4 per cent.; Army class, 11 per cent.; other specialists, 22 per cent.; largely a specialist subject, it appears. Where taught, associated with solid geometry in 50 per cent., taught by a mathematician in 23 per cent., by an art master in 77 per cent.

Other Secondary Schools.—Not done, 64 per cent.; done in upper classes, 9 per cent.; middle classes, 14 per cent.; lower classes, 7 per cent., the rest unclassified. Where taught, associated with solid geometry in 29 per cent., taught by a mathematician in 17 per cent., by an art master in 83 per cent. Mainly an art subject in these schools.

5. *Accurate Drawing Work in connection with Projective Geometry; Construction of Conics by Anharmonic Ratio Properties, etc.*—This is done to some extent in 28 per cent. of public schools, and confined to the few best boys who reach this comparatively advanced work. It hardly enters into the scheme of other secondary schools; 11 per cent. gave an affirmative answer of some sort, but it seemed possible that the term "projective geometry"—the geometry of Chasles—was not always understood in this sense.

6. *Graphical Study of Statistics:*

Public Schools.—91 per cent. do this work, beginning it generally in the lower or middle classes; it is confined to upper classes in only 12 per cent., and to Army class only in 4 per cent.

Other Secondary Schools.—An affirmative answer came from 85 per cent., here again mainly in lower and middle classes, 11 per cent. only confining it to upper classes. In 6 per cent. it is said to be correlated with geography teaching.

This, then, is a very usual feature in English secondary schools. As a rule it does not amount to much more than plotting a few tables of statistics and deducing information by interpolation. Probably even higher percentages really do this; in some cases there was reason to believe that informants were misled by the rather ambitious phrase "study of statistics," taking it to imply excursions into economic science.

Mr A. N. Whitehead has advocated a great extension of this type of work. "There are to our hands statistics of trade, both external and internal, statistics of railway traffic, statistics of crime, of income tax returns, of national expenditure, of weather, of pauperism, and of times of sunrise and sunset throughout the year. The reduction of these to graphs, the careful study of the peculiarity of these graphs, the search for correlations among them, and the study of the public events which corresponded in time to peculiarities of graphical form, would teach more mathematics and more knowledge of modern social forces than all our present methods put together*." In his paper on "Higher Mathematics for the Classical Sixth Form†," Mr W. Newbold has described his experiments on these lines with the senior non-mathematical boys at Tonbridge School; but this is pioneer work, and not attempted generally.

7. *Graphical Representation of Functions:*

Public Schools.—There were only two answers in the negative, and I was doubtful as to the informant's meaning in each case. Practically this work is done in all public schools, and indeed is required for all the ordinary external examinations. In 17 per cent. it is begun in the upper classes, 55 per cent. in middle, 25 per cent. in lower, the rest unclassified.

* *Board of Education. Special Reports on the Teaching of Mathematics in the United Kingdom*, No. 1. (Wyman and Sons, 1911).

† *Journ. of the Assoc. of Teachers of Maths. for S.E. England*, Dec. 1911 (copies from the Hon. Sec., the School, Tonbridge).

Other Secondary Schools.—Here again this work is very general, being done in 97 per cent.; 21 per cent. beginning it in upper classes, 54 per cent. in middle, 17 per cent. in lower, the rest unclassified.

Probably in most cases this work resolves itself into plotting the functions, and using the graphs as an aid to the solution of equations. The somewhat subtle ideas suggested by the word "functionality" are perhaps not so generally appreciated. But at any rate the graph has a firm place in lower and middle teaching, and gives a base from which further advances in teaching may be made.

8. *Graphical Treatment of Vectors* is taught in 80 per cent. of public schools and 41 per cent. of other secondary schools, mainly in the upper classes. There was nothing in the answers to indicate how far graphical treatment of vectors is associated with the study of complex numbers, but it is known to be thus associated in some schools. As a rule, it is probably taken in connection with mechanics (velocity, acceleration, force), which is usually in England a part of the mathematical course. The question might with advantage have been expressed more clearly.

9. *Graphical Statics:*

Public Schools.—95 per cent. do this subject, viz., in upper classes, 69 per cent.; middle, 12 per cent.; lower, 1 per cent.; specialists only, 13 per cent.

Other Secondary Schools.—63 per cent., including 38 per cent. which take the subject in upper classes, 14 per cent. in middle; specialists only, 9 per cent.; the rest unclassified.

This teaching is therefore practically universal in public schools, and common in other secondary schools.

10. *Formal Methods of Contracted Multiplication and Division:*

Public Schools.—In 84 per cent., viz., in upper classes, 19 per cent., middle 38 per cent., lower 27 per cent.

Other Secondary Schools.—In 93 per cent., viz., in upper classes, 24 per cent.; middle, 49 per cent.; lower, 16 per cent.; rest unclassified.

There was some reason to think that the force of the word "formal" was not appreciated in all cases. There are indications that the use of formal rules is diminishing. On the one hand, the omission of useless figures and the presentation of results to a given number of significant figures must become increasingly general as practical and laboratory work in the lower classes gains ground; on the other hand, informants say that logarithms prove more useful than contracted methods; that contracted rules are learnt, but seldom used afterwards; that they have given up the use of such rules after trying them for many years; and so forth.

11. *Use of Slide Rule—(a) Front Scales; (b) Trigonometrical Scales:*

The replies did not discriminate between (*a*) and (*b*).

Public Schools.—Slide rule used in 70 per cent., but mainly by specialists (55 per cent.). In the ordinary course for upper classes, 10 per cent.; for middle, 3 per cent.; for lower, 1 per cent.

Other Secondary Schools.—Used in 33 per cent., including upper classes 15 per cent., specialists 11 per cent.

On the whole, the slide rule is used by mathematical teachers only under compulsion.

12.—(*a*) *Tables of Squares:*

Public Schools.—Used in 61 per cent., viz., upper classes, 23 per cent.; middle, 15 per cent.; lower, 3 per cent.; specialists only, 11 per cent.; not classified, 9 per cent.

Other Secondary Schools.—Used in 33 per cent., viz., in upper classes, 18 per cent.; middle, 8 per cent.; lower, 1 per cent.; not classified, 6 per cent.

12.—(*b*) *Tables of Square Roots.*—Practically the same results as in 12 (*a*).

12.—(*c*) *Tables of Cube Roots:*

Public Schools.—Used in 45 per cent., viz., in upper classes, 15 per cent.; middle, 9 per cent.; lower, 3 per cent.; specialists only, 11 per cent.; not classified, 7 per cent.

Other Secondary Schools.—Used in 30 per cent., viz., in upper classes, 17 per cent.; middle, 6 per cent.; lower, 1 per cent.; not classified, 6 per cent.

12.—(*d*) *Tables of Logarithms:*

Public Schools.—Used in 100 per cent., viz., in upper classes, 49 per cent.; middle, 37 per cent.; lower, 4 per cent.; specialists only, 3 per cent.; not classified, 8 per cent.

Other Secondary Schools.—Used in 97 per cent., viz., in upper classes, 58 per cent.; middle, 30 per cent.; lower, 2 per cent.; not classified, 7 per cent.

12.—(*e*) *Tables of Trigonometrical Functions:*

Public Schools.—Used in 100 per cent., viz., in upper classes, 60 per cent.; middle classes, 25 per cent.; lower, 4 per cent.; specialists only, 3 per cent.; not classified, 8 per cent.

Other Secondary Schools.—Used in 88 per cent., viz., in upper classes, 63 per cent.; middle classes, 17 per cent.; lower classes, 1 per cent.; not classified, 7 per cent.

12.—(*f*) *Number of Figures to which Tables are calculated:*

Various combinations of 4, 5, 6, 7 occurred, of which the following were the most general.

Public Schools.—4-figures, 65 per cent., 4-figures supplemented by 7-figures in upper classes 16 per cent.

Other Secondary Schools.—Reckoning percentages on the total number of schools that use tables, the result is: 4-figures 69 per cent., 4-figures supplemented by 7-figures in upper classes 14 per cent.

12.—(*g*) *Résumé of Returns dealing with Tables:*

Different tables are used in the following descending order of frequency—logarithms, trigonometrical functions, squares and square roots, cube roots. Many schools are of opinion that the use of logarithms obviates the necessity of tables of squares and square roots, but there is no doubt that these latter tables have their place in schoolwork. The extent to which the everyday use of tables has established itself is shown by the prevalent use of 4-figure tables; 7-figure tables are too cumbrous for most school purposes. 4-figure tables used to be used mainly by science masters, but have now been largely adopted by mathematicians as well.

13. *Graphical Work in Solution of Transcendental Equations* (e.g., $\log x = 1 + \frac{1}{x}$), *in conjunction with Newton's Rule, Processes of Trial and Error.*

Public Schools.—Done in 47 per cent. of schools, viz., in upper classes, 29 per cent.; middle, 1 per cent.; specialists only, 17 per cent.

Other Secondary Schools.—In 16 per cent., practically always in upper classes.

This work, then, is not very usual, and seldom goes beyond a few illustrative examples.

14.—(*a*) *Estimation of Area by means of Squared Paper.*

Public Schools.—Done in 93 per cent., viz., in upper classes, 3 per cent.; middle, 35 per cent.; lower, 46 per cent.; specialists only, 9 per cent.

Other Secondary Schools.—Done in 93 per cent., viz., in upper classes, 3 per cent.; middle, 23 per cent.; lower, 55 per cent.; not classified, 12 per cent.

This work appears to be a standard part of the curriculum for the middle and lower classes. A rather curious feature is that it seems to be done later in public schools than in other secondary schools. There is some evidence that this is connected with the greater prevalence of geography teaching in other secondary schools, determination of areas by squared paper often being associated with geography.

14.—(*b*) *Estimation of Area by means of the Planimeter.*—Taught in only 15 per cent. of public schools and 6 per cent. of other secondary schools, and in these cases only to a few boys in the highest class.

15. *General Remarks.*—In framing the questionnaire perhaps I did not keep very closely to my brief, as some of the questions have no particular bearing on methods of intuition and experiment. The questionnaire followed the lines of a circular issued by the Central Committee of the International Commission on Mathematical Education, and in spite of the criticism just made it may fairly be said that all the topics referred to have a somewhat modern flavour, while it is equally true that intuition and experiment is the keynote of modern mathematical education. In a broad sense, then, the questionnaire is appropriate to its purpose.

The object of this report is to state facts, not views. If I speak of the spread of intuitional and experimental methods as an advance, I do so merely to save words, and with all possible respect to the views of those who would prefer another description. With this proviso, it is clear that there has been a marked advance in England during the present century.

The use of graphical methods in elementary algebra teaching is universal and entirely a 20th-century development. Other aspects of the same movement are the adoption of descriptive geometry by the mathematicians, the use of handy 4-figure tables, and of graphical methods in statics, and, though, in these cases, the victory is less complete than that of the "graph," it is remarkable and equally modern. The absence of practical astronomy from the ordinary curriculum is perhaps not surprising, in view of the difficulty of organising a systematic course of observations; but a good deal of interesting work is done in a few schools with sun and shadow, and in determination of latitude; here there is room for further advance, where time permits. Practical fieldwork has not the place that it deserves in English schools.

A close connection will be noticed between the activities of the schools and the demands of examining bodies; these bodies have to bear the chief responsibility for what goes on in the schools. To a great extent, they have considered favourably the suggestions made by representative bodies of teaching; and to this is due the wide adoption of some of the subjects referred to in this report. In other subjects, such as practical fieldwork, which are somewhat unsuited for examination tests, it is easier to measure the advance that teachers have made spontaneously.

LISTE DES PUBLICATIONS DU COMITÉ CENTRAL

ET DES

SOUS-COMMISSIONS NATIONALES

PUBLICATIONS DU COMITÉ CENTRAL

PUBLICATIONS DU COMITÉ CENTRAL, 1re série : 1908 à 1911, rédigées par H. FEHR, Secrétaire-général de la Commission.—1 vol. de 200 p. ; 5 fr., Georg & Cie, Genève.

Elles comprennent :

1. *Rapport préliminaire* sur l'organisation de la Commission et le plan général de ses travaux (*Ens. math.*, n° de nov. 1908). [16 p.]

2. *Circulaire n° 1* : Constitution de la Commission.—Sous-commissions nationales (*E. M.*, n° de mai 1909). [12 p.]

3. *Circulaire n° 2* : Nouveaux membres.—Sous-commissions nationales.—Etat des travaux au commencement de 1910 (*E. M.*, mars 1910). [16 p.]

4. *Circulaire n° 3* : Réunion de Bruxelles. Compte rendu des séances de la Commission et des conférences sur l'enseignement scientifique et sur l'enseignement technique moyens faites à Bruxelles du 10 au 16 août. Conférence de M. C. BOURLET sur la pénétration réciproque des mathématiques pures et des mathématiques appliquées dans l'enseignement secondaire (*E. M.*, n° de sept. 1910). [63 p.]

5. *Circulaire n° 4*: Etat des travaux au 1er mars 1911 (*E. M.*, mars 1911). [16 p.]

6. *Circulaire n° 5*: Compte rendu du Congrès tenu à Milan du 18–21 septembre 1911 (*E.M.*, n° de nov. 1911). [75 p.]:

Rapport de la Sous-commission A : 1. La question de la rigueur dans l'enseignement moyen ; 2. La fusion des différentes branches mathématiques. Rapporteurs : MM. CASTELNUOVO et BIOCHE. — Annexe : Note de M. YOUNG (Chicago).

Rapport de la Sous-commission B : L'enseignement mathématique destiné aux étudiants en sciences physiques, en sciences naturelles, etc. Rapporteur : M. TIMERDING.

Sull' insegnamento matematico nelle scuole per gli ingegneri. Par le Prof. COLOMBO.

Mathématiques et Théorie de la connaissance. Par le Prof. F. ENRIQUES.

SOUS-COMMISSIONS NATIONALES

ALLEMAGNE

A. Berichte und Mitteilungen

veranlasst durch die Internationale Mathematische Unterrichts-Kommission. Herausgegeben von W. LIETZMANN. In zwanglosen Heften. gr. 8. Steif geb. (B. G. Teubner, Leipzig.)

1. FEHR, H., Vorbericht über Organisation und Arbeitsplan der Kommission. Deutsche Uebersetzung von W. LIETZMANN. [S. 1–10.] 1909. M. —.30.

2. NOODT, G., Ueber die Stellung der Mathematik im Lehrplan der höheren Mädchenschule vor und nach der Neuordnung des höheren Mädchenschulwesens in Preussen. [S. 11–32.] 1909. M. —.80.

3. KLEIN, F., und H. FEHR, Erstes Rundschreiben des Hauptausschusses. Deutsch bearbeitet von W. LIETZMANN. [S. 33-38.] 1909. M. —.20.

4. KLEIN, F., und H. FEHR, Zweites Rundschreiben des Hauptausschusses. Deutsch bearbeitet von W. LIETZMANN, sowie P. ZÜHLKE, Mathematiker und Zeichenlehrer im Linearzeichenunterricht der preussischen Realanstalten. [S. 39-54.] 1910. M. —.50.

5. LIETZMANN, W., Die Versammlung in Brüssel. Nach dem von H. FEHR verfassten dritten Rundschreiben des Hauptausschusses. [S. 55-74.] 1911. M. —.60.

6. FEHR, H., Viertes Rundschreiben des Hauptausschusses. Deutsch bearbeitet von W. LIETZMANN. [S. 75-88.] 1911. M. —.50.

7. LIETZMANN, W., Der Kongress in Mailand vom 18. bis 20. Sept. 1911, sowie SCHIMMACK, R., Ueber die Verschmelzung verschiedener Zweige des mathematischen Unterrichts. [S. 89-126.] 1912. M. 1.60.

B. **Abhandlungen**

über den mathematischen Unterricht in Deutschland, veranlasst durch die Internationale Mathematische Unterrichts-Kommission. Herausgegeben von F. KLEIN.—5 Bände, in einzeln käuflichen Heften. gr. 8. Steif geb. (B. G. Teubner, Leipzig.)

ERSTER BAND—**Die höheren Schulen in Norddeutschland.** Mit einem Einführungswort von F. KLEIN.

1. Heft: LIETZMANN, W., Stoff und Methode im mathematischen Unterricht der norddeutschen höheren Schulen auf Grund der vorhandenen Lehrbücher. [XII u. 102 S.] 1909. M. 2.—.

2. Heft: LIETZMANN, W., Die Organisation des mathematischen Unterrichts an den höheren Knabenschulen in Preussen. Mit 18 Figuren. [VIII u. 204 S.] 1910. M. 5.—.

3. Heft: LOREY, W., Staatsprüfung und praktische Ausbildung der Mathematiker an den höheren Schulen in Preussen und einigen norddeutschen Staaten. [V u. 118 S.] 1911. M. 3.20.

4. Heft: THAER, A., N. GEUTHER und A. BÖTTGER, Der mathematische Unterricht an den Gymnasien und Realanstalten der Hansestädte, Mecklenburgs und Oldenburgs. [VI u. 93 S.] 1911. M. 2.—.

5. Heft: SCHRÖDER, J., Die neuzeitliche Entwicklung des mathematischen Unterrichts an den höheren Mädchenschulen, insbes. in Norddeutschland. (Unter der Presse.)

ZWEITER BAND—**Die höheren Schulen in Süd- und Mitteldeutschland.** Mit einem Einführungswort von P. TREUTLEIN.

1. Heft: WIELEITNER, H., Der mathematische Unterricht an den höheren Lehranstalten sowie die Ausbildung und Fortbildung der Lehrkräfte im Königreich Bayern. Mit einem Einführungswort von P. TREUTLEIN. [XII u. 85 S.] 1910. M. 2.40.

2. Heft: WITTING, A., Der mathematische Unterricht an den Gymnasien und Realanstalten nach Organisation, Lehrstoff und Lehrverfahren und die Ausbildung der Lehramtskandidaten im Königreich Sachsen. [XII u. 78 S.] 1910. M. 2.20.

3. Heft: GECK, E., Der mathematische Unterricht an den höheren Schulen nach Organisation, Lehrstoff und Lehrverfahren und die Ausbildung der Lehramtskandidaten im Königreich Württemberg. [IV u. 104 S.] 1910. M. 2.60.

4. Heft: CRAMER, H., Der mathematische Unterricht an den höheren Schulen nach Organisation, Lehrstoff und Lehrverfahren und die Ausbildung der Lehramtskandidaten im Grossherzogtum Baden. [IV u. 48 S.] 1910. M. 1.60.

5. Heft: SCHNELL, H., Der mathematische Unterricht an den höheren Schulen nach Organisation, Lehrstoff und Lehrverfahren und die Ausbildung der Lehramtskandidaten im Grossherzogtum Hessen. [VI u. 51 S.] 1910. M. 1.60.

6. Heft: HOSSFELD, Der mathematische Unterricht an den Gymnasien und Realanstalten Thüringens nach Organisation, Lehrstoff und Lehrverfahren. [IV u. 18 S.] 1912. M. —.80.

7. Heft: WIRZ, J., Der mathematische Unterricht an den höheren Knabenschulen sowie die Ausbildung der Lehramtskandidaten in Elsass-Lothringen. [VI u. 58 S.] 1911. M. 1.80.

DRITTER BAND—**Einzelfragen des höheren mathematischen Unterrichts.** Mit einem Einführungswort von F. KLEIN.

1. Heft: SCHIMMACK, R., Die Entwicklung der mathematischen Unterrichtsreform in Deutschland. Mit einem Einführungswort von F. KLEIN. [VI u. 146 S.] 1911. M. 3.60.

2. Heft: TIMERDING, H. E., Die Mathematik in den physikalischen Lehrbüchern. Mit 22 Figuren. [VI u. 112 S.] 1910. M. 2.80.

3. Heft: ZÜHLKE, P., Der Unterricht im Linearzeichnen und in der darstellenden Geometrie an den deutschen Realanstalten. Mit 14 Figuren. [IV u. 92 S.] 1911. M. 2.60.

4. Heft: HOFFMANN, B., Mathematische Himmelskunde und niedere Geodäsie an den höheren Schulen. Mit 9 Figuren. [VI u. 68 S.] 1912. M. 2.—.

5. Heft: TIMERDING, H. E., Die kaufmännischen Aufgaben im mathematischen Unterricht der höheren Schulen. Mit 5 Figuren im Text. [IV u. 45 S.] 1911. M. 1.60.

6. Heft: GEBHARDT, M., Die Geschichte der Mathematik im mathematischen Unterrichte der höheren Schulen Deutschlands, dargestellt vor allem auf Grund alter und neuer Lehrbücher und der Programmabhandlungen höherer Schulen. [VII u. 157 S.] 1912. M. 4.80.

7. Heft: WERNICKE, A., Mathematik und Philosophische Propädeutik. [VII u. 138 S.] 1912. M. 4.—.

8. Heft: KATZ, D., Psychologie und mathematischer Unterricht. (In Vorbereitung.)

9. Heft: LOREY, W., Das Studium der Mathematik an den deutschen Universitäten. (In Vorbereitung.)

VIERTER BAND—**Die Mathematik an den technischen Schulen.** Mit einem Einführungswort von P. STÄCKEL.

1. Heft: GRÜNBAUM, H., Der mathematische Unterricht an den mittleren technischen Fachschulen der Maschinenindustrie. Mit einem Einführungswort von P. STÄCKEL. [XII u. 99 S.] 1910. M. 2.60.

2. Heft: OTT, Karl, Die angewandte Mathematik an den technischen Mittelschulen der Maschinenindustrie. (Unter der Presse.)

3. Heft: GIRNDT, M., Der mathematische Unterricht an den Baugewerkschulen. (In Vorbereitung.)

4. Heft: SCHILLING, C., und MELDAU, H., Der mathematische Unterricht an den deutschen Navigationsschulen. [VI u. 82 S.] 1912. M. 2.—.

5. Heft: TROST, Die mathematischen Fächer an den gewerblichen Fortbildungsschulen. (In Vorbereitung.)

6. Heft: PENNDORF, B., Rechnen und Mathematik im Unterricht der kaufmännischen Lehranstalten. [VI u. 100 S.] 1912. M. 3.—.

7. Heft: JAHNKE, E., Die Mathematik an Hochschulen für besondere Fachgebiete. [VI u. 55 S.] 1911. M. 1.80.

8. Heft: FURTWÄNGLER, Ph., Die mathematische Ausbildung der Landmesser. (In Vorbereitung.)

9. Heft: STÄCKEL, P., Die mathematische Ausbildung der Architekten, Chemiker und Ingenieure an den deutschen Hochschulen. (In Vorbereitung.)

FÜNFTER BAND—**Der mathematische Elementarunterricht und die Mathematik an den Lehrerbildungsanstalten.** Mit einem Einführungswort von F. KLEIN.

1. Heft: LIETZMANN, W., Stoff und Methode des Rechenunterrichts in Deutschland. Ein Literaturbericht. Mit 20 Figuren im Text. Mit einem Einführungswort von F. KLEIN. [VII u. 125 S.] 1912. M. 3.—.

2. Heft: LIETZMANN, W., Stoff und Methode des Raumlehreunterrichts in Deutschland. Ein Literaturbericht. Mit 38 Figuren im Text. [VIII u. 88 S.] 1912. M. 2.80.

3. Heft: Der mathematische Unterricht an den Volksschulen und Lehrerbildungsanstalten Süddeutschlands mit einem Einführungswort von P. TREUTLEIN und mit den Einzelabhandlungen von H. HENSING über die Verhältnisse in Hessen, von E. GECK über Württemberg, von H. CRAMER über Baden, von KERSCHENSTEINER und BOCK über Bayern. [XIV u. 163 S.] 1912. M. 5.—.

4. Heft: DRESSLER, H., Der mathematische Unterricht an den Volksschulen und Lehrerbildungsanstalten in Sachsen und Thüringen. (In Vorbereitung.)

5. Heft: UMLAUF, K., Der mathematische Unterricht an den Seminaren und Volksschulen der Hansestädte. (In Vorbereitung.)

6. Heft: Lietzmann, W., Die Organisation der Volksschulen, gehobenen Volksschulen, Präparandenanstalten, Seminare usw. in Preussen. (In Vorbereitung.)

AUTRICHE

Berichte über den mathematischen Unterricht in Oesterreich. Veranlasst durch die internationale mathematische Unterrichtskommission. Herausgegeben von E. Czuber, W. Wirtinger, R. Suppantschitsch, E. Dintzl. (Alfred Hölder, Wien.) 1912.

Heft 1.—Begleitwort, von E. Czuber.

Realschulen von F. Bergmann.

Volks- und Bürgerschulen von K. Kraus.—(V u. 81 S.; 1 M. 80.)

Heft 2.—Bildungsanstalten für Lehrer und Lehrerinnen von Th. Konrath.

Höhere Handelsschulen von M. Dolinski.

Höhere Forstlehranstalt Reichstadt von M. Adamicka.—(52 S.; 1 M. 20.)

Heft 3.—Gymnasien von E. Dintzl.—(VIII u. 78 S.; 1 M. 20.)

Heft 4.—Mädchenlyzeen von Th. Konrath.

Die praktische Vorbildung für das höhere Lehramt in Oesterreich von J. Loos.

Gewerbliche Lehranstalten von W. Rulf.—(64 S.; 1 M. 60.)

Heft 5.—Technische Hochschulen von E. Czuber.—(V u. 39 S.; 1 M. 20.)

Heft 6.—Die mathematischen Schulbücher an den Mittelschulen und verwandten Lehranstalten von Ph. Freud.—(53 S.; 1 M. 20.)

Heft 7.—Universitäten von R. v. Sterneck.—(50 S.; 1 M. 20.)

Heft 8.—Bericht über die speziellen Verhältnisse des öffentlichen mathematischen Unterrichtes an den Volks- und Mittelschulen Galiziens von St. Zaremba.—(V u. 25 S.; 1 M. 20.)

Heft 9.—Der Unterricht in der darstellenden Geometrie an den Realschulen und Realgymnasien von A. Adler.

Der Unterricht in der darstellenden Geometrie an den Technischen Hochschulen von E. Müller.—(24 S.; 2 M. 40.)

Heft 10.—Hochschule für Bodenkultur von O. Simony.

Montanistische Hochschulen von E. Kobald.

Militär-Erziehungs- und Bildungsanstalten von A. Mikuta.

Technologisches Gewerbemuseum von K. Reich.—(39 S.; 1 M. 20.)

Heft 11.—Die Mathematik im Physikunterricht der österreichischen Mittelschulen von A. Lanner.—(56 S.; 1 M. 20.)

Heft 12.—Die neuesten Einrichtungen in Oesterreich für die Vorbildung der Mittelschullehrer in Mathematik, Philosophie und Pädagogik von A. Höfler.—(103 S.; 2 M.)

BELGIQUE

1er volume. Rapports sur l'enseignement des mathématiques, du dessin et du travail manuel dans les écoles primaires, les écoles normales primaires, les écoles moyennes, les athénées les collèges belges.—1 vol. de 348 p.; prix: 5 fr.; J. Gœmære, Bruxelles. Ce volume comprend:

Rapport sur l'enseignement des mathématiques dans les *écoles primaires et dans les écoles normales primaires*, par M. Dock (33 p.).

Rapport sur l'enseignement du dessin et du travail manuel dans les écoles primaires, les écoles moyennes, les athénées et les collèges, par M. L. Montfort (154 p.).

Rapport sur l'enseignement des mathématiques dans les écoles moyennes, les athénées et les collèges, par M. H. Ploumen (87 p.).

Les tendances actuelles de l'enseignement mathématique en Belgique et leur influence sur les méthodes et les programmes, par H. Ploumen (67 p.).

En préparation; pour paraître en 1913, *2me volume*:

Les mathématiques dans les écoles industrielles et professionnelles, par M. Rombaut, inspecteur honoraire.

L'enseignement des mathématiques dans les Universités et les Ecoles supérieures, par M. Neuberg.

DANEMARK

Bericht über den Mathematikunterricht in Danemark, par Paul HEEGAARD.—1 vol. de 107 p.; 3 fr. 80; Gyldenalske, Copenhague; Georg & Cie, Genève.

1.—Die Schultypen.
2.—Elementarschulen.
3.—Die höheren allgemeinen Schulen.
4.—Die Volkshochschule.
5.—Elementarschulen für Technik, Handel und Seefahrt.
6.—Militärschulen.
7.—Schulen für Land-, Fortwirtschaft usw.
8.—Die Kunstakademie.
9.—Die Universität und die technische Hochschule.
10.—Die Lehrerausbildung.

ESPAGNE

L'enseignement mathématique en Espagne, rapports de la Sous-commission espagnole, par le délégué Z.-G. de GALDEANO (Saragosse). Travaux préparatoires. 2 fasc., 8 et 18 p.; 1910 et 1911.

Mémoires, Tome I, 139 p., 1912. Sommaire de ce premier volume:

M. Torroja et l'évolution de la Géométrie en Espagne, par Miguel VEGAS.

Enseignement de la Géométrie métrique à la Faculté des Sciences, par Cecilio-Jiménez RUEDA.

Les cours d'Analyse mathématique aux Facultés des Sciences espagnoles, par Luis-Octavio de TOLEDO.

L'enseignement du Calcul infinitésimal aux Facultés des Sciences espagnoles, par Patricio PENALVER.

Les Mathématiques à l'Ecole d'Ingénieurs des Eaux et Forêts, par Jorge TORNER.

L'enseignement des Mathématiques à l'Ecole centrale des Ingénieurs industriels, par Carios MATAIX et Alfonso TORAN.

L'enseignement des Mathématiques à l'Ecole supérieure de Guerre, par Miguel CORREA.

Enseignement des Mathématiques aux Ecoles normales, par Leopoldo FERRERAS.

ETATS-UNIS

Report of the United States of North America (11 fascicules).

COMITÉ I et II: Mathematics in the Elementary Schools of the United States (186 p.), 1911.

COMITÉ III et IV: Mathematics in the Public and Private Secondary Schools of the United States (188 p.), 1911.

COMITÉ V: Training of Teachers of Elementary and Secondary Mathematics (24 p.), 1911.

COMITÉ VI: Mathematics in the Technical Secondary Schools in the United States (36 p.), 1912.

COMITÉ VII: Examinations in Mathematics other than those set by the Teacher for his own classes (72 p.), 1911.

COMITÉ VIII: Influences tending to Improve the Condition of Teachers of Mathematics (47 p.), 1912.

COMITÉ IX: Mathematics in the Technological Schools of Collegiate Grade in the United States (44 p.), 1911.

COMITÉ X: Undergraduate Work in Mathematics in Colleges of Liberal Arts and Universities (30 p.), 1911.

COMITÉ XI: Mathematics at West Point and Annapolis (26 p.), 1912.

COMITÉ XII: Graduate Work in Mathematics in Universities and in other Institutions of like grade in the United States (64 p.), 1911.

General Report of the American Commissioners, with *Index* of all the American Reports (84 p.), 1912.

Ces onze fascicules sont publiés et édités par les soins du Bureau of Education, à Washington.

FRANCE

Rapport de la Sous-commission française, 5 volumes. (Librairie Hachette, Paris.)

Tome I.—Enseignement primaire, publié sous la direction de M. Bioche, prof. de mathématiques au Lycée Louis-le-Grand.—85 p. (3 fr. 50):

Avant-propos.

(*a*) Rapport sur l'ensemble des établissements dans lesquels se donne, en France, un enseignement mathématique, par M. Ch. Bioche.

(*b*) Rapport sur l'enseignement mathématique dans les Ecoles primaires élémentaires, par M. J. Lefebvre.

(*c*) Rapport sur l'enseignement mathématique dans les écoles primaires supérieures, par M. G. Tallent.

(*d*) Rapport sur l'enseignement mathématique dans les écoles normales primaires d'instituteurs, en France, par M. A. Vareil.

(*e*) Rapport sur l'Ecole normale supérieure d'enseignement primaire de Saint-Cloud, par M. Goursat.

Appendice.

Tome II.—Enseignement secondaire, publié sous la direction de M. Bioche, prof. de mathématiques au Lycée Louis-le-Grand.—159 p. (5 fr.):

Avant-propos.

(*a*) Rapport sur la place et l'importance des mathématiques dans l'enseignement secondaire en France, par M. Ch. Bioche.

(*b*) Rapport sur les classes de mathématiques spéciales et de Centrale, par M. E. Blutel.

Pièces annexes.

(*c*) Rapport sur l'arithmétique, par M. A. Lévy.

(*d*) Rapport sur l'algèbre, par M. Guitton.

(*e*) Rapport sur la géométrie, par M. Th. Rousseau.

(*f*) Rapport sur l'enseignement de la mécanique, par M. H. Beghin.

(*g*) Rapport sur l'enseignement de la cosmographie, par M. A. Muxart.

(*h*) Rapport sur l'enseignement des mathématiques dans les écoles nouvelles, par Frank Lombard.

Appendice.

Tome III.—Enseignement supérieur, publié sous la direction de M. Albert de Saint-Germain, Doyen honoraire de la Faculté des sciences de Caen, président de la Sous-commission française.—123 p. (4 fr.):

Aperçu général sur l'enseignement supérieur des mathématiques.

(*a*) Rapport sur l'enseignement du calcul différentiel et intégral, de la mécanique rationnelle, de l'astronomie et des mathématiques générales dans les Facultés des sciences en France, par M. E. Vessiot.

(*b*) Rapport sur les enseignements mathématiques d'ordre élevé dans les Facultés des Sciences d'Universités françaises, par M. Emile Borel.

Annexe.—Faculté des Sciences de Paris: programmes des certificats d'études supérieures pour l'année 1911.

(*c*) Rapport sur les diplômes d'études supérieures de sciences mathématiques, par M. A. de Saint-Germain.

(*d*) Rapport sur l'enseignement mathématique dans les instituts techniques des Facultés des sciences, par M. H. Vogt.

(*e*) Rapport sur l'enseignement des mathématiques à l'Ecole normale supérieure et sur l'agrégation des sciences mathématiques, par M. Jules Tannery.

(*f*) Note sur l'enseignement mathématique au Collège de France, par M. A. de Saint-Germain.

(*g*) Rapport sur l'enseignement mathématique à l'école polytechnique, par M. G. Humbert.

(*h*) Rapport sur l'enseignement mathématique à l'Ecole nationale des Ponts et Chaussées, par M. Maurice d'OCAGNE.

(*i*) Rapport sur l'enseignement des mathématiques à l'Ecole nationale supérieure des Mines, par M. René GARNIER.

(*j*) Rapport sur l'enseignement mathématique à l'Ecole nationale des Mines de Saint-Etienne, par M. FRIEDEL.

(*k*) Note sur l'Ecole d'application du Génie maritime, par M. A. JANET.

Tome IV.—ENSEIGNEMENT TECHNIQUE, publié sous la direction de M. P. ROLLET, Directeur de l'Ecole municipale professionnelle Diderot à Paris.—212 p. (5 fr.):

Introduction.

Ecoles pratiques de commerce et d'industrie. Programmes officiels (28 août 1909). Extraits concernant l'enseignement mathématique.

(*a*) Rapport de M. HARANG.

(*b*) Rapport de M. Ch. LAGNEAUX.

(*c*) Rapport de M. Ch. LAGNEAUX.

Ecoles nationales professionnelles. Programme de l'enseignement technique théorique.

(*d*) Rapport sur l'enseignement des mathématiques dans les écoles nationales professionnelles (E. N. P.), par M. LARIVIÈRE.

(*e*) Rapport sur l'enseignement des mathématiques dans les écoles nationales professionnelles, par M. E. TRIPARD.

Ecoles d'arts et métiers. Programmes officiels du 9 mai 1910.

(*f*) Rapport sur l'enseignement des mathématiques dans les écoles d'arts et métiers (1re année), par M. J. ROUMAJON.

(*g*) Rapport sur l'enseignement des mathématiques dans les écoles d'arts et métiers (2e année), par M. BEZINE.

(*h*) Rapport sur l'enseignement de la mécanique dans les écoles d'arts et métiers (3e année), par M. BAZARD.

Ecoles de commerce.—(*i*) Rapport sur l'enseignement mathématique dans les établissements de la Chambre de Commerce de Paris par M. P. MINEUR.

Conservatoire national d'arts et métiers.—(*j*) Rapport sur l'enseignement des mathématiques au Conservatoire national des arts et métiers, par M. Carlo BOURLET.

Ecole centrale des arts et manufactures.—(*k*) Rapport sur l'enseignement mathématique à l'école centrale des arts et manufactures, par M. P. APPELL.

Appendice.

Tome V.—ENSEIGNEMENT DES JEUNES FILLES, publié sous la direction de Mlle AMIEUX, prof. au Lycée Victor-Hugo à Paris.—95 p. (3 fr. 50):

Aperçu général.

Enseignement secondaire. Introduction.

(*a*) Rapport sur la place des mathématiques dans les plans d'études, l'organisation générale et l'enseignement obligatoire, par Mlle A. AMIEUX.

(*b*) Rapport sur l'enseignement des mathématiques dans la 2e période et sur la préparation au baccalauréat et aux examens de l'enseignement secondaire féminin, par Mme BAUDEUF.

(*c*) Rapport sur l'enseignement des mathématiques à l'Ecole normale de Sèvres, par M. P. APPELL.

Enseignement professionnel. Rapport sur les mathématiques dans l'enseignement professionnel des jeunes filles, par Mme PIVOT et Mlle FREDON.

Enseignement primaire. Introduction.

(*a*) Note sur l'Enseignement mathématique dans les écoles primaires élémentaires.

(*b*) Rapport sur l'enseignement mathématique dans les écoles primaires supérieures de jeunes filles, par M. TALLENT.

(*c*) Sur l'enseignement mathématique dans les écoles normales d'Institutrices primaires.

(*d*) Rapports sur l'enseignement mathématique à l'Ecole normale supérieure d'institutrices de Fontenay-aux-Roses, par MM. G. FONTENÉ et KŒNIGS.

HOLLANDE

Rapport sur l'enseignement mathématique dans les Pays-Bas, publié par la Sous-commission nationale, sous la direction de J. CARDINAAL.—1 vol. de 151 p.; 3 fr.; J. Waltman, Delft.

1. L'enseignement mathématique à l'école primaire.
2. L'enseignement mathématique aux "Burgeravondscholen" (écoles dites bourgeoises), écoles professionnelles, écoles de dessin, écoles professionnelles pour filles et écoles techniques.
3. Ecoles de marine.
4. L'enseignement mathématique aux écoles moyennes (Hoogere Burgerscholen). Ecoles moyennes à 3 années d'études.
5. Ecole moyenne à 5 années d'études.
6. Ecoles moyennes pour jeunes filles.
7. L'enseignement mathématique aux gymnases.
8. Les universités.
9. Académie technique.
10. L'enseignement mathématique aux instituts militaires de l'armée de terre dans les Pays-Bas.
11. Ecole de machinistes pour la marine à Hellevœtsluis.
12. Institut Royal de marine à Willemsoord.
13. Rapport complémentaire sur les propositions de la Commission d'Etat pour la réorganisation de l'enseignement, établie par Arrêté Royal du 21 mars 1903, nº 49.

HONGRIE

1. *Abhandlungen über die Reform des mathematischen Unterrichts in Ungarn.*—Im Auftrage der Mathematischen Reform Kommission des Landesvereins der Mittelschulprofessoren nach dem ungarischen Original deutsch herausgegeben von E. BEKE und S. MIKOLA (1911).—160 S.; 4 M. —B. G. Teubner, Leipzig.

Cette étude d'ensemble se trouve complétée par les dix rapports spéciaux ci-après (fr. 0,50 le fascicule; Librairie Georg, Genève).

2. Die Ausbildung der Mittelschulprofessoren, von J. KÜRSCHAK (1911).—20 S.
3. Der heutige Stand des mathematischen Unterrichts am Königlich ungarischen Josefs-Polytechnikum (Technische Hochschule in Budapest), von G. RADOS (1911).—14 S.
4. Der Unterricht der Mathematik am Uebungsgymnasium, von P. v. SZABO (1912).—17 S.
5. Der mathematische Unterricht an den Lehrerbildungsanstalten, von K. GOLDZIHER (1912).—13 S.
6. Der mathematische Unterricht an den höheren Gewerbeschulen und gewerblichen Fachschulen, von D. ARANY (1912).—15 S.
7. Der mathematische Unterricht an den Handelsschulen, von M. HAVAS und S. BOGYO (1912).—13 S.
8. Der mathematische Unterricht an den Bürgerschulen, von J. VOLENSZKY (1912).—18 S.
9. Der mathematische Unterricht an den Mittelschulen (Gymnasien u. Realschulen), von E. BEKE (1912).—24 S.

In Vorbereitung:

10. Der mathematische Unterricht an den Volksschulen, von M. BITTENBINDER.
11. Der mathematische Unterricht an den höheren Mädchenschulen, von A. VISNYA.

ILES BRITANNIQUES

The following papers on the *Teaching of Mathematics in the United Kingdom* have been published by the Board of Education (Editeurs: Wyman and Sons, London):

1. Higher Mathematics for the Classical Sixth Form. By Mr W. NEWBOLD.—14 p. Price 1*d.*
2. The Relations of Mathematics and Physics. By Dr L. N. G. FILON.—9 p. Price 1*d.*

3. The Teaching of Mathematics in London Public Elementary Schools. By Mr P. B. BALLARD.—28 p. Price 2*d.*

4. The Teaching of Elementary Mathematics in English Public Elementary Schools. By Mr H. J. SPENCER.—32 p. Price 2½*d.*

5. The Algebra Syllabus in the Secondary School. By Mr C. GODFREY.—34 p. Price 2½*d.*

6. The Correlation of Elementary Practical Geometry and Geography. By Miss Helen BARTRAM.—8 p. Price 1*d.*

7. The Teaching of Elementary Mechanics. By Mr W. D. EGGAR.—13 p. Price 1*d.*

8. Geometry for Engineers. By Professor D. A. LOW.—15 p. Price 1½*d.*

9. The Organisation of the Teaching of Mathematics in Public Secondary Schools for Girls. By Miss Louisa STORY.—17 p. Price 1½*d.*

10. Examinations from the School Point of View. By Mr Cecil HAWKINS.—104 p. Price 9*d.*

11. The Teaching of Mathematics to Young Children. By Miss Irene STEPHENS.—19 p. Price 1½*d.*

12. Mathematics with relation to Engineering Work in Schools. By Mr T. S. USHERWOOD. —26 p. Price 2*d.*

13. The Teaching of Arithmetic in Secondary Schools. By Mr G. W. PALMER.—33 p. Price 2½*d.*

14. Examinations for Mathematical Scholarships. By Dr F. S. MACAULAY and Mr W. J. GREENSTREET.—53 p. Price 3*d.*

15. The Educational Value of Geometry. By Mr G. St L. CARSON.—17 p. Price 1½*d.*

16. A School Course in Advanced Geometry. By Mr C. V. DURELL.—14 p. Price 1½*d.*

17. Mathematics at Osborne and Dartmouth. By Mr J. W. MERCER and Mr C. E. ASHFORD. —41 p. Price 2½*d.*

18. Mathematics in the Education of Girls and Women. By Miss E. R. GWATKIN, Miss Sara A. BURSTALL, and Mrs Henry SIDGWICK.—32 p. Price 2½*d.*

19. Mathematics in Scotch Schools. By Professor G. A. GIBSON.—49 p. Price 3*d.*

20. The Calculus as a School Subject. By Mr C. S. JACKSON.—18 p. Price 1½*d.*

21. The Relation of Mathematics to Engineering at Cambridge. By Professor B. HOPKINSON. —13 p. Price 1½*d.*

22. The Teaching of Algebra in Schools. By Mr S. BARNARD.—26 p. Price 1½*d.*

23. Research and Advanced Study as a Training for Mathematical Teachers. By Professor G. H. BRYAN.—21 p. Price 1½*d.*

24. The Teaching of Mathematics in Evening Technical Institutions. By Dr W. E. SUMPNER. —9 p. Price 1*d.*

25. The Undergraduate Course in Pass Mathematics generally, and in relation to Economics and Statistics. By Professor A. L. BOWLEY.—14 p. Price 1½*d.*

26. The Preliminary Mathematical Training of Technical Students. By Mr P. ABBOTT.— 17 p. Price 1½*d.*

27. The Training of Teachers of Mathematics. By Dr T. P. NUNN.—17 p. Price 1½*d.*

28. Recent Changes in the Mathematical Tripos at Cambridge. By Mr Arthur BERRY.— 15 p. Price 1½*d.*

29. Mathematics in the Preparatory School. By Mr E. KITCHENER.—15 p. Price 1½*d.*

30. Course in Mathematics for Municipal Secondary Schools. By Mr L. M. JONES.—15 p. Price 1½*d.*

31. Examinations for Mathematical Scholarships at Oxford. By Mr A. E. JOLIFFE.— Examinations for Mathematical Scholarships at Cambridge. By G. H. HARDY.—22 p. Price 2*d.*

32. Parallel Straight Lines and the Method of Direction. By Mr T. JAMES GARSTANG.—8 p. Price 1*d.*

33. Practical Mathematics at Public Schools: Introduction. By Dr H. H. TURNER.— Practical Mathematics at Clifton College. By Mr R. C. FAWDRY.—Practical Mathematics at Harrow School. By Mr A. W. SIDDONS.—Practical Mathematics at Oundle School. By Mr F. W. SANDERSON.—Practical Mathematics at Winchester College. By Mr G. M. BELL.—36 p. Price 1*d.*

34. Mathematical Examinations at Oxford. By Mr A. L. DIXON.—117 p. Price 16*d.*

Ces rapports ont été réunis en *deux volumes* sous le titre : *The Teaching of Mathematics in the United Kingdom.* Part I & Part II. (*Special Reports on Educational Subjects*, Volumes 26 et 27.) —Prix : T. I, 3 sh ; T. II, 1 sh. 9.

ITALIE

Atti della Sottocommissione italiana. (Les fascicules ne seront mis en vente qu'une fois réunis en volume).

1. Scuole infantili ed elementari, prof. CONTI (Roma).—39 p.
2. Scuole normali, prof. CONTI (Roma), 71 p.
3. Scuole classiche :

(*a*) I successivi programmi dal 1867 al 1910, prof. SCARPIS (Bologna).—11 p.

(*b*) Critiche e proposte, prof. FAZZARI (Palermo).—16 p.

4. Scuole ed istituti tecnici, prof. SCORZA (Cagliari).—34 p.
5. Scuole industriali, professionali e commerciali, prof. LAZZERI (Livorno).—19 p.
6. R. Accademia Navale di Livorno e R. Accademia Militare di Torino, prof. LAZZERI (Livorno).—14 p.
7. Intorno all' ordinamento degli studi matematici nel primo biennio universitario in Italia, prof. SOMIGLIANA (Torino).—11 p.
8. Sugli studi per la laurea in Matematica e sulla sezione di Matematica delle Scuole di Magistero, prof. PINCHERLE (Bologna).—16 p.
9. Osservazioni e proposte circa l'insegnamento della matematica nelle scuole elementari, medie e di magistero, prof. PADOA (Genova).—22 p.
10. Sui libri di testo di geometria per le scuole secondarie superiori, prof. SCORZA (Cagliari). —15 p.

En préparation :

11. Sulla evoluzione degli insegnamenti geometrici nelle Università, prof. SEVERI (Padova).

JAPON

Tome I.—*Report on the teaching of mathematics in Japan*, prepared by the Japanese Sub-commission. 1 vol. de 550 p.

The Teaching of Mathematics :

1. Elementary Schools.—61 p.
2. Middle Schools.—69 p.
3. Higher Middle Schools.—48 p.
4. Faculty of Science of Imperial Universities.—Pour la traduction anglaise, voir Chap. VIII. of the Summary Report.
5. Faculty of Technology of the Tokio Imperial University.—7 p.
6. Normal Schools.—34 p.
7. Training of (male) Teachers for Intermediate Schools.—38 p.
8. Girls' High Schools.—32 p.
9. Normal Schools for Women.—45 p.
10. Higher Normal Schools for Women.—44 p.
11. Commercial Schools and Colleges.—25 p.
12. Technical Schools and Colleges.—42 p.
13. Schools under the Army Department.—48 p.
14. Schools under the Navy Department.—13 p.
15. Schools under the Department of Communications.—30 p.

Tome II.—*Summary Report on the Teaching of Mathematics in Japan*, by R. FUJISAWA.—238 p.

NORVÈGE

Les plans d'études des écoles techniques moyennes norvégiennes étant en revision, M. ALFSEN, délégué, a dû retarder la publication de son rapport. Celui-ci portera le titre *Bericht über den mathematischen Unterricht in Norwegen* et comprendra les objets suivants :

1. Einleitung : Uebersicht über die Organisation des norwegischen Schulwesens.
2. Die Mathematik an den niederen und höheren Volkschulen.
3. Die Mathematik an den höheren Schulen ("Mittelschule" und "Gymnasium").
4. Die Mathematik an den Hochschulen (Universität in Christiania und Technische Hochschule in Drontheim).
5. Die Mathematik an den Spezialschulen.
6. Die Seminare für Volkschullehrer.
7. Die pädagogische Ausbildung der Lehrer der höheren Schulen.

ROUMANIE

L'enseignement mathématique en Roumanie. Enseignement secondaire, par G. TZITZEICA (Bucarest).—16 p.

RUSSIE

1. *L'enseignement mathématique dans les universités, les écoles techniques supérieures et quelques-unes des écoles militaires*, par C. POSSÉ.—100 p. (3 fr.).

2. *L'enseignement mathématique dans les écoles de Finlande*, rédigé par une commission instituée par le Sénat impérial de Finlande.—52 p.

3. *Bericht über den mathematischen Unterricht an den russischen Realschulen*, von K. W. VOGT. —16 p. (fr. 0,60).

4. *L'enseignement mathématique dans les écoles primaires et les écoles normales*, par M. S.

L'enseignement mathématique dans les gymnases de garçons du Ministère de l'Instruction publique et dans les *instituts de jeunes filles* du ressort des établissements de l'Impératrice Marie, par M. KONDRATIEV.—29 p. (fr. 1, —).

5. *L'enseignement mathématique dans les Corps des cadets*, par M. POPRUGENKO.

Notice sur les cours pour la préparation des maîtres des Corps de cadets, par M. MAKCHÉEV.—20 p.

6. (*a*) Sur l'organisation de l'enseignement mathématique dans les *gymnases de jeunes filles* du ressort du Ministère de l'Instruction publique et à l'*Institut pédagogique de jeunes filles*, par M. MICHELSON, prof. à cet institut.

(*b*) Sur l'enseignement mathématique dans les *écoles industrielles* du ressort du Ministère de l'Instruction publique, par MM. KOTOURNITZKY et HATZOUK, professeurs à l'Institut technologique de St Pétersbourg.

(*c*) Sur l'enseignement des mathématiques dans les *gymnases de jeunes filles* dans l'arrondissement scolaire de Varsovie, par M. GORIATCHEV, prof. à l'Université de Varsovie.—37 p.

Rapports déjà rédigés en langue russe ; (en traduction).

(*a*) Les mathématiques dans l'*Institut technologique de St Pétersbourg*, par Boris COÏALOVITSCH, prof. à cet institut.

(*b*) Les mathématiques dans les *cours supérieurs de femmes* (*université de femmes*) à St Pétersbourg, par le même.

(*c*) Les mathématiques dans les *cours supérieurs de femmes à Moscou*, par M. MLODZIEVSKI, anc. prof. à l'Université de Moscou.

(*d*) Les mathématiques dans l'*Institut polytechnique de Varsovie*, par M. MORDOUKHAÏ-BOLTOWSKOÏ, prof. à l'Université de Varsovie.

(*e*) *Sur la préparation des maîtres pour les écoles moyennes secondaires*, par M. KAGAN, prof.-adjoint à l'Université d'Odessa.

(*f*) Les mathématiques dans les *écoles de l'administration générale de l'agriculture*, par M. N. N.

(*g*) Les mathématiques à l'*Institut de Géodésie* de Moscou, par feu B. STRUVE.

SUÈDE

Der mathematische Unterricht in Schweden, herausgegeben von Dr H. von Koch und Dr E. Göransson.—Editeur : C. E. Fritze, Stockholm, 229 p., 4 fr.

Ecoles primaires et écoles normales, par H. Dahlgren : Die Mathematik an den Volksschulen und Volksschullehrerseminaren Schwedens.—52 p.

Ecoles réales, par E. Göransson et E. Hallgren : Die Mathematik an den schwedischen Realschulen.—28 p.

Gymnases, par E. Göransson : Die Mathematik an den schwedischen Gymnasien.—51 p.

Etablissements de jeunes filles, par O. Josephson et Anna Rönström : Die Mathematik an den höheren Mädchenschulen in Schweden.—23 p.

Ecoles professionnelles élémentaires, par K. L. Hagström, G. Erikson et C. Heüman : Die Mathematik an elementartechnischen Gewerbeschulen.—22 p.

Ecoles techniques moyennes, par O. Gallander : Der mathematische Unterricht an den technischen Mittelschulen.—8 p.

Ecoles techniques supérieures, par H. von Koch : Die Mathematik an der technischen Hochschule in Stockholm.—13 p.

Universités, par A. Wiman : Die Mathematik an den schwedischen Universitäten.—18 p.

SUISSE

L'Enseignement mathématique en Suisse. Rapports de la Sous-commission suisse publiés sous la direction de H. Fehr.—1 vol., XVI et 756 p., 18 fr., en 8 fascicules en vente séparément, Georg et Cie, Genève et Bâle.

Fasc. 1.—Les travaux préparatoires : Rapport préliminaire sur l'organisation de la Commission et le plan général de ses travaux, publié au nom du Comité central par H. Fehr, secrétaire-général de la Commission (en français et en allemand).

Organisation des travaux en Suisse.—43 p., 1 fr. 50.

Fasc. 2.—Aperçu général, par H. Fehr.

Der mathematische Unterricht an den schweizerischen Primarschulen, von Just. Stöcklin.

Der mathematische Unterricht an den schweizerischen Sekundarschulen, von Badertscher, Bern.—106 p., 2 fr. 25.

Fasc. 3.—Der mathematische Unterricht an den höheren Mädchenschulen der Schweiz, von E. Gubler, Zürich.

Der mathematische Unterricht an den Lehrer- und Lehrerinnenseminaren der Schweiz, von F. R. Scherrer, Küsnacht.

Organisation und Methodik des mathematischen Unterrichts in den Landerziehungsheimen, von K. Matter, Frauenfeld.—109 p., 2 fr. 25.

Fasc. 4.—Der mathematische Unterricht an den schweizerischen Gymnasien und Realschulen, von K. Brandenberger, Zürich.—167 p., 3 fr. 50.

Fasc. 5.—Les mathématiques dans l'enseignement technique moyen en Suisse, par L. Crelier, Bienne.—112 p., 2 fr. 25.

Fasc. 6.—Les mathématiques dans l'enseignement commercial suisse, par L. Morf, Lausanne. —70 p., 2 fr.

Fasc. 7.—Der mathematische Unterricht an der Eidgenössischen Technischen Hochschule, von M. Grossmann, Zürich.—52 p., 2 fr.

Fasc. 8.—L'Enseignement mathématique à l'Ecole d'Ingénieurs de l'Université de Lausanne, par M. Lacombe, Lausanne.

Der mathematische Unterricht an den schweizerischen Universitäten, von J. H. Graf, Bern.— 72 p., 2 fr. 25.

INDEX TO VOLUME II

COMMUNICATIONS

SECTION II. (GEOMETRY)

SECTIONS III (a) AND III (b). JOINT MEETING. (ASTRONOMY AND STATISTICS)

SECTION III (a). (MECHANICS, PHYSICAL MATHEMATICS, ASTRONOMY)

SECTION III (*b*). (Economics, Actuarial Science, Statistics)

SECTIONS IV (*a*) and IV (*b*). Joint Meeting.

(Philosophy, History, Didactics)

SECTION IV (*a*). (Philosophy and History)

SECTION IV (*b*). (DIDACTICS)

CAMBRIDGE: PRINTED BY JOHN CLAY, M.A. AT THE UNIVERSITY PRESS.

www.ingramcontent.com/pod-product-compliance
Lightning Source LLC
LaVergne TN
LVHW011241110826
845149LV00001B/20
9781418186098